QUADRATIC FORMULA (7-2)

Solutions to $ax^2 + bx + c = 0$, $a \neq 0$:

$$x = \frac{-b \pm \sqrt{b^2 - 4ac}}{2a}$$

GRAPH OF $y = ax^2 + bx + c$ (7-5)

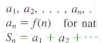

$a > 0$ $a < 0$ Shape is a parabola

$x = -\dfrac{b}{2a}$, $y = -\dfrac{b^2 - 4ac}{4a}$ Vertex

CONIC SECTIONS (7-5, 7-7, 7-8)

$(x - h)^2 + (y - k)^2 = r^2$ Circle, center (h, k), radius r

$y = kx^2$, $x = ky^2$ Parabola

$\dfrac{x^2}{a^2} + \dfrac{y^2}{b^2} = 1$ Ellipse

$\dfrac{x^2}{a^2} - \dfrac{y^2}{b^2} = 1$, $\dfrac{y^2}{b^2} - \dfrac{x^2}{a^2} = 1$ Hyperbola

FUNCTION NOTATION (9-2)

$f(x)$ Value of f at x

$f^{-1}(x)$ Value of inverse function at x

$f^{-1}[f(x)] = f[f^{-1}(x)] = x$ if f^{-1} exists

VARIATION (9-5)

$y = kx$ $k \neq 0$ Direct variation

$y = \dfrac{k}{x}$ $k, x \neq 0$ Inverse variation

$y = kxy$ $k \neq 0$ Joint variation

EXPONENTIAL FUNCTIONS (10-1)

$f(x) = b^x$ Base b

$e \approx 2.718\ 281\ 8$

$f(x) = e^x$ *The* exponential function

LOGARITHMIC FUNCTIONS (10-2, 10-4)

$f(x) = \log_b x$ $b > 0$, $b \neq 1$

$\log_{10} x$ Common logarithm

$\ln x = \log_e x$ Natural logarithm

$y = \log_b x$ is equivalent to $x = b^y$

SEQUENCES AND [...]

$a_1, a_2, \ldots, a_n, \ldots$

$a_n = f(n)$ for nat[...]

$S_n = a_1 + a_2 + \cdots$ [...]ite series

$S_n = \displaystyle\sum_{k=1}^{n} a_k$ Summation notation

ARITHMETIC SEQUENCES (11-2)

$a_n - a_{n-1} = d$ Common difference d

$a_n = a_1 + (n - 1)d$ nth-term formula

$S_n = \dfrac{n}{2}[2a_1 + (n - 1)d]$ Sum of first n terms

$\qquad = \dfrac{n}{2}(a_1 + a_n]$

GEOMETRIC SEQUENCES (11-3)

$\dfrac{a_n}{a_{n-1}} = r$ Common ratio r

$a_n = a_1 r^{n-1}$ nth-term formula

$S_n = \dfrac{a_1 - a_1 r^n}{1 - r}$ $r \neq 1$ Sum of first n terms

$\qquad = \dfrac{a_1 - ra_n}{1 - r}$ $r \neq 1$

$S_\infty = a_1 + a_2 + \cdots$ Sum of infinite sequence

$\qquad = \dfrac{a_1}{1 - r}$

FACTORIAL (11-4)

$n! = 1 \cdot 2 \cdot 3 \cdot \cdots \cdot n$ n factorial, n a natural number

$\dbinom{n}{r} = \dfrac{n!}{r!(n - r)!}$ n, r natural numbers, $r \leq n$

BINOMIAL FORMULA (11-4)

$$(a + b)^n = \sum_{k=0}^{n} \binom{n}{k} a^{n-k} b^k$$

INTERMEDIATE ALGEBRA

Structure and Use

Books by Barnett—Kearns—Ziegler

Barnett—Kearns: *Elementary Algebra: Structure and Use, 6th edition*
Barnett—Kearns: *Algebra: An Elementary Course, 2d edition*
Barnett—Kearns: *Intermediate Algebra: Structure and Use, 5th edition*
Barnett—Kearns: *Algebra: An Intermediate Course, 2d edition*
Barnett—Ziegler: *College Algebra, 5th edition*
Barnett—Ziegler: *College Algebra with Trigonometry, 5th edition*
Barnett—Ziegler: *Precalculus: Functions and Graphs, 3d edition*

Also available from McGraw-Hill

Schaum's Outline Series in Mathematics & Statistics

Most outlines include basic theory, definitions, hundreds of example problems solved in step-by-step detail, and supplementary problems with answers. Titles on the current list include:

Advanced Calculus
Advanced Mathematics
Analytic Geometry
Basic Mathematics for Electricity & Electronics
Basic Mathematics with Applications to Science and Technology
Beginning Calculus
Boolean Algebra & Switching Circuits
Calculus
Calculus for Business, Economics, & the Social Sciences
College Algebra
College Mathematics
Complex Variables
Descriptive Geometry

Differential Equations
Differential Geometry
Discrete Mathematics
Elementary Algebra
Essential Computer Mathematics
Finite Mathematics
Fourier Analysis
General Topology
Geometry
Group Theory
Laplace Transforms
Linear Algebra
Mathematical Handbook of Formulas & Tables
Matrix Operations
Modern Abstract Algebra

Modern Elementary Algebra
Modern Introductory Differential Equations
Numerical Analysis
Partial Differential Equations
Probability
Probability & Statistics
Real Variables
Review of Elementary Mathematics
Set Theory & Related Topics
Statistics
Technical Mathematics
Tensor Calculus
Trigonometry
Vector Analysis

Schaum's Solved Problems Books

Each title in this series is a complete and expert source of solved problems containing thousands of problems with worked out solutions. Titles on the current list include:

3000 Solved Problems in Calculus
2500 Solved Problems in College Algebra and Trigonometry
2500 Solved Problems in Differential Equations
2000 Solved Problems in Discrete Mathematics

3000 Solved Problems in Linear Algebra
2000 Solved Problems in Numerical Analysis
3000 Solved Problems in Precalculus

Bob Miller's Math Helpers

Bob Miller's Calc I Helper

Bob Miller's Calc II Helper

Bob Miller's Precalc Helper

Available at most college bookstores, or for a complete list of titles and prices, write to: Schaum Division, McGraw-Hill, Inc.
Princeton Road, S-1, Hightstown, NJ   08520

5TH EDITION

INTERMEDIATE ALGEBRA

Structure and Use

Raymond A. Barnett
Merritt College

Thomas J. Kearns
Northern Kentucky University

McGraw-Hill WCB McGraw-Hill

Boston, Massachusetts   Burr Ridge, Illinois   Dubuque, Iowa
Madison, Wisconsin   New York, New York   San Francisco, California   St. Louis, Missouri

WCB/McGraw-Hill

A Division of The **McGraw·Hill** *Companies*

INTERMEDIATE ALGEBRA: Structure and Use

This book is printed on acid-free paper.

4   5   6   7   8   9   0   VNH   VNH   9   0   9   8   7

ISBN 0-07-004573-9

This book was set in Times Roman by York Graphic Services, Inc.
The editors were Michael Johnson, Karen M. Minette, and David A. Damstra;
the production supervisor was Leroy A. Young.
The cover was designed by A Good Thing, Inc.
The photo editor was Elyse Rieder.
Von Hoffmann Press, Inc., was printer and binder.

Chapter-Opening Photo Credits
1: Marvin E. Newman/The Image Bank   *2:* Butch Powell/Stock Photos/The Image Bank   *3:* Hank DeLespinasse/The Image Bank   *4:* K. Wothe/The Image Bank   *5:* Stephen Marks/Stock Photos/The Image Bank   *6:* NASA
7: J. Netherton/The Image Bank   *8:* Ulli Seer/The Image Bank   *9:* Steve Proehl/The Image Bank   *10:* Erik Simonsen/The Image Bank   *11:* Loren M. Winters

Library of Congress Cataloging-in-Publication Data

Barnett, Raymond A.
   Intermediate algebra: structure and use / Raymond A. Barnett,
  Thomas J. Kearns.—5th ed.
     p.   cm.
  Includes index.
  ISBN 0-07-004573-9
  1. Algebra.   I. Kearns, Thomas J.   II. Title.
  QA154.2.B38   1994
  512'.9—dc20   93-2364

About the Authors

Raymond A. Barnett is an experienced teacher and author. He received his B.A. in mathematical statistics from the University of California at Berkeley and his M.A. in mathematics from the University of Southern California. He then went on to become a member of the Department of Mathematics at Merritt College and head of that department for 4 years. He is a member of the Mathematical Association of America (MAA), the National Council of Teachers of Mathematics (NCTM), and the American Association for the Advancement of Science. He is the author or coauthor of seventeen books in mathematics that are still in print—all with a reputation for extremely readable prose and high-quality mathematics.

Thomas J. Kearns received his B.S. from Santa Clara University and his M.S. and Ph.D. from the University of Illinois at Urbana-Champaign. After several years of teaching at the University of Delaware, he was appointed to the faculty at Northern Kentucky University, where he served as chairman of the Department of Mathematical Sciences for ten years. He is a member of the MAA, the NCTM, the American Mathematical Society, and the Operations Research Society of America. He has coauthored four texts with Raymond A. Barnett, as well as texts in college algebra and elementary statistics.

Contents

Tables

Preface

This text is written both for students who have studied only elementary algebra and for students who may have studied intermediate algebra in the past, but need a review of intermediate algebra before proceeding further in mathematics. The improvements in this fifth edition have resulted from generous responses from users and reviewers of the fourth edition. Changes have been made to make this new edition even more accessible to students with average background, to provide a better transition from elementary algebra, and to provide a better transition to subsequent courses.

♦ KEY FEATURES AND CHANGES FROM THE FOURTH EDITION

♦ NEW DESIGN

A new, accessible, **full-color** design and an expanded art program make the book more visually appealing and easier to read, and help reinforce mathematical concepts.

Type style for exponents and fractions has been improved for increased clarity. All examples and application problems now have **titles.**

There is more **boxed material** for emphasis. More **schematics** have been added for clarity.

♦ GRAPHING

A greater emphasis is placed on **graphing,** and in particular on the relationship between the graph of an equation

$$y = (\text{an expression in } x)$$

and the solution of the related equation

$$(\text{an expression in } x) = 0$$

(see Chapters 5, 7, 9, and 10).

♦ CALCULATORS

Calculators are assumed to be available to students. Many of the problems in the text lend themselves naturally to calculator use, including graphing calculator use, and are made easier by such use. However, with the exception of very few exercises, which are noted in the instructions, the problems do not *require* the use of a calculator. For those instructors who want to emphasize calculators more, a graphing calculator supplement is available.

◆ EXAMPLES

There are more **worked examples** and **matched problems.** Illustrative examples of some of the more challenging exercises and applications have been added at the request of users.

◆ EXERCISES

Exercise sets have been considerably expanded in almost every section. **Applications** have been added throughout the book, and existing applications have been brought up to date.

◆ SYSTEMATIC REVIEW

Chapter review exercise sets have been expanded. Sample **chapter tests** have been added after every chapter, and comprehensive **cumulative review exercise sets** have been introduced following Chapters 2, 5, 7, and 10.

◆ CAUTIONS

Common student errors are identified by a special **caution** symbol at places where they naturally occur. The number of these has been increased and they are more prominently displayed.

◆ LEARNING SYSTEM

The text is **written for student comprehension.** Each concept or technique is illustrated with an example, followed by a parallel (matched) problem. Answers to the matched problems are provided at the end of each section so students can immediately check their understanding. This example–matched problem structure encourages active learning rather than passive reading.

The order of topics has been chosen to provide a **smooth transition from elementary to intermediate algebra.** The beginning chapters **review** elementary algebraic concepts, manipulations, and applications. This review keeps the material accessible to students with an average background. Subsequent chapters then extend elementary algebraic ideas to intermediate-level topics.

An **informal style** is used for exposition. Definitions are illustrated with simple examples. There are **no formal statements of theorems** in this text.

◆ ORGANIZATIONAL CHANGES

A wider variety of **word problems** has been included earlier in the text (Section 1-8). The use of formal **set notation** has been reduced. **Positive-integer exponents** are now introduced in the first chapter. Material on **solving equations** and particular kinds of **word problems** has been placed in a separate chapter (Chapter 4) to add focus. **Graphing linear equations and systems of linear inequalities** has been moved forward to Chapter 5, now preceding the chapter on exponents and radicals. All material on **solving and graphing quadratic equations** has been placed in one chapter (Chapter 7). **Determinants and Cramer's rule** have been moved from the appendix to an optional section (Section 8-5) in the text. A discussion of the **algebra of functions** has been added to Section 9-2 and **compound interest** to Section 10-1.

♦ **REVISED SECTIONS**

Several sections have been substantially rewritten. In particular, see the sections on **factoring second-degree equations** (Sections 2-3, 2-4, 2-5, 2-6), **graphing linear equations and inequalities** (Sections 5-1, 5-2, 5-5), **complex numbers** (Section 6-8), **square roots, radicals, and simplest radical form** (Sections 6-5, 6-6, 6-7), **graphing quadratic polynomials** (Sections 7-5, 7-6), **functions** (Sections 9-1, 9-2), **inverse functions** (Section 9-4), and **exponential functions** (Section 10-1).

♦ **IMPORTANT FEATURES RETAINED AND IMPROVED IN THE FIFTH EDITION**

Graded Exercise Sets   Graded exercise sets are divided into A, B, and C groupings. The **A problems** are straightforward and representative of the easier examples in the section. The **B problems** represent the more challenging examples in the section, but still emphasize mechanics. The **C problems** provide a mixture of more difficult mechanics, theory, and an extension of the material in the section. The C problems may include some challenging problems that do not parallel worked examples in the section. In short, the exercises are designed so that an average or below-average student will be able to experience success and the more capable students will still be challenged.

Applications   The subject matter is related to the real world through numerous **realistic applications** from the physical sciences, business and economics, the life sciences, and the social sciences.

Major Topic Development   The text continues to use a **spiral technique** for major topics wherever possible; that is, a topic will be introduced in a relatively simple framework and then returned to, reinforced, and developed further in the later sections. For example, consider these topics:

Solving equations: Sections 1-7, 2-7, 4-1, 4-4, 7-1, 7-2, 7-4, 8-1, 8-3, 8-7, 10-5

Word problems: Sections 1-8, 2-7, 4-2, 4-3, 4-5, 5-1, 6-3, 7-3, 8-2, 8-3, 8-6, 9-5, 10-5

Graphing: Chapters 5, 7, 8, 9, 10

Inequalities: Sections 1-2, 5-3, 5-4, 5-5, 7-9, 8-6

Functions: Chapters 9, 10, Section 11-1

The use of this spiral technique continues from the companion text: *Elementary Algebra: Structure and Use,* Sixth Edition.

History   **Historical comments** are included for interest and perspective.

Chapter Summaries   **Chapter summary sections** include a review of the chapter with all important terms and symbols. Also included are a **comprehensive review exercise set** for the chapter and a short **chapter test. Cumulative review exercises** are included after Chapters 2, 5, 7, and 10.

Answers   **Answers** to all chapter review exercises, chapter tests, cumulative review exercises, and all odd-numbered problems from the section exercises are provided in the back of the book. Answers to the exercises in the chapter summaries (review exercises, chapter tests, and cumulative review exercises) are keyed, by numbers in italics, to corresponding sections in the text.

♦ **ADDITIONAL STUDENT AIDS**

Think Boxes   **Think Boxes** (dashed boxes) are used to enclose steps that are usually done mentally (see Sections 1-3, 1-5, 1-6, 1-7, 2-2).

Annotation   **Annotation** of examples and development is found throughout the text to help students through critical steps (see Sections 1-2, 1-3, 1-4, 1-5, 1-6, 1-7).

Color   **Functional use of color** guides students through critical steps (see Sections 1-7, 2-6, 2-7). In this edition, color also has been used to clarify graphs where more than one equation or inequality is graphed (see Sections 8-6 and 8-7).

Formula Summary   **Summaries** of algebraic formulas, symbols, and real-number properties, all keyed to the sections in which they are introduced, are included inside the front cover of the book for convenient reference. Summaries of geometric and other common formulas, and the metric system, are provided inside the back cover of the book.

♦ **STUDENT SUPPLEMENTS**

1. A STUDENT'S SOLUTIONS MANUAL is available through your bookstore. The manual includes key ideas and formulas, solutions to odd-numbered end-of-section exercises, all solutions to the end-of-chapter exercises and chapter tests, and an appendix on setting up word problems.

2. MATHWORKS is a self-paced interactive tutorial specifically linked to the text. It reinforces selected topics and provides unlimited opportunities to review concepts and to practice problem solving. It requires virtually *no* computer training and is available for IBM, IBM-compatible, and Macintosh computers.

3. Course VIDEOTAPES are available.
4. THE GRAPHING CALCULATOR ENHANCEMENT MANUAL presents an integrated approach that utilizes calculator-based graphing to enhance understanding and development. It includes calculator exercises and examples as well as appendixes on how to use the most popular calculators.

♦ **INSTRUCTOR SUPPLEMENTS**

1. An INSTRUCTOR'S RESOURCE MANUAL provides sample tests, transparency masters, an applications index, and additional teaching suggestions and assistance.

2. An INSTRUCTOR'S SOLUTIONS MANUAL contains detailed solutions to even end-of-section exercises, and all cumulative review exercises, as well as the answers to all problems.
3. The PROFESSOR'S ASSISTANT is a unique computerized test generator available to instructors. This system allows the instructor to create tests using algorithmically generated test questions and those from a standard test bank. This testing system enables the instructor to choose questions either manually or randomly by section, question type, difficulty level, and other criteria. This system is available for IBM, IBM-compatible, and Macintosh computers.
4. A PRINTED AND BOUND TEST BANK is also available. This is a hard-copy listing of the questions found in the standard test bank.

For further information, please contact your local McGraw-Hill sales representative.

◆ ERROR CHECK

This text has been carefully and independently checked and proofread by a number of people. Because of this, the authors and publisher believe it to be substantially error-free. However, if errors remain, the authors and publisher would be grateful to be notified and receive corrections. Corrections should be sent to:

Mathematics Editor
McGraw-Hill, Inc.
College Division
1221 Avenue of the Americas
New York, NY 10020

◆ ACKNOWLEDGMENTS

In addition to the authors, many others are involved in the publication of book and the authors have benefited from a great deal of help in preparing this edition. We are grateful to the many users of the fourth edition for their kind remarks and helpful suggestions. We thank the following, in particular, for their detailed reviews:

Jeffrey O. Bauer, Northeast Community College
Brian Buhrman, Iowa Western Community College
Margaret D. Donlan, University of Delaware
Kendall Griggs, Hutchinson Community College
Martin Johnson, Montgomery Community College
Herbert F. Kramer, Longview Community College
Grace Malaney, Donnelly College
Ashley M. Martin, Fairmont State College
Peggy Miller, Kearney State College
Analy Scorsone, Lexington Community College
Mark Serebransky, Camden County College
Jane E. Sieberth, Franklin University
Deborah H. White, College of the Redwoods
Brad Wind, Marshall University
Kelly Wyatt, Umpqua Community College

We especially thank Sr. Margaret Anne Kraemer of Northern Kentucky University and Steven Blasberg of West Valley College, for their careful checking of examples, exercises, and answers; Fred Safier of City College of San Francisco for his accurate checking of exercise sets and his skillful preparation of the student's solutions manual; Karen Minette, associate mathematics editor, for guiding the preparation of the supplements; Phyllis Niklas for her expert manuscript editing; David Damstra, senior editing supervisor, for his conscientious supervision of production; Nancy Evans, marketing manager, for her tireless efforts in promoting the book; and Michael Johnson, mathematics editor, for the services he provided the authors and his guidance of the entire project from beginning to end.

Raymond A. Barnett
Thomas J. Kearns

To the Student

Mastery of the material in this text is important. It will be critical for your success in all college mathematics courses and quantitative courses in other disciplines. There are several things you can do that will help you reach the goal of mastering this material.

First, approach the subject with a **positive attitude.** This text has been written to help you to understand intermediate algebra and to use it to solve problems. Over the past 25 years tens of thousands of students have succeeded in algebra with the help of previous editions of this book. **You, too, can succeed!** But you must want to and believe that you can.

Second, do your **work on a regular basis.** Mathematics is not a spectator sport. You cannot learn to swim, or draw, or speak a foreign language simply by watching someone else. Similarly, you cannot learn mathematics by just reading worked examples or watching your instructor work problems. You must **work problems.** This takes time and effort. Moreover, mathematical learning is cumulative—as you progress through a subject like algebra, you continually need what you have already learned. Thus, it is very important that you keep up with assigned work. **Don't fall behind.**

Third, try the following **study process.** It will help you use this text effectively:

1. **Read** the mathematical development.

   Keep pencil and paper at hand while you read. Make notes. Check any details that aren't provided. Try examples of your own.

2. **Work** through the illustrative example.

   Try to understand each step. There will be a similar problem, called a "Matched Problem," after the example.

3. **Try** the matched problem following the example.

   The answer to the Matched Problem can be found at the end of the section.

4. **Review** the main ideas and any new terminology in the section.

   Pay particular attention to boxed material and any terms in bold type.

5. **Work** the assigned exercises at the end of the section.

   This is the most important part of the learning process!

There are more than enough problems in this text for you to work. Use your assignments as a guide. However, if you are having trouble, you may have to do more of the A problems to get started. If you continue to have trouble with the problems, see your instructor. If you find the assignments too easy, try more of the C problems and check with your instructor—you may be ready for the next course in your curriculum.

Good luck!

Raymond A. Barnett
Thomas J. Kearns

Pedagogical Use of Color

Color in the text figures is used to improve clarity and understanding. Various colors are used in those graphs where different lines are being plotted simultaneously and need to be distinguished.

In addition to the figures, the text has been enhanced with color as well. We have used the following colors to distinguish the various boxes:

RULES/DEFINITIONS

PROPERTIES

STRATEGIES

INTERMEDIATE ALGEBRA

Structure and Use

Preliminaries

Algebra is often referred to as ''generalized arithmetic.'' In arithmetic we deal with basic arithmetic operations: addition, subtraction, multiplication, and division on specific numbers. In algebra we continue to use all that we know in arithmetic, but we also reason and work with *variables,* symbols that represent (or are place-holders for) one or more numbers. The rules for manipulating and reasoning with these symbols depend on certain properties of numbers, since the symbols represent numbers. In this chapter we review important number systems and some of the basic properties that determine how we can manipulate algebraic expressions. The final sections of the chapter review the basic concepts involved in solving equations and word problems.

Photo reference: see Exercise 1-4, Problem 101.

1-1 Basic Concepts

- ◆ Sets
- ◆ Variables and Algebraic Expressions
- ◆ The Set of Real Numbers
- ◆ The Real Number Line

In this section, we recall the familiar number systems of arithmetic and beginning algebra, all contained in the set of real numbers identified with points on the number line. We begin with a brief introduction to the ideas and terminology of *sets,* and also review the concepts of *variables* and *algebraic expressions.*

◆ SETS

Our use of the word ''set'' will not differ appreciably from the way it is used in everyday language. Words such as ''set,'' ''collection,'' ''bunch,'' and ''flock'' all convey the same idea. Thus, we think of a **set** as a collection of objects with the important property that given any object we can tell whether it is or is not in the set. Capital letters, such as A, B, and C, are often used to designate particular sets. For example,

$$A = \{3, 5, 7\} \qquad B = \{4, 5, 6\}$$

specify sets A and B consisting of the numbers enclosed within the braces { }.
A set is usually described in one of two ways:

1. By **listing** the elements between braces:

$$\{3, 5, 7\}$$

where the order in which the elements are listed does not matter. Note that a given element is not listed more than once.
2. By enclosing a **rule** within braces that determines the elements in the set:

Read ''the set of all x such that $x^2 = 81$.''

Example 1
Describing a Set

Let A be the set of all numbers x such that $x^2 = 25$.

(A) Specify A by the listing method.
(B) Specify A by the rule method.

Solution

(A) Listing method: $A = \{-5, 5\}$
(B) Rule method: $A = \{x \mid x^2 = 25\}$ Read ''the set of all x such that $x^2 = 25$.''

Matched Problems

Each example in this text is followed by a similar problem, called a *matched problem*, for you to work. The answers to the matched problems are found at the end of the section, just before the exercise set.

Matched Problem 1 Let B be the set of all x such that $x^2 = 49$.

(A) Specify B by the listing method.
(B) Specify B by the rule method.

If each element of set A is also an element of set B, we say that A is a **subset** of set B. For example, the set of all women in a class is a subset of the whole class. The definition of a subset allows a set to be a subset of itself. If two sets have exactly the same elements, the sets are said to be **equal.** Symbolically:

Subsets and Equality

$A \subset B$	means	''A is a subset of B''	$\{3, 5\} \subset \{3, 5, 7\}$
$A = B$	means	''A is equal to B''	$\{4, 6\} = \{6, 4\}$

Some texts use a variation of this notation where $A \subseteq B$ denotes that A is a subset of B and may possibly be equal to B, while $A \subset B$ denotes that A is a subset of B that is not equal to B. We will not need to make this distinction.

Example 2
Sets and Subsets

Let $A = \{-3, 0, 5\}$, $B = \{0, 5, -3\}$, and $C = \{0, 5\}$. Indicate true or false:

(A) $C \subset A$ **(B)** $A = B$ **(C)** $A \subset B$
(D) $A = C$ **(E)** $B \subset C$

Solution
(A) True: each of the objects in C is also in A.
(B) True: A and B consist of the same three numbers.
(C) True: every object in A is also in B.
(D) False: A and C do not consist of the same numbers; -3 belongs to A but not to C.
(E) False: there is a number, namely -3, that is in set B but not in set C.

Matched Problem 2 Let $M = \{-4, 6\}$, $N = \{6, -4\}$, and $P = \{-4\}$. Indicate true (T) or false (F):

(A) $M = N$ **(B)** $P \subset N$ **(C)** $M \subset P$
(D) $N \subset M$ **(E)** $M = P$

A more complete treatment of sets may be found in Appendix A.

♦ **VARIABLES AND ALGEBRAIC EXPRESSIONS**

The letter x used in $x^2 = 25$ in Example 1 is a variable. In general, a **variable** is a symbol used to represent unspecified elements from a set with two or more elements. This set is called the **replacement set** for the variable. A **constant,** on the other hand, is a symbol that names exactly one object. The symbol "8" is a constant, since it always names the number eight.

An **algebraic expression** is a meaningful symbolic form involving constants, variables, mathematical operations, and grouping symbols. For example,

$$2 + 8 \qquad 4 \cdot 3 - 7 \qquad 16 - 3(7 - 4)$$

$$5x - 3y \qquad 7(x + 2y) \qquad 4\{u - 3[u - 2(u + 1)]\}$$

are all algebraic expressions.

Two or more algebraic expressions, each taken as a single entity and joined by plus or minus signs, are called **terms.** For reasons that will become clear in Section 1-6, a term includes the sign that precedes it. Two or more algebraic expressions joined by multiplication are called **factors.** For example,

has two terms, $3(x - y)$ and $(x + y)(x - y)$, and each of these terms has two factors. The first term has factors 3 and $(x - y)$, and the second term has factors $(x + y)$ and $(x - y)$. A term may contain several factors, and a factor may contain several terms.

Example 3
Terms and Factors

Identify the terms and factors in the following algebraic expressions:

(A) $3xy$ (B) $x + y - z - 3$ (C) $x(x - 1) + 2x$

Solution

(A) The expression has three factors: 3, x, and y.
(B) The expression has four terms: x, y, $-z$, and -3.
(C) The expression consists of two terms: $x(x - 1)$ and $2x$. The first term has two factors, x and $x - 1$, and the second of these factors has two terms, namely x and -1. The second term, $2x$, has two factors, 2 and x.

Matched Problem 3

Identify the terms and factors in the following algebraic expressions:

(A) $a + b - 1$ (B) $23xyz$ (C) $x(x - 1)(x + y + 2)$

To **evaluate** an algebraic expression for particular values of the variables means to replace each variable by a given value and then calculate the resulting arithmetic value. Thus, for example, to evaluate $3xy$ when $x = 5$ and $y = 7$, we calculate

$$3xy = 3(5)(7) = 105$$

The introduction of variables into mathematics in the Western world occurred about A.D. 1600, although the basic idea had been developed earlier in India. A French mathematician, François Vieta (1540–1603), is singled out as the one mainly responsible for this new idea in the West. Many mark this point as the beginning of modern mathematics. Variables allow us to concisely state general properties of numbers, such as

$$a + b = b + a \qquad \text{for all numbers } a \text{ and } b$$

Properties like this are reviewed in Section 1-3. Variables also enable us to write statements about problem situations in compact equations, such as

$$16 + 2x = 1 + 5x \qquad \text{or} \qquad 3x - 2(2x - 5) = 2(x + 3) - 8$$

The ideas behind equality statements and equations are reviewed in Sections 1-2 and 1-7. All the work you will do in this text will relate in some way to algebraic expressions involving variables. You will be asked to *translate* problem situations into algebraic expressions and equations involving them, to *rewrite* such expressions in different forms, to *evaluate* them, to *solve* equations involving them, to *graph* equations and solutions to equations, and to *use* all of these skills to solve applied problems.

♦ **THE SET OF REAL NUMBERS**

The **real number system** is a number system with which you should be familiar, since positive real numbers represent lengths and other similar measurements. Because the symbols used in algebra initially represent real numbers, it is important to review briefly the set of real numbers and some of its important subsets. These subsets are the *natural numbers* (or *positive integers* or *counting numbers*), the *whole numbers,* the *integers,* the *rational numbers,* and the *irrational numbers.*

The **natural numbers** (or **positive integers** or **counting numbers**) are the familiar numbers 1, 2, 3, 4, . . . , with the three dots meaning ''continuing in this way indefinitely.'' The **whole numbers** consist of the natural numbers and the number 0. The **integers** consist of the whole numbers together with a negative number for each positive integer. The **rational numbers** are those numbers that can be represented as quotients of two integers. Rational numbers also can be represented as decimal fractions, which either terminate or repeat in blocks. Thus, for example, the decimal representations of the rational numbers 2, $\frac{4}{3}$, and $\frac{5}{11}$ are, respectively,

$$2 = 2.0000 \ldots \qquad \tfrac{4}{3} = 1.333 \ldots \qquad \tfrac{5}{11} = 0.454545 \ldots$$

The blocks 0, 3, and 45 repeat indefinitely, as indicated by the three dots. The repeating decimals also can be indicated by an overbar, as in

$$2 = 2.\overline{0} \qquad \tfrac{4}{3} = 1.\overline{3} \qquad \tfrac{5}{11} = 0.\overline{45}$$

The decimal representation can be found by long division. The problem of finding a fraction corresponding to a repeating decimal is explored in Exercise 1-1, Problems 77–84.

Irrational numbers are those numbers associated with lengths that cannot be represented as the ratio of two integers or the negative of such a ratio. These can be represented as unending decimal fractions that do not repeat. Examples of irrational numbers include the square root of any integer that is not a perfect square, for example, $\sqrt{2}$, $\sqrt{3}$, $\sqrt{5}$, etc. The number π also is irrational, as is the number represented by e, a symbol you will encounter in Chapter 10. The decimal representations of $\sqrt{2}$ and π are

$$\sqrt{2} = 1.41421356\ldots \qquad \pi = 3.14159265\ldots$$

No block of decimals will repeat indefinitely in either of these numbers.

We will use the letters N, W, Z, Q, and R to represent the natural numbers (N), whole numbers (W), integers (Z), rational numbers (Q), and real numbers (R), respectively. The subset relationships can then be written symbolically as

$$N \subset W \subset Z \subset Q \subset R$$

Figure 1 illustrates these relationships.

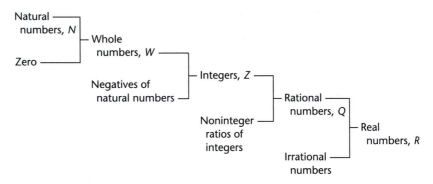

Figure 1   The real number system

Example 4
Subsets of R

Let $A = \{-3, 0, 5\}$, $B = \{0.25, \sqrt{3}, -1\}$, and $C = \{0, 1, \frac{1}{2}\}$. Indicate true (T) or false (F):

(A) $A \subset N$ **(B)** $A \subset Z$ **(C)** $B \subset Q$
(D) $B \subset R$ **(E)** $C \subset W$ **(F)** $C \subset Q$

Solution   **(A)** False: -3 is not a natural number.
(B) True: all the numbers in A are integers.
(C) False: $\sqrt{3}$ is not a rational number.
(D) True: all the numbers in B are real numbers.
(E) False: $\frac{1}{2}$ is not a whole number.
(F) True: all the numbers in C are rational numbers.

Matched Problem 4   Let $A = \{-3, 0, \frac{2}{5}\}$, $B = \{-10.75, \sqrt{2}, 10\}$, and $C = \{-\frac{1}{3}, 0, \frac{1}{3}\}$. Indicate true (T) or false (F):

(A) $A \subset Z$ **(B)** $A \subset Q$ **(C)** $B \subset R$
(D) $B \subset Q$ **(E)** $C \subset Z$ **(F)** $C \subset Q$

♦ THE REAL NUMBER LINE

A one-to-one correspondence exists between the set of real numbers and the set of points on a line; that is, each real number corresponds to exactly one point, and each point to exactly one real number. A line with a real number associated with each point, and vice versa, as shown in Figure 2, is called a **real number line,** or simply a **real line.** Each number associated with a point is called the **coordinate** of the point. The point with coordinate 0 is called the **origin.** The arrow on the number line indicates a positive direction; the coordinates of all points to the right of the origin are called **positive real numbers,** and those to the left of the origin are called **negative real numbers.**

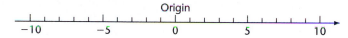

Figure 2   A real number line

| Example 5 | Locate the following points on a number line: $-\sqrt{27}$, $-\frac{4}{3}$, 2, π, 7.64. |

Locating Real Numbers on a Number Line

Solution

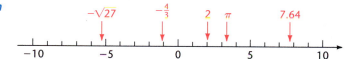

Matched Problem 5   Locate the following points on a number line: -3, $-\sqrt{2}$, $\frac{2}{3}$, $\frac{6}{3}$, 2.75.

Table 1 summarizes the number systems within the set of real numbers.

Table 1   **The Set of Real Numbers**

Symbol	Number System	Description	Examples
N	Natural numbers	Counting numbers, also called positive integers	1, 2, 3, . . .
W	Whole numbers	Natural numbers and 0	0, 1, 2, 3, . . .
Z	Integers	Set of natural numbers, their negatives, and 0	. . . , -2, -1, 0, 1, 2, . . .
Q	Rational numbers	Any number that can be represented as a/b, where a and b are integers and $b \neq 0$	-4; $-\frac{3}{5}$; 0; 1; $\frac{2}{3}$; 3.67
R	Real numbers	Set of all numbers, rational and irrational, corresponding to points on a number line	-4; $-\frac{3}{5}$; 0; 1; $\frac{2}{3}$; 3.67; $\sqrt{2}$; π; $\sqrt[3]{5}$

Answers to
Matched Problems

1. **(A)** $\{-7, 7\}$ **(B)** $\{x \mid x^2 = 49\}$
2. **(A)** T   **(B)** T   **(C)** F   **(D)** T   **(E)** F
3. **(A)** Terms: a, b, -1 **(B)** Factors: 23, x, y, z
   (C) Factors: x, $(x - 1)$, $(x + y + 2)$. The second factor has two terms, x and -1. The third factor has three terms, x, y, and 2
4. **(A)** F   **(B)** T   **(C)** T   **(D)** F   **(E)** F   **(F)** T
5.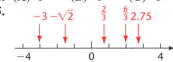

EXERCISE 1-1

The exercise sets in this text are divided according to difficulty into A, B, and C groupings. The A problems are mostly easy and routine. The B problems are less routine but still emphasize the mechanics of the material in the section. The C problems are more challenging and present a mixture of more complicated mechanics and a few problems of a more theoretical nature. Application problems, placed after the C-level exercises in some sections, are grouped differently: they vary from relatively easy (no stars) and moderately difficult problems (marked ★) to more difficult ones (marked ★★).

A Let $A = \{1, 2, 3, 4, 5, 6\}$, $B = \{1, 3, 5\}$, $C = \{2, 4, 6\}$, $D = \{1, 2, 3\}$, $E = \{4, 5, 6\}$, $G = \{1, 2\}$, $H = \{3, 4\}$, $J = \{5\}$, $K = \{6\}$. Indicate true (T) or false (F).

1. $D \subset B$ 2. $J \subset B$ 3. $K \subset B$ 4. $B \subset A$

5. $G \subset D$ 6. $H \subset C$ 7. $J \subset E$ 8. $K \subset B$

9. $H \subset E$ 10. $H \subset J$ 11. $A \subset B$ 12. $E \subset E$

In Problems 13–18, locate the given numbers on a real number line.

13. 3, $\frac{3}{5}$, $\sqrt{5}$, $-\frac{5}{3}$, -2 14. 2, $\frac{2}{5}$, $\sqrt{7}$, $-\frac{5}{2}$, -3

15. 4, $\frac{4}{7}$, $\sqrt{3}$, $-\frac{7}{4}$, -1 16. 5, $\frac{1}{3}$, $-\sqrt{2}$, $-\frac{3}{2}$, 0

17. 0, $\frac{3}{4}$, $-\sqrt{3}$, $-\frac{4}{3}$, -5 18. 1, $\frac{3}{7}$, $-\sqrt{5}$, $-\frac{7}{3}$, -4

19. Give an example of a negative integer, an integer that is neither positive nor negative, and a positive integer.

20. Give an example of a negative rational number, a rational number that is neither positive nor negative, and a positive rational number.

21. Give an example of a rational number that is not an integer.

22. Give an example of an integer that is not a natural number.

Write each set in Problems 23–34 using the listing method; that is, list the elements between braces. Use the notation $\varnothing$ to indicate that the set contains no elements. The set containing no elements is called the **empty set.**

23. $\{x \mid x$ is a counting number between 5 and 10$\}$

24. $\{x \mid x$ is a counting number between 4 and 8$\}$

25. $\{x \mid x$ is a letter in "*status*"$\}$

26. $\{x \mid x$ is a letter in "*Illinois*"$\}$

27. $\{x \mid x$ was a woman elected as president of the United States before 1992$\}$

28. $\{x \mid x$ is a month starting with the letter $B\}$

29. $\{x \mid x - 5 = 0\}$ 30. $\{x \mid x + 3 = 0\}$

31. $\{x \mid x + 9 = x + 1\}$ 32. $\{x \mid x - 3 = x + 2\}$

33. $\{x \mid x^2 = 4\}$ 34. $\{x \mid x^2 = 9\}$

B 35. Indicate whether each of the following is true (T) or false (F):
   (A) All natural numbers are integers.
   (B) All real numbers are irrational.
   (C) All rational numbers are real numbers.

36. Indicate whether each of the following is true (T) or false (F):
   (A) All integers are natural numbers.
   (B) All rational numbers are real numbers.
   (C) All natural numbers are rational numbers.

Each of the following numbers lies between two successive integers. Indicate which two. For example, $\sqrt{10}$ lies between 3 and 4.

37. **(A)** $\frac{15}{4}$ **(B)** $-\frac{4}{3}$ **(C)** $-\sqrt{7}$

38. **(A)** $\frac{17}{3}$ **(B)** $-\frac{11}{3}$ **(C)** $\sqrt{13}$

39. **(A)** $\frac{12}{5}$ **(B)** $-\frac{22}{5}$ **(C)** $\sqrt{45}$

40. **(A)** $\frac{19}{4}$ **(B)** $-\frac{27}{4}$ **(C)** $\sqrt{24}$

41. **(A)** $\frac{33}{2}$ **(B)** $-\frac{41}{2}$ **(C)** $\sqrt{32}$

42. (A) $\frac{53}{7}$ **(B)** $-\frac{38}{7}$ **(C)** $\sqrt{77}$

43. (A) $\frac{48}{9}$ **(B)** $-\frac{32}{9}$ **(C)** $\sqrt{129}$

44. (A) $\frac{47}{3}$ **(B)** $-\frac{31}{3}$ **(C)** $\sqrt{63}$

Identify the terms in the given algebraic expression. You do not need to identify the terms within factors.

45. $3x - 4y + 5xy$ **46.** $x - 2y - 3z$

47. $x(2y - 3z)$ **48.** $5(a - 2b + 3c)$

49. $xyz + w(x + y + z)$ **50.** $a + b(1 + c + d)$

Identify the factors in the given algebraic expression. You do not need to identify factors within separate terms.

51. $2abc$ **52.** $3x(x + 2)$

53. $(x + 2)(x - 3)$ **54.** $4ab(c - 5)$

55. $3x(x + 3y)(3y + x)$ **56.** $(xy + z)(xz + y)(yz + x)$

Let $A = \{-4, \frac{11}{4}, \sqrt{11}\}$, $B = \{-2, 0, 4\}$, $C = \{\frac{1}{8}, 8, -8, -\frac{1}{8}\}$. *Indicate true (T) or false (F).*

57. $A \subset Q$ **58.** $B \subset Q$ **59.** $C \subset Q$ **60.** $A \subset N$

61. $A \subset R$ **62.** $A \subset Z$ **63.** $B \subset W$ **64.** $B \subset Z$

65. $B \subset N$ **66.** $C \subset R$ **67.** $C \subset Z$ **68.** $C \subset N$

69. $Z \subset Q$ **70.** $W \subset N$ **71.** $R \subset Q$ **72.** $R \subset Z$

C *For the sets M and N given, find*
(A) *The set of all elements that belong to M or to N, or to both*[†]
(B) *The set of all elements that belong to both M and N*

73. $M = \{1, 2, 3, 4\}, N = \{2, 4, 6\}$

74. $M = \{-1, 0, 1, 2\}, N = \{-2, 0, 2\}$

75. $M = \{4, 6, 8, 10\}, N = \{2, 4, 6\}$

76. $M = \{1, 3, 5, 7\}, N = \{5, 7, 9\}$

A repeating decimal can be converted to a fraction involving two integers as follows: if c = 0.151515 . . . , then 100c = 15.151515 . . . , and we get

$$100c = 15.151515 \ldots$$
$$\underline{c = 0.151515 \ldots} \quad \text{Subtract.}$$
$$99c = 15$$
$$c = \tfrac{15}{99} = \tfrac{5}{33}$$

[†]Unless otherwise stated, we will use "or" as it is usually used in mathematics—that is, as an inclusive or that includes both possibilities. The phrase "or both" then becomes unnecessary.

We multiplied by 100 to get the repeating decimals to line up so that they were eliminated upon subtracting. For repeating blocks of different lengths, we would use different multipliers. If the number included a nonrepeating part first, such as c = 0.3151515 . . . , we would multiply by 10 and 1,000 to obtain

$$1{,}000c = 315.1515 \ldots$$
$$\underline{10c = 3.1515 \ldots} \quad \text{Subtract.}$$
$$990c = 312$$
$$c = \tfrac{312}{990} = \tfrac{52}{165}$$

Convert the repeating decimals to a fraction involving two integers.

77. $0.090909 \ldots$ **78.** $0.181818 \ldots$

79. $0.123123123 \ldots$ **80.** $0.543543543 \ldots$

81. $0.5676767 \ldots$ **82.** $0.8909090 \ldots$

83. $0.23456456456 \ldots$ **84.** $0.1357242424 \ldots$

Express each number as a decimal fraction to the capacity of your calculator. Observe the repeating decimal representation of the rational numbers and the apparent nonrepeating decimal representation of the irrational numbers.

85. (A) $\frac{8}{9}$ **(B)** $\frac{3}{11}$ **(C)** $\sqrt{5}$ **(D)** $\frac{11}{8}$

86. (A) $\frac{13}{6}$ **(B)** $\sqrt{21}$ **(C)** $\frac{7}{16}$ **(D)** $\frac{29}{111}$

87. (A) $\frac{2}{5}$ **(B)** $-\frac{3}{40}$ **(C)** $\sqrt{3}$ **(D)** $\frac{1}{333}$

88. (A) $\frac{1}{9}$ **(B)** $-\frac{7}{32}$ **(C)** $\sqrt{8}$ **(D)** $\frac{1}{999}$

APPLICATIONS

89. The executive committee of a student council consists of a president, vice president, secretary, and treasurer, and is denoted by the set $\{P, V, S, T\}$. List all the two-person subcommittees that can be formed from the executive committee. That is, list all the two-element subsets of $\{P, V, S, T\}$. There are six of them.

90. List all the three-person subcommittees possible for the executive committee described in Problem 89. There are four of them.

91. List all the sets of three cards that can be formed from a set of five cards denoted A, K, Q, J, and 10. That is, list all the three-element subsets of $\{A, K, Q, J, 10\}$. There are ten of them.

92. List all the two-element subsets that can be formed from the set in Problem 91. There are ten of them.

1-2   Equality and Inequality

♦ Equality Relation
♦ Inequality Relation

Equality and inequality relations are reviewed in this section.

♦ EQUALITY RELATION

The use of an **equality sign,** $=$, between two expressions asserts that the two expressions are names or descriptions of exactly the same object. The symbol $\neq$ means **is not equal to.** Statements involving the use of an equality or inequality sign may be true or they may be false.

$$15 - 3 = 4 \cdot 3 \qquad \text{True statement}$$

$$7 - 2 = \tfrac{8}{2} \qquad \text{False statement}$$

$$15 - 3 \neq 12 \qquad \text{False statement}$$

$$7 - 2 \neq \tfrac{8}{2} \qquad \text{True statement}$$

The equality sign did not appear until rather late in mathematical history—the sixteenth century. It was introduced by the English mathematician Robert Recorde (1510–1558).

If two algebraic expressions involving at least one variable are joined with an equal sign, the resulting form is called an **algebraic equation.** The following are algebraic equations in one or more variables:

$$2x - 3 = 3(x - 5) \qquad a + b = b + a \qquad 3x + 5x = 7$$

Since a variable represents unspecified numbers from a given replacement set, an algebraic equation is neither true nor false as it stands. It does not become true or false until the variables have been replaced by specific values. Formulating algebraic equations is an important first step in solving many practical problems using algebraic methods.

Example 1
Formulating
Algebraic Equations

Translate each statement into an algebraic equation using x as the only variable.

(A) 5 times a number is 3 more than twice the number.
(B) 4 times a number is 5 less than twice the number.
(C) If 5 is added to a certain number, the result is twice the quantity that is 1 less than the original number.

Solution   **(A)** Let $x =$ The unknown number; then the statement translates as follows:

$$\underbrace{5}\;\underbrace{\text{times}}\;\underbrace{\text{a number}}\;\underbrace{\text{is}}\;\underbrace{3}\;\underbrace{\text{more than}}\;\underbrace{\text{twice the number}}$$
$$\;5\quad\cdot\quad\;x\quad=\;2x\quad+\quad\quad 3$$

Thus,

$$5x = 2x + 3$$

(B) Let x = The unknown number; then the statement translates as follows:

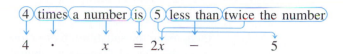

4 · x = 2x − 5

In this context, "less than" means "subtracted from." Thus,

$$4x = 2x - 5 \qquad \text{Not } 4x = 5 - 2x$$

(C) Let x = The unknown number; then the statement translates as follows:

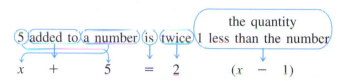

x + 5 = 2 (x − 1)

Matched Problem 1 Translate each statement into an algebraic equation using x as the only variable:

(A) 12 is 9 less than twice a certain number.
(B) 3 times a certain number is 6 less than twice that number.
(C) If 6 is subtracted from a certain number, the difference is twice a number that is 4 less than the original number.

Several important properties of the equality symbol follow directly from its logical meaning. These properties hold any time the symbol is used.

Basic Properties of Equality

If a, b, and c are names of objects, then:

1. $a = a$. **Reflexive property**
2. If $a = b$, then $b = a$. **Symmetric property**
3. If $a = b$ and $b = c$, then $a = c$. **Transitive property**
4. If $a = b$, then either may replace the other in any statement without changing the truth or falsity of the statement. **Substitution principle**

The properties of equality are used extensively throughout mathematics. Using the symmetric property, we may reverse the left and right sides of an equation any time we wish. For example:

$$\text{If} \quad A = P + Prt, \quad \text{then} \quad P + Prt = A.$$

Using the transitive property, we find that, for example:

$$\text{If} \quad 2x + 3x = (2 + 3)x \quad \text{and} \quad (2 + 3)x = 5x, \quad \text{then} \quad 2x + 3x = 5x.$$

And, finally, if we know, for example, that

$$C = \pi D \qquad \text{and} \qquad D = 2R$$

then, using the substitution principle, D in the first formula may be replaced by $2R$ from the second formula to obtain

$$C = \pi D = \pi(2R) = 2\pi R$$

◆ INEQUALITY RELATION

The **inequality,** or **order, relation** arises from the relationships "less than" and "greater than." If a and b are any real numbers, then "a is less than b" means that a is to the left of b on the number line. Similarly, "a is greater than b" means that a is to the right of b on the number line.

Just as we use $=$ to replace the words "is equal to," we use the **inequality symbols** $<$ and $>$ to represent "is less than" and "is greater than," respectively. Thus, we can write the following symbolic forms and their corresponding verbal forms:

Inequality Symbols		
$a < b$	a is less than b	$5 < 8$
$a > b$	a is greater than b	$8 > 5$
$a \leq b$	a is less than b or equal to b	$5 \leq 8;\ 5 \leq 5$
$a \geq b$	a is greater than b or equal to b	$8 \geq 5;\ 5 \geq 5$

Figure 1 illustrates the geometric interpretation of the $<$ and $>$ symbols. Since a is to the left of b on the number line, $a < b$. Since c is to the right of d, $c > d$.

Figure 1 $\quad a < b,\ c > d$

It may be obvious from a number line that statements such as $5 < 8$ and $-8 < -5$ are true. However, graphing will not make it equally obvious that

$$\tfrac{3}{7} < \tfrac{17}{38} \qquad \text{and} \qquad \sqrt{5} < \tfrac{43}{19}$$

are also true. To make the inequality relation precise so that we can interpret it relative to *all* real numbers, we need an algebraic definition of the concept.

Definition of $a < b$ and $b > a$

For a and b real numbers, we say that **a is less than b** or **b is greater than a** and write

$$a < b \qquad \text{or} \qquad b > a$$

if there exists a positive real number p such that $a + p = b$, or equivalently, $b - a = p$.

We would expect that if a positive number is added to *any* real number, the sum will be larger than the original. That is essentially what the definition states.

Example 2
Inequalities

Indicate true (T) or false (F). Letters a, b, c, and d refer to Figure 1 above.

(**A**) $-3{,}000 > 0$ (**B**) $-10 \leq 2$ (**C**) $-5 \geq -5$
(**D**) $0 < -25$ (**E**) $a < c$ (**F**) $d \geq b$

Solution (**A**) F (**B**) T (**C**) T (**D**) F (**E**) T (**F**) F

Matched Problem 2

Indicate true (T) or false (F). Letters a, b, c, and d refer to Figure 1 above.

(**A**) $0 < 25$ (**B**) $-35 \geq 3$ (**C**) $-5 \leq -5$
(**D**) $-25 > 0$ (**E**) $b \leq c$ (**F**) $d < b$

The following two important inequality properties hold for all real numbers:

Basic Inequality Properties

For any real numbers a, b, and c:

1. Exactly one of the following   **Trichotomy property**
   is true: $a < b$, $a = b$, or $a > b$.
2. If $a < b$ and $b < c$, then $a < c$.   **Transitive property**

The **double inequality** $a < x \leq b$ means that $a < x$ and $x \leq b$; that is, x is between a and b, including b but not including a. Similar interpretations are given to forms such as $a \leq x < b$, $a \leq x \leq b$, and $a < x < b$.

Inequality statements involving one variable, such as

$$x > 2 \qquad -2 < x \leq 3$$

$$x \leq -3 \qquad 0 \leq x \leq 5$$

can be graphed on a real number line. To **graph an inequality statement** in one variable on a real number line is to graph the set of all real number replacements of the variable that make the statement true. This set is called the **solution set** of the inequality statement. The solution set usually is an interval on the number line.

Example 3
Graphing Inequalities

Graph on a real number line:

(**A**) $x > 2$ (**B**) $-2 < x \leq 3$

Solution (**A**) The solution set for $x > 2$ is the set of *all* real numbers greater than 2. Graphically, this set includes all the points to the right of 2:

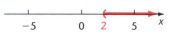

The parenthesis through 2 indicates that 2 is not included.

(**B**) The solution set for $-2 < x \leq 3$ is the set of *all* real numbers between -2 and 3, including 3 but not -2. Graphically:

(indicates -2 is not included;
] indicates 3 is included.

Matched Problem 3   Graph on a real number line:

(A) $x \leq -3$ (B) $0 < x < 5$

Just as formulating algebraic equations is an important step in solving many practical problems, so is formulating appropriate algebraic inequality statements.

Example 4
**Formulating
Algebraic Inequalities**

Translate each statement into an algebraic inequality statement using x as the only variable:

(A) 8 times a number is greater than or equal to 10 more than the number.
(B) 4 less than twice a number is less than 6 times the number.

Solution   (A) Let $x =$ The unknown number(s); then the statement translates as follows:

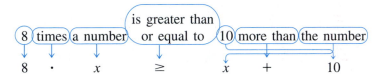

Or, more compactly,

$$8x \geq x + 10$$

(B) Let $x =$ The unknown number(s); then the statement translates as follows:

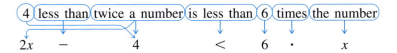

Or, more compactly,

$$2x - 4 < 6x$$

 The phrase "less than" needs to be read in context very carefully. In the first part of the statement in Example 4(B), it means "subtract from." However, "is less than" in the middle of the statement means $<$. The verb "is" helps to make the distinction.

 You must also be careful to distinguish between *algebraic expressions* and *algebraic statements*. Algebraic statements are algebraic expressions related by an equality or inequality. For specific values of the variables involved, an algebraic expression just represents a real-number value. However, for specific values of the variables, an algebraic statement becomes a true or false statement. By way of analogy, statements correspond to declarative sentences, and expressions correspond to phrases within a sentence.

Matched Problem 4   Translate each statement into an algebraic inequality statement using x as the only variable:

(A) 7 less than twice a number is greater than 5 times the number.
(B) 3 more than a number is less than or equal to 5 less than twice the number.

Solving various kinds of inequalities will be considered further in Chapters 5, 7, and 8.

Answers to Matched Problems

1. (A) $12 = 2x - 9$ (B) $3x = 2x - 6$
   (C) $x - 6 = 2(x - 4)$
2. (A) T   (B) F   (C) T   (D) F   (E) T   (F) T
3. (A)

   $$\begin{array}{c}\text{number line with arrow left from } -3 \\ -5 \ -3 \quad 0 \qquad 5 \qquad x\end{array}$$

   (B)

   $$\begin{array}{c}\text{number line segment from } 0 \text{ to } 5 \\ -5 \qquad 0 \qquad 5 \qquad x\end{array}$$
4. (A) $2x - 7 > 5x$ (B) $x + 3 \le 2x - 5$

EXERCISE 1-2

A *Write in symbolic form.*

1. $11 - 5$ is equal to $\frac{12}{2}$.   2. $\frac{36}{2}$ is equal to $3 \cdot 6$.
3. 4 is greater than -18.   4. 3 is greater than -7.
5. -12 is less than -3.   6. -20 is less than 1.
7. x is greater than or equal to -8.
8. x is less than or equal to 12.
9. x is between -2 and 2.
10. x is between 0 and 10.

Replace each question mark with =, <, or > to form a true statement.

11. $-3 \,?\, 7$ 12. $1 \,?\, -6$
13. $-2 \,?\, -5$ 14. $0 \,?\, -100$
15. $-1.35 \,?\, -1$ 16. $-6.33 \,?\, -6$
17. $\frac{55}{5} \,?\, \frac{22}{2}$ 18. $10 + 3 \,?\, \frac{39}{3}$
19. $6 - 4 \,?\, \frac{24}{6}$ 20. $6 \cdot 2 \,?\, 4 + 6$
21. $\frac{3}{2} \,?\, \frac{3}{4}$ 22. $-\frac{3}{2} \,?\, -\frac{3}{4}$
23. $-10 \,?\, -2 \,?\, -1$ 24. $-2 \,?\, 0 \,?\, 2$

Referring to the number line below, replace each question mark in Problems 25–36 with either < or > to form a true statement.

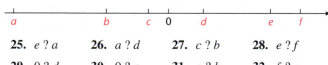

25. $e \,?\, a$ 26. $a \,?\, d$ 27. $c \,?\, b$ 28. $e \,?\, f$
29. $0 \,?\, d$ 30. $0 \,?\, a$ 31. $a \,?\, b$ 32. $f \,?\, a$
33. $d \,?\, c$ 34. $c \,?\, 0$ 35. $d \,?\, 0$ 36. $d \,?\, e$

B 37. If we add a positive real number to any real number, will the sum be greater than or less than the original number?

38. If we add a positive real number to any real number, will the sum be to the right or left of the original number on a number line?

Graph on a real number line for x a real number.

39. $x \le 3$ 40. $x \ge -2$
41. $-5 < x \le -1$ 42. $-4 \le x < 1$
43. $x > -4$ 44. $x > 1$
45. $-1 < x < 3$ 46. $-2 \le x \le 2$
47. $x \ge 0$ 48. $3 < x$
49. $-4 \le x \le -1$ 50. $-3 \le x < 0$
51. $-2 < x \le 5$ 52. $0 < x < 8$

Translate each statement into an algebraic equation or inequality statement using x as the only variable.

53. $x - 8$ is positive.   54. $2x - 8$ is negative.
55. $x + 4$ is not negative.   56. $4x + 1$ is not positive.
57. 80 is 3 more than twice a certain number.
58. 18 is 3 times a certain number.
59. x is greater than or equal to -3 and less than 4.
60. x is greater than 0 and less than 10.
61. 26 is 12 less than a certain number.
62. 32 is 5 less than a certain number.
63. x is less than 6 less than twice x.
64. x is greater than 9 more than 4 times x.
65. 6 times a number is 4 more than 3 times the number.
66. 7 times a number is 12 less than 4 times the number.
67. 6 less than a certain number is 5 times the number that is 7 more than the certain number.
68. 5 more than a certain number is 3 times the number that is 4 less than the certain number.
69. $\frac{9}{5}C + 32$ is between 63 and 72, inclusive.

70. $\frac{5}{9}(F - 32)$ is between 20 and 25, inclusive.

71. The sum of three consecutive natural numbers is 186.

72. The sum of three consecutive natural numbers is 372.

C *Replace each question mark with an appropriate expression or statement to make the statement an illustration of the given property of equality or inequality.*

73. Symmetric property If $-5 = t$, then ?.

74. Transitive property If $x < 3$ and $3 < y$, then ?.

75. Transitive property If $5x + 7x = (5 + 7)x$ and $(5 + 7)x = 12x$, then ?.

76. Trichotomy property If $x \neq 3$, then either $x < 3$ or ?.

77. Reflexive property $3 - x = ?$.

78. Substitution principle If $2x + 3y = 5$ and $y = x - 3$, then $2x + 3(?) = 5$.

In Problems 79–92, identify the property of equality or inequality that is illustrated by the statement.

79. If $x - 5 = 8$, then $8 = x - 5$.

80. If $x \neq 2$, then $x > 2$ or $x < 2$.

81. If $x + 11 = y - 3$ and $y - 3 = z + 14$, then $x + 11 = z + 14$.

82. If $x + 11 = y - 3$ and $y = 8$, then $x + 11 = 5$.

83. For any real number x, one of the following must be true: $x < 0$, $x = 0$, or $x > 0$.

84. Since $\pi < 3.2$, if $x > 3.2$, then $x > \pi$.

85. If $x > 10$ and $x = y - 7$, then $y - 7 > 10$.

86. If $a^2 + b^2 = c^2$, then $c^2 = a^2 + b^2$.

87. If $a^2 + b^2 = c^2$ and $c^2 = d^2 + e^2$, then $a^2 + b^2 = d^2 + e^2$.

88. Since $\frac{1}{3} = \frac{2}{6}$ and $\frac{1}{3} = 0.33333\ldots$, it must be that $\frac{2}{6} = 0.33333\ldots$.

89. If $\sqrt{a^2 + b^2} = 36$ and $a^2 + b^2 = c^2$, then $\sqrt{c^2} = 36$.

90. If $a^2 + b^2 = c^2$ and $c^2 = 25$, then $a^2 + b^2 = 25$.

91. If $\dfrac{x + 3}{12} = y$ and $x = 8$ then $\frac{11}{12} = y$.

92. For $x = 8$ and $y = 9$, $x + y = 8 + 9$.

93. What is wrong with the following argument? Four is an even number and 8 is an even number; hence we can write "4 = Even number" and "8 = Even number." By the symmetric property for equality, we can write "Even number = 8," and we can conclude, using the transitive property for equality

(since 4 = Even number and Even number = 8), that 4 = 8.

94. What is wrong with the following argument? Rod is a human and Jan is a human; hence we can write "Rod = Human" and "Jan = Human." By the symmetric property for equality, we can write "Human = Jan," and we can conclude, using the transitive property for equality (since Rod = Human and Human = Jan), that Rod = Jan.

APPLICATIONS

95. In a rectangle with area 90 square meters, the length is 3 meters less than twice the width. Write an equation relating the area to the length and width, using x as the only variable. [*Note:* Area = (Length)(Width).]

96. In a rectangle with area 500 square inches, the length is 5 inches longer than the width. Write an equation relating the area to the length and the width, using x as the only variable.

★ **97.** In a rectangle with perimeter 210 centimeters, the length is 10 centimeters less than 3 times the width. Write an equation relating the perimeter to the length and width, using x as the only variable.

★ **98.** In a rectangle with perimeter 186 feet, the width is 7 feet less than the length. Write an equation relating the perimeter to the length and the width, using x as the only variable.

★★ **99.** A rectangular lot is to be fenced with 400 feet of wire fencing.
(A) If the lot is x feet wide, write a formula for the area A of the lot in terms of x.
(B) What real number values can x assume? Write the answer in terms of a double inequality statement that also allows the area to be 0.

★★ **100.** A flat sheet of cardboard in the shape of an 18 by 12 inch rectangle is to be used to make an open-topped box by cutting an x by x inch square out of each corner and folding the remaining part appropriately.
(A) Write an equation for the volume of the box in terms of x.
(B) What real number values can x assume? Formulate the answer in terms of a double inequality statement that also allows the volume to be 0.
[*Note:* Volume = (Length)(Width)(Height).]

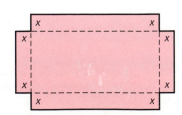

1-3 Real Number Properties

- ♦ Commutative Properties
- ♦ Associative Properties
- ♦ Order of Operations and the Distributive Properties
- ♦ Simplifying Algebraic Expressions
- ♦ Identity Properties

In the previous sections we discussed algebraic expressions and their use in formulating equations and inequality statements. We now review some of the basic properties of the set of real numbers that enable us to convert algebraic expressions into "equivalent forms." These properties become operational rules in the algebra of real numbers.

Basic Properties of the Set of Real Numbers

Let x, y, and z be arbitrary real numbers.

ADDITION PROPERTIES

Closure: $x + y$ is a unique real number; that is, the sum of two real numbers is also a real number.

Associative: $(x + y) + z = x + (y + z)$

Commutative: $x + y = y + x$

Identity: 0 is the additive identity; that is, $0 + x = x + 0 = x$ for all x, and 0 is the only real number with this property.

Inverse: For each x, $-x$ is its unique additive inverse; that is, $x + (-x) = (-x) + x = 0$, and $-x$ is the only real number relative to x with this property.

MULTIPLICATION PROPERTIES

Closure: xy is a unique real number; that is, the product of two real numbers is also a real number.

Associative: $(xy)z = x(yz)$

Commutative: $xy = yx$

Identity: 1 is the multiplicative identity; that is, for all x, $1x = x1 = x$, and 1 is the only real number with this property.

Inverse: For each $x \neq 0$, $1/x$ is its unique multiplicative inverse; that is, $x(1/x) = (1/x)x = 1$, and $1/x$ is the only real number relative to x with this property.

COMBINED PROPERTIES

Distributive: $\quad x(y + z) = xy + xz$

$\qquad\qquad\qquad (x + y)z = xz + yz$

You have been using many of these properties in arithmetic for a long time. The names of the properties may appear unnecessary, but it is useful to be able to refer to them. In this section we will informally consider several of these properties, and we will discuss others later as needed.

♦ COMMUTATIVE PROPERTIES

The commutative properties for addition and multiplication indicate that the order in which addition or multiplication is performed doesn't matter: $2 + 3 = 3 + 2$ and $2 \cdot 3 = 3 \cdot 2$.

Example 1
Using the Commutative Properties

Use the appropriate commutative property to replace each question mark with an appropriate symbol:

(A) $x + 9 = 9 + ?$ (B) $7y = ?7$

(C) $2x + 3y = ? + 2x$ (D) $3 + yx = 3 + ?y$

Solution

(A) $x + 9 = 9 + \mathbf{x}$ (B) $7y = \mathbf{y}7$

(C) $2x + 3y = \mathbf{3y} + 2x$ (D) $3 + yx = 3 + \mathbf{x}y$

Matched Problem 1

Use the appropriate commutative property to replace each question mark with an appropriate symbol:

(A) $x + 3 = 3 + ?$ (B) $3x + 5y = 5y + ?$

(C) $5(y + x) = 5(? + y)$ (D) $y + x4 = y + 4?$

Do the commutative properties hold relative to subtraction and division? That is, are the statements $x - y = y - x$ and $x \div y = y \div x$ true for all real numbers x and y, division by 0 excluded? The answer is no. For example, $5 - 3 \neq 3 - 5$ and $8 \div 4 \neq 4 \div 8$.

♦ ASSOCIATIVE PROPERTIES

When computing

$$4 + 3 + 5 \qquad \text{or} \qquad 4 \cdot 3 \cdot 5$$

we don't need parentheses to show us which two numbers are to be added or multiplied first. The reason for this is found in the associative properties. These properties tell us how parentheses can be moved relative to addition and multiplication, and allow us to write

$$(4 + 3) + 5 = 4 + (3 + 5) \qquad \text{and} \qquad (4 \cdot 3) \cdot 5 = 4 \cdot (3 \cdot 5)$$

So order of grouping doesn't matter relative to either operation.

Example 2 **Using the Associative** **Properties**	Use the appropriate associative property to replace each question mark with an appropriate expression: **(A)** $(x + 7) + 2 = x + (?)$ **(B)** $3(5y) = (?)y$ **(C)** $x + (x + 3) = (?) + 3$ **(D)** $(2y)y = 2(?)$

Solution

(A) $(x + 7) + 2 = x + (7 + 2) = x + 9$
(B) $3(5y) = (3 \cdot 5)y = 15y$
(C) $x + (x + 3) = (x + x) + 3 = 2x + 3$
(D) $(2y)y = 2(yy) = 2y^2$ Recall that $y^2 = y \cdot y$, $y^3 = y \cdot y \cdot y$, etc.

Matched Problem 2

Use the appropriate associative property to replace each question mark with an appropriate expression:

(A) $(x + 3) + 9 = x + (?)$ **(B)** $6(3y) = (?)y$
(C) $(5 + z) + z = 5 + (?)$ **(D)** $(3z)z = 3(?)$

The associative properties do not hold for subtraction and division. For example, $(8 - 4) - 2 \neq 8 - (4 - 2)$ and $(8 \div 4) \div 2 \neq 8 \div (4 \div 2)$. You should evaluate each side of each of these equations to see why. The associative properties of addition and multiplication are referred to as the *associativity* of these operations. The commutative properties are also referred to as *commutativity*. In summary:

Associativity and Commutativity

Relative to addition, commutativity and associativity permit us to change the order of addition at will and insert or remove parentheses as we please. The same is true for multiplication, but not for subtraction and division.

♦ **ORDER OF OPERATIONS AND THE DISTRIBUTIVE PROPERTIES**

The distributive properties allow us to convert a product such as $2(3 + 4)$ to a sum: $2 \cdot 3 + 2 \cdot 4$. To correctly interpret the expression $2 \cdot 3 + 2 \cdot 4$, we need to recall the standard convention that indicates the order of operations in the absence of parentheses in such an expression.

Order of Operations

Unless grouping symbols indicate otherwise, multiplication and division are to be done before addition and subtraction. Operations are done from left to right.

Thus, to correctly interpret the expression $2 \cdot 3 + 2 \cdot 4$, we should first perform both multiplications $2 \cdot 3$ and $2 \cdot 4$, and then add the results to obtain $6 + 8 = 14$.

In general, the distributive properties allow us to convert a product $x(y + z)$ to a sum $xy + xz$, and vice versa. The process of rewriting $x(y + z)$ as $xy + xz$ is usually called **multiplying out,** or just **multiplying.** The process of rewriting the sum $xy + xz$ as the product $x(y + z)$ is usually called **factoring out the common factor** or **taking out the common factor;** in this case, the common factor is x.

Example 3
Using the Distributive Properties

Multiply, using a distributive property:

(A) $3(x + y)$ **(B)** $5(2x^2 + 3)$

(C) $2x(x + 1)$ **(D)** $(3 + x)y$

Solution

(A) $3(x + y) = 3x + 3y$ **(B)** $5(2x^2 + 3) = 10x^2 + 15$

(C) $2x(x + 1) = 2x^2 + 2x$ **(D)** $(3 + x)y = 3y + xy$

Matched Problem 3

Multiply, using a distributive property:

(A) $5(m + n)$ **(B)** $4(3y^3 + 5)$

(C) $3y(2 + y)$ **(D)** $(a + 4)b$

Example 4
Taking Out Common Factors

Take out factors common to both terms:

(A) $6x + 6y$ **(B)** $7x^2 + 14$ **(C)** $ax + ay$

Solution

(A) $6x + 6y = 6x + 6y = 6(x + y)$

(B) $7x^2 + 14 = 7x^2 + 7 \cdot 2 = 7(x^2 + 2)$

(C) $ax + ay = ax + ay = a(x + y)$

Matched Problem 4

Take out factors common to both terms:

(A) $9m + 9n$ **(B)** $6y^3 + 12$ **(C)** $du + dy$

Other useful distributive forms follow from the basic versions given above and from other properties discussed in this chapter:

Additional Forms of the Distributive Property

For real numbers a, b, and c, . . . :

$$a(b - c) = ab - ac = (b - c)a$$

$$a(b + c + d + \cdots + n) = ab + ac + ad + \cdots + an$$

$$= (b + c + d + \cdots + n)a$$

♦ SIMPLIFYING ALGEBRAIC EXPRESSIONS

The commutative, associative, and distributive properties provide us with tools for transforming algebraic expressions into equivalent forms. Two algebraic expressions are **equivalent** over given replacement sets for the variables if both yield the same number for all allowable values of the variables. For example, we can use the associative and commutative properties to formally transform

$$(x + 8) + (y + 2)$$

into a simpler equivalent form. By *formally transforming,* we mean that each step is justified by a basic property or an earlier stated property.

$$
\begin{aligned}
(x + 8) + (y + 2) &= [(x + 8) + y] + 2 && \text{Associative property for } + \\
&= [x + (8 + y)] + 2 && \text{Associative property for } + \\
&= [x + (y + 8)] + 2 && \text{Commutative property for } + \\
&= x + [(y + 8) + 2] && \text{Associative property for } + \\
&= x + [y + (8 + 2)] && \text{Associative property for } + \\
&= x + (y + 10) && \text{Substitution property for } = \\
&= (x + y) + 10 && \text{Associative property for } +
\end{aligned}
$$

Normally, we do most of these steps mentally and simply write

$$
\begin{aligned}
(x + 8) + (y + 2) &= x + 8 + y + 2 \\
&= x + y + 10
\end{aligned}
$$

Even though we did not write each step as in the formal treatment, you should not lose sight of the fact that the associative, commutative, and substitution properties are behind the steps taken in the simpler version.

Example 5
Simplifying

Simplify, using commutative and associative properties:

(A) $(3x + 5) + (2y + 7)$ **(B)** $(6x)(3y)$

Solution   **(A)**
$$
\begin{aligned}
(3x + 5) + (2y + 7) &= 3x + 5 + 2y + 7 \\
&= 3x + 2y + 12
\end{aligned}
$$

(B)
$$
\begin{aligned}
(6x)(3y) &\ \vdots\, = 6x3y \\
&\ \vdots\, = 6 \cdot 3xy \\
&= 18xy
\end{aligned}
$$

The **dashed box** is used here to indicate steps that are usually done mentally. This notation will be used throughout the text.

Matched Problem 5   Simplify, using commutative and associative properties:

(A) $(2a + 3) + (3b + 11)$ **(B)** $(4m)(6n)$

◆ **IDENTITY PROPERTIES**

The identity property for addition states that 0 is the unique real number having the property that when it is added to any real number we get that number back again. Thus,

$$5 + 0 = 5 \qquad 0 + 3x = 3x \qquad (x + y) + 0 = x + y$$

Similarly, the number 1 plays the same role relative to multiplication. That is, when any number is multiplied by 1, we get that number back again. Thus, we can write

$$1 \cdot 8 = 8 \qquad 1x = x \qquad 1xy = xy$$

The closure properties are discussed in Problems 87–94 in Exercise 1-3. The inverse properties will be discussed in Sections 1-4 and 1-5.

Answers to Matched Problems

1. **(A)** x **(B)** $3x$ **(C)** x **(D)** x
2. **(A)** $3 + 9$, or 12 **(B)** $6 \cdot 3$, or 18 **(C)** $z + z$, or $2z$ **(D)** zz, or z^2
3. **(A)** $5m + 5n$ **(B)** $12y^3 + 20$ **(C)** $6y + 3y^2$ **(D)** $ab + 4b$
4. **(A)** $9(m + n)$ **(B)** $6(y^3 + 2)$ **(C)** $d(u + y)$
5. **(A)** $2a + 3b + 14$ **(B)** $24mn$

EXERCISE 1-3

A *All variables represent real numbers.*

Calculate each of the following using the agreed upon order of operations.

1. $3 + 4 \cdot 5$
2. $2 \cdot 3 + 4$
3. $5 \cdot 6 - 3 \cdot 4$
4. $3 \cdot 5 + 4 \cdot 2$
5. $4 \cdot 2 + 3(8 - 3 \cdot 2)$
6. $5 \cdot 2 - 3(9 - 2 \cdot 4)$

Replace each question mark with an appropriate expression that will illustrate the use of the indicated real-number property.

7. **Commutative property (+)** $x + 3 = ?$
8. **Commutative property (+)** $m + n = ?$
9. **Associative property (·)** $5(7z) = ?$
10. **Associative property (·)** $(uv)w = ?$
11. **Commutative property (·)** $nm = ?$
12. **Commutative property (·)** $dc = ?$
13. **Associative property (+)** $9 + (11 + M) = ?$
14. **Associative property (+)** $(x + 7) + 5 = ?$
15. **Distributive property** $3(x + 1) = ?$
16. **Distributive property** $x(x + 1) = ?$

17. **Distributive property** $(2 + x)x = ?$
18. **Distributive property** $(4 + x)3 = ?$
19. **Identity property (+)** $7x + 0 = ?$
20. **Identity property (+)** $0 + (x + z) = ?$
21. **Identity property (·)** $1(x + y) = ?$
22. **Identity property (·)** $1(uv) = ?$

State the justifying property for each statement.

23. $12 + w = w + 12$
24. $2x + 3 = 3 + 2x$
25. $m + (n + 3) = (m + n) + 3$
26. $(3x + y) + 5 = 3x + (y + 5)$
27. $20x = x20$
28. $MN = NM$
29. $2(x + 5) = 2x + 10$
30. $x(y + 1) = xy + x \cdot 1$
31. $4(8y) = (4 \cdot 8)y$
32. $(12u)v = 12(uv)$
33. $3x + 0 = 3x$
34. $0 + (2x + 3) = 2x + 3$
35. $1m = m$
36. $uv = 1uv$
37. $4 + 2x = 2(2 + x)$
38. $3x + 9 = 3(x + 3)$

Remove parentheses and simplify:

39. $(x + 7) + 2$
40. $3 + (5 + m)$
41. $4(5y)$
42. $6(8n)$
43. $12 + (u + 3)$
44. $(4 + x) + 13$

45. $(3x)7$ **46.** $4(y3)$

47. $0 + 1x$ **48.** $(1y + 3) + 0$

B *State the justifying property for each statement.*

49. $2 + (y + 3) = 2 + (3 + y)$

50. $(3m)n = n(3m)$

51. $7(y4) = 7(4y)$

52. $5 + (y + 2) = (y + 2) + 5$

53. $3x + 2y = 2y + 3x$

54. $3x + (2x + 5y) = (3x + 2x) + 5y$

55. $(2x)(x + 3) = 2[x(x + 3)]$

56. $(x + 3) + (2 + y) = x + [3 + (2 + y)]$

Remove parentheses and simplify:

57. $(x + 7) + (y + 4) + (z + 1)$

58. $(7 + m) + (8 + n) + (3 + p)$

59. $(3x + 5) + (4y + 6)$

60. $(3a + 7) + (5b + 2)$

61. $0 + (1x + 3) + (y + 2)$

62. $1(x + 3) + 0 + (y + 2)$

63. $(12m)(3n)(1p)$

64. $(8x)(4y)(2z)$

Use the distributive property to rewrite each sum as a product and each product as a sum.

65. $(x + 5)x$ **66.** $(x + 2) \cdot 3$ **67.** $2x + 18$

68. $7x + 21$ **69.** $(x + 2)8$ **70.** $8x + 2$

71. $6x + 12$ **72.** $6(x + 12)$ **73.** $ax + ay$

74. $x + xy$ **75.** $y + xy$ **76.** $ba + ca$

77. $a(a + b)$ **78.** $(x + y)y$ **79.** $(x + 1)y$

80. $a(b + 1)$

C *Indicate true (T) or false (F). For each false statement, find values for the variables that will illustrate its falseness.*

81. (A) $a + b = b + a$ (B) $a - b = b - a$
    (C) $ab = ba$ (D) $a \div b = b \div a$

82. (A) $(a + b) + c = a + (b + c)$
    (B) $(a - b) - c = a - (b - c)$
    (C) $a(bc) = (ab)c$
    (D) $(a \div b) \div c = a \div (b \div c)$

83. $a \div (b + c) = a \div b + a \div c$

84. $a \div (b \times c) = (a \div b) \times (a \div c)$

85. $a \times (b \div c) = (a \times b) \div (a \times c)$

86. $a + (b \div c) = (a + b) \div (a + c)$

The closure properties listed for the set of real numbers indicate that this set is "closed" under addition and multiplication in the sense that the sum and product of two real numbers are also real numbers. On the other hand, the positive integers are not closed under subtraction, since the difference of two positive integers may not be positive. In Problems 87–94, determine whether the given set is closed under the given operation.

87. $Z; -$ **88.** $Z; \div$ **89.** $N; +$ **90.** $N; -$

91. The positive rational numbers; $\times$

92. The positive rational numbers; $-$

93. The negative integers; $\times$

94. The negative integers; $+$

95. Supply a reason for each step.

STATEMENT

1. $3x + (4y + 5x) = 3x + (5x + 4y)$

2. $= (3x + 5x) + 4y$

3. $= (3 + 5)x + 4y$

4. $= 8x + 4y$

96. Supply a reason for each step.

STATEMENT

1. $3(4y + 5x) = 3(4y) + 3(5x)$

2. $= (3 \cdot 4)y + (3 \cdot 5)x$

3. $= 12y + 15x$

4. $= 15x + 12y$

97. Supply a reason for each step.

STATEMENT

1. $(x + 3) + (y + 4) = (x + 3) + (4 + y)$

2. $= x + [3 + (4 + y)]$

3. $= x + [(3 + 4) + y]$

4. $= x + (7 + y)$

5. $= x + (y + 7)$

6. $= (x + y) + 7$

98. Supply a reason for each step.

STATEMENT

1. $(5x)(2y) = (x5)(2y)$

2. $\qquad = x[5(2y)]$

3. $\qquad = x[(5 \cdot 2)y]$

4. $\qquad = x(10y)$

5. $\qquad = (x10)y$

6. $\qquad = (10x)y$

7. $\qquad = 10(xy)$

1-4 Addition and Subtraction of Real Numbers

- ♦ The Negative of a Number
- ♦ The Absolute Value of a Number
- ♦ Addition of Real Numbers
- ♦ Subtraction of Real Numbers
- ♦ Combined Operations

In this section, we review addition and subtraction of real numbers. Before doing so, we need to recall the operations of finding "the negative of" and "the absolute value of" a real number. These operations are useful in describing operations related to addition, subtraction, multiplication, and division of real numbers.

♦ ## THE NEGATIVE OF A NUMBER

For each real number x, we denote its additive inverse by

$$-x \begin{cases} \text{Additive inverse of } x \\ \text{Opposite of } x \\ \text{Negative of } x \end{cases}$$

All the names on the right describe the same thing and are used interchangeably. Recall that the **opposite of,** or **negative of, a number x** is obtained from x by changing its sign. Thus, the opposite of -5 is $+5$ and the opposite of $+5$ is -5. The opposite of, or negative of, 0 is 0.

Example 1
Evaluating Opposites

Find:

(A) $-(+5)$ **(B)** $-(-8)$ **(C)** $-(0)$ **(D)** $-[-(-4)]$

Solution

(A) $-(+5) = -5$ **(B)** $-(-8) = +8$ or 8
(C) $-(0) = 0$ **(D)** $-[-(-4)] = -(+4) = -4$

Examples 1(A) and (B) illustrate the fact that

$-x$ is not necessarily a negative number.

In particular, $-x$ represents a negative number if x is positive and a positive number if x is negative. Thus, for example, $-(+5) = -5$ and $-(-5) = +5$. More generally, it can be shown that:

Double Negative Property

For a any real number:

$$-(-a) = a$$

Matched Problem 1 Find:

(A) $-(+11)$ **(B)** $-(-12)$ **(C)** $-(0)$ **(D)** $-[-(+6)]$

It is important to note the three distinct uses of the minus sign.

Multiple Uses of the Minus Sign

1. As the operation "subtract": $9 \overset{\downarrow}{-} 3 = 6$

2. As the operation "the negative or opposite of": $\overset{\downarrow}{-}(-8) = 8$

3. As part of a number symbol: $\overset{\downarrow}{-}4$

♦ **THE ABSOLUTE VALUE OF A NUMBER**

The **absolute value** of a number x, denoted by the symbol

$$|x|$$

gives the magnitude of x without its sign.

Note that the absolute-value sign involves vertical bars, not square brackets. That is, you must distinguish between $|x|$ and $[x]$.

The absolute value of a number also can be thought of geometrically as the distance of the number from 0 on the real number line, expressed as a positive number or 0. For example, both 5 and -5 are 5 units from 0. See Figure 1.

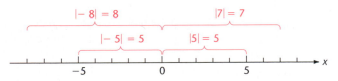

Figure 1 Absolute value

Thus, we can write $|5| = 5$ and $|-5| = 5$. Figure 1 also illustrates the fact that $|-8| = 8$ and $|7| = 7$.

Symbolically, and more formally, we define absolute value as follows:

Absolute Value

$$|x| = \begin{cases} x & \text{if } x \text{ is positive} \\ 0 & \text{if } x \text{ is } 0 \\ -x & \text{if } x \text{ is negative} \end{cases}$$

The first two parts of the definition can be combined to obtain:

$$x = \begin{cases} x & \text{if } x \geq 0 \\ -x & \text{if } x < 0 \end{cases}$$

Note that $-x$ is positive if x is negative. It is important to remember that:

The absolute value of a number is never negative.

Example 2
Evaluating Absolute Values

Evaluate:

(A) $|24|$　　(B) $|-7|$　　(C) $|0|$　　(D) $-(|-8| + |-3|)$

Solution

(A) $|24| = 24$　　(B) $|-7| = 7$　　(C) $|0| = 0$
(D) $-(|-8| + |-3| = -(8 + 3) = -(11) = -11$

Here we have used $-(11)$ to represent the opposite, or negative, of the positive number 11 and used -11 to represent a negative number, but they represent the same number.

Matched Problem 2

Evaluate:

(A) $|-13|$　　(B) $|43|$　　(C) $-|-4|$　　(D) $-(|-6| - |+2|)$

◆ **ADDITION OF REAL NUMBERS**

We now review addition of real numbers. We will assume that you can add positive numbers and focus on what is necessary to include negative numbers in addition.

Addition of Real Numbers: $a + b$

1. If either number is 0, use the identity property for addition:

$$0 + b = b \qquad 0 + 8 = 8$$
$$a + 0 = a \qquad (-13) + 0 = -13$$

2. If a and b are both positive, add as you have learned in arithmetic.
3. If a and b are both negative, take the opposite of the sum of the absolute values.

$(-3) + (-8) = -(3 + 8) = -11$

4. If a and b have opposite signs, subtract the smaller absolute value from the larger; then attach the sign of the original number with the larger absolute value.

$3 + (-8) = -(8 - 3) = -5$

Case 3 can be rephrased as follows: mentally block out the signs of the two numbers, add as in arithmetic, and then attach a minus sign to the result. Mentally blocking out the signs of the numbers amounts to taking their absolute values. For example,

$$(-5) + (-3) = -(|-5| + |3|) = -(5 + 3) = -8$$

Similarly, case 4 can be restated: mentally block out the signs of the two numbers, subtract the smaller from the larger, and then attach the sign of the number with the larger absolute value. For example,

$$(-3) + (+9) = +(9 - 3) = 6$$

$$(+3) + (-9) = -(9 - 3) = -6$$

To add three or more numbers, add all the positive numbers together, add all the negative numbers together, and then add the two resulting sums as above. The commutative and associative properties justify this procedure.

Example 3
Adding Real Numbers

Add: $3 + (-6) + 8 + (-4) + (-5)$

Solution

$$3 + (-6) + 8 + (-4) + (-5) = (3 + 8) + [(-6) + (-4) + (-5)]$$
$$= 11 + (-15) = -(15 - 11) = -4$$

Matched Problem 3   Add: $6 + (-8) + (-4) + 10 + (-3) + 1$

♦ **SUBTRACTION OF REAL NUMBERS**

We define subtraction in terms of addition. We can perform numerical computations as always, but this definition will make algebraic manipulation easier.

> **Subtraction of Real Numbers: $a - b$**
>
> For a and b any real numbers:
>
> $$a - b = a + (-b)$$
>
> That is, to subtract b from a, add the opposite of b to a.

Recall that the opposite of b is the same as the negative of b and the additive inverse of b:

Opposite of -9

$$(-3) - (-9) = (-3) + 9$$

Change to addition

For integers, you should get to the point where you can perform this type of subtraction mentally, and simply write down the answer.

Example 4
Subtracting
Real Numbers

Subtract:

(A) $8 - (-5)$ **(B)** $(-8) - 5$
(C) $(-8) - (-5)$ **(D)** $0 - 5$

Solution

(A) $8 - (-5) \boxed{= 8 + 5} = 13$ Change subtraction to addition and replace -5 by its opposite.

(B) $(-8) - 5 \boxed{= (-8) + (-5)} = -13$

(C) $(-8) - (-5) \boxed{= (-8) + 5} = -3$

(D) $0 - 5 \boxed{= 0 + (-5)} = -5$

Matched Problem 4

Subtract:

(A) $4 - 7$ **(B)** $7 - (-4)$ **(C)** $(-7) - 4$
(D) $(-7) - (-4)$ **(E)** $(-4) - (-7)$ **(F)** $0 - (-4)$

You may be more used to thinking of the difference $a - b$ as representing what must be added to b to get a. That is,

$$b + (a - b) = a$$

This result is proved in Problem 97 in Exercise 1-4. The result can be rephrased as

$$a - b = c \qquad \text{means} \qquad b + c = a$$

This last form provides a convenient way to check subtraction. For example, to check that

$$37 - (-15) = 52$$

we check that this addition holds:

$$(-15) + 52 = 37$$

♦ **COMBINED OPERATIONS**

When three or more terms are combined in addition and subtraction and symbols of grouping are omitted, we mentally convert any subtraction to addition and add. Thus,

$$8 - 5 + 3 \boxed{= 8 + (-5) + 3} = 6$$
$$\text{Think.}$$

Example 5
Adding and Subtracting
Several Terms

Evaluate:

(A) $2 - 3 - 7 + 4$ **(B)** $-4 - 8 + 2 + 9$

Solution **(A)** $2 - 3 - 7 + 4 \boxed{= 2 + (-3) + (-7) + 4} = -4$

Think.

(B) $-4 - 8 + 2 + 9 \boxed{= (-4) + (-8) + 2 + 9} = -1$

Think.

Matched Problem 5 Evaluate: **(A)** $5 - 8 + 2 - 6$ **(B)** $-6 + 12 - 2 - 1$

Recall that to evaluate an algebraic expression for particular values of the variables means to replace each variable by a given number and calculate the resulting arithmetic value.

Example 6
Evaluating Sums and Differences

Evaluate for $x = 2$, $y = -3$, and $z = -9$:

(A) $x + y$ **(B)** $y - z$ **(C)** $y - (z - x)$ **(D)** $\left| (-y) - |z| \right|$

Solution **(A)** $x + y$

$\boxed{(\ ) + (\ )}$ Use of parentheses as indicated prevents many sign errors.

$(2) + (-3) = -1$

(B) $y - z$

$\boxed{(\ ) - (\ )}$

$(-3) - (-9)$
$= (-3) + 9 = 6$

(C) $y - (z - x)$ Replace parentheses with brackets, substitute values, and then evaluate starting inside the square brackets.

$(-3) - [(-9) - 2]$ The use of brackets as an alternate form of parentheses makes this line easier to read.

$\boxed{= (-3) - (-11)}$

$= (-3) + (11) = 8$

(D) $\left| (-y) - |z| \right|$
$\left| [-(-3)] - |-9| \right|$ Evaluate $-(-3)$ and $|-9|$ first.
$= |(3) - (9)|$ Subtract inside absolute value signs.
$= |-6|$ Take the absolute value.
$= 6$

Matched Problem 6 Evaluate for $x = -4$, $y = 5$, and $z = -11$:

(A) $y + z$ **(B)** $x - y$ **(C)** $(z - x) + y$ **(D)** $\left| (x) - |z| \right|$

The above examples have involved only integers in order to focus your attention on the sign properties, not the arithmetic. The same will also be true of most of the problems in Exercise 1-4. However, the definitions of addition and subtraction, and the sign properties, are true for *all* real numbers. The next two examples involve noninteger rational numbers, written as fractions in the next example, and written as decimals in the following one.

Example 7
Calculating Sums and Differences

Evaluate each expression for $x = \frac{1}{3}$ and $y = -\frac{1}{4}$:

(A) $x + y$ **(B)** $x - y$ **(C)** $-x - y$ **(D)** $-x + y$

Solution   **(A)** $x + y = \frac{1}{3} + (-\frac{1}{4}) = \frac{1}{3} - \frac{1}{4} = \frac{4}{12} - \frac{3}{12} = \frac{1}{12}$
(B) $x - y = \frac{1}{3} - (-\frac{1}{4}) = \frac{1}{3} + \frac{1}{4} = \frac{4}{12} + \frac{3}{12} = \frac{7}{12}$
(C) $-x - y = (-\frac{1}{3}) - (-\frac{1}{4}) = (-\frac{1}{3}) + \frac{1}{4} = -(\frac{1}{3} - \frac{1}{4}) = -\frac{1}{12}$
(D) $-x + y = (-\frac{1}{3}) + (-\frac{1}{4}) = -(\frac{1}{3} + \frac{1}{4}) = -\frac{7}{12}$

Matched Problem 7   Evaluate each expression for $x = -\frac{2}{3}$ and $y = \frac{3}{4}$:

(A) $x + y$ **(B)** $x - y$ **(C)** $-x - y$ **(D)** $-x + y$

Example 8
Calculating Sums and Differences

Using a calculator, evaluate each expression for $x = -504.394$, $y = 829.077$, and $z = -1{,}023.998$:

(A) $y - x$ **(B)** $x - (y - z)$

Solution   **(A)** This problem can be entered directly to obtain

$$829.077 - (-504.394) = 1{,}333.471$$

(B) Evaluate

$$y - z = 829.077 - (-1{,}023.998)$$
$$= 1{,}853.075$$

Then evaluate

$$x - (y - z) = x - 1{,}853.075$$

to obtain $-2{,}357.469$. Rearranging the expression $x - (y - z)$ algebraically, as will be done in Section 2-1, or making use of available memory in your calculator can simplify the solution.

Matched Problem 8   Use the values of x, y, and z in Example 8 to evaluate:

(A) $x - y$ **(B)** $x - (z - y)$

Answers to Matched Problems
1. **(A)** -11 **(B)** 12   **(C)** 0   **(D)** 6
2. **(A)** 13   **(B)** 43   **(C)** -4 **(D)** -4
3. 2
4. **(A)** -3 **(B)** 11   **(C)** -11 **(D)** -3 **(E)** 3   **(F)** 4
5. **(A)** -7 **(B)** 3
6. **(A)** -6 **(B)** -9 **(C)** -2 **(D)** 15
7. **(A)** $\frac{1}{12}$ **(B)** $-\frac{17}{12}$ **(C)** $-\frac{1}{12}$ **(D)** $\frac{17}{12}$
8. **(A)** $-1{,}333.471$ **(B)** $1{,}348.681$

EXERCISE 1-4

A *In Problems 1–18, evaluate the expression.*

1. $-(+7)$
2. $-(+12)$
3. $-(-6)$
4. $-(-8)$
5. $|+2|$
6. $|+9|$
7. $|-27|$
8. $|-32|$
9. $|0|$
10. $-(0)$
11. $(-7) + (-3)$
12. $(-7) + (+3)$
13. $(+7) + (-3)$
14. $(-12) + (+8)$
15. $(+3) - (+9)$
16. $(+3) - (-9)$
17. $(+9) - (-3)$
18. $(-9) - (-3)$

19. The negative of a number is (*always, sometimes, never*) a negative number.

20. The absolute value of a number is (*always, sometimes, never*) a positive number.

B *Evaluate.*

21. $-[-(-3)]$
22. $-[-(+6)]$
23. $-|-(+2)|$
24. $-|-(-3)|$
25. $-(|-9| - |-3|)$
26. $-(|-14| - |-8|)$
27. $(-2) + (-6) + 3$
28. $(-2) + (-8) + 5$
29. $5 - 7 - 3$
30. $3 - 2 + 4$
31. $-7 + 6 - 4$
32. $-4 + 7 - 6$
33. $-2 - 3 + 6 - 2$
34. $-4 + 7 - 3 - 2$
35. $6 - [3 - (-9)]$
36. $(-10) - [(-6) + 3]$
37. $[6 - (-8)] - [(-8) - 6]$
38. $[3 - 5] + [(-5) - (-2)]$
39. $\{4 - [5 - (6 - 7)]\}$
40. $\{1 - [3 - (5 - 7)]\}$
41. $|2 - |(3 - |5 - 8|)||$
42. $|1 - |(2 - |4 - 7|)||$
43. $|1 - [3 - (5 - 7)]| + |2 - [4 - (6 - 8)]|$
44. $|1 - (3 - |5 - 7|)| - |2 - (4 - |6 - 8|)|$

Replace each question mark with an appropriate real number.

45. $-(?) = 5$
46. $-(?) = -8$
47. $|?| = 7$
48. $|?| = -4$
49. $(-3) + ? = -8$
50. $? + 5 = -6$
51. $(-3) - ? = -8$
52. $? - (-2) = -4$

Evaluate for $x = 23.417$, $y = -52.608$, and $z = -13.012$.

53. $y + x$
54. $x + z$
55. $x - y$
56. $z - y$
57. $(x - z) + y$
58. $(y + z) - z$
59. $|y - z|$
60. $|x + y + z|$
61. $|z - x - y|$
62. $x - |y|$
63. $|z| - x$
64. $|y + z|$

Evaluate for $x = 3$, $y = -8$, and $z = -2$.

65. $x + y$
66. $y + z$
67. $y - x$
68. $y - z$
69. $(x - z) + y$
70. $y - (z - x)$
71. $|(-z) - |y||$
72. $||-y| - 12|$
73. $-||y| - |x||$
74. $-||-10| - |x||$

Evaluate for $x = \frac{1}{3}$, $y = -\frac{1}{2}$, and $z = -\frac{1}{4}$.

75. $x + y$
76. $y + z$
77. $y - x$
78. $y - z$
79. $(x - z) + y$
80. $y - (z - x)$
81. $|(-z) - |y||$
82. $||-y| - 12|$
83. $-||y| - |x||$
84. $-||-10| - |x||$

C *In Problems 85–96, determine which of the statements hold for all integers a, b, and c. Illustrate each false statement with an example showing that it is false.*

85. $a + b = b + a$
86. $a + (-a) = 0$
87. $a - b = b - a$
88. $a - b = a + (-b)$
89. $(a + b) + c = a + (b + c)$
90. $(a - b) - c = a - (b - c)$
91. $|a + b| = |a| + |b|$
92. $|a - b| = |a| - |b|$
93. $|a| = |-a|$
94. $a + (b - c) = (a + b) - c$
95. $a - (b + c) = (a - b) + c$
96. $|a - b| = |b - a|$

97. Supply the reason for each step.

STATEMENT

1. $b + [a + (-b)] = b + [(-b) + a]$
2. $\qquad = [b + (-b)] + a$
3. $\qquad = 0 + a$
4. $\qquad = a$

98. Supply the reason for each step.

STATEMENT

1. $(a + b) + [(-a) + (-b)] = (b + a) + [(-a) + (-b)]$

2. $\qquad = [(b + a) + (-a)] + (-b)$

3. $\qquad = \{b + [a + (-a)]\} + (-b)$

4. $\qquad = (b + 0) + (-b)$

5. $\qquad = b + (-b)$

6. $\qquad = 0$

7. Therefore, $-(a + b) = (-a) + (-b)$.

99. Supply the reason for each step. You may want to use the result of Problem 98.

STATEMENT

1. $a - (b + c) = a + [-(b + c)])$

2. $\qquad = a + [(-b) + (-c)]$

3. $\qquad = [a + (-b)] + (-c)$

4. $\qquad = (a - b) - c$

100. Supply the reason for each step.

STATEMENT

1. $(a - b) + (b - a) = [a + (-b)] + [b + (-a)]$

2. $\qquad = \{[a + (-b)] + b\} + (-a)$

3. $\qquad = \{a + [(-b) + b]\} + (-a)$

4. $\qquad = (a + 0) + (-a)$

5. $\qquad = a + (-a)$

6. $\qquad = 0$

7. Therefore, $-(a - b) = (b - a)$

APPLICATIONS

101. You own a stock that is traded on a stock exchange. On Monday it closed at $23.50 per share; it fell $3.25 on Tuesday and another $6.75 on Wednesday; it rose $2.50 on Thursday; and it finished strongly on Friday by rising $7.75. Use addition of signed numbers to determine the closing price of the stock on Friday.

102. Find, using subtraction of signed numbers, the difference in the height between the highest point in the United States, Mount McKinley (20,270 feet), and the lowest point in the United States, Death Valley (−280 feet).

103. In golf, scores are usually compared to a standard called par. A score 2 less than par on a hole is called an eagle, and 1 less than par is called a birdie. A score of 1 more than par is called a bogie, 2 more than par a double bogie, and so on. If a golfer plays the first six holes with scores of par, bogie, bogie, birdie, par, par, what is the score at this point relative to par?

104. In one year a taxpayer reported the following sales of stock: four stocks sold for profits of $2,145, $1,800, $640, and $4,756; six stocks sold for losses of $928, $392, $1,240, $64, $4,786, and $2,392. What total net loss or gain does the taxpayer have from these sales?

1-5 Multiplication and Division of Real Numbers

- ♦ Multiplication of Real Numbers
- ♦ Sign Properties for Multiplication
- ♦ Division of Real Numbers
- ♦ Combined Operations

Having discussed addition and subtraction, we now turn to multiplication and division. We will then be in a position to consider problems involving all four arithmetic operations ($+, -, \cdot, \div$).

♦ MULTIPLICATION OF REAL NUMBERS

Multiplication of real numbers is carried out as follows, based upon the multiplication of positive numbers known from arithmetic:

Multiplication of Real Numbers: *ab*

1. If either number is 0, the product is 0:

$$0 \cdot b = 0 \qquad 0 \cdot 7 = 0 \qquad a \cdot 0 = 0 \qquad -3 \cdot 0 = 0$$

2. If a and b are both positive, multiply as you have learned in arithmetic.

3. If a and b are both negative, take the product of the absolute values.
$$(-3)(-7) = 3 \cdot 7 = 21$$

4. If a and b have opposite signs, take the opposite of the product of the absolute values. $\qquad (-3)7 = -(3 \cdot 7) = -21$

Case 1 is a result that can be proved on the basis of the material in the preceding sections (see Problem 99 in Exercise 1-5). To see why numbers with opposite signs are multiplied as indicated in case 4, consider the product $(+2)(-7)$. We start with something we know is true and then proceed through a sequence of logical steps to a conclusion that also must be true:

$$(+2)(-7) = (1 + 1)(-7) \qquad \text{The sum } 1 + 1 = 2 \text{ by addition. Next, use the distributive property.}$$

$$= 1(-7) + 1(-7) \qquad \text{Use the multiplicative identity property.}$$

$$= (-7) + (-7) \qquad \text{Add.}$$

$$= -14$$

Similarly, we can show that $(-2)(-7)$ must be $+14$ as indicated in case 3:

$$(-2)(-7) = (0 - 2)(-7) \qquad \text{The difference } 0 - 2 = -2 \text{ by subtraction. Next, use the distributive property.}$$

$$= 0(-7) - 2(-7) \qquad \text{Use case 1 and the product } 2(-7) = -14 \text{ just established above.}$$

$$= 0 - (-14) \qquad \text{Subtract.}$$

$$= 14$$

From cases 2, 3, and 4, we see that:

The product of two numbers with unlike signs is negative.
The product of two numbers with like signs is positive.

Example 1
Multiplying

Evaluate:

(A) $3(-5)$ **(B)** $(-3)(-5)$ **(C)** $0(-5)$

Solution   **(A)** $3(-5) = -(3 \cdot 5) = -15$ **(B)** $(-3)(-5) = 3 \cdot 5 = 15$

(C) $0(-5) = 0$

Matched Problem 1   Evaluate: **(A)** $4(-3)$ **(B)** $(-4)3$ **(C)** $(-4)(-3)$ **(D)** $0(-3)$

♦ **SIGN PROPERTIES FOR MULTIPLICATION**

Several important sign properties for multiplication are summarized in the following result:

Sign Properties for Multiplication

For any real numbers a and b:

(A) $(-1)a = -a$ **(B)** $(-a)b = -(ab)$ **(C)** $(-a)(-b) = ab$

Example 2
Applying Sign Properties for Multiplication

Evaluate $(-a)b$ and $-(ab)$ for $a = -5$ and $b = 4$.

Solution $(-a)b = [-(-5)]4 = 5 \cdot 4 = 20$
$-(ab) = -[(-5)4] = -(-20) = 20$

Matched Problem 2 Evaluate $(-a)(-b)$ and ab for $a = -5$ and $b = 4$.

Expressions of the form $-ab$ occur frequently, and at first glance may seem confusing. If you were asked to evaluate $-ab$ for $a = -3$ and $b = +2$, how would you proceed? Would you take the negative of a and then multiply it by b, or multiply a and b first and then take the negative of the product? Actually, it doesn't matter! Because of the sign properties, we get the same result either way, since $(-a)b = -(ab)$. Other expressions also are equal to $-ab$:

$$-ab = \begin{cases} (-a)b \\ a(-b) \\ -(ab) \\ (-1)ab \end{cases}$$

We are at liberty to replace any one of these five forms with another from the same group. Note that $-ab$ does not mean that the signs of both a and b are changed: $-ab \neq (-a)(-b)$.

Example 3
Evaluating $-ab$

Evaluate $-ab$ for $a = -7$ and $b = 4$.

Solution Proceed in one of the following ways:

$$-(-7)(4) = -[(-7)(4)] = -(-28) = 28$$

or

$$-(-7)(4) = [-(-7)](4) = (7)(4) = 28$$

Both are correct.

Matched Problem 3   Evaluate $-ab$ for $a = 6$ and $b = -3$ in two different ways.

♦   ## DIVISION OF REAL NUMBERS

Just as subtraction is defined in terms of addition, we define division in terms of multiplication. Recall that every nonzero real number b has a unique multiplicative inverse, denoted by $1/b$. For example:

NUMBER	MULTIPLICATIVE INVERSE
3	$\frac{1}{3}$
$-\frac{1}{2}$	-2
$\frac{3}{5}$	$\frac{5}{3}$
3.159	$0.316\ 558\ 71$ (approximately)

The last inverse is an approximation obtained from a calculator. The inverse property for multiplication states that

$$b \cdot \frac{1}{b} = \frac{1}{b} \cdot b = 1$$

We define division so that $a \div b$ means a times the multiplicative inverse of b, just as $a - b$ means a plus the additive inverse of b.

Definition of Division

For real numbers a and b, with $b \neq 0$,

$$a \div b = a \cdot \frac{1}{b}$$

Notice that division by 0 is not defined since 0 doesn't have a multiplicative inverse. If $\frac{1}{0}$ did exist, then we would have both

$$0 \cdot \frac{1}{0} = 0 \qquad \text{By the multiplication rules}$$

and also

$$0 \cdot \frac{1}{0} = 1 \qquad \text{By the definition of multiplicative inverse}$$

This is impossible. Therefore:

Zero cannot be used as a divisor—ever!
Division by 0 is undefined.

The division symbol $\div$, as well as the alternative $\overline{)}$, from arithmetic are not used very often in algebra and higher mathematics. The horizontal bar ($-$) and slash mark (/) are the symbols most frequently used. Thus,

$$a/b, \qquad \frac{a}{b}, \qquad a \div b, \qquad b\overline{)a} \qquad \text{\textcolor{blue}{In each case, } } b \text{ \textcolor{blue}{is the divisor.}}$$

all name the same number, assuming the quotient is defined. Thus, we can write

$$a/b = \frac{a}{b} = a \div b = b\overline{)a}$$

Division is actually carried out using the following rules, assuming we can do the division of positive numbers from arithmetic:

Division of Real Numbers: *a/b*

1. If $a = 0$ and $b \neq 0$, $\dfrac{a}{b} = \dfrac{0}{b} = 0$ $\qquad \dfrac{0}{3} = 0, \quad \dfrac{0}{-15} = 0$

If $b = 0$, $\dfrac{a}{b} = \dfrac{a}{0}$ is not defined $\qquad \dfrac{3}{0}, \dfrac{-15}{0}, \dfrac{0}{0}$ are not defined.

2. If a and b are both positive, divide as you learned in arithmetic.

3. If a and b are both negative, take the quotient of the absolute values.

$$\frac{-15}{-3} = \frac{15}{3} = 5$$

4. If a and b have opposite signs, take the opposite of the quotient of the absolute values.

$$\frac{-15}{3} = -\left(\frac{15}{3}\right) = -5, \quad \frac{15}{-3} = -\left(\frac{15}{3}\right) = -5$$

From cases 2, 3, and 4, we see that:

The quotient of two numbers with unlike signs is negative.
The quotient of two numbers with like signs is positive.

Example 4
Dividing

Evaluate:

(A) $\dfrac{-22}{11}$ **(B)** $\dfrac{-36}{-12}$ **(C)** $\dfrac{48}{-16}$

Solution **(A)** $\dfrac{-22}{11} = -2$ **(B)** $\dfrac{-36}{-12} = 3$ **(C)** $\dfrac{48}{-16} = -3$

Matched Problem 4 Evaluate: **(A)** $\dfrac{-36}{-9}$ **(B)** $\dfrac{24}{-8}$ **(C)** $\dfrac{-72}{12}$

Several important sign properties for division are summarized in the following result.

Sign Properties for Division

For all real numbers a and b, $b \neq 0$:

(A) $\dfrac{-a}{-b} = \dfrac{a}{b} = -\dfrac{-a}{b} = -\dfrac{a}{-b}$ $\dfrac{-3}{-4} = \dfrac{3}{4} = -\dfrac{-3}{4} = -\dfrac{3}{-4}$

(B) $\dfrac{-a}{b} = \dfrac{a}{-b} = -\dfrac{a}{b}$ $\dfrac{-3}{4} = \dfrac{3}{-4} = -\dfrac{3}{4}$

Example 5
**Applying Sign
Properties for Division**

Evaluate $\dfrac{-a}{b}$, $\dfrac{a}{-b}$, and $-\dfrac{a}{b}$ for $a = -6$ and $b = 2$.

Solution

$$\frac{-a}{b} = \frac{-(-6)}{2} = \frac{6}{2} = 3$$

$$\frac{a}{-b} = \frac{(-6)}{-(2)} = \frac{-6}{-2} = 3$$

$$-\frac{a}{b} = -\frac{-6}{2} = -(-3) = 3$$

Matched Problem 5

Evaluate $\dfrac{-a}{-b}$, $\dfrac{a}{b}$, and $-\dfrac{-a}{b}$ for $a = -6$ and $b = 2$.

Recall from Section 1-4 that we check subtraction by using addition:

$$a - b = c \qquad \text{means} \qquad b + c = a$$

In the same way, we check division using multiplication:

$$a \div b = c \qquad \text{means} \qquad b \cdot c = a$$

For example, to check $\frac{153}{9} = 17$, we verify by multiplying

$$9 \cdot 17 = 153$$

◆ **COMBINED OPERATIONS**

We now consider problems involving various combinations of the arithmetic operations $+$, $-$, $\cdot$, and $\div$, as well as grouping symbols such as parentheses ( ), brackets [ ], braces { }, and fraction bars $-$.

To evaluate an expression such as

$$6 - 4(-3) + \frac{-6}{2}$$

we need to recall the order of operations agreed upon in Section 1-3:

> **Order of Operations**
>
> Unless grouping symbols indicate otherwise, multiplication and division are to be done before addition and subtraction. Operations are done from left to right.

Thus,

$$6 - 4(-3) + \tfrac{-6}{2} = 6 - (-12) + (-3) = 6 + 12 + (-3) = 15$$

Example 6
Applying the Order of Operations

Evaluate:

(A) $5 - [4 - 3(2 - 8) \div 6 - 2]$ (B) $\dfrac{5 - 4 - 3 \cdot 2}{8 \div 4 - 3 \cdot 2}$

Solution   (A) $5 - [4 - 3(2 - 8) \div 6 - 2]$ Simplify within the inner parentheses first.

$= 5 - [4 - 3(-6) \div 6 - 2]$ Inside the brackets, do the multiplications and divisions, left to right.

$= 5 - [4 - (-18) \div 6 - 2]$

$= 5 - [4 - (-3) - 2]$ Do the subtractions inside the brackets.

$= 5 - [4 + 3 - 2]$

$= 5 - 5$

$= 0$

(B) $\dfrac{5 - 4 - 3 \cdot 2}{8 \div 4 - 3 \cdot 2}$ The division bar is a grouping symbol, splitting the problem into a numerator and denominator that should be simplified separately. Do the multiplications and divisions in each first.

$= \dfrac{5 - 4 - 6}{2 - 6}$ Do the subtractions in the numerator and denominator.

$= \dfrac{-5}{-4}$

$= \dfrac{5}{4}$

Matched Problem 6   Evaluate: (A) $1 - [2 - 3(4 - 8) \div 2 \cdot 4]$ (B) $\dfrac{2 + 4 \div 2 - 4}{9 \div 3 - 6 \div 2}$

Example 7
Combining Operations

Evaluate for $x = -24$, $y = 2$, and $z = -3$:

(A) $2x - 3yz + \dfrac{x}{z}$ (B) $\dfrac{x}{y} - \dfrac{16z + xy}{y + z}$ (C) $-x - y(x - 5yz)$

Solution **(A)** $2x - 3yz \quad + \dfrac{x}{z}$ Substitute the given values.

$2(\ ) - 3(\ )(\ ) + \dfrac{(\ )}{(\ )}$ Using parentheses as indicated will help to reduce sign errors.

$2(-24) - 3(2)(-3) + \dfrac{(-24)}{(-3)}$ Multiplication and division precede addition and subtraction.

$= (-48) - (-18) + 8$

$= (-48) + 18 \quad + 8$

$= -22$

(B) $\dfrac{x}{y} - \dfrac{16z + xy}{y + z}$ Substitute the given values.

$\dfrac{(-24)}{2} - \dfrac{16(-3) + (-24)(2)}{2 + (-3)}$

$= (-12) - \dfrac{(-48) + (-48)}{-1}$

$= (-12) - \dfrac{-96}{-1}$

$= (-12) - 96 = -108$

(C) $-x - y(x - 5yz)$

$-(-24) - 2[(-24) - 5(2)(-3)]$ Notice how brackets and parentheses are used.

$= 24 - 2[(-24) - (-30)]$ Brackets are just another kind of parentheses; each may replace the other, as desired.

$= 24 - 2(6)$

$= 24 - 12$

$= 12$

Matched Problem 7   Evaluate for $u = 36$, $v = -4$, and $w = -3$:

(A) $3vw - \dfrac{u}{3w} + 4v$ **(B)** $\dfrac{9w - 8v}{v - w} - \dfrac{u}{v}$ **(C)** $u - [7 - 2(u - 4vw)]$

We have again used integer examples to focus attention on the properties rather than the arithmetic. You should be able to apply the properties using rational numbers as well.

Example 8
Combined Operations with Fractions

Evaluate for $x = \frac{2}{3}$ and $y = -\frac{4}{9}$:

(A) $xy + y$ **(B)** $(-x) + y \div x$ **(C)** $x \div y + x$
(D) $(-x) + x \div y$

Solution   **(A)** $xy + y = \frac{2}{3}(-\frac{4}{9}) + (-\frac{4}{9}) = -(\frac{2}{3} \times \frac{4}{9}) + (-\frac{4}{9})$
$= -\frac{8}{27} + (-\frac{4}{9}) = -\frac{8}{27} - \frac{12}{27} = -\frac{20}{27}$

(B) $(-x) + y \div x = (-\frac{2}{3}) + (-\frac{4}{9}) \div (\frac{2}{3}) = -\frac{2}{3} + (-\frac{4}{9}) \div \frac{2}{3}$

$\qquad = -\frac{2}{3} + (-\frac{4}{9} \times \frac{3}{2}) = -\frac{2}{3} - \frac{2}{3} = -\frac{4}{3}$

(C) $x \div y + x = \frac{2}{3} \div (-\frac{4}{9}) + \frac{2}{3} = -(\frac{2}{3} \times \frac{9}{4}) + \frac{2}{3}$

$\qquad = -\frac{3}{2} + \frac{2}{3} = -\frac{9}{6} + \frac{4}{6} = -\frac{5}{6}$

(D) $(-x) + x \div y = (-\frac{2}{3}) + \frac{2}{3} \div (-\frac{4}{9}) = (-\frac{2}{3}) - (\frac{2}{3} \times \frac{9}{4})$

$\qquad = (-\frac{2}{3}) - \frac{3}{2} = -\frac{4}{6} - \frac{9}{6} = -\frac{13}{6}$

Matched Problem 8 Evaluate for $x = -\frac{3}{4}$ and $y = \frac{5}{8}$:

(A) $xy + y$ **(B)** $(-x) + y \div x$ **(C)** $x \div y + x$ **(D)** $(-x) + x \div y$

Answers to Matched Problems

1. **(A)** -12 **(B)** -12 **(C)** 12 **(D)** 0 2. Both are -20.
3. $-[(6)(-3)] = -(-18) = 18$ and $(-6)(-3) = 18$
4. **(A)** 4 **(B)** -3 **(C)** -6 5. All three are -3.
6. **(A)** -25 **(B)** Not defined. 7. **(A)** 24 **(B)** 4 **(C)** 5
8. **(A)** $\frac{5}{32}$ **(B)** $-\frac{1}{12}$ **(C)** $-\frac{39}{20}$ **(D)** $-\frac{9}{20}$

EXERCISE 1-5

A *Evaluate, performing the indicated operations.*

1. $(-3)(-5)$
2. $(-7)(-4)$
3. $(-18) \div (-6)$
4. $(-20) \div (-4)$
5. $(-2)(+9)$
6. $(+6)(-3)$
7. $\dfrac{-9}{+3}$
8. $\dfrac{+12}{-4}$
9. $0(-7)$
10. $(-6)0$
11. $0/5$
12. $0/(-2)$
13. $3/0$
14. $-2/0$
15. $0 \div 0$
16. $\dfrac{0}{0}$
17. $\dfrac{-21}{3}$
18. $\dfrac{-36}{-4}$
19. $(-4)(-2) + (-9)$
20. $(-7) + (-3)(+2)$
21. $(+5) - (-2)(+3)$
22. $(-7) - (-3)(-4)$
23. $5 - \dfrac{-8}{2}$
24. $7 - \dfrac{-16}{-2}$
25. $(-1)(-8)$ and $-(-8)$
26. $(-1)(+3)$ and $-(+3)$
27. $-12 + \dfrac{-14}{-7}$
28. $\dfrac{-10}{5} + (-7)$
29. $\dfrac{6(-4)}{-8}$
30. $\dfrac{5(-3)}{3}$
31. $\dfrac{22}{-11} - (-4)(-3)$
32. $3(-2) - \dfrac{-10}{-5}$

33. $\dfrac{-16}{2} - \dfrac{3}{-1}$
34. $\dfrac{27}{-9} - \dfrac{-21}{-7}$
35. $(+5)(-7)(+2)$
36. $(-6)(-3)(+4)$
37. $(-22)(+36)(0)$
38. $(+19)(0)(-35)$

B *Evaluate, performing the indicated operations.*

39. $[(+2) + (-7)][(+8) - (+10)]$
40. $[(-3) - (+8)][(+4) + (-2)]$
41. $12 - 7[(-4)(5) - 2(-8)]$
42. $9 - 5[(-2) - 3]$
43. $\dfrac{9}{-3} - \dfrac{3 + 9(-2)}{-2 - (-3)}$
44. $\dfrac{4(-2) - (-5)}{(-9) - (-6)} - \dfrac{-24}{-8}$
45. $\{[8 \div (-2)] - [21 + 5(-3)]\} - (-2)(-4)$
46. $7 - \{9 - [5 - 2(-3)] - (8 \div 2)\}$
47. $[8 - (9 - 7)] \div 6 - 3$ **48.** $[8 - (9 - 7)] \div (6 - 3)$
49. $\dfrac{1 - 4 \div 2 \cdot 3}{6 \div 2 - 4 \div 2}$
50. $1 - [3 - 5(7 - 9) \div 10]$
51. $12 \div \{4 - [3(2 \cdot 9) \div 27 - 1]\}$
52. $\dfrac{2 + 2 \div 2 \cdot 2}{4 \div 4 \cdot 3 + 1}$
53. $[5 \div (4 - 3) - 8] \div (6 - 3)$
54. $\dfrac{5 - 4 \cdot 3 + 2}{8 + 4 \div 2 \cdot 3}$

55. $\dfrac{2 + 4 \div 2 - 4}{1 + 2 \cdot 3 - 4}$

56. $\{5 - [4 - 3(2 - 1) \div 2] \cdot 3\}$

57. $[8 - (9 - 7)] \div (6 - 4) + (4 + 2) \div 2$

58. $[8 - (9 - 7)] \div \{[(6 - 5) + (4 - 3)] \div 2\}$

59. $8 - [(9 - 3) \div 2 - 2 + 4 - 4 \div 2]$

60. $(8 - 1) - 7 \div (6 - 5) + 3 - 4 \div 2$

61. $\dfrac{(8 - 1) - 8 \div (6 - 2)}{4 + 2 \div 2}$ **62.** $\dfrac{8 - [9 - 6 \div (3 - 2)]}{(4 - 2) \div 2}$

63. $\dfrac{1 - 2 \cdot 3 + 4}{6 - 4 \div 2 \cdot 3}$ **64.** $\dfrac{8 + 4 \div 2 - 1}{5 - 6 \div 3 - 3}$

Evaluate Problems 65–88 for $w = 2$, $x = -3$, $y = 0$, and $z = -24$.

65. z/w

66. z/x

67. w/y

68. y/x

69. $\dfrac{z}{x} - wz$

70. $wx - \dfrac{z}{w}$

71. $\dfrac{xy}{w} - xyz$

72. $wxy - \dfrac{y}{z}$

73. $-|w||x|$

74. $|x||z|$

75. $\dfrac{|z|}{|x|}$

76. $-\dfrac{|z|}{|w|}$

77. $(wx - z)(z - 8x)$

78. $(5x - z)(wx - 3w)$

79. $wx + \dfrac{z}{wx} + wz$

80. $xyz + \dfrac{y}{z} + x$

81. $\dfrac{8x}{z} - \dfrac{z - 6x}{wx}$

82. $\dfrac{w - x}{w + x} - \dfrac{z}{2x}$

83. $\dfrac{24}{3w - 2x} - \dfrac{24}{3w + 2x}$ **84.** $\dfrac{48}{z + 8x} - \dfrac{48}{z - 8x}$

85. $\dfrac{z}{wx} - 2[z + 3(2x - w)]$

86. $\dfrac{8wx}{-z} - x[5 + 2(z + 9w)]$

87. $\dfrac{8y - w}{8x - z}$ **88.** $\dfrac{wx(w + x)}{wxy(w + x + y)}$

Evaluate Problems 89–92 for $w = \frac{1}{2}$, $x = -\frac{1}{3}$, $y = \frac{3}{4}$, and $z = -\frac{5}{12}$.

89. $\dfrac{z}{x} - wz$ **90.** $wx - \dfrac{z}{w}$ **91.** $\dfrac{xy}{w} - xyz$

92. $wxy - \dfrac{y}{z}$

93. Any integer divided by 0 is (*always, sometimes, never*) 0.

94. Zero divided by any integer is (*always, sometimes, never*) 0.

95. A product made up of an odd number of negative factors is (*sometimes, always, never*) negative.

96. A product made up of an even number of negative factors is (*sometimes, always, never*) negative.

C 97. If the quotient x/y exists, and neither x nor y is 0, when is it equal to $-|x|/|y|$?

98. If the quotient x/y exists, and neither x nor y is 0, when is it equal to $|x|/|y|$?

99. Provide the reason for each step in the proof that $a0 = 0$ for all real numbers a (case 1, Multiplication of Real Numbers).

STATEMENT

1. $a0 = a(0 + 0)$

2. $a0 = a0 + a0$

3. $a0 + [-(a0)] = (a0 + a0) + [-(a0)]$

4. $0 = a0 + \{a0 + [-(a0)]\}$

5. $0 = a0 + 0$

6. $0 = a0$

7. $a0 = 0$

100. Provide the reason for each step in the proof that $(-1)a = -a$ [part (A), Sign Properties for Multiplication].

STATEMENT

1. $a + (-1)a = 1a + (-1)a$

2. $= a[1 + (-1)]$

3. $= a \cdot 0$

4. $= 0$

5. Therefore, $(-1)a = -a$.

101. Provide the reason for each step in the proof that $(-a)b = -(ab)$ [part (B), Sign Properties for Multiplication].

STATEMENT

1. $(-a)b + ab = [(-a) + a]b$

2. $= 0 \cdot b$

3. $= 0$

4. Therefore, $(-a)b = -(ab)$.

102. Provide the reason for each step in the proof that $\dfrac{-a}{b} = -\dfrac{a}{b}$ for $b \neq 0$ [part (B), Sign Properties for Division].

STATEMENT

1. $\dfrac{a}{b} + \dfrac{-a}{b} = a \cdot \dfrac{1}{b} + (-a) \cdot \dfrac{1}{b}$

2. $\qquad\quad = [a + (-a)] \cdot \dfrac{1}{b}$

3. $\qquad\quad = 0 \cdot \dfrac{1}{b}$

4. $\qquad\quad = 0$

5. Therefore, $\dfrac{-a}{b} = -\dfrac{a}{b}$.

APPLICATIONS

103. For a round of 18 holes, a golfer had 3 birdies, 7 bogies, 2 double bogies, and the rest pars. If par for the round would total 72, what was the golfer's score? (Refer to Problem 103, Section 1-4 for the terminology.)

104. An investor sells 200 shares of one stock at a profit of $6.50 per share, 500 shares of another at a loss of $3.75 per share, and 400 shares of a third stock at a profit of $0.75 per share. What is the investor's net profit?

105. Motion picture film is sometimes used as a model for multiplication of signed numbers. Suppose a film is taken of a person walking at 3 miles per hour. Treat the forward direction as positive and the backward direction as negative.
(A) If the walker is walking forward and the film is run forward at twice the normal speed, how fast and in what direction will the person appear to be walking?
(B) If the walker is walking forward and the film is run backward at twice the normal speed, how fast and in what direction will the person appear to be walking?

106. Continue Problem 105:
(A) If the walker is walking backward and the film is run forward at twice the normal speed, how fast and in what direction will the person appear to be walking?
(B) If the walker is walking backward and the film is run backward at twice the normal speed, how fast and in what direction will the person appear to be walking?

1-6 Exponents and Order of Operations

- Natural Number Exponents
- Order of Operations
- Combining Like Terms

The use of natural number exponents provides a compact way of writing repeated products of the same factor. We review this use in this section and extend the agreement on order of operations to include exponents. Exponents are also used to identify like terms in algebraic expressions. Such terms may then be combined to simplify the expression.

◆ NATURAL NUMBER EXPONENTS

Recall that

$$a^5 = a \cdot a \cdot a \cdot a \cdot a \qquad \text{Five factors of } a$$

and, in general:

Natural Number Exponents

For n a natural number and a a real number:

$$\underset{\text{Base}}{\overset{\text{Exponent}}{a^n}} = a \cdot a \cdot \ \cdots \cdot a \qquad n \text{ factors of } a \qquad 2^3 = 2 \cdot 2 \cdot 2 = 8$$

The expression a^n is read "a raised to the nth power." For $n = 2$ and 3, this is shortened to "a squared" and "a cubed," respectively. Exponent forms are encountered so frequently in algebra that it is essential for you to become completely familiar with their basic properties and uses. The first of these properties is given below. Additional properties will be considered in detail in Chapter 6.

Consider:

$$a^3 a^4 = \overset{\text{3 factors}}{\overbrace{(a \cdot a \cdot a)}} \overset{\text{4 factors}}{\overbrace{(a \cdot a \cdot a \cdot a)}} = \overset{\text{3 + 4 factors}}{\overbrace{a \cdot a \cdot a \cdot a \cdot a \cdot a \cdot a}} = a^{3+4} = a^7$$

This suggests that for any real number a and any positive-integer exponents m and n:

Exponent Property 1

$$a^m a^n = a^{m+n} \qquad a^5 a^2 = a^{5+2} = a^7 \qquad 2^3 \cdot 2^4 = 2^7$$

Example 1
Applying Exponent Property 1

Simplify by rewriting each expression so that the variable x occurs only once:

(A) $x^{12} x^{13}$ **(B)** $(3x^4)(5x^3)$ **(C)** $x^a x^{a+1}$ **(D)** $x^{2a+1} x^{a+3}$

Solution

(A) $x^{12} x^{13} = x^{12+13} = x^{25}$

(B) $(3x^4)(5x^3) = (3 \cdot 5)(x^4 x^3)$
$\qquad\qquad\quad = 15x^{4+3}$
$\qquad\qquad\quad = 15x^7$

(C) $x^a x^{a+1} = x^{a+a+1} = x^{2a+1}$

(D) $x^{2a+1} x^{a+3} = x^{2a+1+a+3} = x^{3a+4}$

Matched Problem 1

Simplify by rewriting each expression so that the variable x occurs only once:

(A) $a^7 a^3$ **(B)** $(2y^2)(6y^3)$ **(C)** $x^{2k} x^{k-1}$ **(D)** $x^{2k-1} x^{k+1}$

♦ **ORDER OF OPERATIONS**

Recall that in the order of operations agreed upon in Sections 1-3 and 1-5, multiplications and divisions are to be done before additions and subtractions, unless grouping symbols indicate otherwise. The scheme is extended to include the taking of powers (and later roots) by agreeing that:

Powers take precedence over multiplication and division.

Thus, for example,

The exponent applies only to the 2.

$$\frac{2^3}{5} = \frac{8}{5} = 1.6 \qquad \text{and} \qquad 5 \cdot 2^3 = 5 \cdot 8 = 40$$

On the other hand,

$$\left(\frac{2}{5}\right)^3 = (0.4)^3 = 0.064 \qquad \text{and} \qquad (5 \cdot 2)^3 = 10^3 = 1{,}000$$

Here, the parentheses indicate that the exponent applies to the expression within.

We summarize the complete agreement on order of operations as follows:

Order of Operations

(A) If no grouping symbols are present:
1. Apply exponents before any multiplication or division is performed.
2. Perform any multiplication and division next, proceeding from left to right.
3. Then perform any addition and subtraction, proceeding from left to right.

(B) If symbols of grouping are present:
1. Simplify above and below any fraction bars using the steps in part (A).
2. Simplify within other symbols of grouping, generally starting with the innermost and working outward, following the steps in part (A).

Particular care is required when applying exponents to expressions involving negative numbers:

$$-4^2 \neq (-4)^2$$

To evaluate -4^2, we first apply the exponent:

$$-4^2 = -(4^2) = -16$$

On the other hand, to evaluate $(-4)^2$, we simply square -4:

$$(-4)^2 = (-4)(-4) = 16$$

That is, in evaluating -4^2, the absence of parentheses indicates that the exponent applies only to the 4 and does not include the negative sign.

Example 2
Applying the Order of Operations

Evaluate for $x = 6$, $y = 3$, and $z = -12$:

(A) $3x^2 - xy^2 - z$ **(B)** $\dfrac{zx^2}{y} + \dfrac{xy - y^2z}{x} - (-z)^2$

Solution   **(A)** $3x^2 - xy^2 - z = 3(6)^2 - 6(3)^2 - (-12)$
$$= 3(36) - 6(9) + 12$$
$$= 66$$

(B) $\dfrac{zx^2}{y} + \dfrac{xy - y^2z}{x} - (-z)^2 = \dfrac{(-12)(6^2)}{3} + \dfrac{6 \cdot 3 - (3^2)(-12)}{6} - (-12)^2$

$$= \dfrac{(-12)(36)}{3} + \dfrac{18 + 108}{6} - 144$$

$$= -108 + 21 - 144 = -231$$

Matched Problem 2   Evaluate for $x = 2$, $y = 3$, and $z = -6$:

(A) $xy + z^2 - 2xy^2 - x^2$ **(B)** $\dfrac{(x - y)z}{x} + \dfrac{xz - y^2}{y} - (y - z)^2$

◆ **COMBINING LIKE TERMS**

A constant that is present as a factor in a term is called the **numerical coefficient,** or simply the **coefficient,** of the term. For example, in the term $4x^2$, 4 is the coefficient, and in the term $-3xy^2$, -3 is the coefficient. If no constant appears in the term, then the coefficient is understood to be 1. The coefficient in the term $-x$ is -1. Two terms are called **like terms** if they have exactly the same variable factors and these variables are raised to the same powers in each term. The numerical coefficients may or may not be the same. Thus, $-5x$ and $7x$ are like terms. So too are $3xy^2$ and $-8xy^2$, but $2x^2y$ and $2xy^2$ are not like terms. Since constant terms involve no variables, all constant terms are like terms. If an algebraic expression contains two or more like terms, these terms can be combined into a single term by making use of the distributive property.

Example 3
Combining Like Terms

Combine like terms:

(A) $3x + 7x$ **(B)** $6m - 9m$
(C) $3z + 5 - z + 2$ **(D)** $5x^3y - 2xy - x^3y - 2x^3y$

Solution   **(A)** $3x + 7x = (3 + 7)x = 10x$

(B) $6m - 9m = (6 - 9)m = -3m$

(C) $3z + 5 - z + 2$ $\boxed{= 3z - z + 5 + 2}$
$\boxed{= (3 - 1)z + (5 + 2)}$
$= 2z + 7$

(D) $5x^3y - 2xy - x^3y - 2x^3y$ $\boxed{= 5x^3y - x^3y - 2x^3y - 2xy}$
$\boxed{= (5 - 1 - 2)x^3y - 2xy}$
$= 2x^3y - 2xy$

Matched Problem 3 Combine like terms:

(A) $5y + 4y$ **(B)** $2u - 6u$
(C) $4x - 1 + 3x + 2$ **(D)** $6mn^2 - m^2n - 3mn^2 - mn^2$

In the example, free use was made of the properties discussed in this chapter. The steps illustrated in the dashed boxes are usually done mentally. The process is quickly mechanized as follows:

> ### Combining Like Terms
>
> Like terms are combined by adding their numerical coefficients.

Example 4
Combining Like Terms

Combine like terms:

(A) $3x - 5y + 6x + 2y$
(B) $x^3y^2 - 2x^2y^3 + 5x^2y^2 - 4x^2y^3 - x^3y^2 - 5x^2y^2$

Solution

(A) $3x - 5y + 6x + 2y = 3x + 6x - 5y + 2y = 9x - 3y$
(B) $x^3y^2 - 2x^2y^3 + 5x^2y^2 - 4x^2y^3 - x^3y^2 - 5x^2y^2$
$\quad = x^3y^2 - x^3y^2 - 2x^2y^3 - 4x^2y^3 + 5x^2y^2 - 5x^2y^2$
$\quad = -6x^2y^3$

Matched Problem 4 Combine like terms:

(A) $7m + 8n - 5m - 10n$
(B) $2u^4v^2 - 3uv^3 - u^4v^2 + 6u^4v^2 + 2uv^3 - 6u^4v^2$

We often have to remove parentheses before combining like terms. Parentheses preceded by a minus sign can lead to sign errors if we are not careful. Remember that

$$-a = (-1)a$$

so that we can think of removing parentheses preceded by a minus sign as multiplying by -1:

$$-(a + b) = (-1)(a + b) = (-1)a + (-1)b = -a - b$$

It is important that *every* term within the parentheses be multiplied by -1, or equivalently, that the sign of every term within the parentheses be reversed.

Example 5
Removing Parentheses and Combining Like Terms

Remove parentheses and combine like terms:

(A) $5x - (2y + 8x)$ (B) $5x - (2y - 8x)$
(C) $5x - 2[8 - 3(x - 4)]$

Solution
(A) $5x - (2y + 8x) = 5x - 2y - 8x$

Remove the parentheses by reversing the sign of each term. It may be helpful to think of this as multiplying each term by -1.

$$= (5 - 8)x - 2y$$
$$= -3x - 2y$$

(B) $5x - (2y - 8x) = 5x - 2y + 8x$
$$= 13x - 2y$$

(C) $5x - 2[8 - 3(x - 4)] = 5x - 2[8 - 3x + 12]$ Remove inner parentheses first.
$$= 5x - 2(20 - 3x)$$
$$= 5x - 40 + 6x$$
$$= 11x - 40$$

Matched Problem 5

Remove parentheses and combine like terms:

(A) $7y - (5x - 3y)$ (B) $7y - (5x + 8y)$
(C) $8x - 3[x - 7(x - 1)]$

Answers to Matched Problems
1. (A) a^{10} (B) $12y^5$ (C) x^{3k-1} (D) x^{3k} 2. (A) 2   (B) -85
3. (A) $9y$ (B) $-4u$ (C) $7x + 1$ (D) $2mn^2 - m^2n$
4. (A) $2m - 2n$ (B) $u^4v^2 - uv^3$
5. (A) $10y - 5x$ (B) $-5x - y$ (C) $26x - 21$

EXERCISE 1-6

A *Replace the question mark with the appropriate symbol.*

1. $3^2 3^5 = 3^?$ 2. $2^2 2^4 = 2^?$
3. $2^5 2^? = 2^8$ 4. $3^? 3^6 = 3^{10}$
5. $y^2 y^7 = y^?$ 6. $x^7 x^5 = x^?$
7. $y^8 = y^3 y^?$ 8. $x^{10} = x^? x^6$
9. $10^5 10^6 10^7 = 10^?$ 10. $10^2 10^4 10^6 = 10^?$
11. $10^{13} 10^? 10 = 10^{19}$ 12. $10^? 10^{17} 10 = 10^{22}$

Simplify by rewriting each expression so that the variable occurs only once.

13. $(5x^2)(2x^9)$ 14. $(2x^3)(3x^7)$
15. $(4y^3)(3y)(y^6)$ 16. $(2x^2)(3x^3)(x^4)$

Simplify.

17. $(4 \times 10^5)(5 \times 10^4)$ 18. $(6 \times 10^3)(2 \times 10^4)$
19. $(5 \times 10^8)(7 \times 10^9)$ 20. $(2 \times 10^3)(3 \times 10^{12})$

Simplify by removing parentheses, if any, and combining like terms.

21. $9x + 8x$ 22. $7x + 3x$
23. $9x - 8x$ 24. $7x - 3x$
25. $5x + x + 2x$ 26. $3x + 4x + x$
27. $4t - 8t - 9t$ 28. $2x - 5x + x$
29. $4y + 3x + y$ 30. $2x + 3y + 5x$
31. $8 + 4x - 4$ 32. $-3 - x + 5$
33. $5m + 3n - m - 9n$ 34. $2x + 8y - 7x - 5y$
35. $3(u - 2v) + 2(3u + v)$
36. $2(m + 3n) + 4(m - 2n)$
37. $4(m - 3n) - 3(2m + 4n)$
38. $2(x - y) - 3(3x - 2y)$
39. $(2u - v) + (3u - 5v)$ 40. $(x + 3y) + (2x - 5y)$
41. $(3x + 2) + (x - 5)$ 42. $(y + 7) - (4 - 2y)$
43. $2(x + 5) - (x - 1)$ 44. $3(z - 2) - 2(z - 3)$

Evaluate for a = 3, b = −2, and c = 6.

45. $-ab^2$ **46.** $-bc^2$ **47.** $-(a-b)^2$

48. $-(b-c)^2$ **49.** $ab^2 - bc^2$ **50.** $ac^2 - ab^2$

B *Replace the question mark with the appropriate symbol or expression.*

51. $x^k x^m = x^?$ **52.** $x^? x^k = x^m$

53. $x^k x^? = x^{k+1}$ **54.** $x^? x^m = x^{2m}$

55. $x^{2m} x^{5m} = x^?$ **56.** $x^{3m} x^? = x^{7m}$

57. $x^{m+1} x^{2m+3} = x^?$ **58.** $x^{3m-1} x^{2m+1} = x^?$

59. $x^? x^{3m+2} = x^{5m+7}$ **60.** $x^? x^{4m-1} = x^{5m+3}$

Simplify by rewriting each expression so that the variable x occurs only once.

61. $(3x^m)(2x^{m+1})$ **62.** $(4x^{3m})(3x^{4m})$

63. $(5x^{2m})(2x^{5m})$ **64.** $(4x^{3m+2})(3x^m)$

Simplify by removing grouping symbols, if any, and combining like terms.

65. $-x^2 y + 3x^2 y - 5x^2 y$

66. $-4r^3 t^3 - 7r^3 t^3 - 7r^3 t^3 + 9r^3 t^3$

67. $y^3 + 4y^2 - 10 + 2y^3 - y + 7$

68. $3x^2 - 2x + 5 - x^2 + 4x - 8$

69. $a^2 - 3ab + b^2 + 2a^2 + 3ab - 2b^2$

70. $2x^2 y + 2xy^2 - 5xy + 2xy^2 - xy - 4x^2 y$

71. $x - 3y - 4(2x - 3y)$ **72.** $a + b - 2(a - b)$

73. $y - 2(x - y) - 3x$ **74.** $x - 3(x + 2y) + 5y$

75. $-2(-3x + 1) - (2x + 4)$

76. $-3(-t + 7) - (t - 1)$

77. $2(x - 1) - 3(2x - 3) - (4x - 5)$

78. $-2(y - 7) - 3(2y + 1) - (-5y + 7)$

79. $4t - 3[4 - 2(t - 1)]$

80. $3x - 2[2x - (x - 7)]$

81. $3[x - 2(x + 1)] - 4(2 - x)$

82. $2(3x + y) - 2[y - (3x + 2)]$

Evaluate for x = 3, y = −2, and z = 6.

83. $x^2 y - xy^2$ **84.** $x(y - z) \div y^2$

85. $-x^2 - yz^2 \div x$ **86.** $12x \div y^2 - z^2$

C *Simplify by removing grouping symbols, if any, and combining like terms.*

87. $2a - 3\{a + 2[a - (a + 5)] + 1\}$

88. $b - \{b - [b - (b - 1)]\}$

89. $a - \{b - [c - (a - b) - c] - (c - a)\} + b$

90. $3a - 2\{a - [a + 4(a - 3)] - 5\}$

91. $\{[(c - 1) - 1] - c\} - 1$

92. $1 - \{1 - [1 - (1 - b)]\}$

93. $a - \{1 - [a - (1 - a)]\}$

94. $b - \{[1 - (b - 1)] - b\}$

Evaluate for $x = \frac{1}{2}$, $y = -\frac{2}{3}$, and $z = \frac{1}{6}$.

95. $x^2 y - xy^2$ **96.** $x(y - z) \div y^2$

97. $-x^2 - yz^2 \div x$ **98.** $12x \div y^2 - z^2$

Evaluate each of the products and use the results to make a conjecture about the given general form.

99. $(x^2)^3 = x^2 x^2 x^2 = x^?$, $(x^3)^4 = x^3 x^3 x^3 x^3 = x^?$, $(x^5)^2 = x^5 x^5 = x^?$; $(x^m)^n = x^?$

100. $(xy)^2 = (xy)(xy) = x^? y^?$, $(xy)^3 = (xy)(xy)(xy) = x^? y^?$, $(xy)^4 = (xy)(xy)(xy)(xy) = x^? y^?$; $(xy)^m = x^? y^?$

1-7   Solving Equations

♦   Solutions to Equations
♦   Solving Linear Equations
♦   Literal Equations

In this section, we consider basic methods of solving equations such as

$$2(3x - 5) - 2 = 5 - (3x + 2) \tag{1}$$
$$2(3x - y) - 2 = y - (3x + 2) \tag{2}$$

We discuss what it means to *solve an equation* and describe the basic strategy for doing so. The strategy is applied to equations of the form (1) and then to literal equations of the form (2).

◆ SOLUTIONS TO EQUATIONS

A **solution,** or **root,** of an equation involving a single variable is a replacement of the variable by a constant that makes the left side of the equation equal to the right side. For example, $x = 4$ is a solution of

$$2x - 1 = x + 3$$

since

$$2(4) - 1 = 4 + 3 \qquad \text{That is, } 7 = 7.$$

The set of all solutions is called the **solution set.** To **solve an equation** is to find its solution set.

Example 1
Verifying Solutions

Verify that $x = -3$ is a solution to each of these equations:

(A) $2x + (5 - x) = -2(x + 2)$ **(B)** $x^2 + 7 = (x - 1)^2$
(C) $x^3 + (x + 8)^2 + 2 = 0$ **(D)** $|x - 3| = 6$

Solution

In each case we substitute $x = -3$ into the left and right sides of the equation, evaluate each side separately, and conclude that $x = -3$ is a solution when we obtain the same value for each side.

(A) The left-hand side (LHS) of the equation becomes

$$2(-3) + [5 - (-3)] = -6 + 8 = 2$$

The right-hand side (RHS) of the equation becomes

$$-2[(-3) + 2] = -2(-1) = 2$$

(B) LHS: $(-3)^2 + 7 = 9 + 7 = 16$

RHS: $[(-3) - 1]^2 = [-4]^2 = 16$

(C) LHS: $(-3)^3 + [(-3) + 8]^2 + 2 = (-27) + 5^2 + 2 = -27 + 25 + 2 = 0$
RHS: 0
(D) LHS: $|(-3) - 3| = |-6| = 6$
RHS: 6

Matched Problem 1

Verify that $x = -5$ is a solution to each of these equations:

(A) $3x - (1 + x) = x + (x + 1)$ **(B)** $(x + 4)^2 = x + 6$
(C) $x^2 + 4x - 5 = 0$ **(D)** $|2x - 3| = |3x + 2|$

Verifying that a particular number is in fact a solution to an equation becomes an arithmetic problem. *Finding* such solutions is a different problem. Our objective is to develop a systematic method of solving equations that is free of guesswork. We start by introducing the idea of *equivalent equations*. We say that two equations are **equivalent** if they both have the same solution set; that is, any solution of one is a solution of the other.

The basic idea in solving equations is to perform operations on equations that produce *simpler* equivalent equations, and to continue the process until we reach an equation whose solutions are obvious—generally, equations such as

$$x = -5 \qquad \text{or} \qquad x = \tfrac{1}{2} \qquad \text{or} \qquad x = 7$$

The following properties of equality produce equivalent equations when applied:

Equality Properties

For a, b, and c any real numbers:

1. If $a = b$, then $a + c = b + c$.   **Addition property**
   If $x - 2 = 3$, then $(x - 2) + 2 = 3 + 2$.
2. If $a = b$, then $a - c = b - c$.   **Subtraction property**
   If $x + 4 = 5$, then $(x + 4) - 4 = 5 - 4$.
3. If $a = b$, then $ca = cb$, $c \neq 0$.   **Multiplication property**
   If $\frac{x}{2} = 3$, then $2 \cdot \frac{x}{2} = 2 \cdot 3$.
4. If $a = b$, then $\dfrac{a}{c} = \dfrac{b}{c}$, $c \neq 0$.   **Division property**
   If $5x = 10$, then $\frac{5x}{5} = \frac{10}{5}$.

These properties of equality follow directly from the basic equality properties discussed in Section 1-2.

We can think of the process of solving an equation as a game. The objective of the game is to isolate the variable (with a coefficient of 1) on one side of the equation, leaving a constant on the other side. It is usual, but not necessary, to isolate the variable on the left. We use the equality properties to solve equations, such as (1) or (2), that involve no powers higher than 1 for the variable for which we are solving:

$$2(3x - 5) - 2 = 5 - (3x + 2) \tag{1}$$

$$2(3x - y) - 2 = y - (3x + 2) \tag{2}$$

Such equations are called **first-degree equations** or **linear equations** in the variable for which we are solving.

◆ SOLVING LINEAR EQUATIONS

To solve equation (1), we first simplify both sides of the equation, removing parentheses using the distributive property:

$$2(3x - 5) - 2 = 5 - (3x + 2)$$

On the right side of the equation it is helpful to think of $-(3x + 2)$ as $(-1)(3x + 2)$. This will help avoid sign errors.

$$6x - 10 - 2 = 5 - 3x - 2$$

$$6x - 12 = 3 - 3x$$

Add 12 to each side, using the addition property.

$$6x = 15 - 3x$$

Add $3x$ to each side, using the addition property.

$$9x = 15$$

Divide each side by 9, using the division property.

$$x = \frac{15}{9} = \frac{5}{3}$$

We check this solution by substituting $\frac{5}{3}$ into each side of the equation and showing that the same value results on each side:

$$\text{LHS: } 2(3 \cdot \tfrac{5}{3} - 5) - 2 = 2(5 - 5) - 2 = -2$$

$$\text{RHS: } 5 - (3 \cdot \tfrac{5}{3} + 2) = 5 - (5 + 2) = 5 - 7 = -2$$

The following strategy will prove helpful in solving first-degree, or linear, equations.

Equation-Solving Strategy for Linear Equations

1. Use the multiplication property to remove fractions, if present.
2. Simplify the left and right sides of the equation by removing grouping symbols and combining like terms.
3. Use the equality properties to get all variable terms on one side, usually the left, and all constant terms on the other side. Combine like terms in the process.
4. Isolate the variable, with a coefficient of 1, using the division or multiplication property of equality.

Example 2
Solving a Linear Equation

Solve $3x - 2(2x - 5) = 2(x + 3) - 8$ and check.

Solution

$$3x - 2(2x - 5) = 2(x + 3) - 8$$

Clear parentheses using the distributive property.

$$3x - 4x + 10 = 2x + 6 - 8$$

Combine like terms.

$$-x + 10 = 2x - 2$$

To isolate x on the left side, we apply the subtraction property, simplify, and then apply the subtraction property again.

$$-x + 10 - 10 = 2x - 2 - 10$$

$$-x = 2x - 12$$

$$-x - 2x = 2x - 12 - 2x$$

$$-3x = -12 \qquad \text{Apply the division property.}$$

$$\frac{-3x}{-3} = \frac{-12}{-3}$$

$$x = 4$$

We have produced a sequence of simpler equivalent equations using equality properties. Since 4 is a solution to the last equation, it must be a solution to the original equation. We *check* the solution by substituting $x = 4$ into the original equation:

Check

$$3x - 2(2x - 5) = 2(x + 3) - 8$$

$$3 \cdot 4 - 2(2 \cdot 4 - 5) \overset{?}{=} 2(4 + 3) - 8$$

$$12 - 2 \cdot 3 \overset{?}{=} 2 \cdot 7 - 8$$

$$6 \overset{\checkmark}{=} 6$$

Matched Problem 2   Solve $8x - 3(x - 4) = 3(x - 4) + 6$ and check.

If all terms in Example 2 had been transferred to the left side, leaving 0 on the right, and like terms combined, we would have obtained

$$-3x + 12 = 0$$

This is a special case of

$$ax + b = 0 \qquad\qquad (3)$$

An equation having only one variable x, and no powers of x higher than 1, will be equivalent to Equation (3) if x is involved only in addition, subtraction, and multiplication. Such an equation is called a **linear equation** in the variable x. Any resulting equation $ax + b = 0$ in which the coefficient of x is not zero has a unique solution, as can be seen here:

$$ax + b = 0 \qquad a \neq 0$$

$$ax = -b$$

$$x = -\frac{b}{a}$$

If $a = 0$ in Equation (3), then the original equation equivalent to (3) may have no solution at all or infinitely many solutions. Consider the following equation:

$$2(x - 2) = 4x - (2x + 3)$$

$$2x - 4 = 4x - 2x - 3$$

$$2x - 4 = 2x - 3$$

$$0x = 1 \qquad \text{0x} - 1 = 0 \text{ is of form } ax + b = 0.$$

$$0 = 1 \qquad \text{Not possible!}$$

Since $0 = 1$ can never be true, the original equivalent equation has no solution. Now consider the following equation:

$$4x + 2 = 2(2x + 1)$$

$$4x + 2 = 4x + 2$$

$$0x = 0 \qquad \text{0x + 0 = 0 is of form } ax + b = 0.$$

$$0 = 0$$

Since $0 = 0$ is always true, the same is true of the original equivalent equation; that is, every real number is a solution, and the equation has infinitely many solutions.

In general, it is important to know under what conditions a particular type of equation has a solution and how many solutions are possible. We have now answered both questions for equations equivalent to

$$ax + b = 0 \qquad a \neq 0$$

Any such equation has exactly one solution, namely, $-b/a$. Other types of equations will be studied later that have more than one solution. For example,

$$x^2 - 4 = 0 \qquad \text{Second-degree equation in one variable}$$

has two solutions: -2 and 2.

Example 3
Solving Linear Equations

Solve:

(A) $\frac{1}{3}x - 2 = \frac{3}{4}x + 8$ **(B)** $3(x + 5) - 2(4 + x) = 1 - (8 - x)$
(C) $2x - 1 + 3(x - 2) = 4x - 3 - (4 - x)$

Solution

(A) $\frac{1}{3}x - 2 = \frac{3}{4}x + 8$ Multiply by 12, using the multiplication property, to clear fractions.

$$4x - 24 = 9x + 96 \qquad \text{Apply the addition and subtraction properties.}$$
$$4x - 9x = 96 + 24$$
$$-5x = 120 \qquad \text{Apply the division property.}$$
$$x = -24$$

Check   LHS: $\frac{1}{3}(-24) - 2 = -8 - 2 = -10$

RHS: $\frac{3}{4}(-24) + 8 = -18 + 8 = -10$

(B) $3(x + 5) - 2(4 + x) = 1 - (8 - x)$ Remove parentheses. Be careful of the signs when removing the parentheses from $-(8 - x)$.

$$3x + 15 - 8 - 2x = 1 - 8 + x \qquad \text{Combine like terms to simplify each side.}$$
$$x + 7 = x - 7 \qquad \text{Subtract } x \text{ from each side.}$$
$$7 = -7 \qquad \text{Since this is impossible, the equation has no solution.}$$

There is no solution.

(C) $2x - 1 + 3(x - 2) = 4x - 3 - (4 - x)$ Remove parentheses.
$$2x - 1 + 3x - 6 = 4x - 3 - 4 + x \qquad \text{Simplify each side.}$$
$$5x - 7 = 5x - 7 \qquad \text{Since this is always true, every real number is a solution.}$$

The solution set is the entire set of real numbers.

Matched Problem 3   Solve:

(A) $\frac{1}{2}x + 3 = 4 - \frac{1}{5}x$ (B) $7(x - 1) - 8x = (x - 5) - 2(x + 1)$
(C) $7x - 6(x - 1) = 5x - 4(x - 3)$

♦ LITERAL EQUATIONS

An equation involving more than one variable can be solved for a specific variable. The strategy developed in this section can be applied provided the equation is linear in that specific variable when the other variables are treated as constants. That is, the specific variable must occur with no powers higher than 1 and must be involved only in additions, subtractions, and multiplications. Because our solution will still involve other variables (letters), such equations are often called **literal equations.**

Example 4
Solving a
Literal Equation

Solve for y in terms of x:

$$2(3x - y) - 2 = y - (3x + 2)$$

Solution

$2(3x - y) - 2 = y - (3x + 2)$	Clear parentheses. Treat $-(3x + 2)$ as $(-1)(3x + 2)$ so that removing parentheses gives $-3x - 2$.
$6x - 2y - 2 = y - 3x - 2$	Apply the addition property.
$6x - 2y - 2 \mathbf{+ 2} = y - 3x - 2 \mathbf{+ 2}$	
$6x - 2y = y - 3x$	Apply the subtraction property.
$6x - 2y \mathbf{- y} = y - 3x \mathbf{- y}$	
$6x - 3y = -3x$	Apply the subtraction property.
$6x - 3y \mathbf{- 6x} = -3x \mathbf{- 6x}$	
$-3y = -9x$	Apply the division property.
$\dfrac{-3y}{\mathbf{-3}} = \dfrac{-9x}{\mathbf{-3}}$	
$y = 3x$	

Note that in solving for y in terms of x, we isolated y on the left with a coefficient of 1 and that the right side does not contain y in any form.

Matched Problem 4   Solve for y in terms of x: $3(x + y) + 1 = 3(3x + 1) + y$

Answers to
Matched Problems

1. (A) Both sides equal -11.   (B) Both sides equal 1.
   (C) Both sides equal 0.   (D) Both sides equal 13.
2. $x = -9$ 3. (A) $x = \frac{10}{7}$ (B) All real numbers   (C) No solution
4. $y = 3x + 1$

EXERCISE 1-7

A *Solve and check.*

1. $3x = 27$
2. $5x = 30$
3. $-2x = 14$
4. $-4x = 20$
5. $5x - 3 = 17$
6. $3x - 9 = 24$
7. $2x + 7 = 19$
8. $4x + 3 = 18$
9. $-3x + 7 = -5$
10. $-4x + 9 = 25$
11. $3(x + 4) = 9$
12. $6(x - 5) = 30$
13. $3(x - 8) = x + 6$
14. $5(x + 2) = 10$
15. $-(x - 2) = 9$
16. $-(x - 3) = 11$
17. $-3(x - 5) = 21$
18. $-2(x - 7) = 12$
19. $-5(3 - x) = 40$
20. $-3(8 - x) = 30$
21. $3(x + 2) = 5(x - 6)$
22. $5x + 10(x - 2) = 40$
23. $4(x - 2) = 4x - 8$
24. $3y + 6 = 3(y + 2)$
25. $5 + 4(t - 2) = 2(t + 7) + 1$
26. $7x - (8x - 4) - 10 = 5 - (4x + 2)$
27. $3x - (x + 2) = 5x - 3(x - 1)$
28. $x - 2(x - 4) = 3x - 2(2x + 1)$
29. $10x + 25(x - 3) = 275$
30. $x + (x + 2) + (x + 4) = 54$
31. $5x - (7x - 4) - 2 = 5 - (3x + 2)$
32. $-3(4 - t) = 5 - (t + 1)$
33. $2(3x + 1) - 8x = 2(1 - x)$
34. $3(4x - 2) - 8x = 6x - 2(x + 3)$
35. $x + 5 = 3x - 4$
36. $4x - 5 = 5 - 4x$
37. $3x + 7 = 1 - (x + 4)$
38. $-2(x - 5) = 3(4 + x)$
39. $\frac{1}{2}x = 5$
40. $\frac{2}{5}x = 7$
41. $\frac{3}{5}x = \frac{1}{2}$
42. $\frac{2}{7}x = \frac{3}{4}$
43. $\frac{2}{3}x + \frac{3}{4} = \frac{5}{6}$
44. $\frac{1}{2}x - \frac{1}{3} = \frac{1}{4}$

B
45. $\frac{1}{2}x + 3 = \frac{3}{4}x - 5$
46. $\frac{1}{2}x - 5 = \frac{1}{5}x - 2$
47. $\frac{2}{5}x + \frac{3}{10} = \frac{1}{2}x - \frac{1}{10}$
48. $\frac{1}{3}x - \frac{1}{2} = \frac{1}{4} - \frac{1}{6}x$
49. $x - \frac{3}{5} = \frac{1}{3}x + \frac{1}{2}$
50. $\frac{2}{3} - \frac{1}{4}x = x - \frac{3}{4}$
51. $x(x - 1) + 5 = x^2 + x - 3$
52. $x(x + 2) = x(x + 4) - 12$
53. $x(x - 4) - 2 = x^2 - 4(x + 3)$
54. $t(t - 6) + 8 = t^2 - 6t - 3$
55. $(1 - x)x = (x - 1)(-x)$
56. $x(x - 4) = (4 + x)x$
57. $(2 + x)(-3) = (3 - x)(-2)$
58. $4x(x - 5) = -(5 - 4x)x - 15$
59. $3(x - 2)x = 1 - x(3 + 2x)2 + 7x^2$
60. $x^2 - 5 = (x - 1)x + (x - 3)$

Solve for the indicated variable in terms of the other variable.

61. $10x + 5y = 150$; solve for x
62. $10x + 5y = 150$; solve for y
63. $3x - 4y + 5 = x - 2y + 3$, solve for x
64. $3x - 4y + 5 = x - 2y + 3$; solve for y
65. $2 - 3(x + 3y) = x - 5(y + 6)$; solve for y
66. $2 - 3(x + 3y) = x - 5(y + 6)$; solve for x
67. $3(x + y) - 1 = 17 - 3y$; solve for x
68. $3(y - 2x) = 9(x - 1)$; solve for y
69. $\frac{1}{3}x = \frac{3}{4}y$; solve for y
70. $\frac{2}{3}x = \frac{4}{5}y$; solve for y
71. $x - \frac{2}{5} = \frac{3}{8}y + \frac{4}{5}$; solve for y
72. $x + \frac{1}{3}y = \frac{1}{4}x - y$; solve for y
73. $\frac{1}{2}(x - 4) = \frac{1}{3}(y - 5)$; solve for y
74. $\frac{3}{4}(1 - x) = \frac{2}{3}(y - 1)$; solve for y

C *Identify the equality property that justifies each statement. The properties that may be used include: the addition, subtraction, multiplication, and division properties listed in this section; the reflexive, symmetric, and transitive properties listed in Section 1-2; and the substitution principle listed in Section 1-2.*

75. If $x + 7 = 11$, then $x + 5 = 9$.
76. If $3x = 18$, then $6x = 36$.
77. If $3x = 4y$ and $4y = 5z$, then $3x = 5z$.
78. If $2x - 5 = 3x + 12$, then $-x - 5 = 12$.
79. If $x + y = 11$, then $11 = x + y$.
80. If $x = 7$ and $3x + 5 = 4y - 9$, then $26 = 4y - 9$.
81. If $12x = 30$, then $4x = 10$.
82. If $8x = 3y - 5$, then $8x + 5 = 3y$.

Determine which, if either, of the equations in (A) and (B) are equivalent to the given equation.

83. $4x - 5 = 6$; **(A)** $3x = 11 - x$ **(B)** $x = \frac{4}{11}$
84. $x + 3 = 4x - 9$; **(A)** $5x = 12$ **(B)** $3x = 6$

85. $2x - 7 = 7 - 5x$; **(A)** $7x = 14$ **(B)** $-14 = -3x$

86. $x + 8 = -9 + 4x$; **(A)** $-3x = 1$ **(B)** $3x = 17$

87. $1 - (x - 1) = 5$; **(A)** $-x = 5$ **(B)** $x = 3$

88. $2 - (3 - x) = 4(x - 5)$; **(A)** $3x = -21$ **(B)** $x = \frac{19}{3}$

Solve.

89. $x - [2x - 3(x - 4)] = 5$

90. $1 - \{x - 2[x - 3(4 - x)]\} = 1 - (1 - x)$

91. $0.25 - [1.75 - (4.25 - x)] = 4x + 3.75$

92. $x - \{1.1 - [2.2 - (3.3 - x)]\} = 4.4$

93. $\frac{1}{2} - [x - (\frac{1}{3} - \frac{1}{4}x)] = \frac{1}{6}$

94. $x - \{\frac{2}{3} - [\frac{3}{4} - (\frac{5}{6} - x)]\} = \frac{1}{12} + x$

95. $1.2 - [\frac{3}{5} - (x - \frac{4}{5})] = \frac{1}{5} - [0.4 - (0.7 - x)]$

96. $\frac{1}{4} - [0.75 - (x - 0.50)] = 1.75 - [\frac{3}{4} - (\frac{1}{4} - x)]$

97. $x - [1 - (y + x)] = y - [1 - (x + y)]$; solve for y

98. $x - [1 - (y + x)] = y - [1 - (x + y)]$; solve for x

99. $\frac{1}{3} - [x - (y - \frac{1}{6})] = x - [y - (x - \frac{1}{2})]$; solve for x

100. $\frac{1}{3} - [x - (y - \frac{1}{6})] = x - [y - (x - \frac{1}{2})]$; solve for y

1-8 Word Problems

- A Strategy for Solving Word Problems
- Number Problems
- Geometric Problems

A great many practical problems can be solved using algebraic techniques—so many, in fact, there is no one method of attack that will work for all. However, in this section, we formulate a strategy that may help you organize your approach. Then we apply this strategy to solve some number and geometric problems.

♦ A STRATEGY FOR SOLVING WORD PROBLEMS

The following strategy will help organize your approach to solving word problems:

A Strategy for Solving Word Problems

1. **Read the problem carefully**—several times if necessary—until you understand the problem, know what is to be found, and know what is given.
2. **If appropriate, draw figures or diagrams** and label known and unknown parts. Look for formulas connecting the known quantities with the unknown quantities.
3. **Let one of the unknown quantities be represented by a variable**, say x, and try to represent all other unknown quantities in terms of x. This is an important step and must be done carefully. Be sure you clearly understand what you are letting x represent.
4. **Form an equation relating the unknown quantities to the known quantities.** This step may involve the translation of an English sentence into an algebraic sentence, the use of relationships in a geometric figure, the use of certain formulas, and so on.
5. **Solve the equation**, and write answers to *all* parts requested in the original problem.
6. **Check all solutions** in the original problem.

Steps 3 and 4 in the strategy are often referred to as "setting up" the word problem. Step 4, in particular, can be difficult, and there is no systematic procedure that will work for all problems.

♦ **NUMBER PROBLEMS**

In Section 1-2 we translated verbal forms into symbolic forms. We now take advantage of that experience to solve a variety of problems about numbers.

Example 1
A Number Problem

Find a number such that 16 more than twice the number is 1 more than 5 times the number.

Solution   Let x = The number. We symbolize each part of the problem as follows:

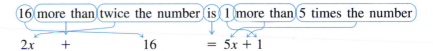

We now solve the equation:

$$2x + 16 = 5x + 1$$
$$15 = 3x$$
$$5 = x$$

That is, $x = 5$.

Check   16 more than twice the number: $2(5) + 16 = 26$

1 more than 5 times the number: $5(5) + 1 = 26$

Matched Problem 1   Find a number such that 12 more than 3 times the number is 5 times the number.

Example 2
A Number Problem

Find a number such that 2 less than twice the number is 5 times the quantity that is 2 more than the number.

Solution   Let x = The number. Symbolize each part:

2 less than twice the number: $2x - 2$ Not $2 - 2x$

is: =

5 times the quantity that is 2 more than the number: $5(x + 2)$ Why would $5x + 2$ be incorrect?

Write an equation and solve:

$$2x - 2 = 5(x + 2)$$
$$2x - 2 = 5x + 10$$
$$-3x = 12$$
$$x = -4$$

Checking is left to you.

Matched Problem 2   Find a number such that 4 times the quantity that is 2 less than the number is 1 more than 3 times the number. Write an equation and solve.

Example 3
A Number Problem

Find three consecutive even numbers such that twice the first plus the third is 10 more than the second.

Solution   Let

$$x = \text{First of three consecutive even numbers}$$
$$x + 2 = \text{Second consecutive even number}$$

The difference between any two consecutive even numbers is 2.

$$x + 4 = \text{Third consecutive even number}$$

Form an equation and solve:

Twice the first plus the third is 10 more than the second

$$2x \qquad + \quad (x + 4) \ = \ (x + 2) \quad + \quad 10$$
$$2x + x + 4 = x + 2 + 10$$
$$3x + 4 = x + 12$$
$$2x = 8$$
$$x = 4$$
$$x + 2 = 6$$
$$x + 4 = 8$$

$\left.\right\}$ Three consecutive even numbers

Thus, the three consecutive even numbers are 4, 6, and 8.

Check   Twice the first plus the third: $2 \cdot 4 + 8 = 8 + 8 = 16$

10 more than the second: $6 + 10 = 16$

Matched Problem 3   Find three consecutive odd numbers such that 4 times the first, minus the third, is the same as the second. Write an equation and solve.

◆   **GEOMETRIC PROBLEMS**

Recall that the perimeter of a triangle or rectangle is the distance around the figure, as illustrated in Figure 1.

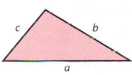   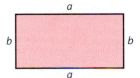

(**A**) Triangle: $P = a + b + c$ (**B**) Rectangle: $P = 2a + 2b$

Figure 1

Example 4
A Geometric Problem

If one side of a triangle is one-fourth the perimeter, the second side is 7 meters, and the third side is two-fifths the perimeter, what is the perimeter?

Solution   Let P = Perimeter. Draw a triangle and label its sides, as shown. Thus,

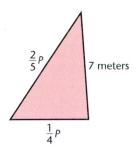

$$P = a + b + c$$

$$P = \tfrac{1}{4}P + 7 + \tfrac{2}{5}P$$

$$\boxed{20 \cdot P = 20 \cdot \tfrac{1}{4}P + 20 \cdot 7 + 20 \cdot \tfrac{2}{5}P}$$ Clear fractions by multiplying by 20.

$$20P = 5P + 140 + 8P$$

$$7P = 140$$

$$P = 20 \text{ meters}$$

Check

$$\text{Side 1} = \tfrac{1}{4}P = \tfrac{1}{4} \cdot 20 = 5 \text{ meters}$$

$$\text{Side 2} = 7 \text{ meters}$$

$$\text{Side 3} = \tfrac{2}{5}P = \tfrac{2}{5} \cdot 20 = \underline{8 \text{ meters}}$$

$$20 \text{ meters} \quad \text{Perimeter}$$

Matched Problem 4   If one side of a triangle is one-third the perimeter, the second side is 7 centimeters, and the third side is one-fifth the perimeter, what is the perimeter of the triangle? Set up an equation and solve.

Example 5
A Geometric Problem

Find the dimensions of a rectangle with perimeter 84 centimeters if its width is two-fifths its length.

Solution   Draw a rectangle and label its sides. If x = Length, then $\frac{2}{5}x$ = Width. Begin with the formula for the perimeter of a rectangle:

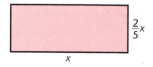

$$2a + 2b = P$$

$$2x + 2 \cdot \tfrac{2}{5}x = 84$$

$$2x + \tfrac{4}{5}x = 84$$

$$\boxed{5 \cdot 2x + 5 \cdot \tfrac{4}{5}x = 5 \cdot 84}$$ Clear fractions.

$$10x + 4x = 420$$

$$14x = 420$$

$$x = 30 \text{ centimeters} \qquad \text{Length}$$

$$\tfrac{2}{5}x = 12 \text{ centimeters} \qquad \text{Width}$$

Checking is left to you.

Matched Problem 5   Find the dimensions of a rectangle with perimeter 176 centimeters if its width is three-eighths its length. Write an equation and solve.

Answers to Matched Problems
1. $x = 6$ 2. $4(x - 2) = 3x + 1$; $x = 9$
3. $4x - (x + 4) = x + 2$; 3, 5, 7   4. $P = \frac{1}{3}P + 7 + \frac{1}{5}P$; $P = 15$ centimeters
5. $2x + 2 \cdot \frac{3}{8}x = 176$; 64 by 24 centimeters

EXERCISE 1-8

A *If x represents a number, write an algebraic expression for each of the following numbers:*

1. Twice x
2. 3 times x
3. 3 less than x
4. 2 more than x
5. 2 more than twice x
6. 5 less than twice x
7. 4 less than 3 times x
8. 5 more than 6 times x
9. 1 more than the negative of $2x$
10. 1 less than the negative of $3x$

Find a number meeting each of the indicated conditions. Write an equation using x and solve.

11. Twice x is 3 less than x.
12. 3 less than x is 2 more than twice x.
13. 2 more than twice x is 4 less than 3 times x.
14. 4 less than 3 times x is 1 more than the negative of $2x$.
15. 1 more than the negative of $2x$ is 3 times x.
16. 3 times x is 2 more than x.
17. 2 more than x is 5 less than twice x.

18. 5 less than twice x is 5 more than 6 times x.
19. 5 more than 6 times x is 1 less than the negative of $3x$.
20. 1 less than the negative of $3x$ is twice x.

Set up appropriate equations and solve.

21. A 12-foot steel rod is cut into two pieces so that one piece is 3 feet less than twice the length of the other piece. How long is each piece?
22. A 32-centimeter string is cut into two pieces so that one piece is 4 centimeters more than 3 times the length of the other piece. How long is each piece?
23. Find the dimensions of a rectangle with perimeter 36 feet if the width is 6 feet shorter than the length.
24. Find the dimensions of a rectangle with perimeter 54 meters if the length is 7 meters longer than the width.

B 25. Find the dimensions of a rectangle with perimeter 66 centimeters if its length is 3 centimeters more than twice its width.
26. Find the dimensions of a rectangle with perimeter 128 meters if its length is 6 meters less than 4 times the width.
27. Find a number such that 2 less than one-sixth the number is 1 more than one-fourth the number.

28. Find a number such that 5 less than half the number is 3 more than one-third the number.

29. Find three consecutive odd numbers such that the sum of the first and second is 5 more than the third.

30. Find three consecutive odd numbers such that the sum of the second and third is 1 more than 3 times the first.

31. Find the dimensions of a rectangle with perimeter 84 meters if its width is one-sixth its length.

32. Find the dimensions of a rectangle with perimeter 72 centimeters if its width is one-third its length.

33. Find a number such that 4 less than three-fifths the number is 8 more than one-third the number.

34. Find a number such that 5 more than two-thirds the number is 10 less than one-fourth the number.

Find a number meeting each of the indicated conditions. Write an equation using x and solve.

35. 7 times a number is 12 less than 4 times the number.

36. 6 times a number is 24 more than 3 times the number.

37. 3 more than twice the number is 12 less than 3 times the number.

38. 1 more than 5 times the number is 6 times 1 more than the number.

The area of a trapezoid is given by the formula $A = \frac{1}{2}h(B + b)$, where h is the height, B is the base, and b is the upper base. See the figure.

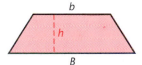

39. Find the upper base of a trapezoid with area 100 square inches, height 10 inches, and base 12 inches.

40. Find the base of a trapezoid with area 120 square inches, height 12 inches, and upper base 6 inches.

C *Find a number meeting each of the indicated conditions. Write an equation using x and solve.*

41. Three consecutive integers whose sum is 96

42. Three consecutive integers whose sum is 78

43. Three consecutive even numbers whose sum is 42

44. Three consecutive even numbers whose sum is 54

Set up appropriate equations and solve.

45. The sum of two numbers is 55 and their quotient is 4. [Recall that $a \div b = 4$ means $a = 4b$.]

46. The sum of two numbers is 36 and their quotient is 3.

47. The sum of two numbers is 48 and their quotient is 5.

48. The sum of two numbers is 49 and their quotient is 6.

49. If one side of a triangle is two-fifths the perimeter P, the second side is 70 centimeters, and the third side is one-fourth the perimeter, what is the perimeter?

50. If one side of a triangle is one-fourth the perimeter P, the second side is 3 meters, and the third side is one-third the perimeter, what is the perimeter?

51. On a trip across the Grand Canyon in Arizona, a group traveled one-third the distance by mule, 6 kilometers by boat, and one-half the distance by foot. How long was the trip?

52. A high-diving tower is located in a lake. If one-fifth the height of the tower is in sand, 6 meters in water, and one-half the total height in air, what is the total height of the tower?

53. Two numbers add to 40 and have the property that 3 times the smaller number is 10 more than twice the larger. Find the smaller number. [Note that if two numbers add to 40, they can be denoted by x and $40 - x$.]

54. Two numbers add to 45 and have the property that 9 times the smaller number is 9 more than twice the larger. Find the smaller number.

55. Two numbers add to 70 and have the property that 3 times the smaller number is as much more than the greater number as 3 times the greater exceeds 7 times the smaller. Find the smaller number.

56. Two numbers add to 60 and have the property that 5 times the smaller number exceeds 92 by the same amount as twice the greater exceeds 44. Find the smaller number.

57. A rectangle has length equal to 3 times its width. If the width is increased by 20 feet and the length decreased by 20 feet, the rectangle becomes a square. Find the width of the rectangle.

58. A rectangle has length equal to 3 times its width. If the width is increased by 20 feet and the length decreased by 20 feet, the rectangle increases in area by 600 square feet. Find the width of the rectangle.

59. If one-half the supplement of an angle is decreased by one-sixth of the angle, the result is an angle of 10°. Find the original angle. (The supplement of an angle is 180° minus the angle.)

60. If one-fourth the supplement of an angle is increased by one-half of the angle, the result is the original angle. Find the original angle.

CHAPTER SUMMARY

1-1   BASIC CONCEPTS

A **set** is a collection of objects. Sets are usually described by **listing** {List of elements} or by a **rule** {x | Rule that determines that x is a member}. If each element of set A is also in set B, we say A is a **subset** of B and write $A \subset B$. A **variable** is a symbol that represents unspecified elements from a **replacement set;** a **constant** is a symbol for one object in a set. An **algebraic expression** is comprised of variables, constants, mathematical operation signs, and grouping symbols. Expressions joined by plus or minus signs are called **terms;** those joined by multiplication are **factors.** To **evaluate** an algebraic expression for particular values of the variables means to replace each variable by its given value, and then calculate the resulting arithmetic value.

The **real number system** is outlined in the diagram:

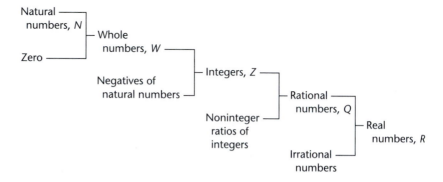

The real numbers can be represented as points on a **real number line,** or **real line,** where each point is associated with a single real number called the **coordinate** of the point. The point with coordinate 0 is called the **origin,** with **positive real numbers** to the right and **negative real numbers** to the left.

1-2   EQUALITY AND INEQUALITY

Algebraic expressions joined by an **equality sign** are called **algebraic equations.** Equality satisfies these **equality properties:**

Reflexive property: $a = a.$

Symmetric property: If $a = b$, then $b = a$.

Transitive property: If $a = b$ and $b = c$, then $a = c$.

Substitution principle: If $a = b$, then either may be substituted for the other.

The **inequality symbols** $<, >, \leq, \geq$ denote **less than, greater than, less than or equal to,** and **greater than or equal to,** respectively; $a < b$ and $b > a$ mean $a + p = b$ for some positive number p. Inequality satisfies these **inequality properties:**

Trichotomy property: Exactly one of $a < b$, $a = b$, and $a > b$ is true.

Transitive property: If $a < b$ and $b < c$, then $a < c$.

For an **inequality statement**—that is, an inequality involving a variable—the numbers that make it true are called the **solution set.** The solution set can be represented by a **graph of the inequality statement.**

1-3   REAL NUMBER PROPERTIES

The real numbers under addition and multiplication satisfy several basic properties that are listed and named in Section 1-3. Among these are the **commutative properties** ($a + b = b + a$; $a \cdot b = b \cdot a$), the **associative properties** [$(a + b) + c = a + (b + c)$; $(ab)c = a(bc)$], the **distributive properties** [$a(b + c) = ab + ac$; $(a + b)c = ac + bc$], and the **identity properties** ($a + 0 = 0 + a = a$; $1 \cdot a = a \cdot 1 = a$). These properties allow us to manipulate algebraic expressions. Two algebraic expressions are **equivalent** if they yield the same number for all possible values of the variables.

1-4   ADDITION AND SUBTRACTION OF REAL NUMBERS

For any real number x, $-x$ denotes its **additive inverse,** the number we add to x to get 0; this is also called the **negative of the number.** For any real number x, its **absolute value** $|x|$ is the distance from x to the origin, a nonnegative quantity. Rules for **addition of real numbers** are given in Section 1-4. **Subtraction** is defined in terms of addition by $a - b = a + (-b)$; $a - b = c$ means $b + c = a$.

1-5   MULTIPLICATION AND DIVISION OF REAL NUMBERS

Rules for **multiplication of real numbers** are given in Section 1-5. For any real number $b \neq 0$, the **multiplicative inverse** of b is denoted by $1/b$. **Division** is defined in terms of multiplication: $a \div b = a \cdot 1/b$; $a \div b = c$ means $bc = a$. Since 0 has no multiplicative inverse, **division by 0 is not defined.** Multiplication and division satisfy these **sign properties:**

$$(-a)b = -(ab) = a(-b) \qquad (-a)(-b) = ab$$

$$\frac{-a}{b} = \frac{a}{-b} = -\frac{a}{b} \qquad \frac{-a}{-b} = \frac{a}{b} = -\frac{-a}{b} = -\frac{a}{-b}$$

1-6   EXPONENTS AND ORDER OF OPERATIONS

For a natural number n, the **base** a raised to the **exponent** n is $a^n = a \cdot a \cdot \cdots \cdot a$ (n factors). Natural number exponents satisfy this **property of exponents:** $a^m a^n = a^{m+n}$. The **order of operations** for arithmetic operations, unless **grouping symbols** indicate otherwise, is first to apply exponents; then do multiplications and divisions, left to right; and lastly, do additions and subtractions, left to right. The distributive properties are used to **combine like terms**—that is, terms with identical variables to the same power.

1-7   SOLVING EQUATIONS

A **solution,** or **root,** of an equation in one variable x is a replacement value for x that makes the equality true. To **solve an equation** means to find all solutions—the **solution set.** Two equations are **equivalent** if they have the same solution set. Equations are generally solved by performing operations that produce simpler equivalent equations. The following **properties of equality** yield equivalent equations:
1. If $a = b$, then $a + c = b + c$ and $a - c = b - c$.
2. If $a = b$ and $c \neq 0$, then $a \cdot c = b \cdot c$ and $a/c = b/c$.

An **equation-solving strategy** is to
1. Clear fractions.
2. Simplify both sides.
3. Get all variable terms to one side, constants to the other, and combine like terms.
4. Isolate the variable.

A **linear equation** in a variable x involves no power of x higher than 1 and involves x only in additions, subtractions, and multiplications. Any linear equation in the variable x is equivalent to $ax + b = 0$ and has the unique solution $x = -b/a$ if $a \neq 0$. If $a = 0$, then

$ax + b = 0$ has either no solution or infinitely many solutions. The same strategy can be applied to solving certain **literal equations,** that is, equations involving more than one variable that are to be solved for one of the variables.

1-8 WORD PROBLEMS

A **general strategy for solving word problems** is, briefly:
1. Read the problem very carefully.
2. Write down the important facts and relationships.
3. Identify the unknown quantities in terms of one variable.
4. Find the equation.
5. Solve the equation. Write down all solutions asked for.
6. Check the solutions.

CHAPTER REVIEW EXERCISE

Work through all the problems in this chapter review and check answers in the back of the book. Answers to all the problems are there, and following each answer is a number in italics indicating the section in which that type of problem is discussed. Where weaknesses show up, review appropriate sections in the text.

A **1.** In the following list of numbers

$3.127\overline{127}$, 143, -1.43, $\sqrt{5}$, $\frac{2}{3}$, π, -1, 0, 12, $-\frac{3}{7}$

    (A) Identify the natural numbers.
    (B) Identify the whole numbers.
    (C) Identify the integers.
    (D) Identify the rational numbers.
    (E) Identify the irrational numbers.

2. Graph on a real number line:
    (A) $x < -1$ **(B)** $-4 \le x < 3$

3. Rewrite $5^6 5^7$ as a power of 5.

4. Rewrite $x^2 x^8$ as a power of x.

Evaluate.

5. $3 \cdot 7 - 4$ **6.** $7 + 2 \cdot 3$

7. $(-8) + 3$ **8.** $(-9) + (-4)$

9. $(-3) - (-9)$ **10.** $4 - 7$

11. $0 - (-3)$ **12.** $(-12) - 0$

13. $(-7)(-4)$ **14.** $3(-6)$

15. $(-16)/4$ **16.** $(-12)/(-2)$

17. $(-6)/0$ **18.** $0/(-3)$

19. $10 - 3(6 - 4)$ **20.** $(-8) - (-2)(-3)$

21. $(-9) - [(-12)/3]$ **22.** $(4 - 8) + 4(-2)$

23. $|-8|$ **24.** $-(-5)$

25. $-[-(-3)]$ **26.** $-|-(-2)|$

27. $-(|-3| + |-2|)$ **28.** $-(|-8| - |3|)$

Remove parentheses and simplify using commutative, associative, and distributive properties.

29. $7 + (x + 3)$ **30.** $(3x)5$

31. $(2x)(4y)$

32. $(y + 7) + (x + 2) + (z + 3)$

33. $0 + (1x + 2)$ **34.** $(x + 0) + 0y$

35. $3(x - 5)$ **36.** $(a + b)7$

37. $2 + 3(x + 4)$ **38.** $(x - 1)2 + 3$

Replace each question mark with $<$ or $>$ to form a true statement.

39. $7 \,?\, 2$ **40.** $-7 \,?\, -2$

41. $-12 \,?\, 0$ **42.** $-342 \,?\, -3$

43. $0 \,?\, -45$ **44.** $-50 \,?\, 20$

Combine like terms.

45. $3a + 8a - 5a$ **46.** $3a - (7a - 9a)$

47. $2x + 3y - 4x + 5y - 6x$

48. $5x - 3y + 2z - 1$

49. $2x^3y - 6xy^3 + 10x^3y + x^2y^2$

50. $2xy - 3(xy - 2xy^2) - 3xy^2$

B *Evaluate.*

51. $2[9 - 3(3 - 1)]$ **52.** $6 - 2 - 3 - 4 + 5$

53. $[(-3) - (-3)] - (-4)$ **54.** $[-(-4)] + (-|-3|)$

55. $[(-16)/2] - (-3)(4)$

56. $(-2)(-4)(-3) - \dfrac{-36}{(-2)(9)}$

57. $2\{9 - 2[(3 + 1) - (1 + 1)]\}$

58. $(-3) - 2\{5 - 3[2 - 2(3 - 6)]\}$

59. $\dfrac{12 - (-4)(-5)}{4 + (-2)} - \dfrac{-14}{7}$

60. $\dfrac{24}{(-4) + 4} - \dfrac{24}{(-4) + 4}$

61. $-3 \cdot 5^2 + 5(-3)^2$

62. $-(3 \cdot 5)^2 + 5(-3)^2$

63. $15 \div (-5) \cdot 6^2$

64. $3x \div [(-x) \cdot (x + 1)]^2$ for $x = 5$

65. $3[14 - x(x + 1)]$ for $x = 3$

66. $-(-x)$ for $x = -2$

67. $-(|x| - |w|)$ for $x = -2$, $w = -10$

68. $(x + y) - z$ for $x = 6$, $y = -8$, $z = 4$

69. $\left(2x - \dfrac{z}{x}\right) - \dfrac{w}{x}$ for $w = -10$, $x = -2$, $z = 0$

70. $\dfrac{(xyz + xz) - z}{z}$ for $x = -6$, $y = 0$, $z = -3$

71. $\dfrac{x - 3y}{z - x} - \dfrac{z}{xy}$ for $x = -3$, $y = 2$, $z = -12$

Solve for x.

72. $2x - 3 = 4x + 5$

73. $x - y = 2x + 3y + 4$

74. $\frac{1}{3}x + \frac{1}{4}x = \frac{5}{6}$

75. $x - (5 - 3x) = 2x - 3 - (2 - 2x)$

76. $3x - 4 - (5 - 6x) = 2x - 1 - (7x - 8)$

77. $x + y = xy$

In Problems 78–83, translate each statement into an algebraic equation or inequality statement using x as the only variable. Do not solve the equation or inequality.

78. $x - 1$ is positive.

79. $2x + 3$ is not negative.

80. 50 is 10 less than twice a certain number.

81. x is less than 12 less than twice x.

82. x is greater than or equal to -5 and less than 5.

83. 8 more than a certain number is 5 times the number that is 6 less than the certain number.

84. In a rectangle with area 1,200 square centimeters, the width is 10 centimeters less than the length x. Write an equation relating the area to the length and width, using x as the only variable.

85. If the length of a rectangle is 5 meters longer than its width x and the perimeter is 43 meters, write an algebraic equation relating the sides and the perimeter.

C *In Problems 86–92, state the real-number property that justifies each statement.*

86. $5 + (x + 3) = 5 + (3 + x)$

87. $5 + (3 + x) = (5 + 3) + x$

88. $5(x3) = 5(3x)$

89. $5(3x) = (5 \cdot 3)x$

90. $(x + y) + 0 = x + y$

91. $(ab) + [-(ab)] = 0$

92. $3(x - 4) = 3x - 12$

93. Evaluate $uv - 3\{x - 2[(x + y) - (x - y)] + u\}$ for $u = -2$, $v = 3$, $x = 2$, and $y = -3$.

94. Evaluate for $w = -4$ and $x = 2$:

$$\dfrac{5w}{x - 7} - \dfrac{wx - 4}{x - w}$$

95. Evaluate for $x = -1$:

$$1 - \dfrac{x}{x + \dfrac{3}{x - \frac{1}{2}}} + \dfrac{x}{\dfrac{5}{x + \frac{1}{2}} - 1} + 1$$

96. Find a number such that 3 times the quantity that is 4 more than the number is 5 times the original number.

97. Three consecutive natural numbers have the property that twice the sum of the first two is 11 more than 3 times the third. Find the numbers.

98. Two consecutive natural numbers have the property that their product is 7 less than the product of the smaller number times the successor of the larger. Find the numbers.

99. A 66-inch board is cut into three pieces in such a way that the longest piece is 3 times the length of the shortest piece and 2 times the length of the remaining piece. What are the three lengths?

100. If the width of a rectangle with perimeter 76 centimeters is 2 centimeters less than three-fifths the length, what are the dimensions of the rectangle?

CHAPTER PRACTICE TEST

The following practice test is provided for you to test your knowledge of the material in this chapter. You should try to complete it in 50 minutes or less. Answers in the back of the book indicate the section in the text where the material in the question is covered. Actual tests in your class may vary from this practice test in difficulty, length, or emphasis, depending on the goals of your course or instructor.

Evaluate.

1. $(-3)(-5) + [(-3) - (-5)] - (-3)[5 - (-3)]$

2. $|-5^2| - |-5|^2 - (-|5|^2)$

3. $(-3)(-6)^2 + (-3)(-6^2) - [(-3)(-6)]^2$

4. $(-6)^2 \div (-3) - [(-6) \div 3]^2 + (-6)^2 \div (-3)^2$

5. $xy^2 - x(x - y) + x \div y^2$ for $x = 12$, $y = -2$

Rewrite the expression in the form indicated.

6. $x^3 x^4 x^2$ as a power of x

7. $a(b + c - d)$ as a sum of three terms

8. $3x - 6xy + 9x^2$ in the form $3x(?)$

9. $4x^2 y + 2xy - 5x^2 y - 3xy + 6x^2 y$ with like terms combined

Graph on a real number line.

10. $x > -2$

11. $-4 < x \leq -1$

Translate each statement into an algebraic expression, equation, or inequality, using x as the only variable. Do not solve any equations or inequalities that result.

12. The sum of 3 times a given number and the product of that number with the number that is 2 greater than the given number.

13. The square of a number is less than the product of the quantity that is 4 less than the number and the quantity that is 1 more than the number.

14. Three more than a given number is 2 times the number that is 3 less than the given number.

Solve.

15. $3x - 4(x + 5) = 6 - (7 - x)$

16. $x(x + 2) = x^2 - 4x + 12$

17. $5x - 4(3 - x) = 4x + 5(x - 2) - 2$

18. $a + b + c = abc$ for a

Set up appropriate equations and solve.

19. Find a number such that 4 times the quantity that is 5 less than the number is 1 more than the original number.

20. In a triangle with perimeter 40 inches, the longest side is 3 times the length of one of the remaining sides and 2 inches longer than the third side. Find the lengths of the sides.

2

Polynomials and Factoring

In this chapter, we consider the operations of addition, subtraction, and multiplication on polynomials. We then look at the process of factoring a polynomial, that is, writing the polynomial as a product; and finally, we use factoring to solve equations.

Photo reference: see Exercise 2-1, Problem 100.

2-1 Addition and Subtraction of Polynomials

- ◆ Polynomials
- ◆ Removing Grouping Symbols
- ◆ Inserting Grouping Symbols
- ◆ Addition and Subtraction

Algebraic expressions such as

$$2x + 3 \qquad 4x^2 - x + 5 \qquad x^2 + 2xy + y^2$$

are called polynomials. In this section, we recall some terminology for describing polynomials and review how to add and subtract them.

◆ POLYNOMIALS

An algebraic expression involving only the operations of addition, subtraction, and multiplication on variables and constants is called a **polynomial.** Phrased differently, a polynomial is a sum of terms, each of which is a product of constants and variables raised to whole-number powers. Here are some examples:

POLYNOMIALS

$$3x - 1 \qquad x \qquad 2x^2 - 3x + 2 \qquad 5$$
$$x^3 - 3x^2y - 4y^2 \qquad 0 \qquad x^2 - \tfrac{2}{3}xy + 2y^2 \qquad (x + y)^2$$

In a polynomial, a variable cannot appear in a denominator, as an exponent, within a radical, or within absolute-value bars. Therefore, the following expressions are *not* polynomials:

NONPOLYNOMIALS

$$\frac{2x + 1}{3x^2 - 5x + 7} \qquad 3^x \qquad x^3 - 2\sqrt{x} + \frac{1}{x^3}$$

$$|2x^3 - 5| \qquad \frac{1}{x} \qquad \sqrt{x}$$

We see that a polynomial in one variable x is constructed by adding or subtracting constants and terms of the form ax^n, where a is a real number and n is a natural number. A polynomial in two variables x and y is constructed by adding or subtracting constants and terms of the form $ax^m y^n$, where again a is a real number and m and n are natural numbers.

It is convenient to identify certain types of polynomials for more efficient study. The concept of *degree* is used for this purpose. The **degree of a term** in a polynomial is the sum of the powers of the variables in the term. Thus, if a term has only one variable, the degree of the term is the power of the variable. For example, the degree of $3x^5$ is 5 and the degree of $3x^5y^2$ is 7. A nonzero constant term—that is, a term without any variables—is assigned degree 0. The **degree of a polynomial** is the degree of its term with the highest degree. The constant 0, either as a term or as a polynomial, is not assigned a degree.

Example 1
Determining Degrees

What is the degree of each term in the following polynomials? What is the degree of the polynomial?

(A) $4x^3$ (B) $3x^3y^2$ (C) $3x^5 - 2x^4 + x^2 - 3$
(D) $x^2 - 2xy + y^2 + 2x - 3y + 2$

Solution

(A) This polynomial and its only term are of degree 3.
(B) This polynomial and its only term are of degree 5.
(C) The degrees of the four terms, in order, are 5, 4, 2, and 0; the degree of the polynomial is 5.
(D) The degrees of the six terms, in order, are 2, 2, 2, 1, 1, and 0; the degree of the polynomial is 2.

Matched Problem 1

What is the degree of each term in the following polynomials? What is the degree of the polynomial?

(A) $7x^5$ (B) $3x^3y^4$ (C) $5x^3 - 7x^2 + x - 9$
(D) $4x^2y - xy^2 + xy + y^2 + x + 6$

We also call a one-term polynomial a **monomial,** a two-term polynomial a **binomial,** and a three-term polynomial a **trinomial.** For example,

Three terms	Two terms	One term	One term
$4x^3 - 3x + 7$	$5x - 2y$	$6x^4y^3$	7
Trinomial	Binomial	Monomial	Monomial
Degree 3	Degree 1	Degree 7	Degree 0

Recall that the constant factor in a term in a polynomial is called the *coefficient.* The constant term in a polynomial is sometimes referred to as a **constant coefficient.** The coefficient of a term in a polynomial includes the sign that precedes it, so a term such as $-3x^2$ should be thought of as $(-3)x^2$.

Example 2
Identifying Coefficients

What is the coefficient of each term in the following polynomial?

$$3x^4 - 2x^3 + x^2 - x + 3$$

Solution

$$3x^4 - 2x^3 + x^2 - x + 3 = 3x^4 + (-2)x^3 + 1 \cdot x^2 + (-1)x + 3$$

Coefficient of x^3 Coefficient of x
Coefficient of x^4 Coefficient of x^2 Constant term, or constant coefficient

The coefficient of x^4 is 3, that of x^3 is -2, that of x^2 is 1, and that of x is -1. The constant coefficient is 3.

Matched Problem 2 What is the coefficient of each term in the following polynomial?

$$5x^4 - x^3 - 3x^2 + x - 7$$

♦ **REMOVING GROUPING SYMBOLS**

We simplify expressions such as

$$2(3x - 5y) - 2(x + 3y)$$

by rewriting, using the various forms of the distributive property:

$$6x - 10y - 2x - 6y$$

and then combining like terms, to obtain

$$4x - 16y$$

 Note that in rewriting $-2(x + 3y)$ as $-2x - 6y$, both terms within the parentheses were multiplied by -2. To rewrite an expression such as $-(x + 3y)$ without parentheses, we think of the term as having coefficient -1:

$$-(x + 3y) = (-1)(x + 3y) = -x - 3y$$
Think

Example 3
Removing Parentheses

Remove parentheses and simplify:

(A) $2(3x^2 - 2x + 5) + (x^2 + 3x - 7)$
(B) $(x^3 - 2x + 6) - (2x^3 - x^2 + 3x - 3)$
(C) $[3x^2 - (2x + 1)] - (x^2 - 1)$
(D) $y - \{x - [2y - (x + y)]\}$

Solution **(A)** $2(3x^2 - 2x + 5) + (x^2 + 3x - 7)$

$$= 2(3x^2 - 2x + 5) + 1(x^2 + 3x - 7)$$
Think

$$= 6x^2 - 4x + 10 + x^2 + 3x - 7 \qquad \text{Combine like terms.}$$

$$= 7x^2 - x + 3$$

(B) $(x^3 - 2x + 6) - (2x^3 - x^2 + 3x - 3)$

$$= 1(x^3 - 2x + 6) + (-1)(2x^3 - x^2 + 3x - 3) \qquad \text{Be careful of the}$$
Think $\qquad\qquad\qquad\qquad\qquad\qquad\qquad\qquad\qquad$ signs here.

$$= x^3 - 2x + 6 - 2x^3 + x^2 - 3x + 3$$

$$= -x^3 + x^2 - 5x + 9$$

(C) $[3x^2 - (2x + 1)] - (x^2 - 1)$

$$= [3x^2 - 2x - 1] - (x^2 - 1) \qquad \text{Remove inner parentheses first.}$$

$$= 3x^2 - 2x - 1 - x^2 + 1$$

$$= 2x^2 - 2x$$

(D) $y - \{x - [2y - (x + y)]\}$ Remove innermost parentheses first.

$= y - \{x - [2y - x - y]\}$ Simplify within brackets.

$= y - \{x - [y - x]\}$ Remove inner brackets.

$= y - \{x - y + x\}$

$= y - x + y - x$

$= 2y - 2x$

Matched Problem 3   Remove parentheses and simplify:

(A) $3(u^2 - 2v^2) + (u^2 + 5v^2)$
(B) $(m^3 - 3m^2 + m - 1) - (2m^3 - m + 3)$
(C) $(x^3 - 2) - [2x^3 - (3x + 4)]$
(D) $\{x - [y - (z + x)] + x\} - [y - (x + z)]$

♦ **INSERTING GROUPING SYMBOLS**

We will sometimes have to insert grouping symbols into an expression such as

$$3x + 4y - 3xz - 4yz$$

We apply the distributive properties to rewrite the expression:

$$\begin{aligned} 3x + 4y - 3xz - 4yz &= 3x + 4y + (-3xz - 4yz) \\ &= 1(3x + 4y) + (-z)(3x + 4y) \\ &= 1(3x + 4y) - z(3x + 4y) \end{aligned}$$

Our work can be checked by removing the parentheses:

$$1(3x + 4y) - z(3x + 4y) = 3x + 4y - 3xz - 4yz$$

Example 4
Inserting Grouping Symbols

Replace each question mark with an appropriate algebraic expression:

(A) $a + b - c = a + (?)$ **(B)** $a + b - c = a - (?)$
(C) $x - 3y + 5z = x + (?)$ **(D)** $x - 3y + 5z = x - (?)$

Solution   **(A)** $a + b - c = a + (b - c)$
(B) $a + b - c = a + (b - c)$ Convert addition to subtraction of the opposite.
$= a - [-(b - c)]$ Take the opposite by multiplying by -1.
$= a - (-1)(b - c)$
$= a - (-b + c)$
(C) $x - 3y + 5z = x + (-3)y + 5z$
$= x + [(-3)y + 5z]$
$= x + (-3y + 5z)$
(D) $x - 3y + 5z = x + (-3)y + 5z$
$= x + (-3y + 5z)$
$= x - [-(-3y + 5z)]$
$= x - (3y - 5z)$

Matched Problem 4   Replace each question mark with an appropriate algebraic expression:

(A) $a - b + c = a + (?)$ **(B)** $a - b + c = a - (?)$
(C) $x + 3y - 5z = x + (?)$ **(D)** $x + 3y - 5z = x - (?)$

◆ ADDITION AND SUBTRACTION

Addition and subtraction of polynomials can be thought of in terms of removing parentheses and combining like terms, as illustrated in Example 3. Horizontal and vertical arrangements are illustrated in the next two examples. You should be able to work either way, letting the situation dictate the choice.

Example 5
Adding Polynomials

Add $x^4 - 3x^3 + x^2$, $-x^3 - 2x^2 + 3x$, and $3x^2 - 4x - 5$.

Solution   Add horizontally by removing parentheses and combining like terms:

$$(x^4 - 3x^3 + x^2) + (-x^3 - 2x^2 + 3x) + (3x^2 - 4x - 5)$$
$$= x^4 - 3x^3 + x^2 - x^3 - 2x^2 + 3x + 3x^2 - 4x - 5$$
$$= x^4 - 4x^3 + 2x^2 - x - 5$$

Add vertically by lining up like terms and adding their coefficients:

$$
\begin{array}{r}
x^4 - 3x^3 + x^2 \\
- x^3 - 2x^2 + 3x \\
3x^2 - 4x - 5 \\
\hline
x^4 - 4x^3 + 2x^2 - x - 5
\end{array}
$$

Matched Problem 5   Add $3x^4 - 2x^3 - 4x^2$, $x^3 - 2x^2 - 5x$, and $x^2 + 7x - 2$ both horizontally and vertically.

Example 6
Subtracting Polynomials

Subtract $4x^2 - 3x + 5$ from $x^2 - 8$.

Solution   Subtract horizontally by removing parentheses and combining like terms:

$$(x^2 - 8) - (4x^2 - 3x + 5) = x^2 - 8 - 4x^2 + 3x - 5$$
$$= -3x^2 + 3x - 13$$

Subtract vertically by changing the signs of the expression to be subtracted, lining up like terms, and adding their coefficients:

$$
\begin{array}{r}
x^2 - 8 \\
-4x^2 + 3x - 5 \\
\hline
-3x^2 + 3x - 13
\end{array}
$$

Change signs and add; be careful of sign errors here.

Matched Problem 6   Subtract $2x^2 - 5x + 4$ from $5x^2 - 6$.

Answers to Matched Problems

1. **(A)** 5   **(B)** 7
   (C) Degrees of terms, in order, are 3, 2, 1, and 0; degree of polynomial is 3.
   (D) Degrees of terms, in order, are 3, 3, 2, 2, 1, and 0; degree of polynomial is 3.
2. The coefficients, in order, are 5, -1, -3, 1, and -7.
3. **(A)** $4u^2 - v^2$ **(B)** $-m^3 - 3m^2 + 2m - 4$
   (C) $-x^3 + 3x + 2$ **(D)** $4x - 2y + 2z$
4. **(A)** $-b + c$ **(B)** $b - c$ **(C)** $3y - 5z$ **(D)** $-3y + 5z$
5. $3x^4 - x^3 - 5x^2 + 2x - 2$
6. $3x^2 + 5x - 10$

EXERCISE 2-1

A *Identify each of the following as a monomial, binomial, or trinomial, and give its degree.*

1. $3x^2 + 7$
2. $1 + 2y + 3y^2$
3. $-u^5 + u^4v^2 - u^3v$
4. $-7a^2 + 2b^5$
5. $-t^3 + 8$
6. $13x^3y^2z$
7. $5p^2r^3st^2$
8. $xyz + x^2z^2 + yz^2$
9. 123
10. $x - y - z$

Given the polynomial $7x^4 - 3x^3 - x^2 + x - 3$, indicate the following, with terms counted from the left:

11. The coefficient of the second term
12. The coefficient of the third term
13. The exponent of the variable in the second term
14. The exponent of the variable in the fourth term
15. The coefficient of the fourth term
16. The coefficient of the first term

Add.

17. $6x + 5$ and $3x - 8$
18. $3x - 5$ and $2x + 3$
19. $7x - 5$, $-x + 3$, and $-8x - 2$
20. $2x + 3$, $-4x - 2$, and $7x - 4$
21. $5x^2 + 2x - 7$, $2x^2 + 3$, and $-3x - 8$
22. $2x^2 - 3x + 1$, $2x - 3$, and $4x^2 + 5$
23. 3, $4 + 5a$, and $6a + 7b$
24. $3p + 2q + 1$, $4p + 5q$, and $6p$
25. $2x + 3y$, $4x - 5y$, and $6x + 7y$
26. $-x - yz$, $y + yz$, and $x - y$

Subtract.

27. $3x$ from $-2x$
28. $-3x$ from $-2x$
29. $4x^2$ from x^2
30. $-2x^2$ from $5x^2$
31. $-3x^2$ from $6x^2$
32. $-6x^2$ from $3x^2$
33. $3x - 8$ from $2x - 7$
34. $4x - 9$ from $2x + 3$
35. $2y^2 - 6y + 1$ from $y^2 - 6y - 1$
36. $x^2 - 3x - 5$ from $2x^2 - 6x - 5$

B *Add or subtract as indicated. Remove any grouping symbols present.*

37. $-x^2y + 3x^2y - 5x^2y$
38. $-4r^3t^3 - 7r^3t^3 - 7r^3t^3 + 9r^3t^3$
39. $y^3 + 4y^2 - 10 + 2y^3 - y + 7$
40. $3x^2 - 2x + 5 - x^2 + 4x - 8$
41. $a^2 - 3ab + b^2 + 2a^2 + 3ab - 2b^2$
42. $2x^2y + 2xy^2 - 5xy + 2xy^2 - xy - 4x^2y$
43. $x - 3y - 4(2x - 3y)$
44. $a + b - 2(a - b)$
45. $y - 2(x - y) - 3x$
46. $x - 3(x + 2y) + 5y$
47. $-2(-3x + 1) - (2x + 4)$
48. $-3(-t + 7) - (t - 1)$
49. $2(x - 1) - 3(2x - 3) - (4x - 5)$
50. $-2(y - 7) - 3(2y + 1) - (-5y + 7)$
51. $4t - 3[4 - 2(t - 1)]$
52. $3x - 2[2x - (x - 7)]$
53. $3[x - 2(x + 1)] - 4(2 - x)$
54. $2(3x + y) - 2[y - (3x + 2)]$

Replace each question mark with an appropriate algebraic expression.

55. $5 + m - 2n = 5 - (?)$
56. $2 + 3x - y = 2 - (?)$

57. $2 - x - y = 2 - x + (?)$

58. $4 - a - 2b = 4 - a + (?)$

59. $2 - x - y = 2 - x - (?)$

60. $4 - a - 2b = 4 - a - (?)$

61. $2 - x - y = 2 + (?)$

62. $4 - a - 2b = 4 + (?)$

63. $2 - x - y = 2 - (?)$

64. $4 - a - 2b = 4 - (?)$

65. $w^2 - y - z = w^2 - (?)$

66. $x - y - z + 5 = x - (?)$

Add.

67. $2x^4 - x^2 - 7$, $3x^3 + 7x^2 + 2x$, and $x^2 - 3x - 1$

68. $3x^3 - 2x^2 + 5$, $3x^2 - x - 3$, and $2x + 4$

69. $2x^3 - 3x^4 + 4x^5$, $1 - x + x^2$, and $-4x^4 - 3x^3 + 2x^2 + x$

70. $x + 2x^3 - 3x^5$, $2 - 4x^2 + 6x^4$, and $5x^5 - 4x^4 - 3x^3 - 2x^2 + x - 6$

71. $xy - x^2y + xy^2 - x^2y^2$, $3xy - 4x^2y + 2xy^2 - 5x^2y^2$, and $6xy + 2x^2y - 3xy^2 + 5x^2y^2$

72. $a^2b^2 - ab + 1$, $2ab^2 - 3a^2b$, and $4a^2b - 3ab - 2ab^2 - 1$

Subtract.

73. $5x^3 - 3x + 1$ from $2x^3 + x^2 - 1$

74. $3x^3 - 2x^2 - 5$ from $2x^3 - 3x + 2$

75. $xy - x^2y + xy^2 - x^2y^2$ from $3xy - 4x^2y + 2xy^2 - 5x^2y^2$

76. $a^2b^2 - ab + 1$ from $2ab^2 - 3a^2b$

77. $1 - x + x^2$ from $-4x^4 - 3x^3 + 2x^2 + x$

78. $2 - 4x^2 + 6x^4$ from $5x^5 - 4x^4 - 3x^3 - 2x^2 + x - 6$

79. $3xy - 4x^2y + 2xy^2 - 5x^2y^2$ from $6xy + 2x^2y - 3xy^2 + 5x^2y^2$

80. $2ab^2 - 3a^2b$ from $4a^2b - 3ab - 2ab^2 - 1$

81. $2x^3 - 3x^4 + 4x^5$ from $-4x^4 - 3x^3 + 2x^2 + x$

82. $x + 2x^3 - 3x^5$ from $5x^5 - 4x^4 - 3x^3 - 2x^2 + x - 6$

83. $xy - x^2y + xy^2 - x^2y^2$ from $6xy + 2x^2y - 3xy^2 + 5x^2y^2$

84. $a^2b^2 - ab + 1$ from $4a^2b - 3ab - 2ab^2 - 1$

C *Remove grouping symbols and combine like terms.*

85. $2t - 3\{t + 2[t - (t + 5)] + 1\}$

86. $x - \{x - [x - (x - 1)]\}$

87. $w - \{x - [z - (w - x) - z] - (x - w)\} + x$

88. $3x^2 - 2\{x - x[x + 4(x - 3)] - 5\}$

89. $\{[(x - 1) - 1] - x\} - 1$

90. $1 - \{1 - [1 - (1 - x)]\}$

91. $x - \{1 - [x - (1 - x)]\}$

92. $x - \{[1 - (x - 1)] - x\}$

APPLICATIONS

93. The width of a rectangle is 5 meters less than its length. If x is the length of the rectangle, write an algebraic expression that represents the perimeter P of the rectangle, and simplify the expression.

94. Repeat Problem 93 if the length of the rectangle is 3 meters more than twice its width.

95. A pile of coins consists of nickels, dimes, and quarters. There are 5 fewer dimes than nickels and 2 more quarters than dimes. If x equals the number of nickels, write an algebraic expression that represents the value of the pile in cents. Simplify the expression. [*Hint:* If x represents the number of nickels, then what do $x - 5$ and $(x - 5) + 2$ represent?]

96. A parking meter contains dimes and quarters only. There are 4 fewer quarters than dimes. If x represents the number of dimes, write an algebraic expression that represents the total value of all coins in the meter in cents. Simplify the expression.

★ **97.** A board is to be cut into four pieces. The largest piece is to be 3 times as long as the smallest; the other two pieces are each to be twice as long as the smallest. If x represents the length of the smallest piece, write an algebraic expression for the total length of the board. Simplify the expression.

★ **98.** A wire is to be cut into three pieces so that the second piece is 3 feet longer than the first and the third is 8 feet longer than the sum of the other two. Let x represent the length of the smallest piece. Write an algebraic expression for the total length of the wire in feet. Simplify the expression.

★ **99.** A jogger runs for some time at a rate of 8 kilometers per hour and then runs twice as long at a rate of 12 kilometers per hour. Let t be the time, in hours, run at the slower pace. Write an algebraic expression for the total distance run. Simplify the expression. (Recall that Distance = Rate × Time.)

★ **100.** A racer drives for 2 hours at one speed and then for another hour and a half at a speed 12 miles per hour slower. Let s be the initial speed. Write an algebraic expression for the total distance driven. Simplify the expression.

2-2   Multiplication of Polynomials

♦   Multiplication of Monomials
♦   Multiplication of Polynomials
♦   Multiplication of Binomials
♦   Squaring Binomials

To develop a method for multiplying polynomials, we first multiply monomials using Property 1 of exponents from Section 1-6. We can then multiply any polynomial by making repeated use of distributive properties. Multiplication and squaring of binomials are both required so often in mathematics that it will be helpful to be able to perform these operations quickly and efficiently. In this section we will show you how.

♦   **MULTIPLICATION OF MONOMIALS**

Monomials were multiplied in Example 1 of Section 1-6 using Property 1 of exponents. We review the process.

Example 1 **Multiplying Monomials**	Multiply: **(A)** x^3x^5 **(B)** $(3m^{12})(5m^{23})$ **(C)** $(-3x^3y^4)(2x^2y^3)$

Solution   **(A)** $x^3x^5 \boxed{= x^{3+5}} = x^8$ $x^3x^5 \neq x^{3 \cdot 5}$

(B) $(3m^{12})(5m^{23}) \boxed{= 3 \cdot 5m^{12+23}} = 15m^{35}$

(C) $(-3x^3y^4)(2x^2y^3) \boxed{= (-3)(2)x^{3+2}y^{4+3}} = -6x^5y^7$

Matched Problem 1   Multiply: **(A)** y^4y^7 **(B)** $(9x^4)(3x^2)$ **(C)** $(4u^3v^2)(-3uv^3)$

♦   **MULTIPLICATION OF POLYNOMIALS**

How do we multiply polynomials with more than one term? The distributive property plays a central role in the process and leads directly to the following mechanical rule:

> **Mechanics of Multiplying Two Polynomials**
>
> To multiply two polynomials, multiply each term of one by each term of the other. Then add like terms.

Example 2 **Multiplying Polynomials**	Multiply: **(A)** $3x^2(2x^2 - 3x + 4)$ **(B)** $(2x - 3)(3x^2 - 2x + 3)$

Solution   (A) $3x^2(2x^2 - 3x + 4) = 6x^4 - 9x^3 + 12x^2$

(B) $(2x - 3)(3x^2 - 2x + 3)$

$$= 2x(3x^2 - 2x + 3) + (-3)(3x^2 - 2x + 3)$$

Apply the distributive property; be careful of the signs.

$$= 6x^3 - 4x^2 + 6x - 9x^2 + 6x - 9$$

$$= 6x^3 - 13x^2 + 12x - 9$$

We can also carry out this multiplication vertically, in much the same way as we multiply large positive integers:

$$
\begin{array}{r}
3x^2 - 2x + 3 \\
2x - 3 \\
\hline
6x^3 - 4x^2 + 6x \\
- 9x^2 + 6x - 9 \\
\hline
6x^3 - 13x^2 + 12x - 9
\end{array}
$$

Note that either way, each term in $3x^2 - 2x + 3$ is multiplied by each term in $2x - 3$. In the vertical arrangement, by multiplying by $2x$ first the like terms line up more conveniently.

Matched Problem 2   Multiply:

(A) $2m^3(3m^2 - 4m - 3)$ (B) $(2x^2 + 3x - 1)(3x - 4)$

(C) $(2x^2 + 3x - 2)(3x^2 - 2x + 1)$

♦   **MULTIPLICATION OF BINOMIALS**

In the following sections, we will consider the problem of writing a polynomial as a product of two factors. To test possibilities and check answers, it is essential that you be able to find products of first-degree polynomials such as $(4x - 3)(2x + 1)$ and $(3x - y)(x + 2y)$ efficiently. We will use a horizontal arrangement and try to discover a method that will enable us to do this. We start by multiplying each term in the first binomial times each term in the second binomial in a particular order:

First terms   Outer terms   Inner terms   Last terms

$(4x - 3)(2x + 1)$ $(4x - 3)(2x + 1)$ $(4x - 3)(2x + 1)$ $(4x - 3)(2x + 1)$

Performing the four multiplications on one line, we obtain

Product of   Product of   Product of   Product of
First terms   Outer terms   Inner terms   Last terms

$$(4x - 3)(2x + 1) = 8x^2 + \quad 4x \quad - \quad 6x \quad - \quad 3$$

The inner and outer products are like terms and hence combine into one term. Thus,

$$(4x - 3)(2x + 1) = 8x^2 - 2x - 3$$

To speed up the process we combine the inner and outer products mentally. The method just described is called the **FOIL method,** for the sequence **F**irst, **O**uter, **I**nner, **L**ast in which the multiplications are carried out. Note that the product of two first-degree polynomials is a second-degree polynomial. A simple three-step process for carrying out the FOIL method is illustrated in Example 3.

<table>
<tr><td>

Example 3
Multiplying Using FOIL

</td><td>

Multiply by the FOIL method:

(A) $(3x + 1)(2x - 5)$ **(B)** $(x - 4y)(x + 4y)$
(C) $(x + a)(x + a)$

</td></tr>
<tr><td align="right">

Solution

</td><td>

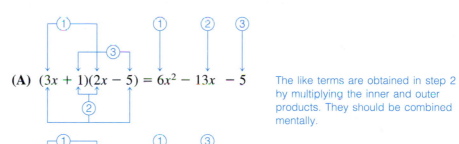

(A) $(3x + 1)(2x - 5) = 6x^2 - 13x - 5$ The like terms are obtained in step 2 by multiplying the inner and outer products. They should be combined mentally.

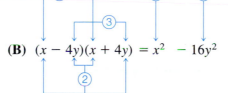

(B) $(x - 4y)(x + 4y) = x^2 - 16y^2$ There is no middle term because its coefficient is 0.

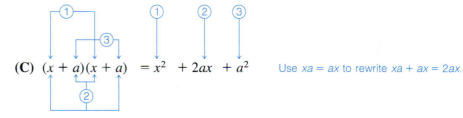

(C) $(x + a)(x + a) = x^2 + 2ax + a^2$ Use $xa = ax$ to rewrite $xa + ax = 2ax$.

</td></tr>
</table>

Matched Problem 3   Multiply by the FOIL method:

(A) $(2x - 1)(x + 4)$ **(B)** $(3x - y)(x + y)$
(C) $(a - b)(a - b)$ **(D)** $(3x + 4)(3x - 4)$

In the next section, we will try to reverse this process. Given a polynomial such as $6x^2 - 13x - 5$, we will want to find first-degree factors that produce this polynomial as a product. To test and check possible factors, it is necessary to be able to multiply factors such as $3x + 1$ and $2x - 5$ quickly and accurately.

The FOIL method works equally well on any binomials, not just those of degree 1. For arbitrary binomials, however, the outer and inner products will not necessarily be like terms.

Example 4
Multiplying Using FOIL

Multiply by the FOIL method:

$$(a^2 + 3b^2)(a - 2b^3)$$

Solution

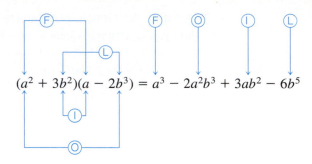

$$(a^2 + 3b^2)(a - 2b^3) = a^3 - 2a^2b^3 + 3ab^2 - 6b^5$$

Matched Problem 4   Multiply by the FOIL method: $(x^3 - 1)(x^2 + 3y)$

Example 5
A Business Application

The revenue from sales of a product is the selling price times the number of items sold. An automobile tire store sells one brand of tire for $80 each and at that price sells 90 per month. The store manager knows from past data that for each dollar by which the price is reduced, the monthly sales will increase by 5. Let x be the number of dollars by which the price is reduced. Write an expression for the resulting monthly revenue using x as the variable. Express the answer as a polynomial.

Solution   If the price is reduced by x dollars, the price is $80 - x$. Each dollar reduction results in 5 more tires sold, so the number of additional tires sold is $5x$, and the total number of tires sold is $90 + 5x$. Therefore, the revenue is

Price Number sold

$$(80 - x)(90 + 5x) = 7,200 + 400x - 90x - 5x^2$$
$$= -5x^2 + 310x + 7,200$$

Matched Problem 5   The store in Example 5 sells another brand of tire for $105 each and at that price sells 50 tires per month. For each dollar the price is reduced, the number of tires sold per month will increase by 3. Let x be the number of dollars by which the price is reduced. Write an expression for the resulting monthly revenue using x as the variable. Express the answer as a polynomial.

♦ SQUARING BINOMIALS

Since

$$(A + B)^2 = (A + B)(A + B) = A^2 + 2AB + B^2$$
$$(A - B)^2 = (A - B)(A - B) = A^2 - 2AB + B^2$$

we can formulate a simple mechanical rule for squaring any binomial directly.

Mechanical Rule for Squaring Binomials

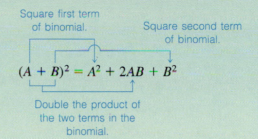

Square first term
of binomial.

Square second term
of binomial.

$$(A + B)^2 = A^2 + 2AB + B^2$$

Double the product of
the two terms in the
binomial.

We do the same thing for $(A - B)^2$, except that the sign of the middle term on the right becomes negative:

$$(A - B)^2 = A^2 - 2AB + B^2$$

Example 6
Squaring Binomials

Square each binomial:

(A) $(x + h)^2$ **(B)** $(2x + 1)^2$ **(C)** $(2x - 3y)^2$

Solution

(A) $(x + h)^2 = x^2 + 2(x)(h) + h^2 = x^2 + 2xh + h^2$
(B) $(2x + 1)^2 = (2x)^2 + 2(2x)(1) + 1^2 = 4x^2 + 4x + 1$
(C) $(2x - 3y)^2 = (2x)^2 - 2(2x)(3y) + (3y)^2$
$\qquad\qquad = 4x^2 - 12xy + 9y^2$

Matched Problem 6

Square each binomial:

(A) $(x + 5y)^2$ **(B)** $(5a - 2b)^2$ **(C)** $(3x + 7)^2$

Note that, in general,

$$(a + b)^2 \neq a^2 + b^2 \qquad (a + b)^2 = a^2 + 2ab + b^2$$

$$(a - b)^2 \neq a^2 - b^2 \qquad (a - b)^2 = a^2 - 2ab + b^2$$

Answers to
Matched Problems

1. **(A)** y^{11} **(B)** $27x^6$ **(C)** $-12u^4v^5$
2. **(A)** $6m^5 - 8m^4 - 6m^3$ **(B)** $6x^3 + x^2 - 15x + 4$
   (C) $6x^4 + 5x^3 - 10x^2 + 7x - 2$
3. **(A)** $2x^2 + 7x - 4$ **(B)** $3x^2 + 2xy - y^2$
   (C) $a^2 - 2ab + b^2$ **(D)** $9x^2 - 16$
4. $x^5 + 3x^3y - x^2 - 3y$ 5. $-3x^2 + 265x + 5,250$
6. **(A)** $x^2 + 10xy + 25y^2$ **(B)** $25a^2 - 20ab + 4b^2$ **(C)** $9x^2 + 42x + 49$

EXERCISE 2-2

A *Multiply.*

1. y^2y^3
2. x^3x^2
3. $(5y^4)(2y)$
4. $(2x)(3x^4)$
5. $(8x^{11})(-3x^9)$
6. $(-7u^9)(5u^7)$
7. $(-3u^4)(2u^5)(-u^7)$
8. $(2x^3)(-3x)(-4x^5)$
9. $(cd^2)(c^2d^2)$
10. $(a^2b)(ab^2)$
11. $(-3xy^2z^3)(-5xyz^2)$
12. $(-2xy^3z)(2x^3yz)$
13. $y(y + 7)$
14. $x(1 + x)$
15. $5y(2y - 7)$
16. $3x(2x - 5)$
17. $3a^2(a^3 + 2a^2)$
18. $2m^2(m^2 + 3m)$
19. $2y(y^2 + 2y - 3)$
20. $3x(2x^2 - 3x + 1)$
21. $7m^3(m^3 - 2m^2 - m + 4)$
22. $3x^2(2x^3 + 3x^2 - x - 2)$
23. $5uv^2(2u^3v - 3uv^2)$
24. $4m^2n^3(2m^3n - mn^2)$
25. $2cd^3(c^2d - 2cd + 4c^3d^2)$
26. $3x^2y(2xy^3 + 4x - y^2)$

B
27. $(3y + 2)(2y^2 + 5y - 3)$
28. $(2x - 1)(x^2 - 3x + 5)$
29. $(m + 2n)(m^2 - 4mn - n^2)$
30. $(x - 3y)(x^2 - 3xy + y^2)$
31. $(2m^2 + 2m - 1)(3m^2 - 2m + 1)$
32. $(x^2 - 3x + 5)(2x^2 + x - 2)$
33. $(a + b)(a^2 - ab + b^2)$
34. $(a - b)(a^2 + ab + b^2)$
35. $(2x^2 - 3xy + y^2)(x^2 + 2xy - y^2)$
36. $(a^2 - 2ab + b^2)(a^2 + 2ab + b^2)$

Multiply mentally.

37. $(x + 3)(x + 2)$
38. $(m - 2)(m - 3)$
39. $(a + 8)(a - 4)$
40. $(m - 12)(m + 5)$
41. $(t + 4)(t - 4)$
42. $(u - 3)(u + 3)$
43. $(m - n)(m + n)$
44. $(a + b)(a - b)$
45. $(4t - 3)(t - 2)$
46. $(3x - 5)(2x + 1)$
47. $(3x + 2y)(x - 3y)$
48. $(2x - 3y)(x + 2y)$
49. $(2m - 7)(2m + 7)$
50. $(3y + 2)(3y - 2)$
51. $(6x - 4y)(5x + 3y)$
52. $(3m + 7n)(2m - 5n)$

53. $(2s - 3t)(3s - t)$
54. $(2x - 3y)(3x - 2y)$
55. $(x^3 + y^3)(x + y)$
56. $(x^2 + y^3)(x^3 + y)$
57. $(2x - y^2)(x^2 + 3y)$
58. $(x^3 - y^2)(2x + y^2)$

Square each binomial.

59. $(3x + 2)^2$
60. $(4x + 3y)^2$
61. $(2x - 5y)^2$
62. $(2x - 7)^2$
63. $(6u + 5v)^2$
64. $(7p + 2q)^2$
65. $(2m - 5n)^2$
66. $(4x - 1)^2$
67. $(x^2 - 1)^2$
68. $(x^2 + 4)^2$
69. $(x^2 + y^2)^2$
70. $(x^2 - y^2)^2$

C *Simplify.*

71. $(x + 2y)^3$
72. $(2m - n)^3$
73. $(3x - 1)(x + 2) - (2x - 3)^2$
74. $(2x + 3)(x - 5) - (3x - 1)^2$
75. $2(x - 2)^3 - (x - 2)^2 - 3(x - 2) - 4$
76. $(2x - 1)^3 - 2(2x - 1)^2 + 3(2x - 1) + 7$
77. $-3x\{x[x - x(2 - x)] - (x + 2)(x^2 - 3)\}$
78. $2\{(x - 3)(x^2 - 2x + 1) - x[3 - x(x - 2)]\}$
79. $(2x - 1)(2x + 1)(3x^3 - 4x + 3)$
80. $(x - 1)(x - 2)(2x^3 - 3x^2 - 2x - 1)$
81. $[(3x - 2) + y]^2$
82. $[y + (x - 1)]^2$
83. $[(x + y) - 2][(x + y) + 1]$
84. $[(x - 1) + y][(x + 1) - y]$

In Problems 85–88, multiply out each side and solve the resulting equation.

85. $(x - 2)(x + 3) = (x + 4)(x - 5)$
86. $(x + 2)(x - 3) = (x - 4)(x + 5)$
87. $(x + 4)(x - 1) = (x - 5)(x + 2)$
88. $(x - 4)(x + 1) = (x + 5)(x - 2)$

89. If you are given two polynomials, one of degree m and another of degree n, and $m > n$, then what is the degree of their product?

90. What is the degree of the sum of the two polynomials in Problem 89?

APPLICATIONS

91. The length of a rectangle is 8 meters more than its width. If y is the length of the rectangle, write an algebraic expression that represents its area. Change the expression to a form without parentheses.

92. Repeat Problem 91 if the length of the rectangle is 3 meters less than twice the width.

93. A toy manufacturer can sell 5,000 of a particular toy if the selling price is $8 each. For each quarter dollar that the seller increases the price, the sales are expected to decrease by 100 units. Let x represent the number of quarters by which the price is increased. Write an expression for the resulting revenue using x as the variable.

★ 94. Repeat Problem 93, but now let x represent the selling price after the price is increased by some number of quarter dollar increments.

★ 95. The total yield from a crop is the number of plants times the yield per plant. A fruit grower can achieve a yield of 40 bushels per tree when 120 trees are planted in a particular orchard. If the number of trees is increased, the yield per tree drops due to

crowding. For each additional tree planted, the yield is expected to decrease by $\frac{1}{2}$ bushel per tree. Let x represent the number of trees planted. Write an expression for the total yield using x as the variable.

★ 96. Repeat Problem 95, but now let x represent the number of trees planted in addition to 120 existing trees.

97. Write an equation to represent the following situation: the product of a certain number and 1 more than the number is equal to the product of 1 less than the number times 3 more than the number.

98. Write an equation to represent the following situation: for three successive integers, the square of the largest is 43 more than the product of the other two. Simplify each side of the equation.

99. Solve the equation found in Problem 97.

100. Solve the equation found in Problem 98.

  ## 2-3   Basic Factoring Methods

♦ Factoring Out Common Factors
♦ Factoring by Grouping
♦ Recognizing Perfect Squares
♦ Sum and Difference of Two Squares

We can view the distributive property

$$a(b + c) = ab + ac$$

from left to right as multiplying out a product. If we view it from right to left, we are rewriting a sum as a product, that is, factoring the right-hand side. In this case, we are taking out the common factor a. This is the basis for the first factoring methods considered in this section. The remaining methods result from reversing common product forms.

♦ FACTORING OUT COMMON FACTORS

The common factors that are easiest to recognize are monomials.

Example 1
Taking Out Common Monomial Factors

Factor out factors common to all terms:

(A) $8x^2 - 14x$ **(B)** $3a^2b - 6ab + 15a^3b^2$

Solution   **(A)** $8x^2 - 14x = \boxed{2x \cdot 4x - 2x \cdot 7}$

$= 2x(4x - 7)$

(B) $3a^2b - 6ab + 15a^3b^2 \boxed{= \mathbf{3ab} \cdot a - \mathbf{3ab} \cdot 2 + \mathbf{3ab} \cdot 5a^2b}$

$$= \mathbf{3ab}(a - 2 + 5a^2b)$$

Matched Problem 1 Factor out factors common to all terms:

(A) $20x^2 + 4x$ **(B)** $4u^3v^2 - 10u^2v^4 + 18u^2v$

Common factors do not have to be monomials. Compare the following:

$$ax + ay = a(x + y)$$

$$(w + 1)x + (w + 1)y = (w + 1)(x + y)$$

In essence, the factoring is the same. We must be ready to recognize that $w + 1$ is a single number, just as a is in the first example. Similarly, the following two examples are also essentially the same:

$$3xz + 4yz = (3x + 4y)z$$

$$3x(2x + y) + 4y(2x + y) = (3x + 4y)(2x + y)$$

Common factors can be taken out on either side, left or right, because multiplication is commutative.

Example 2
Taking Out Common Factors

Factor out factors common to all terms:

(A) $3y(y + 2) - 5(y + 2)$ **(B)** $x^2(x - 1) + 2(x - 1)$

Solution **(A)** $3y(y + 2) - 5(y + 2) \boxed{= 3y(y + 2) - 5(y + 2)}$

$$= (3y - 5)(y + 2)$$

(B) $x^2(x - 1) + 2(x - 1) = (x^2 + 2)(x - 1)$

Matched Problem 2 Factor out factors common to all terms:

(A) $x(x^2 + y^2) - y(x^2 + y^2)$ **(B)** $2x(x - 5) - 3(x - 5)$

♦ **FACTORING BY GROUPING**

In some situations we may be able to take a polynomial with no apparent common factor and find one when the terms are properly grouped. In the next example, grouping the first two and last two terms will yield a common factor in two steps.

Example 3
Factoring by Grouping

Factor by grouping:

(A) $x^2 + x + 3x + 3$ **(B)** $3y^2 + 6y - 5y - 10$

(C) $2x^2 - 2xy - xy^2 + y^3$

Solution   **(A)** $x^2 + x + 3x + 3$ Group the first two and last two terms.

$\qquad = (x^2 + x) + (3x + 3)$ Remove common factors from each group. Think of $x^2 + x$ as $x \cdot x + x \cdot 1$.

$\qquad = x(x + 1) + 3(x + 1)$ Take out the common factor $x + 1$ to complete the factoring.

$\qquad = (x + 3)(x + 1)$

(B) $3y^2 + 6y - 5y - 10$ Group the first two and last two terms.

$\qquad = (3y^2 + 6y) + (-5y - 10)$ Remove common factors from each group.

$\qquad = 3y(y + 2) + (-5)(y + 2)$ Take out the common factor $y + 2$.

$\qquad = (3y - 5)(y + 2)$

(C) $2x^2 - 2xy - xy^2 + y^3$

$\qquad = (2x^2 - 2xy) + (-xy^2 + y^3)$

$\qquad = 2x(x - y) + y^2(-x + y)$ Recognize $-x + y = (-1)(x - y)$.

$\qquad = 2x(x - y) + (y^2)(-1)(x - y)$

$\qquad = (2x - y^2)(x - y)$

Matched Problem 3   Factor by grouping:

(A) $x^3 + 3x^2 + x + 3$ **(B)** $2x^2 + 4x - 3x - 6$

(C) $3ab^2 + b^3 + 3a^2 + ab$

Not all polynomials can be regrouped for factoring as easily as those in Example 3. You may have to rearrange the terms in order to regroup them successfully, and even this may not always work. Many polynomials with integer coefficients cannot be factored at all using integer coefficients.

Example 4
Factoring by Grouping

Factor $2x^2 + y^3 - xy^2 - 2xy$ by grouping.

Solution

$2x^2 + y^3 - xy^2 - 2xy$

$\qquad = (2x^2 + y^3) + (-xy^2 - 2xy)$ If we group the first two and last two terms, no common factor appears.

$\qquad = (2x^2 + y^3) + (-xy)(y + 2)$

$2x^2 + y^3 - xy^2 - 2xy$ Rearrange the terms and try again.

$\qquad = 2x^2 - xy^2 + y^3 - 2xy$

$\qquad = (2x^2 - xy^2) + (y^3 - 2xy)$

$\qquad = x(2x - y^2) + y(y^2 - 2x)$

$\qquad = x(2x - y^2) + y(-1)(2x - y^2)$ $y^2 - 2x = (-1)(2x - y^2)$

$\qquad = (x - y)(2x - y^2)$

Still another rearrangement of the terms,

$$2x^2 + y^3 - xy^2 - 2xy = 2x^2 - 2xy - xy^2 + y^3$$

yields the polynomial as it appeared in Example 3(C), where grouping also worked.

Matched Problem 4   Factor $y^3 + 3x^2 + xy + 3xy^2$ by grouping.

♦ **RECOGNIZING PERFECT SQUARES**

Consider the product

$$(x + b)^2 = (x + b)(x + b) = x^2 + 2bx + b^2$$

The role played by b identifies the expression $x^2 + 2bx + b^2$ as a **perfect square:**

$$x^2 + 2bx + b^2 = (x + b)^2$$

The last term is the square
of one-half the coefficient
of x.

One-half the
coefficient of x

**Example 5
Recognizing
Perfect Squares**

Identify the perfect squares among these second-degree polynomials; factor those that are.

(A) $x^2 + 10x + 25$ **(B)** $x^2 + 4x + 16$
(C) $x^2 - 6x + 9$ **(D)** $4x^2 + 20x + 25$

Solution **(A)** $x^2 + 10x + 25$ is a perfect square, since the constant term is the square of one-half the coefficient of x: $25 = (\frac{1}{2} \cdot 10)^2$. Thus, $x^2 + 10x + 25 = (x + 5)^2$.

(B) $x^2 + 4x + 16$ is not a perfect square, since $16 \neq (\frac{1}{2} \cdot 4)^2$.

(C) $x^2 - 6x + 9$ is a perfect square, since $9 = [\frac{1}{2}(-6)]^2$. To factor, recognize that $b = \frac{1}{2}(-6) = -3$, so

$$x^2 - 6x + 9 = [x + (-3)]^2 = (x - 3)^2$$

(D) Since the coefficient of x^2 in $4x^2 + 20x + 25$ is not 1, we must modify our approach somewhat. Notice that the first term, $4x^2$, is $(2x)^2$. Rewrite the first-degree term, $20x$, as $10(2x)$ and take one-half the coefficient of $2x$ rather than that of x, that is, 5. The constant is 5^2, so this is a perfect square:

$$4x^2 + 20x + 25 = (2x + 5)^2$$

A substitution may make this process easier to see. Let $y = 2x$. Then

$$\begin{aligned} 4x^2 + 20x + 25 &= y^2 + 10y + 25 \\ &= (y + 5)^2 \\ &= (2x + 5)^2 \end{aligned}$$

Matched Problem 5

Identify the perfect squares among these second-degree polynomials; factor those that are.

(A) $x^2 + 8x + 8$ **(B)** $x^2 + 4x + 4$
(C) $x^2 - 12x + 36$ **(D)** $9x^2 + 6x + 1$

♦   **SUM AND DIFFERENCE OF TWO SQUARES**

If we multiply $(A - B)$ and $(A + B)$, we obtain

$$(A - B)(A + B) = A^2 - B^2$$

which is a **difference of two squares.** Writing this result in reverse order, we obtain a very useful factoring formula. If we try to factor the **sum of two squares,** $A^2 + B^2$, we find that it cannot be factored using integer coefficients unless A and B have common factors. This will be shown in Section 2-5.

Sum and Difference of Two Squares

$A^2 + B^2$ cannot be factored using integer coefficients unless
             A and B have common factors

$A^2 - B^2 = (A - B)(A + B)$

Example 6
Factoring $A^2 + B^2$ and $A^2 - B^2$

Factor, if possible, using integer coefficients:

(A) $x^2 - 16$ **(B)** $4x^2 - y^2$ **(C)** $x^2 + 9y^2$ **(D)** $8x^2 - 50y^2$

Solution   **(A)** $x^2 - 16 = x^2 - 4^2 = (x - 4)(x + 4)$

(B) $4x^2 - y^2 = (2x)^2 - y^2 = (2x - y)(2x + y)$

(C) $x^2 + 9y^2 = x^2 + (3x)^2$ is the sum of two squares and, therefore, is not factorable using integer coefficients.

(D) $8x^2 - 50y^2 = 2(4x^2 - 25y^2)$
$$= 2[(2x)^2 - (5y)^2]$$
$$= 2(2x - 5y)(2x + 5y)$$

Matched Problem 6   Factor, if possible, using integer coefficients.

(A) $x^2 + 16$ **(B)** $x^2 - 9y^2$ **(C)** $3x^2 - 48$ **(D)** $9x^2 - 49y^2$

Answers to Matched Problems
1. **(A)** $4x(5x + 1)$ **(B)** $2u^2v(2uv - 5v^3 + 9)$
2. **(A)** $(x - y)(x^2 + y^2)$ **(B)** $(2x - 3)(x - 5)$
3. **(A)** $(x^2 + 1)(x + 3)$ **(B)** $(2x - 3)(x + 2)$ **(C)** $(b^2 + a)(3a + b)$
4. $(y + 3x)(x + y^2)$
5. **(A)** Not a perfect square   **(B)** $(x + 2)^2$ **(C)** $(x - 6)^2$ **(D)** $(3x + 1)^2$
6. **(A)** Not factorable using integer coefficients
   (B) $(x - 3y)(x + 3y)$ **(C)** $3(x - 4)(x + 4)$ **(D)** $(3x - 7y)(3x + 7y)$

EXERCISE 2-3

A *Factor out factors common to all terms.*

1. $3xz - 6x$

2. $2ab + 4$

3. $8x + 2y$

4. $5xy - y$

5. $6x^2 + 9x$

6. $8xy - 20y^2$

7. $14x^2y - 7xy^2$

8. $2x^2y^3 + 3x^2y^2$

9. $2x(x - 3) + z(x - 3)$

10. $a(c + d) + b(c + d)$

11. $(a + b)c - (a + b)d$

12. $3u(v + 1) - u(v + 1)$

13. $x(x - y) - y(x - y)$

14. $a(x + h) + 3(x + h)$

15. $ab(c + d) - (c + d)$

16. $x(pq - r) + y(pq - r)$

17. $x^5 + x^4 + x^3$

18. $y^2 - y^4 - y^8$

19. $a^2b - a^3b^2 - a^4b$

20. $rst - rs + rt$

Identify the perfect squares among the second-degree polynomials in Problems 21–32; factor those that are.

21. $x^2 - 10x + 25$

22. $x^2 + 8x + 16$

23. $x^2 + 12x + 36$

24. $x^2 - 4x + 4$

25. $x^2 - 12x - 36$

26. $x^2 - 8x - 16$

27. $x^2 - 20x + 100$

28. $x^2 - 16x + 64$

29. $x^2 + 24x + 144$

30. $x^2 + 20x + 100$

31. $x^2 + 36x + 288$

32. $x^2 + 15x + 75$

Factor.

33. $abc + bcd + cde$

34. $2xy + 8xz - 4yz$

35. $xy^2z^3 - x^2y^3z + x^3yz^2$

36. $2xy^2 + 4x^2y + 6x^2y^2$

37. $v^2 - 25$

38. $x^2 - 81$

39. $9x^2 - 4$

40. $4m^2 - 1$

41. $x^2 + 49$

42. $y^2 + 64$

43. $9x^2 - 16y^2$

44. $25u^2 - 4v^2$

B **45.** $4x^2 - 9y^2$

46. $x^2 - 16y^2$

47. $25x^2 - 64y^2$

48. $x^2 + 144y^2$

49. $121x^2 + 1$

50. $36x^2 - 49y^2$

Factor out factors common to all terms.

51. $a(a - 2) + 3(2 - a)$

52. $(2x - 3) + x(3 - 2x)$

53. $x(x - 1) - (1 - x)$

54. $x(x - y) - y(y - x)$

55. $(x - 2) - 5(2 - x)$

56. $2(a - b) - 8(b - a)$

57. $3(x - 1) + 6(1 - x)$

58. $4(x^2 - 2) + 10(2 - x^2)$

Factor out common factors from each group and, if possible, complete the factoring.

59. $(2x^3 + x^2) + (6x + 3)$

60. $(x^3 - 5x) + (20x^2 - 100)$

61. $(abc - a^2b^2) - (c^3 - abc^2)$

62. $(4x^2 - 8x) - (6x - 12)$

63. $(3xy + 6x) - (y^2 + 2y)$

64. $(2x + 4y) - (xy + y^2)$

Factor by grouping, if possible. (Compare these problems to Problems 59–64.)

65. $2x^3 + x^2 + 6x + 3$

66. $x^3 - 5x + 20x^2 - 100$

67. $abc - a^2b^2 - c^3 + abc^2$

68. $4x^2 - 8x - 6x + 12$

69. $3xy + 6x - y^2 - 2y$

70. $2x + 4y - xy - y^2$

Factor.

71. $3x^2 - 27$

72. $4x^2 + 16$

73. $5x^2 - 125$

74. $6x^2 - 24$

75. $3x^2 - 12y^2$

76. $4x^2 - 36y^2$

77. $x^2y^2 - 1$

78. $x^2y^2 - z^2$

79. $a^2b^2 - c^2d^2$

80. $4 - a^2b^2$

C *Factor by grouping, if possible.*

81. $2x^4 - 3x^2 + 6x^2 - 9$

82. $x^3 - x^2 + 3x - 3$

83. $x^3y^2 - x^2y^3 + 3x - 3y$

84. $xy^3 + x^3y + x^2 + y^2$

85. $4ab - b^2 - 4a^2 + ab$

86. $2x^2 - x^3 - 2y^2 + xy^2$

87. $xy^2z - xyz^2 - xy + xz$

88. $4b^2 - ab + a^2 - 4ab$

89. $2x - 6y - 9yz + 3xz$

90. $x - xy - z + zy$

91. $3a^2 + bc + ab + 3ac$

92. $x^2 + 6y + x^2y + 6$

93. $12 + xy^2 - 4x - 3y^2$

94. $5xy - z^2 + xyz - 5z$

95. $ac + bd + ab + cd$

96. $ab + cd - bd - ac$

97. $yz + 2xy - 2xz - y^2$

98. $2xy + 9 - 6y - 3x$

2-4 Factoring Second-Degree Polynomials

Products of the form

$$(ax + b)(cx + d) = acx^2 + (ad + bc)x + bd$$

or

$$(ax + by)(cx + dy) = acx^2 + (ad + bc)xy + bdy^2$$

are easily obtained by multiplying out. We would like to be able to reverse the process and factor expressions of the form

$$Ax^2 + Bx + C \quad \text{or} \quad Ax^2 + Bxy + Cy^2$$

For example, can we factor $2x^2 - x - 6$ into two first-degree polynomials with integer coefficients a, b, c, and d:

$$2x^2 - x - 6 = (ax + b)(cx + d)$$

We can easily check that

$$2x^2 - x - 6 = (2x + 3)(x - 2)$$

but if the factors were not supplied, how could we find them? We now develop a trial-and-error method for factoring any second-degree polynomial.

Consider the polynomial $x^2 - 5x + 6$. We would like to factor

$$x^2 - 5x + 6 = (x + a)(x + b)$$

If we multiply out $(x + a)(x + b)$, we obtain

$$x^2 - 5x + 6 = x^2 + (a + b)x + ab$$

so we must have $ab = 6$ and $a + b = -5$. At this stage, we might try all possible a and b that multiply to 6 and find that $a = -2$ and $b = -3$ work; that is,

$$x^2 - 5x + 6 = (x - 2)(x - 3)$$

Some observations, however, will limit the number of possible a and b to try and thus shorten our work:

1. a and b must have the same sign since the product is positive.
2. The sign must be negative to make the sum negative.
3. The order does not matter, since

$$(x + a)(x + b) = (x + b)(x + a)$$

That is, we would not have to check both $a = -1$, $b = -6$ and also $a = -6$, $b = -1$.

We therefore look for two negative numbers that multiply to 6 and add to -5. The only pairs that multiply to 6 are $-2, -3$ and $-1, -6$. Of these, only the first pair, $-2, -3$ add to -5.

The trial-and-error method is somewhat more complicated when the coefficient of x^2 is not 1. Consider again the polynomial $2x^2 - x - 6$. We begin by comparing the polynomial and the multiplied-out form of $(ax + b)(cx + d)$:

$$2x^2 - x - 6 = acx^2 + (ad + bc)x + bd$$

Thus, we need to look for coefficients a, b, c, and d such that

$$ac = 2$$
$$bd = -6$$
$$ad + bc = -1$$

There are only a limited number of integer pairs that meet the first two conditions:

$ac = 2$:		$bd = -6$:	
a	c	b	d
$1 \cdot 2$		$2(-3)$	
$2 \cdot 1$		$(-2)3$	
$(-1)(-2)$		$3(-2)$	
$(-2)(-1)$		$(-3)2$	

We need only try the first combination here for a and c, since the other three will be related to it in this way:

$$
\left.
\begin{array}{l}
(1 \cdot x + b)(2 \cdot x + d) \\
(2 \cdot x + d)(1 \cdot x + b) \\
(-x - b)(-2x - d) \\
(-2x - d)(-x - b)
\end{array}
\right\}
$$

All multiply to $2x^2 + (2b + d)x + bd$.

Next, we try each pair for bd until we either find one that works or exhaust all the possibilities. In this case, $b = -2$ and $d = 3$ give the correct middle term. Thus,

$$2x^2 - x - 6 = (ax + b)(cx + d)$$
$$= (x - 2)(2x + 3)$$

If checking all possibilities does not yield a factorization, we will say that the polynomial is not factorable using integers.

Example 1
Trial-and-Error
Factoring

Factor each polynomial, if possible, using integer coefficients:

(A) $x^2 - x - 12$ **(B)** $2x^2 - 5xy + 2y^2$ **(C)** $x^2 + x + 4$
(D) $12x^2 + x - 6$

Solution **(A)** $x^2 - x - 12$ Put in what we know.

         $= (x - \quad)(x + \quad)$ The signs must be opposite.

Now test all possible factors of 12 to see if any produce the correct middle term:

$$\begin{array}{c} 12 \\ \hline 1 \cdot 12 \\ 12 \cdot 1 \\ 2 \cdot 6 \\ 6 \cdot 2 \\ 3 \cdot 4 \\ 4 \cdot 3 \end{array}$$

The last choice, $4 \cdot 3$, yields the desired results, so

$$x^2 - x - 12 = (x - 4)(x + 3)$$

(B) $2x^2 - 5xy + 2y^2$

       $= (2x - \quad)(x - \quad)$ Both signs must be negative to give a positive coefficient to y^2 and a negative in the middle term.

Try all possible factors for 2.

$$\begin{array}{c} 2 \\ \hline 1 \cdot 2 \\ 2 \cdot 1 \end{array}$$

The first works:

$$(2x - y)(x - 2y) = 2x^2 - 5xy + 2y^2$$

(C) $x^2 + x + 4 = (x + \quad)(x + \quad)$

We try all ways of factoring 4 with positive factors:

$$\begin{array}{c} 4 \\ \hline 1 \cdot 4 \\ 4 \cdot 1 \\ 2 \cdot 2 \end{array}$$

None of these provides the correct middle term, so $x^2 + x + 4$ is not factorable using integer coefficients.

(D) $12x^2 + x - 6 = (\quad x + \quad)(\quad x - \quad)$

                              ↑   ↑   ↑   ↑

                              ?   ?   ?   ?

The signs must be opposite to produce the negative constant and positive coefficient of x.

We now consider all positive factors for 12 and 6:

12	6
$1 \cdot 12$	$1 \cdot 6$
$2 \cdot 6$	$2 \cdot 3$
$3 \cdot 4$	$3 \cdot 2$
$4 \cdot 3$	$6 \cdot 1$
$6 \cdot 2$	
$12 \cdot 1$	

We check each choice for 12 with each possibility of 6—there are 24 such combinations—until we find one that produces the correct middle term. Eventually, we obtain

$$12x^2 + x - 6 = (4x + 3)(3x - 2)$$

If none of the combinations had worked, we would have concluded the polynomial is not factorable using integer coefficients.

Matched Problem 1 Factor each polynomial, if possible, using integer coefficients.

(A) $x^2 + 7x + 12$ (B) $x^2 + x - 5$
(C) $x^2 - xy - 6y^2$ (D) $4x^2 + 11x - 3$

In problems like Example 1(D), if the coefficients of the first and last terms get larger and larger with more and more factors, the number of combinations that need to be checked increases very rapidly. And it is quite possible in most practical situations that none of the combinations will work. However, the approach presented here will work for many of the factoring problems you will encounter. The next section introduces a systematic approach to the problem of factoring that will reduce the amount of trial and error substantially and even tell you whether a polynomial can be factored before you proceed too far.

In conclusion, we point out that if a, b, and c are selected at random out of the integers, the probability that $ax^2 + bx + c$ is not factorable in the integers is much greater than the probability that it is.

Answers to
Matched Problem

1. (A) $(x + 4)(x + 3)$ (B) Not factorable using integers
(C) $(x - 3y)(x + 2y)$ (D) $(4x - 1)(x + 3)$

EXERCISE 2-4

A *Factor in the integers, if possible. If the expression is not factorable, say so.*

1. $x^2 + 3x + 2$ **2.** $x^2 + 8x + 15$ **3.** $x^2 + 7x + 10$

4. $x^2 + 4x + 3$ **5.** $x^2 - 2x - 3$ **6.** $x^2 - 2x - 8$

7. $x^2 + 3x - 4$ **8.** $x^2 + 5x - 6$

9. $x^2 - 9x + 20$ **10.** $x^2 - 9x + 18$

11. $x^2 - 6x + 5$ **12.** $x^2 - 6x + 8$

13. $x^2 - 6x + 9$ **14.** $x^2 + 8x + 16$

15. $x^2 + 14x + 49$ **16.** $x^2 - 4x - 4$

17. $x^2 - 6xy + 9y^2$ **18.** $x^2 + 8xy + 16y^2$

19. $x^2 + 4x - 5$ **20.** $x^2 + 10x + 24$

21. $x^2 - 3x - 18$ **22.** $x^2 + 5x - 24$

23. $x^2 + 4x - 12$ **24.** $x^2 + 2x - 4$

25. $x^2 - 9x + 8$

26. $x^2 + 9x - 10$

27. $x^2 + 2x + 4$

28. $x^2 - 8x + 16$

29. $2x^2 - x - 1$

30. $2x^2 + 7x + 6$

31. $2x^2 + 7x + 5$

32. $2x^2 + x - 3$

33. $3x^2 + 7x - 4$

34. $3x^2 - 11x + 10$

35. $9x^2 + 8$

36. $3x^2 - 15$

37. $4x^2 - 40$

38. $6x^2 + 12$

39. $2x^2 - 200$

40. $5x^2 - 4$

B **41.** $2x^2 - xy + y^2$

42. $2x^2 + xy - 6y^2$

43. $2x^2 - 9xy + 10y^2$

44. $2x^2 - 7xy + 3y^2$

45. $4x^2 + 4x - 3$

46. $4x^2 + 11x - 3$

47. $4x^2 + 3x - 1$

48. $4x^2 + 4x - 15$

49. $2x^2 + 4x + 3$

50. $4x^2 + 4x + 1$

51. $4x^2 + 12x + 9$

52. $2x^2 - 4x - 5$

53. $x^2 - x - 2$

54. $9x^2 - 6x + 1$

55. $9x^2 + 12x + 4$

56. $x^2 + x - 1$

57. $6x^2 - 7x + 2$

58. $6x^2 - 5x - 4$

59. $6x^2 - xy - y^2$

60. $6x^2 - xy - 2y^2$

61. $x^2 + 24$

62. $x^2 + 60$

63. $x^2 + y^2$

64. $x^2 - a^2y^2$

C **65.** $9x^2 + 6xy + y^2$

66. $9x^2 + 12xy - 4y^2$

67. $4x^2 - 4xy - y^2$

68. $4x^2 - 12xy + 9y^2$

69. $8x^2 + 14x + 7$

70. $16x^2 + 7x + 6$

71. $8x^2 - 14x - 15$

72. $16x^2 + 2x - 3$

73. $6x^2 + 10xy + 5y^2$

74. $9x^2 + 11xy + 4y^2$

75. $6x^2 + 21xy - 12y^2$

76. $9x^2 + 26xy - 3y^2$

A polynomial such as $x^4 - x^2 - 12$ is not second-degree but still can be factored using the techniques of this section. We can think of x^2 as the variable, rewrite the polynomial as $(x^2)^2 - (x^2) - 12$, and factor it as

$$(x^2 - 4)(x^2 + 3) = (x - 2)(x + 2)(x^2 + 3)$$

as in Example 1(A). You may find it easier to use a substitution: let $u = x^2$ so that $x^4 - x^2 - 12 = u^2 - u - 12 = (u - 4)(u + 3) = (x^2 - 4)(x^2 + 3)$. Factor the following polynomials:

77. $x^4 + 4x^2 + 3$

78. $x^4 - x^2 - 2$

79. $x^4 - 1$

80. $x^4 + 4x^2 + 4$

81. $x^4 + 2x^2 + 1$

82. $x^4 - x^2 - 9$

83. $4x^4 - 1$

84. $4x^4 + 4x^2 + 1$

85. $9x^4 - 12x^2 + 4$

86. $9x^4 - 4$

87. $x^8 - 1$

88. $x^8 + 1$

89. $x^6 - 2x^3 + 1$

90. $4x^6 - 1$

91. $x^6 + 36$

92. $x^4 - 16$

93. $x^6 - 2x^3 - 8$

94. $x^6 - x^3 - 2$

95. $x^6 + 5x^3 + 6$

96. $x^6 - 10x^3 + 24$

97. $2x^4 - x^2 - 3$

98. $6x^4 - 7x^2 - 3$

99. $x^6 + x^3 - 6$

100. $2x^6 - 7x^3 - 4$

2-5 Factoring with the *ac* Test

In Example 1(D) of Section 2-4, the polynomial $12x^2 + x - 6$ was factored by trial and error as

$$12x^2 + x - 6 = (4x + 3)(3x - 2)$$

The following process leads to the same result:

$12x^2 + x - 6$ Rewrite x as $9x - 8x$; we'll see in a moment what leads to this choice.

$= 12x^2 + 9x - 8x - 6$ Group the first two and last two terms.

$= (12x^2 + 9x) - (8x + 6)$ Factor out common factors.

$= 3x(4x + 3) - 2(4x + 3)$ Factor out common factors again.

$= (3x - 2)(4x + 3)$

The first step in the process, rewriting x as $9x - 8x$, is what makes it work. To see why this choice works, consider the process again:

$$12x^2 + x - 6$$

Rewrite x as $px + qx$; note that $p + q$ must be 1.

$$= 12x^2 + px + qx - 6$$
$$= (12x^2 + px) + (qx - 6)$$
$$= x(12x + p) + (qx - 6)$$

We would like to take out a common factor of degree 1 from $(12x + p)$ and $(qx - 6)$. Since both these expressions are of degree 1, for there to be a common factor, one must be a multiple of the other. This means that the ratio of 12 to q must be the same as the ratio of p to -6; that is,

$$\frac{12}{q} = \frac{p}{-6}$$

So $pq = -72$. Therefore, we need two integers p and q with $p + q = 1$ and $pq = -72$. The numbers 9 and -8 work, and this is what lead to our rewriting x as $9x - 8x$ above.

    This process can be generalized and applied to second-degree polynomials of the following types:

$$ax^2 + bx + c \qquad \text{and} \qquad ax^2 + bxy + cy^2 \tag{1}$$

The process provides a test, called the ***ac* test for factorability,** that not only tells us if the polynomials of type (1) can be factored using integer coefficients, but, in addition, leads to a direct way of factoring those that are factorable.

The *ac* Test for Factorability

If, in a polynomial of type (1), the product ac has two integer factors p and q whose sum is the coefficient of the middle term b; that is, if integers p and q exist so that

$$pq = ac \qquad \text{and} \qquad p + q = b \tag{2}$$

then the polynomial has first-degree factors with integer coefficients. If no integers p and q exist that satisfy Equations (2), then the polynomial does not have first-degree factors with integer coefficients.

    Once we find integers p and q in the ac test, if they exist, our work is almost finished. We can then write polynomials of type (1), splitting the middle term, in forms that can be factored by grouping.

Applying the *ac* Test

When the *ac* test produces integers p and q such that

$$pq = ac \quad \text{and} \quad p + q = b \qquad (2)$$

rewrite the polynomial

$$ax^2 + bx + c \quad \text{or} \quad ax^2 + bxy + cy^2 \qquad (1)$$

as

$$ax^2 + px + qx + c \quad \text{or} \quad ax^2 + pxy + qxy + cy^2 \qquad (3)$$

The results in (3) can always be factored by grouping.

Example 1
Using the *ac* Test

Factor, if possible, using integer coefficients:

(A) $2x^2 + 7x - 4$ **(B)** $4x^2 + 5x - 6$ **(C)** $8x^2 + 12xy + 9y^2$

Solution **(A)** *Step 1: Test.* Test $2x^2 + 7x - 4$ for factorability using the *ac* test:

$$ac = 2(-4) = -8$$

We need two integers p and q such that $pq = -8$ and $p + q = 7$. We write (or think of) all integer pairs that multiply to -8:

$$\frac{pq = -8}{\begin{array}{c} 2(-4) \\ (-2)4 \\ 1(-8) \\ (-1)8 \end{array}}$$

Then we test to see if any of these add to 7, the coefficient of the middle term. The last pair works; that is,

$$\overset{p \quad q \quad ac}{(-1)(8) = -8} \quad \text{and} \quad \overset{p \qquad q \quad b}{(-1) + (8) = 7}$$

We conclude that the polynomial can be factored.
Step 2: Factor. We now split the middle term, $7x$, of the original expression using $p = -1$ and $q = 8$:

$$2x^2 + \overset{b}{7x} - 4 = 2x^2 - \overset{p}{1x} + \overset{q}{8x} - 4$$

Factor the resulting polynomial by grouping. This will always work if we can get to this step, and it won't matter which order of p and q is used—that is, the process will work if they are reversed.

$$
\begin{aligned}
2x^2 + 7x - 4 \\
= 2x^2 - x + 8x - 4 \qquad &\text{Group the first two and last two terms.} \\
= (2x^2 - x) + (8x - 4) \qquad &\text{Factor out common factors.} \\
= x(2x - 1) + 4(2x - 1) \\
= (x + 4)(2x - 1)
\end{aligned}
$$

This process can be reduced to a few key operational steps when all the commentary is eliminated and some of the process is done mentally. The only trial and error occurs in step 1, and with a little practice this step can be done fairly quickly.

(B) Compute ac for $4x^2 + 5x - 6$:

$$ac = 4(-6) = -24$$

We need two integers, p and q, whose product is -24 and whose sum is 5, the coefficient of the middle term.

$pq = -24$	
$1(-24)$	$8(-3)$
$(-1)24$	$(-8)3$
$12(-2)$	$6(-4)$
$(-12)2$	$(-6)4$

We find 8 and -3 will work, so we split the middle term, $5x = 8x - 3x$, and factor by grouping:

$$
\begin{aligned}
4x^2 + 5x - 6 &= 4x^2 + 8x - 3x - 6 \\
&= (4x^2 + 8x) - (3x + 6) \\
&= 4x(x + 2) - 3(x + 2) \\
&= (4x - 3)(x + 2)
\end{aligned}
$$

(C) Compute ac for $8x^2 + 12xy + 9y^2$:

$$ac = 8 \cdot 9 = 72$$

Thus, we need p and q with $pq = 72$ and $p + q = 12$. In this case, both p and q must be positive, so we look only at these possibilities:

$pq = 72$	
$1 \cdot 72$	$4 \cdot 18$
$2 \cdot 36$	$6 \cdot 12$
$3 \cdot 24$	$8 \cdot 9$

None of these add up to 12, so we can conclude that $8x^2 + 12xy + 9y^2$ is not factorable using integer coefficients.

Matched Problem 1   Factor, if possible, using integer coefficients:

(A) $3x^2 - 2xy - y^2$ (B) $6x^2 - 7x - 5$ (C) $4x^2 + 10x + 9$

If we apply the *ac* test to the sum of squares $x^2 + y^2$, we have

$$ac = 1 \cdot 1 = 1$$

Thus, we need p and q with $pq = 1$ and $p + q = 0$. No such integers exist, so $x^2 + y^2$ is not factorable using integer coefficients. It follows that $A^2 + B^2$ is not factorable using integer coefficients unless A and B have a common factor, as claimed in Section 2-3.

Answers to   **1. (A)** $(3x + y)(x - y)$ **(B)** $(3x - 5)(2x + 1)$
Matched Problem   **(C)** Not factorable

EXERCISE 2-5

A *In Problems 1–74, factor, if possible, using integer coefficients. Use the ac test and proceed as in Example 1.*

1. $4x^2 + 11x - 3$
2. $4x^2 - x - 3$
3. $4x^2 + 4x - 3$
4. $4x^2 - 8x - 5$
5. $4x^2 + 11x + 8$
6. $4x^2 + 21x + 5$
7. $4x^2 - 25x + 6$
8. $6x^2 + 12x + 9$
9. $6x^2 - 25x + 4$
10. $6x^2 - 29x - 5$
11. $4x^2 + 4x - 15$
12. $4x^2 - 4x - 3$
13. $5x^2 - 4x + 3$
14. $2x^2 + 7x + 5$
15. $2x^2 - 7x - 15$
16. $6x^2 - 5x + 4$
17. $3x^2 + 8x - 3$
18. $3x^2 - 5x + 7$
19. $4x^2 - 3x + 2$
20. $3x^2 + 14x - 5$
21. $7x^2 + 26x - 8$
22. $3x^2 - x - 10$
23. $3x^2 + 11x + 10$
24. $7x^2 + 12x - 4$

B 25. $4x^2 + 4x + 3$
26. $6x^2 - 7x - 10$
27. $6x^2 - 29x + 20$
28. $8x^2 - 12x + 5$
29. $6x^2 + 7x - 3$
30. $6x^2 + 5x - 4$
31. $6x^2 + 7x - 20$
32. $6x^2 + 14x + 9$
33. $6x^2 + x - 15$
34. $6x^2 - 43x + 7$
35. $6x^2 + 20x + 5$
36. $6x^2 + 37x + 6$
37. $8x^2 + 2x - 3$
38. $8x^2 - 6x - 9$
39. $4x^2 - 7x + 10$
40. $10x^2 - 23x - 5$
41. $4x^2 + 7x - 10$
42. $10x^2 + 5x - 1$

43. $8x^2 - 34x - 9$
44. $8x^2 - 6x + 7$
45. $4x^2 - 13xy + 10y^2$
46. $4x^2 - xy - 14y^2$
47. $4x^2 - 9xy - 5y^2$
48. $6x^2 - xy - 2y^2$
49. $6x^2 + 14xy - 12y^2$
50. $10x^2 + 24xy + 25y^2$

C 51. $8x^2 + 6x - 9$
52. $8x^2 - 6x - 9$
53. $6x^2 - 31x - 30$
54. $6x^2 + 31x - 30$
55. $24x^2 + 38x + 15$
56. $24x^2 - 2x - 15$
57. $6x^2 + 2x - 15$
58. $6x^2 - 38x + 15$
59. $24x^2 + 106x - 9$
60. $24x^2 + 110x + 9$
61. $24x^2 + 30x - 9$
62. $24x^2 + 42x + 9$
63. $12x^2 + 19xy + 5y^2$
64. $6x^2 - 16xy + 3y^2$
65. $6x^2 - 20xy - 25y^2$
66. $9x^2 - 14xy + 4y^2$
67. $4x^2 - 17xy - 15y^2$
68. $4x^2 - 25xy + 12y^2$
69. $6x^2 + 28x - 5$
70. $6x^2 - 28x + 15$
71. $18x^2 + 37x - 20$
72. $18x^2 + 5x - 2$
73. $18x^2 + 9xy - 20y^2$
74. $18x^2 + 3xy - 28y^2$

75. Find all integers b such that $x^2 + bx - 18$ can be factored using integer coefficients.

76. Find all integers b such that $x^2 + bx + 20$ can be factored using integer coefficients.

77. Find all integers c between 0 and 10 such that $x^2 + 5x + c$ can be factored using integer coefficients.

78. Find all integers c between 0 and 10 such that $x^2 - 6x + c$ can be factored using integer coefficients.

Refer to the approach introduced in Exercise 2-4 (Problems 77–100) for factoring certain polynomials of degree higher than 2. Apply the ac test and factor, if possible, using integer coefficients.

79. $6x^4 + 7x^2 + 2$ **80.** $5x^4 - 11x^2 + 2$

81. $2x^4 + 5x^2y^2 + 3y^4$ **82.** $3x^4 + 5x^2y^2 - 2y^4$

83. $2x^6 - x^3 - 3$ **84.** $3x^6 - 10x^3 - 8$

85. $12x^4 + 25x^2 + 12$ **86.** $12x^4 + 51x^2 + 12$

87. $12x^4 + 70x^2 - 12$ **88.** $12x^4 + 10x^2 - 12$

89. $24x^6 - 69x^3 - 9$ **90.** $24x^6 - 6x^3 - 9$

     # 2-6 More Factoring

♦ Sum and Difference of Two Cubes
♦ Combined Forms
♦ Higher-Degree Polynomials
♦ General Strategy

In this section, we consider a variety of factoring methods, including methods for polynomials of degree greater than 2. We conclude by suggesting a general strategy for applying the various factoring processes.

♦ **SUM AND DIFFERENCE OF TWO CUBES**

We can verify, by direct multiplication of the right sides, the following factoring formulas for the **sum and difference of two cubes:**

> ### Sum and Difference of Two Cubes
>
> $$A^3 + B^3 = (A + B)(A^2 - AB + B^2) \qquad (1)$$
> $$A^3 - B^3 = (A - B)(A^2 + AB + B^2) \qquad (2)$$

You should learn these formulas. Observing the pattern of signs in them may help you remember them:

$$A^3 + B^3 = (A + B)(A^2 - AB + B^2)$$

Same sign
Opposite sign
Always plus

$$A^3 - B^3 = (A - B)(A^2 + AB + B^2)$$

Same sign
Opposite sign
Always plus

Neither $A^2 - AB + B^2$ nor $A^2 + AB + B^2$ can be factored using integer coefficients. (See Problems 107 and 108, Exercise 2-6.)

To factor $y^3 - 27$, we first note that this expression can be written in the form

$$y^3 - 3^3$$

Thus, we are dealing with the difference of two cubes. If we let $A = y$ and $B = 3$ in factoring formula (2), we obtain

$$y^3 - 27 = y^3 - 3^3 = (y - 3)(y^2 + 3y + 9)$$

Example 1
Factoring $A^3 + B^3$
and $A^3 - B^3$

Factor as far as possible using integer coefficients:

(A) $x^3 + 64$ **(B)** $8x^3 - 1$ **(C)** $16x^3 + 2y^3$
(D) $8x^3 - 27y^3$

Solution   **(A)** $x^3 + 64 \boxed{= x^3 + 4^3} = (x + 4)(x^2 - 4x + 16)$

(B) $8x^3 - 1 \boxed{= (2x)^3 - 1^3} = (2x - 1)[(2x)^2 + 1 \cdot 2x + 1^2]$
$$= (2x - 1)(4x^2 + 2x + 1)$$

(C) $16x^3 + 2y^3 = 2(8x^3 + y^3) \boxed{= 2[(2x)^3 + y^3]}$
$$= 2(2x + y)[(2x)^2 - 2xy + y^2]$$
$$= 2(2x + y)(4x^2 - 2xy + y^2)$$

(D) $8x^3 - 27y^3 = (2x)^3 - (3y)^3$
$$= (2x - 3y)[(2x)^2 + (2x)(3y) + (3y)^2]$$
$$= (2x - 3y)(4x^2 + 6xy + 9y^2)$$

Matched Problem 1   Factor as far as possible using integer coefficients:

(A) $27x^3 + 1$ **(B)** $x^3 - 64$ **(C)** $81x^3 - 3y^3$
(D) $8x^3 + 125y^3$

♦ **COMBINED FORMS**

In general, factoring will be simpler if we first remove common factors before proceeding:

> **General Factoring Principle**
>
> Remove common factors first before proceeding further.

Example 2
Factoring

Factor as far as possible using integer coefficients:

(A) $3x^2 + 12x + 12$ **(B)** $4x^2 - 100$
(C) $5x^2 - 5x - 60$ **(D)** $11x^3 - 297y^3$

Solution   **(A)** $3x^2 + 12x + 12$ Factor out the common factor.
$$= 3(x^2 + 4x + 4)$$ Recognize the perfect square.
$$= 3(x + 2)^2$$

(B) $4x^2 - 100$ Factor out the common factor 4.
$$= 4(x^2 - 25)$$ Factor the difference of two squares.
$$= 4(x - 5)(x + 5)$$

Since both $4x^2$ and 100 are perfect squares, this problem also could have been treated as the difference of two squares initially.

$$4x^2 - 100 = (2x)^2 - 10^2 = (2x - 10)(2x + 10)$$
$$= 2(x - 5)2(x + 5)$$
$$= 4(x - 5)(x + 5)$$

(C) $5x^2 - 5x - 60$ Factor out the common factor 5.
$= 5(x^2 - x - 12)$ Factor $x^2 - x - 12$ by trial and error.
$= 5(x - 4)(x + 3)$
(D) $11x^3 - 297y^3$ Factor out the common factor 11.
$= 11(x^3 - 27y^3)$ Factor the difference of two cubes.
$= 11(x - 3y)(x^2 + 3xy + 9y^2)$

Matched Problem 2   Factor as far as possible using integer coefficients:

(A) $2x^3 + 250$ **(B)** $6x^2 + 18x - 24$
(C) $12x^2 - 108$ **(D)** $4x^2 + 40x + 100$

Example 3
Polynomial Common Factors

Factor:

$$(x - 1)^2(x + 2) - (x - 1)(x + 2)^2$$

Solution   Note that $(x - 1)(x + 2)$ is a factor in both terms. Therefore,

$$(x - 1)^2(x + 2) - (x - 1)(x + 2)^2$$
$= (x - 1)(x + 2)[(x - 1) - (x + 2)]$ Simplify within the brackets.
$= (x - 1)(x + 2)[x - 1 - x - 2]$
$= (x - 1)(x + 2)(-3)$
$= -3(x - 1)(x + 2)$

Matched Problem 3   Factor: $(x + 3)^2(x - 3) - (x + 3)(x - 3)^2$

Occasionally, polynomial forms of a more general nature than those considered in Sections 2-3, 2-4, and 2-5 can be factored by appropriate grouping of terms. The following example illustrates the process.

Example 4
Factoring by Grouping

Factor by grouping terms:

(A) $x^2 + 3xy - 2x - 6y$ **(B)** $x^3 + 3x^2 + 4x + 12$

Solution   **(A)** $x^2 + 3xy - 2x - 6y$ Group the first two and last two terms.
$= (x^2 + 3xy) + (-2x - 6y)$ Remove common factors.
$= x(x + 3y) - 2(x + 3y)$ Remove the common factor $x + 3y$.
$= (x - 2)(x + 3y)$

(B) $x^3 + 3x^2 + 4x + 12$ Group the first two and last two terms.
$= (x^3 + 3x^2) + (4x + 12)$ Remove the common factors.
$= x^2(x + 3) + 4(x + 3)$ Remove the common factor $x + 3$.
$= (x^2 + 4)(x + 3)$

Matched Problem 4   Factor by grouping terms:

(A) $x^2 - xy + 3x - 3y$ (B) $2x^3 - x^2 + 4x - 2$

♦ **HIGHER-DEGREE POLYNOMIALS**

The techniques and examples of factoring introduced so far have dealt mainly with polynomials of degree 2 and certain polynomials of degree 3. Some higher-degree polynomials can be factored in the same ways.

Example 5
Factoring Higher-Degree Polynomials

Factor:

(A) $x^4 - 81$ (B) $x^4 + 6x^2 + 8$ (C) $x^6 + y^3$

Solution   (A) $x^4 - 81$ Think of x^4 as $(x^2)^2$.
$= (x^2)^2 - 9^2$ Factor as the difference of two squares.
$= (x^2 - 9)(x^2 + 9)$
$= (x - 3)(x + 3)(x^2 + 9)$
(B) $x^4 + 6x^2 + 8$ Think of x^4 as $(x^2)^2$.
$= (x^2)^2 + 6(x^2) + 8$ Factor as a second-degree polynomial in x^2.
$= (x^2 + 4)(x^2 + 2)$
(C) $x^6 + y^3$ Think of x^6 as $(x^2)^3$.
$= (x^2)^3 + y^3$ Factor as the sum of two cubes.
$= (x^2 + y)[(x^2)^2 - (x^2)y + y^2]$ Simplify the second factor.
$= (x^2 + y)(x^4 - x^2y + y^2)$

It may be easier to see the structure of a polynomial with a substitution. In Example 5(B), we think of x^4 as $(x^2)^2$. If we write $u = x^2$, then $u^2 = x^4$ and the factoring process becomes

$$x^4 + 6x^2 + 8$$ Replace x^4 by u^2 and x^2 by u.
$$= u^2 + 6u + 8$$ Factor.
$$= (u + 4)(u + 2)$$ Replace u by x^2.
$$= (x^2 + 4)(x^2 + 2)$$

Matched Problem 5   Factor: (A) $x^4 - y^4$ (B) $x^4 - 3x^2 + 2$ (C) $8x^6 - 27$

♦ **GENERAL STRATEGY**

There is no standard procedure (**algorithm**) for factoring a polynomial. However, the following strategy may be helpful.

General Strategy for Factoring

1. Remove common factors (Section 2-3).
2. If the polynomial has two terms, look for a difference of two squares or a sum or difference of two cubes (Sections 2-3 and 2-6).
3. If the polynomial has three terms:
   (A) See if it is a perfect square (Section 2-3).
   (B) Try trial and error (Section 2-4).
   (C) Or use the *ac* test (Section 2-5).
4. If the polynomial has more than three terms, try grouping (Sections 2-3 and 2-6).
5. Check that each factor has been completely factored.

Factoring requires skill, creativity, and perseverance. Often the appropriate technique is not immediately apparent, and practice is necessary to develop your recognition of what might work on a given problem. Exercise 2-6 contains numerous problems for this purpose.

Answers to Matched Problems

1. **(A)** $(3x + 1)(9x^2 - 3x + 1)$ **(B)** $(x - 4)(x^2 + 4x + 16)$
   (C) $3(3x - y)(9x^2 + 3xy + y^2)$ **(D)** $(2x + 5y)(4x^2 - 10xy + 25y^2)$
2. **(A)** $2(x + 5)(x^2 - 5x + 25)$ **(B)** $6(x + 4)(x - 1)$
   (C) $12(x - 3)(x + 3)$ **(D)** $4(x + 5)^2$
3. $6(x - 3)(x + 3)$ 4. **(A)** $(x + 3)(x - y)$ **(B)** $(x^2 + 2)(2x - 1)$
5. **(A)** $(x^2 + y^2)(x - y)(x + y)$ **(B)** $(x^2 - 2)(x - 1)(x + 1)$
   (C) $(2x^2 - 3)(4x^4 + 6x^2 + 9)$

EXERCISE 2-6

A *In Problems 1–106, factor as far as possible using integer coefficients.*

1. $5v^2 - 125$
2. $9x^2 - 81$
3. $v^3 - 125$
4. $3x^3 - 81$
5. $84m^2 - 21$
6. $5x^2 - 245$
7. $y^3 + 64$
8. $y^3 - 64$
9. $x^3 + 1$
10. $y^3 - 1$
11. $m^3 - n^3$
12. $p^3 + q^3$
13. $8x^3 + 27$
14. $u^3 - 8v^3$
15. $6u^2v^2 - 3uv^3$
16. $2x^3y - 6x^2y^3$
17. $2x^2 - 8$
18. $3y^2 - 27$
19. $2x^3 + 8x$
20. $3x^4 + 27x^2$
21. $12x^3 - 3xy^2$
22. $2u^3v - 2uv^3$
23. $2x^4 + 2x$
24. $xy^3 + x^4$
25. $6x^2 + 36x + 48$
26. $4x^2 - 4x - 24$

27. $3x^3 - 6x^2 + 15x$
28. $2x^3 - 2x^2 + 8x$

B 29. $x^4 - 9$
30. $x^4 - 4y^4$
31. $x^6 + 27y^6$
32. $x^6 + 8$
33. $x^4 + 4x^2 + 4$
34. $x^4 + 6x^2 + 9$
35. $x^6 + 6x^3 + 9$
36. $x^8 - 8x^4 + 16$
37. $x^2y^2 - 16$
38. $m^2n^2 - 36$
39. $a^3b^3 + 8$
40. $27 - x^3y^3$
41. $4x^3y + 14x^2y^2 + 6xy^3$
42. $3x^3y - 15x^2y^2 + 18xy^3$
43. $4u^3 + 32v^3$
44. $54x^3 - 2y^3$
45. $60x^2y^2 - 200xy^3 - 35y^4$
46. $60x^4 + 68x^3y - 16x^2y^2$
47. $xy + 2x + y^2 + 2y$
48. $x^2 + 3x + xy + 3y$
49. $x^2 - 5x + xy - 5y$
50. $x^2 - 3x - xy + 3y$
51. $ax - 2bx - ay + 2by$
52. $mx + my - 2nx - 2ny$

53. $15ac - 20ad + 3bc - 4bd$

54. $2am - 3an + 2bm - 3bn$

55. $x^3 - 2x^2 - x + 2$ 56. $x^3 - 2x^2 + x - 2$

57. $(y - x)^2 - y + x$ 58. $x^2(x - 1) - x + 1$

59. $x^2y^2 - xy - 6$ 60. $a^2b^2 - 7ab + 12$

61. $z^4 - z^2 - 6$ 62. $x^4 + 4x^2 + 4$

C 63. $x(x + 1)^2 - x^2(x + 1)$

64. $(x + 1)^3(x + 2)^2 - (x + 1)^2(x + 2)^3$

65. $x^3(x - 1) - x(x - 1)^3$

66. $(x - 2)(x + 1)^3 - (x - 2)^2(x + 1)^2$

67. $(x + 4)^2(x - 1)^4 - (x + 4)^3(x - 1)^3$

68. $(x - 5)^5(x + 2)^2 - (x - 5)^4(x + 2)^3$

69. $(x + 2)^4(x - 4)^4 - (x + 2)^3(x - 4)^5$

70. $(x - 3)^3(x + 1)^3 - (x - 3)^2(x + 1)^4$

71. $x^8 - 4$ 72. $a^6 + 8a^3 - 20$

73. $r^4 - s^4$ 74. $16a^4 - b^4$

75. $x^4 - 3x^2 - 4$ 76. $x^4 - 7x^2 - 18$

77. $(x - 3)^2 - 16y^2$ 78. $(x + 2)^2 - 9y^2$

79. $(a - b)^2 - 4(c - d)^2$

80. $(x^2 - x)^2 - 9(y^2 - y)^2$

81. $25(4x^2 - 12xy + 9y^2) - 9a^2b^2$

82. $18a^3 - 8a(x^2 + 8x + 16)$

83. $x^6 - 1$ 84. $a^6 - 64b^6$

85. $2x^3 - x^2 - 8x + 4$ 86. $4y^3 - 12y^2 - 9y + 27$

87. $25 - a^2 - 2ab - b^2$ 88. $x^2 - 2xy + y^2 - 9$

89. $x^4 - 1$ 90. $x^4 - y^4$

91. $x^6 - y^6$ 92. $x^4 - 16$

93. $x^6 - 64$ 94. $x^6 + y^6$

95. $x^8 - y^8$ 96. $x^8 - 1$

97. $16x^4 - x^2 + 6xy - 9y^2$

98. $x^4 - x^2 + 4x - 4$

99. $x^3 - 2x^2 + 3x - 6$

100. $x^3 + 2x^2 - 5x - 10$

101. $x^5 - x^4 + x - 1$

102. $x^5 + x^3 + 2x^2 + 2$

103. $3x^3 - x^2 + 12x - 4$

104. $2x^3 + x^2 + 4x + 2$

105. $x^2 + 4x - y^2 + 4$

106. $x^2 + y^2 - z^2 + 2xy$

107. Use the *ac* test to show that $A^2 - AB + B^2$ cannot be factored using integer coefficients.

108. Use the *ac* test to show that $A^2 + AB + B^2$ cannot be factored using integer coefficients.

2-7 Solving Equations by Factoring

♦ Zero Factor Property
♦ Solving Equations by Factoring

An equation in which a polynomial is equated to 0 can be solved readily if the polynomial can be completely factored. We solve the original equation by setting each factor equal to 0. A property of real numbers, called the *zero factor property*, justifies this process.

♦ **ZERO FACTOR PROPERTY**

When nonzero real numbers are multiplied, the product will never be 0. Equivalently, if a product is 0, then at least one of the factors must be 0. This is the **zero factor property.**

> **Zero Factor Property**
>
> For real numbers a and b:
>
> $$a \cdot b = 0 \qquad \text{if and only if} \qquad a = 0 \text{ or } b = 0 \text{ (or both)}$$

We apply this property to solving equations of the form $a \cdot b = 0$. The property allows us to solve the simpler equations $a = 0$ or $b = 0$.

♦ **SOLVING EQUATIONS BY FACTORING**

Example 1
Solving by Factoring

Solve by factoring, if possible:

(A) $x^2 + 2x - 15 = 0$ **(B)** $4x^2 = 6x$ **(C)** $2x^2 - 8x + 3 = 0$

Solution **(A)** $x^2 + 2x - 15 = 0$

$(x - 3)(x + 5) = 0$ *$(x - 3)(x + 5) = 0$ if and only if $(x - 3) = 0$ or $(x + 5) = 0$.*

$x - 3 = 0 \qquad \text{or} \qquad x + 5 = 0$

$x = 3 \qquad\qquad\qquad x = -5$

(B) $\qquad 4x^2 = 6x$ Simplify the equation by dividing both sides by 2, a common factor of both coefficients.

$\qquad 2x^2 = 3x$ Rewrite the equation as a polynomial equal to 0.

$2x^2 - 3x = 0$ Factor the polynomial.

$x(2x - 3) = 0$ *$x(2x - 3) = 0$ if and only if $x = 0$ or $2x - 3 = 0$.*

$x = 0 \qquad \text{or} \qquad 2x - 3 = 0$

$x = \tfrac{3}{2}$

Note that if we had begun by dividing both sides of $4x^2 = 6x$ by x, we would have lost the solution $x = 0$.

(C) The polynomial cannot be factored using integer coefficients; hence, another method must be used to find the solution. This will be discussed later.

Matched Problem 1

Solve by factoring, if possible:

(A) $x^2 - 2x - 8 = 0$ **(B)** $9t^2 = 6t$ **(C)** $x^2 - 3x = 3$

Example 2
A Number Problem

The product of two successive positive even integers is 48. Find the numbers.

Solution Let

$$x = \text{First number}$$

so that

$$x + 2 = \text{Second number}$$

Then

$$x(x + 2) = 48$$
$$x^2 + 2x = 48$$
$$x^2 + 2x - 48 = 0$$
$$(x + 8)(x - 6) = 0$$

$$x + 8 = 0 \qquad \text{or} \qquad x - 6 = 0$$
$$x = -8 \qquad\qquad\qquad x = 6$$

Since we are seeking a positive solution, $x = -8$ is discarded. The solution is $x = 6$. That is, the successive integers are 6 and 8.

Matched Problem 2   The square of a positive integer is equal to the sum of 14 and the quantity 5 times the integer. Find the integer.

Example 3
A Geometry Problem

The length of a rectangle is 1 inch more than twice its width. If the area is 21 square inches, find its dimensions.

Solution   Draw a figure and label the sides, as shown.

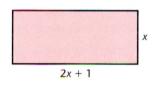

$$x(2x + 1) = 21$$
$$2x^2 + x - 21 = 0$$
$$(2x + 7)(x - 3) = 0$$

$$2x + 7 = 0 \qquad \text{or} \qquad x - 3 = 0$$
$$2x = -7 \qquad\qquad\qquad x = 3$$
$$\cancel{x = -\tfrac{7}{2}} \qquad$$ A negative length is not possible, so must be discarded.

The solution is $x = 3$. That is,

$$x = 3 \text{ inches} \qquad \text{Width}$$
$$2x + 1 = 7 \text{ inches} \qquad \text{Length}$$

In practical problems involving second-degree equations, one of two solutions often must be discarded because it will not make sense in the problem.

Matched Problem 3   The base of a triangle is 6 meters longer than its height. If the area is 20 square meters, find the base and height. The formula for the area of a triangle is $A = \frac{1}{2}bh$.

Answers to
Matched Problems

1. **(A)** $x = 4, -2$ **(B)** $t = 0, \frac{2}{3}$ **(C)** Cannot be factored using integer coefficients
2. 7   3. Height = 4 meters, Base = 10 meters

EXERCISE 2-7

A *Solve Problems 1–78 by factoring.*

1. $x^2 + 3x + 2 = 0$
2. $x^2 + 8x + 15 = 0$
3. $x^2 + 7x + 10 = 0$
4. $x^2 + 4x + 3 = 0$
5. $x^2 - 2x - 3 = 0$
6. $x^2 - 2x - 8 = 0$
7. $x^2 + 3x - 4 = 0$
8. $x^2 + 5x - 6 = 0$
9. $x^2 - 9x + 20 = 0$
10. $x^2 - 9x + 18 = 0$
11. $x^2 - 6x + 5 = 0$
12. $x^2 - 6x + 8 = 0$
13. $x^2 - 6x + 9 = 0$
14. $x^2 + 8x + 16 = 0$
15. $x^2 + 14x + 49 = 0$
16. $x^2 - 4x - 4 = 0$
17. $2x^2 - x - 1 = 0$
18. $2x^2 + 7x + 6 = 0$
19. $2x^2 + 7x + 5 = 0$
20. $2x^2 + x - 3 = 0$
21. $u^2 + 5u = 0$
22. $v^2 - 3v = 0$
23. $3A^2 = -12A$
24. $4u^2 = 8u$
25. $x^2 - 11x - 12 = 0$
26. $y^2 - 6y + 5 = 0$
27. $x^2 + 4x - 5 = 0$
28. $x^2 - 4x - 12 = 0$
29. $3Q^2 - 10Q - 8 = 0$
30. $2d^2 + 15d - 8 = 0$

B 31. $4x^2 + 4x - 3 = 0$
32. $4x^2 + 11x - 3 = 0$
33. $4x^2 + 3x - 1 = 0$
34. $4x^2 + 4x - 15 = 0$
35. $2x^2 + 4x + 3 = 0$
36. $4x^2 + 4x + 1 = 0$
37. $4x^2 + 12x + 9 = 0$
38. $2x^2 - 4x - 5 = 0$
39. $x^2 - x - 2 = 0$
40. $9x^2 - 6x + 1 = 0$
41. $9x^2 + 12x + 4 = 0$
42. $x^2 + x - 1 = 0$
43. $6x^2 - 7x + 2 = 0$
44. $6x^2 - 5x - 4 = 0$
45. $u^2 = 2u + 3$
46. $m^2 + 2m = 15$
47. $3x^2 = x + 2$
48. $2x^2 = 3 - 5x$
49. $y^2 = 5y - 2$
50. $3 = t^2 + 7t$
51. $2x(x - 1) = 3(x + 1)$
52. $3x(x - 2) = 2(x - 2)$
53. $t^2 = 4$
54. $y^2 = 9$
55. $m^2 + 4m = 12$
56. $A^2 = 2A + 8$
57. $2y^2 = 2 + 3y$
58. $L^2 - 2L = 15$
59. $2x^2 + 2 = 5x$
60. $x^2 - 3x = 10$
61. $x^2 + x = 6$
62. $x^2 + x = 2$

C 63. $8x^2 + 14x + 7 = 0$
64. $16x^2 + 7x + 6 = 0$
65. $8x^2 - 14x - 15 = 0$
66. $16x^2 + 2x - 3 = 0$
67. $x^4 - 81 = 0$
68. $2x^4 - 32 = 0$
69. $(x^3 - 1)(x^2 - 4) = 0$
70. $(x^4 - 1)(x^3 - 8) = 0$
71. $x^4 - 18x^2 + 81 = 0$
72. $x^4 - 8x^2 + 16 = 0$
73. $x^6 + 7x^3 - 8 = 0$
74. $x^6 - 26x^3 - 27 = 0$
75. $x^2 + 3x = 1 + 3x$
76. $x(x + 1) - (x + 1) = x^2$
77. $(x + 1)(x + 2) - 6x = (2x - 1)(x - 1)$
78. $2x(x - 1) = (x - 2) + x$

APPLICATIONS

79. The product of 1 more than a positive integer and 2 more than the integer is 132. Find the integer.

80. The product of 1 less than a positive integer and 2 more than the integer is 54. Find the integer.

81. Two successive negative integers have product 72. Find the integers.

82. Two successive positive integers have product 42. Find the integers.

83. Twice the square of a number is 21 more than the number. Find the number.

84. Twice the square of a number is 15 more than the number. Find the number.

85. A triangle with base equal to half the height of the triangle has area 36 square centimeters. Find the base and height of the triangle $[A = \frac{1}{2}bh]$.

86. A rectangle with length equal to 4 times its width has area 144 square meters. Find the dimensions of the rectangle.

87. The width of a rectangle is 8 inches less than its length. If its area is 33 square inches, find its dimensions.

88. Find the base and height of a triangle with area 2 square feet if its base is 3 feet longer than its height $[A = \frac{1}{2}bh]$.

89. A rectangle with area 108 square inches has length and width that add to 21. Find the dimensions of the rectangle.

90. A triangle with area 24 square feet has base and height that add to 11. Find the base and height of the triangle $[A = \frac{1}{2}bh]$.

CHAPTER SUMMARY

2-1 ADDITION AND SUBTRACTION OF POLYNOMIALS

A **polynomial** is an algebraic expression involving only the operations of addition, subtraction, and multiplication on variables and constants. The constant factor in a term is called the **coefficient.** The **degree of a term** in a polynomial is the sum of the powers of the variables; nonzero constants are assigned degree 0. The **degree of a polynomial** is the highest degree of its terms. Polynomials with one, two, and three terms are called **monomials, binomials,** and **trinomials,** respectively. Addition and subtraction of polynomials are accomplished by **grouping** and combining like terms.

2-2 MULTIPLICATION OF POLYNOMIALS

Monomials are multiplied by using the property of exponents: $x^m x^n = x^{m+n}$. **Polynomials** are multiplied by multiplying each term of one by each term of the other and adding. **Binomials** can be multiplied mentally by the **FOIL method,**

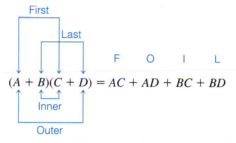

$$(A + B)(C + D) = AC + AD + BC + BD$$

and squared by the formulas

$$(A + B)^2 = A^2 + 2AB + B^2 \qquad (A - B)^2 = A^2 - 2AB + B^2$$

2-3 BASIC FACTORING METHODS

To **factor** a polynomial means to find factors whose product is the given polynomial. Factors common to all terms can be factored out by the distributive property. **Grouping** may lead to factoring out common factors. The polynomial $x^2 + bx + c$ is a **perfect square** if $c = (b/2)^2$. In this case,

$$x^2 + bx + \left(\frac{b}{2}\right)^2 = \left(x + \frac{b}{2}\right)^2$$

The **sum of two squares,** $A^2 + B^2$, cannot be factored unless there are common factors. The **difference of two squares** is factored as

$$A^2 - B^2 = (A - B)(A + B)$$

2-4 FACTORING SECOND-DEGREE POLYNOMIALS

Second-degree polynomials can be factored by **trial and error.**

2-5 FACTORING WITH THE *ac* TEST

The ***ac* test** to factor $ax^2 + bx + c$ or $ax^2 + bxy + cy^2$ involves finding two integers p and q such that $pq = ac$, $p + q = b$, rewriting the middle term as $px + qx$ or $pxy + qxy$, and then factoring by grouping.

2-6   MORE FACTORING

The **sum and difference of two cubes** are factored as

$$A^3 + B^3 = (A + B)(A^2 - AB + B^2)$$

$$A^3 - B^3 = (A - B)(A^2 + AB + B^2)$$

The factoring techniques outlined above also can be applied to some higher-degree polynomials. A substitution may make the factoring clearer. A **general strategy for factoring** is:
1. Remove common factors.
2. If the polynomial has two terms, look for a difference of two squares or a sum or difference of two cubes.
3. If the polynomial has three terms:
   (A) See if it is a perfect square.
   (B) Try trial and error.
   (C) Or use the *ac* test.
4. If a polynomial has four or more terms, try grouping.
5. Check that each factor has been completely factored.

2-7   SOLVING EQUATIONS BY FACTORING

Equations where a polynomial is set equal to 0 can be solved by factoring the polynomial and using the **zero factor property:** $ab = 0$ if and only if $a = 0$ or $b = 0$.

CHAPTER REVIEW EXERCISE

Work through all the problems in this chapter review and check answers in the back of the book. Answers to all the problems are there, and following each answer is a number in italics indicating the section in which that type of problem is discussed. Where weaknesses show up, review appropriate sections in the text.

A   **1.** What is the degree of the term $-8x^2$?

2. In the polynomial $x^5y^3 + 3x^3y^4 - 4x^6y^2 + 5x^5y^4$, what is the smallest degree of a term?

3. Given the polynomial $3x^5 - 2x^3 + 7x^2 - x + 2$:
   (A) What is its degree?
   (B) What is the degree of the second term?

4. Consider the product $(3x^3y^2)(-2xy^2z^3)$ of two monomials.
   (A) What is the degree of the first factor?
   (B) What is the degree of the product?

Add, subtract, or multiply as indicated.

5. $(2x + 5) + (x^2 - 4)$

6. $(2x + 5) - (x^2 - 4)$

7. $(2x + 5)(x^2 - 4)$

8. $(x - 3)(x + 3)$

9. $(x - 3) - (x + 3)$

10. $(x - 3) + (x + 3)$

11. $(x + 4) - (x^2 - 5x + 6)$

12. $(x + 4)(x^2 - 5x + 6)$

13. $(x + 4) + (x^2 - 5x + 6)$

14. $(3x^2 + 2x + 1) - (2x^2 - 3x + 4)$

15. $(3x^2 + 2x + 1) + (2x^2 - 3x + 4)$

16. $(3x^2 + 2x + 1)(2x^2 - 3x + 4)$

17. $[x^2 - (1 - x - x^2)] - (x + 1)$

18. $-2x\{(x^2 + 2)(x - 3) - x[x - x(3 - x)]\}$

19. $\{x - [x - (x - 1)]\} - \{1 - [1 - (1 - x)]\}$

20. $(x - 2)^2(x - 3) - (x - 2)(x - 3)^2$

Factor as far as possible in the integers.

21. $3x^4 + 9x^3$

22. $2x^2y^3 + x^3y$

23. $6x^2 + 15x$

24. $x^2 - 16$

25. $x^2 + 25$

26. $14x^2 - 56$

27. $6x^2y + 9xy + 12xy^2$

28. $6x^2y + 9xy + 8x + 12$

B   **29.** $2x^2 - 10x + x - 5$

30. $4x^2 + 11x - 3$

31. $4x^2 - 25$

32. $x^2 + 3xy + 2y^2$

33. $a^2 + 9$

34. $x^2 + 2x + 4 + 2x$

35. $2x^2 - 5x - 3$

36. $x^2 - 3x + 9 - 3x$

37. $3x^2 + 2x - 3x - 2$

38. $a^3 - 8$

39. $x^2 + 6x + 8$

40. $2x^2 - 5xy + 2y^2$

41. $x^3 + 64$

42. $a^2 - 9b^2$

43. $x^2 + 6x + 9$

44. $x^2 - 8x + 16$

45. $x^2 - 8x + 12$ **46.** $x^2 + 9x + 18$

47. $x^2 - 10x - 11$ **48.** $x^2 + 11x - 26$

49. $x^2 - 12x + 36$ **50.** $x^2 + x + 6$

51. $8x^3 - 125$ **52.** $x^4 - 1$

53. $x^4 + 1$ **54.** $4x^2 + 12x + 9$

55. $9x^2 - 12x + 4$ **56.** $2x^2 + 3x + 1$

57. $2x^2 - 5x + 3$ **58.** $6x^2 + x - 1$

59. $x^3 - y^3$ **60.** $x^2 + 4xy + 4y^2$

61. $x^2 + 3x + 4$ **62.** $x^2 - 3x - 6$

Solve by factoring.

63. $x^2 + 7x + 12 = 0$ **64.** $x^4 - 16 = 0$

65. $x^2 + 3x - 10 = 0$ **66.** $x^2 - 5x = 6$

67. $x^2 + x = 20$ **68.** $x^2 - 4x - 12 = 0$

69. $x^3 - 1 = 0$ **70.** $x^2 - 1 = 0$

71. $x^2 + 8 = 0$ **72.** $x^3 + 8 = 0$

73. $x^2 + 2x = 1$ **74.** $x^2 - 2x = -1$

75. $2x - x^2 = 1$ **76.** $x^2 = 2x + 1$

C *Factor as far as possible in the integers.*

77. $x^6 + 1$ **78.** $2x^4 + 6x^3 + 3x^2 + x^3$

79. $x^2a^3 - 4xa^3 + 4a^3$ **80.** $3x^3y - 3xy^3$

81. $x^3 - 3x^2 + x - 3$ **82.** $x^6 + x^5 + x^2 + x$

83. $2x^3 - 2x^2 - 4x$ **84.** $x^2 + 2xy + y^2 - 1$

85. $x^6 - 1$ **86.** $8x^3 + 125$

87. $3x^3 + 2x^2 - 15x - 10$ **88.** $x^4 + 2x^3 - 3x^2$

89. $-x^5y^3 - 2x^4y^2 - x^3y$ **90.** $2a^4 + 2a$

In Problems 91–94, solve by factoring.

91. $2x^2 - x = 3$ **92.** $x^2 - x = 20$

93. $6x^2 + 5x - 6 = 0$ **94.** $12x^2 + 29x - 8 = 0$

95. The product of a positive integer and the number that is 4 greater than the integer is 96. Find the integer.

96. Twice the square of a positive number is 10 more than 8 times the number. Find the number.

97. Twice the square of a positive integer is 7 more than the square of the number that is 3 greater than the integer. Find the integer.

98. A rectangle is 5 times as long as it is wide. Find the dimensions of the rectangle if its area is 180 square inches.

99. A triangle with area 150 square centimeters has height equal to one-third the base. Find the dimensions of the triangle.

100. A rectangle has length 2 feet longer than the width. If both length and width are increased by 1 foot, the resulting area is 48 square feet. Find the original dimensions of the rectangle.

CHAPTER PRACTICE TEST

The following practice test is provided for you to test your knowledge of the material in this chapter. You should try to complete it in 50 minutes or less. Answers in the back of the book indicate the section in the text where the material in the question is covered. Actual tests in your class may vary from this practice test in difficulty, length, or emphasis, depending on the goals of your course or instructor.

1. For the polynomial $3x^4 - 4x^3 + x^2 - 2x - 5$, counting terms from the left, identify:
   (A) The degree of the second term
   (B) The coefficient of the second term
   (C) The degree of the polynomial

2. For the product $(-2x^3y^3)(3x^2y^5)$, what is:
   (A) The degree of the first factor?
   (B) The degree of the product?
   (C) The product?

Add, subtract, or multiply, as indicated.

3. $(3x^2 - x + 2) - (2x^2 - 3x - 4)$

4. $(4x + 5)(x^2 + 3x + 5)$

5. $(2x - 7)(x - 4)$

6. $(x^3 + x + 2) + (x^2 - 4x + 6) + (x^3 - x^2 - 9)$

7. $\{3 - [x - (3 - x)]\} - \{x + [3 - (x - 3)]\}$

Factor as far as possible in the integers.

8. $2ab^2 + 4a^2b^2 + 6a^2b$ **9.** $5x^2 - 80$

10. $x^2 + 12x + 36$ **11.** $x^2 - 12x + 32$

12. $x^2 + 12x - 45$ **13.** $x^3 - 64$

14. $x^2 + 4xy - 3xy - 12y^2$ **15.** $6x^2 - 7x - 20$

16. $x^4 + 5x^2 - 24$

In Problems 17–19, solve by factoring, if possible.

17. $x^2 - 7x + 10 = 0$ **18.** $2x^2 - 3x = 2$

19. $x^2 + 3x + 18 = 0$

20. Two more than twice the square of a positive number is 10 times the quantity that is 3 more than the number. Find the number.

21. The length of a rectangle is 2 feet less than 3 times its width. Find the dimensions if the rectangle has area 5 square feet.

CUMULATIVE REVIEW EXERCISE, CHAPTERS 1–2

This set of exercises reviews the major concepts and techniques of Chapters 1–2. Work through all the problems and check answers in the back of the book. Answers to all the problems are there, and following each answer is a number in italics indicating the section in which that type of problem is discussed. Where weaknesses show up, review appropriate sections in the text.

A *Translate each statement into an algebraic equation or inequality statement using x as the only variable. Do not solve the equation or inequality.*

1. Three times the quantity 2 less than a number is 4 more than the number.

2. The product of 2 less than a number and 3 more than the number is 84.

3. Two more than 3 times a number is less than 4 more than one-half the number.

Find, evaluate, or calculate the quantity requested.

4. $(-5) + (-8)$ **5.** $(-5)(-8)$

6. $(-12) - (-6)$ **7.** $(-12) \div (-6)$

8. $(-3)[4 + (-5)6]$ **9.** $(-3)4 - 5 \cdot 6$

10. $-3^2 \cdot 4 - 5$ **11.** $|4 - 3^2|$

12. $1 - |2 - 3| - 4|5 - 6|$

13. (A) The coefficient of x^2 in $5x^3 - 4x^2 + 3x - 2$
    (B) The coefficient of xy in $x^2y - xy + xy^2$

14. (A) The degree of the polynomial $2 + 5x - 7x^2 + 9x^3$
    (B) The degree of the polynomial $13x^7y + 14x^4y^2 + 15x^2y^3$

Rewrite in the form indicated.

15. x^5x^{11} as a power of x

16. $3(x + 8)$ without parentheses

17. $3x - 4y - 5x + 6y$ with like terms combined

18. $x(yz) + (xy)z$ without parentheses and with like terms combined

19. $(x^2 - 5x + 9) + (2x^2 + 4x + 6)$ without parentheses and with like terms combined

20. $(3x^2 + 5x + 7) - (4x^2 - 5x - 6)$ without parentheses and with like terms combined

21. $(x + 1)(x^2 + 2x + 1)$ without parentheses and with like terms combined

22. $(x + 2)(3x - 4)$ without parentheses and with like terms combined

23. $x^2 + 4x + 4$ in factored form

24. $x^2 + 5x + 6$ in factored form

25. $x^2 - 10x - 24$ in factored form

26. $3x^4 - 9x^3$ in factored form

27. $x^3 - x$ in factored form

28. $x^3 - 1$ in factored form

29. $x^2 + 25$ in factored form

30. $x^2 - 10x + 25$ in factored form

Solve for x.

31. $3x - 5 = 7x + 9$

32. $-x - 1 = -2 + 3x$

33. $3[x - 5(x + 2)] = 7x + 9[4 - 6(x - 8)]$

34. $\dfrac{x}{2} + \dfrac{1}{2} = x - \dfrac{1}{10}$

35. $x(x - 3) = (x - 1)(x + 2)$

36. $x(x - 1)(x + 2) = 0$ **37.** $y = 2x + 3$

38. $\dfrac{x}{2} + \dfrac{y}{3} = \dfrac{1}{4}$ **39.** $x^2 - 3x - 4 = 0$

Set up an appropriate equation or inequality and solve.

40. Three more than the quantity 2 less than a number is 4 less than the quantity 5 times the number. Find the number.

41. The sum of three consecutive integers is 111. Find the smallest of the three integers.

42. Find three consecutive integers such that the first integer plus twice the third is equal to 1 more than three times the second.

43. A rectangle with perimeter 32 inches has length equal to 1 more than twice the width. Find the dimensions.

44. A 2 by 4 piece of lumber is 16 feet long. It is cut into three pieces in such a way that the longest piece is 3 times the length of the shortest and the other piece is 1 foot longer than the shortest. Find the lengths of the three pieces.

B *Translate each statement into an algebraic equation or inequality statement using x as the only variable. Do not solve the equation or inequality.*

45. One less than the square of a quantity that is 2 less than a number is 3 more than the number.

46. Twice the square of a positive integer is 14 less than the square of a number that is 5 more than the integer.

47. The absolute value of 5 less than a number is 9 less than the number.

Find, evaluate, or calculate the quantity requested.

48. $x^2 - 3xy$ for $x = -1$, $y = -2$

49. $2x - 3xy^2$ for $x = 3$, $y = -1$

50. $\dfrac{x}{3 - \dfrac{4}{x - \dfrac{5}{x + 6}}}$ for $x = -1$

51. $ax^2 + bx + c$ for $a = 3$, $b = 4$, $c = 5$, $x = -2$

52. $|x - 3y|$ for $x = -2$, $y = 3$

53. $xy \div z^2$ for $x = 3$, $y = 4$, $z = 2$

Rewrite in the form indicated.

54. $x^4 x^7 x^{10}$ as a power of x

55. $x(2x^2 - 3x + 4)$ without parentheses and with like terms combined

56. $x^2(x - 1)^2$ without parentheses and with like terms combined

57. $(xy + x^2 y^4 - x^4 y^6) - (xy - 2x^2 y^4 - 3x^4 y^6)$ without parentheses and with like terms combined

58. $(x^2 + 3)(x^4 - x^2 + 5)$ without parentheses and with like terms combined

59. $x^2 + 7x + \frac{49}{4}$ in factored form

60. $x^2 - 11x - 60$ in factored form

61. $x^3 - 8$ in factored form

62. $x^3 + x^2 + x + 1$ in factored form

63. $x^3 + 4x$ in factored form

64. $2x^2 + x - 3$ in factored form

65. $2x^2 - 5x - 12$ in factored form

66. $x^2 - 7x - 60$ in factored form

67. $4x^2 + 12x + 9$ in factored form

68. $x^3 + 6x^2 + 9x$ in factored form

Solve for x.

69. $x - 2(x - 3) = 3 - 4(x - 5)$

70. $6 + x[5 - 4(x - 3)] = 2 + x[3 - 4(x + 5)]$

71. $x^2 - 8x + 16 = 0$

72. $x^2 + 5x - 24 = 0$

73. $\dfrac{2}{3}x - \dfrac{4}{5}(x - 6) = \dfrac{x}{10} + 7$

74. $x(x^2 - 5) = x^3 + 7x + 9$

Set up an appropriate equation or inequality and solve.

75. The perimeter of a triangle is 72 centimeters. The longest side is 6 centimeters less than twice the length of the shortest side and also equal to $\frac{5}{8}$ of twice the length of the remaining side. Find the three sides.

76. The area of a rectangle is 175 square feet. The length is 7 times the width. Find the dimensions.

77. The area of a rectangle is 60 square feet. The length is 7 feet more than the width. Find the dimensions.

78. The product of 2 more than a positive number and 2 less than the number is 4 less than the square of the number. Find the number.

79. The product of a number and the quantity 5 less than the opposite of the number is 6. Find the number.

80. Twice the square of a positive integer is 14 less than the square of a number that is 5 more than the integer. Find the integer.

C *Rewrite in the form indicated.*

81. $(x - 1)[(x^2 + 2) + (x^2 - 3)]$ without parentheses and with like terms combined

82. $-2x[3(x + 4) - x(x - 5)]$ without parentheses and with like terms combined

83. $(x^2 + x + 1)(2x^2 - 3x + 4)$ without parentheses and with like terms combined

84. $2x^2 - 7x - 30$ in factored form

85. $4x^2 - 37x - 30$ in factored form

86. $2x^2 - 11x - 30$ in factored form

87. $ax - by - ay + bx$ in factored form

88. $x^4 + 4x^2 + 4$ in factored form

89. $x^5 - 3x^3 - 4x$ in factored form

90. $8x^3 - a^3y^3$

91. $x^6 - 64$

Solve for x.

92. $x^3 + 64 = 0$

93. $x^3 - 64 = x(x^2 + x + 16)$

94. $3 + 5(x + 7) = 7 + 5(x - 3)$

95. $x(5 - x) = x(5x - 1) - 5x$

96. $x - y = xy + 1$

Set up an appropriate equation or inequality and solve.

97. A rectangle has length equal to 3 times its width. A triangle has base equal to the width of the rectangle and height equal to the length of the rectangle. The area of the rectangle is equal to twice the area of the triangle. Find the dimensions of each.

98. A rectangle has length equal to 3 times its width. The area of the rectangle is 6 times its perimeter. Find the dimensions.

99. The average of a set of numbers is the sum of the numbers divided by the number of numbers in the set. The average of four consecutive even integers is equal to $\frac{7}{8}$ of the largest integer of the four. Find the integers.

100. A 102-problem exercise set in an algebra text is divided into A, B, and C problems. The number of A problems is 1 less than twice the number of C problems. The number of A and B problems combined is 10 more than 3 times the number of C problems. How many problems are there of each type?

3

Algebraic Fractions

Polynomials are built up from constants and variables using the operations of addition, subtraction, and multiplication. If we allow division as well, we obtain algebraic forms called *rational expressions.* In this chapter we consider these expressions, and how to add, subtract, multiply, and divide them.

Photo reference: see Exercise 3-5, Problem 80.

3-1 Rational Expressions

- ◆ Rational Expressions
- ◆ Fundamental Principle of Fractions
- ◆ Reducing to Lowest Terms
- ◆ Raising to Higher Terms

The concept of reducing fractions to lower terms is familiar from arithmetic. In this section, the concept is extended to algebraic fractional forms.

◆ RATIONAL EXPRESSIONS

Fractional forms in which the numerator and denominator are polynomials are called **rational expressions.** For example,

$$\frac{1}{x} \qquad \frac{-6}{y-3} \qquad \frac{x-2}{x^2-2x+5} \qquad \frac{x^2-3xy+y^2}{3x^3y^4} \qquad \frac{3x^2-8x-4}{5}$$

are all rational expressions. In the first two examples, the numerator is a polynomial since a nonzero constant is a polynomial of degree 0. More generally, a rational expression is an algebraic expression involving only the operations of addition, subtraction, multiplication, and division on variables and constants.

Each rational expression names a real number for real number replacements of the variables, with division by 0 excluded. Hence, all properties of the real numbers apply to these expressions. In particular, the basic properties of fractions apply. In this section we recall the *fundamental principle of fractions,* which allows us to rewrite a fraction in higher or lower terms, and we use this principle to change the form of rational expressions.

◆ FUNDAMENTAL PRINCIPLE OF FRACTIONS

You will recall from arithmetic that

$$\frac{8}{12} = \frac{8 \div 4}{12 \div 4} = \frac{2}{3} \qquad \text{and} \qquad \frac{3}{5} = \frac{2 \cdot 3}{2 \cdot 5} = \frac{6}{10}$$

The first example illustrates reducing a fraction to lowest terms, while the second example illustrates raising a fraction to higher terms. Both illustrate the use of the **fundamental principle of fractions.**

Fundamental Principle of Fractions

For all polynomials P, Q, and K, with Q, $K \neq 0$:

$$\frac{PK}{QK} = \frac{P}{Q}$$
We may divide out a common factor K from both the numerator and denominator. This is called **reducing to lower terms.**

$$\frac{P}{Q} = \frac{PK}{QK}$$
We may multiply the numerator and denominator by the same nonzero factor K. This is called **raising to higher terms.**

♦ REDUCING TO LOWEST TERMS

To **reduce a fraction to lowest terms** is to divide out *all* common factors from the numerator and denominator.

It is important to keep in mind when reducing fractions to lowest terms that it is only common *factors* in products that can be divided out. Common *terms* in sums or differences cannot be divided out. For example:

Common factors may be divided out: $\dfrac{2y}{3y} = \dfrac{2\overset{1}{\cancel{y}}}{3\underset{1}{\cancel{y}}} = \dfrac{2}{3}$

Common terms may not be divided out: $\dfrac{2+y}{3+y} \neq \dfrac{2+\cancel{y}}{3+\cancel{y}} \neq \dfrac{2}{3}$

Example 1
Reducing to Lowest Terms

Reduce to lowest terms:

(A) $\dfrac{8x^2y}{12xy^2}$ **(B)** $\dfrac{6x^2 - 3x}{3x}$ **(C)** $\dfrac{x^2y - xy^2}{x^2 - xy}$ **(D)** $\dfrac{3x^3 - 2x^2 - 8x}{x^4 - 8x}$

Solution   **(A)** $\dfrac{8x^2y}{12xy^2} = \dfrac{(4xy)(2x)}{(4xy)(3y)}$ Divide out the common factor $4xy$.

$\qquad\qquad = \dfrac{2x}{3y}$

(B) $\dfrac{6x^2 - 3x}{3x} = \dfrac{\overset{1}{\cancel{3x}}(2x - 1)}{\underset{1}{\cancel{3x}}}$ Factor the numerator. Divide out the common factor $3x$.

$\qquad\qquad = 2x - 1$ Note that we cannot divide out the *term* $3x$ in $6x^2 - 3x$. We must factor first in order to divide out common factors.

(C) $\dfrac{x^2y - xy^2}{x^2 - xy} = \dfrac{\overset{1}{\cancel{xy}}(\overset{1}{x - \cancel{y}})}{\underset{1}{\cancel{x}}(\underset{1}{x - \cancel{y}})}$ Factor numerator and denominator. Then divide out the common factors.

$\qquad\qquad = y$

(D) $\dfrac{3x^3 - 2x^2 - 8x}{x^4 - 8x} = \dfrac{\overset{1}{\cancel{x}}(3x^2 - 2x - 8)}{\underset{1}{\cancel{x}}(x^3 - 8)}$

$\qquad\qquad = \dfrac{(3x + 4)\overset{1}{(x - \cancel{2})}}{\underset{1}{(x - \cancel{2})}(x^2 + 2x + 4)}$

$\qquad\qquad = \dfrac{3x + 4}{x^2 + 2x + 4}$

Matched Problem 1   Reduce to lowest terms:

(A) $\dfrac{16u^5v^3}{24u^3v^5}$ **(B)** $\dfrac{4x}{8x^2 - 4x}$ **(C)** $\dfrac{x^2 - 3x}{x^2y - 3xy}$ **(D)** $\dfrac{2x^3 - 8x}{2x^4 + 16x}$

Example 2
Simplifying and Reducing

Simplify the numerator and reduce to lowest terms:

(A) $\dfrac{(x + h)^2 - x^2}{h}$ **(B)** $\dfrac{(2 + h)^3 - 8}{h}$

(C) $\dfrac{4(x + 1)^3(x + 2)^3 - 3(x + 1)^4(x + 2)^2}{(x + 2)^6}$

Solution **(A)** $\dfrac{(x + h)^2 - x^2}{h} = \dfrac{x^2 + 2xh + h^2 - x^2}{h}$

$$= \dfrac{2xh + h^2}{h} = \dfrac{\cancel{h}(2x + h)}{\cancel{h}}$$

$$= 2x + h$$

(B) First simplify the numerator. Note that it is the difference of two cubes.

$$(2 + h)^3 - 8 = (2 + h)^3 - 2^3 \qquad \text{Use } A^3 - B^3 = (A - B)(A^2 + AB + B^2).$$
$$= [(2 + h) - 2][(2 + h)^2 + 2(2 + h) + 2^2]$$
$$= h[4 + 4h + h^2 + 4 + 2h + 4]$$
$$= h(12 + 6h + h^2)$$

Now reduce the fraction:

$$\dfrac{(2 + h)^3 - 8}{h} = \dfrac{\cancel{h}(12 + 6h + h^2)}{\cancel{h}}$$

$$= 12 + 6h + h^2$$

(C) First factor the numerator:

$$4(x + 1)^3(x + 2)^3 - 3(x + 1)^4(x + 2)^2$$
$$= (x + 1)^3(x + 2)^2[4(x + 2) + 3(x + 1)]$$
$$= (x + 1)^3(x + 2)^2[4x + 8 + 3x + 3]$$
$$= (x + 1)^3(x + 2)^2(7x + 11)$$

Now reduce the fraction:

$$\dfrac{4(x + 1)^3(x + 2)^3 - 3(x + 1)^4(x + 2)^2}{(x + 2)^6} = \dfrac{(x + 1)^3(x + 2)^2(7x + 11)}{(x + 2)^6}$$

$$= \dfrac{(x + 1)^3(7x + 11)}{(x + 2)^4}$$

Matched Problem 2   Simplify the numerator and reduce to lowest terms:

(A) $\dfrac{(2 + h)^2 - 2^2}{h}$ **(B)** $\dfrac{(x + h)^3 - x^3}{h}$

(C) $\dfrac{5(x - 1)^4(x + 3)^2 - 2(x - 1)^5(x + 3)}{(x + 3)^4}$

♦ **RAISING TO HIGHER TERMS**

The following example shows how to use the fundamental principle of fractions to raise fractions to higher terms. We will need to use this technique later when we add and subtract rational expressions.

Example 3
Raising to
Higher Terms

Complete the process of raising to higher terms by replacing each question mark with an appropriate expression.

(A) $\dfrac{3}{2x} = \dfrac{?}{16xy^2}$ (B) $\dfrac{2x}{3y} = \dfrac{?}{3xy - 3y^2}$ (C) $\dfrac{x - 2y}{2x + y} = \dfrac{2x^2 - 5xy + 2y^2}{?}$

Solution (A) $\dfrac{3}{2x} = \dfrac{(8y^2)(3)}{(8y^2)(2x)} = \dfrac{24y^2}{16xy^2}$ By what must $2x$ be multiplied to obtain $16xy^2$?

(B) $\dfrac{2x}{3y} = \dfrac{(x - y)(2x)}{(x - y)(3y)} = \dfrac{2x^2 - 2xy}{3xy - 3y^2}$

(C) $\dfrac{x - 2y}{2x + y} = \dfrac{(2x - y)(x - 2y)}{(2x - y)(2x + y)} = \dfrac{2x^2 - 5xy + 2y^2}{4x^2 - y^2}$

Matched Problem 3

Complete the process of raising to higher terms by replacing each question mark with an appropriate expression.

(A) $\dfrac{3u^2}{4v} = \dfrac{?}{20u^2v^2}$ (B) $\dfrac{3m}{2n} = \dfrac{3m^3 - 3m}{?}$ (C) $\dfrac{2x - 3}{x - 4} = \dfrac{?}{3x^2 - 14x + 8}$

Answers to
Matched Problems

1. (A) $\dfrac{2u^2}{3v^3}$ (B) $\dfrac{1}{2x - 1}$ (C) $\dfrac{1}{y}$ (D) $\dfrac{x - 2}{x^2 - 2x + 4}$

2. (A) $4 + h$ (B) $3x^2 + 3xh + h^2$ (C) $\dfrac{(x - 1)^4(3x + 17)}{(x + 3)^3}$

3. (A) $15u^4v$ (B) $2m^2n - 2n$ (C) $6x^2 - 13x + 6$

EXERCISE 3-1

A *Reduce to lowest terms.*

1. $\dfrac{36}{90}$ 2. $\dfrac{48}{120}$ 3. $\dfrac{75}{45}$ 4. $\dfrac{210}{150}$

5. $\dfrac{3x^3}{6x^5}$ 6. $\dfrac{9u^5}{3u^6}$ 7. $\dfrac{14x^3y}{21xy^2}$ 8. $\dfrac{20m^4n^6}{15m^5n^2}$

9. $\dfrac{-x^4y^2}{xy^6}$ 10. $\dfrac{-a^5b^2}{a^3b^4}$

11. $\dfrac{(-a)^5b^2}{a^3(-b)^4}$ 12. $\dfrac{(-x)^4y^2}{x(-y)^6}$

13. $\dfrac{15y^3(x - 9)^3}{5y^4(x - 9)^2}$ 14. $\dfrac{2x^2(x + 7)}{6x(x + 7)}$

15. $\dfrac{(2x - 1)(2x + 1)}{3x(2x + 1)}$ 16. $\dfrac{(x + 3)(2x + 5)}{2x^2(2x + 5)}$

17. $\dfrac{(x - 1)^2(x + 3)^5}{x(x - 1)^4}$ 18. $\dfrac{x^4(x + 2)^3}{x(x + 2)^4}$

19. $\dfrac{x^3(x + 5)^4}{x^5(x + 5)}$ 20. $\dfrac{(x + 2)^3(x - 3)^2}{(x + 2)^4(x - 1)^3}$

21. $\dfrac{x^2 - 2x}{2x - 4}$ 22. $\dfrac{2x^2 - 10x}{4x - 20}$

23. $\dfrac{m^2 - mn}{m^2n - mn^2}$ 24. $\dfrac{a^2b + ab^2}{ab + b^2}$

25. $\dfrac{x^2 - 3x}{x^3 + 2x^2}$ 26. $\dfrac{x^2 + 5x}{x^3 + 10x^2 + 25x}$

27. $\dfrac{x^3 + 4x^2 + 4x}{x^2(x + 2)}$

28. $\dfrac{x^3 + x^2}{x^2 + 3x}$

Complete the process of raising to higher terms by replacing each question mark with an appropriate expression.

29. $\dfrac{3}{2x} = \dfrac{?}{8x^2y}$

30. $\dfrac{5x}{3} = \dfrac{10x^3y^2}{?}$

31. $\dfrac{7}{3y} = \dfrac{?}{6x^3y^2}$

32. $\dfrac{5u}{4v^2} = \dfrac{20u^3v}{?}$

33. $\dfrac{6xy}{16x^2y^2} = \dfrac{?}{8xy^2}$

34. $\dfrac{18a^2b^3}{6ab^2} = \dfrac{6ab^2}{?}$

35. $\dfrac{30a^4b^2}{20ab^3} = \dfrac{6a^3b}{?}$

36. $\dfrac{24x^3y^3}{40x^2y^4} = \dfrac{?}{20xy}$

B *Reduce to lowest terms.*

37. $\dfrac{x^2 + 6x + 8}{3x^2 + 12x}$

38. $\dfrac{x^2 + 5x + 6}{2x^2 + 6x}$

39. $\dfrac{x^2 - 9}{x^2 + 6x + 9}$

40. $\dfrac{x^2 - 4}{x^2 + 4x + 4}$

41. $\dfrac{4x^2 - 9y^2}{4x^2y + 6xy^2}$

42. $\dfrac{a^2 - 16b^2}{4ab - 16b^2}$

43. $\dfrac{x^2 + 2x + 1}{x^2 + 3x + 2}$

44. $\dfrac{x^2 - 2x + 1}{x^2 - 1}$

45. $\dfrac{x^2 + 4x + 4}{x^2 - 4}$

46. $\dfrac{x^2 - 8x + 16}{x^2 - 2x - 8}$

47. $\dfrac{x^2 + 3x + 2}{x^2 + 5x + 6}$

48. $\dfrac{x^2 + 7x + 12}{x^2 + 5x + 6}$

49. $\dfrac{x^2 - xy + 2x - 2y}{x^2 - y^2}$

50. $\dfrac{u^2 + uv - 2u - 2v}{u^2 + 2uv + v^2}$

51. $\dfrac{6x^3 + 28x^2 - 10x}{12x^3 - 4x^2}$

52. $\dfrac{12x^3 - 78x^2 - 42x}{16x^4 + 8x^3}$

53. $\dfrac{x^3 - 8}{x^2 - 4}$

54. $\dfrac{y^3 + 27}{2y^3 - 6y^2 + 18y}$

55. $\dfrac{2x^3 + 4x^2 - 6x}{6x^3 - 6x^2}$

56. $\dfrac{4x^3 - 4x}{8x^3 - 16x^2 + 8x}$

57. $\dfrac{3x^3 - 9x^2}{6x^3 - 36x^2 + 54x}$

58. $\dfrac{6x^3 + 6x^2 - 6x}{8x^3 + 32x^2 + 32x}$

59. $\dfrac{3x + x^2 + 15 + 5x}{x^2 + 13x + 30}$

60. $\dfrac{4x^2 - 4x - 8}{4x - 8 - 2x + x^2}$

61. $\dfrac{x^3 + 1}{x^2 - 1}$

62. $\dfrac{x^3 - 1}{x^2 - 2x + 1}$

63. $\dfrac{x^2 + 10x + 21}{x^2 + 5x - 14}$

64. $\dfrac{x^2 - 4x + 3}{x^2 + 2x - 15}$

65. $\dfrac{6x^2z + 4xyz}{3xyz + 2y^2z}$

66. $\dfrac{2x^2 + x - 3}{4x^2 + x - 5}$

67. $\dfrac{x^3 - 3x^2 + 2x - 6}{x^2 - 5x + 6}$

68. $\dfrac{x^3 - x^2 + x - 1}{x^2 - 1}$

69. $\dfrac{6a^2b + 3ab^2}{8a^3b + 4a^2b^2}$

70. $\dfrac{2x^2y^2 - 4xy^3}{4x^3y - 8x^2y^2}$

71. $\dfrac{2x^2 - x - 1}{2x^2 + 3x + 1}$

72. $\dfrac{3x^2 + 8x - 3}{3x^2 - 8x + 3}$

73. $\dfrac{2x^2 - 3x + 2x - 3}{2x + 3 + 2x^2 + 3x}$

74. $\dfrac{3x - 2 - 3x^2 + 2x}{3x - 3x^2 + 2 - 2x}$

Complete the process of raising to higher terms by replacing each question mark with an appropriate expression.

75. $\dfrac{3x}{4y} = \dfrac{?}{4xy + 4y^2}$

76. $\dfrac{4m}{5n} = \dfrac{4m^2 - 4mn}{?}$

77. $\dfrac{x - 2y}{x + y} = \dfrac{x^2 - 3xy + 2y^2}{?}$

78. $\dfrac{2x + 5}{x - 3} = \dfrac{?}{3x^2 - 8x - 3}$

79. $\dfrac{4x^2 + 8x}{12x^2 + 48x + 48} = \dfrac{?}{6x + 12}$

80. $\dfrac{6x^3 + 6x^2}{10x^2 + 10x} = \dfrac{?}{5x + 5}$

81. $\dfrac{x^3 + 3x^2 + 2x}{x^3 + x^2} = \dfrac{?}{x^2}$

82. $\dfrac{x^3 - x}{x^2 + x} = \dfrac{?}{x + 1}$

C *Reduce to lowest terms.*

83. $\dfrac{x^3 - y^3}{3x^3 + 3x^2y + 3xy^2}$

84. $\dfrac{2u^3v - 2u^2v^2 + 2uv^3}{u^3 + v^3}$

85. $\dfrac{ux + vx - uy - vy}{2ux + 2vx + uy + vy}$

86. $\dfrac{mx - 2my + nx - 2ny}{mx - 2my - nx + 2ny}$

87. $\dfrac{x^4 - y^4}{(x^2 - y^2)(x + y)^2}$

88. $\dfrac{x^4 - 2x^2y^2 + y^4}{x^4 - y^4}$

89. $\dfrac{x^4 + x}{x^3 - x^2 + x}$

90. $\dfrac{x^5 + x^3 + x}{x^4 + x^2 + 1}$

91. $\dfrac{x^2 + y^2}{x^3 + y^3}$

92. $\dfrac{z^3 - z}{z^3 - z^2 - z + 1}$

93. $\dfrac{x^2u + uxy + vxy + vy^2}{xu^2 + uvy + uvx + yv^2}$

94. $\dfrac{x^2y^2z - xyz^2 - xy + z}{x^2yz^2 - xz - xy^2z + y}$

95. $\dfrac{(1 + h)^3 - 1}{h}$

96. $\dfrac{3(x + 1)^2(x - 1)^3 - 3(x + 1)^3(x - 1)^2}{(x - 1)^6}$

97. $\dfrac{2(x + 3)(x - 2)^4 - 4(x + 3)^2(x - 2)^3}{(x - 2)^8}$

98. $\dfrac{(2 - h)^3 - 8}{h}$

99. $\dfrac{(x + h)^2 + 2(x + h) - (x^2 + 2x)}{h}$

100. $\dfrac{(x + h)^2 - 3(x + h) - (x^2 - 3x)}{h}$

3-2 Multiplication and Division

- Multiplication of Rational Expressions
- Division of Rational Expressions

Multiplication and division of rational expressions are based upon the corresponding properties of fractions.

♦ MULTIPLICATION OF RATIONAL EXPRESSIONS

We start with the process of multiplying rational forms:

> **Multiplication**
>
> If P, Q, R, and S are polynomials (Q, $S \neq 0$), then
>
> $$\frac{P}{Q} \cdot \frac{R}{S} = \frac{P \cdot R}{Q \cdot S}$$

Note that this is exactly the same rule you would use to multiply numerical fractions with P, Q, R, and S representing integers.

Example 1
Multiplying Rational Expressions

Multiply and reduce to lowest terms:

(A) $\dfrac{3a^2b}{4c^2d} \cdot \dfrac{8c^2d^3}{9ab^2}$

(B) $(x^2 - 4) \cdot \dfrac{2x - 3}{x + 2}$

(C) $\dfrac{4a^2 - 9b^2}{4a^2 + 12ab + 9b^2} \cdot \dfrac{6a^2b}{8a^2b^2 - 12ab^3}$

Solution

(A) $\dfrac{3a^2b}{4c^2d} \cdot \dfrac{8c^2d^3}{9ab^2} = \dfrac{(3a^2b)(8c^2d^3)}{(4c^2d)(9ab^2)} = \dfrac{24a^2bc^2d^3}{36ab^2c^2d}$

$= \dfrac{(12\cancel{abc^2d})(2ad^2)}{(12\cancel{abc^2d})(3b)} = \dfrac{2ad^2}{3b}$

It is not necessary to multiply out the numerator and denominator completely before dividing out common factors. Any factor in either numerator may be divided out with any like factor in either denominator. Thus, the process is easily shortened to the following:

$$\frac{\overset{1}{\cancel{3}}\cdot \overset{a}{\cancel{a^2}}\cdot \overset{1}{\cancel{b}}}{\underset{1}{\cancel{4}}\cdot \underset{1}{\cancel{e^2}}\cdot \underset{1}{\cancel{d}}}\cdot \frac{\overset{2}{\cancel{8}}\cdot \overset{1}{\cancel{e^2}}\cdot \overset{d^2}{\cancel{d^3}}}{\underset{3}{\cancel{9}}\cdot \underset{1}{\cancel{a}}\cdot \underset{b}{\cancel{b^2}}}=\frac{2ad^2}{3b}$$

(B) $(x^2-4)\cdot \dfrac{2x-3}{x+2}$ Factor where possible. Divide out common factors; then multiply and write the answer.

$$=\frac{\overset{1}{\cancel{(x+2)}}(x-2)}{1}\cdot \frac{(2x-3)}{\underset{1}{\cancel{(x+2)}}}$$

$$=(x-2)(2x-3)$$

(C) $\dfrac{4a^2-9b^2}{4a^2+12ab+9b^2}\cdot \dfrac{6a^2b}{8a^2b^2-12ab^3}$

$$=\frac{\overset{1}{\cancel{(2a-3b)}}\overset{1}{\cancel{(2a+3b)}}}{\underset{(2a+3b)}{\cancel{(2a+3b)^2}}}\cdot \frac{\overset{3a}{\cancel{6a^2b}}}{\underset{2b}{\cancel{4ab^2}}\underset{1}{\cancel{(2a-3b)}}}$$

$$=\frac{3a}{2b(2a+3b)}$$

Matched Problem 1   Multiply and reduce to lowest terms:

(A) $\dfrac{4x^2y^3}{9w^2z}\cdot \dfrac{3wz^2}{2xy^4}$ **(B)** $\dfrac{x+5}{x^2-9}\cdot (x+3)$

(C) $\dfrac{x^2-9y^2}{x^2-6xy+9y^2}\cdot \dfrac{6x^2y}{2x^2+6xy}$

♦ ## DIVISION OF RATIONAL EXPRESSIONS

The rule for dividing rational expressions follows from the definition of division: $A\div B=A\cdot \left(\dfrac{1}{B}\right)$, where $\dfrac{1}{B}$ is the multiplicative inverse of B. For the quotient $\dfrac{R}{S}$ of two polynomials, the multiplicative inverse is the reciprocal $\dfrac{S}{R}$, since $\dfrac{R}{S}\cdot \dfrac{S}{R}=1$. Thus,

Division

If P, Q, R, and S are polynomials with Q, R, $S \neq 0$, then

Divisor $\dfrac{R}{S}$ Reciprocal of divisor $= \dfrac{S}{R}$

$$\frac{P}{Q} \div \frac{R}{S} = \frac{P}{Q} \cdot \frac{S}{R}$$

That is, to divide one rational expression by another, multiply by the reciprocal of the divisor.

Example 2
Dividing Rational
Expressions

Divide and reduce to lowest terms:

(A) $\dfrac{6a^2b^3}{5cd} \div \dfrac{3a^2c}{10bd}$ **(B)** $(4 - x) \div \dfrac{2x^2 - 32}{6xy}$

(C) $\dfrac{10x^3y}{3xy + 9y} \div \dfrac{4x^2 - 12x}{x^2 - 9}$

Solution   **(A)** $\dfrac{6a^2b^3}{5cd} \div \dfrac{3a^2c}{10bd} = \dfrac{\overset{2}{\cancel{6a^2b^3}}}{\underset{1}{\cancel{5cd}}} \cdot \dfrac{\overset{1}{\cancel{10bd}}}{\underset{1}{\cancel{3a^2c}}} = \dfrac{4b^4}{c^2}$

(B) $(4 - x) \div \dfrac{2x^2 - 32}{6xy} = (4 - x) \cdot \dfrac{6xy}{2(x - 4)(x + 4)}$

$$= (-1)(\cancel{x - 4}) \cdot \dfrac{\overset{3}{\cancel{6xy}}}{\underset{1}{\cancel{2}}(\cancel{x - 4})(x + 4)}$$

$$= -\dfrac{3xy}{x + 4}$$

(C) $\dfrac{10x^3y}{3xy + 9y} \div \dfrac{4x^2 - 12x}{x^2 - 9} = \dfrac{\overset{5\ \ x^2}{\cancel{10x^3y}}}{\underset{1}{3y}\underset{1}{(x + 3)}} \cdot \dfrac{\overset{1}{(x + 3)}\overset{1}{(x - 3)}}{\underset{2\ 1}{4x}\underset{1}{(x - 3)}}$

$$= \dfrac{5x^2}{6}$$

Note that in part (B) we rewrote

$$4 - x = (-1)(x - 4)$$

Recall that more generally

$$(a - b) = (-1)(b - a) = -(b - a)$$

This gives us the following sign rule for dividing out opposites:

Dividing Out Opposites

$$\frac{a - b}{b - a} = \frac{a - b}{-(a - b)} = -1$$

Matched Problem 2 Divide and reduce to lowest terms:

(A) $\dfrac{8w^2z^2}{9x^2y} \div \dfrac{4wz}{6xy^2}$ (B) $\dfrac{2x^2 - 8}{4x} \div (x + 2)$

(C) $\dfrac{x^2 - 4x + 4}{4x^2y - 8xy} \div \dfrac{x^2 + x - 6}{6x^2 + 18x}$

CAUTION   Equality between rational expressions must be interpreted carefully. For example, we can rewrite

$$\frac{x^2 - x}{x} = \frac{\overset{1}{\cancel{x}}(x - 1)}{\underset{1}{\cancel{x}}} = x - 1$$

That is,

$$\frac{x^2 - x}{x} = x - 1 \tag{1}$$

However, if we replace x by 0, the right side of this equation equals -1 while the left side is undefined. That is,

$$\frac{x^2 - x}{x} \neq x - 1 \qquad \text{for } x = 0$$

Therefore, to be strictly correct, we should write Equation (1) as

$$\frac{x^2 - x}{x} = x - 1 \qquad \text{for } x \neq 0$$

For convenience, we will continue to write equations such as Equation (1) without listing the exception "for $x \neq 0$." The equal sign in such equations should then be interpreted to mean that the two expressions are equal only if both are defined.

Answers to Matched Problems

1. (A) $\dfrac{2xz}{3wy}$ (B) $\dfrac{x + 5}{x - 3}$ (C) $\dfrac{3xy}{x - 3y}$

2. (A) $\dfrac{4wzy}{3x}$ (B) $\dfrac{x - 2}{2x}$ (C) $\dfrac{3}{2y}$

EXERCISE 3-2

Do not change improper fractions in your answers to mixed fractions; that is, write $\frac{7}{2}$, not $3\frac{1}{2}$.

A *Multiply and reduce to lowest terms.*

1. $\dfrac{10}{9} \cdot \dfrac{12}{15}$

2. $\dfrac{3}{7} \cdot \dfrac{14}{9}$

3. $\dfrac{18}{35} \cdot \dfrac{45}{24}$

4. $\dfrac{36}{25} \cdot \dfrac{40}{54}$

5. $\dfrac{2a}{3bc} \cdot \dfrac{9c}{a}$

6. $\dfrac{2x}{3yz} \cdot \dfrac{6y}{4x}$

7. $\dfrac{3x^2}{4} \cdot \dfrac{16y}{12x^3}$

8. $\dfrac{2x^2}{3y^2} \cdot \dfrac{9y}{4x}$

9. $\dfrac{6xy}{z} \cdot \dfrac{x}{9yz}$

10. $\dfrac{xy}{4z} \cdot \dfrac{10yz}{3x}$

11. $\dfrac{8x^2}{5y} \cdot \dfrac{15xy^2}{16}$

12. $\dfrac{6xy^2}{7} \cdot \dfrac{21x}{2y^2}$

13. $12xy \cdot \dfrac{x}{4yz}$

14. $2x^2 \cdot \dfrac{3}{4xy}$

Divide and reduce to lowest terms.

15. $\dfrac{18}{35} \div \dfrac{45}{24}$

16. $\dfrac{36}{25} \div \dfrac{40}{54}$

17. $\dfrac{11}{5} \div \dfrac{22}{25}$

18. $\dfrac{9}{4} \div \dfrac{36}{15}$

19. $\dfrac{9m}{8n} \div \dfrac{3m}{4n}$

20. $\dfrac{6x}{5y} \div \dfrac{3x}{10y}$

21. $\dfrac{a}{4c} \div \dfrac{a^2}{12c^2}$

22. $\dfrac{2x}{3y} \div \dfrac{4x}{6y^2}$

23. $\dfrac{x}{3y} \div 3y$

24. $2xy \div \dfrac{x}{y}$

25. $\dfrac{6xy}{z} \div \dfrac{x}{9yz}$

26. $\dfrac{xy}{4z} \div \dfrac{10yz}{3x}$

27. $\dfrac{8x^2}{5y} \div \dfrac{15xy^2}{16}$

28. $\dfrac{6xy^2}{7} \div \dfrac{21x}{2y^2}$

29. $12xy \div \dfrac{x}{4yz}$

30. $2x^2 \div \dfrac{3}{4xy}$

31. $\dfrac{x}{4yz} \div 12xy$

32. $\dfrac{3}{4xy} \div 2x^2$

Perform the indicated operations and reduce to lowest terms.

33. $\dfrac{8x^2}{3xy} \cdot \dfrac{12y^3}{6y}$

34. $\dfrac{6a^2}{7c} \cdot \dfrac{21cd}{12ac}$

35. $\dfrac{21x^2y^2}{12cd} \div \dfrac{14xy}{9d}$

36. $\dfrac{3uv^2}{5w} \div \dfrac{6u^2v}{15w}$

37. $\dfrac{9u^4}{4v^3} \div \dfrac{-12u^2}{15v}$

38. $\dfrac{-6x^3}{5y^2} \div \dfrac{18x}{10y}$

39. $\dfrac{3c^2d}{a^3b^3} \div \dfrac{3a^3b^3}{cd}$

40. $\dfrac{uvw}{5xyz} \div \dfrac{5vy}{uwxz}$

41. $\dfrac{3(-x)^2y}{2z} \cdot \dfrac{4z}{x}$

42. $\dfrac{2x^2(-y)}{3z} \cdot \dfrac{4z}{-x}$

43. $\dfrac{3(-x)y^2}{2z} \div 6xy$

44. $\dfrac{2(-x)^2y}{3z} \div 6xy$

45. $3xy \div \dfrac{2x(-y)}{3z}$

46. $-4x \div \dfrac{6xy}{5z}$

47. $\dfrac{-3abc^2}{2d} \div \dfrac{3a^2b}{4cd^2}$

48. $\dfrac{4(-a)b^2}{3cd^2} \div \dfrac{2ab}{3(-c)d}$

B 49. $\dfrac{3x^2y}{x-y} \cdot \dfrac{x-y}{6xy}$

50. $\dfrac{x+3}{2x^2} \cdot \dfrac{4x}{x+3}$

51. $\dfrac{x+3}{x^3+3x^2} \cdot \dfrac{x^3}{x-3}$

52. $\dfrac{a^2-a}{a-1} \cdot \dfrac{a+1}{a}$

53. $\dfrac{x-2}{4y} \div \dfrac{x^2+x-6}{12y^2}$

54. $\dfrac{4x}{x-4} \div \dfrac{8x^2}{x^2-6x+8}$

55. $\dfrac{6x^2}{4x^2y-12xy} \cdot \dfrac{x^2+x-12}{3x^2+12x}$

56. $\dfrac{2x^2+4x}{12x^2y} \cdot \dfrac{6x}{x^2+6x+8}$

57. $(t^2-t-12) \div \dfrac{t^2-9}{t^2-3t}$

58. $\dfrac{2y^2+7y+3}{4y^2-1} \div (y+3)$

59. $\dfrac{x^2-4}{x+3} \cdot \dfrac{x^2-9}{x+2}$

60. $\dfrac{x^2-1}{x-2} \cdot \dfrac{x^2-4}{x-1}$

61. $\dfrac{x^2+3x+2}{x+3} \cdot \dfrac{x^2+5x+6}{x+1}$

62. $\dfrac{x^2-3x+2}{x} \cdot \dfrac{x^2-x}{x-2}$

63. $\dfrac{x^2 + 3x + 2}{x + 3} \div \dfrac{x^2 + 5x + 6}{x + 1}$

64. $\dfrac{x^2 - 3x + 2}{x} \div \dfrac{x^2 - x}{x - 2}$

65. $\dfrac{x + 1}{x^2 + 8x + 15} \cdot \dfrac{x^2 + 6x + 5}{x + 3}$

66. $\dfrac{x^2 + 8x + 12}{x + 4} \cdot \dfrac{x + 6}{x^2 + 6x + 8}$

67. $\dfrac{x + 1}{x^2 + 8x + 15} \div \dfrac{x^2 + 6x + 5}{x + 3}$

68. $\dfrac{x^2 + 8x + 12}{x + 4} \div \dfrac{x + 6}{x^2 + 6x + 8}$

69. $\dfrac{m + n}{m^2 - n^2} \cdot \dfrac{m^2 - mn}{m^2 - 2mn + n^2}$

70. $\dfrac{x^2 - 6x + 5}{x^2 - x - 9} \cdot \dfrac{x^2 + 2x - 15}{x^2 + 2x}$

71. $\dfrac{m + n}{m^2 - n^2} \div \dfrac{m^2 - mn}{m^2 - 2mn + n^2}$

72. $\dfrac{x^2 - 6x + 9}{x^2 - x - 6} \div \dfrac{x^2 + 2x - 15}{x^2 + 2x}$

73. $-(x^2 - 3x) \cdot \dfrac{x - 2}{x - 3}$ **74.** $-(x^2 - 4) \cdot \dfrac{3}{x + 2}$

75. $\left(\dfrac{d^5}{3a} \div \dfrac{d^2}{6a^2}\right) \cdot \dfrac{a}{4d^3}$ **76.** $\dfrac{d^5}{3a} \div \left(\dfrac{d^2}{6a^2} \cdot \dfrac{a}{4d^3}\right)$

77. $\dfrac{2x^2}{3y^3} \cdot \dfrac{-6yz}{2x} \cdot \dfrac{y}{-xz}$ **78.** $\dfrac{-a}{-b} \cdot \dfrac{12b^2c}{15ac} \cdot \dfrac{-10}{4b}$

79. $\dfrac{3xy}{2z} \cdot \dfrac{-4xz}{5y} \cdot \dfrac{-5yz}{6x}$

80. $\dfrac{-abc}{4d} \cdot \dfrac{3ac}{2bd} \cdot \dfrac{12ad^2}{a^2}$

81. $\dfrac{x}{2y} \cdot \dfrac{3z}{4w} \cdot \dfrac{6y}{x} \cdot \dfrac{-w}{3z}$

82. $\dfrac{3a}{4c} \cdot \dfrac{-d}{6b} \cdot \dfrac{2b}{a} \cdot \dfrac{2c}{d}$

C 83. $\dfrac{9 - x^2}{x^2 + 5x + 6} \cdot \dfrac{x + 2}{x - 3}$

84. $\dfrac{2 - m}{2m + m^2} \cdot \dfrac{m^2 + 4m + 4}{m^2 - 4}$

85. $\dfrac{x^2 + 2x + 1}{x - 2} \cdot \dfrac{x^2 - 4}{x^2 - 2x - 3} \cdot \dfrac{x^2 - 5x + 6}{x + 1}$

86. $\dfrac{x^2 - 1}{x + 2} \cdot \dfrac{x^2 - 4}{x - 3} \cdot \dfrac{x^2 - 9}{x^2 - 3x + 2}$

87. $\dfrac{x^2 - 4x + 3}{x - 2} \cdot \dfrac{x^2 - 5x + 6}{x^2 - 3x + 2} \cdot \dfrac{x^2 - x - 2}{x^2 - 6x + 9}$

88. $\dfrac{x^2 - 2x + 1}{x^2 + 4x + 4} \cdot \dfrac{x^2 + 5x + 6}{x^2 + 2x - 3} \cdot \dfrac{x^2 + 6x + 8}{x^2 + 3x - 4}$

89. $\dfrac{x^2 - x - 2}{x^2 + 6x + 9} \div \left(\dfrac{x^2 - 4x + 3}{x^2 + x - 6} \cdot \dfrac{x^2 - 4x + 4}{x^2 - 9}\right)$

90. $\dfrac{x^2 + 2x - 3}{x^2 + 4x + 4} \div \left(\dfrac{x^2 - 5x + 4}{x^2 - x - 6} \cdot \dfrac{x^2 - 4x + 3}{x^2 + 6x + 8}\right)$

91. $\dfrac{x^2 - xy}{xy + y^2} \div \left(\dfrac{x^2 - y^2}{x^2 + 2xy + y^2} \div \dfrac{x^2 - 2xy + y^2}{x^2y + xy^2}\right)$

92. $\left(\dfrac{x^2 - xy}{xy + y^2} \div \dfrac{x^2 - y^2}{x^2 + 2xy + y^2}\right) \div \dfrac{x^2 - 2xy + y^2}{x^2y + xy^2}$

For each equation, what is the exception that will make it true in the strict sense? For example, to make

$$\frac{x^2 - x}{x} = x - 1$$

true, we should add ''for $x \neq 0$.''

93. $\dfrac{x^2 - 1}{x - 1} = x + 1$ **94.** $\dfrac{x^2 - 4}{x + 2} = x - 2$

95. $\dfrac{x^2 + x - 6}{x + 3} = x - 2$

96. $\dfrac{x^2 + 3x + 2}{x + 2} = x + 1$

97. $\dfrac{x^3 - 2x^2 + x}{x^2 - x} = x - 1$

98. $\dfrac{x^3 - x}{x^2 - 1} = x$

99. $\dfrac{x^4 + x^3}{x^4 - x^3} = \dfrac{x + 1}{x - 1}$

100. $\dfrac{x^3 - 2x^2}{x^2 - 3x + 2} = \dfrac{x}{x - 1}$

3-3 Addition and Subtraction

Like multiplication and division, addition and subtraction of rational expressions are based on the corresponding properties of fractions. Thus:

Addition and Subtraction

If P, D, and Q are polynomials ($D \neq 0$), then

$$\frac{P}{D} + \frac{Q}{D} = \frac{P + Q}{D} \tag{1}$$

$$\frac{P}{D} - \frac{Q}{D} = \frac{P - Q}{D} \tag{2}$$

In words: if the denominators of two rational expressions are the same, we may either add or subtract the expressions by adding or subtracting the numerators and placing the result over the common denominator. If the denominators are not the same, we use the fundamental principle of fractions to change the form of each fraction so they have a common denominator, and then use either (1) or (2).

Even though any common denominator will do, the problem generally will be less complicated if the least common denominator (LCD) is used. Recall that the least common denominator is the least common multiple (LCM) of all the denominators; that is, it is the "smallest" quantity exactly divisible by each denominator.

If the LCM is not obvious, then it is found as follows:

Finding the Least Common Multiple (LCM)

Step 1. Factor each expression completely using integer coefficients.
Step 2. The LCM must contain each *different* factor that occurs in any of the expressions to the highest power it occurs in any one expression. The LCM is the product of these powers.

Example 1
Finding the LCM

Find the least common multiple for

(A) $18x^3$, $15x$, $10x^2$ **(B)** $6(x - 3)$, $x^2 - 9$, $4x^2 + 24x + 36$

Solution **(A)** Write each expression in completely factored form, including coefficients:

$$18x^3 = 2 \cdot 3^2 x^3 \qquad 15x = 3 \cdot 5x \qquad 10x^2 = 2 \cdot 5x^2$$

The LCM must contain each of the different factors, 2, 3, 5, and x, to the highest power it occurs in any one expression. Thus,

$$LCM = 2 \cdot 3^2 \cdot 5x^3 = 90x^3$$

(B) Factor each expression completely:

$$6(x - 3) = 2 \cdot 3(x - 3) \qquad x^2 - 9 = (x - 3)(x + 3)$$

$$4x^2 + 24x + 36 = 4(x^2 + 6x + 9) = 2^2(x + 3)^2$$

The different factors of these expressions are listed below, along with the highest power to which each occurs:

2	2^2 in the third expression
3	3 in the first expression
$x - 3$	$x - 3$ in the first and second expressions
$x + 3$	$(x + 3)^2$ in the third expression

Thus,

$$\text{LCM} = 2^2 \cdot 3(x - 3)(x + 3)^2 = 12(x - 3)(x + 3)^2$$

Matched Problem 1   Find the LCM for

(A) $15y^2, 12y, 9y^4$ **(B)** $3x^2 - 12, x^2 - 4x + 4, 12(x + 2)$

Example 2
Adding and Subtracting Rational Expressions

Combine into a single fraction and simplify:

(A) $\dfrac{x + 1}{x - 2} + \dfrac{3x - 2}{x - 2}$ **(B)** $\dfrac{1}{x - 3} - \dfrac{x - 2}{x - 3}$

Solution **(A)** $\dfrac{x + 1}{x - 2} + \dfrac{3x - 2}{x - 2}$

When a numerator has more than one term, place this numerator in parentheses before proceeding. Since denominators are the same, use (1) to add.

$$= \frac{(x + 1) + (3x - 2)}{x - 2}$$

Simplify the numerator.

$$= \frac{x + 1 + 3x - 2}{x - 2}$$

$$= \frac{4x - 1}{x - 2}$$

(B) $\dfrac{1}{x - 3} - \dfrac{x - 2}{x - 3}$

Use (2) to subtract.

$$= \frac{1 - (x - 2)}{x - 3}$$

Simplify the numerator. Watch signs.

$$= \frac{1 - x + 2}{x - 3}$$

Sign errors are frequently made here.

$$= \frac{3 - x}{x - 3}$$

$3 - x \neq x - 3$; $3 - x = -(x - 3)$

$$= \frac{-(x - 3)}{(x - 3)}$$

Reduce to lowest terms.

$$= -1$$

Matched Problem 2   Combine into a single fraction and simplify:

$$\frac{x + 3}{2x - 5} - \frac{3x - 2}{2x - 5}$$

Example 3
Adding and Subtracting Rational Expressions

Combine into a single fraction and simplify:

$$\frac{3}{2y} - \frac{1}{3y^2} + 1$$

Solution

$$\frac{3}{2y} - \frac{1}{3y^2} + 1 \qquad \text{LCD} = 6y^2$$

$$= \frac{3y \cdot 3}{3y \cdot 2y} - \frac{2 \cdot 1}{2 \cdot 3y^2} + \frac{6y^2}{6y^2} \qquad \text{Use the fundamental principle of fractions to make each denominator } 6y^2.$$

$$= \frac{9y}{6y^2} - \frac{2}{6y^2} + \frac{6y^2}{6y^2}$$

$$= \frac{9y - 2 + 6y^2}{6y^2} \qquad \text{Arrange the numerator in descending powers of } y.$$

$$= \frac{6y^2 + 9y - 2}{6y^2}$$

Matched Problem 3   Combine into a single fraction and simplify:

$$\frac{5}{4x^3} - \frac{1}{3x} + 2$$

We can summarize the process of adding and subtracting rational expressions as follows:

Summary: Adding and Subtracting Rational Expressions

Step 1.   Find the LCD of the terms. That is, find the LCM of the de-nominators.

Step 2.   Use the fundamental principle of fractions to convert each term to a fraction with the LCD as denominator.

Step 3.   Add and subtract the numerators of the fractions formed in step 2, as indicated, to find the numerator of the resulting fraction. Use the LCD as the denominator.

Step 4.   Reduce the answer if possible.

Example 4
Adding and Subtracting Rational Expressions

Combine into a single fraction and simplify:

$$\frac{4}{3x^2 - 27} - \frac{x - 1}{4x^2 + 24x + 36}$$

Solution $\dfrac{4}{3x^2 - 27} - \dfrac{x - 1}{4x^2 + 24x + 36}$ Factor each denominator completely.

$$= \dfrac{4}{3(x - 3)(x + 3)} - \dfrac{(x - 1)}{2^2(x + 3)^2}$$

LCD $= 12(x - 3)(x + 3)^2$. Use the fundamental principle of fractions to make each denominator $12(x - 3)(x + 3)^2$.

$$= \dfrac{\mathbf{4(x + 3)} \cdot 4}{\mathbf{4(x + 3)} \cdot 3(x - 3)(x + 3)} - \dfrac{\mathbf{3(x - 3)} \cdot (x - 1)}{\mathbf{3(x - 3)} \cdot 2^2(x + 3)^2}$$

$$= \dfrac{16(x + 3)}{12(x - 3)(x + 3)^2} - \dfrac{3(x - 3)(x - 1)}{12(x - 3)(x + 3)^2}$$

$$= \dfrac{16(x + 3) - 3(x - 3)(x - 1)}{12(x - 3)(x + 3)^2}$$

Note that neither $x + 3$ nor $x - 3$ can be removed as a common factor here; the whole numerator must be written as a product before any factors can be divided out.

$$= \dfrac{16x + 48 - 3(x^2 - 4x + 3)}{12(x - 3)(x + 3)^2}$$

$$= \dfrac{16x + 48 - 3x^2 + 12x - 9}{12(x - 3)(x + 3)^2}$$

$$= \dfrac{-3x^2 + 28x + 39}{12(x - 3)(x + 3)^2}$$

Matched Problem 4   Combine into a single fraction and simplify:

$$\dfrac{3}{2x^2 - 8x + 8} - \dfrac{x + 1}{3x^2 - 12}$$

Answers to Matched Problems

1. (A) $180y^4$ (B) $12(x - 2)^2(x + 2)$

2. -1 **3.** $\dfrac{24x^3 - 4x^2 + 15}{12x^3}$ **4.** $\dfrac{-2x^2 + 11x + 22}{6(x - 2)^2(x + 2)}$

EXERCISE 3-3

A *Find the least common multiple (LCM) for each group of expressions.*

1. $3, x$

2. $4, y$

3. $x, 1$

4. $y, 1$

5. v^2, v, v^3

6. x, x, x^2

7. $3x, 6x^2, 4$

8. $8u^3, 6u, 4u^2$

9. $x + 1, x - 2$

10. $x - 2, x + 3$

11. $y + 3, 3y$

12. $x - 2, 2x$

Combine into a single fraction and simplify.

13. $\dfrac{7x}{5x^2} + \dfrac{2}{5x^2}$

14. $\dfrac{3m}{2m^2} + \dfrac{1}{2m^2}$

15. $\dfrac{4x}{2x - 1} - \dfrac{2}{2x - 1}$

16. $\dfrac{5a}{a - 1} - \dfrac{5}{a - 1}$

17. $\dfrac{y}{y^2 - 9} - \dfrac{3}{y^2 - 9}$

18. $\dfrac{2x}{4x^2 - 9} + \dfrac{3}{4x^2 - 9}$

19. $\dfrac{5}{3k} - \dfrac{6x - 4}{3k}$

20. $\dfrac{1}{2a^2} - \dfrac{2b - 1}{2a^2}$

21. $\dfrac{3x}{y} + \dfrac{1}{4}$

22. $\dfrac{2}{x} - \dfrac{1}{3}$

23. $\dfrac{2}{y} + 1$

24. $x + \dfrac{1}{x}$

25. $\dfrac{u}{v^2} - \dfrac{1}{v} + \dfrac{u^3}{v^3}$

26. $\dfrac{1}{x} - \dfrac{y}{x^2} + \dfrac{y^2}{x^3}$

27. $\dfrac{2}{3x} - \dfrac{1}{6x^2} + \dfrac{3}{4}$

28. $\dfrac{1}{8u^3} + \dfrac{5}{6u} - \dfrac{3}{4u^2}$

29. $\dfrac{2}{x+1} + \dfrac{3}{x-2}$

30. $\dfrac{1}{x-2} + \dfrac{1}{x+3}$

31. $\dfrac{3}{y+3} - \dfrac{2}{3y}$

32. $\dfrac{2}{x-2} - \dfrac{3}{2x}$

33. $\dfrac{x}{y}\left(\dfrac{1}{x} + \dfrac{1}{y}\right)$

34. $\left(\dfrac{x}{y} - \dfrac{y}{x}\right)\dfrac{x}{x+y}$

35. $\dfrac{x-y}{x} \div \left(\dfrac{y}{x} - \dfrac{x}{y}\right)$

36. $\dfrac{3}{x} \div \left(\dfrac{2}{x} - \dfrac{1}{2x}\right)$

37. $\dfrac{x}{y} \div \dfrac{y}{z} - \dfrac{y}{x} \div \dfrac{y}{z}$

38. $\dfrac{x}{y} - \dfrac{y}{z} \div \dfrac{y}{x} - \dfrac{y}{z}$

B *Find the LCM for each group of expressions.*

39. $12x^3,\ 8x^2y^2,\ 3xy^2$ **40.** $9u^3v^2,\ 6uv,\ 12v^3$

41. $15x^2y,\ 25xy,\ 5y^2$ **42.** $18m^4n^2,\ 12m^2n^4,\ 9mn$

43. $6(x-1),\ 9(x-1)^2$ **44.** $8(y-3)^2,\ 6(y-3)$

45. $6(x-7)(x+7),\ 8(x+7)^2$

46. $3(x-5)^2,\ 4(x+5)(x-5)$

47. $x^2 - 4,\ x^2 + 4x + 4$

48. $x^2 - 6x + 9,\ x^2 - 9$

49. $3x^2 + 3x,\ 4x^2,\ 3x^2 + 6x + 3$

50. $3m^2 - 3m,\ m^2 - 2m + 1,\ 5m^2$

Combine into a single fraction and simplify.

51. $\dfrac{2}{9u^3v^2} - \dfrac{1}{6uv} + \dfrac{1}{12v^3}$

52. $\dfrac{1}{12x^3} + \dfrac{3}{8x^2y^2} - \dfrac{2}{3xy^2}$

53. $\dfrac{4t-3}{18t^3} + \dfrac{3}{4t} - \dfrac{2t-1}{6t^2}$

54. $\dfrac{3y+8}{4y^2} - \dfrac{2y-1}{y^3} - \dfrac{5}{8y}$

55. $\dfrac{t+1}{t-1} - 1$ **56.** $2 + \dfrac{x+1}{x-3}$

57. $5 + \dfrac{a}{a+1} - \dfrac{a}{a-1}$

58. $\dfrac{1}{y+2} + 3 - \dfrac{2}{y-2}$

59. $\dfrac{2}{3(x-5)^2} - \dfrac{1}{4(x+5)(x-5)}$

60. $\dfrac{1}{6(x-7)(x+7)} + \dfrac{3}{8(x+7)^2}$

61. $\dfrac{5}{6(x-1)} + \dfrac{2}{9(x-1)^2}$

62. $\dfrac{3}{8(y-3)^2} - \dfrac{1}{6(y-3)}$

63. $\dfrac{3}{x+3} - \dfrac{3x+1}{(x-1)(x+3)}$

64. $\dfrac{4}{2x-3} - \dfrac{2x+1}{(2x-3)(x+2)}$

65. $\dfrac{3s}{3s^2-12} + \dfrac{1}{2s^2+4s}$

66. $\dfrac{2t}{3t^2-48} + \dfrac{t}{4t+t^2}$

67. $\dfrac{3}{x^2-4} - \dfrac{1}{x^2+4x+4}$

68. $\dfrac{2}{x^2-6x+9} - \dfrac{1}{x^2-9}$

69. $\dfrac{2}{x+3} - \dfrac{1}{x-3} + \dfrac{2x}{x^2-9}$

70. $\dfrac{2x}{x^2-y^2} + \dfrac{1}{x+y} - \dfrac{1}{x-y}$

71. $\dfrac{1}{x^2}\left(\dfrac{x}{x+1} + \dfrac{x}{x-1}\right)$

72. $\dfrac{1}{x^2} \div \left(\dfrac{x}{x+1} + \dfrac{x}{x-1}\right)$

73. $\dfrac{(x+1)^2}{x} \div \left(\dfrac{x+1}{x} - \dfrac{x}{x-1}\right)$

74. $\dfrac{(x+1)^2}{x}\left(\dfrac{x+1}{x} - \dfrac{x}{x-1}\right)$

75. $\dfrac{x}{(x+1)^2}\left(\dfrac{2}{x} + \dfrac{x}{x+\frac{1}{2}}\right)$

76. $\dfrac{(x-1)^2}{x} \div \left(\dfrac{-2}{x} + \dfrac{x}{x-\frac{1}{2}}\right)$

Combine into a single fraction and simplify. Recall that
$b - a = -(a - b)$; *thus,* $3 - y = -(y - 3)$, $1 - x = -(x - 1)$, *and so on.*

77. $\dfrac{5}{y - 3} - \dfrac{2}{3 - y}$ **78.** $\dfrac{3}{x - 1} + \dfrac{2}{1 - x}$

79. $\dfrac{3}{x - 3} + \dfrac{x}{3 - x}$ **80.** $\dfrac{-2}{2 - y} - \dfrac{y}{y - 2}$

81. $\dfrac{1}{5x - 5} - \dfrac{1}{3x - 3} + \dfrac{1}{1 - x}$

82. $\dfrac{x + 7}{ax - bx} + \dfrac{y + 9}{by - ay}$

C 83. $\dfrac{x}{x^2 - x - 2} - \dfrac{1}{x^2 + 5x - 14} - \dfrac{2}{x^2 + 8x + 7}$

84. $\dfrac{m^2}{m^2 + 2m + 1} + \dfrac{1}{3m + 3} - \dfrac{1}{6}$

85. $\dfrac{1}{3x^2 + 3x} + \dfrac{1}{4x^2} - \dfrac{1}{3x^2 + 6x + 3}$

86. $\dfrac{1}{3m(m - 1)} + \dfrac{1}{m^2 - 2m + 1} - \dfrac{1}{5m^2}$

87. $\dfrac{xy^2}{x^3 - y^3} - \dfrac{y}{x^2 + xy + y^2}$

88. $\dfrac{x}{x^2 - xy + y^2} - \dfrac{xy}{x^3 + y^3}$

89. $\dfrac{x + 1}{x + 3}\left(\dfrac{2}{x + 1} - \dfrac{1}{x + 2}\right)$

90. $\dfrac{x - 4}{x + 2}\left(\dfrac{2}{x - 4} - \dfrac{1}{x - 1}\right)$

91. $\dfrac{x^2 + 3x + 2}{x + 3}\left(\dfrac{x + 3}{x + 1} - \dfrac{x + 3}{x + 2}\right)$

92. $\dfrac{x^2 + 2x}{x + 4}\left(\dfrac{x + 4}{x} - \dfrac{x + 4}{x + 2}\right)$

93. $\dfrac{x}{x^2 - 1} \div \left(\dfrac{1}{x + 1} + \dfrac{1}{x - 1}\right)$

94. $\dfrac{x}{x^2 - 1} \div \left(\dfrac{1}{x - 1} - \dfrac{1}{x + 1}\right)$

95. $\dfrac{1}{x} + \dfrac{x}{x + 1} - \dfrac{x + 1}{x + 2}$

96. $\dfrac{3}{x} - \dfrac{x + 1}{x - 1} + \dfrac{x - 1}{x + 1}$

97. $\left(\dfrac{1}{a} - \dfrac{1}{b}\right)\left(\dfrac{1}{a} + \dfrac{1}{b}\right)$

98. $\left(\dfrac{a}{b} + \dfrac{b}{a}\right)\left(\dfrac{a}{b} - \dfrac{b}{a}\right)$

99. $\left(\dfrac{a}{b} + \dfrac{b}{a}\right) \div \left(\dfrac{a}{b} - \dfrac{b}{a}\right)$

100. $\left(\dfrac{1}{a} - \dfrac{1}{b}\right) \div \left(\dfrac{1}{a} + \dfrac{1}{b}\right)$

3-4 Quotients of Polynomials

- ♦ Algebraic Long Division
- ♦ Synthetic Division (Optional)

There are times when it is useful to find the quotients of polynomials by a long-division process similar to that used in arithmetic. For some division problems the process also can be shortened by a technique called *synthetic division.*

♦ ALGEBRAIC LONG DIVISION

If we divide 33 by 7, we obtain a quotient of 4 with remainder 5. This means that $33 = 4 \cdot 7 + 5$. Similarly, if we divide a polynomial A by a polynomial B and obtain a *quotient* of Q with *remainder R,* it means that $A = QB + R$. We also interpret the remainder just as we would with the arithmetic problem. Since $33 = 4 \cdot 7 + 5$, $\frac{33}{7} = 4 + \frac{5}{7}$. Similarly, if $A = QB + R$, then

$$\frac{A}{B} = Q + \frac{R}{B}$$

The actual division can be accomplished by **algebraic long division,** which is analogous to the same process in arithmetic. Some examples will illustrate the process.

Example 1
Long Division of Polynomials

Divide using the long-division process, and check:

(A) $(3x^2 + 11x - 20) \div (x + 5)$ (B) $(x^3 - 27) \div (x - 3)$
(C) $(x^3 + 2x^2 - 3x + 4) \div (x - 2)$

Solution   (A) $x + 5 \overline{)3x^2 + 11x - 20}$ Arrange both polynomials in descending powers of the variable if this is not already done.

$$\begin{array}{r} 3x \\ x + 5 \overline{)3x^2 + 11x - 20} \end{array}$$

Divide the first term of the divisor into the first term of the dividend. That is, what must x be multiplied by so that the product is exactly $3x^2$? Answer: $3x$.

$$\begin{array}{r} 3x \\ x + 5 \overline{)3x^2 + 11x - 20} \\ \underline{3x^2 + 15x} \\ - 4x - 20 \end{array}$$

Multiply the divisor by $3x$, line up like terms as indicated, subtract, and bring down -20 from above.

$$\begin{array}{r} 3x - 4 \\ x + 5 \overline{)3x^2 + 11x - 20} \\ \underline{3x^2 + 15x} \\ - 4x - 20 \\ \underline{- 4x - 20} \\ 0 \end{array}$$

Repeat the process until the degree of the remainder is less than that of the divisor, or the remainder is 0.

Remainder.

Check $(x + 5)(3x - 4) = 3x^2 + 11x - 20$

$$\begin{array}{r} x^2 + 3x + 9 \\ \textbf{(B)} \quad x - 3 \overline{)x^3 + 0x^2 + 0x - 27} \\ \underline{x^3 - 3x^2} \\ 3x^2 + 0x \\ \underline{3x^2 - 9x} \\ 9x - 27 \\ \underline{9x - 27} \\ 0 \end{array}$$

Note that the terms $0x^2$ and $0x$ need to be included. Now proceed as in part (A).

Remainder.

Check   Note that as the difference of two cubes,

$$x^3 - 27 = (x - 3)(x^2 + 3x + 9)$$

(C)

$$
\begin{array}{r}
x^2 + 4x + 5 \\
x - 2\overline{)x^3 + 2x^2 - 3x + 4} \\
\underline{x^3 - 2x^2} \\
4x^2 - 3x \\
\underline{4x^2 - 8x} \\
5x + 4 \\
\underline{5x - 10} \\
14
\end{array}
$$

Proceed as above until the degree of the remainder is less than the degree of the divisor.

Remainder.

Check   Note that if we have

$$(x^3 + 2x^2 - 3x + 4) \div (x - 2) = (x^2 + 4x + 5), \text{ remainder } 14$$

then

$$x^3 + 2x^2 - 3x + 4 = (x - 2)(x^2 + 4x + 5) + 14$$

or

$$\frac{x^3 + 2x^2 - 3x + 4}{x - 2} = x^2 + 4x + 5 + \frac{14}{x - 2}$$

We check the division by verifying either of these last equations. The first is an easier equation to verify: just multiply out and simplify the right side.

Matched Problem 1   Divide using the long-division process, and check:

(A) $(4x^3 - 12x^2 - x + 3) \div (x - 3)$ **(B)** $(x^3 + 125) \div (x + 5)$
(C) $(4x^3 + 3x^2 - 2x + 1) \div (x - 1)$

   Each of the quotients in Example 1 involved division by a linear polynomial. The same process works for divisors of degrees larger than 1. When the divisor is linear in the form $x - a$, however, there is a relationship between the remainder and the value of the polynomial at a. Consider the polynomials in Example 1, as summarized in Table 1.

Table 1

Polynomial	Divisor	a	Remainder	Value of Polynomial at $x = a$
$3x^2 + 11x - 20$	$x + 5$	-5	0	0
$x^3 - 27$	$x - 3$	3	0	0
$x^3 + 2x^2 - 3x + 4$	$x - 2$	2	14	14

If we write $P = (x - a)Q + R$, we can see that R is the value of P when $x = a$ because $(x - a)Q$ becomes $(0)Q = 0$.

Example 2
Comparing the Remainder and the Polynomial Value

Divide $x^5 - 3x^3 + x^2 - 8$ by $x - 2$, and compare the remainder with the value of $x^5 - 3x^3 + x^2 - 8$ at $x = 2$.

Solution

$$
\begin{array}{r}
x^4 + 2x^3 + x^2 + 3x + 6 \\
x - 2\overline{)x^5 - 0x^4 - 3x^3 + x^2 - 0x - 8} \\
\underline{x^5 - 2x^4} \\
2x^4 - 3x^3 \\
\underline{2x^4 - 4x^3} \\
x^3 + x^2 \\
\underline{x^3 - 2x^2} \\
3x^2 - 0x \\
\underline{3x^2 - 6x} \\
6x - 8 \\
\underline{6x - 12} \\
4 = R
\end{array}
$$

The value of the polynomial at $x = 2$ is

$$2^5 - 3 \cdot 2^3 + 2^2 - 8 = 32 - 24 + 4 - 8 = 4$$

which is equal to R, as claimed.

Matched Problem 2

Divide $x^5 - 3x^3 + x^2 - 8$ by $x + 1$, and compare the remainder with the value of $x^5 - 3x^3 + x^2 - 8$ at $x = -1$.

♦ **SYNTHETIC DIVISION (OPTIONAL)**

Any polynomial can be divided by a first-degree polynomial of the form $x - r$ using the algebraic long-division process described earlier in this section. In some circumstances such divisions have to be done repeatedly and a quicker, more concise method, called **synthetic division,** is useful. The method is most easily understood through an example. Let us start by dividing $P = 2x^4 + 3x^3 - x - 5$ by $x + 2$, using ordinary long division. The critical parts of the process are indicated in color:

$$
\begin{array}{r}
\textbf{2x}^3 - \textbf{1x}^2 + \textbf{2x} - \textbf{5} \qquad \text{Quotient} \\
x + 2\overline{)\textbf{2x}^4 + 3x^3 + 0x^2 - \textbf{1x} - 5} \qquad \text{Dividend} \\
\underline{2x^4 + \textbf{4x}^3} \\
- \textbf{1x}^3 + 0x^2 \\
\underline{- 1x^3 - \textbf{2x}^2} \\
\textbf{2x}^2 - 1x \\
\underline{2x^2 + \textbf{4x}} \\
- \textbf{5x} - 5 \\
\underline{- 5x - \textbf{10}} \\
\textbf{5} \qquad \text{Remainder}
\end{array}
$$

Divisor

Thus,

$$2x^4 + 3x^3 - x - 5 = (x + 2)(2x^3 - x^2 + 2x - 5) + 5$$

or

$$\frac{2x^4 + 3x^3 - x - 5}{x + 2} = 2x^3 - x^2 + 2x - 5 + \frac{5}{x + 2}$$

The essential parts of the division process, which were printed in color above, are arranged more conveniently here:

$$
\begin{array}{c}
\text{Dividend coefficients} \\
\begin{array}{r|rrrrr}
 & 2 & 3 & 0 & -1 & -5 \\
 &   & 4 & -2 & 4 & -10 \\
\hline
2 & 2 & -1 & 2 & -5 & 5 \\
\end{array}
\end{array}
$$

Quotient coefficients   Remainder

We see that the second and third rows of numerals are generated as follows. The first coefficient 2 of the dividend is brought down and multiplied by 2 from the divisor, and the product 4 is placed under the second dividend coefficient 3 and subtracted. The difference -1 is again multiplied by the 2 from the divisor, and the product is placed under the third coefficient from the dividend and subtracted. This process is repeated until the remainder is reached. The process can be made a little faster, and less prone to sign errors, by changing $+2$ from the divisor to -2 and adding instead of subtracting. Thus,

$$
\begin{array}{c}
\text{Dividend coefficients} \\
\begin{array}{r|rrrrr}
 & 2 & 3 & 0 & -1 & -5 \\
 &   & -4 & 2 & -4 & 10 \\
\hline
-2 & 2 & -1 & 2 & -5 & 5 \\
\end{array}
\end{array}
$$

Quotient coefficients   Remainder

We outline the steps as follows:

Key Steps in the Synthetic Division Process: $P \div (x - a)$

Step 1. Arrange the coefficients of P in order of descending powers of x. Write 0 as the coefficient for each missing power.

Step 2. After writing the divisor in the form $x - a$, use a to generate the second and third rows of numbers as follows. Bring down the first coefficient of the dividend and multiply it by a; then add the product to the second coefficient of the dividend. Multiply this sum by a, and add the product to the third coefficient of the dividend. Repeat the process until a product is added to the constant term of P.

Step 3. The last number in the third row of numbers is the remainder; the other numbers in the third row are the coefficients of the quotient, which is of degree 1 less than P.

This process is well-suited to hand calculator use. Store a; then proceed from left to right recalling a and using it as indicated.

Example 3
Synthetic Division

Use synthetic division to find the quotient and remainder resulting from dividing $P = 4x^5 - 30x^3 - 50x - 2$ by $x + 3$.

Solution $x + 3 = x - (-3)$; therefore, $a = -3$.

$$
\begin{array}{r|rrrrrr}
 & 4 & 0 & -30 & 0 & -50 & -2 \\
 &   & -12 & 36 & -18 & 54 & -12 \\
\hline
-3 & 4 & -12 & 6 & -18 & 4 & -14 \\
\end{array}
$$

The quotient is $4x^4 - 12x^3 + 6x^2 - 18x + 4$ with a remainder of -14.

Matched Problem 3
Use synthetic division to find the quotient and remainder resulting from dividing $P = 3x^4 - 11x^3 - 18x + 8$ by $x - 4$.

Answers to Matched Problems
1. **(A)** $4x^2 - 1$ **(B)** $x^2 - 5x + 25$ **(C)** $4x^2 + 7x + 5$, $R = 6$
2. $R = -5$
3. Quotient $3x^3 + x^2 + 4x - 2$; remainder 0

EXERCISE 3-4

A *Divide using the long-division process. Check the answers.*

1. $(3x^2 - 5x - 2) \div (x - 2)$

2. $(2x^2 + x - 6) \div (x + 2)$

3. $(2y^3 + 5y^2 - y - 6) \div (y + 2)$

4. $(x^3 - 5x^2 + x + 10) \div (x - 2)$

5. $(3x^2 - 11x - 1) \div (x - 4)$

6. $(2x^2 - 3x - 4) \div (x - 3)$

7. $(8x^2 - 14x + 13) \div (2x - 3)$

8. $(6x^2 + 5x - 6) \div (3x - 2)$

9. $(6x^2 + x - 13) \div (2x + 3)$

10. $(6x^2 + 11x - 12) \div (3x - 2)$

11. $(x^2 - 4) \div (x - 2)$

12. $(y^2 - 9) \div (y + 3)$

13. $(6x^2 + 13x + 6) \div (3x + 2)$

14. $(12x^2 + 23x + 10) \div (3x + 2)$

15. $(6x^3 + 17x^2 + 12x) \div (2x + 3)$

16. $(2x^3 - 3x^2 - 9x) \div (2x + 3)$

17. $(4x^4 + 3x^3 - x^2) \div (x + 1)$

18. $(4x^4 + x^3 - 3x^2) \div (x + 1)$

Divide. Write the quotient and indicate the remainder. Write the divisor in the form $x - a$. Check that the remainder is the same as the value of the polynomial at $x = a$.

19. $(x^3 + 2x^2 + 3x + 4) \div (x - 2)$

20. $(x^3 + 2x^2 + 3x + 4) \div (x - 1)$

21. $(x^3 + 2x^2 + 3x + 4) \div (x + 1)$

22. $(x^3 + 2x^2 + 3x + 4) \div (x + 2)$

23. $(2x^3 - x^2 + x - 2) \div (x - 3)$

24. $(2x^3 - x^2 + x - 2) \div (x - 1)$

25. $(2x^3 - x^2 + x - 2) \div (x + 1)$

26. $(2x^3 - x^2 + x - 2) \div (x + 3)$

27. $(x^4 + x^3 + 3x^2 + 3x + 5) \div (x - 4)$

28. $(x^4 + x^3 + 3x^2 + 3x + 5) \div (x - 2)$

29. $(x^4 + x^3 + 3x^2 + 3x + 5) \div (x + 2)$

30. $(x^4 + x^3 + 3x^2 + 3x + 5) \div (x + 4)$

B *Divide using the long-division process. Check the answers.*

31. $(12x^2 + 11x - 2) \div (3x + 2)$

32. $(8x^2 - 6x + 6) \div (2x - 1)$

33. $(8x^2 + 7) \div (2x - 3)$

34. $(9x^2 - 8) \div (3x - 2)$

35. $(-7x + 2x^2 - 1) \div (2x + 1)$

36. $(13x - 12 + 3x^2) \div (3x - 2)$

37. $(x^3 - 1) \div (x - 1)$

38. $(a^3 + 27) \div (a + 3)$

39. $(x^4 - 81) \div (x - 3)$

40. $(x^4 - 16) \div (x + 2)$

41. $(4a^2 - 22 - 7a) \div (a - 3)$

42. $(8c + 4 + 5c^2) \div (c + 2)$

43. $(x + 5x^2 - 10 + x^3) \div (x + 2)$

44. $(5y^2 - y + 2y^3 - 6) \div (y + 2)$

45. $(3 + x^3 - x) \div (x - 3)$

46. $(3y - y^2 + 2y^3 - 1) \div (y + 2)$

Divide. Write the quotient and indicate the remainder. Write the divisor in the form $x - a$. Check that the remainder is the same as the value of the polynomial at $x = a$.

47. $(x^3 - 2x + 4) \div (x - 2)$

48. $(x^3 - 2x + 4) \div (x - 1)$

49. $(x^3 - 2x + 4) \div (x + 1)$

50. $(x^3 - 2x + 4) \div (x + 2)$

51. $(x^4 - 3x + 5) \div (x - 4)$

52. $(x^4 - 3x + 5) \div (x - 2)$

53. $(x^4 - 3x + 5) \div (x + 2)$

54. $(x^4 - 3x + 5) \div (x + 4)$

Use division to evaluate the polynomial $P = x^5 - 3x^3 + x^2 - 1$ for the following values of x.

55. 3   **56.** 2   **57.** 1

58. −1   **59.** −2   **60.** −3

Use division to evaluate the polynomial $P = 5x^4 - 4x^3 + 2x^2 - 2x + 1$ for the following values of x.

61. −1   **62.** −2   **63.** −3

64. 3   **65.** 2   **66.** 1

C *Divide using the long-division process. Check the answers.*

67. $(9x^4 - 2 - 6x - x^2) \div (3x - 1)$

68. $(4x^4 - 10x - 9x^2 - 10) \div (2x + 3)$

69. $(8x^2 - 7 - 13x + 24x^4) \div (3x + 5 + 6x^2)$

70. $(16x - 5x^3 - 8 + 6x^4 - 8x^2) \div (2x - 4 + 3x^2)$

71. $(9x^3 - x + 2x^5 + 9x^3 - 2 - x) \div (2 + x^2 - 3x)$

72. $(12x^2 - 19x^3 - 4x - 3 + 12x^5) \div (4x^2 - 1)$

73. $(6x^4 + x^3 - 7x^2 - x + 1) \div (2x^2 + 3x + 1)$

74. $(6x^4 + x^3 - 9x^2 - 3x + 1) \div (3x^2 - 4x + 3)$

75. $(2x^4 + 3x^3 - 2x - 3) \div (2x^2 + 5x + 3)$

76. $(3x^4 - 8x^3 + 4x^2 + 3x - 2) \div (3x^2 - 5x + 2)$

77. $(3x^5 + x^4 - 48x - 16) \div (3x^2 - 5x - 2)$

78. $(3x^5 + x^4 - 48x - 16) \div (x^3 + 2x^2 + 4x + 8)$

Let P be the polynomial $x^3 - 6x^2 + 12x - 4$. For the polynomial D given, find polynomials Q and R such that $P = QD + R$ and either the degree of R is less than the degree of D, or $R = 0$.

79. $D = x^2 - 3x + 2$ **80.** $D = x^2 + 2$

81. $D = x^3 - 4$ **82.** $D = x - 4$

Let P be the polynomial $x^4 - 4x^2 + 7x + 2$. For the polynomial D given, find polynomials Q and R such that $P = QD + R$ and either the degree of R is less than the degree of D, or $R = 0$.

83. $D = x^2 - x + 1$ **84.** $D = x + 2$

85. $D = x^3 + 2$ **86.** $D = x - 2$

Synthetic division can be used with any values for a, not just integer values. Use synthetic division and a calculator to evaluate (to six decimal places) the polynomial $P = x^3 - 2x^2 - 3x + 4$ for the following values of x.

87. 1.1   **88.** 1.01   **89.** 1.35   **90.** 2.11

91. −3.3   **92.** 1.001   **93.** 3.102   **94.** −3.141

Use synthetic division and a calculator to evaluate (to six decimal places) the polynomial $P = 2x^3 - x^2 + x - 1$ for the following values of x.

95. −1.2   **96.** −3.45   **97.** −1.001   **98.** 1.23

99. 2.01   **100.** 0.95

3-5 Complex Fractions

A fractional form with fractions in its numerator or denominator, such as

$$\frac{x+1}{\frac{2}{3}} \quad \text{or} \quad \frac{1 + \dfrac{1}{x}}{\dfrac{x}{2}}$$

is called a **complex fraction.** It is often necessary to represent a complex fraction as a **simple fraction**—that is, as the quotient of two polynomials. The process does not involve any new concepts. It is a matter of applying what we already know in the right way. In particular, we will find the fundamental principle of fractions,

$$\frac{PK}{QK} = \frac{P}{Q} \qquad Q, K \neq 0$$

of considerable use. A numerical problem will show the basic processes.

Example 1
Simplifying Complex Fractions

Express as simple fractions:

(A) $\dfrac{\frac{3}{5}}{\frac{9}{10}}$ **(B)** $\dfrac{1\frac{1}{2}}{3\frac{2}{3}}$

Solution **(A)** The problem can be solved by treating the complex fraction as the quotient of two fractions and dividing:

$$\frac{\frac{3}{5}}{\frac{9}{10}} = \frac{3}{5} \div \frac{9}{10} = \frac{\overset{1}{\cancel{3}}}{\underset{1}{\cancel{5}}} \cdot \frac{\overset{2}{\cancel{10}}}{\underset{3}{\cancel{9}}} = \frac{2}{3}$$

We also can simplify the complex fraction by multiplying the numerator and denominator by a number that will clear both of fractions. The LCD of the numerator and denominator, in this case 10, will do this:

$$\frac{\frac{3}{5}}{\frac{9}{10}} = \frac{10 \cdot \frac{3}{5}}{10 \cdot \frac{9}{10}} = \frac{2 \cdot 3}{1 \cdot 9} = \frac{6}{9} = \frac{2}{3}$$

Although the first approach appears to be easier on the surface, the second can be very helpful when dealing with complex algebraic fractions.

(B) Recall that $1\frac{1}{2}$ and $3\frac{2}{3}$ represent sums and not products; that is, $1\frac{1}{2} = 1 + \frac{1}{2}$ and $3\frac{2}{3} = 3 + \frac{2}{3}$. Thus,

$$\frac{1\frac{1}{2}}{3\frac{2}{3}} = \frac{1 + \frac{1}{2}}{3 + \frac{2}{3}} \qquad \text{Write mixed fractions as sums.}$$

$$= \frac{6\left(1 + \frac{1}{2}\right)}{6\left(3 + \frac{2}{3}\right)} \qquad \text{Multiply the numerator and denominator by 6,}$$
the LCD of all fractions within the main fraction.

$$= \frac{6 \cdot 1 + 6 \cdot \frac{1}{2}}{6 \cdot 3 + 6 \cdot \frac{2}{3}} \qquad \text{The denominators 2 and 3 divide out.}$$

$$= \frac{6 + 3}{18 + 4} = \frac{9}{22} \qquad \text{A simple fraction.}$$

We also could have rewritten $1\frac{1}{2}$ as $\frac{3}{2}$ and $3\frac{2}{3}$ as $\frac{11}{3}$, and proceeded as in part (A). This would be easier in this case, but we will need to apply the idea illustrated above to algebraic fractions.

Matched Problem 1   Express as simple fractions: **(A)** $\dfrac{\frac{3}{5}}{\frac{1}{4}}$ **(B)** $\dfrac{2\frac{3}{4}}{4\frac{1}{3}}$

We now apply the same techniques to complex algebraic fractions.

Example 2
Simplifying Complex Algebraic Fractions

Express as simple fractions:

(A) $\dfrac{1 - \dfrac{1}{x^2}}{1 + \dfrac{1}{x}}$ **(B)** $\dfrac{\dfrac{a}{b} - \dfrac{b}{a}}{\dfrac{a}{b} + 2 + \dfrac{b}{a}}$

Solution

(A) $\dfrac{1 - \dfrac{1}{x^2}}{1 + \dfrac{1}{x}}$ Multiply the numerator and denominator by x^2, the LCD of all fractions within the complex fraction.

$$= \frac{x^2\left(1 - \dfrac{1}{x^2}\right)}{x^2\left(1 + \dfrac{1}{x}\right)}$$

$$= \frac{x^2 \cdot 1 - x^2 \cdot \dfrac{1}{x^2}}{x^2 \cdot 1 + x^2 \cdot \dfrac{1}{x}}$$

$$= \frac{x^2 - 1}{x^2 + x} \qquad \text{Factor the numerator and denominator to reduce to lowest terms.}$$

$$= \frac{(x - 1)\overset{1}{\cancel{(x + 1)}}}{x\underset{1}{\cancel{(x + 1)}}}$$

$$= \frac{x - 1}{x}$$

(B) $\dfrac{\dfrac{a}{b} - \dfrac{b}{a}}{\dfrac{a}{b} + 2 + \dfrac{b}{a}}$ LCD = ab

$$= \dfrac{ab\left(\dfrac{a}{b} - \dfrac{b}{a}\right)}{ab\left(\dfrac{a}{b} + 2 + \dfrac{b}{a}\right)}$$

$$= \dfrac{ab \cdot \dfrac{a}{b} - ab \cdot \dfrac{b}{a}}{ab \cdot \dfrac{a}{b} + ab \cdot 2 + ab \cdot \dfrac{b}{a}}$$

$$= \dfrac{a^2 - b^2}{a^2 + 2ab + b^2} \qquad \text{Reduce to lowest terms.}$$

$$= \dfrac{(a - b)(a + b)}{(a + b)^2}$$

$$= \dfrac{a - b}{a + b} \qquad \text{A simple fraction.}$$

Matched Problem 2 Express as simple fractions:

(A) $\dfrac{1 - \dfrac{1}{3x}}{1 - \dfrac{1}{9x^2}}$ **(B)** $\dfrac{\dfrac{x}{y} + 1 - \dfrac{2y}{x}}{\dfrac{x}{y} - \dfrac{y}{x}}$

In some cases, it may be easier to simplify the numerator and denominator separately by carrying out the indicated operations. For example, in Example 2(B), we also could have proceeded in this way:

$$\dfrac{\dfrac{a}{b} - \dfrac{b}{a}}{\dfrac{a}{b} + 2 + \dfrac{b}{a}} = \dfrac{\dfrac{a^2}{ab} - \dfrac{b^2}{ab}}{\dfrac{a^2}{ab} + \dfrac{2ab}{ab} + \dfrac{b^2}{ab}} = \dfrac{\dfrac{a^2 - b^2}{ab}}{\dfrac{a^2 + 2ab + b^2}{ab}}$$

$$= \dfrac{(a - b)(a + b)}{ab} \div \dfrac{(a + b)^2}{ab}$$

$$= \dfrac{(a - b)(a + b)}{ab} \cdot \dfrac{ab}{(a + b)^2}$$

$$= \dfrac{(a - b)\overset{1}{\cancel{(a + b)}}}{\cancel{ab}} \cdot \dfrac{\overset{1}{\cancel{ab}}}{\underset{a + b}{\cancel{(a + b)^2}}} = \dfrac{a - b}{a + b}$$

This approach will be used in the next example.

Example 3
Simplifying Complex Algebraic Fractions

Express as simple fractions:

(A) $\dfrac{\dfrac{1}{x+h} - \dfrac{1}{x}}{h}$ **(B)** $\dfrac{\dfrac{1}{a^2} - \dfrac{1}{b^2}}{a-b}$

Solution **(A)** $\dfrac{\dfrac{1}{x+h} - \dfrac{1}{x}}{h} = \dfrac{\dfrac{x}{(x+h)x} - \dfrac{x+h}{(x+h)x}}{h}$

Add the fractions in the numerator using $(x+h)x$ as common denominator.

$= \dfrac{\dfrac{x-(x+h)}{(x+h)x}}{h}$ Simplify.

$= \dfrac{\dfrac{-h}{(x+h)x}}{h}$ Divide.

$= \dfrac{\overset{-1}{\cancel{h}}}{(x+h)x} \cdot \dfrac{1}{\underset{1}{\cancel{h}}}$ Simplify.

$= \dfrac{-1}{(x+h)x}$

(B) $\dfrac{\dfrac{1}{a^2} - \dfrac{1}{b^2}}{a-b} = \dfrac{\dfrac{b^2}{a^2b^2} - \dfrac{a^2}{a^2b^2}}{a-b}$

Subtract the fractions in the numerator using a^2b^2 as common denominator.

$= \dfrac{\dfrac{b^2-a^2}{a^2b^2}}{a-b}$ Factor $b^2 - a^2$, divide, and simplify.

$= \dfrac{\overset{-1}{\cancel{(b-a)}}(b+a)}{a^2b^2} \cdot \dfrac{1}{\underset{1}{\cancel{a-b}}}$

$= -\dfrac{a+b}{a^2b^2}$

Problems in later courses will make extensive use of fractional forms such as these.

Matched Problem 3   Express as simple fractions:

(A) $\dfrac{\dfrac{1}{(x+h)^2} - \dfrac{1}{x^2}}{h}$ **(B)** $\dfrac{\dfrac{1}{a} - \dfrac{1}{b}}{a-b}$

Answers to Matched Problems

1. (A) $\frac{12}{5}$ (B) $\frac{33}{52}$

2. (A) $\dfrac{3x}{3x+1}$ (B) $\dfrac{x+2y}{x+y}$

3. (A) $-\dfrac{2x+h}{(x+h)^2 x^2}$ (B) $\dfrac{-1}{ab}$

EXERCISE 3-5

In Problems 1–78, rewrite the expression as a simple fraction reduced to lowest terms.

A

1. $\dfrac{\frac{1}{2}}{\frac{2}{3}}$ 2. $\dfrac{\frac{1}{4}}{\frac{2}{3}}$ 3. $\dfrac{\frac{3}{8}}{\frac{5}{12}}$ 4. $\dfrac{\frac{4}{15}}{\frac{5}{6}}$

5. $\dfrac{1\frac{1}{3}}{2\frac{1}{6}}$ 6. $\dfrac{3\frac{1}{10}}{2\frac{1}{5}}$ 7. $\dfrac{1\frac{2}{9}}{2\frac{5}{6}}$ 8. $\dfrac{2\frac{4}{15}}{1\frac{7}{10}}$

9. $\dfrac{\frac{3}{4}}{\frac{x}{y}}$ 10. $\dfrac{\frac{-1}{5}}{\frac{a}{b}}$ 11. $\dfrac{\frac{-x}{10}}{\frac{y}{2}}$ 12. $\dfrac{\frac{4}{x}}{\frac{y}{2}}$

13. $\dfrac{\frac{-ab}{c}}{\frac{-bc}{a}}$ 14. $\dfrac{\frac{a^2}{b}}{\frac{-a}{b^2}}$ 15. $\dfrac{\frac{2x^2}{y^2}}{\frac{4x}{y}}$ 16. $\dfrac{\frac{-x}{3y^2}}{\frac{6x^2}{5y}}$

17. $\dfrac{\frac{x}{y}}{\frac{1}{y^2}}$ 18. $\dfrac{\frac{1}{b^2}}{\frac{a}{b}}$ 19. $\dfrac{\frac{y}{2x}}{\frac{1}{3x^2}}$ 20. $\dfrac{\frac{2x}{5y}}{\frac{1}{3x}}$

21. $\dfrac{\frac{1}{1+x} - 1}{x}$ 22. $\dfrac{1 - \frac{1}{x}}{x-1}$ 23. $\dfrac{\frac{1}{3} - \frac{1}{x}}{x-3}$

24. $\dfrac{\frac{1}{2+x} - \frac{1}{2}}{x}$ 25. $\dfrac{\frac{a}{b} - 1}{1 - \frac{b}{a}}$ 26. $\dfrac{1 + \frac{x}{y}}{1 + \frac{y}{x}}$

27. $\dfrac{3 - \frac{a}{b}}{b - \frac{a}{3}}$ 28. $\dfrac{a - \frac{b}{c}}{c - \frac{b}{a}}$ 29. $\dfrac{\frac{x}{y} - z}{\frac{x}{z} - y}$

30. $\dfrac{2 - \frac{x}{y}}{1 - \frac{2y}{x}}$

B

31. $\dfrac{\frac{2}{x} + \frac{3}{y}}{\frac{9}{y} + \frac{6}{x}}$ 32. $\dfrac{\frac{x}{2} - \frac{y}{3}}{\frac{y}{6} - \frac{x}{4}}$ 33. $\dfrac{\frac{a}{3} - \frac{2}{b}}{\frac{b}{2} - \frac{3}{a}}$

34. $\dfrac{\frac{1}{a} + \frac{2}{b}}{\frac{6}{a} + \frac{3}{b}}$ 35. $\dfrac{1 + \frac{3}{x}}{x - \frac{9}{x}}$ 36. $\dfrac{1 - \frac{2}{x}}{x - \frac{4}{x}}$

37. $\dfrac{1 - \frac{y^2}{x^2}}{1 - \frac{y}{x}}$ 38. $\dfrac{\frac{a^2}{b^2} - 1}{\frac{a}{b} - 1}$ 39. $\dfrac{\frac{1}{x} + \frac{1}{y}}{\frac{y}{x} - \frac{x}{y}}$

40. $\dfrac{b - \frac{a^2}{b}}{\frac{1}{a} - \frac{1}{b}}$ 41. $\dfrac{\frac{1}{x^2} + x}{\frac{1}{x} + 1}$ 42. $\dfrac{\frac{8}{x} - x^2}{\frac{2}{x} - 1}$

43. $\dfrac{\frac{9}{x^2} - \frac{x}{3}}{\frac{3}{x} - 1}$ 44. $\dfrac{\frac{x^2}{4} + \frac{2}{x}}{\frac{x}{2} + 1}$

45. $\dfrac{\frac{1}{(1+x)^2} - 1}{x}$ 46. $\dfrac{\frac{3}{x^2} - \frac{3}{y^2}}{x - y}$

47. $\dfrac{\frac{-4}{a^2} + \frac{4}{b^2}}{a - b}$ 48. $\dfrac{\frac{1}{(2+h)^2} - \frac{1}{4}}{h}$

49. $\dfrac{\frac{4a}{b} - \frac{b}{a}}{\frac{4a}{b} - 4 + \frac{b}{a}}$ 50. $\dfrac{\frac{x}{y} - \frac{9y}{x}}{\frac{x}{y} + 6 + \frac{9y}{x}}$

51. $\dfrac{\frac{x}{y} - 2 + \frac{y}{x}}{\frac{x}{y} - \frac{y}{x}}$ 52. $\dfrac{\frac{1}{x^2} + \frac{2}{x} + 1}{1 - \frac{1}{x^2}}$

53. $\dfrac{1 + \dfrac{4}{x} + \dfrac{4}{x^2}}{\dfrac{1}{x} + \dfrac{2}{x^2}}$

54. $\dfrac{a + \dfrac{6}{a} + 9}{\dfrac{3}{a} + 1}$

74. $\dfrac{\left(x - \dfrac{1}{x}\right)\left(1 - \dfrac{1}{x}\right)}{\left(\dfrac{1}{x} - \dfrac{1}{x^2}\right)\left(\dfrac{1}{x^3} - \dfrac{1}{x^2}\right)}$

55. $\dfrac{x + 3 - \dfrac{4}{x}}{1 + \dfrac{1}{x} - \dfrac{2}{x^2}}$

56. $\dfrac{1 + \dfrac{2}{x} - \dfrac{15}{x^2}}{1 + \dfrac{4}{x} - \dfrac{5}{x^2}}$

75. $1 - \dfrac{x - \dfrac{1}{x}}{1 - \dfrac{1}{x}}$

76. $\dfrac{t - \dfrac{1}{1 + \dfrac{1}{t}}}{t + \dfrac{1}{t - \dfrac{1}{t}}}$

57. $\dfrac{\dfrac{a^2}{a - b} - a}{\dfrac{b^2}{a - b} + b}$

58. $\dfrac{n - \dfrac{n^2}{n - m}}{1 + \dfrac{m^2}{n^2 - m^2}}$

77. $1 + \dfrac{1}{1 + \dfrac{1}{1 + \dfrac{1}{1 + x}}}$

78. $1 - \dfrac{1}{1 - \dfrac{1}{1 - \dfrac{1}{1 - x}}}$

59. $\dfrac{\dfrac{1}{x} + \dfrac{5}{x^2} + \dfrac{6}{x^3}}{\dfrac{8}{x^2} + \dfrac{6}{x} + 1}$

60. $\dfrac{x + 4 - \dfrac{5}{x}}{1 + \dfrac{6}{x} + \dfrac{5}{x^2}}$

APPLICATIONS

79. A formula for the average rate r for a round-trip between two points, where the rate going is r_G and the rate returning is r_R, is given by the complex fraction

$$r = \dfrac{2}{\dfrac{1}{r_G} + \dfrac{1}{r_R}}$$

Express r as a simple fraction.

61. $\dfrac{x - 3y + \dfrac{2y^2}{x}}{1 + \dfrac{y}{x} - \dfrac{2y^2}{x^2}}$

62. $\dfrac{\dfrac{x^2}{y^2} + \dfrac{2x}{y} + 1}{\dfrac{x^2}{y} + 3x + 2y}$

63. $\dfrac{\dfrac{1}{2} + \dfrac{3}{x} + \dfrac{4}{x^2}}{x + 1 - \dfrac{2}{x}}$

64. $\dfrac{\dfrac{1}{4} + \dfrac{1}{2x} - \dfrac{3}{4x^2}}{1 + \dfrac{1}{x} - \dfrac{2}{x^2}}$

80. The airspeed indicator on a jet aircraft registers 500 miles per hour. If the plane is traveling with an airstream moving at 100 miles per hour, then the plane's ground speed would be 600 miles per hour—or would it? According to Einstein, velocities must be added according to the following formula:

$$v = \dfrac{v_1 + v_2}{1 + \dfrac{v_1 v_2}{c^2}}$$

where v is the resultant velocity, c is the speed of light, and v_1 and v_2 are the two velocities to be added. Convert the right side of the equation into a simple fraction.

65. $\dfrac{\dfrac{m}{m + 2} - \dfrac{m}{m - 2}}{\dfrac{m + 2}{m - 2} - \dfrac{m - 2}{m + 2}}$

66. $\dfrac{\dfrac{y}{x + y} - \dfrac{x}{x - y}}{\dfrac{x}{x + y} + \dfrac{y}{x - y}}$

67. $\dfrac{\dfrac{1}{x - 1} + \dfrac{1}{x + 1}}{\dfrac{1}{x^2 - 1} + \dfrac{1}{x^2 + 1}}$

68. $\dfrac{\dfrac{1}{1 + x} - \dfrac{1}{1 - x}}{\dfrac{1}{1 + x^2} - \dfrac{1}{1 - x^2}}$

C 69. $1 - \dfrac{1}{1 - \dfrac{1}{x}}$

70. $2 - \dfrac{1}{1 - \dfrac{2}{x + 2}}$

71. $1 - \dfrac{2}{1 - \dfrac{2}{1 + x}}$

72. $2 - \dfrac{1}{2 - \dfrac{1}{2 - x}}$

73. $\dfrac{\left(1 + \dfrac{1}{x}\right)\left(1 - \dfrac{1}{x^2}\right)}{\left(x + 1\right)\left(1 + \dfrac{1}{x}\right)}$

CHAPTER SUMMARY

3-1   RATIONAL EXPRESSIONS

A **rational expression** is an algebraic expression involving only the operations of addition, subtraction, multiplication, and division on variables and constants. A rational expression A/B can be **reduced to lowest terms** or **raised to higher terms** by using the **fundamental principle of fractions:**

$$\frac{A}{B} = \frac{AK}{BK}$$

3-2   MULTIPLICATION AND DIVISION

Rational expressions are multiplied and divided as follows:

$$\frac{A}{B} \cdot \frac{C}{D} = \frac{A \cdot C}{B \cdot D} \qquad \frac{A}{B} \div \frac{C}{D} = \frac{A}{B} \cdot \frac{D}{C} = \frac{A \cdot D}{B \cdot C}$$

The relation $\dfrac{a-b}{b-a} = -1$ may help reduce the resulting expression.

3-3   ADDITION AND SUBTRACTION

Rational expressions with a common denominator are added and subtracted as follows:

$$\frac{A}{D} + \frac{B}{D} = \frac{A+B}{D} \qquad \frac{A}{D} - \frac{B}{D} = \frac{A-B}{D}$$

Rational expressions with different denominators are converted to expressions with a common denominator and then added or subtracted. Using the **least common denominator (LCD)** as the common denominator simplifies the addition or subtraction. The LCD is the **least common multiple (LCM)** of the denominators. It is the product of the highest powers of each factor that occurs in the denominators.

3-4   QUOTIENTS OF POLYNOMIALS

A polynomial P can be divided by a polynomial **divisor** D by using **algebraic long division** to yield a **quotient** Q and **remainder** R, so that $P = D \cdot Q + R$ and either R is 0 or the degree of R is smaller than that of D. When the divisor is of the form $x - a$, the remainder is the value of the polynomial P at $x = a$. When the divisor is $x - a$, the division process can be done quickly using **synthetic division.**

3-5   COMPLEX FRACTIONS

A **complex fraction** is a fraction in which the numerator and/or denominator contain a fraction. The fundamental principle of fractions and the operations of addition, subtraction, multiplication, and division are used to convert complex fractions to **simple fractions**—that is, to quotients of two polynomials.

CHAPTER REVIEW EXERCISE

Work through all the problems in this chapter review and check answers in the back of the book. Answers to all the problems are there, and following each answer is a number in italics indicating the section in which that type of problem is discussed. Where weaknesses show up, review appropriate sections in the text.

A *Reduce to lowest terms.*

1. $\dfrac{42}{105}$

2. $\dfrac{a^4b^3c^2}{a^2b^2c^2}$

3. $\dfrac{x^2 - 4x}{x^2 + x - 20}$

4. $\dfrac{x^3 - 4x}{x^3 + 4x^2 + 4x}$

Replace the question mark with the appropriate algebraic expression.

5. $\dfrac{6xy^2}{14x^2y} = \dfrac{?}{7xy}$

6. $\dfrac{3xy^2}{8xyz} = \dfrac{?}{16x^2yz^2}$

Find the least common multiple (LCM).

7. $4x,\ 6x^2,\ 9x^3$

8. $x^3y,\ 2x^2y^2,\ 3xy^3$

9. $x^2,\ x + 1,\ x^2 - 1$

Perform the indicated operations and simplify. Express each answer as a simple fraction in lowest terms.

10. $\dfrac{18x^3y^2(z + 3)}{12xy^2(z + 3)^2}$

11. $\dfrac{x^2 + 2x + 1}{x^2 - 1}$

12. $1 + \dfrac{2}{3x}$

13. $\dfrac{2}{x} - \dfrac{1}{6x} + \dfrac{1}{3}$

14. $\dfrac{1}{6x^3} - \dfrac{3}{4x} - \dfrac{2}{3}$

15. $\dfrac{4x^2y^3}{3a^2b^2} \div \dfrac{2xy^2}{3ab}$

16. $\dfrac{6x^2}{3(x - 1)} - \dfrac{6}{3(x - 1)}$

17. $1 - \dfrac{m - 1}{m + 1}$

18. $\dfrac{3}{x - 2} - \dfrac{2}{x + 1}$

19. $(d - 2)^2 \div \dfrac{d^2 - 4}{d - 2}$

20. $\dfrac{x + 1}{x + 2} - \dfrac{x + 2}{x + 3}$

21. $\dfrac{1}{x} - \dfrac{1}{x + 1}$

22. $\dfrac{2}{x + 4} - \dfrac{1}{2}$

23. $\dfrac{x + 2}{3} \cdot \dfrac{6x}{x^2 + 2x}$

24. $x + \dfrac{x}{x - 1}$

25. $\dfrac{2x^3}{9}\left(\dfrac{3x}{4y^2} \cdot \dfrac{6y}{x^2}\right)$

26. $\dfrac{2x^3}{9}\left(\dfrac{3x}{4y^2} + \dfrac{6y}{x^2}\right)$

27. $\dfrac{\frac{1}{4}}{\frac{2}{3}}$

28. $\dfrac{2\frac{3}{4}}{1\frac{1}{2}}$

29. $\dfrac{1 - \dfrac{2}{y}}{1 + \dfrac{1}{y}}$

30. $\dfrac{\dfrac{5x}{6}}{\dfrac{25x^2}{24}}$

31. $\dfrac{\dfrac{2}{3 + x} - \dfrac{2}{3}}{x}$

32. $\dfrac{x - \dfrac{y}{3}}{3 - \dfrac{y}{x}}$

33. $\dfrac{\dfrac{x + 1}{y}}{\dfrac{x^2 - 1}{y^2}}$

B *Reduce to lowest terms.*

34. $\dfrac{x^2 + 3x - 4}{x^2 - 5x + 4}$

35. $\dfrac{x^2 - 1}{x^3 - 1}$

36. $\dfrac{x^3 - 3x^2 + 2x}{x^3 + 2x^2 - 3x}$

37. $\dfrac{x^2 + 1}{x^3 + 1}$

Replace the question mark with the appropriate algebraic expression.

38. $\dfrac{x + 3}{x - 3} = \dfrac{?}{x^2 - 9}$

39. $\dfrac{x^2 + 2x + 1}{x^4 + x} = \dfrac{?}{x^3 - x^2 + x}$

40. $\dfrac{3x^2 + 6x}{4x^3 + 4x^2} = \dfrac{?}{8x^4 + 24x^3 + 16x^2}$

Find the least common multiple (LCM).

41. $(x + 1)^2,\ (x + 1)(x - 2),\ (x - 2)^2$

42. $3(x - 2),\ 4(x + 3),\ 5(x^2 + x - 6)$

43. $x^2 - 9,\ x^2 + 6x + 9,\ x^2 - 6x + 9$

Divide to find the quotient and remainder.

44. $(x^3 + 8) \div (x + 2)$

45. $(x^2 + 9) \div (x - 3)$

46. $(x^3 - 3x^2 + x - 3) \div (x - 1)$

47. $(x^3 + x) \div (x^2 + 1)$

48. $(x^4 + 2x^3 + 3x^2 + 4x + 5) \div (x^2 + 2)$

49. $(x^4 + x^2 - 1) \div (x + 2)$

50. $(x^4 - 1) \div (x - 1)$

51. $(x^4 + x + x^3) \div (1 + x + x^2)$

Perform the indicated operations and simplify. Express each answer as a simple fraction in lowest terms.

52. $\dfrac{2}{5b} - \dfrac{4}{3b^3} - \dfrac{1}{6a^2b^2}$

53. $\dfrac{2}{2x - 3} - 1$

54. $\dfrac{4x^2y}{3ab^2} \div \left(\dfrac{2a^2x^2}{b^2y} \cdot \dfrac{6a}{2y^2} \right)$

55. $\dfrac{x}{x^2 + 4x} + \dfrac{2x}{3x^2 - 48}$

56. $\dfrac{x^3 - x}{x^2 - x} \div \dfrac{x^2 + 2x + 1}{x}$

57. $\dfrac{\dfrac{x}{y} - \dfrac{y}{x}}{\dfrac{x}{y} + 1}$

58. $\dfrac{x}{x^3 - y^3} - \dfrac{1}{x^2 + xy + y^2}$

59. $\dfrac{\dfrac{y^2}{x^2 - y^2} + 1}{\dfrac{x^2}{x - y} - x}$

60. $\dfrac{x^3 - 1}{x^2 + x + 1} \div \dfrac{x^2 - 1}{x^2 + 2x + 1}$

61. $\dfrac{1}{3x^2 - 27} - \dfrac{x - 1}{4x^3 + 24x^2 + 36x}$

62. $1 + \dfrac{1}{x} + \dfrac{1}{1 + x}$

63. $\left(\dfrac{1 + x}{x} \div \dfrac{x}{1 - x} \right) \div \dfrac{1 - x^2}{x^2}$

64. $\dfrac{x + 2}{x^2 - 9} \cdot \dfrac{x^2 - 2x - 3}{x^2 + 3x + 2}$

65. $\dfrac{x}{x^2 + 3x + 2} \cdot \dfrac{x^2 + 5x + 6}{x^2 + 7x + 12} \cdot \dfrac{x^2 + 5x + 4}{x^2}$

66. $x - \dfrac{x + 1}{\dfrac{1}{x} + 1}$

67. $x + \dfrac{1}{x} - \dfrac{1}{\dfrac{1}{x}} - \dfrac{1}{\dfrac{1}{x}}$

C *In Problems 68–74, perform the indicated operations and simplify. Express each answer as a simple fraction in lowest terms.*

68. $\dfrac{4}{s^2 - 4} + \dfrac{1}{2 - s}$

69. $\dfrac{y^2 - y - 6}{(y + 2)^2} \cdot \dfrac{2 + y}{3 - y}$

70. $\dfrac{y}{x^2} \div \left(\dfrac{x^2 + 3x}{2x^2 + 5x - 3} \div \dfrac{x^3y - x^2y}{2x^2 - 3x + 1} \right)$

71. $\dfrac{1 - \dfrac{1}{1 + \dfrac{x}{y}}}{1 - \dfrac{1}{1 - \dfrac{x}{y}}}$

72. $\left(x - \dfrac{1}{1 - \dfrac{1}{x}} \right) \div \left(\dfrac{x}{x + 1} - \dfrac{x}{1 - x} \right)$

73. $\dfrac{1}{x^2 + x} - \dfrac{1}{x^2 + 3x + 2} + \dfrac{1}{x^2 + 2x}$

74. $\dfrac{x}{x^2 - 1} - \dfrac{x}{x^2 - 2x + 1} - \dfrac{2}{x^2 + 2x + 1}$

75. Evaluate $x^4 - x^3 + x^2 - x + 1$ for $x = -2$ by dividing the polynomial by $x + 2$.

CHAPTER PRACTICE TEST

The following practice test is provided for you to test your knowledge of the material in this chapter. You should try to complete it in 50 minutes or less. Answers in the back of the book indicate the section in the text where the material in the question is covered. Actual tests in your class may vary from this practice test in difficulty, length, or emphasis, depending on the goals of your course or instructor.

Reduce to lowest terms.

1. $\dfrac{6xy^2z^3}{15x^2yz^2}$

2. $\dfrac{x(x^2 + x - 2)}{x^3 + 5x^2 + 6x}$

Find the least common multiple (LCM).

3. $6x^2yz, \ 3xy^2, \ 4y^2z^4$

4. $x + 2, \ x^2 + 3x + 2, \ x^2 + 2x + 1$

Replace the question mark with the appropriate algebraic expression.

5. $\dfrac{3x^2y}{8z^3} = \dfrac{15x^2y^2z^3}{?}$

6. $\dfrac{x + 1}{x + 2} = \dfrac{?}{x^2 + 5x + 6}$

Perform the indicated operations and simplify. Express each answer as a simple fraction reduced to lowest terms.

7. $3 - \dfrac{x}{x-1}$

8. $\dfrac{2}{x} + \dfrac{3}{x^2} + \dfrac{4}{x^3}$

9. $\dfrac{xy^2}{3z} \cdot \dfrac{12xz}{7y^3}$

10. $\dfrac{4x^2y}{5z^2} \div \dfrac{8xy^2}{15z}$

11. $\dfrac{x(x-1)}{x^2+2x} \cdot \dfrac{x+2}{x^2-2x+1}$

12. $\dfrac{x}{x+3} - \dfrac{x}{x-2}$

13. $\dfrac{1}{x} - \dfrac{1}{x-1} + \dfrac{1}{x^2}$

14. $\dfrac{\dfrac{1}{x-2}}{\dfrac{x}{x^2-4x+4}}$

15. $\dfrac{x}{x+1}\left(\dfrac{x}{2} - \dfrac{\frac{1}{2}}{x}\right)$

16. $\dfrac{x - \dfrac{x}{\dfrac{1}{x}}}{x}$

17. $\dfrac{\dfrac{1}{x+1} - \dfrac{1}{x-1}}{\dfrac{2}{x^2-1}}$

In Problems 18 and 19, divide to find the quotient and remainder.

18. $(x^3 - x^2 + x - 1) \div (x - 1)$

19. $\dfrac{x^3 + 2x^2 + 3x + 4}{x^2 + x + 1}$

20. Find the value of the polynomial $x^3 + 4x^2 + 5x + 6$ for $x = 7$ by dividing the polynomial by $x - 7$.

4

Solving Equations and Applications

In Chapters 1–3, you have been introduced, or reintroduced, to several major themes that occur in the study of algebra. These themes are reflected in the kinds of instructions you have carried out in the exercise sets. They might be grouped as follows:

Translate

Rewrite

Solve

Evaluate

Use

For example, you have *translated* verbal statements and relationships into algebraic expressions and equations. Both expressions and equations have been *rewritten,* or simplified. The process of rewriting was done for a purpose: to get the expression into a more useful form or to change the equation into one that is easier to solve. You have then *solved* equations in a variety of contexts. We discussed how to solve linear equations in Chapter 1, equations with a factorable polynomial set equal to 0 in Chapter 2, and equations involving fractions in Chapter 3. Since algebraic expressions represent numbers, you have *evaluated* expressions for particular values of the variables involved. This is done every time the solution to an equation is checked. Finally, you *used* these techniques to solve applied problems.

In this chapter, we bring together all the techniques we have developed thus far for solving equations and then consider more applications that require equation solving.

Photo reference: see Exercise 4-3, Problem 3.

145

4-1   Solving Equations Involving Fractions

♦ No Variables in Denominators
♦ Variables in Some Denominators

Recall the strategy for solving equations given in Section 1-7:

> ## Equation-Solving Strategy
>
> 1. Use multiplication to remove fractions if present.
> 2. Simplify the left and right sides of the equation by removing grouping symbols and combining like terms.
> 3. Use equality properties to get all variable terms on one side, usually the left, and all constant terms on the other side. Combine like terms in the process.
> 4. Isolate the variable with a coefficient of 1, using the division or multiplication property of equality.

In this section we apply this strategy to equations involving fractions, and in particular to equations with variables in the denominator of a fraction.

♦ NO VARIABLES IN DENOMINATORS

In Chapter 1, we applied the equation-solving strategy to some equations involving fractions, but only when no variables occurred in denominators. The process is reviewed in the following examples. We start by using the multiplication property of equality to clear fractions. What do we multiply both sides by to clear the fractions? We use any common multiple, preferably the LCM, of all denominators present in the equation.

Example 1
Clearing Fractions and Solving

Solve:

(A) $\dfrac{x}{3} - \dfrac{1}{2} = \dfrac{5}{6}$ **(B)** $0.2x + 0.3(x - 5) = 13$ **(C)** $5 - \dfrac{2x - 1}{4} = \dfrac{x + 2}{3}$

Solution   **(A)**

$$\frac{x}{3} - \frac{1}{2} = \frac{5}{6}$$

Clear fractions by multiplying both sides by 6, the LCM of all the denominators.

$$6 \cdot \left(\frac{x}{3} - \frac{1}{2}\right) = 6 \cdot \frac{5}{6}$$

Clear parentheses.

$$6 \cdot \frac{x}{3} - 6 \cdot \frac{1}{2} = 6 \cdot \frac{5}{6}$$

$$2x - 3 = 5$$ The equation is now free of fractions.

$$2x = 8$$

$$x = 4$$

To check this answer, verify that $\frac{4}{3} - \frac{1}{2} = \frac{5}{6}$. The work is left to you, as is the checking in the remaining examples.

(B) The decimals in an equation such as this are fractions that can be cleared by multiplying by an appropriate power of 10:

$$0.2x + 0.3(x - 5) = 13 \qquad \text{Multiply by 10 to clear decimals.}$$

$$2x + 3(x - 5) = 130$$

$$2x + 3x - 15 = 130$$

$$5x = 145$$

$$x = 29$$

(C) Before multiplying both sides by 12, the LCM of the denominators, enclose any numerator with more than one term in parentheses:

$$5 - \frac{(2x - 1)}{4} = \frac{(x + 2)}{3} \qquad \text{Multiply both sides by 12.}$$

$$\overset{3}{12} \cdot 5 - \overset{3}{\cancel{12}} \cdot \frac{(2x - 1)}{\underset{1}{\cancel{4}}} = \overset{4}{\cancel{12}} \cdot \frac{(x + 2)}{\underset{1}{\cancel{3}}} \qquad \begin{array}{l}\text{12 is exactly divisible by each}\\\text{denominator.}\end{array}$$

$$60 - 3(2x - 1) = 4(x + 2)$$

$$60 - 6x + 3 = 4x + 8$$

$$-6x + 63 = 4x + 8$$

$$-10x = -55$$

$$x = \tfrac{11}{2} \quad \text{or} \quad 5.5$$

Matched Problem 1   Solve:

(A) $\frac{1}{4}x - \frac{2}{3} = \frac{5}{12}x$ **(B)** $0.3(x + 2) + 0.5x = 3$ **(C)** $\dfrac{x + 3}{4} - \dfrac{x - 4}{2} = \dfrac{3}{8}$

Be careful not to confuse *algebraic expressions* involving fractions with *algebraic equations* involving fractions. Consider the two problems:

(A) Solve: $\dfrac{x}{2} + \dfrac{x}{3} = 10$ **(B)** Add: $\dfrac{x}{2} + \dfrac{x}{3} + 10$

The problems look very much alike but are actually very different. To solve the *equation* in (A) we multiply both sides by 6, the LCM of 2 and 3, to clear the fractions. This works so well for equations, you may want to do the same thing for problems like (B). However, (B) is an algebraic *expression*, not an equation, and the multiplication property of equality does not apply. If we multiply expression (B) by 6, we obtain an expression 6 times as large as the original. To add the terms in the expression in (B), we find the LCD and proceed as described in Section 3-3.

Compare the following correct procedures for (A) and (B):

(A) $\dfrac{x}{2} + \dfrac{x}{3} = 10$ **(B)** $\dfrac{x}{2} + \dfrac{x}{3} + 10$

$$6 \cdot \dfrac{x}{2} + 6 \cdot \dfrac{x}{3} = 6 \cdot 10$$ $$= \dfrac{3 \cdot x}{3 \cdot 2} + \dfrac{2 \cdot x}{2 \cdot 3} + \dfrac{6 \cdot 10}{6 \cdot 1}$$

$$3x + 2x = 60$$ $$= \dfrac{3x}{6} + \dfrac{2x}{6} + \dfrac{60}{6}$$

$$5x = 60$$ $$= \dfrac{5x + 60}{6}$$

$$x = 12$$

◆ VARIABLES IN SOME DENOMINATORS

If an equation involves a variable in one or more denominators, such as

$$\frac{2}{3} - \frac{2}{x} = \frac{4}{x}$$

we proceed in essentially the same way as above,

but we must avoid any value of x that makes a denominator 0.

Example 2
Clearing Variables
from Denominators
and Solving

Solve:

$$\frac{2}{3} - \frac{2}{x} = \frac{4}{x}$$

Solution $$\frac{2}{3} - \frac{2}{x} = \frac{4}{x} \qquad x \neq 0$$ We note that $x \neq 0$; then multiply both sides by $3x$, the LCM of the denominators. If 0 turns up later as a "solution," it must be discarded.

$$3x \cdot \frac{2}{3} - 3x \cdot \frac{2}{x} = 3x \cdot \frac{4}{x}$$ $3x$ is exactly divisible by each denominator.

$$2x - 6 = 12$$

$$2x = 18$$

$$x = 9$$

Check $$\frac{2}{3} - \frac{2}{9} \stackrel{?}{=} \frac{4}{9}$$

$$\frac{6}{9} - \frac{2}{9} \stackrel{\checkmark}{=} \frac{4}{9}$$

It is especially important to check answers when there is a variable in the denominator. You must be sure no denominator is 0.

Matched Problem 2   Solve: $\dfrac{3}{x} - \dfrac{1}{2} = \dfrac{4}{x}$

Multiplying both sides of an equation by an expression involving a variable may change the solution set. The new equation may have additional solutions, called **extraneous solutions,** that are not solutions of the original equation. For example, if we multiply both sides of $x = 1$ by x, we obtain $x^2 = x$. The number 0 is a solution to the new equation, but not to the original one. Thus, it is important that we check solutions.

Example 3
Extraneous Solutions

Solve:

$$\frac{3x}{x-2} - 4 = \frac{14 - 4x}{x-2}$$

Solution

$$\frac{3x}{x-2} - 4 = \frac{14 - 4x}{x-2} \qquad x \neq 2 \qquad \text{If 2 turns up as a "solution," it must be discarded.}$$

$$(x-2)\frac{3x}{x-2} - (x-2) \cdot 4 = (x-2)\frac{14-4x}{x-2}$$

$$3x - 4(x-2) = 14 - 4x$$

$$3x - 4x + 8 = 14 - 4x$$

$$-x + 8 = 14 - 4x$$

$$3x = 6$$

$$x = 2 \qquad \text{Recall that 2 cannot be a solution to the original equation.}$$

Since $x = 2$ cannot be a solution, the equation has no solution.

Matched Problem 3   Solve: $\dfrac{5x}{x-2} + \dfrac{10}{2-x} = 3$

Example 4
Clearing Variables from Denominators and Solving

Solve:

$$2 - \frac{3x}{1-x} = \frac{8}{x-1}$$

Solution

$$2 - \frac{3x}{1-x} = \frac{8}{x-1} \qquad \text{Recall that } 1 - x = -(x-1).$$

$$2 - \frac{3x}{-(x-1)} = \frac{8}{x-1} \qquad \text{Recall that } -\frac{a}{-b} = \frac{a}{b}.$$

$$2 + \frac{3x}{x - 1} = \frac{8}{x - 1}$$ Multiply both sides by $(x - 1)$, keeping in mind that $x \neq 1$.

$$(x - 1)2 + (x - 1)\frac{3x}{x - 1} = (x - 1)\frac{8}{x - 1}$$

$$2x - 2 + 3x = 8$$

$$5x = 10$$

$$x = 2$$

Check $2 - \dfrac{3 \cdot 2}{1 - 2} \stackrel{?}{=} \dfrac{8}{2 - 1}$

$$2 - (-6) \stackrel{?}{=} 8$$

$$8 \stackrel{\checkmark}{=} 8$$

Matched Problem 4   Solve: $\dfrac{2x}{x - 1} - 3 = \dfrac{7 - 3x}{x - 1}$

Example 5
Solving a Literal Equation

Solve for x:

$$\frac{x}{a} + \frac{x}{b} = 1$$

Solution

$$\frac{x}{a} + \frac{x}{b} = 1$$ Clear fractions by multiplying by ab. Note that $a \neq 0$, $b \neq 0$.

$$ab\frac{x}{a} + ab\frac{x}{b} = ab \cdot 1$$

$$bx + ax = ab$$ Factor out x on the left side.

$$(b + a)x = ab$$

$$x = \frac{ab}{b + a}$$

Matched Problem 5   Solve for x: $\dfrac{x}{a} - \dfrac{x}{b} = 1$

Example 6
A Number Problem

A fraction has numerator 3 greater than its denominator. If the numerator and denominator are both increased by 6, the resulting fraction is equal to $\frac{5}{4}$. Find the original fraction.

Solution   Let x represent the denominator of the original fraction, so that $x + 3$ represents the numerator. The original fraction is then

$$\frac{x + 3}{x}$$

Increasing both the numerator and denominator by 6 yields the equation

$$\frac{x+3+6}{x+6} = \frac{5}{4}$$

$$\frac{x+9}{x+6} = \frac{5}{4}$$ Clear fractions by multiplying by $4(x+6)$.

$$4(x+6)\frac{x+9}{x+6} = 4(x+6)\frac{5}{4} \quad x \neq -6$$

$$4(x+9) = 5(x+6)$$

$$4x+36 = 5x+30$$

$$6 = x$$

The original fraction is therefore

$$\frac{6+3}{6} = \frac{9}{6}$$

Check   9 is 3 more than 6, as required. When the numerator and denominator are both increased by 6, the resulting fraction is

$$\frac{9+6}{6+6} = \frac{15}{12} = \frac{5}{4}$$

as desired.

Matched Problem 6   A fraction has numerator 3 less than its denominator. If both the numerator and denominator are increased by 5, the resulting fraction is equal to $\frac{3}{4}$. Find the original fraction.

Answers to Matched Problems

1. **(A)** $x = -4$ **(B)** $x = 3$ **(C)** $x = \frac{19}{2}$ or 9.5   2. $x = -2$

3. No solution   4. $x = 2$ 5. $x = \dfrac{ab}{b-a}$ 6. $\frac{4}{7}$

EXERCISE 4-1

A *Solve.*

1. $\dfrac{x}{5} - 2 = \dfrac{3}{5}$

2. $\dfrac{x}{7} - 1 = \dfrac{1}{7}$

3. $\dfrac{x}{3} + \dfrac{x}{6} = 4$

4. $\dfrac{y}{4} + \dfrac{y}{2} = 9$

5. $\dfrac{m}{4} - \dfrac{m}{3} = \dfrac{1}{2}$

6. $\dfrac{n}{5} - \dfrac{n}{6} = \dfrac{6}{5}$

7. $\dfrac{5}{12} - \dfrac{m}{3} = \dfrac{4}{9}$

8. $\dfrac{2}{3} - \dfrac{x}{8} = \dfrac{5}{6}$

9. $0.7x = 21$

10. $0.9x = 540$

11. $0.7x + 0.9x = 32$

12. $0.3x + 0.5x = 24$

13. $\dfrac{1}{2} - \dfrac{2}{x} = \dfrac{3}{x}$

14. $\dfrac{2}{x} - \dfrac{1}{3} = \dfrac{5}{x}$

15. $\dfrac{1}{m} - \dfrac{1}{9} = \dfrac{4}{9} - \dfrac{2}{3m}$

16. $\dfrac{1}{2t} + \dfrac{1}{8} = \dfrac{2}{t} - \dfrac{1}{4}$

17. $\dfrac{x-2}{3} + 1 = \dfrac{x}{7}$

18. $\dfrac{x+3}{2} - \dfrac{x}{3} = 4$

19. $\dfrac{2x-3}{9} - \dfrac{x+5}{6} = \dfrac{3-x}{2} - 1$

20. $\dfrac{3x+4}{3} - \dfrac{x-2}{5} = \dfrac{2-x}{15} - 1$

21. $0.1(x-7) + 0.05x = 0.8$

22. $0.4(x+5) - 0.3x = 17$

23. $0.02x - 0.5(x-2) = 5.32$

24. $0.3x - 0.04(x+1) = 2.04$

25. $\dfrac{1}{2x} - \dfrac{3}{5} = -\dfrac{7}{6x}$

26. $\dfrac{5}{2x} - \dfrac{2}{x} = \dfrac{1}{6}$

27. $\dfrac{3}{x} + \dfrac{5}{2x} = \dfrac{1}{2} + \dfrac{7}{2x}$

28. $\dfrac{1}{2} - \dfrac{2}{3x} = \dfrac{1}{2x} + \dfrac{1}{3x}$

29. $\dfrac{8}{3x} - \dfrac{1}{15} = \dfrac{1}{x} + \dfrac{4}{3x}$

30. $\dfrac{5}{2x} + \dfrac{1}{3} = \dfrac{1}{6} + \dfrac{7}{2x}$

31. $\dfrac{1}{2x} + \dfrac{1}{4} = \dfrac{1}{x} + \dfrac{1}{4x}$

32. $\dfrac{2}{3x} + \dfrac{1}{x} = \dfrac{4}{3x} + \dfrac{1}{3}$

33. $\dfrac{1}{4} - \dfrac{1}{4x} + \dfrac{1}{2} = \dfrac{2}{x}$

34. $\dfrac{1}{2} - \dfrac{5}{2x} - \dfrac{1}{x} = -\dfrac{3}{2x}$

35. $\dfrac{3}{2x} + \dfrac{2}{x} = 1 - \dfrac{5}{2x}$

36. $\dfrac{2}{3x} + \dfrac{1}{6} = \dfrac{5}{2x} - \dfrac{1}{x}$

B **37.** $\dfrac{7}{y-2} - \dfrac{1}{2} = 3$

38. $\dfrac{9}{A+1} - 1 = \dfrac{12}{A+1}$

39. $\dfrac{3}{2x-1} + 4 = \dfrac{6x}{2x-1}$

40. $\dfrac{5x}{x+5} = 2 - \dfrac{25}{x+5}$

41. $\dfrac{2E}{E-1} = 2 + \dfrac{5}{2E}$

42. $\dfrac{3N}{N-2} - \dfrac{9}{4N} = 3$

43. $\dfrac{n-5}{6n-6} = \dfrac{1}{9} - \dfrac{n-3}{4n-4}$

44. $\dfrac{1}{3} - \dfrac{s-2}{2s+4} = \dfrac{s+2}{3s+6}$

45. $5 + \dfrac{2x}{x-3} = \dfrac{6}{x-3}$

46. $\dfrac{6}{x-2} = 3 + \dfrac{3x}{x-2}$

47. $\dfrac{x^2+2}{x^2-4} = \dfrac{x}{x-2}$

48. $\dfrac{5}{x-3} = \dfrac{33-x}{x^2-6x+9}$

49. $\dfrac{3}{x+1} + \dfrac{4}{x-1} = \dfrac{5}{x^2-1}$

50. $\dfrac{2}{x-3} - \dfrac{1}{x+3} = \dfrac{4}{x^2-9}$

51. $\dfrac{1}{x+2} - \dfrac{1}{x-3} = \dfrac{1}{x^2-x-6}$

52. $\dfrac{1}{x-1} - \dfrac{2}{2x+1} = \dfrac{3}{2x^2-x-1}$

53. $\dfrac{2}{x-3} + \dfrac{1}{x+4} = \dfrac{5}{x^2+x-12}$

54. $\dfrac{3}{x-2} + \dfrac{6}{x+4} = \dfrac{9}{x^2+2x-8}$

55. $\dfrac{3}{x+5} = \dfrac{2}{x+4} - \dfrac{1}{x^2+9x+20}$

56. $\dfrac{8}{x-3} = \dfrac{4}{x+2} - \dfrac{2}{x^2-x-6}$

57. $\dfrac{4}{2x^2+3x+1} = \dfrac{2}{x^2+x-2}$

58. $\dfrac{3}{6x^2+7x+1} = \dfrac{1}{2x^2-x-1}$

59. $\dfrac{3}{x^2+5x+6} = \dfrac{6}{2x^2+8x-10}$

60. $\dfrac{9}{3x^2-x-2} = \dfrac{3}{x^2+x-2}$

61. $\dfrac{3x^2+5x+7}{x^2+3x+5} = 3$

62. $\dfrac{2x^2-4x+9}{x^2+4x-9} = 2$

63. $\dfrac{14x^2+15x+13}{2x^2+5x+3} = 7$

64. $\dfrac{10x^2+9x-8}{2x^2-8x+9} = 5$

In Problems 65–70, solve for x.

65. $\dfrac{x+a}{x+b} = 2$

66. $\dfrac{x+1}{x+2} = a$

67. $\dfrac{x+1}{x-1} = b$

68. $3 = \dfrac{x-a}{x-b}$

69. $\dfrac{x+a}{x+b} = c$

70. $\dfrac{x-a}{x-b} = c$

71. A fraction has numerator 6 less than its denominator. If both the numerator and denominator are decreased by 4, the resulting fraction is equal to $\frac{1}{3}$. Find the original fraction.

72. A fraction has numerator 9 less than its denominator. If both the numerator and denominator are increased by 5, the resulting fraction is equal to $\frac{1}{2}$. Find the original fraction.

73. A fraction has numerator 7 less than its denominator. If both the numerator and denominator are increased by 4, the resulting fraction is equal to $\frac{1}{2}$. Find the original fraction.

74. A fraction has numerator 12 less than its denominator. If both the numerator and denominator are decreased by 5, the resulting fraction is equal to $\frac{1}{3}$. Find the original fraction.

C *Solve.*

75. $\dfrac{3x}{24} - \dfrac{2-x}{10} = \dfrac{5+x}{40} - \dfrac{1}{15}$

76. $\dfrac{2x}{10} - \dfrac{3-x}{14} = \dfrac{2+x}{5} - \dfrac{1}{2}$

77. $\dfrac{5t-22}{t^2-6t+9} - \dfrac{11}{t^2-3t} - \dfrac{5}{t} = 0$

78. $\dfrac{x-33}{x^2-6x+9} + \dfrac{5}{x-3} = 0$

79. $5 - \dfrac{2x}{3-x} = \dfrac{6}{x-3}$ **80.** $\dfrac{3x}{2-x} + \dfrac{6}{x-2} = 3$

81. $\dfrac{1}{c^2-c-2} - \dfrac{3}{c^2-2c-3} = \dfrac{1}{c^2-5c+6}$

82. $\dfrac{5t-22}{t^2-6t+9} - \dfrac{11}{t^2-3t} = \dfrac{5}{t}$

83. $\dfrac{x+3}{(x-1)(x-2)} - \dfrac{x+1}{(x-2)(x-3)} = \dfrac{2}{(x-1)(x-3)}$

84. $\dfrac{3}{x^2+x} - \dfrac{2}{x^2-x} = \dfrac{1}{x^2-1}$

85. $\dfrac{1}{x} - \dfrac{1}{x^2} = \dfrac{12}{x^3}$

86. $\dfrac{3}{x^2+2x+1} - \dfrac{5}{x+1} = 2$

87. $\dfrac{1}{(x-1)^2} + \dfrac{1}{(x+1)^2} = \dfrac{2}{x^2-1}$

88. $\dfrac{1}{x^2+3x+2} + \dfrac{1}{x^2-3x+2} = \dfrac{2}{x^2-x-2}$

89. $\dfrac{2x+5}{x^2-1} = \dfrac{6x+5}{3x^2+3x}$

90. $\dfrac{2x-7}{x^2-2x+1} = \dfrac{10x+7}{5x^2-5x}$

91. $\dfrac{3x}{x+2} + \dfrac{4x}{x-3} = \dfrac{7x^2+6x+5}{x^2-x-6}$

92. $\dfrac{7x}{x-1} - \dfrac{4x}{x+2} = \dfrac{3x^2+7x-4}{x^2+x-2}$

93. $\dfrac{x-1}{x+2} + \dfrac{x-2}{x+3} = 2 + \dfrac{x-3}{x^2+5x+6}$

94. $\dfrac{x+2}{x-4} + \dfrac{x+1}{x-3} = \dfrac{x}{x^2-7x+12} + 2$

In Problems 95–98, solve for x.

95. $\dfrac{x}{ab} + \dfrac{x}{bc} + \dfrac{x}{ac} = 1$

96. $\dfrac{x}{a+b} + \dfrac{x}{a-b} = 1$

97. $\dfrac{ax}{a-b} - \dfrac{bx}{a+b} = 1$

98. $\dfrac{ax}{x-a} - \dfrac{bx}{x-b} = a - b$

99. The denominator of a fraction is 3 less than its numerator. If the numerator is decreased by 5 and the denominator is increased by 4, the sum of the resulting fraction and the original fraction is equal to 2. Find the original fraction.

100. The denominator of a fraction is 2 less than its numerator. If the numerator is decreased by 1 and the denominator is increased by 5, the sum of the resulting fraction and the original fraction is equal to 2. Find the original fraction.

  # 4-2 **Application: Ratio and Proportion Problems**

- Ratios
- Proportions

In this section we apply the equation-solving strategy to problems involving *ratio* and *proportion*.

♦ RATIOS

The **ratio** of two quantities is the first quantity divided by the second quantity. Symbolically:

The Ratio of *a* to *b*

For quantities *a* and *b*, $b \neq 0$, the ratio of *a* to *b*, is

$$\frac{a}{b}$$

The ratio $\dfrac{a}{b}$ is also written as *a/b* or *a:b* and is read ''*a* to *b*.''

Example 1
Expressing Ratios

If a parking meter has 45 nickels and 15 quarters, what is the ratio of nickels to quarters?

Solution   The ratio is

$$\frac{\text{Number of nickels}}{\text{Number of quarters}} = \frac{45}{15} = \frac{3}{1}$$

That is, the ratio of nickels to quarters is $3:1$, or 3 to 1, so that for every 3 nickels there is 1 quarter.

Matched Problem 1

If a parking meter has 45 nickels, 30 dimes, and 15 quarters, what is the ratio of quarters to nickels? Of quarters to dimes? Which ratio is greater?

   In addition to providing a way of comparing known quantities, ratios also provide a way of finding unknown quantities.

Example 2
Using Ratios

Suppose you are told that the ratio of quarters to dimes in a parking meter is $\frac{3}{5}$ and that there are 40 dimes in the meter. How many quarters are in the meter?

Solution   Let

$$q = \text{Number of quarters}$$

Then the ratio of quarters to dimes is $q/40$. We also know that this ratio is $\frac{3}{5}$. Thus,

$$\frac{q}{40} = \frac{3}{5} \qquad \textcolor{blue}{\text{To isolate } q, \text{ multiply both sides by 40.}}$$

$$q = 40 \cdot \frac{3}{5}$$

$$= 24 \text{ quarters}$$

We also could have solved this equation by multiplying both sides by $40 \cdot 5$. Then

$$5q = 40 \cdot 3$$

$$q = \frac{120}{5} = 24$$

The equation $5q = 40 \cdot 3$ results from multiplying the numerator of each side by the denominator of the other. This process is called **cross multiplying** and can be visualized as

$$\frac{q}{40} \diagup\!\!\!\!\!\diagdown \frac{3}{5}$$

$$5q = 40 \cdot 3$$

 Cross multiplying applies *only* to equations of the particular form

$$\frac{A}{B} = \frac{C}{D}$$

from which we conclude that $AD = BC$, when $B, D \neq 0$.

Matched Problem 2   If the ratio of dimes to quarters in a meter is $\frac{3}{2}$ and there are 24 quarters in the meter, how many dimes are there?

Ratios are sometimes expressed using the word "per" or as "averages." For example, the ratio of miles driven to gallons of gasoline consumed, that is,

$$\frac{\text{Number of miles driven}}{\text{Number of gallons of gasoline consumed}}$$

is expressed as miles *per* gallon. In baseball, a player's batting *average* is the ratio

$$\frac{\text{Number of hits}}{\text{Number of times at bat}}$$

In this case, the ratio is usually converted to a decimal number with three decimal places.

Example 3
A Baseball Problem

A baseball player has 77 hits in 280 times at bat, for a batting average of $\frac{77}{280} = .275$. How many consecutive hits does the player need to raise the average to .300?

Solution   Let x denote the number of consecutive hits needed. Then

$$\frac{77 + x}{280 + x} = .300$$

$$\frac{77 + x}{280 + x} = \frac{3}{10} \qquad \text{Cross multiply to clear the fractions.}$$

$$770 + 10x = 840 + 3x$$

$$7x = 70$$

$$x = 10$$

Check   After 10 consecutive hits, the player has 87 hits in 290 times at bat, for an average of $\frac{87}{290} = .300$.

Matched Problem 3

A baseball player has 51 hits in 170 times at bat, for a batting average of .300. How many consecutive hits does the player need to raise the average to .320?

The ratio of hits to times at bat is very seldom a ''nice'' number like .275, .300, or .320. For example, 34 hits in 117 times at bat yields an average of .290 598 . . . . For this reason, situations like Example 3 are better modeled as inequality problems. A more appropriate question would be ''how many hits does the player need to raise the average to *at least* .300?'' We will explore such questions in Chapter 5.

Ratios also occur frequently in gambling situations. The terms ''odds'' and ''probabilities'' reflect ratios. In situations where all outcomes are viewed as *equally likely* to occur, we have these definitions:

$$\frac{\text{Number of favorable outcomes possible}}{\text{Total number of possible outcomes}} = \text{Probability of winning}$$

$$\frac{\text{Number of unfavorable outcomes possible}}{\text{Total number of possible outcomes}} = \text{Probability of losing}$$

$$\frac{\text{Number of favorable outcomes possible}}{\text{Number of unfavorable outcomes possible}} = \text{Odds in favor of winning}$$

$$\frac{\text{Number of unfavorable outcomes possible}}{\text{Number of favorable outcomes possible}} = \text{Odds against winning}$$

Probabilities are usually interpreted as representing ratios of success in the long run. For example,

$$\text{Probability of winning} = \frac{\text{Number of wins}}{\text{Number of trials}}$$

in a very large number of trials.

Example 4
A Gambling Problem

An American roulette wheel has 38 numbers, each of which is equally likely to occur when a fair wheel is spun. Of the 38 numbers, 18 are red. If a red number occurring is viewed as a favorable, or winning, outcome, what are the following ratios?

(A) Probability of winning   **(B)** Probability of losing
(C) Odds in favor   **(D)** Odds against

Solution   **(A)** Probability of winning $= \dfrac{\text{Number of favorable outcomes possible}}{\text{Total number of possible outcomes}} = \dfrac{18}{38}$

(B) Probability of losing $= \dfrac{\text{Number of unfavorable outcomes possible}}{\text{Total number of possible outcomes}} = \dfrac{20}{38}$

(C) Odds in favor of winning $= \dfrac{\text{Number of favorable outcomes possible}}{\text{Number of unfavorable outcomes possible}}$

$= \dfrac{18}{20}$ Read "18 to 20."

(D) Odds against winning $= \dfrac{\text{Number of unfavorable outcomes possible}}{\text{Number of favorable outcomes possible}}$

$= \dfrac{20}{18}$ Read "20 to 18."

Matched Problem 4

Suppose that in American roulette the numbers 1, 2, 3, 4, 5, and 6 are viewed as favorable, or winning. What are the following ratios?

(A) Probability of winning   **(B)** Probability of losing
(C) Odds in favor   **(D)** Odds against

♦   **PROPORTIONS**

A statement of equality between two ratios, as in Example 2, is called a **proportion;** that is,

Proportion

$$\frac{a}{b} = \frac{c}{d} \qquad b, d \neq 0$$

Example 5
A Proportion Problem

If a car can travel 192 kilometers on 32 liters of gas, how far will it go on 60 liters?

Solution   Here, the ratio $\frac{192}{32}$ represents kilometers per liter. Let

$$x = \text{Distance traveled on 60 liters}$$

Then,

$$\frac{x}{60} = \frac{192}{32}$$

$\dfrac{\text{kilometers}}{\text{liter}} = \dfrac{\text{kilometers}}{\text{liter}}$ (kilometers per liter)

$$x = 60 \cdot \frac{192}{32}$$

We isolate x by multiplying both sides by 60. (We do not need to use the LCM of 60 and 32.)

$$= 360 \text{ kilometers}$$

Matched Problem 5   If there are 24 milliliters of sulfuric acid in 64 milliliters of solution, how many milliliters of sulfuric acid are in 48 milliliters of the same solution? Set up a proportion and solve.

Proportions can be used to convert metric units to English units, and vice versa. A summary of metric units is located inside the back cover of the text.

Example 6
Metric Conversion

If there is 0.45 kilogram in 1 pound, how many pounds are in 90 kilograms?

Solution   Let x = Number of pounds in 90 kilograms. Set up a proportion. It is usually preferable to put x in the numerator on the left side whenever you have a choice. That is, set up a proportion of the form

$$\frac{\text{Pounds}}{\text{Kilograms}} = \frac{\text{Pounds}}{\text{Kilograms}}$$

Each ratio represents pounds per kilogram.

$$\frac{x}{90} = \frac{1}{0.45}$$

$$x = 90 \cdot \frac{1}{0.45}$$

$$= 200 \text{ pounds}$$

Matched Problem 6   If there are 2.2 pounds in 1 kilogram, how many kilograms are in 100 pounds? Set up a proportion and solve to two decimal places.

Example 7
Metric Conversion

If there are 3.76 liters in 1 gallon, how many gallons are in 50 liters? Set up a proportion and solve to two decimal places.

Solution   Let x = Number of gallons in 50 liters. We set up a proportion of the form

$$\frac{\text{Gallons}}{\text{Liters}} = \frac{\text{Gallons}}{\text{Liters}}$$

Each side gives gallons per liter.

$$\frac{x}{50} = \frac{1}{3.76}$$

$$x = 50 \cdot \frac{1}{3.76}$$

$$= 13.30 \text{ gallons}$$

Matched Problem 7   If there are 1.09 yards in 1 meter, how many meters are in 80 yards? Set up a proportion and solve to two decimal places.

The same technique used here to convert between metric and English measurements also can be used to convert between different currency systems. See Problems 55–58 in Exercise 4-2.

When the ratio of two quantities, *A* and *B*, is constant, we say that *A* **is proportional to** *B*, or that the quantities are **directly proportional**, or that *A* **varies directly as** *B*. If *A* is proportional to *B* and *A* = 5 when *B* = 12, then we can find the value of *A* for any value of *B* using a proportion. For example, to find the value of *A* corresponding to *B* = 20, we solve

$$\frac{5}{12} = \frac{A}{20}$$

to obtain $A = \frac{100}{12} = 8\frac{1}{3}$. Examples of direct variation can be found in Exercise 4-2, Problems 93–96. This topic is considered in more detail in Section 9-5.

If you are having trouble with word problems—and many people do—return to the worked-out examples. Cover the solutions with a piece of paper, proceed with your own solution until you get stuck, and then uncover only enough of the solution to get you started again. After completing an example in this way, immediately work the matched problem following the example. Then work similar problems in the exercise set.

Answers to
Matched Problems

1. $\frac{1}{3}$; $\frac{1}{2}$; the ratio of quarters to dimes is the greater
2. 36 dimes
3. 5 hits

4. **(A)** $\frac{6}{38}$ **(B)** $\frac{32}{38}$ **(C)** $\frac{6}{32}$ **(D)** $\frac{32}{6}$ 5. $\frac{x}{48} = \frac{24}{64}$; $x = 18$ milliliters

6. $\frac{x}{100} = \frac{1}{2.2}$; $x = 45.45$ kilograms   7. $\frac{x}{80} = \frac{1}{1.09}$; $x = 73.39$ meters

---

EXERCISE 4-2

A *Write as a ratio.*

1. 33 dimes to 22 nickels

2. 17 quarters to 51 dimes

3. 25 centimeters to 10 centimeters

4. 30 meters to 18 meters

5. 300 kilometers to 24 liters

6. 320 miles to 12 gallons

Solve each proportion.

7. $\dfrac{m}{16} = \dfrac{5}{4}$ 8. $\dfrac{n}{12} = \dfrac{2}{3}$

9. $\dfrac{x}{13} = \dfrac{21}{39}$ 10. $\dfrac{x}{12} = \dfrac{27}{18}$

11. $\dfrac{7}{36} = \dfrac{n}{9,000}$ 12. $\dfrac{5}{36} = \dfrac{n}{12,600}$

13. $\dfrac{2}{38} = \dfrac{m}{1,900}$ 14. $\dfrac{5}{38} = \dfrac{m}{5,700}$

Set up appropriate proportions and solve. Compute decimal answers to two decimal places.

15. If in a pay telephone the ratio of quarters to dimes is $\frac{5}{8}$ and there are 96 dimes, how many quarters are there?

16. If in a parking meter the ratio of pennies to nickels is $\frac{13}{6}$ and there are 78 nickels, how many pennies are there?

17. If the ratio of the length of a rectangle to its width is $\frac{5}{3}$ and its width is 24 meters, how long is it?

18. If the ratio of the width of a rectangle to its length is $\frac{4}{7}$ and its length is 56 centimeters, how wide is it?

19. If a car can travel 108 kilometers on 12 liters of gas, how far will it go on 18 liters?

20. If a boat can travel 72 miles on 18 gallons of diesel fuel, how far will it travel on 15 gallons?

21. For a particular modeling agency, the ratio of male to female models is 3:8. If the agency has 216 female models, how many male models does it have?

22. A race horse has a ratio of races won to races started of 2:13. If the horse has started 117 races, how many has it won?

23. At a certain medical school the ratio of applicants accepted to the total number of applicants is 4:15. If the school has 705 applicants, how many will be accepted?

24. If a recipe that serves 6 calls for $1\frac{1}{2}$ teaspoons of salt, how much salt is needed if the recipe is made to serve 8?

25. If the numbers 1, 2, 3, . . . , 12 are viewed as favorable, or winning, occurrences in American roulette, what is the probability of winning?

26. There are 18 even numbers on an American roulette wheel. If an even number occurring is viewed as favorable, or winning, what is the probability of winning?

If a pair of fair dice, one red and one blue, are rolled, there are 36 equally likely possible ways the dice can come up. For example, the red die can show 3 dots, and the blue die 5 dots. The 36 equally likely outcomes are shown below. The problems that follow concern the total number of dots on the two dice together. For example, of the 36 possibilities, there are 6 with a total of 7 dots on the two dice.

BLUE DIE

RED DIE						
•	(1, 1)	(1, 2)	(1, 3)	(1, 4)	(1, 5)	(1, 6)
⠢	(2, 1)	(2, 2)	(2, 3)	(2, 4)	(2, 5)	(2, 6)
⠒	(3, 1)	(3, 2)	(3, 3)	(3, 4)	(3, 5)	(3, 6)
⠲	(4, 1)	(4, 2)	(4, 3)	(4, 4)	(4, 5)	(4, 6)
⠵	(5, 1)	(5, 2)	(5, 3)	(5, 4)	(5, 5)	(5, 6)
⠶	(6, 1)	(6, 2)	(6, 3)	(6, 4)	(6, 5)	(6, 6)

27. If winning means that a sum of 5 occurs, what is the probability of winning?

28. If winning means that a sum of 10 occurs, what is the probability of winning?

29. If winning means that a sum of 6 occurs, what are the odds in favor of winning?

30. If winning means that a sum of 8 occurs, what are the odds in favor of winning?

31. If winning means that a sum of 7 or 11 occurs, what are the odds in favor of winning?

32. If winning means that a sum of 7 or 11 occurs, what are the odds against winning?

33. If winning means that a sum of 7 or 11 occurs, what is the probability of winning?

34. If winning means that a sum of 7 or 11 occurs, what is the probability of losing?

*When a proportion is written in the form $a:b = c:d$, the first and fourth terms, that is, a and d, are called the **extremes** of the proportion. The second and third terms, b and c, are called the **means** of the proportion. Use the given information to find the remaining term in the proportion.*

35. The product of the extremes is 72, one mean is 18; find the other mean.

36. The product of the extremes is 180, one mean is 24; find the other mean.

37. The product of the means is 120, one extreme is 24; find the other extreme.

38. The product of the means is 60, one extreme is 6; find the other extreme.

In **similar triangles,** *that is, triangles with the same shape, the ratio of corresponding sides is constant. In the triangles drawn in the figure,*

$$\frac{a}{A} = \frac{b}{B} = \frac{c}{C}$$

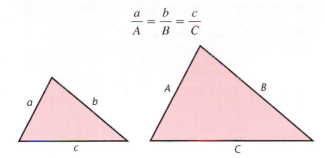

Use the information provided to find the measurement requested.

39. $A = 6$ inches, $B = 11$ inches, $a = 8$ inches; find b

40. $b = 4$ centimeters, $B = 15$ centimeters, $a = 10$ centimeters; find A

41. $A = 8$ meters, $a = 18$ meters, $b = 30$ meters; find B

42. $A = 4$ feet, $B = 20$ feet, $b = 12$ feet; find a

B *Set up appropriate proportions and solve. Compute decimal answers to two decimal places.*

43. If there are 9 milliliters of hydrochloric acid in 46 milliliters of solution, how many milliliters will be in 52 milliliters of solution?

44. If 0.75 cup of flour is needed in a recipe that will feed 6 people, how much flour will be needed in the recipe that will feed 9 people?

45. A 35- by 23-millimeter colored slide is used to make an enlargement whose longer side is 10 inches. How wide will the enlargement be if all of the slide is used?

46. A 3.25- by 4.25-inch negative is used to produce an enlargement whose shortest side is 12 inches. How long will the enlargement be if all of the negative is used?

47. If there is 0.26 gallon in 1 liter, how many liters are in 5 gallons?

48. If there is 0.94 liter in 1 quart, how many quarts are in 10 liters?

49. If there are 1.0567 quarts in 1 liter, how many liters are in 1 gallon? Recall that 1 gallon equals 4 quarts.

50. If there are 2.205 pounds in 1 kilogram, how many kilograms are in 10 pounds?

51. If there is 0.6215 mile in 1 kilometer, how many kilometers are in 1 mile?

52. If there is 0.9144 meter in 1 yard, how many yards are in 1 meter?

53. If there are 2.54 centimeters in 1 inch, how many inches are in 100 centimeters?

54. If there is 0.0353 gram in 1 ounce, how many ounces are in 1,000 grams?

55. If there are 1.54 U.S. dollars in 1 British pound, how many British pounds are there in $100?

56. If there are 2.22 U.S. dollars in 1 Dutch guilder, how many Dutch guilders are there in $100?

57. If there are 1.5 U.S. dollars in 10 Swedish krona, how many U.S. dollars are there in 65 Swedish krona?

58. If there is 0.8418 U.S. dollar in 1 Canadian dollar, how many U.S. dollars are there in 50 Canadian dollars?

59. If a commission of $240 is charged on the purchase of 200 shares of a stock, how much commission would be charged for 500 shares of the same stock?

60. If the commission on the sale of a $120,000 property is $7,500, what would be the commission on the sale of a $225,000 property?

61. If the price/earnings ratio of a common stock is 12:1 and the price of the stock is $66 per share, what does the stock earn per share?

62. If the price/earnings ratio of a common stock is 8.4 (that is, 8.4:1), and the stock earns $23.50 per share, what is the price of the stock per share?

A **pie chart** *is a common way of displaying parts of a total represented by various categories. For example, in a voting population that consists of 50% registered Democrats, 40% registered Republicans, and 10% registered Independents, the three categories and their relative sizes may be displayed as shown in the pie chart in the figure on page 162. The size of each slice is proportional to the relative size of the category. Thus, the size of the central angle that determines the slice is proportional to the relative size of the category. For example, the angle for the Democratic slice should be 180°. This is obtained from the proportion*

$$\frac{50}{100} = \frac{x}{360}$$

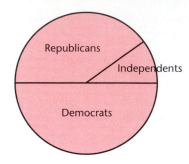

Republicans

Independents

Democrats

Determine the size of the central angle corresponding to each category in the following problems.

63. Total 150 items; categories: *A* 25 items, *B* 60 items, *C* 65 items

64. Total 200 items; categories: *A* 40 items, *B* 70 items, *C* 90 items

65. Total 100 items; categories: *A* 35 items, *B* 45 items, remaining items in *C*

66. Total 450 items; categories: *A* 125 items, *B* 100 items, remaining items in *C*

67. Total 1,000 items; categories: *A* 200 items, *B* 250 items, *C* 400 items, remaining items in *D*

68. Total 1,000 items; categories: *A* 300 items, *B* 100 items, *C* 350 items, remaining items in *D*

C *Fractional parts of angle measurements are often expressed in* **minutes** *and* **seconds;** *60 minutes equals 1 degree, 60 seconds equals 1 minute. To convert 0.4° to minutes, solve the proportion*

$$\frac{x}{0.4} = \frac{60}{1}$$

to obtain x = 24 minutes. To convert 0.41° to minutes, solve

$$\frac{x}{0.41} = \frac{60}{1}$$

to get x = 24.6. Then convert 0.6 minute to seconds:

$$\frac{x}{0.6} = \frac{60}{1}$$

$$x = 36$$

Thus, 3.41° equals 3 degrees, 24 minutes, 36 seconds. Minutes are denoted by ' and seconds by ", so the angle of 3 degrees, 24 minutes, 36 seconds is denoted 3°24'36". Convert each of the following angle measurements to degree-minute-second form.

69. 40.65°   **70.** 50.45°   **71.** 50.43°

72. 40.52°   **73.** 70.605°   **74.** 60.705°

Convert each of the following angle measurements to decimal form. Round answers to four decimal places.

75. 30°15'45"   **76.** 40°24'15"   **77.** 60°48'6"

78. 20°45'30"   **79.** 10°42'18"   **80.** 80°12'54"

Latitude *is the angle, measured in degrees, minutes, and seconds, between a point on the earth's surface, the center of the earth, and the equator. In Problems 81–84, convert the given latitude into decimal form. Round answers to four decimal places.*

81. 15°12'20"   **82.** 25°15'40"

83. 60°24'45"   **84.** 23°30'12"

85. A difference of 10" in latitude corresponds to a north-to-south distance of approximately 0.2 mile. What distance corresponds to a change in latitude of 1°?

86. What north-to-south distance corresponds to a change in latitude of 1'?

87. What change in latitude corresponds to a north-to-south distance of 25 miles?

88. What change in latitude corresponds to a north-to-south distance of 40 miles?

Set up appropriate proportions and solve. Compute decimal answers to two decimal places.

89. Estimate the total number of trout in a lake if a sample of 300 is netted, marked, and released, and after a suitable period for mixing, a second sample of 250 produces 25 marked trout. (Assume that the ratio of the marked trout in the second sample to the total number of the sample is the same as the ratio of those marked in the first sample to the total lake population.)

90. Repeat Problem 89 with a first (marked) sample of 400 and a second sample of 264 with only 24 marked trout.

91. If in the figure the diameter of the smaller pipe is 12 millimeters and the diameter of the larger pipe is 24 centimeters, how much force would be required to lift a 1,200-kilogram car? (Neglect the weight of the hydraulic lift equipment and use the proportion shown in the figure.)

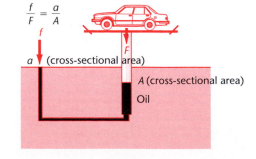

$$\frac{f}{F} = \frac{a}{A}$$

a (cross-sectional area)

A (cross-sectional area)

Oil

92. Do you have any idea how one might measure the circumference of the earth? In 240 B.C. Eratosthenes measured the size of the earth from its curvature. At Syene, Egypt (lying on the Tropic of Cancer), the sun was directly overhead at noon on June 21. At the same time in Alexandria, a town 500 miles directly north, the sun's rays fell at an angle of 7.5° to the vertical. Using this information and a little knowledge of geometry (see the figure), Eratosthenes was able to approximate the circumference of the earth using the following proportion: the circumference of the earth C is to 500 as 360 is to 7.5. Compute Eratosthenes' estimate.

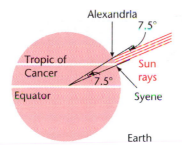

93. The ratio of the volume of a sphere to the cube of its radius is a constant. If a sphere of radius 10 inches has a volume of 4,188.79 cubic inches, what is the volume of a sphere with radius 5 inches?

94. The cost of a cubical box is directly proportional to the cube of its side. If a cubical box with side length 24 inches costs $4.20, how much will a cubical box with side 20 inches cost?

95. The simple interest earned on an investment of P dollars is directly proportional to the time it is invested. If the investment earns $234 in 3 years, how much will it earn in 8 years?

96. The ratio of the force stretching a spring to the distance stretched is constant. If a force of 20 grams stretches the spring 8 centimeters, how far will a force of 75 grams stretch the spring?

When two sides of a triangle are connected by a line segment parallel to the third side, similar triangles result. In the figure, if side DE is parallel to side AC, then triangle ABC is similar to triangle DBE. As a result, the following proportions hold for the sides:

$$\frac{DB}{AB} = \frac{EB}{CB} = \frac{DE}{AC}$$

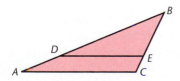

Use this result to find the measurements requested in the following problems.

97. The height of the tree shown in this figure:

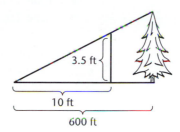

98. The altitude of the balloon shown in this figure:

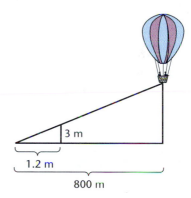

99. The width of the river shown in this figure:

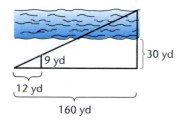

100. The length of the lake shown in this figure:

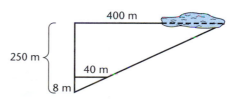

4-3 **Application: Rate-Time Problems**

- ♦ Rate-Time Formulas
- ♦ Rate-Time Problems

There are many types of rate-time problems in addition to the distance-rate-time problems with which you are probably familiar. In this section we will look at rate-time problems as a general class of problems that includes distance-rate-time problems as one of many special cases.

♦ RATE-TIME FORMULAS

If a runner travels 20 kilometers in 2 hours, then the ratio

$$\frac{20 \text{ kilometers}}{2 \text{ hours}} \quad \text{or} \quad 10 \text{ kilometers per hour}$$

is called a **rate.** It is the average number of kilometers produced (traveled) in each unit of time (each hour). Similarly, if an automatic bottling machine bottles 3,200 bottles of soft drinks in 20 minutes, then the ratio

$$\frac{3,200 \text{ bottles}}{20 \text{ minutes}} = 160 \text{ bottles per minute}$$

is also a rate. It is the average number of bottles produced (filled) in each unit of time (each minute).

In general, if Q is the quantity of something produced (kilometers, words, parts, and so on) in T units of time (hours, years, minutes, seconds, and so on) and R is the average rate (quantity produced in one unit of time), then Q, T, and R are related by formula (1).

Quantity-Rate-Time Formula

$Q = RT$ Quantity = (Rate)(Time)   (1)

In the special case where Q is distance D, then

$D = RT$ Distance = (Rate)(Time)   (2)

You should learn Formulas (1) and (2) and recognize that Formula (2) is simply a special case of (1). Two other forms of Formulas (1) and (2), one for rate R and the other for time T, are easily derived from (1) and (2) by dividing both sides of each by either T or R:

$$Q = RT$$
$$RT = Q$$

$$T = \frac{Q}{R} \qquad R = \frac{Q}{T}$$

$$D = RT$$
$$RT = D$$

$$T = \frac{D}{R} \qquad R = \frac{D}{T}$$

◆ RATE-TIME PROBLEMS

Example 1

A Quantity-Rate-Time Problem

A gas pump in a service station can deliver 36 liters in 4 minutes.

(A) What is its rate (liters per minute)?
(B) How much gas can be delivered in 5 minutes?
(C) How long will it take to deliver 54 liters?

Solution

(A) $R = \dfrac{Q}{T} = \dfrac{36 \text{ liters}}{4 \text{ minutes}} = 9$ liters per minute

(B) $Q = RT = (9 \text{ liters per minute})(5 \text{ minutes}) = 45$ liters

(C) $T = \dfrac{Q}{R} = \dfrac{54 \text{ liters}}{9 \text{ liters per minute}} = 6$ minutes

Matched Problem 1

A woman jogs 18 kilometers in 2 hours.

(A) What is her rate (kilometers per hour)?
(B) How far will she jog in 1.5 hours?
(C) How long will it take her to jog 12 kilometers?

Example 2

A Distance-Rate-Time Problem

A jet plane leaves San Francisco and travels at 650 kilometers per hour toward Los Angeles. At the same time another plane leaves Los Angeles and travels at 800 kilometers per hour toward San Francisco. If the cities are 570 kilometers apart, how long will it take the jets to pass each other, and how far from San Francisco will they be at that time?

Solution

Let T = Number of hours until both planes meet. Draw a diagram and label known and unknown parts. Both planes will have traveled the same amount of time when they meet.

$$\begin{pmatrix} \text{Distance plane} \\ \text{from SF travels} \\ \text{to meeting point} \end{pmatrix} + \begin{pmatrix} \text{Distance plane} \\ \text{from LA travels} \\ \text{to meeting point} \end{pmatrix} = \begin{pmatrix} \text{Total distance} \\ \text{from} \\ \text{SF to LA} \end{pmatrix}$$

$$D_1 \qquad + \qquad D_2 \qquad = \qquad 570$$

$$650T \qquad + \qquad 800T \qquad = \qquad 570$$

$$1{,}450T = 570$$

$$T = \frac{570}{1{,}450} \approx 0.39 \text{ hour}$$

$$\text{Distance from SF} = (\text{Rate from SF})(\text{Time from SF})$$

$$\approx (650)(0.39) = 253.5 \text{ kilometers}$$

You may find it helpful to summarize the given information in a table:

	Rate	Time	Distance
From SF	650	T	$650T$
From LA	800	T	$800T$
Total			570

The sum of the distances is 570, so we obtain the equation $650T + 800T = 570$, as above.

Matched Problem 2 If a printing press can print 45 fliers per minute and a newer press can print 80, how long will it take both presses together to print 4,500 fliers if they start simultaneously? How many will the older press have printed by then?

Example 3
A Quantity-Rate-Time Problem

A printing press can print 45 fliers per minute and a newer press can print 80. How long will it take both presses together to print 4,500 fliers if the newer press is brought on the job 10 minutes later than the older press and both continue until the job is completed? Note that here $Q = RT$ takes the following form: the number of fliers printed (Q) = the number of fliers per minute (R) times the number of minutes (T).

Solution Let

$$x = \text{Time to complete whole job}$$

Then

$$x = \text{Time old press is on the job}$$

$$x - 10 = \text{Time new press is on the job}$$

$$\begin{pmatrix} \text{Quantity} \\ \text{printed} \\ \text{by old press} \end{pmatrix} + \begin{pmatrix} \text{Quantity} \\ \text{printed} \\ \text{by new press} \end{pmatrix} = \begin{pmatrix} \text{Total} \\ \text{needed} \end{pmatrix}$$

Remember: Quantity = (Rate)(Time)

$$45x + 80(x - 10) = 4,500$$

$$45x + 80x - 800 = 4,500$$

$$125x = 5,300$$

$$x = 42.4 \text{ minutes}$$

Matched Problem 3 A car leaves a city traveling at 60 kilometers per hour. How long will it take a second car traveling at 80 kilometers per hour to catch up to the first car if it leaves 2 hours later? Set up an equation and solve. [*Hint:* Both cars will have traveled the same distance when the second car catches up to the first.]

Example 4
A Distance-Time-Rate Problem

A speedboat takes 1.5 times longer to go 120 miles up a river than to return. If the boat cruises at 25 miles per hour in still water, what is the rate of the current?

Solution Let

$$x = \text{Rate of current in miles per hour}$$
$$25 - x = \text{Rate of boat upstream}$$
$$25 + x = \text{Rate of boat downstream}$$

$$\text{Time upstream} = (1.5)(\text{Time downstream})$$

$$\frac{\text{Distance upstream}}{\text{Rate upstream}} = (1.5)\frac{\text{Distance downstream}}{\text{Rate downstream}} \qquad \text{Recall that } T = \frac{D}{R}.$$

$$\frac{120}{25 - x} = (1.5)\frac{120}{25 + x}$$

$$\frac{120}{25 - x} = \frac{180}{25 + x}$$

$$(25 + x)120 = (25 - x)180$$

$$3{,}000 + 120x = 4{,}500 - 180x$$

$$300x = 1{,}500$$

$$x = 5 \text{ miles per hour} \qquad \text{Rate of current}$$

Check
$$\text{Time upstream} = \frac{\text{Distance upstream}}{\text{Rate upstream}} = \frac{120}{20} = 6 \text{ hours}$$

$$\text{Time downstream} = \frac{\text{Distance downstream}}{\text{Rate downstream}} = \frac{120}{30} = 4 \text{ hours}$$

Thus, time upstream is 1.5 times longer than time downstream.

Matched Problem 4 A fishing boat takes twice as long to go 24 miles up a river than to return. If the boat cruises at 9 miles per hour in still water, what is the rate of the current?

Example 5
A Quantity-Rate-Time Problem

An advertising company has an automated mailing machine that can fold, stuff, and address a particular mailing in 6 hours. With the help of a newer machine the job can be completed in 2 hours. How long would it take the new machine to do the job alone?

Solution   Let

$$x = \text{Time for new machine to do the job alone}$$

If a job can be completed by the old machine in 6 hours, then that machine's rate of completion is $\frac{1}{6}$ job per hour. That is, from the rate-time formula $Q = RT$, 1 job (Q) = the number of jobs per hour (R) times the number of hours (T), so $1 = R \cdot 6$ and $R = \frac{1}{6}$. Similarly, if a job can be completed by both machines together in 2 hours, then the rate of completion is $\frac{1}{2}$ job per hour. If a job can be completed by the new machine alone in x hours, then the rate of completion for that machine is $1/x$ job per hour. Thus,

$$\left(\begin{array}{c}\text{Rate of old}\\\text{machine}\end{array}\right) + \left(\begin{array}{c}\text{Rate of new}\\\text{machine}\end{array}\right) = \left(\begin{array}{c}\text{Rate}\\\text{together}\end{array}\right)$$

$$\frac{1}{6} \quad + \quad \frac{1}{x} \quad = \quad \frac{1}{2} \qquad \text{Multiply by } 6x.$$

$$x + 6 = 3x$$

$$2x = 6$$

$$x = 3 \text{ hours} \qquad \text{New machine alone}$$

Check $\frac{1}{6} + \frac{1}{3} = \frac{1}{6} + \frac{2}{6} = \frac{3}{6} = \frac{1}{2}$

Matched Problem 5   At a family cabin, water is pumped and stored in a large water tank. Two pumps are used for this purpose. One can fill the tank by itself in 6 hours, and the other can do the job in 9 hours. How long will it take both pumps operating together to fill the tank?

Exercise 4-3 contains not only rate-time problems, but also additional applied problems unclassified as to type, including rate problems that do not directly involve time.

Answers to Matched Problems
1. **(A)** 9 kilometers per hour   **(B)** 13.5 kilometers   **(C)** $1\frac{1}{3}$ hours
2. 36 minutes; 1,620 fliers   3. $60(x + 2) = 80x$; $x = 6$ hours
4. 3 miles per hour   5. 3.6 hours

EXERCISE 4-3

A 1. A worker in a fast-food restaurant earns a daily gross pay of $45.20 for 8 hours of work.
   (A) What is the pay rate, in dollars per hour?
   (B) How much will the worker earn for 52 hours of work?
   (C) How long will it take the worker to earn $500 gross pay?

2. A recreational vehicle traveling at normal highway speeds uses 30 gallons of fuel in 4 hours.
   (A) At what rate is it using fuel, in gallons per hour?
   (B) How much fuel will it consume in 18 hours of driving?
   (C) How long will it take to use an entire 44 gallon tank of fuel?

3. In the 10 years prior to the 1989 international ban on ivory trade, the African elephant population declined by 691,000.
   (A) At what rate did the population decline, in elephants per year?
   (B) At the same rate, how much would the population decline in 15 years?

(C) At the same rate of decline, how long would the remaining population of elephants in Africa, estimated at 609,000, have lasted?

4. In the first 11 years of his career, baseball player Cal Ripken, Jr., had 1762 base hits.
   (A) At what average rate did he accumulate hits, in hits per year?
   (B) At the same rate, how many hits would he accumulate in 15 years?
   (C) At the same rate, how long would it take him to get 3000 hits?

Set up appropriate equations and solve.

5. Two cars leave Chicago at the same time and travel in opposite directions. If one travels at 62 kilometers per hour and the other at 88 kilometers per hour, how long will it take them to be 750 kilometers apart?

6. Two airplanes leave Miami at the same time and fly in opposite directions. If one flies at 840 kilometers per hour and the other at 510 kilometers per hour, how long will it take them to be 3,510 kilometers apart?

7. The distance between towns A and B is 750 kilometers. If a passenger train leaves town A and travels toward town B at 90 kilometers per hour at the same time a freight train leaves town B and travels toward A at 35 kilometers per hour, how long will it take the two trains to meet?

8. Repeat Problem 7 using 630 kilometers for the distance between the two towns, 100 kilometers per hour as the rate for the passenger train, and 40 kilometers per hour as the rate for the freight train.

9. Two cruise ships leaving different ports at the same time head toward each other at speeds of 18 and 24 knots. (A knot is a speed of 1 nautical mile per hour.) If the ports are 630 nautical miles apart, how long does it take the cruise ships to meet and pass?

10. A coal barge leaves its Ohio River loading facility and travels downstream at a speed of 18 miles per hour. At the same time, an empty barge 350 miles downriver is heading upstream at a rate of 10 miles an hour. How long does it take for the barges to meet and pass?

11. An office worker can fold and stuff 14 envelopes per minute. If another office worker can fold and stuff 10 envelopes per minute, how long will it take them working together to fold and stuff 1,560 envelopes?

12. One file clerk can file 12 folders per minute and a second clerk 9. How long will it take them working together to file 672 folders?

13. Three workers can do the same task working alone in times of $3\frac{1}{3}$, $2\frac{1}{2}$, and 2 hours, respectively. How long will it take them to perform the task working together?

14. Three workers can do the same task working alone in times of $3\frac{1}{3}$, $2\frac{1}{2}$, and $1\frac{1}{4}$ hours, respectively. How long will it take them to perform the task working together?

15. A car leaves a town traveling at 50 kilometers per hour. How long will it take a second car traveling at 60 kilometers per hour to catch up to the first car if it leaves 1 hour later?

16. Repeat Problem 15 if the first car travels at 45 kilometers per hour and the second car leaves 2 hours later traveling at 75 kilometers per hour.

17. Pipe A can fill a tank in 8 hours and pipe B can fill the same tank in 6 hours. How long will it take both pipes together to fill the tank?

18. A typist can complete a mailing in 5 hours. If another typist requires 7 hours, how long will it take both working together to complete the mailing?

B 19. Find the total time to complete the job in Problem 11 if the second (slower) office worker is brought on the job 15 minutes after the first person has started.

20. Find the total time to complete the job in Problem 12 if the faster file clerk is brought on the job 14 minutes after the slower clerk has started.

21. A painter can paint a house in 5 days. With the help of another painter, the house can be painted in 3 days. How long would it take the second painter to paint the house alone?

22. One worker from a landscape service can maintain a customer's yard in 3 hours per week. If the service sends a second worker, the maintenance requires only $1\frac{2}{3}$ hours per week. How long would it take the second worker doing it alone?

23. The average speed of an express bus is 1.5 times the average speed of the local bus. If the express travels 60 miles in 40 minutes less time than the local, what are the two speeds?

24. The average speed of an express bus is $1\frac{1}{3}$ times the average speed of the local bus. If the express travels 40 miles in 25 minutes less time than the local, what are the two speeds?

25. A man can walk 12 miles to town and then return via bus at 6 times his walking rate. If the walk to town takes $2\frac{1}{2}$ hours more than the return bus trip, at what rate does the man walk?

26. A hiker walks 6 miles into a national forest on a trail that is mainly uphill. She walks back down at a rate 1.2 times as fast as she walked up. If the return half of the trip takes 40 minutes less than the hike up, at what rates did the hiker walk uphill and then back down?

27. Two clerical workers process tax forms. One can process 8 forms per hour more than the other. If the faster worker can process 286 forms in the same time the slower worker can process 260, what are their rates in forms per hour?

28. Two long-distance runners would be separated by $\frac{1}{2}$ mile if both ran for 1 hour at their individual average speeds. The faster runner can run $2\frac{2}{3}$ miles in the same time it takes the other to run $2\frac{1}{2}$ miles. What are their individual average speeds?

29. A hiker can average a walking speed of 1.6 miles per hour on the uphill trail and 2.4 miles per hour on the downhill return. How far up the trail should he go if he must return to the starting point 10 hours after beginning the hike?

30. You are at a river resort and rent a motorboat for 5 hours at 7 A.M. You are told that the boat will travel at 8 kilometers per hour upstream and 12 kilometers per hour returning. You decide that you would like to go as far up the river as you can and still be back at noon. At what time should you turn back, and how far from the resort will you be at that time?

31. A plane with a normal airspeed of 165 miles per hour flies 425 miles against a head wind and then returns with the same wind as a tail wind. If the outbound trip takes 1.2 times as long as the return, what is the wind speed?

32. A canoeist who can paddle at 4.6 miles per hour in still water, paddles 12 miles downstream and then returns upstream. If the trip back takes 1.3 times as long as the downstream portion, what is the current in the stream?

C 33. Three seconds after firing a rifle at a target, a sharpshooter hears the sound of impact. If sound travels at 335 meters per second and the bullet at 670 meters per second, how far away is the target?

34. An explosion is set off on the surface of the water 11,000 feet from a ship. If the sound reaches the ship through the water 7.77 seconds before it arrives through the air and if sound travels through water 4.5 times faster than through air, how fast (to the nearest foot per second) does sound travel in air and in water?

Problems 35–38 lead to second-degree equations that can be solved by factoring. The factors, however, involve larger numbers than usual. It may be helpful to observe that $(x + 10a)(x + 10b) = x^2 + 10(a + b) + 100ab$. Thus, to factor a second-degree expression like $x^2 + 30x - 10,800$, it is easier to look for two integers that multiply to -108 and add to 3 than to look for two integers that multiply to $-10,800$ and add to 30.

35. An office's new copy machine can print 30 copies per minute more than their old machine. It takes the new machine 5 minutes less to print 1,800 copies. What are the two rates?

36. The faster of a team's two race cars averages 10 miles per hour faster than their slower car and finishes a 500-mile race $12\frac{1}{2}$ minutes ahead of the slower car. What are the cars' average speeds?

37. Taking a 600-mile trip at an average speed 10 miles per hour faster than she usually drives reduces a woman's travel time by 2 hours. What is her normal driving speed?

38. A data entry operator must enter 7,200 items into a computer. If he increases his normal entry rate by 10 items per minute, he can do the job in 10 minutes less time. What is his normal entry rate?

  # 4-4 **Formulas and Literal Equations**

♦ Solving Equations: A Summary
♦ Formulas and Literal Equations

In this section we summarize the equation-solving techniques developed thus far. We then apply these techniques to solve formulas, or literal equations, for one variable in terms of the others.

♦ **SOLVING EQUATIONS: A SUMMARY**

Recall the strategy we have developed for solving equations:

> ## Strategy for Solving Linear Equations
>
> 1. Clear fractions.
> 2. Simplify both sides.
> 3. Get all variable terms to one side, constants to the other, and combine like terms.
> 4. Isolate the variable.

The same strategy also can be applied to equations with variables in the denominator as long as a linear equation results after step 1. When variables are cleared from a denominator, extraneous solutions may be introduced. It is always necessary to check solutions.

Recall also the approach to solving certain equations by factoring:

> ## Strategy for Solving Polynomial Equations
>
> To solve an equation of the form
>
> $$(\text{Polynomial in } x) = 0$$
>
> 1. Factor the polynomial into the product of first-degree factors.
> 2. Set each factor equal to 0.

If step 1 cannot be done, other methods must be used. A general method for solving second-degree polynomials is considered in Chapter 7. Approximation methods for solving higher-degree polynomials are considered in more advanced courses.

Example 1
Solving Equations

Solve:

(A) $\dfrac{1}{2}x - \dfrac{2}{3} = \dfrac{1}{6}$ **(B)** $\dfrac{1}{x} + \dfrac{3}{5} = \dfrac{1}{2}$ **(C)** $x - \dfrac{2}{x} = 1$

Solution **(A)** $\dfrac{1}{2}x - \dfrac{2}{3} = \dfrac{1}{6}$ Clear fractions by multiplying by 6.

$$3x - 4 = 1$$
$$3x = 5$$
$$x = \dfrac{5}{3}$$

Check $\frac{1}{2} \cdot \frac{5}{3} - \frac{2}{3} = \frac{5}{6} - \frac{4}{6} = \frac{1}{6}$

(B) $\dfrac{1}{x} + \dfrac{3}{5} = \dfrac{1}{2}$ Clear fractions by multiplying by $10x$; remember that if $x = 0$ turns up as a solution, it must be discarded.

$$10 + 6x = 5x$$
$$x = -10$$

Check $\dfrac{1}{-10} + \dfrac{3}{5} = \dfrac{-1}{10} + \dfrac{6}{10} = \dfrac{5}{10} = \dfrac{1}{2}$

(C) $\qquad x - \dfrac{2}{x} = 1$ Clear fractions by multiplying by x; remember that if $x = 0$ turns up as a solution in this problem, it must be discarded.

$$x^2 - 2 = x$$
$$x^2 - x - 2 = 0 \qquad \text{Factor.}$$
$$(x - 2)(x + 1) = 0$$

$$x - 2 = 0 \qquad \text{or} \qquad x + 1 = 0$$
$$x = 2 \qquad\qquad\qquad x = -1$$

Check $\quad 2 - \frac{2}{2} = 2 - 1 = 1; \; -1 - \dfrac{2}{-1} = -1 - (-2) = 1$

Matched Problem 1   Solve:

(A) $\dfrac{3}{x} + 2 = x$ **(B)** $\dfrac{1}{3} + \dfrac{3}{x} = \dfrac{5}{7}$ **(C)** $x + \dfrac{2}{3} = 1 - \dfrac{x}{4}$

♦ FORMULAS AND LITERAL EQUATIONS

The process of changing a formula into an equivalent one by solving for a different variable is necessary in many applied situations. As long as we are solving for a variable that does not occur to a higher power than 1, the process is no different than solving a linear equation in one variable. This process was first considered in Section 1-7. The same strategy will apply:

Solving a Literal Equation for a Particular Variable

1. Clear fractions.
2. Simplify both sides.
3. Get all terms with the variable in question to one side, all remaining terms to the other, and combine like terms.
4. Isolate the variable by dividing by its coefficient.

For example, to solve $D = RT$ for R, steps 1–3 are unnecessary. We simply divide by the coefficient T to obtain

$$\frac{D}{T} = R$$

and write this as

$$R = \frac{D}{T}$$

by using the symmetric property of equality from Section 1-2.

Example 2
Solving for a
Particular Variable

Solve $F = \frac{9}{5}C + 32$ (the Celsius-Fahrenheit conversion formula) for C.

Solution

$$F = \tfrac{9}{5}C + 32 \qquad \text{Multiply by 5 to clear fractions.}$$

$$5F = 9C + 160 \qquad \text{Leave terms involving } C \text{ on the right, shift all others to the left.}$$

$$5F - 160 = 9C \qquad \text{Divide by 9.}$$

$$\tfrac{5}{9}F - \frac{160}{9} = C \qquad \text{To obtain the usual form of this equation, factor out } \tfrac{5}{9} \text{ on the left side.}$$

$$\tfrac{5}{9}(F - 32) = C \quad \text{or} \quad C = \tfrac{5}{9}(F - 32)$$

Matched Problem 2

Solve $A = P + Prt$ (the simple interest formula) for r.

Example 3
Solving for a
Particular Variable

Solve $a_n = a_1 + (n-1)d$ (the formula for the nth term of an arithmetic sequence) for n. Treat a_n as a fixed number.

Solution

$$a_n = a_1 + (n-1)d \qquad \text{Simplify the right side.}$$

$$a_n = a_1 + nd - d \qquad \text{Move all terms not involving } n \text{ to the left.}$$

$$a_n - a_1 + d = nd \qquad \text{Divide by the coefficient of } n.$$

$$\frac{a_n - a_1 + d}{d} = n \quad \text{or} \quad n = \frac{a_n - a_1 + d}{d}$$

Matched Problem 3

Solve

$$S = \frac{n}{2}[2a + (n-1)d]$$

(the formula for the sum of an arithmetic sequence) for d.

Example 4
Solving for a
Particular Variable

Solve the formula $A = P + Prt$ for P.

Solution

$$A = P + Prt$$

$$P + Prt = A$$

Since P is a common factor to both terms on the left, we factor P out and complete the problem:

$$P(1 + rt) = A$$

$$\frac{P(1 + rt)}{1 + rt} = \frac{A}{1 + rt}$$

$$P = \frac{A}{1 + rt}$$ Note that P appears only on the left side.

 CAUTION   If we write $P = A - Prt$, we have not solved for P. To solve for P is to isolate P on the left side with a coefficient of 1. In general, if the variable for which we are solving appears on both sides of the equation, we have not solved for it!

Matched Problem 4   Solve $A = xy + xz$ for x.

Answers to Matched Problems   **1. (A)** $x = -1, 3$ **(B)** $x = \frac{63}{8}$ **(C)** $x = \frac{4}{15}$ **2.** $r = \frac{A - P}{Pt}$

3. $d = \frac{2S - 2an}{n(n - 1)}$ **4.** $x = \frac{A}{y + z}$

EXERCISE 4-4

A *Solve.*

1. $\frac{1}{3}x + \frac{1}{4} = \frac{1}{6}$

2. $\frac{1}{2}x - \frac{1}{3} = \frac{1}{4}$

3. $1 - \frac{1}{2}x = \frac{2}{3}$

4. $\frac{4}{5}x - \frac{3}{4} = \frac{1}{2}$

5. $\frac{1}{x} + 3 = \frac{5}{3}$

6. $2 - \frac{3}{x} = \frac{4}{5}$

7. $5 - \frac{6}{x} = \frac{7}{8}$

8. $\frac{4}{x} + 5 = \frac{6}{7}$

9. $\frac{1}{x} + \frac{1}{2x} = \frac{1}{3}$

10. $1 - \frac{2}{x} = \frac{3}{4x}$

11. $\frac{5}{x} - \frac{4}{3x} = \frac{1}{2}$

12. $\frac{1}{x} + \frac{3}{5x} = \frac{7}{9}$

13. $\frac{x + 1}{3} - \frac{1}{2} = \frac{x + 1}{12}$

14. $\frac{3(x - 1)}{5} - \frac{1}{4} = \frac{x - 1}{20}$

15. $\frac{1}{x - 1} + \frac{2x}{x - 1} = 3$

16. $4 - \frac{1}{x + 1} = \frac{5}{x + 1}$

17. $\frac{2}{x} + \frac{3}{x + 1} = \frac{4}{x^2 + x}$

18. $\frac{7}{x + 2} - \frac{6}{x + 1} = \frac{1}{(x + 2)(x + 1)}$

19. $\frac{1}{x + 1} - \frac{1}{x - 1} = \frac{x}{x^2 - 1}$

20. $\frac{3x}{x^2 - 4} - \frac{2}{x + 2} = \frac{1}{x - 2}$

Solve for the indicated variable.

21. $A = \frac{1}{2}bh$ for h.   *Area of a triangle*

22. $V = \frac{1}{3}\pi r^2 h$ for h.   *Volume of a cone*

23. $A = \frac{1}{2}(a + b)h$ for a.   *Area of a trapezoid*

24. $A = \frac{1}{2}(a + b)h$ for h.   *Area of a trapezoid*

25. $y = 3x + 7$ for x.   *Equation of a line*

26. $y = -7x + 3$ for x.   *Equation of a line*

27. $l = \frac{\pi}{180}r\theta$ for θ.   *Arc length on a circle*

28. $A = \frac{\pi}{360}r^2\theta$ for θ.   *Area of a circular sector*

29. $P = S(1 - dt)$ for t.   *Simple discount*

30. $V = V_0(1 + Bt)$ for t.   *Gas expansion*

31. $V = \frac{4}{3}\pi ab^2$ for a.   *Volume of ellipsoid*

32. $S_n = \frac{n}{2}(a_1 + a_n)$ for n.   *Arithmetic progression*

B 33. $P = \frac{p}{p + q}$ for p.   *Probability*

34. $P = \dfrac{p}{p + q}$ for q.   *Probability*

35. $(ERA) = \dfrac{9R}{I}$ for I.   *Baseball—earned run average*

36. $(IQ) = \dfrac{100(MA)}{(CA)}$ for (CA).   *Psychology— intelligence quotient*

37. $F = m\dfrac{v - v_0}{t}$ for v.   *Momentum*

38. $z = \dfrac{x - \mu}{\sigma}$ for x.   *Statistics*

39. $y = Q(P - V) - F$ for V.   *Profit analysis*

40. $F_2 = \alpha Y_1 + (1 - \alpha)F_1$ for F_1.   *Forecasting*

Solve.

41. $x(x - 2)(x - 4) = 0$

42. $x(x + 3)(x + 5) = 0$

43. $1 = \dfrac{7}{x} - \dfrac{12}{x^2}$

44. $\dfrac{14}{x^2} = 1 + \dfrac{5}{x}$

45. $x(3x + 7) = x(5x + 2)$

46. $x(4x - 3) = x(3x - 4)$

47. $1 = \dfrac{2}{x - 1} + \dfrac{6}{(x - 1)^2}$

48. $\dfrac{15}{(x + 2)^2} = \dfrac{8}{x + 2} - 1$

49. $6x^2 + 5x - 6 = 0$

50. $12x^2 - 7x - 12 = 0$

C *Solve for the indicated variable.*

51. $\dfrac{1}{f} = \dfrac{1}{a} + \dfrac{1}{b}$ for f.   *Optics—focal length*

52. $\dfrac{1}{R} = \dfrac{1}{R_1} + \dfrac{1}{R_2}$ for R.   *Electric circuits*

53. $C = \tfrac{1}{2}QC_h + \dfrac{D}{Q}C_0$ for D.   *Inventory*

54. $H = \dfrac{kA(t_2 - t_1)}{L_1}$ for A.   *Heat flow*

55. $\dfrac{P_1 V_1}{T_1} = \dfrac{P_2 V_2}{T_2}$ for T_2.   *Gas law*

56. $\dfrac{P_1 V_1}{T_1} = \dfrac{P_2 V_2}{T_2}$ for V_1.   *Gas law*

57. $y = \dfrac{2x - 3}{3x - 5}$ for x.   *Rational equation*

58. $y = \dfrac{3x + 2}{2x - 4}$ for x.   *Rational equation*

    ## 4-5   Miscellaneous Applications

We have used this general strategy for solving word problems:

A Strategy for Solving Word Problems

1. Read the problem carefully—several times if necessary—until you understand the problem, know what is to be found, and know what is given.
2. If appropriate, draw figures or diagrams and label known and unknown parts. Look for formulas connecting the known quantities with the unknown quantities.
3. Let one of the unknown quantities be represented by a variable, say x, and try to represent all other unknown quantities in terms of x. This is an important step and must be done carefully. Be sure you clearly understand what you are letting x represent.

4. Form an equation relating the unknown quantities to the known quantities. This step may involve the translation of an English sentence into an algebraic sentence, the use of relationships in a geometric figure, the use of certain formulas, and so on.
5. Solve the equation and write answers to *all* questions in the problem.
6. Check all solutions in the original problem.

Thus far, when given an applied problem, you have usually been told what type of problem you were dealing with—number, geometric, ratio and proportion, rate-time, and so forth. Another specific type of problem, mixture problems, will be considered in detail in Section 8-2. Having information about the problem type at the beginning provides a suggestion of how to set up and solve the problem. We now consider some problems without this extra information. A large number of such problems are provided in the exercise set for this section.

Example 1
A Word Problem

Five people form a glider club and decide to share the cost of a glider equally. They find, however, that if they let three more join the club, the share for each of the original five will be reduced by $480. What is the total cost of the glider?

Solution   Let x be the cost of the glider. With five shares, each share of the cost of the glider is $x/5$. With an additional three shares, each share is $x/8$. Each share for eight people is $480 less than it is for five; that is,

$$\frac{x}{8} = \frac{x}{5} - 480$$

Solve by clearing fractions first:

$$5x = 8x - 40 \cdot 480$$

$$3x = 19{,}200$$

$$x = 6{,}400 \text{ dollars}$$

Check $6,400 divided into five shares is $1,280 each; divided into eight shares it is $800 each. The difference is $480 per share.

Matched Problem 1

Three people bought a sailboat together. If they had taken in a fourth person, the cost for each would have been reduced by $400. What was the total cost of the boat?

Applied problems very commonly involve the use of percentages. Recall that a **percent** is just a fraction expressed in hundredths. For example,

$$35\% = \tfrac{35}{100} = 0.35$$

Also, when a percentage of something is calculated, this means to multiply. Thus, 35% of an amount A means 0.35 times A, or $0.35A$.

Example 2
A Word Problem

If a stock that you bought on Monday went up 10% on Tuesday and fell 10% on Wednesday, how much did you pay for the stock on Monday if you sold it on Wednesday for $99?

Solution

Let x be the Monday price. Then the Tuesday price is $1.1x$. Remember that a 10% increase means the new price is $100\% + 10\% = 110\%$ of the old, that is, 1.1 times the old. The Wednesday price is 0.9 times the Tuesday price, or $(0.9)(1.1x)$. Thus,

$$(0.9)(1.1)x = 99$$
$$0.99x = 99$$
$$x = 100$$

Check

Monday price $100, Tuesday price $110, Wednesday price $99.

Matched Problem 2

A company's sales decreased 5% in 1987 and increased 5% in 1988. What were the sales in 1986 if sales for 1988 were $9,975,000?

Answers to
Matched Problems

1. $4,800
2. $10,000,000

EXERCISE 4-5

The problems in this set of exercises contain a variety of applications, including number, geometric, ratio and proportion, and rate-time problems. The more difficult problems are marked with two stars (★★), the moderately difficult problems with one star (★), and the easier problems are not marked.

NUMBER

1. Find three consecutive even integers with the sum of the first two numbers equal to 8 more than the third number.

2. Find three consecutive integers with the sum of the second and third numbers equal to 12 less than 3 times the first number.

3. Find two numbers whose sum is $\frac{11}{4}$ and whose product is $\frac{5}{8}$.

4. The sum of two numbers is equal to 8 times their product. One of the numbers is $\frac{1}{3}$. What is the other number?

SPORTS

5. Baseball player Wade Boggs had 1,965 hits in the first 5,699 times at bat in his career.
   (A) At what rate is he hitting, in hits per time at bat? (This rate is called his batting average.)

   (B) At the same rate, how many hits will he have in his first 9,000 times at bat?
   (C) At the same rate, how many times at bat will he need to reach 3,000 hits?

★6. A major league baseball player finished the season with 122 singles. Five percent of his hits were triples. Two out of every 15 hits was a home run, and he hit 1 more double than home run. How many hits did he have?

GEOGRAPHY

Problems 7–12 refer to the following. The temperature in degrees Celsius at a depth of x meters beneath the surface of the earth is given by the formula

$$T = S + 2.5\left(\frac{x - 3{,}000}{100}\right) \qquad \text{for } x \geq 3{,}000$$

where S is the surface temperature.

7. If the surface temperature is 20°C, what is the temperature at a depth of 4,000 meters?

8. If the surface temperature is 40°C, what is the temperature at a depth of 3,600 meters?

★9. If the surface temperature is 25°C, at what depth is the temperature 60°C?

★10. If the surface temperature is 30°C, at what depth is the temperature 50°C?

★**11.** At what depth is the temperature 60°C higher than the temperature at the surface?

★**12.** At what depth is the temperature 100°C higher than the temperature at the surface?

LIFE SCIENCES

13. A good approximation for the normal weight w (in kilograms) of a person over 150 centimeters tall is given by the formula $w = 0.98h - 100$, where height h is in centimeters. What would be the normal height of a person with a normal weight of 76 kilograms?

14. Find the normal height of a person in Problem 13 with a normal weight of 55 kilograms.

15. A scuba diver knows that 1 atmosphere of pressure is the weight of a column of air 1 square inch extending straight up from the surface of the earth without end (14.7 pounds per square inch). Also, the water pressure below the surface increases 1 atmosphere for each 33 feet of depth. In terms of a formula,

$$P = 1 + \frac{D}{33}$$

where P is pressure in atmospheres and D is depth of the water in feet. At what depth will the pressure be 3.6 atmospheres?

16. A company selling water-resistant watches advertises that they are waterproof to 3 atmospheres of pressure. How deep could a diver go (see Problem 15) and safely use the watch?

★**17.** A wildlife management group approximated the number of chipmunks in a wildlife preserve by using the popular capture-mark-recapture technique. Using live traps, they captured and marked 600 chipmunks and then released them. After a period for mixing, they captured another 500 and found 60 marked ones among them. Assuming that the ratio of the total chipmunk population to the chipmunks marked in the first sample is the same as the ratio of all chipmunks in the second sample to those found marked, estimate the chipmunk population in the preserve.

★**18.** A naturalist for a fish-and-game department estimated the total number of rainbow trout in a lake by using the method described in Problem 17. The naturalist netted, marked, and released 200 rainbow trout. A week later, after thorough mixing, 200 more were netted and 8 marked trout were found among them. Estimate the total rainbow trout population in the lake.

DOMESTIC

19. A 2-gallon can of concrete seal will cover 220 square feet. How many gallons are required to cover a garage measuring 20 by 27 feet?

20. A 50-pound bag of lawn fertilizer treats 2,250 square feet of lawn. If a homeowner has a yard measuring 7,800 square feet, how many bags of fertilizer will be needed to treat the lawn?

21. The annual tax on a home valued at $90,000 is $840. What is the value of a house taxed at the same rate if the tax is $2,226?

22. An automobile can travel 518 miles on 18.5 gallons of gasoline.
   (A) At what rate does the car consume gas, in miles per gallon?
   (B) How far can the car go on 30 gallons of gasoline?
   (C) How many gallons are required on a trip of 700 miles?

23. If 40 gallons of an insect spray solution contains 12 ounces of insecticide, how much insecticide will be in 140 gallons of the same spray?

24. Sixteen ounces of pole bean seed are needed to plant a 200-foot row.
   (A) At what rate are the seeds being planted, in ounces per foot?
   (B) How many ounces of seed will be required to plant a 44-foot row of beans?
   (C) How long a row can be planted with 24 ounces of seed?

25. A student needs at least 80% of all points on the tests in a class to get a B. There are three 100-point tests and a 250-point final. The student's test scores are 72, 85, and 78 for the 100-point tests. What is the least score the student can make on the final and still get a B?

26. Repeat Problem 25, but this time suppose the student scores 95, 87, and 66 on the three 100-point tests.

★**27.** The cruising speed of an airplane is 150 miles per hour (relative to ground). You wish to hire the plane for a 3-hour sightseeing trip. You instruct the pilot to fly north as far as possible and still return to the airport at the end of the allotted time.
   (A) How far north should the pilot fly if there is a 30-mile-an-hour wind blowing from the north?
   (B) How far north should the pilot fly if there is no wind blowing?

★**28.** Repeat Problem 27 for an airplane with a cruising speed of 350 kilometers per hour (in still air) and a wind blowing at 70 kilometers per hour from the north.

MUSIC

29. Starting with a string tuned to a given note, one can move up and down the scale simply by decreasing or increasing its length (while maintaining the same tension) according to simple whole-number ratios (see the figure). For example, $\frac{8}{9}$ of the C string gives the next higher note D, $\frac{2}{3}$ of the C string gives G, and $\frac{1}{2}$ of the C string gives C 1 octave higher. (The reciprocals of these fractions, $\frac{9}{8}$, $\frac{3}{2}$, and 2, respectively, are proportional to the frequencies of these notes.) Find the lengths of seven strings (each less than 30 inches) that will produce the following seven chords when paired with a 30-inch string:

(A) Octave   1:2   (B) Fifth   2:3
(C) Fourth   3:4   (D) Major third   4:5
(E) Minor third 5:6   (F) Major sixth   3:5
(G) Minor sixth 5:8

	C	D	E	F	G	A	B	C	D	E	F	G	A	B	C
Relative string length	2	$\frac{16}{9}$	$\frac{8}{5}$	$\frac{3}{2}$	$\frac{4}{3}$	$\frac{6}{5}$	$\frac{16}{15}$	1	$\frac{8}{9}$	$\frac{4}{5}$	$\frac{3}{4}$	$\frac{2}{3}$	$\frac{3}{5}$	$\frac{8}{15}$	$\frac{1}{2}$
Scale ratios (proportional to frequencies)	$\frac{1}{2}$	$\frac{9}{16}$	$\frac{5}{8}$	$\frac{2}{3}$	$\frac{3}{4}$	$\frac{5}{6}$	$\frac{15}{16}$	1	$\frac{9}{8}$	$\frac{5}{4}$	$\frac{4}{3}$	$\frac{3}{2}$	$\frac{5}{3}$	$\frac{15}{8}$	2
Frequencies	132	149	165	176	198	220	248	264	297	330	352	396	440	495	528

30. The three major chords in music are composed of notes whose frequencies are in the ratio 4:5:6. If the first note of a chord has a frequency of 264 hertz (middle C on the piano), find the frequencies of the other two notes. [*Hint:* Set up two proportions using 4:5 and 4:6, respectively.]

BUSINESS

31. The price of a women's blouse after a 30% discount is $25.06. What was the original price?

32. A company that marks all items up 40% offers a color television for sale at $364. What was the company's cost?

★33. A company that markets calendars can produce x calendars at a cost of $12,000 + 1.3x$ dollars. The $12,000 represents a fixed setup cost for beginning production, and it costs $1.30 to produce each calendar. If each calendar sells for $2.50, how many calendars must the company sell to break even on one production run?

★34. Refer to Problem 33. If the company can reduce its setup cost to $10,000 and its per calendar cost to $1.25, how many calendars must be produced and sold to break even?

★35. If $10,000 is invested, part at an annual interest rate of 4.5% and the rest at a rate of 7%, how much should be invested at each rate to yield $600 total for the year?

★36. If $10,000 is invested, part at an annual interest rate of 4% and the rest at a rate of 8.5%, how much should be invested at each rate to yield $600 total for the year?

37. The sticker price for a 1992 Saturn included a destination charge of $300 in addition to the manufacturer's suggested retail price (MSRP). The cost to the dealer was 86% of the MSRP. If the sticker price was $11,200, what was the dealer's cost?

38. The sticker price for a 1992 Honda included a destination charge of $275 in addition to the manufacturer's suggested retail price (MSRP). The cost to the dealer was 90% of the MSRP. If the sticker price was $14,800, what was the dealer's cost?

39. A piece of industrial machinery is depreciated $3,500 per year. If the equipment is valued at $6,400 after 8 years, what was its original value?

40. A business computer originally valued at $140,000 is depreciated by $16,000 per year. If the value is now $36,000, how long has the company had the computer?

41. If you paid $160 for a camera after receiving a discount of 20%, what was the price of the camera before the discount?

42. A car rental company charges $21 per day and 10¢ per mile. If a car was rented for 2 days, how far was it driven if the total bill came to $53.20?

43. It costs a book publisher $74,200 to prepare a book for publication (typesetting, art, editing, and so on); printing and binding costs are $5.50 per book. If the book is sold to bookstores for $19.50 per copy, how many copies must be sold for the publisher to break even?

44. A woman borrowed a sum of money from a bank at 18% simple interest. At the end of 10 months she repaid the bank $1,380. How much was borrowed from the bank? [*Hint:* $A = P + Prt$, where A is the amount repaid, P is the amount borrowed, r is the interest rate expressed as a decimal, and t is time in years.]

Problems 45–48 refer to the following business situation: the cost of producing x items usually involves a fixed setup cost S and a cost per item C, so that the total production cost is S + Cx. The revenue from selling the x items at a price P is Px. The **break-even point** *is the value of x where revenue equals cost, that is, where S + Cx = Px. Find the break-even point for the given values of S, C, and P.*

45. $S = \$8{,}000$, $C = \$18$, $P = \$28$

46. $S = \$2{,}000$, $C = \$4$, $P = \$6$

47. $S = \$26{,}000$, $C = \$42$, $P = \$55$

48. $S = \$40{,}000$, $C = \$220$, $P = \$300$

49. Find the break-even point if the revenue from x units is equal to $x^2 + 40x$ and the cost is $x^2 - 600x + 22{,}400$.

50. Find the break-even point if the revenue from x units is equal to $x^2 + 180x$ and the cost is $x^2 - 720x + 55{,}800$.

GEOMETRY

51. A rectangle with perimeter 52 inches has length equal to 2 less than 3 times the width. Find the dimensions.

52. A rectangle with area 36 square feet has width 5 feet less than its length. Find the length.

The usual standard of measurement for angles uses a complete revolution of 360° as the reference. A second system for measuring angles uses the **radian** *as the unit of measurement and assigns 2π radians as the measurement of a full revolution. The proportion*

$$\frac{Degrees}{360°} = \frac{Radians}{2\pi}$$

allows conversion between the two systems. Convert the following angle measurements as indicated.

53. 30° to radians

54. 240° to radians

55. $\dfrac{\pi}{4}$ radian to degrees

56. $\dfrac{2\pi}{3}$ radians to degrees

57. 120° to radians

58. 15° to radians

59. $\dfrac{7\pi}{4}$ radians to degrees

60. $\dfrac{5\pi}{8}$ radians to degrees

The length of an arc on a circle has the same ratio to the full circumference as the corresponding angle does to a full revolution. (See the figure.) That is,

$$\frac{Arc\ length}{Circumference} = \frac{Angle}{360°}$$

The circumference of a circle with radius r is given by 2πr. Find the arc lengths corresponding to the following angle and radius measurements. Express the answer as a multiple of π.

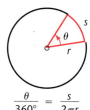

$$\frac{\theta}{360°} = \frac{s}{2\pi r}$$

61. $r = 5$, $\theta = 36°$

62. $r = 8$, $\theta = 72°$

63. $r = 100$, $\theta = 100°$

64. $r = 50$, $\theta = 50°$

65. $r = 6$, $\theta = 40°$

66. $r = 10$, $\theta = 120°$

67. $r = 10$, $\theta = 225°$

68. $r = 1$, $\theta = 300°$

PHYSICS AND ENGINEERING

69. If a small steel ball is thrown downward from a tower with an initial velocity of 15 meters per second, its velocity in meters per second after t seconds is given approximately by

$$v = 15 + 9.75t$$

How many seconds are required for the object to attain a velocity of 93 meters per second?

70. How long would it take the ball in Problem 69 to reach a velocity of 120 meters per second?

71. If the large cross-sectional area in a hydraulic lift (see the figure) is approximately 630 square centimeters and a person wants to lift 2,250 kilograms with a 25-kilogram force, how large should the small cross-sectional area be?

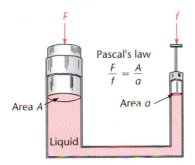

Pascal's law

$$\frac{F}{f} = \frac{A}{a}$$

Area A Area a

Liquid

72. If the large cross-sectional area of the hydraulic lift shown in the figure is 560 square centimeters and the small cross-sectional area is 8 square centimeters, how much force f will be required to lift 2,100 kilograms?

73. A type of physics problem with wide applications is the *lever problem*. For a lever to be in static equilibrium (balanced), relative to a fulcrum, the sum of the downward forces times their respective distances on one side of the fulcrum must equal the sum of the downward forces times their respective distances on the other side of the fulcrum (see the figure). If a person has a 200-centimeter steel wrecking bar and places a fulcrum 20 centimeters from one end, how much can be lifted with a force of 50 kilograms on the long end?

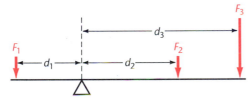

74. Two people decided to move a 1,920-pound rock by use of a 9-foot steel bar (see the figure). If they place the fulcrum 1 foot from the rock and one of them applies a force of 150 pounds on the other end, how much force will the second person have to apply 2 feet from that end to lift the rock?

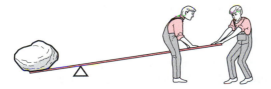

★75. In 1849, during a celebrated experiment, the French mathematician Fizeau made the first accurate approximation of the speed of light. By using a rotating disk with notches equally spaced on the circumference and a reflecting mirror 5 miles away (see the figure), he was able to measure the elapsed time for the light traveling to the mirror and back. Calculate his estimate for the speed of light (in miles per second) if his measurement for the elapsed time was $\frac{1}{20,000}$ second ($d = rt$).

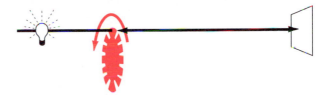

★★76. An earthquake emits a primary wave and a secondary wave. Near the surface of the earth, the primary wave travels at about 5 miles per second and the secondary wave at about 3 miles per second. From the time lag between the two waves arriving at a given seismic station, it is possible to estimate the distance to the quake. (The *epicenter* can be located by getting distance bearings at three or more stations.) Suppose a station measured a time difference of 12 seconds between the arrival of the two waves. How far would the earthquake be from the station?

★77. The ratio of the length of an auto's skid marks to the square of the speed when the brakes were applied is a constant. If a speed of 30 miles per hour produces skid marks 20 feet long, what speed will account for 80-foot-long skid marks?

78. If 90 milliliters of acid is contained in 240 milliliters of an acid solution, how many milliliters of acid will be in 50 milliliters of solution?

GOVERNMENT

Property tax rates are often expressed in **mills,** *a mill being a tenth of a cent, or $0.001. A rate of 1 mill corresponds to a tax of $1 per every $1,000 of assessed property value.*

79. A school district proposes a 0.4 mill tax increase. How much additional tax would be levied on property valued at $78,000?

80. A county library system receives the revenue from a 0.2 mill property tax. How much does the tax on a home valued at $85,000 contribute to the library?

81. A city taxes real estate at the rate of 2.8 mills. What will be the tax on a house valued at $220,000?

82. If the city in Problem 81 raises a total of $1,450,000 in taxes, what is the total value of taxed real estate in the city?

The United States Constitution calls for the House of Representatives to be apportioned on the basis of population, with every state having at least one representative, and with reapportionment to follow the census taken every 10 years. The total number of seats in the House is set by law at 435. The problem of rounding off the apportioned numbers has been a source of political controversy for two centuries. The population of the United States according to the 1990 census is 248,700,000.

83. Census figures show the 1990 population of California to be 29,760,000 and that of Delaware to be 666,000. To how many seats in the House of Representatives is each of these states entitled? (There will be two integer answers, depending on whether the exact proportion is rounded up or down.)

84. Repeat Problem 83 for Texas, with a population of 16,987,000, and Montana, with a population of 799,000.

85. Refer to Problem 83. What will be the average size of a congressional district in California and Delaware if their apportioned numbers are rounded up? If the numbers are rounded down?

86. Refer to Problem 84. What will be the average size of

a congressional district in Texas and Montana if their apportioned numbers are rounded up? If the numbers are rounded down?

PUZZLES

87. A pole is located in a pond. One-fifth of the length of the pole is in the sand, 4 meters is in water, and two-thirds of the length is in the air. How long is the pole?

★**88.** Diophantus, an early Greek algebraist (A.D. 280), was the subject for a famous ancient puzzle. See if you can find Diophantus' age at death from the following information: Diophantus was a boy for one-sixth of his life; after one-twelfth more he grew a beard; after one-seventh more he married, and after 5 years of marriage he was granted a son; the son lived one-half as long as his father; and Diophantus died 4 years after his son's death.

★★**89.** A classic problem is the courier problem. If a column of soldiers 3 miles long is marching at 5 miles per hour, how long will it take a courier on a motorcycle traveling at 25 miles per hour to deliver a message from the end of the column to the front and then return?

★★**90.** After 12:00 noon exactly, what time will the hands of a clock be together again?

CHAPTER SUMMARY

4-1   SOLVING EQUATIONS INVOLVING FRACTIONS

The strategy for solving linear equations is:

1. Clear fractions.
2. Simplify both sides.
3. Get all variable terms to one side, constants to the other, and combine like terms.
4. Isolate the variable.

In particular, equations involving fractions are solved by clearing fractions and then proceeding as usual on the resulting equation with integer coefficients. Clearing variables from denominators may introduce **extraneous solutions** which are not solutions to the original equation.

4-2   APPLICATION: RATIO AND PROPORTION PROBLEMS

The **ratio** of a to b is a/b. A **proportion** is a statement $a/b = c/d$ that two ratios are equal.

4-3   APPLICATION: RATE-TIME PROBLEMS

The equation-solving strategy can be applied to any word problem, including quantity-rate-time problems. The formula **Quantity = Rate × Time** involves average rates over time.

4-4 FORMULAS AND LITERAL EQUATIONS

The equation-solving strategy also can be applied to formulas or literal equations to solve for one variable in terms of the others.

4-5 MISCELLANEOUS APPLICATIONS

The strategy for solving word problems is:

1. Read the problem very carefully.
2. Write down the important facts and relationships.
3. Identify the unknown quantities in terms of one variable.
4. Find the equation.
5. Solve the equation. Write down answers to all questions in the problem.
6. Check the solutions.

This strategy is applied to any word problem, regardless of the particular type of problem.

CHAPTER REVIEW EXERCISE

Work through all the problems in this chapter review and check answers in the back of the book. Answers to all the problems are there, and following each answer is a number in italics indicating the section in which that type of problem is discussed. Where weaknesses show up, review appropriate sections in the text.

A *In Problems 1–13, solve the equation.*

1. $3x + 8 = 1 - (5 - 9x)$
2. $\dfrac{x}{3} + \dfrac{x}{4} = \dfrac{1}{5}$

3. $1 - [x - (1 - x)] = x - [1 - (x - 1)]$

4. $x - \dfrac{1}{2} = \dfrac{x}{3}$

5. $0.4x + 0.3x = 6.3$
6. $-\frac{3}{5}y = \frac{2}{3}$

7. $\dfrac{x}{4} - 3 = \dfrac{x}{5}$
8. $\dfrac{x}{4} - \dfrac{x - 3}{3} = 2$

9. $\dfrac{x}{3} + \dfrac{1}{2} = \dfrac{x}{4} + \dfrac{x + 6}{12}$

10. $\dfrac{x}{5} + \dfrac{1}{2} = \dfrac{4x + 7}{10} - \dfrac{x}{5}$

11. $\dfrac{1}{5}x + \dfrac{3}{10} = \dfrac{3}{4}$

12. $0.5x - 9.9 = 0.005x$

13. $0.05n + 0.1(n - 3) = 1.35$

14. Solve $A = \dfrac{bh}{2}$ for b.   *Area of a triangle*

15. Solve $3x + 5y = 15$ for y.

Set up appropriate equations and solve. Compute any decimal answers to two decimal places.

16. Six less than two-thirds of a number is 1 more than one-fifth of the number. Find the number.

17. Ten more than half a number is 60% of the number. Find the number.

18. A population grows by 6% one year and 8.5% the next. If the population after the two years is 253,022, what was it originally?

19. A gardener wishes to add 300 pounds of sand to each 40 square feet of a garden. How large a portion of the garden can be treated with a ton of sand?

20. For a particular type of paint, 2.5 gallons are required to cover 1,175 square feet.
    (A) At what rate, in gallons per square foot, is the paint being applied?
    (B) How many gallons will be needed to cover 4,000 square feet?
    (C) How many square feet can be covered with 4 gallons of paint?

21. An office supply company sells large mailing envelopes at 11¢ each for the first 1,000 ordered and 9.5¢ apiece for each additional envelope ordered. How many envelopes can be purchased for $224?

22. A salesperson earns a commission of $360 on the first $3,000 in sales each month, and $15 for each additional $1,000 in sales. How much did the salesperson sell if the commission for the month was $1,065?

23. If you paid $210 for a stereo that was on sale for 30% off list price, what was the price before the sale?

24. If there are 2.54 centimeters in 1 inch, how many inches are in 127 centimeters?

25. If 50 milliliters of a solution contains 18 milliliters of alcohol, how many milliliters of alcohol are in 70 milliliters of the same solution?

26. If the ratio of all the squirrels in a forest to the ones that were captured, marked, and released is $\frac{55}{6}$, and there are 360 marked squirrels, how many squirrels are in the forest?

27. If 100 milliliters is equivalent to 3.4 fluid ounces, how many milliliters are in 32 fluid ounces?

28. If 1 inch is equivalent to 25.4 millimeters, how many millimeters are in $\frac{3}{4}$ inch?

29. If 1 CI (a Cayman Islands, British West Indies, dollar) is worth 1.25 U.S. dollars, what is the cost in CI of a meal for which the charge is 36 U.S. dollars?

30. What central angle will correspond to a slice of pie that amounts to $\frac{55}{100}$ of the total pie?

31. A small private university rejects 7 applicants for every 2 it accepts. If 148 students are accepted, how many applied?

32. An investor divides investments into three types, putting $16,000 in bonds, $\frac{1}{3}$ of the total in stocks, and $\frac{6}{13}$ of the total in money market funds. How much is invested?

33. The price of a winter coat is first reduced by 20% and then by another 19%. If the resulting sale price is $158.40, what was the original price?

34. A rectangle with perimeter 19 meters has width equal to 2 more than $\frac{1}{4}$ of the length. Find the dimensions.

35. A rectangle with area 60 square meters has length 7 meters more than its width. Find the width.

36. Yaks are used to remove debris from the climbing camps on Mt. Everest. To remove 33 tons of debris requires 550 yak loads. How many pounds are in each yak load?

37. A telephone solicitation company experiences 4 successful sales for every 150 calls made. How many calls must it make to make 150 sales?

B *In Problems 38–51, solve the equation.*

38. $\dfrac{2}{3m} - \dfrac{1}{4m} = \dfrac{1}{12}$

39. $\dfrac{3x}{x-5} - 8 = \dfrac{15}{x-5}$

40. $\dfrac{2}{x} + \dfrac{1}{x-2} = 1$

41. $2 - \dfrac{3}{x} = \dfrac{4}{5}$

42. $\dfrac{6}{x} - \dfrac{5}{x-1} = \dfrac{4}{x^2-x}$

43. $1 + \dfrac{2}{x} = \dfrac{3}{x^2}$

44. $x(2x-1)(2x+3) = 0$

45. $\dfrac{5}{2x+3} - 5 = \dfrac{-5x}{2x+3}$

46. $\dfrac{3}{x} - \dfrac{2}{x+1} = \dfrac{1}{2x}$

47. $\dfrac{11}{9x} - \dfrac{1}{6x^2} = \dfrac{3}{2x}$

48. $\dfrac{u-3}{2u-2} = \dfrac{1}{6} - \dfrac{1-u}{3u-3}$

49. $\dfrac{x}{x^2-6x+9} - \dfrac{1}{x^2-9} = \dfrac{1}{x+3}$

50. $\dfrac{1}{x+1} + \dfrac{2}{x} = \dfrac{11}{4x}$

51. $\dfrac{x-3}{x-2} = \dfrac{x}{x-2} + \dfrac{x+1}{2-x}$

52. Solve $S = \dfrac{n(a+L)}{2}$ for L. *Arithmetic progression*

53. Solve $P = M - Mdt$ for M. *Mathematics of finance*

Set up appropriate equations and solve. Compute any decimal answers to two decimal places.

54. If a fourth-grade class with 28 students has 4 students who are retained in the same grade the next year, what is the ratio of students promoted to students retained?

55. A bag of candy of various colors contains 40 pieces. If 12 of the pieces are red, what is the ratio of red pieces to the total number of pieces?

56. On a college campus, the ratio of female to male varsity athletes is 3:2. If the campus has 270 varsity athletes, how many are female?

57. A triangle has sides of lengths 5, 10, and 14 inches. In a similar triangle the shortest side has length 7 inches. What are the lengths of the other two sides?

58. If one car leaves a town traveling at 56 kilometers per hour, how long will it take a second car traveling at 76 kilometers per hour to catch up, if the second car leaves 1.5 hours later?

59. Suppose one printing press can print 45 brochures per minute and a newer press can print 55. How long will it take to print 3,000 brochures if the newer press is brought on the job 10 minutes after the older press has started and both continue until finished?

60. A student received grades of 65 and 80 on two tests. What grade on a third test will give the student an average of 75 for the three tests?

61. What Fahrenheit temperature corresponds to 15°C (Celsius)? Use $C = \frac{5}{9}(F - 32)$.

C *In Problems 62–65, solve the equation.*

62. $\dfrac{x - 3}{12} - \dfrac{x + 2}{9} = \dfrac{1 - x}{6} - 1$

63. $\dfrac{7}{2 - x} = \dfrac{10 - 4x}{x^2 + 3x - 10}$

64. $\dfrac{1}{x^2} + \dfrac{1}{(x + 1)^2} = \dfrac{2}{x^2 + x}$

65. $\dfrac{1}{x^2 + x} + \dfrac{1}{x^2 + 2x + 1} = \dfrac{2x + 1}{x(x + 1)^2}$

66. Solve $\dfrac{x - a}{a} - \dfrac{x - b}{b} = c$ for x.

67. Solve $y = \dfrac{4x + 3}{2x - 5}$ for x in terms of y.

68. Solve $\dfrac{1}{f} = \dfrac{1}{f_1} + \dfrac{1}{f_2}$ for f_1.   *Optics*

Set up appropriate equations and solve. Compute any decimal answers to two decimal places.

69. The length of time it takes fruit to ripen is related to altitude. The ratio between the difference in ripening time and the difference in altitude is a constant. If an increase in altitude of 500 feet produces a difference of 4 more days in the ripening time, how much longer will it take fruit to ripen at an altitude 2,500 feet higher?

70. A basketball player has been successful on 75% of free throws, making 72 of 96. How many consecutive free throws must the player make to raise the percentage to 80%?

71. The numerator of a fraction is 1 less than the denominator. If $\frac{1}{2}$ is added to the fraction, the result is 1 less than 3 times the original fraction. Find the fraction.

72. The average speed of a passenger train on a cross-country route is 1.5 times as fast as a freight train. The passenger train travels the 1,040 miles from Chicago to Denver in $6\frac{2}{3}$ hours less than it takes the freight train. What are the two speeds?

73. An express cross-country bus travels at an average speed of 66 miles per hour, while a local bus averages 48 miles per hour. If the express bus leaves 3 hours after the local, how long will it take it to catch up?

74. The simple interest formula $A = P + Prt$ gives the amount A accumulated from a principal P after t years at interest rate r. How long does it take for the amount to equal twice the original principal when $r = 6\%$?

75. Recall the Fahrenheit-Celsius conversion formula $F = \frac{9}{5}C + 32$. At what temperature are the Fahrenheit and Celsius readings the same?

CHAPTER PRACTICE TEST

The following practice test is provided for you to test your knowledge of the material in this chapter. You should try to complete it in 50 minutes or less. Answers in the back of the book indicate the section in the text where the material in the question is covered. Actual tests in your class may vary from this practice test in difficulty, length, or emphasis, depending on the goals of your course or instructor.

Solve.

1. $1 - 2x = 3(x - 4)$

2. $1 + \dfrac{1}{2}x = \dfrac{3}{4}$

3. $1 + \dfrac{1}{2x} = \dfrac{3}{4}$

4. $\dfrac{1}{x} + \dfrac{1}{2x} = \dfrac{3}{4}$

5. $\dfrac{2}{x - 1} + \dfrac{3}{x - 2} = \dfrac{3}{2}$

6. $\dfrac{2x}{x - 1} + 3 = 1 + \dfrac{2}{x - 1}$

7. $\dfrac{1}{x - 1} = \dfrac{x - 2}{(x - 3)(x - 4)}$

Solve for the indicated variable.

8. $3x + 4y = 5$ for y

9. $\dfrac{3}{x} + \dfrac{4}{y} = 5$ for x

10. $3x^4 = 5y + 6$ for y

Set up an equation and solve.

11. A copying machine can complete a particular job in 40 minutes. If a newer machine is also used on the same job, the two machines together can complete the job in 60% less time. How long would it take the newer machine alone to complete the job?

12. The ratio of all bass in a lake to those that have been caught, tagged, and released is 35:3. If 60 bass have been tagged, how many bass are in the lake?

13. A tablespoon is $\frac{1}{16}$ cup, and a pint is 2 cups. How many tablespoons are there in $3\frac{1}{2}$ pints?

14. If 50 milliliters of a solution contains 14 milliliters of acid, how much acid is there in 85 milliliters of the same solution?

15. Yaks are used to remove debris from the climbing camps on Mt. Everest. To remove 33 tons of debris requires 550 yak loads. How many yak loads are needed to remove $\frac{1}{2}$ ton of debris?

16. A car leaves a city traveling at 45 kilometers per hour. How long will it take a second car traveling at 60 kilometers per hour to catch up to the first car if it leaves 1 hour and 20 minutes later?

5

Linear Equations, Inequalities, and Graphs

In the previous chapter, we brought together five themes that recur throughout the study of algebra: translate, rewrite, solve, evaluate, use. In this chapter, we add another theme that we have considered only briefly thus far: *graphing*. In Chapter 1, we graphed the relationships for inequality statements involving only one variable on a number line. The number line is a one-dimensional figure with each point uniquely named by a single real number. In this chapter, we begin to graph relationships between two variables. To do so, we must identify points on a plane with numbers. For this purpose, we introduce a coordinate system whereby each point in a plane is uniquely named by a pair of real numbers. Graphing the points corresponding to pairs of numbers related by an equation allows us to visualize the relationship of the two variables in the equation.

To conclude the chapter, we also develop techniques for solving linear inequalities similar to those already developed for linear equations. We then consider the graphs of inequalities in both one and two variables.

Photo reference: see Exercise 5-3, Problem 91.

5-1   Graphing Linear Equations

♦ Cartesian Coordinate System
♦ Graphing a First-Degree Equation in Two Variables
♦ Vertical and Horitzontal Lines
♦ Using Different Scales for the Coordinate Axes
♦ Interpreting Graphs

The expressions and relationships we have considered thus far have, for the most part, involved only one variable. A notable exception has been the solution of literal equations. Literal equations, or formulas, describe relationships between two or more variables. Such relationships are at the heart of the mathematical analysis of many practical and theoretical problems. In this section we consider relationships between two variables graphically.

To graph relationships involving one variable, such as the inequalities we graphed in Section 1-2, we identify numbers with points on a number line, a one-dimensional figure. To graph relationships involving two variables, we must extend this identification of numbers to points on a plane, a two-dimensional figure. To do this, we introduce a coordinate system whereby each point in a plane is uniquely named by a pair of real numbers.

♦ CARTESIAN COORDINATE SYSTEM

To form a *cartesian coordinate system* we select two real number lines, one vertical and one horizontal, and let them cross through their origins (that is, at **0**), as indicated in Figure 1a. Up and to the right are the traditional choices for the positive directions. Initially, we use the same scale for each number line. Later, we will consider problems where it is useful to use different scales.

These two number lines are called the **vertical axis** and the **horizontal axis** and, together, the **coordinate axes.** The coordinate axes divide the plane into four parts called **quadrants.** The quadrants are numbered counterclockwise from I to IV, starting in the upper right quadrant. All points in the plane lie in one of the four quadrants, except the points on the coordinate axes.

We assign coordinates to a point P in the plane as indicated in Figure 1b. Pass horizontal and vertical lines through the point. The vertical line will intersect the horizontal axis at a point with coordinates a, and the horizontal line will intersect the vertical axis at a point with coordinate b. These two numbers form the **coordinates** (a, b) of the point P in the plane. In particular, the coordinates of the point Q shown in Figure 1b are $(-10, 5)$ and those of the point R are $(5, 10)$.

The first coordinate a of the coordinates of point P is also called the **abscissa** of P; the second coordinate b of the coordinates of point P is also called the **ordinate** of P. The abscissa for Q in Figure 1b is -10 and the ordinate for Q is 5. The point with coordinates $(0, 0)$ is called the **origin.**

We know that coordinates (a, b) exist for each point in the plane since every point on each axis has a real number associated with it. Hence, by the procedure described, each point located in the plane can be labeled with a unique pair of real numbers. Conversely, by reversing the process, each pair of real numbers can be associated with a unique point in the plane.

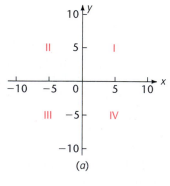

(a)

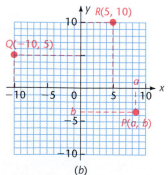

(b)

Figure 1

The system that we have just defined is called a **cartesian coordinate system,** also referred to as a **rectangular coordinate system.**

| **Example 1** Finding Coordinates of Points | Find the coordinates of each of the points A, B, C, and D. |

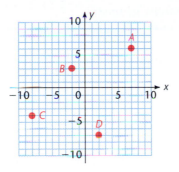

Solution $A(7, 6)$ $B(-2, 3)$ $C(-8, -4)$ $D(2, -7)$

Matched Problem 1   Find the coordinates, using the figure in Example 1, for each of the following points:

(A) 2 units to the right and 1 unit up from A
(B) 2 units to the left and 2 units down from C
(C) 1 unit up and 1 unit to the left of D
(D) 2 units to the right of B

Given the coordinates of a point, we can locate it in the plane. The process is called **plotting** the point.

Example 2
Plotting Points

Plot the point associated with each of the following coordinate pairs:

$(2, 7)$ $(7, 2)$ $(-8, 4)$ $(4, -8)$ $(-4, -8)$ $(3, 0)$ $(0, 3)$ $(-8, -4)$

Solution

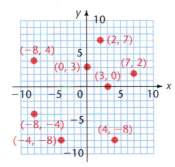

To plot $(-8, 4)$, for example, start at the origin and count 8 units to the left; then go straight up 4 units.

It is very important to note that the ordered pair $(2, 7)$ and the pair $(7, 2)$ are not the same. The order of the coordinates is important.

Matched Problem 2   Plot the point associated with each of the following coordinate pairs: $(3, 4)$, $(-3, 2)$, $(-2, -2)$, $(4, -2)$, $(0, 1)$, $(-4, 0)$.

The development of the cartesian coordinate system represented a very important advance in mathematics. It was through the use of this system that René Descartes (1596–1650), a French philosopher–mathematician, was able to transform geometric problems requiring long, tedious reasoning into algebraic problems that could be solved almost mechanically. This joining of algebra and geometry is now known as **analytic geometry.**

Two fundamental problems of analytic geometry are the following:

1. Given an equation, find its graph.
2. Given a geometric figure, such as a straight line or circle, find an equation that has this figure as its graph.

In this section, we will be interested mainly in the first problem. In the next section, we will consider the second problem for straight lines. Before we take up the first problem, however, let us recall what is meant by the solution of an equation.

A **solution of an equation in two variables** is an ordered pair of real numbers that satisfies the equation. We usually agree that the first element in the ordered pair will replace x and the second element will replace y. For example, if $y = 2x - 3$, then $(4, 5)$ is a solution of the equation since

$$5 = 2 \cdot 4 - 3$$

This allows us to define the graph of an equation as follows: the **graph of an equation** is the set of all the solutions of the equation plotted as points in a cartesian coordinate system.

The Graph of an Equation

The graph of an equation in two variables in a rectangular coordinate system meets the following two conditions:

1. If an ordered pair of numbers is a solution to the equation, the corresponding point must be on the graph of the equation.
2. If a point is on the graph of an equation, its coordinates must satisfy the equation.

◆ GRAPHING A FIRST-DEGREE EQUATION IN TWO VARIABLES

Suppose we are interested in graphing

$$y = 2x - 4$$

We start by finding some of its solutions. To do so, we assign any convenient value to x in $y = 2x - 4$, and then solve for y. For example, if $x = 3$, then

$$y = 2(3) - 4 = 2$$

Hence, $(3, 2)$

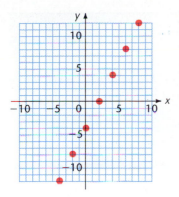

Figure 2

is a solution of $y = 2x - 4$. By proceeding in this manner we can get as many solutions to this equation as we desire. Thus, the solution set is infinite. Table 1 lists some solutions, and we have plotted these solutions in a cartesian coordinate system in Figure 2. The horizontal axis is identified with x and the vertical axis with y.

Table 1

Choose x	Compute $2x - 4 = y$	Write Ordered Pair (x, y)
-4	$2(-4) - 4 = -12$	$(-4, -12)$
-2	$2(-2) - 4 = -8$	$(-2, -8)$
0	$2(0) - 4 = -4$	$(0, -4)$
2	$2(2) - 4 = 0$	$(2, 0)$
3	$3(2) - 4 = 2$	$(3, 2)$
4	$2(4) - 4 = 4$	$(4, 4)$
6	$2(6) - 4 = 8$	$(6, 8)$
8	$2(8) - 4 = 12$	$(8, 12)$

It appears in Figure 2 that the graph of the equation is a straight line. If we knew this for a fact, then graphing $y = 2x - 4$ would be easy. We would simply find two solutions of the equation, plot them, and then graph as much of $y = 2x - 4$ as we like by drawing a straight line through the two points. In fact, the graph of $y = 2x - 4$ is a straight line. More generally, we have the following result, which we state without proof:

Graphs of First-Degree Equations in Two Variables

The graph of any equation of the form

$$y = mx + b \qquad \text{or} \qquad Ax + By = C$$

where m, b, A, B, and C are constants, with A and B not both 0, and where x and y are variables, is a straight line.

For this reason, equations that can be rewritten in either of these forms are called **linear equations in x and y.**

For example, the graphs of

$$y = \tfrac{2}{3}x - 5 \qquad \text{and} \qquad 2x - 3y = 12$$

are straight lines, since the first is of the form $y = mx + b$ and the second is of the form $Ax + By = C$.

Graphing Linear Equations

To graph $y = mx + b$ or $Ax + By = C$:

Step 1. Find two solutions of the equation. A third solution is sometimes useful as a checkpoint.

Step 2. Plot the solutions in a coordinate system.

Step 3. Draw a straight line through the points plotted in step 2.

The third solution provides a checkpoint, since if the line does not pass through all three points, a mistake has been made in finding the solutions.

Example 3
Graphing Linear Equations

(A) Graph $y = 2x - 4$.　　**(B)** Graph $x + 3y = 6$.

Solution　**(A)** Make up a table of at least two solutions, plot them, and then draw a straight line through these points. We can use values from Table 1 since the equation is the same one considered there.

x	$2x - 4 = y$	(x, y)
0	$2(0) - 4 = -4$	$(0, -4)$
2	$2(2) - 4 = 0$	$(2, 0)$
4	$2(4) - 4 = 4$	$(4, 4)$

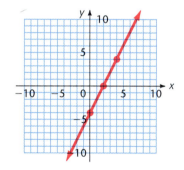

(B) To graph $x + 3y = 6$, assign to either x or y any convenient value and solve for the other variable. If we let $x = 0$, a convenient value, then

$$0 + 3y = 6$$
$$3y = 6$$
$$y = 2$$

Thus, $(0, 2)$ is a solution.
If we let $y = 0$, another convenient choice, then

$$x + 3(0) = 6$$
$$x + 0 = 6$$
$$x = 6$$

Thus, $(6, 0)$ is a solution.

To find a checkpoint, choose another value for x or y, say $x = -6$. Then

$$-6 + 3y = 6$$
$$3y = 12$$
$$y = 4$$

Thus, $(-6, 4)$ is also a solution.

We summarize these results in a table and then draw the graph.

(x, y)	
$(0, 2)$	y intercept
$(6, 0)$	x intercept
$(-6, 4)$	Checkpoint

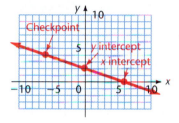

The first two solutions found in Example 3(B) indicate where the graph crosses the coordinate axes and are called the y and x intercepts, respectively. That is, the **y intercept** is the y value where the graph crosses the y axis; the **x intercept** is the x value where the graph crosses the x axis. The intercepts are often the easiest points to find. To find the y intercept we let $x = 0$ and solve for y; to find the x intercept we let $y = 0$ and solve for x. This is called the **intercept method** of graphing a straight line.

Remember that if a straight line does not pass through all three points plotted, then we know we have made a mistake and must go back and check our work.

Matched Problem 3   Graph: **(A)** $y = 2x - 6$ **(B)** $3x + y = 6$

♦ **VERTICAL AND HORIZONTAL LINES**

Linear equations in two variables where one of the variables is absent result in vertical or horizontal lines, as we see in the next example.

Example 4
Equations of Vertical and Horizontal Lines

Graph the equations $y = 4$ and $x = 3$ in a rectangular coordinate system.

Solution   To graph $y = 4$ or $x = 3$ in a rectangular coordinate system, each equation must be provided with the missing variable. This can be done mentally as follows:

$y = 4$	is equivalent to	$0x + y = 4$
$x = 3$	is equivalent to	$x + 0y = 3$

In the first case, we see that no matter what value is assigned to x, $0x = 0$. Thus, as long as $y = 4$, x can assume any value:

$$
\begin{array}{c|c}
x & y \\
\hline
5 & 4 \\
0 & 4 \\
-3 & 4
\end{array}
$$

We conclude that the graph of $y = 4$ is a horizontal line crossing the y axis at 4. Similarly, in the second case, y can assume any value as long as $x = 3$:

$$
\begin{array}{c|c}
x & y \\
\hline
3 & -2 \\
3 & 0 \\
3 & 6
\end{array}
$$

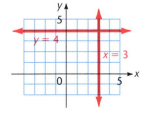

Figure 3

Thus, the graph of $x = 3$ is a vertical line crossing the x axis at 3. The graphs of $y = 4$ and $x = 3$ are shown in Figure 3:

Matched Problem 4   Graph $y = -3$ and $x = -4$ in a rectangular coordinate system.

We can state the following more general result about the graphs of linear equations in two variables with only one variable present:

Horizontal and Vertical Lines

1. The graph of the equation $ax = b$ in a rectangular coordinate system is the vertical line through the point

$$\left(\frac{b}{a}, 0 \right)$$

2. The graph of the equation $cy = d$ in a rectangular coordinate system is the horizontal line through the point

$$\left(0, \frac{d}{c} \right)$$

♦ **USING DIFFERENT SCALES FOR THE COORDINATE AXES**

Equations arising from applications often have restrictions put on the variables. They also may involve significant differences in the magnitude of the variables. To graph such equations we may want to use different scales for the two axes.

Example 5
Using Different Scales

Graph $y = 50 + 5x$ for $0 \le x \le 10$.

Solution We first note that x is restricted to values from 0 to 10. Let us find y for three values in this interval. We choose each end value and the middle value:

x	y
0	50
5	75
10	100

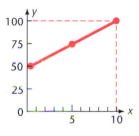

Figure 4

We see that as x varies from 0 to 10, y varies from 50 to 100. To keep the graph on the paper, we choose a different scale on the vertical axis representing y than that used on the horizontal axis representing x. Thus, we obtain Figure 4. The graph is a line segment joining the two points (0, 50) and (10, 100). The dashed lines are intended to guide one's eyes to the endpoint and are not part of the graph.

Matched Problem 5 Graph $y = 100 - 6x$ for $2 \le x \le 15$.

♦ ## INTERPRETING GRAPHS

Graphs convey information, often in a way that tables of values cannot. For example, the depreciation of a \$175,000 piece of equipment over 20 years, at a rate of \$8,000 per year, can be graphed as shown in Figure 5. The yearly values might be displayed as in Table 2. The graph and the table both reflect *straight-line depreciation,* that is, depreciation at a constant rate.

Table 2

Year	Value ($)
0	175,000
1	167,000
2	159,000
3	151,000
...	
18	31,000
19	23,000
20	15,000

Figure 5

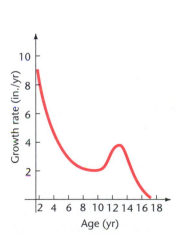

Figure 6

Of course, not all graphs are linear. For example, the graph in Figure 6, displaying the rates at which female children grow, on the average, from birth through their late teens, is not a straight line.

Visually verify the following from the graph in Figure 6: infants grow quite rapidly when newborn, but the rate of growth decreases as the child gets older; at about age 12 or 13, there is a growth spurt where the rate of growth increases rapidly; after this spurt, the growth rate decreases again, approaching 0 by about age 18.

We will continue to look at graphical representations as we consider additional relationships among two variables, and more nonlinear graphs will be considered in later chapters. For the rest of this chapter, however, we will restrict ourselves to linear equations and inequalities.

Answers to Matched Problems

1. (A) $(9, 7)$ **(B)** $(-10, -6)$ **(C)** $(1, -6)$ **(D)** $(0, 3)$

2.

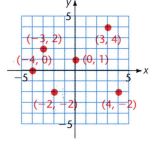

3. (A)

(x, y)
$(0, -6)$
$(3, 0)$
$(2, -2)$

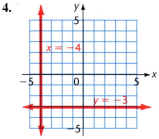

(B)

(x, y)
$(0, 6)$
$(2, 0)$
$(1, 3)$

4.

5.

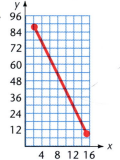

EXERCISE 5-1

A *Problems 1 and 2 refer to the figure below. Write down the coordinates of each point.*

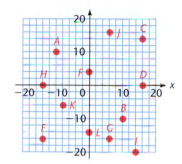

1. A, B, C, D, E, F   **2.** G, H, I, J, K, L

Problems 3 and 4 refer to the figure below. Write down the coordinates of each point.

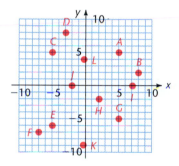

3. A, B, C, D, E, F   **4.** G, H, I, J, K, L

Plot each set of ordered pairs of numbers on the same coordinate system.

5. (4, 4), (−4, 1), (−3, −3), (5, −1), (0, 2), (−2, 0)

6. (3, 1), (−2, 3), (−5, −1), (2, −1), (4, 0), (0, −5)

7. (2, 7), (7, 2), (−6, 3), (−4, −7), (2, 3), (0, −8), (9, 0)

8. (−9, 8), (8, −9), (0, 5), (4, −8), (−3, 0), (7, 7), (−6, −6)

Problems 9 and 10 refer to the figure below. Write down the coordinates of each labeled point to the nearest half unit.

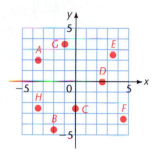

9. A, B, C, D   **10.** E, F, G, H

Problems 11 and 12 refer to the figure below. Write down the coordinates of each labeled point to the nearest half unit.

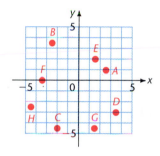

11. A, B, C, D   **12.** E, F, G, H

Plot each set of ordered pairs of numbers on the same coordinate system.

13. $A(3.5, 2.5)$, $B(−4.5, 3)$, $C(0, −3.5)$

14. $A(1.5, 3.5)$, $B(−3.5, 0)$, $C(3, −2.5)$

15. $A(−2.5, −3.5)$, $B(4.5, −1.5)$, $C(−3.5, 0)$

16. $A(−4.5, 1.5)$, $B(−2.5, −4.5)$, $C(1, −1.5)$

Find the x and y intercepts for each equation.

17. $y = 3x − 7$ **18.** $y = −2x + 6$

19. $y = −5x + 2$ **20.** $y = 7x + 4$

21. $2x − 3y = 12$ **22.** $4x − y = 10$

23. $3x + 4y = −20$ **24.** $2x + 3y = −6$

Graph in a rectangular coordinate system.

25. $y = 2x$

26. $y = x$

27. $y = 2x - 2$

28. $y = x - 1$

29. $x + y = -4$

30. $x + y = 6$

31. $x - y = 3$

32. $x - y = 5$

33. $3x + 4y = 12$

34. $2x + 3y = 12$

35. $8x - 3y = 24$

36. $3x - 5y = 15$

37. $y = 3$

38. $x = 2$

39. $x = -4$

40. $y = -3$

41. $y = -x + 2$

42. $y = -2x + 6$

43. $3x + 2y = 10$

44. $2x + y = 7$

45. $5x - 6y = 15$

46. $7x - 4y = 21$

47. $y = 0$

48. $x = 0$

B 49. $y = \frac{1}{3}x + 2$

50. $y = \frac{1}{2}x + 1$

51. $y - \frac{1}{2}x = 0$

52. $y - \frac{1}{4}x = 0$

53. $y - \frac{1}{2}x = -1$

54. $y + \frac{1}{2}x = 1$

55. $\frac{1}{2}x + \frac{1}{3}y = 2$

56. $\frac{1}{3}x + \frac{1}{5}y = 1$

57. $\frac{1}{4}x - \frac{1}{3}y = 1$

58. $\frac{1}{2}x - \frac{1}{4}y = 1$

59. $\frac{1}{2}x + \frac{1}{3}y - \frac{1}{6} = 0$

60. $\frac{1}{2}x - \frac{1}{4}y + \frac{3}{8} = 0$

Write in the form $y = mx + b$ and graph.

61. $3x + 4y = 6$

62. $5x - 2y = 10$

63. $x + 6 = 3x + 2 - y$

64. $y - x - 2 = x + 1$

Write in the form $Ax + By = C$, $A > 0$, and graph.

65. $y = 3x - 4$

66. $y = -2x - 5$

67. $y + 8 = 2 - x - y$

68. $6x - 3 + y = 2y + 4x + 5$

Graph each of the following using a different scale on the vertical axis to keep the size of the graph within reason.

69. $I = 6t,\ 0 \le t \le 10$

70. $d = 60t,\ 0 \le t \le 10$

71. $v = 10 + 32t,\ 0 \le t \le 5$

72. $A = 100 + 10t,\ 0 \le t \le 10$

73. $y = -80x + 250,\ 0 \le x \le 3$

74. $y = -150x + 900,\ 0 \le x \le 6$

75. $y = -125x + 1{,}000,\ 0 \le x \le 8$

76. $y = -60x + 200,\ 0 \le x \le 3$

C *Graph both equations on the same coordinate system. Determine by inspection the coordinates of the point where the two graphs meet, and show that the coordinates of the intersection point satisfy both equations. In these problems, the coordinates of the intersection points will be integers.*

77. $x + y = 3,\ 2x - y = 0$

78. $2x - 3y = 8,\ x + 2y = 11$

79. $x - y = 5,\ 2x - 5y = 4$

80. $x + y = 8,\ x - y = -10$

81. $y = 3x + 2,\ y = -x + 10$

82. $y = x - 4,\ y = -x$

83. $y = x,\ x + y = 6$

84. $y = 3x - 1,\ y = -x - 5$

85. Graph $y = mx - 2$ for $m = 2$, $m = \frac{1}{2}$, $m = 0$, $m = -\frac{1}{2}$, and $m = -2$, all on the same coordinate system.

86. Graph $y = -\frac{1}{2}x + b$ for $b = -6$, $b = 0$, and $b = 6$, all on the same coordinate system.

87. Graph $y = |x|$. [*Hint:* Graph $y = x$ for $x \ge 0$, and graph $y = -x$ for $x < 0$.]

88. Graph $y = |2x|$ and $y = |\frac{1}{2}x|$ on the same coordinate system (see Problem 87.)

Many equations can be graphed by plotting appropriate points and joining these points with a smooth curve. In Problems 89–92, plot the points corresponding to the given values of x and sketch the graph of the equation.

89. $y = x^2$, $x = -3, -2, -1, -0.5, 0, 0.5, 1, 2, 3$

90. $y = -x^2$, $x = -3, -2, -1, -0.5, 0, 0.5, 1, 2, 3$

91. $y = -\sqrt{x}$, $x = 0, 0.25, 1, 2, 4, 6, 9$

92. $y = \sqrt{x}$, $x = 0, 0.25, 1, 2, 4, 6, 9$

APPLICATIONS

93. *Finance* The amount of money in an account that starts with $1,000 and earns simple interest at a rate of 6% for t years is given by $A = 1{,}000 + 60t$. Graph this equation for $0 \le t \le 10$.

94. *Finance* The depreciated value of a piece of equipment that is originally worth $2,200 and is depreciated uniformly by $450 per year for t years is given by $V = 2{,}200 - 450t$. Graph this equation for $0 \le t \le 4$.

95. *Business* The demand for a particular product when it is priced at x dollars per unit is given by the equation $d = 400 - 0.5x$. Graph this equation for $1 \leq x \leq 50$.

96. *Business* The cost of manufacturing x units of a particular product is given by $C = 1,200 + 10x$. Graph this equation for $0 \leq x \leq 1,000$.

97. *Psychology* In 1948 Professor Brown, a psychologist, trained a group of rats (in an experiment on motivation) to run down a narrow passage in a cage to receive food in a box. A harness was put on each rat and the harness was then connected to an overhead wire that was attached to a scale. In this way the rat could be placed at different distances (measured in centimeters) from the food and Professor Brown could then measure the pull (in grams) of the rat toward the food. It was found that the relationship between motivation (pull) and position was given approximately by the equation $p = -\frac{1}{5}d + 70$, $30 \leq d \leq 175$. Graph this equation for the indicated values of d.

98. *Electronics* In a simple electric circuit, such as found in a flashlight, if the resistance is 30 ohms, the current in the circuit I (in amperes) and the electromotive force E (in volts) are related by the equation $E = 30I$. Graph this equation for $0 \leq I \leq 1$.

99. *Biology* In biology there is an approximate rule, called the *bioclimatic rule*, for temperate climates. This rule states that in spring and early summer, periodic phenomena such as blossoming for a given species, appearance of certain insects, and ripening of fruit usually come about 4 days later for each 500 feet of altitude. Stated as a formula,

$$d = 4\left(\frac{h}{500}\right)$$

where d = Change in days and h = Change in altitude in feet. Graph the equation for $0 \leq h \leq 4,000$.

100. *Physics* A small steel ball thrown downward with an initial velocity of 15 meters per second has a velocity after t seconds given by $v = 15 + 9.75t$. Graph this equation for $0 \leq t \leq 8$.

5-2 Slope and Equations of a Line

- ♦ Slope of a Line
- ♦ Slope-Intercept Form
- ♦ Point-Slope Form
- ♦ Vertical and Horizontal Lines
- ♦ Parallel and Perpendicular Lines

In the preceding section we considered this problem: given a linear equation of the form

$$Ax + By = C \qquad \text{or} \qquad y = mx + b$$

find its graph. Now we will consider the reverse problem: given certain information about a straight line in a rectangular coordinate system, find its equation. We start by introducing a measure of the steepness of a line called *slope*.

♦ SLOPE OF A LINE

If we take two points $P_1(x_1, y_1)$ and $P_2(x_2, y_2)$ on a line, then the ratio of the change in y to the change in x as we move from point P_1 to P_2 is called the **slope** of the line.

Slope Formula

If a line passes through $P_1(x_1, y_1)$ and $P_2(x_2, y_2)$, then its slope is given by the formula

$$m = \frac{y_2 - y_1}{x_2 - x_1} \qquad x_1 \neq x_2$$

$$= \frac{\text{Vertical change (rise)}}{\text{Horizontal change (run)}}$$

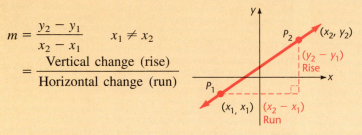

The slope is a measure of the direction and steepness of a line. Table 1 summarizes how the slope indicates direction.

Table 1   Going from Left to Right

Line	Slope	Example
Horizontal	0	
Vertical	Not defined	
Rising	Positive	
Falling	Negative	

For a horizontal line, y does not change as x changes; hence, its slope is 0. On the other hand, for a vertical line, x does not change as y changes; hence, $x_1 = x_2$ and the slope is not defined:

$$m = \frac{y_2 - y_1}{x_2 - x_1} = \frac{y_2 - y_1}{0} \qquad \text{Vertical-line slope is not defined.}$$

Suppose the slope is a positive number m. If x increases by 1 unit, then

$$m = \frac{\text{Change in } y}{\text{Change in } x} = \frac{\text{Change in } y}{1} = \text{Change in } y$$

That is, y must increase by m units, so the line rises as we move from left to right. On the other hand, suppose the slope is a negative number m. If x is increased by 1 unit, then the change in y is still m, but this represents a decrease since m is negative. Thus, the line is falling as we move from left to right.

Example 1
Finding the Slope

Find the slope of the line passing through $(-3, -2)$ and $(3, 4)$. Graph the line.

Solution Let $(x_1, y_1) = (-3, -2)$ and $(x_2, y_2) = (3, 4)$; then

$$m = \frac{y_2 - y_1}{x_2 - x_1} = \frac{4 - (-2)}{3 - (-3)} = \frac{6}{6} = 1$$

It does not matter which point we call P_1 or P_2 as long as we stick to the choice once it is made. If we reverse the choice above, we obtain the same value for the slope, since the sign of the numerator and the denominator both change:

$$m = \frac{(-2) - 4}{(-3) - 3} = \frac{-6}{-6} = 1$$

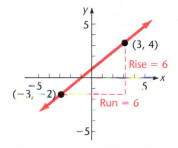

Note that the positive slope here indicates the line is rising, as shown in the graph of the line in Figure 1.

Figure 1

Matched Problem 1

Find the slope of the line passing through $(-2, 7)$ and $(3, -3)$. Graph the line.

♦ SLOPE-INTERCEPT FORM

Any equation of the form $Ax + By = C$, $B \neq 0$, always can be rewritten in the form

$$y = mx + b$$

where m and b are constants. For example, starting with

$$2x + 3y = 6 \qquad Ax + By = C.$$

we solve for y to obtain

$$3y = -2x + 6 \qquad By = -Ax + C$$

$$\boxed{\frac{1}{3}(3y) = \frac{1}{3}(-2x + 6)} \qquad \frac{1}{B}By = \frac{1}{B}(-Ax) + \frac{1}{B}C$$

$$\text{or } y = -\frac{A}{B}x + \frac{C}{B}$$

$$y = -\frac{2}{3}x + 2 \qquad \text{Form } y = mx + b.$$

The constants m and b in $y = mx + b$ have special geometric meaning. For $y = -\frac{2}{3}x + 2$, two points on the line are $(0, 2)$ and $(1, \frac{4}{3})$, since for $x = 0$, $y = 2$, and for $x = 1$, $y = \frac{4}{3}$. The point $(0, 2)$ is the y intercept since this point is on the y axis. The slope can be computed using these two points:

$$\frac{\frac{4}{3} - 2}{1 - 0} = \frac{-\frac{2}{3}}{1} = -\frac{2}{3}$$

Thus, in this case, $m = -\frac{2}{3}$ is the slope and $b = 2$ is the y intercept. In general, we have:

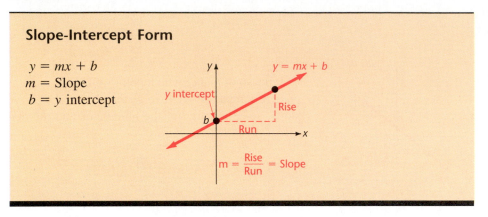

Slope-Intercept Form

$y = mx + b$
$m = $ Slope
$b = y$ intercept

To see why this result is true, consider $y = mx + b$ and let $x = 0$. Then

$$y = m \cdot 0 + b$$
$$= 0 + b = b$$

Thus, b is the intercept, the y coordinate of the point where the graph crosses the y axis.

Next, choose two points (x_1, y_1) and (x_2, y_2) on the graph of $y = mx + b$ (Figure 2). Since the two points are on the graph, they are solutions to the equation. Thus,

$$y_1 = mx_1 + b \qquad \text{and} \qquad y_2 = mx_2 + b$$

Substituting these values into the slope formula, the slope of the line is given by

$$m = \frac{y_2 - y_1}{x_2 - x_1} = \frac{(mx_2 + b) - (mx_1 + b)}{x_2 - x_1} = \frac{mx_2 - mx_1}{x_2 - x_1}$$

$$= \frac{m(x_2 - x_1)}{x_2 - x_1} = m$$

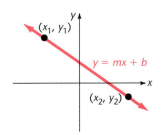

Figure 2

Thus, m is the slope of the graph of $y = mx + b$.

In summary, if an equation of a line is written in the form $y = mx + b$, then b is the y intercept and m is the slope. Conversely, if we know the slope m and y intercept b of a line, we can write its equation in the form $y = mx + b$.

Example 2
Using the Slope-Intercept Form

(A) Find the slope and y intercept of the line $y = \frac{1}{3}x + 2$.
(B) Find the equation of a line with slope -2 and y intercept 3.

Solution

Slope — y intercept

(A) $y = \frac{1}{3}x + 2$
(B) Since $m = -2$ and $b = 3$, then $y = mx + b = -2x + 3$ is the equation.

Matched Problem 2 (A) Find the slope and y intercept of the line $y = \dfrac{x}{2} - 7$.

(B) Find the equation of a line with slope $-\frac{1}{3}$ and y intercept 6.

The slope-intercept form allows us to graph a linear equation very efficiently.

Example 3
Graphing a Line in Slope-Intercept Form

Graph $y = -\frac{1}{2}x + 3$.

Solution   The y intercept is 3, so the point $(0, 3)$ is on the graph. The slope of the line is $-\frac{1}{2}$, so if the x coordinate is increased by 2 units, the y coordinate changes by -1. The resulting point is easily graphed, and the two points yield the graph of the line.

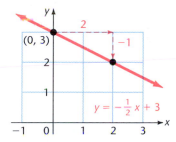

Matched Problem 3 Graph $y = \frac{2}{3}x - 2$.

♦ **POINT-SLOPE FORM**

In Example 2(B) we found the equation of a line given its slope and y intercept. Often it is necessary to find the equation of a line given its slope and the coordinates of a different point through which it passes, or to find the equation of a line given the coordinates of two points through which it passes.

Let a line have slope m and pass through the fixed point (x_1, y_1). If the variable point, (x, y) is to be a point on the line, the slope of the line passing through (x, y) and (x_1, y_1) must be m (see Figure 3).

Thus, the equation

$$\frac{y - y_1}{x - x_1} = m$$

restricts the variable point (x, y) so that only those points in the plane lying on the line will have coordinates that satisfy the equation, and vice versa. This equation is usually written in the following form, referred to as the **point-slope form of the equation of a line:**

Variable point
(x, y)

Fixed point
(x_1, y_1)

$\dfrac{y - y_1}{x - x_1} = m$

Figure 3

Point-Slope Form

The equation of a line passing through (x_1, y_1) with slope m is given by

$$y - y_1 = m(x - x_1)$$

Using this equation in conjunction with the slope formula, we also can find the equation of a line knowing only the coordinates of two points through which it passes.

Example 4
Using the Point-Slope Form

(A) Find an equation of a line with slope $-\frac{1}{3}$ that passes through $(6, -3)$. Write the resulting equation in the form $y = mx + b$.

(B) Find an equation of a line that passes through the two points $(-2, -6)$ and $(2, 2)$. Write the resulting equation in the form $y = mx + b$.

Solution   **(A)**
$$\begin{aligned}
y - y_1 &= m(x - x_1) \qquad &\text{Substitute } y_1 = -3, \, m = -\tfrac{1}{3}, \, x_1 = 6. \\
y - (-3) &= -\tfrac{1}{3}(x - 6) \qquad &\text{Solve for } y \text{ in terms of } x. \\
y + 3 &= -\tfrac{1}{3}(x - 6) \\
y + 3 &= -\tfrac{1}{3}x + 2 \\
y &= -\tfrac{1}{3}x - 1 \qquad &\text{Form } y = mx + b.
\end{aligned}$$

(B) First find the slope of the line using the slope formula

$$m = \frac{y_2 - y_1}{x_2 - x_1} = \frac{2 - (-6)}{2 - (-2)} = 2$$

Now proceed as in part (A), using the coordinates of either point for (x_1, y_1).

Using $(x_1, y_1) = (-2, -6)$:

$$\begin{aligned}
y - y_1 &= m(x - x_1) \\
y - (-6) &= 2[x - (-2)] \\
y + 6 &= 2(x + 2) \\
y + 6 &= 2x + 4 \\
y &= 2x - 2
\end{aligned}$$

Using $(x_1, y_1) = (2, 2)$:

$$\begin{aligned}
y - y_1 &= m(x - x_1) \\
y - 2 &= 2(x - 2) \\
y - 2 &= 2x - 4 \\
y &= 2x - 2
\end{aligned}$$

Either point yields the same equation. Choosing a point with only positive coordinates, if possible, will cut down on sign errors.

Matched Problem 4   **(A)** Find an equation of a line with slope $\frac{2}{3}$ that passes through $(-3, 4)$. Write the resulting equation in the form $y = mx + b$.

(B) Find an equation of a line that passes through the two points $(6, -1)$ and $(-2, 3)$. Write the resulting equation in the form $y = mx + b$.

The slope-intercept form also can be used to find the equation of the line given the slope and one point. In Example 4(A), for instance, given the slope $-\frac{1}{3}$, we know the equation has the form

$$y = -\tfrac{1}{3}x + b$$

Since $(6, -3)$ must satisfy the equation, we can solve for b:

$$-3 = -\tfrac{1}{3}(6) + b$$
$$-3 = -2 + b$$
$$-1 = b$$

Thus, the equation of the line is $y = -\tfrac{1}{3}x - 1$.

♦ ### VERTICAL AND HORIZONTAL LINES

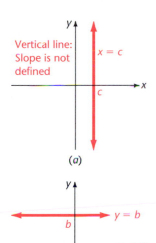

Vertical line:
Slope is not
defined

$x = c$

c

(a)

$y = b$

b

Horizontal
line:
Slope = 0

(b)

Figure 4

Recall from Section 5-1 that linear equations with only one of the variables x or y present represent a vertical or horizontal line. If a line is vertical, its slope is not defined. Since points on a vertical line have constant x coordinates and arbitrary y coordinates, the equation of a vertical line is of the form

$$x = c$$

where c is the x coordinate of each point on the line. See Figure 4a. Similarly, if a line is horizontal, the slope is 0. Then every point on the line has constant y coordinates and arbitrary x coordinates. Thus, the equation of a horizontal line is of the form

$$y = b$$

where b is the y coordinate of each point on the line. See Figure 4b. Also, since a horizontal line has slope 0, then, using the slope-intercept form, we obtain

$$y = mx + b$$
$$= 0x + b$$
$$= b$$

Example 5
Equations for Horizontal and Vertical Lines

What are the equations of vertical and horizontal lines through $(-2, -4)$?

Solution The equation of a vertical line through $(-2, -4)$ is $x = -2$, and the equation of a horizontal line through the same point is $y = -4$ (see the figure).

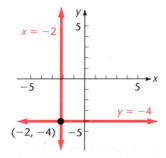

$x = -2$

$y = -4$

$(-2, -4)$

Matched Problem 5   What are the equations of vertical and horizontal lines through $(3, -8)$?

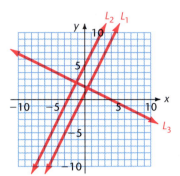

Figure 5

◆ **PARALLEL AND PERPENDICULAR LINES**

Consider the three lines L_1, L_2, and L_3 with equations

$$L_1: \quad y = 2x + 1 \qquad \text{Slope} = 2$$
$$L_2: \quad y = 2x + 5 \qquad \text{Slope} = 2$$
$$L_3: \quad y = -\tfrac{1}{2}x + 2 \qquad \text{Slope} = -\tfrac{1}{2}$$

The lines are graphed in Figure 5.

It appears that lines L_1 and L_2 are *parallel* and that line L_3 is *perpendicular* to both of them. Notice that lines L_1 and L_2 have the same slope, namely 2. This should be expected of parallel lines since they "are in the same direction." Notice also that the slope of line L_3, namely $-\tfrac{1}{2}$, is the negative reciprocal of the slope 2. This illustrates the following general result:

Parallel and Perpendicular Lines

Given nonvertical lines L_1 and L_2 with slopes m_1 and m_2, respectively, then

$$L_1 \| L_2 \qquad \text{if and only if} \qquad m_1 = m_2$$

$$L_1 \perp L_2 \qquad \text{if and only if} \qquad m_1 m_2 = -1, \text{ or } m_2 = -\frac{1}{m_1}$$

Here, $\|$ means "is parallel to" and $\perp$ means "is perpendicular to."

For example, the lines determined by

$$y = \tfrac{2}{3}x - 5 \qquad \text{and} \qquad y = \tfrac{2}{3}x + 8$$

are parallel, since both have the same slope, $\tfrac{2}{3}$. Confirm this visually by graphing both lines. Also, the lines that are determined by

$$y = \frac{x}{3} + 5 \qquad \text{and} \qquad y = -3x - 7$$

are perpendicular, since the product of their slopes is -1; that is, $(\tfrac{1}{3})(-3) = -1$. You should also confirm this visually by graphing both lines.

Example 6
Equations of Parallel and Perpendicular Lines

Given the line $x - 2y = 4$, find the equation of a line that passes through $(2, -3)$ and is:

(A) Parallel to the given line   **(B)** Perpendicular to the given line

Write final equations in the form $y = mx + b$.

Solution First find the slope of the given line by writing $x - 2y = 4$ in the form $y = mx + b$:

$$x - 2y = 4$$

$$-2y = 4 - x$$

$$\boxed{-\tfrac{1}{2}(-2y) = -\tfrac{1}{2}(4 - x)}$$

$$y = -2 + \tfrac{1}{2}x$$

$$y = \tfrac{1}{2}x - 2 \quad \text{↙ Slope}$$

The slope of the given line is $\tfrac{1}{2}$.

(A) The slope of a line parallel to the given line is also $\tfrac{1}{2}$. We have only to find the equation of a line through $(2, -3)$ with slope $\tfrac{1}{2}$ to solve part (A):

$$y - y_1 = m(x - x_1) \qquad m = \tfrac{1}{2} \text{ and } (x_1, y_1) = (2, -3)$$

$$y - (-3) = \tfrac{1}{2}(x - 2)$$

$$y + 3 = \tfrac{1}{2}x - 1$$

$$y = \tfrac{1}{2}x - 4$$

(B) The slope of the line perpendicular to the given line is the negative reciprocal of $\tfrac{1}{2}$, that is, -2. We have only to find the equation of a line through $(2, -3)$ with slope -2 to solve part (B):

$$y - y_1 = m(x - x_1) \qquad m = -2 \text{ and } (x_1, y_1) = (2, -3)$$

$$y - (-3) = -2(x - 2)$$

$$y + 3 = -2x + 4$$

$$y = -2x + 1$$

Matched Problem 6 Given the line $2x = 6 - 3y$, find the equation of a line that passes through $(-3, 9)$ and is:

(A) Parallel to the given line   **(B)** Perpendicular to the given line

Write final equations in the form $y = mx + b$.

Answers to Matched Problems

1. $m = -2$

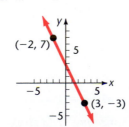

2. (A) $m = 1/2$, $b = -7$ **(B)** $y = -\frac{1}{3}x + 6$

3.

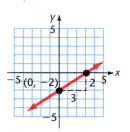

4. (A) $y - 4 = \frac{2}{3}(x + 3)$, or $y = \frac{2}{3}x + 6$ **(B)** $y = -\frac{1}{2}x + 2$

5. $x = 3$, $y = -8$ **6. (A)** $y = -\frac{2}{3}x + 7$ **(B)** $y = \frac{3}{2}x + \frac{27}{2}$

EXERCISE 5-2

A *Find the slope and y intercept, and graph each equation.*

1. $y = 2x - 3$ **2.** $y = x + 1$

3. $y = -x + 2$ **4.** $y = -2x + 1$

Write an equation of the line with slope and y intercept as indicated.

5. Slope $= 5$ **6.** Slope $= 3$
  y intercept $= -2$ y intercept $= -5$

7. Slope $= -2$ **8.** Slope $= -1$
  y intercept $= 4$ y intercept $= 2$

Write an equation of the line that passes through the given point with the indicated slope.

9. $m = 2$; $(5, 4)$ **10.** $m = 3$; $(2, 5)$

11. $m = -2$; $(2, 1)$ **12.** $m = -3$; $(1, 3)$

13. $m = 1$; $(-1, -1)$ **14.** $m = 3$; $(-1, -2)$

15. $m = 2$; $(-2, -1)$ **16.** $m = 4$; $(-1, -1)$

Find the slope of the line that passes through the given points.

17. $(3, 2)$ and $(5, 6)$ **18.** $(1, 3)$ and $(2, 4)$

19. $(2, 1)$ and $(10, 5)$ **20.** $(1, 3)$ and $(7, 5)$

21. $(3, 6)$ and $(5, 2)$ **22.** $(1, 4)$ and $(2, 3)$

23. $(2, 5)$ and $(10, 1)$ **24.** $(1, 5)$ and $(7, 3)$

Write an equation of the line through each indicated pair of points.

25. $(3, 2)$ and $(5, 6)$ **26.** $(1, 3)$ and $(2, 4)$

27. $(2, 1)$ and $(10, 5)$ **28.** $(1, 3)$ and $(7, 5)$

29. $(3, 6)$ and $(5, 2)$ **30.** $(1, 4)$ and $(2, 3)$

31. $(2, 5)$ and $(10, 1)$ **32.** $(1, 5)$ and $(7, 3)$

B *Find the slope and y intercept, and graph each equation.*

33. $y = -\dfrac{x}{3} + 2$ **34.** $y = -\dfrac{x}{4} - 1$

35. $x + 2y = 4$ **36.** $x - 3y = -6$

37. $2x + 3y = 6$ **38.** $3x + 4y = 12$

39. $y = \frac{2}{5}x - 3$ **40.** $y = \frac{3}{4}x - 1$

41. $y = \frac{1}{6}x + 2$ **42.** $y = \frac{1}{2}x + 3$

43. $2x + 3y = 5$ **44.** $3x - 2y = 1$

45. $x - 2y = 3$ **46.** $2x - 3y = 4$

47. $\frac{1}{2}x - \frac{2}{3}y = 1$ **48.** $\frac{2}{3}x + \frac{3}{4}y = 1$

49. $\frac{1}{2}x + \frac{1}{3}y = \frac{1}{4}$ **50.** $\frac{1}{2}x + \frac{1}{5}y = \frac{1}{10}$

Write an equation of the line with slope and y intercept as indicated.

51. Slope $= -\frac{1}{2}$ **52.** Slope $= -\frac{1}{3}$
  y intercept $= -2$ y intercept $= -5$

53. Slope $= \frac{2}{3}$ **54.** Slope $= -\frac{3}{2}$
  y intercept $= \frac{3}{2}$ y intercept $= \frac{5}{2}$

Write an equation of the line that passes through the given point with the indicated slope. Transform the equation into the form $y = mx + b$.

55. $m = -2$; $(-3, 2)$ **56.** $m = -3$; $(4, -1)$

57. $m = \frac{1}{2}$; $(-4, 3)$ **58.** $m = \frac{2}{3}$; $(-6, -5)$

59. $m = -3$; $(-1, 3)$ **60.** $m = -5$; $(2, -1)$

61. $m = -2$; $(-2, 2)$ **62.** $m = -1$; $(-2, -2)$

63. $m = \frac{1}{3}$; $(-1, -1)$ **64.** $m = \frac{1}{5}$; $(-1, -5)$

65. $m = -\frac{1}{2}$; $(-3, 6)$ **66.** $m = -\frac{2}{3}$; $(-3, -2)$

67. $m = \frac{1}{2}$; $(\frac{1}{3}, \frac{2}{3})$ **68.** $m = \frac{1}{5}$; $(\frac{1}{2}, \frac{3}{2})$

69. $m = -\frac{3}{2}$; $(\frac{3}{4}, \frac{3}{2})$ **70.** $m = -\frac{3}{4}$; $(\frac{1}{5}, \frac{1}{2})$

Find the slope of the line that passes through the given points.

71. $(3, 7)$ and $(-6, 4)$ **72.** $(-5, -2)$ and $(-5, -4)$

73. $(4, -2)$ and $(-4, 0)$ **74.** $(-3, 0)$ and $(3, -2)$

75. $(\frac{1}{2}, \frac{2}{3})$ and $(\frac{3}{4}, \frac{1}{3})$ **76.** $(\frac{2}{3}, \frac{3}{4})$ and $(\frac{1}{3}, \frac{5}{4})$

77. $(\frac{1}{2}, -\frac{2}{3})$ and $(-\frac{3}{2}, \frac{1}{6})$ **78.** $(-\frac{1}{4}, \frac{2}{5})$ and $(\frac{3}{4}, -\frac{1}{5})$

Write an equation of the line through each indicated pair of points. Transform the equation into the form $y = mx + b$.

79. $(3, 7)$ and $(-6, 4)$ **80.** $(-5, -2)$ and $(5, -4)$

81. $(4, -2)$ and $(-4, 0)$ **82.** $(-3, 0)$ and $(3, -4)$

83. $(\frac{1}{2}, \frac{2}{3})$ and $(\frac{3}{4}, \frac{1}{3})$ **84.** $(\frac{2}{3}, \frac{3}{4})$ and $(\frac{1}{3}, \frac{5}{4})$

85. $(\frac{1}{2}, -\frac{2}{3})$ and $(-\frac{3}{2}, \frac{1}{6})$ **86.** $(-\frac{1}{4}, \frac{2}{5})$ and $(\frac{3}{4}, -\frac{1}{5})$

Write the equations of the vertical and horizontal lines through each point.

87. $(-3, 5)$ **88.** $(6, -2)$ **89.** $(-1, 22)$ **90.** $(5, 0)$

For each pair of lines, use slopes to determine whether the lines are parallel, perpendicular, or neither.

91. $y = 3x + 4$, $x - 3y = 5$

92. $2x + 3y = 1$, $y = 3x - 1$

93. $4x - 5y = 6$, $y = -\frac{5}{4}x + \frac{3}{2}$

94. $y = 2x - 3$, $\frac{1}{4}x + \frac{1}{2}y = \frac{3}{4}$

C *Given the indicated equation of a line and the indicated point, find the equation of the line through the given point that is (A) parallel to the given line and (B) perpendicular to the given line. Write the answers in the form $y = mx + b$.*

95. $y = \frac{3}{5}x + 1$; $(1, 4)$

96. $y = -\frac{2}{3}x + 3$; $(-1, -2)$

97. $y = -3x + \frac{1}{3}$; $(0, 2)$ **98.** $y = 5x - \frac{1}{4}$; $(-2, 0)$

99. $x + y = 3$; $(1, 1)$ **100.** $x - y = \frac{1}{7}$; $(2, 3)$

101. $y = 3$; $(-2, 5)$ **102.** $x = 4$; $(3, -2)$

APPLICATIONS

103. *Business* A sporting goods store sells a pair of cross-country ski boots costing $20 for $33 and a pair of cross-country skis costing $60 for $93.
    (A) If the markup policy of the store for items costing over $10 is assumed to be linear and is reflected in the pricing of these two items, write an equation that relates retail price R to cost C.

    (B) Graph this equation for $10 \le C \le 300$.
    (C) Use the equation to find the cost of a surfboard retailing for $240.

104. *Business* The management of a company manufacturing ballpoint pens estimates costs for running the company to be $200 per day at zero output and $700 per day at an output of 1,000 pens.
    (A) Assuming total cost per day C is linearly related to total output per day x, write an equation relating these two quantities.
    (B) Graph the equation for $0 \le x \le 2,000$.

105. *Physics* Water freezes at 32°F and 0°C and boils at 212°F and 100°C. Find the linear relationship between the two scales.

106. *Physics* It is known from physics (Hooke's law) that the relationship between the stretch s of a spring and the weight w causing the stretch is linear (a principle upon which all spring scales are constructed). A 10-pound weight stretches a spring 1 inch, and with no weight the stretch of the spring is 0.
    (A) Find a linear equation, $s = mw + b$, that represents this relationship. [*Hint:* Both points $(10, 1)$ and $(0, 0)$ are on its graph.]
    (B) Find the stretch of the spring for 15-pound and 30-pound weights.
    (C) What is the slope of the graph? (The slope indicates the increase in stretch for each pound increase in weight.)
    (D) Graph the equation for $0 \le w \le 40$.

107. *Business* An electronic computer was purchased by a company for $20,000 and is assumed to have a salvage value of $2,000 after 10 years (for tax purposes). Its value is depreciated linearly from $20,000 to $2,000.
    (A) Find the linear equation, $V = mt + b$, that relates value V in dollars to time t in years.
    (B) Find the values of the computer after 4 and 8 years, respectively.
    (C) Find the slope of the graph. (The slope indicates the decrease in value per year.)
    (D) Graph the equation for $0 \le t \le 10$.

108. *Biology* A biologist needs to prepare a special diet for a group of experimental animals. Two food mixes, M and N, are available. If mix M contains 20% protein and mix N contains 10% protein, what combinations of each mix will provide exactly 20 grams of protein? Let x be the amount of M used and y the amount of N used. Then write a linear equation relating x, y, and 20. Graph this equation for $x \ge 0$ and $y \ge 0$.

5-3 Solving Inequalities

♦ Interval Notation
♦ Solving Linear Inequalities
♦ Applications

In Section 1-2 we introduced inequality statements of the form

$$x > 2 \qquad -4 < x \le 3 \qquad x \le -1$$

These have solutions that can be readily graphed as intervals on a real number line. In this section we will consider inequality statements such as

$$3(x - 2) + 1 < 3x - (x + 7)$$

We want to solve this type of inequality by changing it to an equivalent form that represents an interval. First, however, we introduce a convenient notation for intervals.

♦ **INTERVAL NOTATION**

In Section 1-2 we used parentheses ( ) and brackets [ ] in the graphical representation of certain inequality statements. For example,

$-3 < x \le 4$ and

are two ways of indicating that x is between -3 and 4, including 4 but excluding -3. Another convenient way of representing this fact is in terms of the *interval notation*

$$(-3, 4]$$

This notation tells us the endpoints -3 and 4 for the interval and indicates that -3 is excluded but 4 is included in the interval. The use of parentheses and brackets corresponds to the earlier use to mean the endpoint is excluded or included, respectively. Table 1 shows the use of this **interval notation** in its most common forms.

The symbols ∞ and $-\infty$ that appear in Table 1 are not numbers. Instead, they are indicators that an interval continues indefinitely. The symbol ∞, read "infinity," indicates that the intervals (b, ∞) and $[b, \infty)$ extends to the right indefinitely. Similarly, the symbol $-\infty$, read "minus infinity" or "negative infinity," indicates that the intervals $(-\infty, a)$ and $(-\infty, a]$ continue to the left indefinitely. The symbol $+\infty$, read "plus infinity," is sometimes used instead of ∞. The interval $(-\infty, \infty)$ represents the whole real line. Since $-\infty$ and ∞ are not numbers, they are never included in an interval, so they are always accompanied by a parenthesis rather than a bracket.

Table 1

Interval Notation	Inequality Notation	Line Graph
$[a, b]$	$a \leq x \leq b$	
$[a, b)$	$a \leq x < b$	
$(a, b]$	$a < x \leq b$	
(a, b)	$a < x < b$	
$[b, \infty)$	$x \geq b$	
(b, ∞)	$x > b$	
$(-\infty, a]$	$x \leq a$	
$(-\infty, a)$	$x < a$	
$(-\infty, \infty)$		

Example 1
Interval Notation

Write each of the following intervals in inequality notation, and graph on a real number line:

(A) $[-2, 3)$ **(B)** $(-4, 2)$ **(C)** $[-2, \infty)$ **(D)** $(-\infty, 3)$

Solution

(A) $-2 \leq x < 3$

(B) $-4 < x < 2$

(C) $x \geq -2$

(D) $x < 3$

Matched Problem 1

Write each of the following in interval notation, and graph on a real number line:

(A) $-3 < x \leq 3$ **(B)** $-1 \leq x \leq 2$ **(C)** $x > 1$ **(D)** $x \leq 2$

◆ SOLVING LINEAR INEQUALITIES

Interval notation provides a convenient notation for describing the solutions of linear inequalities in one variable, such as

$$2(2x + 3) < 6(x - 2) + 10 \qquad \text{and} \qquad -3 < 2x + 3 \leq 9$$

We now turn to the problem of actually solving such inequalities.

The **solution set** for an inequality is the set of elements from its replacement set that make the inequality a true statement. Any element of the solution set is called a

solution of the inequality. To **solve an inequality** is to find its solution set. Two inequalities are **equivalent** if they have the same solution set. Just as with equations, we try to perform operations on inequalities that produce simpler equivalent inequalities. We continue the process until an inequality is reached whose solution is obvious. The six properties of inequalities listed in the box produce equivalent inequalities when applied. These properties are similar to the addition, subtraction, multiplication, and division properties that we use to solve equations, except when we multiply or divide both sides by a negative number.

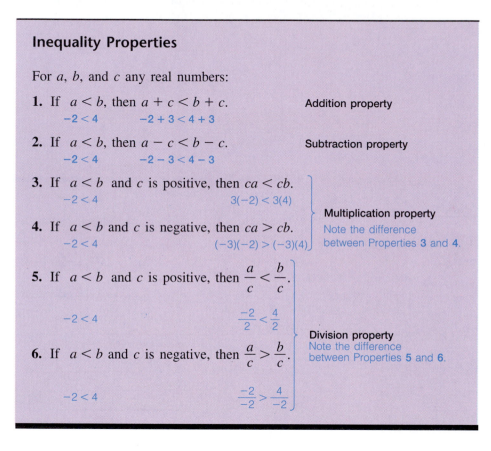

Inequality Properties

For a, b, and c any real numbers:

1. If $a < b$, then $a + c < b + c$.   Addition property
   $-2 < 4$ $-2 + 3 < 4 + 3$

2. If $a < b$, then $a - c < b - c$.   Subtraction property
   $-2 < 4$ $-2 - 3 < 4 - 3$

3. If $a < b$ and c is positive, then $ca < cb$.
   $-2 < 4$ $3(-2) < 3(4)$

Multiplication property
Note the difference
between Properties **3** and **4**.

4. If $a < b$ and c is negative, then $ca > cb$.
   $-2 < 4$ $(-3)(-2) > (-3)(4)$

5. If $a < b$ and c is positive, then $\dfrac{a}{c} < \dfrac{b}{c}$.

   $-2 < 4$ $\dfrac{-2}{2} < \dfrac{4}{2}$

Division property
Note the difference
between Properties **5** and **6**.

6. If $a < b$ and c is negative, then $\dfrac{a}{c} > \dfrac{b}{c}$.

   $-2 < 4$ $\dfrac{-2}{-2} > \dfrac{4}{-2}$

Similar properties hold if each inequality sign is reversed or if $<$ is replaced with $\leq$ and $>$ is replaced with $\geq$. Thus, we find that we can perform essentially the same operations on inequalities that we perform on equations. When working with inequalities, however, we have to be particularly careful of the use of the multiplication and division properties:

The sense of the inequality reverses if we multiply or divide both sides of an inequality statement by a negative number.

The justification of these properties depends on the definition of $<$ given in Chapter 1: $a < b$ if and only if $b - a$ is a positive number. Thus, for the addition property, we begin with $a < b$, that is, $b - a$ is positive. Then

$$(b + c) - (a + c) = b + c - a - c = b - a$$

is also positive, so $a + c < b + c$. The subtraction property is justified in an almost identical way. The multiplication and division properties are considered in Exercise 5-3, Problems 87–90.

The inequality properties allow us to solve linear inequalities in much the same way we solve linear equations. Some examples will illustrate the process.

Example 2
Solving a Linear Inequality

Solve and graph:

$$2(2x + 3) - 10 < 6(x - 2)$$

Solution

$$2(2x + 3) - 10 < 6(x - 2)$$ Simplify left and right sides.

$$4x + 6 - 10 < 6x - 12$$

$$4x - 4 < 6x - 12$$ Use the addition property.

$$\boxed{4x - 4 + 4 < 6x - 12 + 4}$$

$$4x < 6x - 8$$ Use the subtraction property.

$$\boxed{4x - 6x < 6x - 8 - 6x}$$

$$-2x < -8$$ Use the division property. Note that the sense of the inequality must be reversed.

$$\boxed{\dfrac{-2x}{-2} > \dfrac{-8}{-2}}$$

$$x > 4 \quad \text{or} \quad (4, \infty)$$

Matched Problem 2   Solve and graph: $3(x - 1) \geq 5(x + 2) - 5$

Example 3
Solving a Linear Inequality

Solve and graph:

$$\frac{2x - 3}{4} + 6 \geq 2 + \frac{4x}{3}$$

Solution

$$\frac{2x - 3}{4} + 6 \geq 2 + \frac{4x}{3}$$ Multiply both sides by 12, the LCM of 4 and 3. Note that the sense of the inequality will not reverse.

$$12 \cdot \frac{2x - 3}{4} + 12 \cdot 6 \geq 12 \cdot 2 + 12 \cdot \frac{4x}{3}$$

$$3(2x - 3) + 72 \geq 24 + 4 \cdot 4x$$

$$6x - 9 + 72 \geq 24 + 16x$$

$$6x + 63 \geq 24 + 16x$$

$$-10x \geq -39$$ Use the division property. Note that the sense will reverse.

$$x \leq 3.9 \quad \text{or} \quad (-\infty, 3.9]$$

Matched Problem 3   Solve and graph: $\dfrac{4x - 3}{3} + 8 < 6 + \dfrac{3x}{2}$

Example 4
Solving a Double Inequality

Solve and graph:

$$-3 \leq 4 - 7x < 18$$

Solution   The inequality here represents two separate inequalities:

$$-3 \leq 4 - 7x \quad \text{and} \quad 4 - 7x < 18$$

Recall that in chapter 1 we called such an inequality a **double inequality.** We proceed as in the preceding examples, except that we try to isolate x in the middle with a coefficient of 1. That is, we try to solve both inequalities at the same time:

$$-3 \leq 4 - 7x < 18$$ Subtract 4 from each member.

$$-3 - 4 \leq 4 - 7x - 4 < 18 - 4$$

$$-7 \leq -7x < 14$$ Divide each member by -7; note that the sense will reverse.

$$\frac{-7}{-7} \geq \frac{-7x}{-7} > \frac{14}{-7}$$

$$1 \geq x > -2 \quad \text{or} \quad -2 < x \leq 1 \quad \text{or} \quad (-2, 1]$$

Matched Problem 4   Solve and graph: $-3 < 7 - 2x \leq 7$

♦   **APPLICATIONS**

In Section 4-2 we considered the following example:

> A baseball player has 77 hits in 280 times at bat, for a batting average of .275. How many consecutive hits does the player need to raise the average to .300?

At that point we noted that the ratio of hits to times at bat is very seldom a "nice" number like .275 or .300, and that a more appropriate question would be "how many hits does the player need to raise the average to *at least* .300?" The next example addresses this problem.

Example 5
An Application from Baseball

A baseball player has 48 hits in 170 times at bat, for a batting average of approximately .282. How many consecutive hits does the player need to raise the average to at least .300?

Solution Let x be the number of consecutive hits needed. Then

$$\frac{48 + x}{170 + x} \geq .300$$

$$48 + x \geq (.3)170 + .3x$$

$$48 + x \geq 51 + .3x$$

$$48 + .7x \geq 51$$

$$.7x \geq 3$$

$$x \geq \frac{3}{.7}$$

$$x \geq 4.285 \ldots$$

Therefore, since the number of hits must be a natural number, 5 or more consecutive hits are necessary.

Check After 4 consecutive hits, the player's batting average is $\frac{52}{174} = .2988 \ldots$, but after 5 consecutive hits, it is $\frac{53}{175} = .3028 \ldots$.

Matched Problem 5 A basketball player has made 71 out of 92 free-throw attempts, for a free-throw percentage of approximately 77% ($\frac{71}{92} = 0.7717 \ldots$). How many of the next 25 free throws does the player need to make to raise the percentage to at least 80%?

Example 6
An Application in Chemistry

In a chemistry experiment a solution of hydrochloric acid is to be kept between 30°C and 35°C, that is, $30 \leq C \leq 35$. What is the range in temperature in degrees Fahrenheit? Use the formula $C = \frac{5}{9}(F - 32)$.

Solution

$30 \leq C \leq 35$ Replace C with $\frac{5}{9}(F - 32)$.

$30 \leq \frac{5}{9}(F - 32) \leq 35$ Multiply each member by $\frac{9}{5}$.

$$\frac{9}{5} \cdot 30 \leq \frac{9}{5} \cdot \frac{5}{9}(F - 32) \leq \frac{9}{5} \cdot 35$$

$54 \leq F - 32 \leq 63$ Add 32 to each member.

$54 + 32 \leq F - 32 + 32 \leq 63 + 32$

$86 \leq F \leq 95$ or [86, 95]

Matched Problem 6　A film developer is to be kept between 68°F and 77°F, that is, $68 \leq F \leq 77$. What is the range in temperature in degrees Celsius? Use the formula $F = \frac{9}{5}C + 32$.

Answers to Matched Problems

1. **(A)** $(-3, 3]$

(B) $[-1, 2]$

(C) $(1, \infty)$

(D) $(-\infty, 2]$

2. $x \leq -4$　or　$(-\infty, -4]$

3. $x > 6$　or　$(6, \infty)$

4. $5 > x \geq 0$　or　$0 \leq x < 5$ or $[0, 5]$

5. 23 or more　　　6. $20 \leq C \leq 25$　or　$[20, 25]$

EXERCISE 5-3

The replacement set for all variables is the set of real numbers.

A *Write in inequality notation and graph on a real number line.*

1. $[-8, 7]$
2. $(-4, 8)$
3. $[-6, 6)$
4. $(-3, 3]$
5. $[-6, \infty)$
6. $(-\infty, 7)$
7. $(-1, \infty)$
8. $(-\infty, 0]$

Write in interval notation and graph on a real number line.

9. $-2 < x \leq 6$
10. $-5 \leq x \leq 5$
11. $-7 < x < 8$
12. $-4 \leq x < 5$
13. $x \leq -2$
14. $x > 3$
15. $x < 0$
16. $x \geq -5$

Write in interval and inequality notation.

17.

18.

19.

20.

21.

22.

23.

24.

Solve and graph.

25. $7x - 8 < 4x + 7$
26. $4x + 8 \geq x - 1$
27. $3 - x \geq 5(3 - x)$
28. $2(x - 3) + 5 < 5 - x$
29. $-2(1 - x) \leq 3(x - 1)$
30. $3 - x \geq -2(x - 3)$
31. $2 + 3(x - 4) > 4(x - 5)$
32. $1 - (2 - x) < 3(x - 4)$
33. $2(3 - 4x) + 5 \leq 6 - 5(x - 4)$
34. $4 - 5(6 - x) > 7(x - 6) - 5$
35. $3 - [2 - (1 - x)] < 1 - [2 - (3 - x)]$
36. $4 - 5[6 - (x - 7)] \geq 1 - [2(3 - x) - 4]$
37. $\dfrac{N}{-2} > 4$
38. $\dfrac{M}{-3} \leq -2$

39. $-5t < -10$

40. $-7n \geq 21$

41. $3 - m < 4(m - 3)$

42. $2(1 - u) \geq 5u$

43. $-x < x - 1$

44. $1 + x \geq 1 - x$

45. $x + 1 > x - 1$

46. $1 - x < 1 + x$

47. $-2 - \dfrac{B}{4} \leq \dfrac{1 + B}{3}$

48. $\dfrac{y - 3}{4} - 1 > \dfrac{y}{2}$

49. $-4 < 5t + 6 \leq 21$

50. $2 \leq 3m - 7 < 14$

51. $2 < 3 - x \leq 4$

52. $-1 \leq 1 - x < 1$

53. $0 \leq 3 - 4x \leq 5$

54. $-5 < 4 - 3x < -2$

55. $16 < 7 - 3x \leq 31$

56. $-1 \leq 9 - 2x < 5$

B *Translate into inequality statements and solve using inequality methods.*

57. Three less than twice a number is greater than or equal to -6.

58. Five more than twice a number is less than or equal to 7.

59. Fifteen reduced by 3 times a number is less than 6.

60. Five less than 3 times a number is less than or equal to 4 times the number.

61. The sum obtained from adding a number, 1 more than the number, and 2 more than the number is less than 75.

62. The sum obtained from adding a number, 1 less than the number, and 2 less than the number is at least 75.

63. The sum of $\frac{1}{2}$ a number and $\frac{2}{3}$ the number is at most 5 more than the number.

64. The sum of $\frac{1}{3}$ a number and $\frac{3}{4}$ of the number is at least 5 more than the number.

65. One less than the product of a number and $-\frac{1}{2}$ is greater than the number.

66. Negative one-third times the quantity 1 less than a number is greater than the number.

In Problems 67–80, solve and graph.

67. $\dfrac{q}{7} - 3 > \dfrac{q - 4}{3} + 1$

68. $\dfrac{p}{3} - \dfrac{p - 2}{2} \leq \dfrac{p}{4} - 4$

69. $\dfrac{2x}{5} - \dfrac{1}{2}(x - 3) \leq \dfrac{2x}{3} - \dfrac{3}{10}(x + 2)$

70. $\dfrac{2}{3}(x + 7) - \dfrac{x}{4} > \dfrac{1}{2}(3 - x) + \dfrac{x}{6}$

71. $-4 \leq \frac{9}{5}x + 32 \leq 68$

72. $-1 \leq \frac{2}{3}A + 5 \leq 11$

73. $-12 < \frac{3}{4}(2 - x) \leq 24$

74. $24 \leq \frac{2}{3}(x - 5) < 36$

75. $-6 < -\frac{2}{5}(1 - x) \leq 4$

76. $15 \leq 7 - \frac{2}{8}x \leq 21$

77. $\frac{1}{3} \leq \frac{1}{2}x + \frac{1}{4} < \frac{5}{6}$

78. $\frac{1}{4} < \frac{1}{2} + \frac{3}{8}x < \frac{3}{4}$

79. $\frac{1}{2} < \frac{2}{3} - \frac{1}{6}x \leq \frac{2}{3}$

80. $\frac{1}{5} \leq \frac{3}{10} - \frac{2}{5}x \leq \frac{9}{10}$

C 81. If both a and b are negative numbers and b/a is greater than 1, then is $a - b$ positive or negative?

82. If both a and b are positive numbers and b/a is greater than 1, then is $a - b$ positive or negative?

83. Indicate true (T) or false (F):
(A) If $p > q$ and $m > 0$, then $mp < mq$.
(B) If $p < q$ and $m < 0$, then $mp > mq$.
(C) If $p > 0$ and $q < 0$, then $p + q > q$.

84. Indicate true (T) or false (F):
(A) If $p > q$ and $m < 0$, then $\dfrac{p}{m} > \dfrac{q}{m}$.
(B) If $p < q$ and $m > 0$, then $\dfrac{p}{m} > \dfrac{q}{m}$.
(C) If $p < 0$ and $q > 0$, then $q + p < q$.

85. Assume $0 < a < 1$; then

$$a^2 < a$$
$$a^2 - 1 < a - 1$$
$$(a - 1)(a + 1) < a - 1$$
$$a + 1 < 1$$
$$a < 0$$

But it was assumed that $a > 0$. Can you find the error in the argument?

86. Assume that $m > n > 0$; then

$$mn > n^2$$
$$mn - m^2 > n^2 - m^2$$
$$m(n - m) > (n + m)(n - m)$$
$$m > n + m$$
$$0 > n$$

But it was assumed that $n > 0$. Can you find the error in the argument?

87. Supply a reason for each step in the following argument justifying inequality Property 3:

STATEMENT

Assume $a < b$ and $c > 0$.

1. $b - a > 0$

2. $c(b - a) > 0$

3. $cb - ca > 0$

4. $ca < cb$

88. Supply a reason for each step in the following argument justifying inequality Property 6:

STATEMENT

Assume $a < b$ and $c < 0$.

1. $b - a > 0$

2. $\dfrac{b - a}{c} < 0$

3. $\dfrac{b}{c} - \dfrac{a}{c} < 0$

4. $\dfrac{a}{c} > \dfrac{b}{c}$

89. Justify inequality Property 4 using an argument similar to that used in Problem 87 or 88.

90. Justify inequality Property 5 using an argument similar to that used in Problem 87 or 88.

APPLICATIONS

Set up inequalities and solve.

91. *Earth science*   As dry air moves upward it expands, and in so doing cools at a rate of about 5.5°F for each 1,000-foot rise up to about 40,000 feet. If the ground temperature is 70°F, then the temperature T at height h is given approximately by $T = 70 - 0.0055h$. For what range in a balloon's altitude will the temperature be between 26°F and −40°F?

★ **92.** *Energy*   If the power demands in a 110-volt electric circuit in a home vary between 220 and 2,750 watts, what is the range of current flowing through the circuit? ($W = EI$, where W = Power in watts, E = Pressure in volts, and I = Current in amperes.)

93. *Business and economics*   For a business to make a profit it is clear that revenue R must be greater than cost C; in short, a profit will result only if $R > C$. If a company manufactures records and its cost equation for a week is $C = 300 + 1.5x$ and its revenue equation is $R = 2x$, where x is the number of records sold in a week, how many records must be sold for the company to realize a profit?

94. *Psychology*   IQ is given by the formula

$$IQ = \frac{MA}{CA} \cdot 100$$

where MA is mental age and CA is chronological age. If

$$80 \le IQ \le 140$$

for a group of 12-year-old children, find the range of their mental ages.

95. *Politics*   In a county with 44,000 registered voters, 20,000 of them belong to the minority party. How many new voters must the minority register to in-

crease their share of the registered voters to at least 48%, assuming no gain in voters for the other parties?

96. *Medicine*   One measurement of a desirable weight is given by Quetelet's index, which is defined to be the ratio of a person's weight in kilograms to the square of their height in meters. An index between 20 and 25 is considered desirable. A person 1.83 meters tall (about 6 feet) currently weighs 100 kilograms (about 220 pounds). How much weight should the person lose to achieve a desirable index?

97. *Baseball*   In baseball, a pitcher's earned run average (ERA) is defined by the formula

$$ERA = 9 \cdot \frac{\text{Earned runs allowed}}{\text{Innings pitched}}$$

If a pitcher has given up 65 earned runs in 180 innings, the ERA is 3.25. How many consecutive innings without allowing an earned run must the pitcher then pitch to reduce the ERA to 3.00 or less?

98. *Education*   A student's grade point average (GPA) is defined by the formula

$$GPA = \frac{\text{Quality points}}{\text{Credit hours attempted}}$$

A student who has 208 quality points in 74 credit hours attempted has a GPA of 2.81. If the student receives straight A's the next term, 4 quality points will be received for each credit hour attempted. How many credit hours must the student attempt and earn straight A grades in order to raise the GPA to at least 3.00?

99. *Puzzle*   A railroad worker is walking through a train tunnel (see the figure) when he notices an unscheduled train approaching him. If he is three-quarters of the way through the tunnel and the train is one tunnel length ahead of him, which way should he run to maximize his chances of escaping?

100. Repeat Problem 99 for the situation where the worker is 70% of the way through the tunnel when he sees the train.

5-4 Absolute Value in Equations and Inequalities

♦ Absolute Value and Distance
♦ Absolute Value in Equations and Inequalities

Equations and inequalities also may involve absolute values of algebraic expressions. Such problems can be solved by geometric and algebraic approaches. We consider both in this section, after reviewing the absolute-value concept.

♦ ABSOLUTE VALUE AND DISTANCE

Geometric and algebraic definitions of absolute value were introduced in Section 1-4. We review them here.

If a is the coordinate of a point on a real number line, then the nondirected distance from the origin to a, a nonnegative quantity, is represented by $|a|$ and is referred to as the **absolute value** of a. Thus, if $|x| = 5$, then x can be either -5 or 5 (Figure 1).

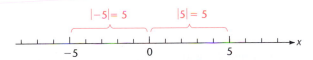

Figure 1

Algebraically, recall that we defined absolute value in Chapter 1 as follows:

Absolute Value

$$|x| = \begin{cases} x & \text{if } x \text{ is positive} \\ 0 & \text{if } x \text{ is } 0 \\ -x & \text{if } x \text{ is negative} \end{cases}$$

Basically, this definition says that if x is a positive number or 0, the absolute value of x is just the number itself. However, if x is negative, the absolute value of x is its opposite, $-x$, which is then a positive number. Thus:

1. **The absolute value of a number is never negative.**
2. **A number and its opposite have the same absolute value.**

Both the geometric and algebraic definitions of absolute value are useful, as will be seen in the material that follows.

Example 1
Evaluating $|x|$

Write without the absolute-value sign:

(A) $|7|$ **(B)** $|\pi - 3|$ **(C)** $|-7|$ **(D)** $|3 - \pi|$

Solution (A) $|7| = 7$

(B) $|\pi - 3| = \pi - 3$ Since $\pi > 3$, $\pi - 3$ is positive.

(C) $|-7| = -(-7) = 7$

(D) $|3 - \pi| = -(3 - \pi) = \pi - 3$ $3 - \pi$ is negative.

Matched Problem 1 Write without the absolute-value sign:

(A) $|8|$ (B) $|\sqrt{5} - 2|$ (C) $|-\sqrt{2}|$ (D) $|2 - \sqrt{5}|$

Since $b - a = -(a - b)$, $b - a$ and $a - b$ have the same absolute value. That is, for all real numbers a and b:

$$|b - a| = |a - b| \qquad |7 - 4| = |3| = 3 = |-3| = |4 - 7|$$

We use this result in defining the distance between two points on a real number line. Since the distance between the two points with coordinates a and b is either $a - b$ or $b - a$, whichever is positive, the distance is $|a - b|$. That is:

Distance Between Points *A* and *B*

Let A and B be two points on a real number line with coordinates a and b, respectively. The **distance between *A* and *B*,** also called the **length of the line segment joining *A* and *B*,** is given by

$$d(A, B) = |b - a|$$

Example 2
Finding Distances on the Real Number Line

Find the distance between points A and B with coordinates a and b, respectively, as given:

(A) $a = 4, b = 9$ (B) $a = 9, b = 4$

(C) $a = 0, b = 6$ (D) $a = -3, b = 5$

Solution (A) $d(A, B) = |9 - 4| = |5| = 5$

(B) $d(A, B) = |4 - 9| = |-5| = 5$

(C) $d(A, B) = |6 - 0| = |6| = 6$

(D) $d(A, B) = |5 - (-3)| = |8| = 8$

Since $|b - a| = |a - b|$, it is clear that

$$d(A, B) = d(B, A)$$

Hence, in computing the distance between two points on a real number line, it does not matter how the two points are labeled—point A can be to the left or to the right of point B. Note also that if A is at the origin O, then

$$d(O, B) = |b - 0| = |b|$$

That is, $|b|$ is the distance from the point with coordinate b to 0, just as we defined it.

Matched Problem 2   Find the indicated distances, given the following points:

(A)	$d(C, D)$	**(B)**	$d(D, C)$	**(C)**	$d(A, B)$
(D)	$d(A, C)$	**(E)**	$d(O, A)$	**(F)**	$d(D, A)$

◆ ABSOLUTE VALUE IN EQUATIONS AND INEQUALITIES

Absolute value is frequently encountered in equations and inequalities. Some of these forms have an immediate geometric interpretation that may help us see the solution.

Example 3
Solving Geometrically

Solve geometrically and graph:

(A)	$|x - 3| = 5$	**(B)**	$|x - 3| < 5$
(C)	$0 < |x - 3| < 5$	**(D)**	$|x - 3| > 5$

Solution   **(A)** Geometrically, $|x - 3|$ represents the distance between x and 3. Thus, in $|x - 3| = 5$, x is a number whose distance from 3 is 5. That is, x is 5 units to the left of 3 or 5 units to the right of 3, as shown:

$x = 3 - 5$ or $x = 3 + 5$ More compactly: $x = 3 \pm 5 = -2$ or 8.
$x = -2$ $x = 8$

(B) Geometrically, in $|x - 3| < 5$, x is a number whose distance from 3 is less than 5; that is, x is within 5 units of 3:

(C) The form $0 < |x - 3| < 5$ is encountered in calculus and advanced mathematics. Geometrically, x is a number whose distance from 3 is less than 5. In addition, since $|x - 3| > 0$, x cannot equal 3. Thus,

(D) Geometrically, in $|x - 3| > 5$, x is a number whose distance from 3 is greater than 5; that is,

$$x < -2 \quad \text{or} \quad x > 8 \qquad \text{Note that this cannot be written as a double inequality.}$$

We summarize the preceding results in Table 1, which allows us to interpret certain algebraic statements about absolute values geometrically.

Table 1   (*d* represents a positive number)

Form	Geometric Interpretation	Graph
$|x - c| = d$	Distance between x and c is equal to d.	
$|x - c| < d$	Distance between x and c is less than d.	
$0 < |x - c| < d$	Distance between x and c is less than d, but $x \neq c$.	
$|x - c| > d$	Distance between x and c is greater than d.	

Matched Problem 3   Solve geometrically and graph. Use $|x + 2| = |x - (-2)|$, so that $|x + 2|$ represents the distance from x to -2.

(A) $|x + 2| = 6$ **(B)** $|x + 2| < 6$
(C) $0 < |x + 2| < 6$ **(D)** $|x + 2| > 6$

Reasoning geometrically, as before, and noting that $|x| = |x + 0|$, we can establish the following result:

Equivalent Statements for Absolute Values

For $p > 0$:

1. $|x| = p$ is equivalent to $x = \pm p$
2. $|x| < p$ is equivalent to $-p < x < p$
3. $|x| > p$ is equivalent to $x < -p$ or $x > p$ *Not* $-p > x > p$.

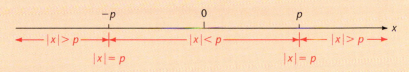

If we replace x in the above box with $ax + b$, we obtain a more general result:

For $p > 0$:

1. $|ax + b| = p$ is equivalent to $ax + b = \pm p$
2. $|ax + b| < p$ is equivalent to $-p < ax + b < p$
3. $|ax + b| > p$ is equivalent to $ax + b < -p$ or $ax + b > p$

In fact, if we replace x in the first box with *any* algebraic expression, we will obtain other variations of these results. Using these results, we often can replace equations or inequalities that involve absolute values by expressions that do not. The resulting equations or inequalities then can be solved by the methods we have already developed.

Example 4
Solving Algebraically

Solve:

(A) $|3x + 5| = 4$ (B) $|x| < 5$
(C) $|2x - 1| < 3$ (D) $|7 - 3x| \leq 2$

Solution

(A) $|3x + 5| = 4$
$$3x + 5 = \pm 4$$
$$3x = -5 \pm 4$$
$$x = \frac{-5 \pm 4}{3}$$
$$= -3, -\tfrac{1}{3}$$

(B) $|x| < 5$
$$-5 < x < 5$$

(C) $|2x - 1| < 3$
$$-3 < 2x - 1 < 3$$
$$-2 < 2x < 4$$
$$-1 < x < 2$$

(D) $|7 - 3x| \leq 2$
$$-2 \leq 7 - 3x \leq 2$$
$$-9 \leq -3x \leq -5$$
$$3 \geq x \geq \tfrac{5}{3}$$
$$\tfrac{5}{3} \leq x \leq 3$$

Matched Problem 4

Solve:

(A) $|2x - 1| = 8$ (B) $|x| \leq 7$
(C) $|3x + 3| \leq 9$ (D) $|5 - 2x| < 9$

<hr>

Example 5
Solving Algebraically

Solve:

(A) $|x| > 3$ (B) $|2x - 1| \geq 3$ (C) $|7 - 3x| > 2$

Solution

(A) $|x| > 3$
$$x < -3 \quad \text{or} \quad x > 3$$
(B) $|2x - 1| \geq 3$

$2x - 1 \leq -3$	or	$2x - 1 \geq 3$
$2x \leq -2$		$2x \geq 4$
$x \leq -1$		$x \geq 2$

(C) $|7 - 3x| > 2$

$7 - 3x < -2$	or	$7 - 3x > 2$
$-3x < -9$		$-3x > -5$
$x > 3$		$x < \frac{5}{3}$

Matched Problem 5

Solve:

(A) $|x| \geq 5$ (B) $|4x - 3| > 5$ (C) $|6 - 5x| > 16$

Since $|a - b| < c$ means that the distance between a and b is less than c, we can rephrase this as "a is within c units of b." This concept occurs frequently in industrial and manufacturing problems, where tolerances such as "within 0.001 centimeter of 20 centimeters" or "within $\frac{1}{100}$ ounce of 16 ounces" may be set on items produced. It also occurs in reports of polls and other statistical surveys, where the statement "60% in favor, with a possible error of 3 points" means "within 3% of 60%" or "60% ± 3%."

Answers to Matched Problems

1. (A) 8   (B) $\sqrt{5} - 2$ (C) $\sqrt{2}$ (D) $\sqrt{5} - 2$
2. (A) 4   (B) 4   (C) 6   (D) 11   (E) 8   (F) 15
3. (A) $x = -8, 4$

(B) $-8 < x < 4$ or $(-8, 4)$

(C) $-8 < x < 4, x \neq -2$

(D) $x < -8$ or $x > 4$

4. (A) $-\frac{7}{2}, \frac{9}{2}$ (B) $-7 \leq x \leq 7$ (C) $-4 \leq x \leq 2$ (D) $-2 < x < 7$
5. (A) $x \leq -5$ or $x \geq 5$ (B) $x < -\frac{1}{2}$ or $x > 2$ (C) $x < -2$ or $x > \frac{22}{5}$

<hr>

EXERCISE 5-4

A *Simplify, and write without absolute-value signs. Leave radicals in radical form and π as π.*

1. $|6 - 2\pi|$ 2. $|13 - 4\pi|$ 3. $|5 - \sqrt{24}|$

4. $|5 - \sqrt{26}|$ 5. $|\sqrt{5}|$ 6. $|-\frac{3}{4}|$

7. $|(-6) - (-2)|$ 8. $|(-2) - (-6)|$ 9. $|5 - \sqrt{5}|$

10. $|\sqrt{7} - 2|$ 11. $|\sqrt{5} - 5|$ 12. $|2 - \sqrt{7}|$

Find the distance between points A and B with coordinates a and b, respectively, as given.

13. $a = -7, b = 5$ 14. $a = 3, b = 12$

15. $a = 5$, $b = -7$

16. $a = 12$, $b = 3$

17. $a = -16$, $b = -25$

18. $a = -9$, $b = -17$

19. $a = 11$, $b = 14$

20. $a = -11$, $b = 14$

21. $a = -11$, $b = -14$

22. $a = 11$, $b = -14$

Find the indicated distances, given the following points:

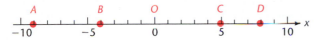

23. $d(B, O)$

24. $d(A, B)$

25. $d(O, B)$

26. $d(B, A)$

27. $d(B, C)$

28. $d(D, C)$

29. $d(A, D)$

30. $d(B, D)$

31. $d(C, A)$

32. $d(O, A)$

Solve and graph.

33. $|x| = 7$

34. $|x| = 5$

35. $|x| \le 7$

36. $|t| \le 5$

37. $|x| \ge 7$

38. $|x| \ge 5$

39. $|y - 5| = 3$

40. $|t - 3| = 4$

41. $|y - 5| < 3$

42. $|t - 3| < 4$

43. $|y - 5| > 3$

44. $|t - 3| > 4$

45. $|u + 8| = 3$

46. $|x + 1| = 5$

47. $|u + 8| \le 3$

48. $|x + 1| \le 5$

49. $|u + 8| \ge 3$

50. $|x + 1| \ge 5$

51. $|1 - x| > 1$

52. $|3 - x| \ge 4$

53. $|4 - x| \le 3$

54. $|2 - x| < 2$

B *Solve.*

55. $|3x + 4| = 8$

56. $|2x - 3| = 5$

57. $|5x - 3| \le 12$

58. $|2x - 3| \le 5$

59. $|2y - 8| > 2$

60. $|3u + 4| > 3$

61. $|5t - 7| = 11$

62. $|6m + 9| = 13$

63. $|9 - 7u| < 14$

64. $|7 - 9M| < 15$

65. $|1 - \frac{2}{3}x| \ge 5$

66. $|\frac{3}{4}x + 3| \ge 9$

67. $|\frac{9}{5}C + 32| < 31$

68. $|\frac{5}{9}(F - 32)| < 40$

Write the following statements in a concise form using absolute values.

69. $x = 2$ or $x = -2$

70. $x = -5$ or $x = 5$

71. $x < -2$ or $x > 2$

72. $x < -5$ or $x > 5$

73. $x \ge -2$ and $x \le 2$

74. $x \ge -5$ and $x \le 5$

75. x is within 0.01 of 10.

76. x is within 0.02 of 100.

77. x is within 0.1 of -5.

78. x is within 0.03 of -10.

79. The distance from x to -8 is less than 5.

80. The distance from x to -5 is less than 8.

81. x and 5 differ by less than 0.01.

82. x and 8 differ by less than 0.1.

C *For what values of x does each of the following hold?*

83. $|x| = x$

84. $|x| = -x$

85. $|x - 5| = x - 5$

86. $|x + 7| = x + 7$

87. $|x + 8| = -(x + 8)$

88. $|x - 11| = -(x - 11)$

89. $|4x + 3| = 4x + 3$

90. $|5x - 9| = (5x - 9)$

91. $|5x - 2| = -(5x - 2)$

92. $|3x + 7| = -(3x + 7)$

93. Find values of x and y that show that $|x - y| = |x| - |y|$ is not always true.

94. Find values of x and y that show that $|x + y| = |x| + |y|$ is not always true.

95. Prove that $|xy| = |x| \cdot |y|$ by showing that it is true for all cases: x positive, negative, or 0; y positive, negative, or 0.

96. Prove that $\left|\dfrac{x}{y}\right| = \dfrac{|x|}{|y|}$ by showing that it is true for all cases: x positive, negative, or 0; y positive, negative, or 0.

5-5 Graphing Linear Inequalities

We now know how to graph first-degree equations such as

$$y = 2x - 3 \qquad \text{or} \qquad 2x - 3y = 5$$

The graphs are straight lines. What happens if we replace the equal signs by inequality signs to obtain the corresponding first-degree inequalities below?

$$y \leq 2x - 3 \qquad \text{or} \qquad 2x - 3y > 5$$

We will find that the graphs of the solution sets are *regions* in the plane, bounded on one side by a straight line, and possibly including the line.

A line in a cartesian coordinate system divides the plane into two **half-planes.** A vertical line divides the plane into left and right half-planes; a nonvertical line divides the plane into upper and lower half-planes. See Figure 1.

The half-plane to the right of the vertical line $x = a$, but excluding the line itself, satisfies the inequality $x > a$. That is, for every point in the half-plane, the x coordinate is greater than a since the half-plane is to the right of a. The solution set to the inequality $x \geq a$ is the half-plane to the right of the line $x = a$ together with the boundary line itself. Similarly, the solution to the inequality $x < a$ is the half-plane to the left of the line $x = a$, and the solution to $x \leq a$ is the same half-plane together with the boundary line. In graphing these solutions, we draw the boundary line with a solid line if it is included or with a dashed line if it is excluded. See Figure 2.

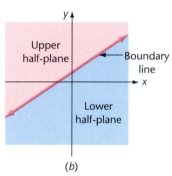

(a)

(b)

Figure 1

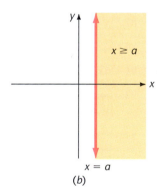 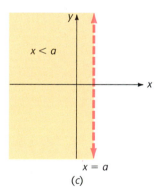 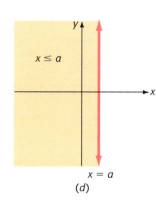

(a) (b) (c) (d)

Figure 2

Graphing $x < a$, $x \leq a$, $x > a$, **or** $x \geq a$

1. The graph of $x < a$ or $x \leq a$ is the half-plane to the left of the boundary line $x = a$, together with the boundary line if the inequality is $\leq$.
2. The graph of $x > a$ or $x \geq a$ is the half-plane to the right of the boundary line $x = a$, together with the boundary line if the inequality is $\geq$.

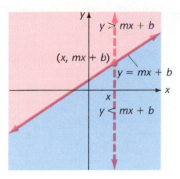

Figure 3

Consider the nonvertical line $y = mx + b$. For a fixed value of x, the point $(x, mx + b)$ lies on the line. Points with y values greater than $mx + b$ lie above the line and those with smaller y values lie below the line. See Figure 3.

Thus, the solution to the inequality $y > mx + b$ is the half-plane above the boundary line. The solution to $y < mx + b$ is the half-plane below the boundary line. The solutions to $y \geq mx + b$ and $y \leq mx + b$ are the corresponding half-planes together with the boundary line. See Figure 4.

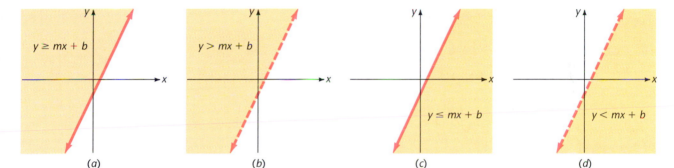

   (a)   (b)   (c)   (d)

Figure 4

Graphing $y < mx + b$, $y \leq mx + b$, $y > mx + b$, or $y \geq mx + b$

1. The graph of $y < mx + b$ or $y \leq mx + b$ is the half-plane below the boundary line $y = mx + b$, together with the boundary line if the inequality is $\leq$.
2. The graph of $y > mx + b$ or $y \geq mx + b$ is the half-plane above the boundary line $y = mx + b$, together with the boundary line if the inequality is $\geq$.

Example 1
Graphing Inequalities

Graph in a rectangular coordinate system:

(A) $2x < 5$ **(B)** $y > -3$ **(C)** $y \geq \frac{3}{4}x - 3$

Solution

(A) Rewrite the inequality as $x < \frac{5}{2}$. The solution set is the half-plane to the left of the vertical line $x = \frac{5}{2}$ as shown in Figure 5a on page 228.

(B) Recognize that $y > -3$ is in the form $y > mx + b$. Here, $m = 0$ and $b = -3$. The boundary line is the horizontal line $y = -3$. The solution set is the half-plane above the boundary line. See Figure 5b.

(C) The solution set is the half-plane above the boundary line $y = \frac{3}{4}x - 3$. In this case, the inequality is $\geq$ so the boundary line is included in Figure 5c.

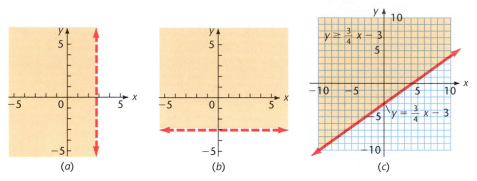

Figure 5

Matched Problem 1   Graph in a rectangular coordinate system:

(A) $y < 2$ **(B)** $3x > -8$ **(C)** $y \leq -x + \frac{5}{2}$

If the inequality is written in the form $Ax + By < C$, or in the same form using $>$, $\leq$, or $\geq$, the solution set is still the half-plane on one side of the boundary line $Ax + By = C$, including the boundary line if the inequality is $\leq$ or $\geq$. However, in this form the correct side of the line cannot be determined so readily from the inequality symbol used. For example, the solution set to $x - y < -1$ is the half-plane *above* the line $x - y = -1$, not below as the symbol $<$ might suggest. To see this, rewrite the inequality:

$$x - y < -1$$
$$-y < -x - 1 \qquad \text{Remember that multiplying by } -1 \text{ will reverse the sense of the inequality.}$$
$$y > x + 1 \qquad \text{The graph of this inequality is the half-plane } above \text{ the boundary line.}$$

It is not necessary to rewrite the inequality in the form $y = mx + b$ to determine the solution. The correct side of the boundary line can be determined by testing one point. For example, with $x - y < -1$, we can try the point $(0, 0)$:

$$0 - 0 \overset{?}{<} -1$$

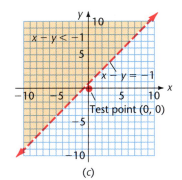

(c)

Figure 6

Since $0 < -1$ is not true, the point $(0, 0)$ does not lie in the correct half-plane. The solution set is, therefore, the half-plane not containing $(0, 0)$. See Figure 6.

This procedure of graphing the boundary line and checking one point works for any first-degree inequality, regardless of its form.

Steps in Graphing Linear Inequalities

1. Graph the corresponding equation obtained by replacing the inequality by $=$. Show this boundary line as a broken line if equality is not included in the original statement, or as a solid line if equality is included in the original statement.

> **2.** Choose a test point in the plane that is not on the line—the origin is the best choice if it is not on the line—and substitute the coordinates into the inequality.
> **3.** Determine which half-plane is the graph of the original inequality and shade it. The graph includes:
>   **(A)** The half-plane containing the test point if the inequality is satisfied by that point.
>   **(B)** The half-plane not containing the test point if the inequality is not satisfied by that point.

Example 2
Graphing a Linear Inequality

Graph $3x - 4y \leq 12$.

Solution   *Step 1.*   First graph the line $3x - 4y = 12$ as a solid line, since equality is included in $3x - 4y \leq 12$. See Figure 7a.

Step 2.   Choose a convenient test point—any point not on the line will do. In this case, the origin results in the simplest computation. See Figure 7b.

Test to see if the origin $(0, 0)$ satisfies the original inequality:

$$3x - 4y \leq 12$$
$$3 \cdot 0 - 4 \cdot 0 \overset{?}{\leq} 12$$
$$0 \overset{\checkmark}{\leq} 12$$

Step 3.   The origin does satisfy the original inequality. Hence, all other points on the same side as the origin are also part of the graph. Thus, the final graph in Figure 7c is the upper half-plane and the line.

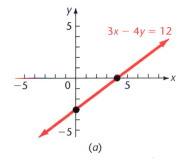

(a)

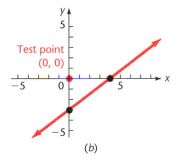

(b)

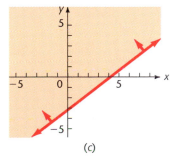

(c)

Figure 7

Matched Problem 2   Graph $2x + 2y \leq 5$.

If we are to graph an inequality involving only one variable, it is important to know the context in which the graph is to occur. In Examples 1(A) and 1(B), the graphs are to occur in a rectangular coordinate system; hence, in each case the presence of a second variable is assumed, but with a 0 coefficient. For example, $y > -3$ is assumed to mean $0x + y > -3$, and the graph is the upper half-plane above the line $y = -3$. If we were to graph $y > -3$ on a real number line, however, then we would obtain

Also, recall that a double inequality statement, such as $-1 \le x < 1$, really indicates two inequalities: $-1 \le x$ and $x < 1$. To graph such a statement in the plane means to plot the points that satisfy both inequalities. In this example, it would be the points between the two boundary lines $x = -1$ and $x = 1$, including the line on the left:

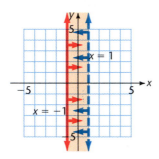

Example 3

Graphing a Double Inequality

Graph in a rectangular coordinate system:

(A) $-2 \le x \le 4$ **(B)** $1 < y \le 3$
(C) $-2 \le x \le 4$ and $1 < y \le 3$

Solution   **(A)** The solution to $-2 \le x$ is the half-plane to the right of $x = -2$. The solution to $x \le 4$ is the half-plane to the left of $x = 4$. The solution to the double inequality is the region common to both half-planes, as shown in Figure 8a.
(B) See Figure 8b.
(C) The solution is the set of points common to *both* the solution to (A) and the solution to (B). This set is graphed in Figure 8c.

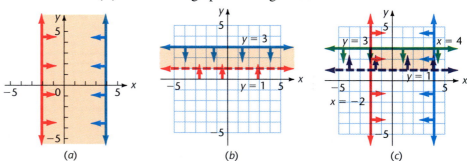

(a)   (b)   (c)

Figure 8

Matched Problem 3   Graph in a rectangular coordinate system:

(A) $1 \le x < 4$ **(B)** $-2 \le y \le 2$
(C) $1 \le x < 4$ and $-2 \le y \le 2$

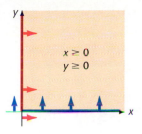

Figure 9

In applied problems, the variables are often restricted to represent only nonnegative numbers, that is, positive numbers or 0. For two variables x and y such a restriction takes the form

$$x \geq 0$$
$$y \geq 0$$

The points that meet these conditions are in the first quadrant or on the axes that serve as boundaries for the first quadrant. See Figure 9.

Example 4

Graphing an Inequality in the First Quadrant

Graph the solution to $2x + 3y \leq 18$ in the first quadrant. That is, graph the simultaneous solution to these three inequalities:

$$2x + 3y \leq 18$$
$$x \geq 0$$
$$y \geq 0$$

Solution This is the set of all ordered pairs of real numbers (x, y) such that x and y are both nonnegative and satisfy $2x + 3y \leq 18$. The result is the portion of the half-plane $2x + 3y \leq 18$ that lies in the first quadrant:

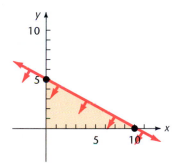

Matched Problem 4 Graph the solution to $2x + 3y \geq 18$ in the first quadrant. That is, graph the simultaneous solution to these three inequalities:

$$2x + 3y \geq 18$$
$$x \geq 0$$
$$y \geq 0$$

An expression like $2x + 3y$, or more generally, $ax + by$, is called a **linear combination of x and y.** Such linear combinations occur often in applied situations. For example, purchasing x items at \$2 per item and y items at \$3 per item leads to a total cost of $2x + 3y$ dollars. Producing x items that require 2 hours each of production time and y items that require 3 hours each, uses a total of $2x + 3y$ hours. Mixing x pounds of an animal feed that contains 2 ounces of protein per pound with y pounds of feed that contains 3 ounces of protein per pound results in a mix containing $2x + 3y$ ounces of protein. The applied problems in this section make use of such combinations.

Answers to Matched Problems

1. **(A)** $y < 2$

(B) $3x > -8$

(C) $y \le -x + \frac{5}{2}$

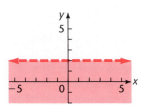

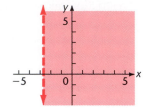

2. $2x + 2y \le 5$

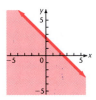

3. **(A)** $1 \le x \le 4$

(B) $-2 \le y \le 2$

(C) $1 \le x \le 4$ and $-2 \le y \le 2$

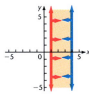

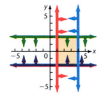

4. $2x + 3y \ge 18$, $x \ge 0$, $y \ge 0$

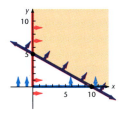

EXERCISE 5-5

A *Graph each inequality in a rectangular coordinate system.*

1. $y < 2x - 3$

2. $y \le 3x + 3$

3. $y \ge 4x + 2$

4. $y > 2x - 4$

5. $x + y \le 6$

6. $x + y \ge 4$

7. $x - y > 3$

8. $x - y < 5$

9. $y \ge x - 2$

10. $y \le x + 1$

11. $y \ge \dfrac{x}{3} - 2$

12. $y \le \dfrac{x}{2} - 4$

13. $y \le \frac{2}{3}x + 5$

14. $y > \dfrac{x}{3} + 2$

15. $x \ge -5$

16. $y \le 8$

17. $y < 0$

18. $x \ge 0$

19. $2x - 3y < 6$

20. $3x + 4y < 12$

21. $3y - 2x \ge 24$

22. $3x + 2y \ge 18$

23. $2x + 5y > 10$

24. $2x - 3y \leq 12$

25. $3x + y \leq 9$

26. $x - 3y \geq 6$

27. $5x - 2y \leq 10$

28. $3x - 2y < 18$

29. $4x - y \geq 8$

30. $5x + 2y < 10$

B *Graph each double inequality in a rectangular coordinate system.*

31. $-1 < x \leq 3$

32. $-3 \leq y < 2$

33. $-2 \leq y \leq 2$

34. $-1 \leq x \leq 4$

35. $2 \leq x < 3$

36. $-1 < x < 6$

37. $-5 < x \leq -2$

38. $-5 \leq x \leq 2$

39. $-3 < y < 1$

40. $2 \leq y \leq 6$

41. $-5 \leq y < -2$

42. $2 < y \leq 7$

Graph the simultaneous solution to each pair of double inequalities in a rectangular coordinate system.

43. $-3 \leq x \leq 3$ and $-1 \leq y \leq 5$

44. $-1 \leq x \leq 5$ and $-2 \leq y \leq 2$

45. $2 \leq x < 3$ and $-1 < y < 6$

46. $-5 < x \leq -2$ and $-5 \leq y \leq 2$

47. $-3 < x < 1$ and $2 \leq y \leq 6$

48. $-5 \leq x < -2$ and $2 < y \leq 7$

49. $-3 \leq x \leq 3$ and $-3 \leq y \leq 3$

50. $-5 < x < 5$ and $-5 < y < 5$

Graph the solution to each inequality in the first quadrant.

51. $x - 3y \geq 6$

52. $5x - 2y \leq 10$

53. $3x - 3y < 18$

54. $4x - y \geq 8$

55. $5x + 2y < 10$

56. $x - 5y \leq 5$

Graph the simultaneous solution to all three inequalities.

57. $3x + 4y \leq 12$
$x \geq 0$
$y \geq 0$

58. $3x + 2y \leq 18$
$x \geq 0$
$y \geq 0$

59. $3x + 4y \geq 12$
$x \geq 0$
$y \geq 0$

60. $3x + 2y \geq 12$
$x \geq 0$
$y \geq 0$

C *Graph each inequality in a rectangular coordinate system.*

61. $\frac{1}{5}x + \frac{2}{5}y \leq 1$

62. $\frac{3}{8}x + \frac{1}{6}y < 1$

63. $\frac{1}{8}x - \frac{1}{6}y < 1$

64. $\frac{5}{2}x - \frac{1}{9}y \leq 1$

65. $-\frac{1}{6}x + \frac{1}{3}y > 1$

66. $-\frac{1}{5}x + \frac{2}{9}y > 1$

67. $\frac{1}{3}x - \frac{3}{5}y \geq 1$

68. $\frac{2}{9}x - \frac{1}{3}y \geq 1$

APPLICATIONS

In each of the following problems, you will be asked to write an inequality and graph it. Restrict your graph to the first quadrant, since in each case, we must have $x \geq 0$ and $y \geq 0$ for the problem to make sense.

69. *Business* A tea shop blends two teas, one worth \$5 per kilogram and the other worth \$6.50 per kilogram.
(A) If the blend contains x kilograms of the first tea and y kilograms of the second, write the total value of the blend as a linear combination of x and y.
(B) Write an inequality to express the fact that the total value of the blend should be at least \$133.
(C) Graph the inequality obtained in part (B).

70. *Business* A gourmet coffee shop blends two kinds of coffee, one worth \$7 per kilogram and the other worth \$9.50 per kilogram.
(A) If the blend contains x kilograms of the first coffee and y kilograms of the second, write the total value of the blend as a linear combination of x and y.
(B) Write an inequality to express the fact that the total value of the blend should not exceed \$133.
(C) Graph the inequality obtained in part (B).

71. *Finance* An investor places x dollars in an account that earns 6% interest per year and y dollars in an account that earns 9% per year.
(A) Express the total earnings for the year as a linear combination of x and y.
(B) Write an equality to express the fact that total earnings are at least \$180.
(C) Graph the inequality obtained in part (B). You may have to use a large scale on the axes.

72. *Finance* An investor places x dollars in an account that earns 4% interest per year and y dollars in an account that earns 8% per year.
(A) Express the total earnings for the year as a linear combination of x and y.
(B) Write an equality to express the fact that total earnings are less than \$180.
(C) Graph the inequality obtained in part (B). You may have to use a large scale of the axes.

73. *Chemistry* A chemist mixes x milliliters of a 25% acid solution with y milliliters of a 60% acid solution.
(A) Express the total amount of acid in the mixture as a linear combination of x and y.
(B) Write an inequality to express the fact that the amount of acid must be at least 300 milliliters.
(C) Graph the inequality obtained in part (B).

74. *Chemistry* A chemist mixes x milliliters of a 40% acid solution with y milliliters of a 75% acid solution.
(A) Express the total amount of acid in the mixture as a linear combination of x and y.

(B) Write an inequality to express the fact that the amount of acid must be at least 300 milliliters.

(C) Graph the inequality obtained in part (B).

75. **Business** A manufacturer of sailboards makes a standard model and a competition model. The pertinent manufacturing data are summarized in the following table:

	Standard Model (Workhours per Board)	Competition Model (Workhours per Board)	Maximum Workhours Available per Week
Fabricating	6	8	120
Finishing	1	3	30

Note that the total number of hours used for fabricating boards is restricted to at most 120 per week. Similarly, the number of hours used in finishing cannot exceed 30. Let x represent the number of standard models and y the number of competition models produced in a week.

(A) Represent the number of fabricating hours used in a week as a linear combination of x and y.

(B) Write an inequality representing the limitation on the number of fabricating hours available per week.

(C) Graph the inequality obtained in part (B) in a rectangular coordinate system.

76. **Business** Refer to Problem 75.

(A) Represent the number of finishing hours used in a week as a linear combination of x and y.

(B) Write an inequality representing the limitation on the number of finishing hours available per week.

(C) Graph the inequality obtained in part (B) in a rectangular coordinate system.

77. **Diet** In an experiment involving guinea pigs, a zoologist needs a food mix with at least 23 grams of protein and at most 6.2 grams of fat. Two commercial mixes are available, with the compositions as shown in the table:

Mix	Protein (%)	Fat (%)
A	20	2
B	10	6

Suppose x grams of mix A and y grams of mix B are used.

(A) Represent the number of grams of protein in the resulting mix as a linear combination of x and y.

(B) Write an inequality representing the limitation on the number of grams of protein.

(C) Graph the inequality obtained in part (B) in a rectangular coordinate system.

78. **Diet** Refer to Problem 77.

(A) Represent the number of grams of fat in the resulting mix as a linear combination of x and y.

(B) Write an inequality representing the limitation on the number of grams of fat.

(C) Graph the inequality obtained in part (B) in a rectangular coordinate system.

CHAPTER SUMMARY

5-1 GRAPHING LINEAR EQUATIONS

A **cartesian coordinate system** is formed with two perpendicular real number lines—one as a **horizontal axis,** the other as a **vertical axis**—intersecting at their origins. These axes divide the plane into four **quadrants.** Every point in the plane corresponds to its **coordinates,** a pair (a, b) where the **abscissa a** is the coordinate of the point projected to the horizontal axis and the **ordinate b** is the coordinate of the point projected to the vertical axis. The point $(0, 0)$ is the **origin.**

The **graph of an equation** in two variables x and y is the set of all points whose coordinates (x, y) are **solutions** of the equation. The graph of a **first-degree, or linear, equation** in x and y is a straight line. The point or value where a graph crosses the y axis is called the **y intercept;** the point or value where it crosses the x axis is called the **x intercept.**

5-2 SLOPE AND EQUATIONS OF A LINE

The **slope** of a nonvertical line is given by

$$m = \frac{y_2 - y_1}{x_2 - x_1}$$

where (x_1, y_1) and (x_2, y_2) are any two points on the line. The equation $y = mx + b$ represents a line in **slope-intercept form;** m is the slope of the line and b is the y intercept. The equation $y - y_1 = m(x - x_1)$ represents a line in **point-slope form;** m is the slope and (x_1, y_1) is any point on the line. A **vertical line** has equation of the form $x = c$, and its slope is not defined. A **horizontal line** has equation of the form $y = b$ and slope 0. Two nonvertical lines are **parallel** when their slopes are equal, and **perpendicular** when the product of their slopes is -1.

5-3 SOLVING INEQUALITIES

The **solution set of a linear inequality** (an inequality in one variable with no exponents and no variable in a denominator) is the set of all values (**solutions**) that make the inequality true. Inequalities are **equivalent** if they have the same solution set. These **inequality properties** produce equivalent inequalities:

$$\text{If } a < b, \text{ then } a + c < b + c \text{ and } a - c < b - c.$$

$$\text{If } a < b \text{ and } c > 0, \text{ then } a \cdot c < b \cdot c \text{ and } \frac{a}{c} < \frac{b}{c}.$$

$$\text{If } a < b \text{ and } c < 0, \text{ then } a \cdot c > b \cdot c \text{ and } \frac{a}{c} > \frac{b}{c}.$$

Note that the sense of an inequality reverses when both sides are multiplied or divided by a negative number. The properties remain valid when $<$ is replaced by $>$, $\leq$, or $\geq$. Solutions to inequalities can be written in **interval notation.**

5-4 ABSOLUTE VALUE IN EQUATIONS AND INEQUALITIES

The distance between two points A and B on the real number line is given by $d(A, B) = |A - B| = |B - A|$. Equations and inequalities involving **absolute value** can thus sometimes be solved geometrically. They are solved algebraically by changing the statement to one without absolute values using the following translations:

1. $|x| = p$ means $x = \pm p$
2. $|x| < p$ means $-p < x < p$
3. $|x| > p$ means $x < -p \text{ or } x > p$

5-5 GRAPHING LINEAR INEQUALITIES

A nonvertical line divides the plane into two **half-planes,** the upper half-plane above the line and the lower half-plane below the line. The graph of a linear inequality in two variables is a half-plane, possibly including the **boundary line.** If the equation of the line is written in the form $y = mx + b$, the solution to $y > mx + b$ is the upper half-plane and the solution to $y < mx + b$ is the lower half-plane. With the boundary line written in other forms, the correct half-plane can be determined by using a test point.

CHAPTER REVIEW EXERCISE

Work through all the problems in this chapter review and check answers in the back of the book. Answers to all the problems are there, and following each answer is a number in italics indicating the section in which that type of problem is discussed. Where weaknesses show up, review appropriate sections in the test.

A *Rewrite in interval notation.*

1. $-1 < x \le 8$ **2.** $x > 2$

3. $x \le 10$ **4.** $-5 \le x < 10$

Rewrite in inequality form and graph on a real number line.

5. $[-4, 8)$ **6.** $(-\infty, 3)$ **7.** $[6, \infty)$ **8.** $(2, 5]$

In Problems 9–11, translate each statement into an inequality.

9. Five less than the quantity 3 times a number is more than half the number.

10. The length x of a rod is within 0.01 centimeter of 20 centimeters.

11. The value of x quarters and y dimes is at least $4.60.

12. Rewrite $|x + 2| < 3$ as an inequality, or a pair of inequalities, without using absolute value.

13. Suppose a student earns x credits of A grades and y credits of B grades. Each credit of A is worth 4 quality points and each credit of B is worth 3. Express the total quality points earned as a linear combination of A and B.

14. Express the distance between the point $a = -8$ and the point $b = 11$ as an absolute value, and evaluate the distance.

15. Give the coordinates of the points A, B, C, and D plotted in the figure.

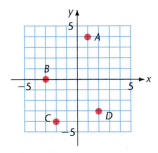

16. Plot the points $(-3, 2)$, $(4, 0)$, $(2, -4)$, $(0, -3)$, and $(-3, -3)$, in a rectangular coordinate system.

In Problems 17–28, solve the equation or inequality.

17. $4x - 9 = x - 15$

18. $2x + 3(x - 1) = 5 - (x - 4)$

19. $4x - 9 < x - 15$ **20.** $-3 < 2x - 5 < 7$

21. $|x| = 6$ **22.** $|x| < 6$

23. $|x| > 6$ **24.** $|y + 9| = 5$

25. $|y + 9| < 5$ **26.** $|y + 9| > 5$

27. $\frac{x}{4} - 1 \ge \frac{x}{3}$ **28.** $\frac{1}{6} \le \frac{1}{4}x + \frac{1}{5} < \frac{1}{3}$

29. What are the slope and y intercept for the graph of $y = -2x - 3$?

30. Write an equation of the line that passes through $(2, 4)$ with slope -2. Write the answer in the form $Ax + By = C, A > 0$.

31. What is the slope of the line that passes through $(1, 3)$ and $(3, 7)$?

32. What is an equation of the line that passes through $(1, 3)$ and $(3, 7)$? Write the answer in the form $y = mx + b$.

In Problems 33–43, graph in a rectangular coordinate system.

33. $y = 2x - 3$ **34.** $2x + y = 6$

35. $x - y \ge 6$ **36.** $y > x - 1$

37. $3x - 2y = 9$ **38.** $y = \frac{1}{3}x - 2$

39. $x = -3$ **40.** $4x - 5y \le 20$

41. $y < \frac{x}{2} + 1$ **42.** $x \ge -3$

43. $-4 \le y < 3$

B *In Problems 44–52, solve and graph the statement on a real number line.*

44. $-14 \le 3x - 2 < 7$ **45.** $-3 \le 5 - 2x < 3$

46. $3(2 - x) - 2 \le 2x - 1$ **47.** $0.4x - 0.3(x - 3) = 5$

48. $|4x - 7| = 5$ **49.** $|4x - 7| \le 5$

50. $|4x - 7| > 5$ **51.** $\frac{x + 3}{8} \le 5 - \frac{2 - x}{3}$

52. $-6 < \frac{3}{5}(x - 4) \le -3$

53. What are the slope and y intercept for the graph of $x + 2y = -6$?

54. Write an equation of the line that passes through $(-3, 2)$ with slope $-\frac{1}{3}$. Write the final answer in the form $y = mx + b$.

55. Write an equation of the line that passes through $(-3, 2)$ and $(3, -2)$. Write the final answer in the form $Ax + By = C$, $A > 0$.

56. Find the equation of the line that passes through $(3, -4)$ and is perpendicular to $x + 2y = -6$? Write the final answer in the form $y = mx + b$.

57. Write the equation of the vertical and horizontal lines that pass through $(5, -2)$.

58. Write an equation of the line that passes through $(-6, 2)$ and is parallel to $3x - 2y = 5$. Write your final answer in the form $y = mx + b$.

In Problems 59–64, graph in a rectangular coordinate system.

59. $-2 < x \leq 4$ and $0 \leq y < 3$

60. $y - \frac{1}{2} = \frac{2}{3}(x + \frac{3}{4})$

61. $2x + 5y \leq 10$, $x \geq 0$, $y \geq 0$

62. $y \geq \frac{3}{5}x + 3$

63. $|x| \leq 4$ **64.** $\frac{1}{10}x + \frac{1}{20}y = \frac{1}{2}$

65. A baseball team has a record of 81 wins and 61 losses. How many of their last 20 games must they win to have a winning rate of at least .585?

66. The number of tickets, y, that are sold for a concert series is related to the price, x, by a linear equation. If the price is set at $40 per ticket, then 1,000 will be sold; if the price is $30, then 1,500 will be sold. Find the linear equation relating x and y. Write your answer in the form $y = mx + b$.

C *In Problems 67–70, solve and graph each statement on a real number line.*

67. $|3 - 2x| \leq 5$

68. $\frac{3x}{5} - \frac{1}{2}(x - 3) \leq \frac{1}{3}(x + 2)$

69. $-4 \leq \frac{2}{3}(6 - 2x) \leq 8$ **70.** $|2x - 3| < -2$

71. $|2x - 3| = 2x - 3$ for what values of x?

72. $|2x - 3| = -(2x - 3)$ for what values of x?

73. Classify each pair of lines as parallel, perpendicular, or neither:
(A) $y = 2x + 3$, $y = \frac{1}{2}x + 8$
(B) $y = -3x - 11$, $3x + y = -1$

74. Graph the lines $y = 3x + 4$ and $y = -x + 8$ on the same rectangular coordinate system. Estimate the coordinates of the point where the lines intersect.

75. An investor places x dollars in a money market fund returning 6% per year and y dollars in a bond fund returning 9%. Graph the set of all points (x, y) that will return the investor at least $5,400 per year.

CHAPTER PRACTICE TEST

The following practice test is provided for you to test your knowledge of the material in this chapter. You should try to complete it in 50 minutes or less. Answers in the back of the book indicate the section in the text where the material in the question is covered. Actual tests in your class may vary from this practice test in difficulty, length, or emphasis, depending on the goals of your course or instructor.

Solve.

1. $|1 - \frac{1}{2}x| \leq \frac{3}{4}$ **2.** $1 - \frac{1}{2}x \geq \frac{3}{4}$ **3.** $-1 \leq \frac{1}{2}x < \frac{3}{4}$

Rewrite the expression in the form requested.

4. The interval $[-3, 6)$ in inequality notation

5. The line $\frac{1}{2}x + \frac{3}{4}y = 5$ in slope-intercept form

6. $|1 - 2x| > 3$ without using absolute value

Find the information requested

7. The slope of the line passing through the points $(-1, 2)$ and $(3, -4)$

8. The equation of the line, in slope-intercept form, that has slope -4 and passes through the point $(5, 6)$

9. The equation of the line, in slope-intercept form, that passes through the points $(3, -2)$ and $(-1, 0)$

10. All pairs of perpendicular or parallel lines among these lines: $x = 2$, $y = 3$, $y = 4x + 5$, $x - 6 = 0$, $12x - 3y = 6$

In Problems 11–17, graph in a rectangular coordinate system.

11. The points $(2, 5)$, $(-4, 0)$, and $(3, -3)$

12. The line with slope 3 and y intercept 4

13. $2x - 3y = 9$ **14.** $4x + 5y \leq 20$

15. $\frac{1}{2}x + \frac{1}{3}y = 2$ **16.** $0.5x - 0.6y \leq 15$

17. The solution in the first quadrant to $3x + 4y \geq 12$

18. Graph $-2 \leq x \leq 3$:
(A) On a number line
(B) In a rectangular coordinate system

19. A work force of 1,370 currently includes 210 minority employees. How many additional minority employees must be hired to raise the minority percentage of the work force to at least 20%.

20. A company produces two products, A and B. Each unit of A requires \$4 worth of materials and \$14 in labor costs. Each unit of B requires \$6 in materials and \$12 in labor. The company's budget permits at most \$1,000 per day to be spent on labor. The company produces x units of Product A and y of product B daily. Write the restriction(s) on labor costs in inequality form and graph the set of coordinates (x, y) that satisfy the restriction(s).

CUMULATIVE REVIEW EXERCISE, CHAPTERS 1–5

This set of exercises reviews the major concepts and techniques of Chapters 1–5. Work through all the problems and check answers in the back of the book. Answers to all the problems are there, and following each answer is a number in italics indicating the section in which that type of problem is discussed. Where weaknesses show up, review appropriate sections in the text.

A *Translate each statement into an algebraic equation or inequality statement using x as the only variable. Do not solve the equation or inequality.*

1. Three less than twice a number is positive.

2. The sum of a number and its reciprocal is at least 1.

3. The ratio of a number to 5 more than itself is 3 to 5.

4. The product of two consecutive integers is 812.

5. The value of a certain number of quarters and 8 fewer dimes is \$13.90.

Find, evaluate, or calculate the quantity requested.

6. $\dfrac{2^2}{3} - \dfrac{2}{3^2} + \left(\dfrac{2}{3}\right)^2$

7. $x^2 + x + \dfrac{1}{x} + \dfrac{1}{x^2}$ for $x = 2$

8. The LCM of $x^2 - 1$, x, and $x - 1$

9. The quotient and remainder when $x^4 - x^3 + 2x^2 - 3$ is divided by $x + 1$.

10. $|-x^2 - 3x - 4|$ for $x = -5$

11. The slope of the line through the points $(2, 0)$ and $(-4, -2)$

12. The equation of the line, in slope-intercept form, with slope 5 passing through $(6, 7)$

13. The coordinates of the points A and B shown in the figures below.

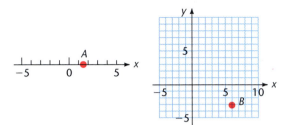

Rewrite in the form indicated.

14. $\dfrac{15x^2y^3z^4}{10x^5y^4z^3}$ in lowest terms

15. $\dfrac{6ab^2c}{a^2bc}$ as a fraction with denominator $a^2b^3c^4$

16. $3ab + 6ac + 9bc + 12ab + 15bc + 18ab$ with like terms combined

17. $(x + 5)(x + 3)$ as a polynomial with like terms combined

18. $\dfrac{1}{x} + \dfrac{2}{x + 1}$ as a fraction reduced to lowest terms

19. $\dfrac{\dfrac{1}{x}}{\dfrac{x}{x + 1}}$ as a fraction reduced to lowest terms

20. The sum of $x^3 + 2x^2 + 3x + 4$ and $x^2 - 2x - 3$ as a polynomial with like terms combined

21. The difference $x^3 + 2x^2 + 3x + 4$ minus $x^2 - 2x - 3$ as a polynomial with like terms combined

22. $x - [3 - 2(x - 1)]$ without grouping symbols and with like terms combined

23. The interval $0 < x \le 5$ in interval notation

24. $x^2 - 9$ in factored form

25. $x^2 + 8x + 16$ in factored form

26. $x^2 - 3x + 2$ in factored form

27. $x^3 + x$ in factored form

28. $x^3 - 27$ in factored form

29. The product $\dfrac{x + 1}{x}$ times $\dfrac{x^2}{x^2 - 1}$ as a fraction reduced to lowest terms

30. The quotient $\dfrac{x + 1}{x}$ divided by $\dfrac{x}{x + 1}$ as a fraction reduced to lowest terms

31. $\dfrac{x^2 + 5x + 4}{x^2 + 4x + 3}$ as a fraction reduced to lowest terms

Solve for x.

32. $3x + 7 = 9 - 11x$

33. $(3x + 7)(9 - 11x) = 0$

34. $\dfrac{3x + 7}{9 - 11x} = 2$

35. $3x + 7 < 9 - 11x$

36. $|3x + 7| < 9$

37. $x^2 + 4x + 4 = 0$

38. $\dfrac{x}{4} + \dfrac{1}{4} = \dfrac{x}{5} + \dfrac{x + 7}{6}$

Graph in a rectangular coordinate system.

39. The points $(-1, 2)$, $(3, -4)$, $(5, 6)$, $(-7, 0)$

40. The line $y = -3x + 2$

41. The solution to the inequality $y > -3x + 2$

42. The line $3x - 4y = 12$

43. The solution to the inequality $3x - 4y > 12$

Set up an appropriate equation or inequality and solve.

44. A rectangle has its longer sides equal to 1.75 times the length of the shorter sides. What are the dimensions if the perimeter is 110 centimeters?

45. An 11-foot-long piece of pipe is to be cut into two pieces, with the longer piece 1.75 times the length of the short. What lengths are to be cut?

46. A baseball pitcher has a strikeout-to-walk ratio of 7:4. If the pitcher has struck out 203 batters, how many batters have been walked?

47. The sum of a number and 7 less than the product of 4 and the original number is 74. Find the number.

48. In a rural school district, 132 students, or 24% of the total enrollment, scored below national norms on a standardized test. How many students are in the district?

49. A rectangle with length equal to 3 times the width is to have perimeter at most 96 meters. What widths are possible?

50. One family leaves Cincinnati, Ohio, at 6 A.M., driving south on Interstate Highway 75 at an average speed of 60 miles per hour. Their neighbors leave Sarasota, Florida, which is 966 miles from Cincinnati, at 10 A.M. and drive north on the same road at an average speed of 61 miles per hour. What time will the cars pass? How far will they be from Cincinnati when they pass?

B *Find, evaluate, or calculate the quantity requested.*

51. The slope of the line through the points $(\frac{1}{2}, \frac{3}{4})$ and $(1, \frac{1}{4})$

52. The equation of the line through the points $(0, 1)$ and $(-2, 3)$

53. The coordinates of the point 3 units to the right and 2 units down from the point $(-1, 4)$

54. The quotient and remainder resulting from dividing $x^2 - 2$ by $x^2 + 1$

55. The LCM of $x^2 + 3x - 4$ and $x^2 - 4x + 3$

Rewrite in the form indicated.

56. $2x + 3y = 4$ in slope-intercept form

57. $\dfrac{1}{x - 1} + \dfrac{1}{x + 1} + \dfrac{1}{x^2 - 1}$ as a fraction reduced to lowest terms

58. $\dfrac{x^2 + 4x + 4}{x^3 + x} \cdot \dfrac{x^3 + 2x^2 + x}{x^2 + 3x + 2}$ as a fraction reduced to lowest terms

59. $\left(\dfrac{1}{x} - x\right) \div \left(\dfrac{1 + x}{x^2}\right)$ as a fraction reduced to lowest terms

60. $\dfrac{1 + \dfrac{1}{x}}{x - \dfrac{1}{x}}$ as a fraction reduced to lowest terms

61. $2x^2 - 5x - 3$ in factored form

62. $x^4 - x$ in factored form

63. $ax^2 - 2a + bx^2 - 2b$ in factored form

64. $(x^2 + 2x + 3)(x^2 - x - 2)$ as a polynomial with like terms combined

Solve.

65. $\dfrac{2}{x^2 - 2x} - \dfrac{1}{x - 2} = 1$

66. $1 - \dfrac{1}{x} - \dfrac{2}{x^2} = 0$

67. $1 + \dfrac{1}{x} = \dfrac{2}{x(2 - x)}$

68. $\dfrac{3}{x + 4} - 5 = \dfrac{6x}{x + 4}$

69. $\dfrac{x}{x^2 - 2x + 1} - \dfrac{1}{x^2 - 1} = \dfrac{1}{x + 1}$

70. $(x - 1)(x + 2) = (x - 1)(2x - 3)$

71. $|2x + 5| \geq 7$

72. $3x + 4y = 5$ for y

Graph in rectangular coordinates.

73. $\frac{2}{3}x - \frac{1}{4}y = 1$

74. $0.5x + 0.75y \geq 1$

75. $2x + 5y \leq 10$
$\quad\ x \geq 0$
$\quad\ y \geq 0$

Set up an appropriate equation or inequality and solve.

76. The Kentucky Derby is a $1\frac{1}{4}$-mile horse race for which the record time of 1 minute 59.2 seconds was set by Secretariat in 1973. The Preakness is a $1\frac{3}{16}$-mile horse race. The record time for the Preakness is 1 minute 53.2 seconds, set by Tank's Prospect in 1975. If Tank's Prospect had run the additional $\frac{1}{16}$ mile at the same pace, would he have bettered Secretariat's time for the $1\frac{1}{4}$-mile race?

77. The live birth rate in the state of Maine for 1990 was 15.5 births for each 1,000 residents. If the number of births was 16,211, what was the population? Round your answer to the nearest hundred.

78. A baseball player has 45 hits in the first 200 times at bat. How many consecutive hits must the player get to raise the batting average (hits/times at bat) to at least .300?

79. The product of 2 less than a positive number times 3 more than the number is 150. Find the number.

80. Rods, furlongs, and chains are measures of length, with a furlong equal to $\frac{1}{8}$ mile. A rod is $\frac{1}{40}$ furlong and a chain is equal to 4 rods. A mile has 5,280 feet. How many feet are there in a chain?

81. The temperature in Las Vegas, Nevada, in July ranges, on average, between 76°F and 105°F. What is this range on the Celsius scale? Recall that $F = \frac{9}{5}C + 32$.

82. The ratio of the longer sides to the shorter sides in a rectangle is 7:4. What are the dimensions if the rectangle has perimeter 83.6 centimeters?

C *Find, evaluate, or calculate the quantity requested.*

83. $1 - \dfrac{x}{1 + \dfrac{x}{1 + x}} - x \div (x + 1)^2$ for $x = 3$

84. The slope of a line perpendicular to the line through the points $(-1, 3)$ and $(3, 1)$

85. The slope-intercept form of the equation of the line parallel to the line through the points $(-1, 3)$ and $(3, 1)$, but passing through the point $(0, 8)$

Rewrite in the form indicated.

86. $x^6 - 1$ in factored form

87. $x^3 - 3x^2 - 4x + 12$ in factored form

88. $3x^3y - 12xy^3$ in factored form

89. $\dfrac{x^2}{x^2 - 1} + \dfrac{x}{x^2 - 4x + 3}$ as a fraction reduced to lowest terms

90. $\left(\dfrac{x}{x + 1} + \dfrac{1}{x - 1}\right) \cdot \dfrac{x + 1}{x^2 + 1}$ as a fraction reduced to lowest terms.

91. $\dfrac{1 + \dfrac{a}{b - a}}{1 + \dfrac{b}{a - b}}$ as a fraction reduced to lowest terms

92. $\dfrac{3x^2(x + 2)^4 - 4x^3(x + 2)^3}{(x + 2)^6}$ as a fraction reduced to lowest terms

93. $\dfrac{1 + \dfrac{b}{a - b}}{b^2} \cdot \left(\dfrac{a}{a - b} - \dfrac{a + b}{a}\right)$ as a fraction reduced to lowest terms

Solve.

94. $\dfrac{x - y}{z} + \dfrac{x - z}{y} = 1$ for x

95. $\dfrac{3}{(x + 2)^2} - \dfrac{2}{x^2} = \dfrac{1}{x^2 + 2x}$

96. $\dfrac{3x}{x - 1} - 4 = \dfrac{4}{1 - x}$

Set up an appropriate equation or inequality and solve.

97. A rectangle has length 3 times the width. If both dimensions are increased by 3, the area becomes 45 square meters. What are the original dimensions?

98. The numerator of a fraction is 1 greater than its denominator. When its reciprocal is subtracted from the fraction, the result is equal to the sum of the original numerator and denominator divided by the product of the original numerator and denominator. Find the original denominator.

99. One brand of fertilizer contains 12% nitrogen. A second brand contains 18% nitrogen. A landscaper will use x pounds of the first brand and y pounds of the second to apply a total of at least 180 pounds of nitrogen to the ball fields in a city park. What values of x and y are allowable? Express your answer as a graph in rectangular coordinates.

100. A cross-country cyclist begins the day at 6 A.M. and pedals at an average speed of 18 miles per hour. The cyclist's family follows in a support van that averages 45 miles per hour, but does not leave until 9:30 A.M. At what time will the van overtake the cyclist?

6

Exponents, Radicals, and Complex Numbers

The concept of exponents was first introduced in Chapter 1. There we considered only positive-integer exponents:

$$\underbrace{a^n = a \cdot a \cdot \cdots \cdot a}_{n \text{ factors}} \qquad 2^4 = 2 \cdot 2 \cdot 2 \cdot 2$$

These were the only exponents needed to work with polynomials. In this chapter, we will extend the exponent concept to include any integer (whether positive, zero, or negative) as an exponent, and then any rational number as an exponent. Rational exponents require a consideration of roots and radicals, and these in turn will lead us to investigate a new number system, the *complex numbers*.

Photo reference: see Exercise 6-3, Problem 51.

6-1 Positive-Integer Exponents

♦ Positive-Integer Exponents and Exponent Properties
♦ Use of the Exponent Properties

In Section 1-6, we introduced the concept of a positive-integer or natural number exponent and one exponent property, which we used repeatedly when dealing with polynomials:

Property 1: Product of Powers

$$a^m a^n = a^{m+n} \qquad a^5 a^2 = a^{5+2} = a^7$$

In this section, we discuss four more exponent properties. The five properties taken together will determine how we extend the exponent concept beyond positive integers.

♦ POSITIVE-INTEGER EXPONENTS AND EXPONENT PROPERTIES

Property 1 for positive-integer exponents was based upon the following observation:

$$a^3 a^4 = \underbrace{(a \cdot a \cdot a)}_{\substack{3 \\ \text{factors}}} \underbrace{(a \cdot a \cdot a \cdot a)}_{\substack{4 \\ \text{factors}}} = \underbrace{a \cdot a \cdot a \cdot a \cdot a \cdot a \cdot a}_{\substack{7 \\ \text{factors}}}$$

$$= a^{3+4} = a^7$$

Other properties can be obtained similarly.

Consider how we might take a power of a power:

$$(a^3)^4 = (a^3)(a^3)(a^3)(a^3)$$

$$\overbrace{= (a \cdot a \cdot a)(a \cdot a \cdot a)(a \cdot a \cdot a)(a \cdot a \cdot a)}^{\text{4 groups of 3 factors each}}$$

$$\overbrace{= a \cdot a \cdot a \cdot a \cdot a \cdot a \cdot a \cdot a \cdot a \cdot a \cdot a \cdot a}^{4 \cdot 3 \text{ factors}}$$

$$= a^{4 \cdot 3} = a^{12}$$

This suggests Property 2:

Property 2: Power of a Power

$$(a^n)^m = a^{mn} \qquad (a^2)^5 = a^{2(5)} = a^{10}$$

Now consider how we would take a power of a product:

$$\overbrace{(ab)^4 = \overbrace{(ab)(ab)(ab)(ab)}^{\text{4 factors of }(ab)} = \overbrace{(a \cdot a \cdot a \cdot a)}^{\text{4 factors of }a}\overbrace{(b \cdot b \cdot b \cdot b)}^{\text{4 factors of }b} = a^4 b^4}$$

This suggests Property 3:

Property 3: Power of a Product

$$(ab)^m = a^m b^m \qquad (ab)^7 = a^7 b^7$$

Next, consider how we would take a power of a quotient:

$$\left(\frac{a}{b}\right)^5 = \overbrace{\left(\frac{a}{b} \cdot \frac{a}{b} \cdot \frac{a}{b} \cdot \frac{a}{b} \cdot \frac{a}{b}\right)}^{\text{5 factors of }a/b} = \underbrace{\overbrace{\frac{a \cdot a \cdot a \cdot a \cdot a}{b \cdot b \cdot b \cdot b \cdot b}}^{\text{5 factors of }a}}_{\text{5 factors of }b} = \frac{a^5}{b^5}$$

This suggests Property 4:

Property 4: Power of a Quotient

$$\left(\frac{a}{b}\right)^m = \frac{a^m}{b^m}, \; b \neq 0 \qquad \left(\frac{a}{b}\right)^3 = \frac{a^3}{b^3}$$

Finally, consider how we can write the quotient of two powers:

(A) $\dfrac{a^7}{a^3} = \dfrac{a \cdot a \cdot a \cdot a \cdot a \cdot a \cdot a}{a \cdot a \cdot a}$

$\qquad = \dfrac{(a \cdot a \cdot a)(a \cdot a \cdot a \cdot a)}{(a \cdot a \cdot a)} = a^4 = a^{7-3}$

(B) $\dfrac{a^3}{a^3} = \dfrac{a \cdot a \cdot a}{a \cdot a \cdot a} = 1$

(C) $\dfrac{a^4}{a^7} = \dfrac{a \cdot a \cdot a \cdot a}{a \cdot a \cdot a \cdot a \cdot a \cdot a \cdot a}$

$\qquad = \dfrac{(a \cdot a \cdot a \cdot a)}{(a \cdot a \cdot a \cdot a)(a \cdot a \cdot a)} = \dfrac{1}{a^3} = \dfrac{1}{a^{7-4}}$

This suggests Property 5:

Property 5: Quotient of Powers

$$\frac{a^m}{a^n} = \begin{cases} a^{m-n} & \text{if } m \text{ is greater than } n \\ 1 & \text{if } m = n \\ \dfrac{1}{a^{n-m}} & \text{if } n \text{ is greater than } m \end{cases} \qquad a \neq 0$$

$$\frac{a^8}{a^3} = a^{8-3} = a^5$$

$$\frac{a^8}{a^8} = 1$$

$$\frac{a^3}{a^8} = \frac{1}{a^{8-3}} = \frac{1}{a^5}$$

 CAUTION

Remember that the properties of exponents apply to products and quotients, not to sums and differences. Many mistakes are made in algebra by applying a property of exponents to the wrong algebraic form. For example,

$$(ab)^2 = a^2b^2 \qquad \text{but} \qquad (a + b)^2 \neq a^2 + b^2$$

The exponent properties for positive-integer exponents are summarized here:

Exponent Properties

For positive integers m and n, the real numbers a and b:

1. $a^m a^n = a^{m+n}$
2. $(a^n)^m = a^{mn}$
3. $(ab)^m = a^m b^m$
4. $\left(\dfrac{a}{b}\right)^m = \dfrac{a^m}{b^m} \qquad b \neq 0$
5. $\dfrac{a^m}{a^n} = \begin{cases} a^{m-n} & \text{if } m \text{ is greater than } n \\ 1 & \text{if } m = n \\ \dfrac{1}{a^{n-m}} & \text{if } n \text{ is greater than } m \end{cases} \qquad a \neq 0$

♦ USE OF THE EXPONENT PROPERTIES

Example 1
Applying the Exponent Properties

Simplify to a form that uses only positive-integer exponents; that is, rewrite the expression so that each variable occurs only once and has only a single positive-integer exponent applied to it.

(A) $x^{12}x^{13}$ **(B)** $(t^7)^5$ **(C)** $(xy)^5$

(D) $\left(\dfrac{u}{v}\right)^3$ **(E)** $\dfrac{x^{12}}{x^4}$ **(F)** $\dfrac{t^4}{t^9}$

Solution (A) $x^{12}x^{13} \boxed{= x^{12+13}} = x^{25}$ (B) $(t^7)^5 \boxed{= t^{7\cdot5}} = t^{35}$

(C) $(xy)^5 = x^5y^5$ (D) $\left(\dfrac{u}{v}\right)^3 = \dfrac{u^3}{v^3}$

(E) $\dfrac{x^{12}}{x^4} \boxed{= x^{12-4}} = x^8$ (F) $\dfrac{t^4}{t^9} \boxed{= \dfrac{1}{t^{9-4}}} = \dfrac{1}{t^5}$

Matched Problem 1 Simplify to a form that uses only positive-integer exponents.

(A) x^8x^6 (B) $(u^4)^5$ (C) $(xy)^9$

(D) $\left(\dfrac{x}{y}\right)^4$ (E) $\dfrac{x^{10}}{x^3}$ (F) $\dfrac{x^3}{x^{10}}$

Example 2
Applying the Exponent Properties

Simplify to a form that uses only positive-integer exponents.

(A) $(x^2y^3)^4$ (B) $\left(\dfrac{u^3}{v^4}\right)^3$ (C) $\dfrac{2x^9y^{11}}{4x^{12}y^7}$

Solution (A) $(x^2y^3)^4 \boxed{= (x^2)^4(y^3)^4} = x^8y^{12}$ (B) $\left(\dfrac{u^3}{v^4}\right)^3 \boxed{= \dfrac{(u^3)^3}{(v^4)^3}} = \dfrac{u^9}{v^{12}}$

(C) $\dfrac{2x^9y^{11}}{4x^{12}y^7} \boxed{= \dfrac{2}{4} \cdot \dfrac{x^9}{x^{12}} \cdot \dfrac{y^{11}}{y^7} = \dfrac{1}{2} \cdot \dfrac{1}{x^3} \cdot \dfrac{y^4}{1}} = \dfrac{y^4}{2x^3}$

Matched Problem 2 Simplify to a form that uses only positive-integer exponents.

(A) $(u^3v^4)^2$ (B) $\left(\dfrac{x^4}{y^3}\right)^2$ (C) $\dfrac{9x^7y^2}{3x^5y^3}$ (D) $\dfrac{(2x^2y)^3}{(4xy^3)^2}$

Knowing the rules of the game of chess doesn't make you skilled at playing chess; similarly, memorizing the properties of exponents doesn't necessarily make you skilled at using these properties. To acquire skill in their use, you must use these properties in a fairly large variety of problems. The following exercises should help you acquire this skill.

Answers to Matched Problems

1. (A) x^{14} (B) u^{20} (C) x^9y^9 (D) $\dfrac{x^4}{y^4}$ (E) x^7 (F) $\dfrac{1}{x^7}$

2. (A) u^6v^8 (B) $\dfrac{x^8}{y^6}$ (C) $\dfrac{3x^2}{y}$ (D) $\dfrac{x^4}{2y^3}$

EXERCISE 6-1

A *Replace each question mark with the appropriate symbol.*

1. $y^2y^8 = y^?$ 2. $x^8x^5 = x^?$ 3. $y^9 = y^3y^?$

4. $x^{11} = x^?x^6$ 5. $(u^4)^3 = u^?$ 6. $(v^2)^3 = ?$

7. $x^{10} = (x^?)^5$ 8. $y^{12} = (y^6)^?$ 9. $(uv)^7 = ?$

10. $(xy)^5 = x^5y^?$ 11. $p^4q^4 = (pq)^?$ 12. $m^3n^3 = (mn)^?$

13. $\left(\dfrac{a}{b}\right)^8 = ?$ 14. $\left(\dfrac{x}{y}\right)^4 = \dfrac{x^?}{y^4}$ 15. $\dfrac{m^3}{n^3} = \left(\dfrac{m}{n}\right)^?$

16. $\dfrac{x^7}{y^7} = \left(\dfrac{x}{y}\right)^?$ **17.** $\dfrac{n^{14}}{n^8} = n^?$ **18.** $\dfrac{x^7}{x^3} = x^?$

19. $m^6 = \dfrac{m^8}{m^?}$ **20.** $x^3 = \dfrac{x^?}{x^4}$ **21.** $\dfrac{x^4}{x^{11}} = \dfrac{1}{x^?}$

22. $\dfrac{a^5}{a^9} = \dfrac{1}{a^?}$ **23.** $\dfrac{1}{x^8} = \dfrac{x^4}{x^?}$ **24.** $\dfrac{1}{u^2} = \dfrac{u^?}{u^9}$

Simplify to a form that uses only positive-integer exponents.

25. $(4x^2)(2x^{10})$ **26.** $(2x^4)(4x^7)$ **27.** $\dfrac{9x^6}{3x^4}$

28. $\dfrac{4x^8}{2x^6}$ **29.** $\dfrac{6m^5}{8m^7}$ **30.** $\dfrac{4u^3}{2u^7}$

31. $(xy)^{10}$ **32.** $(cd)^{12}$ **33.** $\left(\dfrac{m}{n}\right)^5$

34. $\left(\dfrac{x}{y}\right)^6$ **35.** $(4y^3)(5y)(y^7)$ **36.** $(4x^2)(3x^5)(x^4)$

37. $(6 \times 10^8)(8 \times 10^9)$ **38.** $(3 \times 10^3)(4 \times 10^{13})$

39. $(10^7)^2$ **40.** $(10^4)^5$ **41.** $(x^3)^2$

42. $(y^4)^5$

B **43.** $(m^2n^5)^3$ **44.** $(x^2y^3)^4$ **45.** $\left(\dfrac{c^2}{d^5}\right)^3$

46. $\left(\dfrac{a^3}{b^2}\right)^4$ **47.** $\dfrac{9u^8v^6}{3u^4v^8}$ **48.** $\dfrac{2x^3y^8}{6x^7y^2}$

49. $(2s^2t^4)^4$ **50.** $(3a^3b^2)^3$ **51.** $6(xy^3)^5$

52. $2(x^2y)^4$ **53.** $\left(\dfrac{mn^3}{p^2q}\right)^4$ **54.** $\left(\dfrac{x^2y}{2w^2}\right)^3$

55. $\dfrac{(4u^3v)^3}{(2uv^2)^6}$ **56.** $\dfrac{(2xy^3)^2}{(4x^2y)^3}$ **57.** $\dfrac{(9x^3)^2}{(-3x)^2}$

58. $\dfrac{(-2x^2)^3}{(2^2x)^4}$ **59.** $\dfrac{(-ab^2)^3}{-(a^2b)^2}$ **60.** $\dfrac{-(ab^3)^2}{(-a^2b)^3}$

61. $\dfrac{-2a^2b^3}{(-2a^2b)^2}$ **62.** $\dfrac{-(3a^3b)^2}{(-3a^3b)^3}$ **63.** $\dfrac{-3^3}{(-3)^3}$

64. $\dfrac{(-a)^3}{-a^3}$ **65.** $\dfrac{-x^2}{(-x)^2}$ **66.** $\dfrac{-2^2}{(-2)^2}$

67. $\dfrac{(-x^2)^2}{(-x^3)^3}$ **68.** $\dfrac{-2^4}{(-2a^2)^4}$

69. $\left(-\dfrac{x}{y}\right)^3\left(\dfrac{y^2}{w}\right)^2\left(\dfrac{w}{x^2}\right)^3$ **70.** $\left(-\dfrac{a^2b}{c}\right)^2\left(\dfrac{c}{b^2}\right)^3\left(\dfrac{1}{a^3}\right)^2$

71. $-(a^2bc^3)\dfrac{-ab^2c}{(abc)^3}$ **72.** $\dfrac{-a^2b^4c^6}{(-ab^2c^3)^2}(a^3b^2c)$

73. $\dfrac{(a+b)^2(a-b)^2}{a^2 - b^2}$ **74.** $\dfrac{a^2(b-c)^2}{a^2b^2 - a^2c^2}$

75. $\dfrac{a^3b^4 - a^4b^3}{a^2b^2}$ **76.** $\dfrac{a^2b^4 + a^3b^3}{a^4b^2}$

77. $\dfrac{5(x+1)^4x^7 - 7(x+1)^5x^6}{x^{14}}$

78. $\dfrac{8x^7(x-1)^4 - 4x^8(x-1)^3}{(x-1)^8}$

79. $\dfrac{3(x+y)^3(x-y)^4}{(x-y)^26(x+y)^5}$ **80.** $\dfrac{10(u-v+w)^8}{5(u-v+w)^{11}}$

C *In Problems 81–96, simplify, assuming n is restricted so that each exponent represents a positive integer.*

81. $x^{5-n}x^{n+3}$ **82.** $y^{2n+2}y^{n-3}$ **83.** $\dfrac{x^{2n}}{x^n}$

84. $\dfrac{x^{n+2}}{x^n}$ **85.** $(x^{n+1})^2$ **86.** $(x^{n+1})^n$

87. $\dfrac{u^{n+3}v^n}{u^{n+1}v^{n+4}}$ **88.** $\dfrac{(x^ny^{n+1})^2}{x^{2n+1}y^{2n}}$ **89.** $\dfrac{x^nx^{n+1}}{x^{2n}}$

90. $\dfrac{x^{n-1}x^{n+1}}{x^n}$ **91.** $\dfrac{x^ny^n}{(xy)^{n-1}}$ **92.** $\dfrac{x^{n+1}y^{n-1}}{(xy)^n}$

93. $\left(\dfrac{x^n}{y^{n+1}}\right)^2\left(\dfrac{y^n}{x^{n+1}}\right)^2$ **94.** $\left(\dfrac{x^{n-1}}{y^{n+1}}\right)^2 \div \left(\dfrac{x^{n+1}}{y^{n-1}}\right)$

95. $\dfrac{(x+y)^n(x-y)^n}{(x^2-y^2)^n}$ **96.** $\dfrac{(x+y)^{2n}}{(x-y)^{2n}}(x^2-y^2)^n$

97. Factor $x^{2n} - 1$ into a product of two factors.

98. Factor $x^{3n} - 1$ into a product of two factors.

99. Factor $x^{3n} + 1$ into a product of two factors.

100. Factor $x^{4n} - 1$ into a product of three factors.

6-2 Integer Exponents

♦   Zero Exponents
♦   Negative-Integer Exponents

In this section we extend the exponent concept to include 0 and negative integers as exponents. Typical scientific expressions such as the following will then make sense:

The diameter of a red corpuscle is approximately 8×10^{-5} centimeter.

The amount of water found in the air as vapor is about 9×10^{-6} times the amount of water found in the sea.

The focal length of a thin lens is given by $f^{-1} = a^{-1} + b^{-1}$.

In extending the concept of exponent beyond the natural numbers, we will require that any new exponent symbol be defined in such a way that all five properties of exponents for natural numbers continue to hold. We want only one set of properties for all types of exponents, rather than a new set for each new type of exponent.

♦ ZERO EXPONENTS

We start by defining the 0 exponent. We want all the exponent properties to hold, even if some of the exponents are 0. Applying Property 1, with $a \neq 0$, we would get

$$a^0 a = a^0 a^1 = a^{0+1} = a^1 = a$$

Dividing both sides of $a^0 a = a$ by a yields $a^0 = 1$. This suggests that a^0 should be defined as 1 for all nonzero real numbers a. If we were to try to define 0^0, it would have to be either 0 or 1, since Property 1 would give

$$0^0 \cdot 0^0 = 0^{0+0} = 0^0 \qquad \text{{\small \color{blue} The only real numbers satisfying }} x \cdot x = x \text{ {\small \color{blue}are 0 and 1.}}$$

However, neither of these alternatives works satisfactorily. One reason for this will be given later, in Section 10-1. For now, we will simply state that 0^0 is undefined.

Definition of 0 Exponents

For all real numbers $a \neq 0$:

$$a^0 = 1$$

0^0 is not defined.

Example 1
Evaluating Expressions with 0 Exponents

Simplify:

(A) 5^0 **(B)** $(\frac{1}{3})^0$ **(C)** t^0, $t \neq 0$ **(D)** $(x^2y^3)^0$, x, $y \neq 0$

(E) $\left(\dfrac{x^2y^3}{z^4}\right)^0$, x, y, $z \neq 0$

Solution   All are equal to 1.

Matched Problem 1   Simplify:

(A) 12^0 **(B)** 999^0 **(C)** $(\frac{2}{7})^0$
(D) x^0, $x \neq 0$ **(E)** $(m^3n^3)^0$, m, $n \neq 0$

In Section 2-1, a nonzero constant was given degree 0 when treated as a polynomial. You can now see why this was done: for $a \neq 0$, $a = ax^0$.

♦ **NEGATIVE-INTEGER EXPONENTS**

To get an idea of how a negative-integer exponent should be defined, we can proceed as we did in considering how to define the 0 exponent. Again, we want all the exponent properties to hold, even for negative-integer exponents. Using Property 1, for n a positive integer, we would get

$$a^n a^{-n} = a^{n-n} = a^0 = 1$$

That is, a^{-n} must be the multiplicative inverse, or reciprocal, of a^n. In other words,

$$a^{-n} = \frac{1}{a^n} \qquad \text{Since } a^3a^{-3} = a^0 = 1, \ a^{-3} \text{ must be } \frac{1}{a^3}.$$

This kind of reasoning leads us to the following definition:

Definition of Negative-Integer Exponents

If n is a positive integer and a is a nonzero real number, then

$$a^{-n} = \frac{1}{a^n}$$

If n is a positive integer and $a \neq 0$, then

$$\frac{1}{a^{-n}} = \frac{1}{\dfrac{1}{a^n}} = 1 \cdot \frac{a^n}{1} = a^n$$

Thus, for example, $a^{-7} = 1/a^7$ and $a^7 = 1/a^{-7}$. That is, changing the sign of the exponent corresponds to moving the expression between the numerator and denominator.

Example 2
Rewriting Expressions Having Negative Exponents

Rewrite in a form using positive exponents or no exponents:

(A) a^{-7} **(B)** $\dfrac{1}{x^{-8}}$ **(C)** 10^{-3} **(D)** $\dfrac{x^{-3}}{y^{-5}}$

Solution   **(A)** $a^{-7} = \dfrac{1}{a^7}$ **(B)** $\dfrac{1}{x^{-8}} = x^8$

(C) $10^{-3} = \dfrac{1}{10^3}$ or $\dfrac{1}{1{,}000}$ or 0.001

(D) $\dfrac{x^{-3}}{y^{-5}} = \dfrac{x^{-3}}{1} \cdot \dfrac{1}{y^{-5}} = \dfrac{1}{x^3} \cdot \dfrac{y^5}{1} = \dfrac{y^5}{x^3}$

Matched Problem 2   Rewrite in a form using positive exponents or no exponents:

(A) x^{-5} **(B)** $\dfrac{1}{y^{-4}}$ **(C)** 10^{-2} **(D)** $\dfrac{m^{-2}}{n^{-3}}$

We can now replace Property 5 of exponents with a simpler form that does not have any restrictions on the relative size of the exponents:

$$\frac{a^m}{a^n} = a^{m-n} = \frac{1}{a^{n-m}} \qquad a \neq 0$$

For example,

$$\frac{a^8}{a^{10}} = a^{8-10} = a^{-2} = \frac{1}{a^2} = \frac{1}{a^{10-8}}$$

Example 3
Changing the Exponent Form

Rewrite in a form using only negative exponents:

(A) $\dfrac{2^5}{2^8}$ **(B)** $\dfrac{a^{-3}}{a^6}$, $a \neq 0$

Rewrite in a form using only positive exponents:

(C) $\dfrac{2^5}{2^8}$ **(D)** $\dfrac{a^{-3}}{a^6}$, $a \neq 0$

Solution   **(A)** $\dfrac{2^5}{2^8} = 2^{5-8} = 2^{-3}$ **(B)** $\dfrac{a^{-3}}{a^6} = a^{-3-6} = a^{-9}$

(C) $\dfrac{2^5}{2^8} = \dfrac{1}{2^{8-5}} = \dfrac{1}{2^3}$ **(D)** $\dfrac{a^{-3}}{a^6} = \dfrac{1}{a^{6-(-3)}} = \dfrac{1}{a^9}$

Matched Problem 3   Rewrite in a form using only negative exponents:

(A) $\dfrac{3^4}{3^9}$ **(B)** $\dfrac{x^{-2}}{x^3}$

Rewrite in a form using only positive exponents:

(C) $\dfrac{3^4}{3^9}$ **(D)** $\dfrac{x^{-2}}{x^3}$

Table 1 provides a summary of all our work on exponents to this point.

Table 1   **Integer Exponents and Their Properties (Summary)**

Definition of a^n (n an Integer, a a Real Number)	Properties of Exponents (m and n Integers, a and b Real Numbers)
1. If n is a positive integer, then $a^n = \underbrace{a \cdot a \cdot \dots \cdot a}_{n \text{ factors}}$	**1.** $a^m a^n = a^{m+n}$ **2.** $(a^n)^m = a^{mn}$ **3.** $(ab)^m = a^m b^m$
2. If $n = 0$, then $a^n = 1 \qquad a \neq 0$	**4.** $\left(\dfrac{a}{b}\right)^m = \dfrac{a^m}{b^m} \qquad b \neq 0$
3. If n is a positive integer, then $a^{-n} = \dfrac{1}{a^n} \qquad a \neq 0$	**5.** $\dfrac{a^m}{a^n} = a^{m-n} = \dfrac{1}{a^{n-m}} \qquad a \neq 0$

It is important to realize that the situation determines whether positive or negative exponents are the more useful form in a given expression. You will see an example in Section 6-3 where it is desirable to leave negative exponents in the expression. However, for the remainder of this section we will concentrate on rewriting expressions in forms that use only positive exponents. This is commonly done to make comparisons easier.

Example 4
Changing the Form of an Expression

Rewrite in a form using only positive exponents:

(A) $a^5 a^{-2}$ **(B)** $(a^{-3}b^2)^{-2}$ **(C)** $\left(\dfrac{a^{-5}}{a^{-2}}\right)^{-1}$

(D) $\dfrac{4x^{-3}y^{-5}}{6x^{-4}y^3}$ **(E)** $\dfrac{10^{-4} \cdot 10^2}{10^{-3} \cdot 10^5}$ **(F)** $\left(\dfrac{m^{-3}m^3}{n^{-2}}\right)^{-2}$

Solution   **(A)** $a^5 a^{-2} \boxed{= a^{5+(-2)} = a^{5-2}} = a^3$

(B) $(a^{-3}b^2)^{-2} \boxed{= (a^{-3})^{-2}(b^2)^{-2}} = a^6 b^{-4} = \dfrac{a^6}{b^4}$

(C) $\left(\dfrac{a^{-5}}{a^{-2}}\right)^{-1} \boxed{= \dfrac{(a^{-5})^{-1}}{(a^{-2})^{-1}}} = \dfrac{a^5}{a^2} = a^3$

(D) $\dfrac{4x^{-3}y^{-5}}{6x^{-4}y^3} \boxed{= \dfrac{2x^{-3-(-4)}}{3y^{3-(-5)}} = \dfrac{2x^{-3+4}}{3y^{3+5}}} = \dfrac{2x}{3y^8}$

It may be easier in problems like this to change to positive exponents first:

$$\frac{4x^{-3}y^{-5}}{6x^{-4}y^3} = \frac{2x^4}{3x^3y^3y^5} = \frac{2x}{3y^8}$$

(E) $\dfrac{10^{-4} \cdot 10^2}{10^{-3} \cdot 10^5} = \dfrac{10^{-4+2}}{10^{-3+5}} = \dfrac{10^{-2}}{10^2} = \dfrac{1}{10^4} = \dfrac{1}{10,000} = 0.0001$

(F) $\left(\dfrac{m^{-3}m^3}{n^{-2}}\right)^{-2} = \left(\dfrac{m^{-3+3}}{n^{-2}}\right)^{-2} = \left(\dfrac{m^0}{n^{-2}}\right)^{-2} = \left(\dfrac{1}{n^{-2}}\right)^{-2} = \dfrac{1^{-2}}{(n^{-2})^{-2}} = \dfrac{1}{n^4}$

Matched Problem 4   Rewrite in a form using only positive exponents:

(A) $x^{-2}x^6$ **(B)** $(x^3y^{-2})^{-2}$ **(C)** $\left(\dfrac{x^{-6}}{x^{-2}}\right)^{-1}$

(D) $\dfrac{8m^{-2}n^{-4}}{6m^{-5}n^2}$ **(E)** $\dfrac{10^{-3} \cdot 10^5}{10^{-2} \cdot 10^6}$

 As stated earlier, properties of exponents involve products and quotients, not sums and differences. For example,

$$\frac{a^{-2}y}{b} = \frac{y}{a^2b} \qquad \text{but} \qquad \frac{a^{-2}+y}{b} \neq \frac{y}{a^2b}$$

The plus sign in the numerator $a^{-2} + y$ makes a big difference:

$$\frac{a^{-2}+y}{b}$$

represents a compact way of writing a complex fraction. To simplify it, we replace a^{-2} with $1/a^2$, and then proceed as in Section 3-5:

$$\frac{a^{-2}+y}{b} = \frac{\frac{1}{a^2}+y}{b} = \frac{a^2\left(\frac{1}{a^2}+y\right)}{a^2b} = \frac{1+a^2y}{a^2b}$$

 Also, consider

$$(a^{-1}b^{-1})^2 = a^{-2}b^{-2}$$
$$= \frac{1}{a^2b^2}$$

If the product of a^{-1} and b^{-1} is replaced by their sum, the expression is not simplified so easily. In particular, note that

$$(a^{-1} + b^{-1})^2 \neq a^{-2} + b^{-2}$$

The expression $(a^{-1} + b^{-1})^2$ is correctly simplified in Example 5.

Moreover, note that

$$a^{-2} + b^{-2} \neq \frac{1}{a^2 + b^2}$$

You are asked to simplify $a^{-2} + b^{-2}$ in Matched Problem 5.

Example 5
Simplifying

Rewrite in a simplified form using only positive exponents:

(A) $\dfrac{3^{-2} + 2^{-1}}{11}$ **(B)** $(a^{-1} + b^{-1})^2$

Solution **(A)** $\dfrac{3^{-2} + 2^{-1}}{11} = \dfrac{\frac{1}{3^2} + \frac{1}{2}}{11} = \dfrac{\frac{2}{18} + \frac{9}{18}}{11} = \boxed{\dfrac{\frac{11}{18}}{ } = \dfrac{11}{18} \div 11} = \dfrac{11}{18} \cdot \dfrac{1}{11} = \dfrac{1}{18}$

(B) $(a^{-1} + b^{-1})^2 = \left(\dfrac{1}{a} + \dfrac{1}{b}\right)^2 = \left(\dfrac{b + a}{ab}\right)^2 = \dfrac{b^2 + 2ab + a^2}{a^2b^2}$

Alternatively,

$$(a^{-1} + b^{-1})^2 = a^{-2} + 2a^{-1}b^{-1} + b^{-2} = \dfrac{1}{a^2} + \dfrac{2}{ab} + \dfrac{1}{b^2}$$

$$= \dfrac{b^2 + 2ab + a^2}{a^2b^2}$$

Matched Problem 5 Rewrite in a simplified form using only positive exponents:

(A) $\dfrac{2^{-2} + 3^{-1}}{5}$ **(B)** $a^{-2} + b^{-2}$

Sometimes taking out common factors involving negative exponents can make rewriting an expression easier. For example, x^{-4} can be viewed as a common factor in the terms of the expression $x^{-3} + x^{-4}$, by recognizing that

$$x^{-3} = x^{-4}x$$

The expression then can be rewritten as

$$x^{-3} + x^{-4} = x^{-4}(x + 1)$$

Example 6
**Simplifying by Taking
Out a Common Factor**

Rewrite $x^{-3}y^2 + x^{-4}y^3$ as a single fraction involving only positive exponents.

Solution $x^{-3}y^2 + x^{-4}y^3$ Rewrite x^{-3} as $x^{-4}x$.
$\quad = x^{-4}xy^2 + x^{-4}y^3$ Factor out x^{-4}.
$\quad = x^{-4}(xy^2 + y^3)$ Change to positive exponents.
$\quad = \dfrac{xy^2 + y^3}{x^4}$

Matched Problem 6 Rewrite $x^3y^{-2} + x^4y^{-3}$ as a single fraction involving only positive exponents.

Answers to
Matched Problems

1. All are equal to 1.

2. (A) $\dfrac{1}{x^5}$ (B) y^4 (C) $\dfrac{1}{10^2}$ or $\dfrac{1}{100}$ or 0.01   (D) $\dfrac{n^3}{m^2}$

3. (A) 3^{-5} (B) x^{-5} (C) $\dfrac{1}{3^5}$ (D) $\dfrac{1}{x^5}$

4. (A) x^4 (B) $\dfrac{y^4}{x^6}$ (C) x^4 (D) $\dfrac{4m^3}{3n^6}$ (E) $\dfrac{1}{10^2}$ or $\dfrac{1}{100}$ or 0.01

5. (A) $\dfrac{7}{60}$ (B) $\dfrac{a^2 + b^2}{a^2b^2}$ 6. $\dfrac{x^3y + x^4}{y^3}$

EXERCISE 6-2

A *Rewrite in a simplified form using only positive exponents.*

1. 23^0 2. 10^0 3. y^0

4. x^0 5. 3^{-3} 6. 2^{-2}

7. m^{-7} 8. x^{-4} 9. $\dfrac{1}{4^{-3}}$

10. $\dfrac{1}{3^{-2}}$ 11. $\dfrac{1}{y^{-5}}$ 12. $\dfrac{1}{x^{-3}}$

13. $10^7 \cdot 10^{-5}$ 14. $10^{-4} \cdot 10^6$ 15. $y^{-3}y^4$

16. x^6x^{-2} 17. u^5u^{-5} 18. $m^{-3}m^3$

19. $\dfrac{10^3}{10^{-7}}$ 20. $\dfrac{10^8}{10^{-3}}$ 21. $\dfrac{x^9}{x^{-2}}$

22. $\dfrac{a^8}{a^{-4}}$ 23. $\dfrac{z^{-2}}{z^3}$ 24. $\dfrac{b^{-3}}{b^5}$

25. $\dfrac{10^{-1}}{10^6}$ 26. $\dfrac{10^{-4}}{10^2}$ 27. $(10^{-4})^{-3}$

28. $(2^{-3})^{-2}$ 29. $(y^{-2})^{-4}$ 30. $(x^{-5})^{-2}$

31. $(u^{-5}v^{-3})^{-2}$ 32. $(x^{-3}y^{-2})^{-1}$ 33. $(x^2y^{-3})^2$

B 34. $(x^{-2}y^3)^2$ 35. $(x^{-2}y^3)^{-1}$ 36. $(x^2y^{-3})^{-1}$

37. $(m^2)^0$ 38. $1{,}231^0$ 39. $\dfrac{10^{-3}}{10^{-5}}$

40. $\dfrac{10^{-2}}{10^{-4}}$ 41. $\dfrac{y^{-2}}{y^{-3}}$ 42. $\dfrac{x^{-3}}{x^{-2}}$

43. $\dfrac{10^{-13} \cdot 10^{-4}}{10^{-21} \cdot 10^3}$ 44. $\dfrac{10^{23} \cdot 10^{-11}}{10^{-3} \cdot 10^{-2}}$ 45. $\dfrac{18 \times 10^{12}}{6 \times 10^{-4}}$

46. $\dfrac{8 \times 10^{-3}}{2 \times 10^{-5}}$ 47. $\left(\dfrac{y}{y^{-2}}\right)^3$ 48. $\left(\dfrac{x^2}{x^{-1}}\right)^2$

49. $\dfrac{1}{(3mn)^{-2}}$ 50. $(2cd^2)^{-3}$ 51. $(2mn^{-3})^3$

52. $(3x^3y^{-2})^2$ 53. $(m^4n^{-5})^{-3}$ 54. $(x^{-3}y^2)^{-2}$

55. $(2^23^{-3})^{-1}$ 56. $(2^{-3}3^2)^{-2}$

57. $(10^{12} \cdot 10^{-12})^{-1}$ 58. $(10^2 \cdot 3^0)^{-2}$

59. $\dfrac{8x^{-3}y^{-1}}{6x^2y^{-4}}$ 60. $\dfrac{9m^{-4}n^3}{12m^{-1}n^{-1}}$

61. $\dfrac{2a^6b^{-2}}{16a^{-3}b^2}$ 62. $\dfrac{4x^{-2}y^{-3}}{2x^{-3}y^{-1}}$

63. $\left(\dfrac{x^{-1}}{x^{-8}}\right)^{-1}$ 64. $\left(\dfrac{n^{-3}}{n^{-2}}\right)^{-2}$

65. $\left(\dfrac{m^{-2}n^3}{m^4n^{-1}}\right)^2$ 66. $\left(\dfrac{x^4y^{-1}}{x^{-2}y^3}\right)^2$

67. $\left(\dfrac{6nm^{-2}}{3m^{-1}n^2}\right)^{-3}$ 68. $\left(\dfrac{2x^{-3}y^2}{4xy^{-1}}\right)^{-2}$

69. $\left(\dfrac{ab^2}{a^{-2}b^{-3}}\right)^{-1}$ 70. $\left(\dfrac{a^2b}{a^{-1}b^{-2}}\right)^{-3}$

71. $\left(\dfrac{3x^{-1}y^2}{4xy^{-2}}\right)^{-2}$ 72. $\left(\dfrac{2xy^{-1}}{3x^{-1}y}\right)^{-1}$

73. $\left(\dfrac{3x^{-3}y^2}{2x^2y^{-3}}\right)^{-3}$ 74. $\left(\dfrac{4x^3y^{-2}}{3x^{-2}y^{-3}}\right)^{-2}$

75. $\left[\left(\dfrac{x^{-2}y^3t}{x^{-3}y^{-2}t^2}\right)^2\right]^{-1}$ 76. $\left[\left(\dfrac{u^3v^{-1}w^{-2}}{u^{-2}v^{-2}w}\right)^{-2}\right]^2$

77. $\left(\dfrac{2^2x^2y^0}{8x^{-1}}\right)^{-2}\left(\dfrac{x^{-3}}{x^{-5}}\right)^3$

78. $\left(\dfrac{3^3x^0y^{-2}}{2^3x^3y^{-5}}\right)^{-1}\left(\dfrac{3^3x^{-1}y}{2^2x^2y^{-2}}\right)^2$

C 79. $(a^2 - b^2)^{-1}$ 80. $(x + 2)^{-2}$

81. $\dfrac{x^{-1} + y^{-1}}{x + y}$ 82. $\dfrac{2^{-1} + 3^{-1}}{25}$

83. $\dfrac{c - d}{c^{-1} - d^{-1}}$ 84. $\dfrac{12}{2^{-2} + 3^{-1}}$

85. $(x^{-1} + y^{-1})^{-1}$

86. $(2^{-2} + 3^{-2})^{-1}$

87. $(x^{-1} - y^{-1})^2$

88. $(10^{-2} + 10^{-3})^{-1}$

89. $\left(\dfrac{x^{-1}}{x^{-1} - y^{-1}}\right)^{-1}$

90. $\left[\dfrac{u^{-2} - v^{-2}}{(u^{-1} - v^{-1})^2}\right]^{-1}$

95. $x^{-2}y^2 - x^2y^{-2} = \dfrac{?}{x^2y^2}$

96. $x^{-5}y^3 - x^3y^{-5} = \dfrac{?}{x^5y^5}$

Replace each question mark by an appropriate expression.

91. $a^{-4}b + a^{-3}c = \dfrac{?}{a^4}$

92. $a^{-1}b + ac = \dfrac{?}{a}$

93. $a^{-1}b^{-2} + a^{-2}b^{-1} = \dfrac{?}{a^2b^2}$

94. $a^{-2}b^{-4} + a^{-3}b^{-3} = \dfrac{?}{a^3b^4}$

Factor so that one factor is a single term and the other is a polynomial with no common factor. For example,
$x^{-2}y^{-3} + x^{-3}y^{-1} = x^{-3}y^{-3}(x + y^2)$.

97. $a^{-3}b^{-5} + a^{-4}b^{-2}$

98. $x^{-4}y^{-1} + x^{-6}y^2$

99. $x^{-1}y^{-2}z^{-3} + x^{-3}y^{-4}z^{-2}$

100. $a^{-5}b^{-4}c^{-3} + a^{-2}b^{-3}c^{-4}$

6-3 Scientific Notation and Applications

Work in science and engineering often involves the use of very, very large numbers:

The estimated free oxygen of the earth weighs approximately
1,500,000,000,000,000,000,000 grams.

Also involved in the use of very, very small numbers:

The probable mass of a hydrogen atom is
0.000 000 000 000 000 000 000 001 7 gram.

Writing and working with numbers of this type in standard decimal notation is generally awkward. It is often convenient to represent such numbers in **scientific notation**—that is, as the product of a number in the interval [1, 10) and a power of 10. Any positive number, however large or small, that has a finite decimal expansion can be represented in scientific notation.

Here are some examples of numbers in standard notation and scientific notation:

$$
\begin{aligned}
100 &= 1 \times 10^2 & 0.001 &= 1 \times 10^{-3} \\
1{,}000{,}000 &= 1 \times 10^6 & 0.000\ 01 &= 1 \times 10^{-5} \\
5 &= 5 \times 10^0 & 0.7 &= 7 \times 10^{-1} \\
35 &= 3.5 \times 10 & 0.083 &= 8.3 \times 10^{-2} \\
430 &= 4.3 \times 10^2 & 0.004\ 3 &= 4.3 \times 10^{-3} \\
5{,}870 &= 5.87 \times 10^3 & 0.000\ 687 &= 6.87 \times 10^{-4} \\
8{,}910{,}000 &= 8.91 \times 10^6 & 0.000\ 000\ 36 &= 3.6 \times 10^{-7}
\end{aligned}
$$

From these examples you can discover a simple rule that relates the number of decimal places the decimal is moved to the power of 10 that is used:

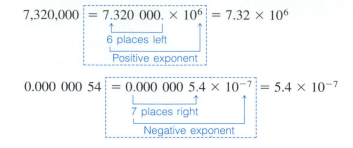

In summary:

> ### Converting Positive Numbers to Scientific Notation
>
> 1. If the number is greater than or equal to 10, the number of places the decimal point is shifted left appears as a positive exponent.
> 2. If the number is less than 10 but greater than or equal to 1, the decimal place is not shifted at all, and the exponent used is 0.
> 3. If the number is less than 1, the number of places the decimal point is shifted right appears as a negative exponent.

Example 1
Converting to Scientific Notation

Write in scientific notation:

(A) 123   (B) 0.001 23   (C) 1.23

Solution   (A) $123 = 1.23 \times 10^2 = 1.23 \times 10^2$

2 places left

Positive exponent

(B) $0.001\,23 = 0001.23 \times 10^{-3} = 1.23 \times 10^{-3}$

3 places right

Negative exponent

(C) $1.23 = 1.23 \times 10^0 = 1.23 \times 10^0 = 1.23$ Numbers in the interval [1, 10) are usually left in standard notation.

0 places moved

0 exponent

Matched Problem 1   Write in scientific notation:

(A) 450   (B) 27,000   (C) 0.05
(D) 0.000 006 3   (E) 0.000 1   (F) 10,000

The process is reversed to convert from scientific notation to standard notation. Numbers in scientific notation with a positive power of 10 can be converted to standard notation simply by multiplying. If the power of 10 involved is negative, we move the power to a denominator and divide.

Example 2
Converting to
Standard Notation

Write in standard notation:

(A) 4.32×10^6 **(B)** 4.32×10^{-5}

Solution   **(A)** $4.32 \times 10^6 = 4,320,000$

6 places right

Positive exponent 6

(B) $4.32 \times 10^{-5} = \dfrac{4.32}{10^5} = 0.000\ 043\ 2$

5 places left

Negative exponent −5

Matched Problem 2   Write in standard notation:

(A) 1.59×10^7 **(B)** 1.59×10^{-6}

Scientific notation may be used to evaluate complicated arithmetic problems.

Example 3
Evaluating Complicated
Expressions

Convert to scientific notation and evaluate:

$$\frac{(0.000\ 000\ 000\ 000\ 026)(720)}{(48,000,000,000)(0.001\ 3)}$$

Solution

$$\frac{(0.000\ 000\ 000\ 000\ 026)(720)}{(48,000,000,000)(0.001\ 3)}$$ Write each number in scientific notation.

$$= \frac{2.6 \times 10^{-14})(7.2 \times 10^2)}{(4.8 \times 10^{10})(1.3 \times 10^{-3})}$$ Collect all powers of 10 together.

$$= \frac{(2.6)(7.2)}{(4.8)(1.3)} \cdot \frac{(10^{-14})(10^2)}{(10^{10})(10^{-3})}$$ Calculate the power of 10 and numerical part separately.

$$= 3 \times 10^{-19}$$

If you try to work Example 3 directly using a hand calculator, you will probably find that some of the numbers cannot be entered unless they are first converted to scientific notation. If you have a calculator, try it. Some calculators can compute directly in scientific notation and read out in scientific notation.

Matched Problem 3   Convert to scientific notation and evaluate:

$$\frac{(42,000)(0.000\ 000\ 000\ 09)}{(600,000,000,000)(0.000\ 21)}$$

Figure 1 on page 258 shows the relative sizes of several familiar objects on a power-of-10 scale. Note that 10^{10} is not just double 10^5. If a normal scale were used here, each interval on the axis would be 100,000 times as long as the one immediately below it.

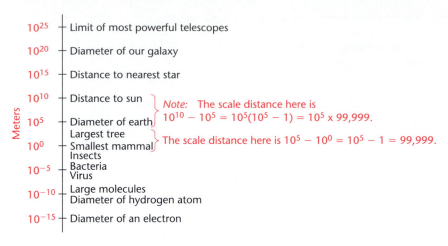

Figure 1

We are able to look back into time by looking out into space. Since light travels at a fast but finite rate, we see heavenly bodies not as they exist now, but as they existed some time in the past. If the distance between the sun and the earth is approximately 9.3×10^7 miles and if light travels at the rate of approximately 1.86×10^5 miles per second, how long does it take the image of the sun to reach us? Using the rate-time formula,

$$t = \frac{d}{r} \approx \frac{9.3 \times 10^7}{1.86 \times 10^5} = 5 \times 10^2 = 500 \text{ seconds or } \frac{500}{60} \approx 8.3 \text{ minutes}$$

Hence, we always see the sun as it was approximately 8.3 minutes ago.

Answers to Matched Problems

1. **(A)** 4.5×10^2 **(B)** 2.7×10^4 **(C)** 5×10^{-2}
   (D) 6.3×10^{-6} **(E)** 1.0×10^{-4} **(F)** 1.0×10^4
2. **(A)** 15,900,000   **(B)** 0.000 001 59   **3.** 3×10^{-14}

EXERCISE 6-3

A *Write in scientific notation.*

1. 70
2. 50
3. 800
4. 600
5. 80,000
6. 600,000
7. 0.008
8. 0.06
9. 0.000 000 08
10. 0.000 06
11. 52
12. 35
13. 0.63
14. 0.72
15. 340
16. 270
17. 0.085
18. 0.032
19. 6,300
20. 5,200
21. 0.000 006 8
22. 0.000 72

Write in standard notation.

23. 8×10^2
24. 5×10^2
25. 4×10^{-2}
26. 8×10^{-2}
27. 3×10^5
28. 6×10^6
29. 9×10^{-4}
30. 2×10^{-5}
31. 5.6×10^4
32. 7.1×10^3
33. 9.7×10^{-3}
34. 8.6×10^{-4}
35. 4.3×10^5
36. 8.8×10^6
37. 3.8×10^{-7}
38. 6.1×10^{-6}

B *Write each number in scientific notation.*

39. 5,460,000,000
40. 42,700,000
41. 0.000 000 072 9
42. 0.000 072 3
43. 0.000 000 000 012 3
44. 0.000 000 000 000 012
45. 6,789,000,000,000
46. 3,450,000,000,000,000
47. 0.102 003 004
48. 0.001 234 000 000 005
49. 1,234,000,567,000
50. 3,210,000,000,456

51. The maximum distance of the planet Jupiter from the sun is 507,000,000 miles.

52. The World Health Organization expects 40,000,000 persons to be infected with the HIV virus by the end of the decade.

53. The energy of a laser beam can go as high as 10,000,000,000,000 watts.

54. The distance that light travels in 1 year is called a light-year. It is approximately 5,870,000,000,000 miles.

55. The nucleus of an atom has a diameter of a little more than $\frac{1}{100,000}$ that of the whole atom.

56. The mass of one water molecule, in grams, is 0.000 000 000 000 000 000 000 03.

Write each number in standard notation.

57. 8.35×10^{10}

58. 3.46×10^9

59. 6.14×10^{-12}

60. 6.23×10^{-7}

61. $2.000\ 01 \times 10^9$

62. $3.010\ 001 \times 10^4$

63. $1.002\ 003 \times 10^{-5}$

64. $7.068\ 09 \times 10^{-6}$

65. The probability of drawing a royal flush in poker is approximately 1.539×10^{-6}.

66. The probability of matching 6 numbers drawn at random from among 48 numbers is approximately 8.149×10^{-8}.

67. The number of possible winning combinations in a state lottery that selects 6 numbers at random from among 48 is approximately 1.227×10^7.

68. The number of possible 5-card poker hands is approximately 2.597×10^6.

69. The diameter of the sun is approximately 8.65×10^5 miles.

70. The distance from the earth to the sun is approximately 9.3×10^7 miles.

71. The probable mass of a hydrogen atom is 1.7×10^{-24} gram.

72. The diameter of a red corpuscle is approximately 7.5×10^{-5} centimeter.

Simplify and express the answer in scientific notation.

73. $(3 \times 10^{-6})(3 \times 10^{10})$

74. $(4 \times 10^5)(2 \times 10^{-3})$

75. $(2 \times 10^3)(3 \times 10^{-7})$

76. $(4 \times 10^{-8})(2 \times 10^5)$

77. $\dfrac{6 \times 10^{12}}{2 \times 10^7}$

78. $\dfrac{9 \times 10^8}{3 \times 10^5}$

79. $\dfrac{15 \times 10^{-2}}{3 \times 10^{-6}}$

80. $\dfrac{12 \times 10^3}{4 \times 10^{-4}}$

Convert each numeral to scientific notation and simplify. Express your answer in scientific notation and in standard notation.

81. $\dfrac{(90,000)(0.000\ 002)}{0.006}$

82. $\dfrac{(0.0006)(4,000)}{0.000\ 12}$

83. $\dfrac{(60,000)(0.000\ 003)}{(0.0004)(1,500,000)}$

84. $\dfrac{(0.000\ 039)(140)}{(130,000)(0.000\ 21)}$

C *Write each answer in scientific notation and in standard notation.*

85. If the mass of the earth is 6×10^{27} grams and each gram is 1.1×10^{-6} ton, find the mass of the earth in tons.

86. In 1929 Vernadsky, a biologist, estimated that all the free oxygen of the earth weighs 1.5×10^{21} grams and is produced by life alone. If 1 gram is approximately 2.2×10^{-3} pound, what is the amount of free oxygen in pounds?

87. It takes a high-speed computer approximately 2.5×10^{-7} second to perform a single addition. How many additions can such a computer perform in 1 second? A nanosecond is 10^{-9} second. How many nanoseconds does it take such a computer to perform a single addition?

88. If electricity in a computer circuit travels at the speed of light (1.86×10^5 miles per second), how far will it travel in the time the computer in Problem 87 takes to perform a single addition? (Size of circuits is a critical problem in computer design.) Give your answer in miles and in feet.

89. India had a 1990 population of 844 million people and a land area of 1,269,000 square miles. What was the population density, expressed in persons per square mile?

90. The United States had a 1990 population of 249.6 million people and a land area of 3,539,000 square miles. What was the population density?

91. In 1990, the United States had a violent crime rate of 731.8 per 100,000 people and a population of 249.6 million people. How many violent crimes occurred that year? Round your answer to the nearest thousand.

92. In 1990, the United Republic of Tanzania had a population of 26.07 million and a growth rate of 37 persons per 1,000 of population. By how much did the population grow that year? Round your answer to the nearest thousand.

6-4 Rational Exponents

♦ Roots of Real Numbers
♦ Rational Exponents

In this section we extend the definition of exponents to include rational numbers as exponents. Expressions such as $7^{1/3}$ will then be meaningful. Again, we want the five exponent properties to continue to be true. If the expression $7^{1/3}$ is to make sense, we would have

$$(7^{1/3})^3 = 7^{(1/3)\cdot 3} = 7^1 = 7$$

so $7^{1/3}$ is a number whose cube is 7; that is, $7^{1/3}$ must name a *cube root* of 7. Thus, we must first consider cube roots and other roots of real numbers before developing rational exponents further.

♦ ROOTS OF REAL NUMBERS

Recall that a square root of a number a is a number b such that $b^2 = a$, and a cube root of a number a is a number c such that $c^3 = a$.

Definition of an *n*th Root

For n a natural number:

a is an nth root of b if $a^n = b$.   2 is a 4th root of 16, since $2^4 = 16$.
  −3 is a cube root of −27, since $(-3)^3 = -27$.

Any positive number has two square roots. For example, 4 has both 2 and −2 as square roots. Zero has only one square root, namely 0 itself. Negative numbers have no real square roots, since no real number squared can be negative.

For cube roots the situation is simpler. Every number has exactly one real cube root. More generally, we have the following result:

Number of Real *n*th Roots of a Real Number *a*

	n EVEN	n ODD
a positive	Two real nth roots	One real nth root
	−2 and 2 are both 4th roots of 16.	2 is the only real cube root of 8.
a negative	No real nth root	One real nth root
	−4 has no real square root.	−2 is the only real cube root of −8.

In Section 6-8, we will see that additional roots can be considered in an extended number system. Until then, we will interpret root to mean a real number root.

We will use the **nth root radical, $\sqrt[n]{a}$,** to denote an nth root of a. The symbol $\sqrt{}$ is called a **radical;** n is called the **index** of the radical, and a is the **radicand.**

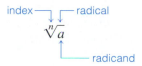

When the index is 2, it is usually omitted. That is, when dealing with square roots, we continue to use $\sqrt{a}$ rather than $\sqrt[2]{a}$. If there are two nth roots, $\sqrt[n]{a}$ denotes the positive root, called the **principal nth root.** The nth root radical is completely described as follows:

nth Root Radical, $\sqrt[n]{a}$		
RADICAND a	INDEX n EVEN	INDEX n ODD
Positive	$\sqrt[n]{a}$ is the positive nth root of a.	$\sqrt[n]{a}$ is the unique nth root of a.
Negative	$\sqrt[n]{a}$ is not a real number.	$\sqrt[n]{a}$ is the unique nth root of a.

An nth root also can be represented using exponents, as we shall see next.

♦ RATIONAL EXPONENTS

If all exponent properties are to continue to hold, even if some of the exponents are not integers, then we would have

$$(5^{1/2})^2 = 5^{2/2} = 5 \qquad \text{and} \qquad (x^{1/3})^3 = x^{3/3} = x$$

That is, $5^{1/2}$ is a number whose square is 5, so $5^{1/2}$ must name a square root of 5. Similarly, $x^{1/3}$ is a number whose cube is x, so $x^{1/3}$ must name a cube root of x.

More generally, if $a^{1/n}$ is to be defined, it must represent an nth root of a, since

$$(a^{1/n})^n = a^{n/n} = a^1 = a$$

We need to be careful in our definition when n is an even number because, from the result above, a will have either two real nth roots or none.

Definition of $a^{1/n}$

For n a positive integer,

$$a^{1/n} = \sqrt[n]{a} \qquad 5^{1/2} = \sqrt{5}; \;\; 5^{1/3} = \sqrt[3]{5}$$

The following table summarizes the definition of $a^{1/n}$ as $\sqrt[n]{a}$:

$a^{1/n}$		
BASE a	n EVEN	n ODD
Positive	$a^{1/n}$ is the positive nth root of a.	$a^{1/n}$ is the unique nth root of a.
Negative	$a^{1/n}$ is not a real number.	$a^{1/n}$ is the unique nth root of a.

You should note that this is exactly the same information given in the definition of $\sqrt[n]{a}$. Only the notation is changed.

Example 1
Evaluating nth Roots

Find each of the following:

(A) $4^{1/2}$ (B) $-4^{1/2}$ (C) $(-4)^{1/2}$
(D) $8^{1/3}$ (E) $(-8)^{1/3}$ (F) $0^{1/5}$

Solution

(A) $4^{1/2} = \sqrt{4} = 2$
(B) $-4^{1/2} = -\sqrt{4} = -2$ Remember that exponents precede subtraction in the order of operations. The minus sign here applies to all of $4^{1/2}$.

(C) $(-4)^{1/2}$ is not a real number.   Note the difference between parts (B) and (C). The minus sign here is applied to the 4 before the exponent is applied.

(D) $8^{1/3} = \sqrt[3]{8} = 2$
(E) $(-8)^{1/3} = \sqrt[3]{-8} = -2$
(F) $0^{1/5} = \sqrt[5]{0} = 0$

Matched Problem 1

Find each of the following:

(A) $9^{1/2}$ (B) $-9^{1/2}$ (C) $(-9)^{1/2}$
(D) $27^{1/3}$ (E) $(-27)^{1/3}$ (F) $0^{1/4}$

How should an expression such as $5^{2/3}$ be defined? Once more, if the properties of exponents are to continue to hold for all rational exponents, then $5^{2/3} = (5^{1/3})^2$; that is, $5^{2/3}$ must represent the square of the cube root of 5. That is, $5^{2/3}$ must be $(\sqrt[3]{5})^2$. This suggests that we define $a^{m/n}$ to be $(\sqrt[n]{a})^m$. This will, in fact, be our definition. However, we need to note that it does not make sense for every a, m, and n. To take an nth root, we need $n > 0$. Moreover, when $a < 0$, we need n to be an odd number. With these exceptions noted, we define $a^{m/n}$ to be $(\sqrt[n]{a})^m$, *whenever the root makes sense.*

Definition of $a^{m/n}$ and $a^{-m/n}$

For m and n natural numbers, $n \geq 2$, and a any real number for which $\sqrt[n]{a}$ is defined,

$$a^{m/n} = (a^{1/n})^m = (\sqrt[n]{a})^m \quad \text{and} \quad a^{-m/n} = \frac{1}{a^{m/n}} = \frac{1}{(\sqrt[n]{a})^m} \qquad a \neq 0$$

$$4^{3/2} = (4^{1/2})^3 = 2^3 = 8 \qquad 4^{-3/2} = \frac{1}{4^{3/2}} = \frac{1}{8} \qquad (-4)^{3/2} \text{ is not defined.}$$

$$(-32)^{3/5} = [(-32)^{1/5}]^3 = (-2)^3 = -8$$

It can be shown, though we will not do so, that all five properties of exponents discussed in Section 6-1 continue to hold.

The following useful relationship is an immediate consequence of the exponent properties:

$$a^{m/n} = (a^{1/n})^m = (a^m)^{1/n} \qquad \begin{aligned} 27^{2/3} &= (27^{1/3})^2 = 3^2 = 9 \\ 27^{2/3} &= (27^2)^{1/3} = (729)^{1/3} = 9 \end{aligned}$$

These equalities hold as long as all the quantities are defined.

Example 2

Evaluating Expressions with Rational Exponents

Simplify, and express your answers using only positive exponents.

(A) $8^{2/3}$ (B) $(-8)^{5/3}$ (C) $(3x^{1/3})(2x^{1/2})$

(D) $(2x^{1/3}y^{-2/3})^3$ (E) $\left(\dfrac{4x^{1/3}}{x^{1/2}}\right)^{1/2}$ (F) $(2a^{1/2} + b^{1/2})(a^{1/2} + 3b^{1/2})$

Solution

(A) $8^{2/3} = (8^{1/3})^2 = 2^2 = 4$ or $8^{2/3} = (8^2)^{1/3} = 64^{1/3} = 4$

(B) $(-8)^{5/3} = [(-8)^{1/3}]^5 = (-2)^5 = -32$ Taking the root first is easier than computing $[(-8)^5]^{1/3}$.

(C) $(3x^{1/3})(2x^{1/2}) = 6x^{1/3+1/2} = 6x^{5/6}$

(D) $(2x^{1/3}y^{-2/3})^3 = 2^3 x^{(1/3)3} y^{(-2/3)3} = 8xy^{-2} = \dfrac{8x}{y^2}$

(E) $\left(\dfrac{4x^{1/3}}{x^{1/2}}\right)^{1/2} = \dfrac{4^{1/2}x^{1/6}}{x^{1/4}} = \dfrac{2}{x^{1/4-1/6}} = \dfrac{2}{x^{1/12}}$

(F) $(2a^{1/2} + b^{1/2})(a^{1/2} + 3b^{1/2}) = 2a^{1/2}a^{1/2} + 6a^{1/2}b^{1/2} + a^{1/2}b^{1/2} + 3b^{1/2}b^{1/2}$
$$= 2a + 7a^{1/2}b^{1/2} + 3b$$

Matched Problem 2

Simplify, and express your answers using only positive exponents:

(A) $9^{3/2}$ (B) $(-27)^{4/3}$ (C) $(5y^{3/4})(2y^{1/3})$

(D) $(2x^{-3/4}y^{1/4})^4$ (E) $\left(\dfrac{8x^{1/2}}{x^{2/3}}\right)^{1/3}$ (F) $(x^{1/2} - 2y^{1/2})(3x^{1/2} + y^{1/2})$

The properties of exponents and the definition of $a^{m/n}$ can be used only as long as we are dealing with symbols that name real numbers. Can you resolve the following contradiction?

$$-1 = (-1)^1 = (-1)^{2/2} = [(-1)^2]^{1/2} = 1^{1/2} = 1$$

The third number listed, $(-1)^{2/2}$, must be $(\sqrt[2]{-1})^2$, which involves the even root of a negative number. Such a root is not defined. In evaluating or manipulating expressions involving rational exponents, we need to make sure that we are not dealing with undefined quantities. As long as the base in question is positive, we can use all the exponent properties. If a negative base is involved, restrictions must be imposed. One such restriction is to require all rational exponents to be reduced to lowest terms. To avoid such complications, we will usually restrict variables in problems with rational exponents to represent positive real numbers.

Answers to
Matched Problems

1. (A) 3   (B) -3 (C) Not a real number   (D) 3   (E) -3 (F) 0

2. (A) 27   (B) 81   (C) $10y^{13/12}$ (D) $\dfrac{16y}{x^3}$

(E) $\dfrac{2}{x^{1/18}}$ (F) $3x - 5x^{1/2}y^{1/2} - 2y$

EXERCISE 6-4

In Problems 1–92, all variables represent positive real numbers.

A *Most of the following are integers. Find them.*

1. $25^{1/2}$ **2.** $36^{1/2}$ **3.** $(-25)^{1/2}$

4. $(-36)^{1/2}$ **5.** $64^{1/3}$ **6.** $125^{1/3}$

7. $(-64)^{1/3}$ **8.** $(-125)^{1/3}$ **9.** $-64^{1/3}$

10. $-125^{1/3}$ **11.** $16^{3/2}$ **12.** $25^{3/2}$

13. $8^{2/3}$ **14.** $27^{2/3}$

Simplify, and express the answer using only positive exponents.

15. $x^{1/4}x^{3/4}$ **16.** $y^{1/5}y^{2/5}$ **17.** $\dfrac{x^{2/5}}{x^{3/5}}$

18. $\dfrac{a^{2/3}}{a^{1/3}}$ **19.** $(x^4)^{1/2}$ **20.** $(y^{1/2})^4$

21. $(a^3b^9)^{1/3}$ **22.** $(x^4y^2)^{1/2}$ **23.** $\left(\dfrac{x^9}{y^{12}}\right)^{1/3}$

24. $\left(\dfrac{m^{12}}{n^{16}}\right)^{1/4}$ **25.** $(x^{1/3}y^{1/2})^6$ **26.** $\left(\dfrac{u^{1/2}}{v^{1/3}}\right)^{12}$

B *Most of the following are rational numbers. Find them.*

27. $(\frac{4}{25})^{1/2}$ **28.** $(\frac{9}{4})^{1/2}$ **29.** $(\frac{4}{25})^{3/2}$

30. $(\frac{9}{4})^{3/2}$ **31.** $(\frac{1}{8})^{2/3}$ **32.** $(\frac{1}{27})^{2/3}$

33. $36^{-1/2}$ **34.** $25^{-1/2}$ **35.** $25^{-3/2}$

36. $16^{-3/2}$ **37.** $5^{3/2} \cdot 5^{1/2}$ **38.** $7^{2/3} \cdot 7^{4/3}$

39. $(3^6)^{-1/3}$ **40.** $(4^{-8})^{3/16}$ **41.** $(-\frac{8}{27})^{2/3}$

42. $(-\frac{8}{27})^{-2/3}$ **43.** $(-\frac{64}{125})^{-4/3}$ **44.** $(-\frac{64}{125})^{4/3}$

45. $(-16)^{-3/2}$ **46.** $-16^{-3/2}$

Simplify, and express the answer using only positive exponents.

47. $x^{1/4}x^{-3/4}$ **48.** $\dfrac{d^{2/3}}{d^{-1/3}}$

49. $n^{3/4}n^{-2/3}$ **50.** $m^{1/2}m^{-1/3}$

51. $(x^{-2/3})^{-6}$ **52.** $(y^{-8})^{1/16}$

53. $(4u^{-2}v^4)^{1/2}$ **54.** $(8x^3y^{-6})^{1/3}$

55. $(x^4y^6)^{-1/2}$ **56.** $(4x^{1/3}y^{3/2})^2$

57. $\left(\dfrac{x^{-2/3}}{y^{-1/2}}\right)^{-6}$ **58.** $\left(\dfrac{m^{-3}}{n^2}\right)^{-1/6}$

59. $(8x^6y^3)^{1/3}$ **60.** $(9x^6y^4)^{1/2}$

61. $(4x^{4/3}y^4)^{1/2}$ **62.** $(27x^{1/2}y^{2/3})^{1/2}$

63. $(8 \times 10^{-6})^{1/3}$ **64.** $(4 \times 10^{12})^{1/2}$

65. $\left(\dfrac{3xy^{-4}}{12x^{-1}y^2}\right)^{1/2}$ **66.** $\left(\dfrac{18x^2y^5}{3x^{-4}y^{-3}}\right)^{-1/2}$

67. $\left(\dfrac{x^{-1/5}y^{1/4}}{z^{1/2}}\right)^{10}$ **68.** $\left(\dfrac{x^{-1/3}y^{-1/2}}{z^{-4}}\right)^{-6}$

69. $\left(\dfrac{25x^5y^{-1}}{16x^{-3}y^{-5}}\right)^{1/2}$

70. $\left(\dfrac{8a^{-4}b^3}{27a^2b^{-3}}\right)^{1/3}$

71. $\left(\dfrac{8y^{1/3}y^{-1/4}}{y^{-1/12}}\right)^2$

72. $\left(\dfrac{9x^{1/3}x^{1/2}}{x^{-1/6}}\right)^{1/2}$

91. $(x^{m/4}x^{m/4})^{-2}$, $m > 0$

92. $(y^{m^2+1}y^{2m})^{1/(m+1)}$, $m > 0$

93. Rewrite the equation $x^{1/n}y^{1/n} = (xy)^{1/n}$ in radical notation.

94. Rewrite the equation $\dfrac{x^{1/n}}{y^{1/n}} = \left(\dfrac{x}{y}\right)^{1/n}$ in radical notation.

C In Problems 73–92, multiply, and express the answer using positive exponents only.

73. $x^{1/2}(x^4 - x^{1/5})$

74. $x^{1/3}(x^{3/2} + x^6)$

75. $3m^{3/4}(4m^{1/4} - 2m^8)$

76. $2x^{1/3}(3x^{2/3} - x^6)$

77. $(2x^{1/2} + y^{1/2})(x^{1/2} + y^{1/2})$

78. $(x^{1/2} + y^{1/2})(x^{1/2} - y^{1/2})$

79. $(x^{1/2} + y^{1/2})^2$

80. $(x^{1/2} - y^{1/2})^2$

81. $(x^{1/3} - 1)(x^{2/3} + x^{1/3} + 1)$

82. $(a^{1/3} + 1)(a^{2/3} - a^{1/3} + 1)$

83. $(x^{3/2} - y^{3/2})(x^{3/2} + y^{3/2})$

84. $(x^{1/4} + y^{1/4})(x^{1/4} - y^{1/4})$

85. $(a^{-1/2} + 3b^{-1/2})(2a^{-1/2} - b^{-1/2})$

86. $(x^{-1/2} - y^{-1/2})^2$

87. $(a^{n/2}b^{n/3})^{1/n}$, $n > 0$

88. $(a^{3/n}b^{3/m})^{1/3}$, $n > 0$, $m > 0$

89. $\left(\dfrac{x^{m+2}}{x^m}\right)^{1/2}$, $m > 0$

90. $\left(\dfrac{a^m}{a^{m-2}}\right)^{1/2}$, $m > 0$

95. Rewrite the equations $(x^{1/n})^n = x$ and $(x^n)^{1/n} = x$ in radical notation.

96. Rewrite the equation $x^{m/n} = x^{mk/nk}$ in radical notation.

97. Is $x^{1/n}y^{1/n} = (xy)^{1/n}$ always true? Test for $x = y = -2$ and $n = 2$. Describe the circumstances under which the equation will be false.

98. Is $\dfrac{x^{1/n}}{y^{1/n}} = \left(\dfrac{x}{y}\right)^{1/n}$ always true? Test for $x = -8$, $y = -2$, and $n = 2$. Describe the circumstances under which the equation will be false.

99. Are $(x^{1/n})^n = x$ and $(x^n)^{1/n} = x$ always true? Test for $x = -2$ and $n = 2$. Describe the circumstances under which the equations will be false.

100. Is $x^{m/n} = x^{mk/nk}$ always true? Test for $x = -8$, $n = 3$, $m = 1$, and $k = 2$. Describe the circumstances under which the equation will be false.

6-5   Radical Forms and Rational Exponents

♦ Converting Forms
♦ Properties of Radicals

In the preceding section we introduced two notations for the nth root of a,

$$\sqrt[n]{a} \qquad \text{Radical notation}$$
$$a^{1/n} \qquad \text{Rational exponent notation}$$

and for powers of an nth root,

$$(\sqrt[n]{a})^m \qquad \text{Radical notation}$$
$$a^{m/n} \qquad \text{Rational exponent notation}$$

It is important to be able to convert from one form to another so that we can choose the more convenient form for a particular purpose. In this section we illustrate such conversions and obtain properties of radicals corresponding to the five exponent properties.

◆ **CONVERTING FORMS**

The relationships

$$a^{m/n} = (a^{1/n})^m = (a^m)^{1/n}$$

convert to the following in radical notation:

$$a^{m/n} = (\sqrt[n]{a})^m = \sqrt[n]{a^m}$$ *Remember:* When *a* is negative, we require that *n* be odd.

The second radical form is usually the more useful. Examples 1 and 2 illustrate how to shift back and forth between exponent and radical forms. All variables represent positive real numbers.

Example 1
Converting to
Radical Form

Convert to radical form:

(A) $5^{1/2}$ **(B)** $x^{1/7}$ **(C)** $7m^{2/3}$
(D) $(3u^2v^3)^{3/5}$ **(E)** $y^{-2/3}$ **(F)** $(x^2 + y^2)^{1/2}$

Solution   **(A)** $5^{1/2} = \sqrt{5}$ *Remember:* $\sqrt{5}$ is the *positive* square root of 5.
(B) $x^{1/7} = \sqrt[7]{x}$
(C) $7m^{2/3} = 7(\sqrt[3]{m})^2 = 7\sqrt[3]{m^2}$ Second radical form is usually more useful.
(D) $(3u^2v^3)^{3/5} = (\sqrt[5]{3u^2v^3})^3$
$\qquad\qquad\quad = \sqrt[5]{(3u^2v^3)^3}$ Second radical form is usually more useful.
$\qquad\qquad\quad = \sqrt[5]{27u^6v^9}$
(E) $y^{-2/3} = \dfrac{1}{y^{2/3}} = \dfrac{1}{(\sqrt[3]{y})^2} = \dfrac{1}{\sqrt[3]{y^2}}$
(F) $(x^2 + y^2)^{1/2} = \sqrt{x^2 + y^2}$ $\sqrt{x^2 + y^2} \neq x + y$,
$\qquad\qquad\qquad\qquad\qquad\qquad$ since $(x + y)^2 = x^2 + 2xy + y^2 \neq x^2 + y^2$

Note in Example 1(E) that the exponent $-2/3$ was treated as $\dfrac{-2}{3}$. The denominator in the fraction must be positive since it represents the index of a radical.

Matched Problem 1

Convert to radical form:

(A) $7^{1/2}$ **(B)** $u^{1/5}$ **(C)** $3x^{3/5}$
(D) $(2x^3y^2)^{2/3}$ **(E)** $x^{-3/4}$ **(F)** $(x^3 + y^3)^{1/3}$

Example 2
Converting to
Exponent Form

Convert to rational exponent form:

(A) $\sqrt{13}$ **(B)** $\sqrt[5]{x}$ **(C)** $\sqrt[4]{w^3}$
(D) $\sqrt[5]{(3x^2y^2)^4}$ **(E)** $\dfrac{1}{\sqrt[3]{x^2}}$ **(F)** $\sqrt[4]{x^4 + y^4}$

Solution   **(A)** $\sqrt{13} = 13^{1/2}$ **(B)** $\sqrt[5]{x} = x^{1/5}$
(C) $\sqrt[4]{w^3} = w^{3/4}$ **(D)** $\sqrt[5]{(3x^2y^2)^4} = (3x^2y^2)^{4/5}$

(E) $\dfrac{1}{\sqrt[3]{x^2}} = \dfrac{1}{x^{2/3}} = x^{-2/3}$

(F) $\sqrt[4]{x^4 + y^4} = (x^4 + y^4)^{1/4}$ Note that $(x^4 + y^4)^{1/4} \neq x + y$.

Matched Problem 2   Convert to rational exponent form:

(A) $\sqrt{17}$ (B) $\sqrt[7]{m}$ (C) $\sqrt[5]{x^2}$

(D) $(\sqrt[7]{5m^3n^4})^3$ (E) $\dfrac{1}{\sqrt[6]{u^5}}$ (F) $\sqrt[5]{x^5 - y^5}$

♦ **PROPERTIES OF RADICALS**

We can use the exponent properties to simplify square root expressions. Consider the following examples of square roots that can be written in simpler form:

1. $\sqrt{2^2} = (2^2)^{1/2} = 2^{2/2} = 2^1 = 2$ and $(\sqrt{2})^2 = (2^{1/2})^2 = 2^1 = 2$
2. $\sqrt{4 \cdot 9} = \sqrt{36} = 6$ and $\sqrt{4}\sqrt{9} = 2 \cdot 3 = 6$, so $\sqrt{4 \cdot 9} = \sqrt{4}\sqrt{9}$
3. $\sqrt{\dfrac{36}{4}} = \sqrt{9} = 3$ and $\dfrac{\sqrt{36}}{\sqrt{4}} = \dfrac{6}{2} = 3$, so $\sqrt{\dfrac{36}{4}} = \dfrac{\sqrt{36}}{\sqrt{4}}$

Each of these statements generalizes to give us the following square root properties:

Square Root Properties

For positive real numbers a and b,

1. $\sqrt{a^2} = a$ and $(\sqrt{a})^2 = a$
2. $\sqrt{ab} = \sqrt{a}\sqrt{b}$
3. $\sqrt{\dfrac{a}{b}} = \dfrac{\sqrt{a}}{\sqrt{b}}$

Example 3
Applying Square Root Properties

Simplify using square root properties:

(A) $\sqrt{90}$ (B) $\sqrt{\frac{5}{16}}$

Solution   (A) $\sqrt{90} = \sqrt{9 \cdot 10} = \sqrt{9}\sqrt{10} = 3\sqrt{10}$

(B) $\sqrt{\dfrac{5}{16}} = \dfrac{\sqrt{5}}{\sqrt{16}} = \dfrac{\sqrt{5}}{4}$

Matched Problem 3   Simplify using square root properties:

(A) $\sqrt{80}$ (B) $\sqrt{\frac{10}{72}}$

We also can use the properties of exponents to obtain properties of radicals in general:

1. $\sqrt[n]{a^n} = (a^n)^{1/n} = a^{n/n} = a$
2. $\sqrt[n]{ab} = (ab)^{1/n} = a^{1/n}b^{1/n} = \sqrt[n]{a}\sqrt[n]{b}$
3. $\sqrt[n]{\dfrac{a}{b}} = \left(\dfrac{a}{b}\right)^{1/n} = \dfrac{a^{1/n}}{b^{1/n}} = \dfrac{\sqrt[n]{a}}{\sqrt[n]{b}}$
4. $\sqrt[kn]{a^{km}} = (a^{km})^{1/kn} = a^{km/kn} = a^{m/n} = \sqrt[n]{a^m}$

In summary:

Properties of Radicals

For positive integers n, m, and k, with $n \geq 2$; and positive real numbers a and b:

1. $\sqrt[n]{a^n} = a$ and $(\sqrt[n]{a})^n = a$
2. $\sqrt[n]{ab} = \sqrt[n]{a}\sqrt[n]{b}$
3. $\sqrt[n]{\dfrac{a}{b}} = \dfrac{\sqrt[n]{a}}{\sqrt[n]{b}}$
4. $\sqrt[kn]{a^{km}} = \sqrt[n]{a^m}$

It is important to remember that these four properties hold in general only if all the roots are defined. This will be true if the bases a and b are restricted to positive real numbers.

Therefore, for the remainder of this section, unless otherwise stated, all variables are assumed to represent positive real numbers.

The following example illustrates how these properties are used. Properties 2 and 3 are used from right to left as well as from left to right.

Example 4
Applying the Properties of Radicals

Simplify by using the properties of radicals:

(A) $\sqrt[5]{(3x^2y)^5}$ (B) $\sqrt[3]{25}\sqrt[3]{5}$ (C) $\sqrt[3]{\dfrac{x}{27}}$ (D) $\sqrt[6]{x^4}$

Solution

(A) $\sqrt[5]{(3x^2y)^5} = 3x^2y$ Property 1

(B) $\sqrt[3]{25}\sqrt[3]{5} = \sqrt[3]{125} = \sqrt[3]{5^3} = 5$ Property 2
                                    Property 1

(C) $\sqrt[3]{\dfrac{x}{27}} = \dfrac{\sqrt[3]{x}}{\sqrt[3]{27}} = \dfrac{\sqrt[3]{x}}{3}$ or $\dfrac{1}{3}\sqrt[3]{x}$ Property 3

(D) $\sqrt[6]{x^4} = \sqrt[2\cdot3]{x^{2\cdot2}} = \sqrt[3]{x^2}$ Property 4

Matched Problem 4

Simplify by using the properties of radicals:

(A) $\sqrt[7]{(u^2 + v^2)^7}$ (B) $\sqrt[5]{6}\sqrt[5]{16}$ (C) $\sqrt[3]{\dfrac{x^2}{8}}$

(D) $\dfrac{\sqrt[3]{54x^8}}{\sqrt[3]{2x^2}}$ (E) $\sqrt[8]{y^6}$

Answers to Matched Problems

1. **(A)** $\sqrt{7}$ **(B)** $\sqrt[5]{u}$ **(C)** $3\sqrt[5]{x^3}$ **(D)** $\sqrt[3]{(2x^3y^2)^2}$ or $(\sqrt[3]{2x^3y^2})^2$
   (E) $\dfrac{1}{\sqrt[4]{x^3}}$ or $\dfrac{1}{(\sqrt[4]{x})^3}$ **(F)** $\sqrt[3]{x^3 + y^3}$

2. **(A)** $17^{1/2}$ **(B)** $m^{1/7}$ **(C)** $x^{2/5}$ **(D)** $(5m^3n^4)^{3/7}$
   (E) $u^{-5/6}$ **(F)** $(x^5 - y^5)^{1/5}$

3. **(A)** $4\sqrt{5}$ **(B)** $\dfrac{\sqrt{5}}{6}$

4. **(A)** $u^2 + v^2$ **(B)** $2\sqrt[5]{3}$ **(C)** $\frac{1}{2}\sqrt[3]{x^2}$ **(D)** $3x^2$ **(E)** $\sqrt[4]{y^3}$

EXERCISE 6-5

In Problems 1–96, all variables are restricted to avoid even roots of negative numbers.

A *Change to radical form. (Do not simplify.)*

1. $11^{1/2}$ 2. $7^{1/2}$ 3. $5^{1/3}$ 4. $6^{1/4}$

5. $u^{3/5}$ 6. $x^{3/4}$ 7. $4y^{3/7}$ 8. $5m^{2/3}$

9. $(4y)^{3/7}$ 10. $(5m)^{2/3}$ 11. $(4ab^3)^{2/5}$

12. $(7x^2y)^{2/3}$ 13. $(a + b)^{1/2}$ 14. $(a^2 + b^2)^{1/2}$

Change to rational exponent form. (Do not simplify.)

15. $\sqrt{6}$ 16. $\sqrt{3}$ 17. $\sqrt[4]{m}$

18. $\sqrt[7]{m}$ 19. $\sqrt[5]{y^3}$ 20. $\sqrt[3]{a^2}$

21. $\sqrt[4]{(xy)^3}$ 22. $\sqrt[5]{(7m^3n^3)^4}$ 23. $\sqrt{x^2 - y^2}$

24. $\sqrt{1 + y^2}$

Simplify using properties of radicals.

25. $\sqrt{72}$ 26. $\sqrt{75}$ 27. $\sqrt{76}$ 28. $\sqrt{84}$

29. $\dfrac{\sqrt{44}}{\sqrt{11}}$ 30. $\dfrac{\sqrt{45}}{\sqrt{5}}$ 31. $\dfrac{\sqrt{48}}{\sqrt{3}}$ 32. $\dfrac{\sqrt{50}}{\sqrt{2}}$

33. $\sqrt{2^6}$ 34. $\sqrt{3^8}$ 35. $\sqrt{3^4}$ 36. $\sqrt{2^{10}}$

37. $\sqrt{(-3)^2}$ 38. $\sqrt{(-5)^2}$ 39. $\sqrt[4]{3^2}$ 40. $\sqrt[4]{5^2}$

41. $\sqrt{x^5}$ 42. $\sqrt{x^3}$ 43. $\sqrt{x^3y^5}$ 44. $\sqrt{x^4y^3}$

45. $\dfrac{\sqrt{x^5}}{\sqrt{x^3}}$ 46. $\dfrac{\sqrt{x^7}}{\sqrt{x^3}}$ 47. $\dfrac{\sqrt{x^3}}{\sqrt{x^9}}$ 48. $\dfrac{\sqrt{x}}{\sqrt{x^7}}$

49. $\sqrt{\dfrac{x^3}{24}}$ 50. $\sqrt{\dfrac{x}{18}}$

B *Change to radical form. (Do not simplify.)*

51. $-5y^{2/5}$ 52. $-3x^{1/2}$

53. $(1 + m^2n^2)^{3/7}$ 54. $(x^2y^2 - w^3)^{4/5}$

55. $w^{-2/3}$ 56. $y^{-3/5}$

57. $(3m^2n^3)^{-3/5}$ 58. $(2xy)^{-2/3}$

59. $a^{1/2} + b^{1/2}$ 60. $x^{-1/2} + y^{-1/2}$

61. $(a^3 + b^3)^{2/3}$ 62. $(x^{1/2} + y^{-1/2})^{1/3}$

Change to rational exponent form. (Do not simplify.)

63. $\sqrt[3]{(a + b)^2}$ 64. $\sqrt[5]{(x - y)^2}$

65. $-3x\sqrt[4]{a^3b}$ 66. $-5\sqrt[3]{2x^2y^2}$

67. $\sqrt[9]{-2x^3y^7}$ 68. $\sqrt[5]{-4m^2n^3}$

69. $\dfrac{3}{\sqrt[3]{y}}$ 70. $\dfrac{2x}{\sqrt{y}}$

71. $\dfrac{-2x}{\sqrt{x^2 + y^2}}$ 72. $\dfrac{2}{\sqrt{x}} + \dfrac{3}{\sqrt{y}}$

73. $\sqrt[3]{m^2} - \sqrt{n}$ 74. $\dfrac{-5u^2}{\sqrt{u} + \sqrt[5]{v^3}}$

In Problems 75–96, simplify using properties of radicals.

75. $\sqrt[3]{40}$ 76. $\sqrt[4]{80}$ 77. $\sqrt[5]{64}$ 78. $\sqrt[3]{54}$

79. $\sqrt[3]{375}$ 80. $\sqrt[3]{250}$ 81. $\sqrt[3]{256}$ 82. $\sqrt[3]{135}$

83. $\dfrac{\sqrt[3]{54}}{\sqrt[3]{2}}$ 84. $\dfrac{\sqrt[3]{500}}{\sqrt[3]{4}}$ 85. $\sqrt[3]{5^6}$ 86. $\sqrt[4]{3^8}$

87. $\sqrt[4]{(-2)^4}$ 88. $\sqrt[6]{(-3)^6}$ 89. $\sqrt[3]{x^4}$ 90. $\sqrt[4]{x^7}$

91. $\dfrac{\sqrt[4]{x^9}}{\sqrt[4]{x}}$ 92. $\dfrac{\sqrt[3]{x^8}}{\sqrt[3]{x^2}}$

C 93. $\sqrt[3]{(3xy^2)^3}$ 94. $\sqrt[4]{(7x^2y^3)^4}$

95. $\sqrt[4]{(7x^7y^3)^5}$ 96. $\sqrt[3]{(3xy^2)^4}$

97. Is $\sqrt{x^2} = x$ always true? Test for $x = -2$. Describe the circumstances under which the equation will be true.

98. Is $\sqrt[3]{x^3} = x$ always true? Test for $x = -2$. Describe the circumstances under which the equation $\sqrt[n]{x^n} = x$ will be true.

99. Is $\sqrt[n]{x^{2n}} = x^2$ always true?

100. Is $\sqrt[n]{x^{3n}} = x^3$ always true?

6-6 Simplest Radical Form

♦ Properties of Radicals
♦ Simplest Radical Form for Square Roots
♦ Simplest Radical Form for More General Radicals
♦ Simplifying $\sqrt[n]{x^n}$ for All Real x

The properties of square roots and other radicals developed in Section 6-5 allow us to rewrite radical expressions in alternative forms. In this section we focus on one form, called *simplest radical form,* that is particularly useful for comparing radical expressions.

♦ **PROPERTIES OF RADICALS**

We list the properties of square roots and other radicals from Section 6-5 for reference:

Properties of Radicals

For positive integers n, m, and k, with $n \geq 2$; and positive real numbers a and b:

SQUARE ROOT PROPERTIES

1. $\sqrt{a^2} = a$, $(\sqrt{a})^2 = a$
2. $\sqrt{ab} = \sqrt{a}\sqrt{b}$
3. $\sqrt{\dfrac{a}{b}} = \dfrac{\sqrt{a}}{\sqrt{b}}$

GENERAL RADICAL PROPERTIES

1. $\sqrt[n]{a^n} = a$, $(\sqrt[n]{a})^n = a$
2. $\sqrt[n]{ab} = \sqrt[n]{a}\sqrt[n]{b}$
3. $\sqrt[n]{\dfrac{a}{b}} = \dfrac{\sqrt[n]{a}}{\sqrt[n]{b}}$
4. $\sqrt[kn]{a^{km}} = \sqrt[n]{a^m}$

These properties of radicals hold only if all roots are defined. For example,

$$\sqrt{4} = \sqrt{(-2)(-2)} \neq \sqrt{-2}\sqrt{-2}$$

To avoid the possibility of undefined roots such as $\sqrt{-2}$ occurring in expressions, we can restrict a and b to positive integers.

Once again, in this section, unless otherwise stated, all variables are assumed to represent positive real numbers.

Near the end of the section we will discuss what happens if we relax this restriction.

♦ **SIMPLEST RADICAL FORM FOR SQUARE ROOTS**

We want to establish a standard form, called **simplest radical form,** for expressions involving radicals. Other forms may sometimes be more useful, but putting expressions in such a standard form makes comparisons easier. We begin by considering simplest radical form for expressions where the only radicals present are square roots. Such as expression is in simplest radical form if the following conditions are met:

Simplest Radical Form for Square Roots

1. No radicand contains a polynomial factor to a power greater than or equal to 2. $\sqrt{x^3}$ violates this condition.

2. No radical appears in a denominator. $\dfrac{3}{\sqrt{5}}$ violates this condition.

3. No fraction appears within a radical. $\sqrt{\tfrac{2}{3}}$ violates this condition.

Example 1

Putting Square Roots in Simplest Radical Form

Change to simplest radical form:

(A) $\sqrt{12x^3}$ **(B)** $\dfrac{4x}{\sqrt{8x}}$ **(C)** $\sqrt{\dfrac{y}{2x}}$

Solution   **(A)** $\sqrt{12x^3}$

The expression violates Condition 1. Factor as many perfect squares from $12x^3$ as possible.

$= \sqrt{4x^2 \cdot 3x}$ Use Property 2 for square roots.

$= \sqrt{4x^2}\sqrt{3x}$

$= 2x\sqrt{3x}$

(B) $\dfrac{4x}{\sqrt{8x}}$

The expression violates Condition 2. Multiply the numerator and denominator by a quantity that will make the denominator a perfect square.

$= \dfrac{4x\sqrt{2x}}{\sqrt{8x}\sqrt{2x}}$ Apply Property 2 for square roots to the denominator.

$= \dfrac{4x\sqrt{2x}}{\sqrt{16x^2}}$

$= \dfrac{4x\sqrt{2x}}{4x} = \sqrt{2x}$

We also could have multiplied the numerator and denominator by $\sqrt{8x}$ rather than $\sqrt{2x}$, but the resulting fraction would have more common factors to eliminate from the numerator and denominator. Try it to see why.

(C) $\sqrt{\dfrac{y}{2x}}$

The expression violates Condition 3. Multiply the numerator and denominator of the fraction inside the radical by a quantity that will make the denominator a perfect square.

$= \sqrt{\dfrac{y \cdot 2x}{2x \cdot 2x}}$

$= \sqrt{\dfrac{2xy}{4x^2}}$ Apply Property 3 for square roots.

$= \dfrac{\sqrt{2xy}}{\sqrt{4x^2}} = \dfrac{\sqrt{2xy}}{2x}$

Matched Problem 1   Change to simplest radical form:

(A) $\sqrt{18y^5}$ (B) $\dfrac{6a}{\sqrt{4b}}$ (C) $\sqrt{\dfrac{3y}{8x}}$

As mentioned above, the simplest radical form may not be the most useful form. In a product such as

$$\frac{4}{\sqrt{8x}} \cdot \frac{\sqrt{8x}}{2}$$

it is easier to multiply directly rather than to change either factor to simplest form. On the other hand, if $4x/\sqrt{8x}$ is to be evaluated for several values of x, the work will be much easier if the expression is first changed to $\sqrt{2x}$ as in Example 1(B).

♦ **SIMPLEST RADICAL FORM FOR MORE GENERAL RADICALS**

For more general expressions involving radicals with index greater than 2, we will say that an expression is in simplest radical form if the following conditions are met:

Simplest Radical Form

1. No radicand contains a polynomial factor to a power greater than or equal to the index of the radical. $\sqrt[3]{x^4}$ violates this condition.

2. No radical appears in a denominator. $\dfrac{3}{\sqrt[3]{5}}$ violates this condition.

3. No fraction appears within a radical. $\sqrt[3]{\frac{2}{3}}$ violates this condition.
4. The power of the radicand and the index of the radical have no common factor other than 1. $\sqrt[6]{x^4}$ violates this condition.

Example 2
Putting Expressions in Simplest Radical Form

Change to simplest radical form:

(A) $\sqrt[3]{54x^5}$ (B) $\dfrac{2}{\sqrt[3]{x}}$ (C) $\sqrt[3]{\dfrac{y}{4x}}$ (D) $\sqrt[9]{x^6}$

Solution   (A) $\sqrt[3]{54x^5}$ The expression violates Condition 1. Factor as many cubes as possible from $54x^5$.

$= \sqrt[3]{27x^3 \cdot 2x^2}$ Apply radical Property 2.

$= \sqrt[3]{27x^3}\sqrt[3]{2x^2}$

$= 3x\sqrt[3]{2x^3}$

(B) $\dfrac{2}{\sqrt[3]{x}}$ The expression violates Condition 2. Multiply the numerator and denominator by a quantity that makes the denominator a perfect cube.

$$= \dfrac{2\sqrt[3]{x^2}}{\sqrt[3]{x}\sqrt[3]{x^2}}$$ Apply radical Property 2 to the denominator.

$$= \dfrac{2\sqrt[3]{x^2}}{\sqrt[3]{x^3}} = \dfrac{2\sqrt[3]{x^2}}{x}$$

(C) $\sqrt[3]{\dfrac{y}{4x}}$ The expression violates Condition 3. Multiply the numerator and denominator inside the radical by a quantity that makes the denominator a perfect cube.

$$= \sqrt[3]{\dfrac{y \cdot 2x^2}{4x \cdot 2x^2}}$$ Apply radical Property 3.

$$= \dfrac{\sqrt[3]{2x^2y}}{\sqrt[3]{8x^3}} = \dfrac{\sqrt[3]{2x^2y}}{2x}$$

(D) $\sqrt[9]{x^6}$ The expression violates Condition 4, since the index 9 and the power 6 in the radicand have a common factor.

$$= \sqrt[3\cdot3]{x^{3\cdot2}}$$ Apply radical Property 4.

$$= \sqrt[3]{x^2}$$

   The process of removing radicals from a denominator, as in Example 2(B), is called **rationalizing the denominator.** In general, both the numerator and denominator are multiplied by an expression so that the denominator becomes free of radicals. For example, if the denominator is $\sqrt{8x}$, multiply by $\sqrt{2x}$ as in Example 1(B):

$$\sqrt{8x} \cdot \sqrt{2x} = \sqrt{16x^2} = 4x$$

If the denominator is $\sqrt[3]{x}$, multiply by $\sqrt[3]{x^2}$ as in Example 2(B):

$$\sqrt[3]{x} \cdot \sqrt[3]{x^2} = \sqrt[3]{x^3} = x$$

Matched Problem 2   Change to simplest radical form:

(A) $\sqrt[3]{32m^8}$ **(B)** $\sqrt[12]{y^8}$ **(C)** $\dfrac{6u}{\sqrt[3]{4x}}$ **(D)** $\sqrt[3]{\dfrac{x}{2y^2}}$

Example 3
Changing to Simplest Radical Form

Change to simplest radical form:

(A) $\sqrt{12x^3y^5z^2}$ **(B)** $\sqrt[6]{16x^4y^2}$ **(C)** $\dfrac{6x^2}{\sqrt[3]{9x}}$ **(D)** $\sqrt[3]{\dfrac{2a^2}{3b^2}}$

Solution   **(A)** $\sqrt{12x^3y^5z^2}$ Factor as many perfect squares as possible out of the radicand.

$$= \sqrt{(2^2x^2y^4z^2)(3xy)}$$ Apply Property 1.
$$= \sqrt{2^2x^2y^4z^2}\sqrt{3xy}$$
$$= 2xy^2z\sqrt{3xy}$$

(B) $\sqrt[6]{16x^4y^2}$ Notice that $16 = 4^2$, so the radicand is $(4x^2y)^2$. The index 6 and the power in the radicand have common factor 2.

$$= \sqrt[6]{(4x^2y)^2}$$ Apply Property 4.
$$= \sqrt[3]{4x^2y}$$

(C) $\dfrac{6x^2}{\sqrt[3]{9x}}$ Rationalize the denominator.

$$= \frac{6x^2}{\sqrt[3]{9x}} \cdot \frac{\sqrt[3]{3x^2}}{\sqrt[3]{3x^2}}$$

$$= \frac{6x^2\sqrt[3]{3x^2}}{\sqrt[3]{3^3x^3}} = \frac{6x^2\sqrt[3]{3x^2}}{3x}$$

$$= 2x\sqrt[3]{3x^2}$$

(D) $\sqrt[3]{\dfrac{2a^2}{3b^2}}$ Multiply the numerator and denominator by a quantity that makes the denominator a perfect cube.

$$= \sqrt[3]{\frac{(2a^2)(3^2b)}{(3b^2)(3^2b)}}$$

$$= \sqrt[3]{\frac{18a^2b}{3^3b^3}} = \frac{\sqrt[3]{18a^2b}}{\sqrt[3]{3^3b^3}} = \frac{\sqrt[3]{18a^2b}}{3b}$$

Matched Problem 3   Change to simplest radical form:

(A) $\sqrt[3]{16}$ **(B)** $\sqrt[3]{16x^7y^4z^3}$ **(C)** $\sqrt[9]{8x^6y^3}$

(D) $\dfrac{6}{\sqrt{2x}}$ **(E)** $\dfrac{10x^3}{\sqrt[3]{4x^2}}$ **(F)** $\sqrt[3]{\dfrac{3y^2}{2x^4}}$

◆ SIMPLIFYING $\sqrt[n]{x^n}$ FOR ALL REAL x

In the preceding discussion we restricted variables to represent positive numbers to avoid taking undefined roots. An expression such as $\sqrt{x^2}$ makes sense for any value of x, since $x^2 \geq 0$, but it is not necessarily equal to x. For example, if $x = 2$, then $\sqrt{2^2} = 2$. On the other hand, if $x = -2$, then $\sqrt{(-2)^2} = 2$ but not -2. In general, if $x \geq 0$, then $\sqrt{x^2} = x$, but if $x < 0$, $\sqrt{x^2} = -x$. We can therefore write

$$\sqrt{x^2} = \begin{cases} x & \text{if } x > 0 \\ 0 & \text{if } x = 0 \\ -x & \text{if } x < 0 \end{cases}$$

This corresponds exactly to the definition of $|x|$, the absolute value of x. That is, $\sqrt{x^2}$ and $|x|$ actually are the same:

For x any real number:

$$\sqrt{x^2} = |x|$$

Only if x is restricted to nonnegative real numbers can we drop the absolute-value sign.

For expressions such as $\sqrt[3]{x^3}$, we do not have the same kind of problem. For *all* real numbers,

$$\sqrt[3]{x^3} = x$$

and we do not need an absolute-value sign on the right. You should test this for $x = 2$ and $x = -2$.

Following the same reasoning, we can obtain the following more general result:

For *any* real number x and positive integer n:

$$\sqrt[n]{x^n} = \begin{cases} |x| & \text{if } n \text{ is even} \\ x & \text{if } n \text{ is odd} \end{cases} \qquad \begin{array}{l} \sqrt[4]{x^4} = |x| \\ \sqrt[5]{x^5} = x \end{array}$$

Example 4
Simplifying Expressions Involving $\sqrt[n]{x^n}$

Simplify $\sqrt[3]{x^3} + \sqrt{x^2}$:

(A) For x a positive number   **(B)** For x a negative number

Solution   **(A)** For x a positive number:

$$\sqrt[3]{x^3} + \sqrt{x^2} = x + |x| = x + x = 2x$$

(B) For x a negative number:

$$\sqrt[3]{x^3} + \sqrt{x^2} = x + |x| = x + (-x) = 0$$

Matched Problem 4   Simplify $2\sqrt[3]{x^3} - \sqrt{x^2}$:

(A) For x a positive number   **(B)** For x a negative number

Answers to Matched Problems

1. **(A)** $3y^2\sqrt{2y}$ **(B)** $\dfrac{3a\sqrt{b}}{b}$ **(C)** $\dfrac{\sqrt{6xy}}{4x}$

2. **(A)** $2m^2\sqrt[3]{4m^2}$ **(B)** $\sqrt[3]{y^2}$ **(C)** $\dfrac{3u\sqrt[3]{2x^2}}{x}$
   (D) $\dfrac{\sqrt[3]{4xy}}{2y}$

3. **(A)** $2\sqrt[3]{2}$ **(B)** $2x^2yz\sqrt[3]{2xy}$ **(C)** $\sqrt[3]{2x^2y}$
   (D) $\dfrac{3\sqrt{2x}}{x}$ **(E)** $5x^2\sqrt[3]{2x}$ **(F)** $\dfrac{\sqrt[3]{12x^2y^2}}{2x^2}$

4. **(A)** x **(B)** $3x$

EXERCISE 6-6

In Problems 1–94, simplify, and write in simplest radical form. All variables represent positive real numbers.

A

1. $\sqrt{y^2}$ 2. $\sqrt{x^2}$ 3. $\sqrt{4u^2}$

4. $\sqrt{9m^2}$ 5. $\sqrt{49x^4y^2}$ 6. $\sqrt{25x^2y^4}$

7. $\sqrt{18}$ 8. $\sqrt{8}$ 9. $\sqrt{m^3}$ 10. $\sqrt{x^3}$

11. $\sqrt{8x^3}$ 12. $\sqrt{18y^3}$ 13. $\sqrt{\frac{1}{9}}$ 14. $\sqrt{\frac{1}{4}}$

15. $\dfrac{1}{\sqrt{y^2}}$ 16. $\dfrac{1}{\sqrt{x^2}}$ 17. $\dfrac{1}{\sqrt{5}}$ 18. $\dfrac{1}{\sqrt{3}}$

19. $\sqrt{\frac{1}{5}}$ 20. $\sqrt{\frac{1}{3}}$ 21. $\dfrac{1}{\sqrt{y}}$ 22. $\dfrac{1}{\sqrt{x}}$

23. $\sqrt{\dfrac{1}{y}}$ 24. $\sqrt{\dfrac{1}{x}}$ 25. $\sqrt{9x^3y^5}$ 26. $\sqrt{4x^5y^3}$

27. $\sqrt{18x^8y^5}$ 28. $\sqrt{8x^7y^6}$ 29. $\dfrac{1}{\sqrt{2x}}$ 30. $\dfrac{1}{\sqrt{3y}}$

31. $\dfrac{6x^2}{\sqrt{3x}}$ 32. $\dfrac{4xy}{\sqrt{2y}}$ 33. $\dfrac{3a}{\sqrt{2ab}}$ 34. $\dfrac{2x^2y}{\sqrt{3xy}}$

35. $\sqrt{\dfrac{6x}{7y}}$ 36. $\sqrt{\dfrac{3m}{2n}}$

B

37. $\sqrt{\dfrac{9m^5}{2n}}$ 38. $\sqrt{\dfrac{4a^3}{3b}}$ 39. $\sqrt[4]{16x^8y^4}$

40. $\sqrt[5]{32m^5n^{15}}$ 41. $\sqrt[3]{2^4x^4y^7}$ 42. $\sqrt[4]{2^4a^5b^8}$

43. $\sqrt[3]{32x^4y^6}$ 44. $\sqrt[4]{32x^4y^6}$ 45. $\sqrt{32x^4y^6}$

46. $\sqrt[5]{32x^4y^6}$ 47. $\sqrt[5]{24x^3y^4z^5}$ 48. $\sqrt{24x^3y^4z^5}$

49. $\sqrt[4]{24x^3y^4z^5}$ 50. $\sqrt[3]{24x^3y^4z^5}$ 51. $\sqrt[4]{x^2}$

52. $\sqrt[10]{x^6}$ 53. $\sqrt{2}\sqrt{8}$ 54. $\sqrt[3]{3}\sqrt[3]{9}$

55. $\sqrt{18m^3n^4}\sqrt{2m^3n^2}$

56. $\sqrt[3]{9x^2y}\sqrt[3]{3xy^2}$

57. $\dfrac{6}{\sqrt[3]{3}}$ 58. $\dfrac{2}{\sqrt[3]{2}}$

59. $\dfrac{\sqrt{4a^3}}{\sqrt{3b}}$ 60. $\dfrac{\sqrt{9m^5}}{\sqrt{2n}}$

61. $\sqrt{a^2+b^2}$ 62. $\sqrt[3]{x^3+y^3}$

63. $\sqrt[3]{\dfrac{8x^3}{27y^6}}$ 64. $\sqrt[4]{\dfrac{a^8b^4}{16c^{12}}}$

65. $-m\sqrt[5]{36m^7n^{11}}$ 66. $-2x\sqrt[3]{8x^8y^{13}}$

67. $\sqrt[6]{x^4(x-y)^2}$ 68. $\sqrt[8]{2^6(x+y)^6}$

69. $\sqrt[3]{2x^2y^3}\sqrt[3]{3x^5y}$ 70. $\sqrt[4]{6u^3v^4}\sqrt[4]{4u^5v}$

71. $\sqrt[3]{4x^2y}\sqrt[3]{2x^4y^2}$ 72. $\sqrt[3]{3x^5y^2}\sqrt[3]{9xy^4}$

73. $\sqrt[3]{9x^2y^2}\sqrt[3]{24xy^2}$ 74. $\sqrt[3]{16xy^2}\sqrt[3]{4xy^2}$

75. $\dfrac{\sqrt[3]{9x^2y^2}}{\sqrt[3]{72xy^2}}$ 76. $\dfrac{\sqrt[3]{40xy^5}}{\sqrt[3]{5xy}}$

77. $\dfrac{4x^3y^2}{\sqrt[3]{2xy^2}}$ 78. $\dfrac{8u^3v^5}{\sqrt[3]{4u^2v^2}}$

79. $-2x\sqrt[3]{\dfrac{3y^2}{4x}}$ 80. $6c\sqrt[3]{\dfrac{2ab}{9c^2}}$

C

81. $\dfrac{x-y}{\sqrt[3]{x-y}}$ 82. $\dfrac{1}{\sqrt[3]{(x-y)^2}}$

83. $\sqrt[4]{\dfrac{3y^3}{4x}}$ 84. $\sqrt[5]{\dfrac{4n^2}{16m^3}}$

85. $-\sqrt{x^4+2x^2}$ 86. $\sqrt[4]{m^4+4m^6}$

87. $\sqrt[4]{16x^4}\sqrt[3]{16x^{24}y^4}$ 88. $\sqrt[3]{8\sqrt{16x^6y^4}}$

89. $\sqrt[3]{3m^2n^2}\sqrt[4]{3m^3n^2}$ 90. $\sqrt{2x^5y^3}\sqrt[3]{16x^7y^7}$

91. $\sqrt[3]{x^{3n}(x+y)^{3n+6}}$ 92. $\sqrt[n]{x^{2n}y^{n^2+n}}$

93. $\sqrt[4]{x^ny^{n^3+n}}$ 94. $\sqrt[4]{x^{4n}(x-y)^{8n-4}}$

*Simplify each of the following for (**A**) x a positive number and (**B**) x a negative number.*

95. $2\sqrt[3]{x^3}+4\sqrt{x^2}$ 96. $\sqrt[3]{8x^3}-\sqrt{16x^2}$

97. $\sqrt[5]{x^5}+\sqrt[4]{x^4}$ 98. $\sqrt[6]{x^6}+\sqrt[3]{x^3}$

99. $3\sqrt[4]{x^4}-2\sqrt[5]{x^5}$ 100. $5\sqrt[7]{x^7}-3\sqrt[6]{x^6}$

6-7 Basic Operations Involving Radicals

♦ Sums and Differences
♦ Products
♦ Rationalizing Denominators in Quotients

The algebraic expressions we can work with are becoming more and more complex. Polynomials were built up from addition, subtraction, and multiplication.

Since rational expressions allow division, we can now add, subtract, multiply, and divide such expressions. In this section, we discuss how to perform the same operations on some algebraic expressions that involve radicals.

All variables in this section will again represent positive real numbers.

◆ SUMS AND DIFFERENCES

Algebraic expressions involving radicals often can be simplified by adding and subtracting those terms that contain exactly the same radical expressions. We proceed in essentially the same way as we do when we combine like terms in polynomials.

Example 1
Adding and Subtracting Radical Expressions

Combine as many terms as possible:

(A) $5\sqrt{3} + 4\sqrt{3}$ (B) $2\sqrt[3]{xy^2} - 7\sqrt[3]{xy^2}$
(C) $3\sqrt{xy} - 2\sqrt[3]{xy} + 4\sqrt{xy} - 7\sqrt[3]{xy}$

Solution

(A) $5\sqrt{3} + 4\sqrt{3} \;\boxed{= (5 + 4)\sqrt{3}} = 9\sqrt{3}$ $5\sqrt{3}$ and $4\sqrt{3}$ are treated as like terms, just as we would treat $5x$ and $4x$.

(B) $2\sqrt[3]{xy^2} - 7\sqrt[3]{xy^2} \;\boxed{= (2 - 7)\sqrt[3]{xy^2}} = -5\sqrt[3]{xy^2}$

(C) $3\sqrt{xy} - 2\sqrt[3]{xy} + 4\sqrt{xy} - 7\sqrt[3]{xy}$

$\boxed{= 3\sqrt{xy} + 4\sqrt{xy} - 2\sqrt[3]{xy} - 7\sqrt[3]{xy}}$

$= 7\sqrt{xy} - 9\sqrt[3]{xy}$

Matched Problem 1

Combine as many terms as possible:

(A) $6\sqrt{2} + 2\sqrt{2}$ (B) $3\sqrt[5]{2x^2y^3} - 8\sqrt[5]{2x^2y^3}$
(C) $5\sqrt[3]{mn^2} - 3\sqrt{mn} - 2\sqrt[3]{mn^2} + 7\sqrt{mn}$

Thus, if two terms contain exactly the same radical—having the same index and the same radicand—they can be combined into a single term.

Occasionally, terms containing different radicals can be combined after they have been expressed in simplest radical form.

Example 2
Adding and Subtracting Radical Expressions

Express terms in simplest radical form and combine where possible:

(A) $4\sqrt{8} - 2\sqrt{18}$ (B) $2\sqrt{12} - \sqrt{\tfrac{1}{3}}$ (C) $\sqrt[3]{81} - \sqrt[3]{\tfrac{1}{9}}$

Solution

(A) $4\sqrt{8} - 2\sqrt{18}$ Change both radicals to simplest radical form.
$= 4\sqrt{4 \cdot 2} - 2\sqrt{9 \cdot 2}$
$= 8\sqrt{2} - 6\sqrt{2}$
$= 2\sqrt{2}$

(B) $2\sqrt{12} - \sqrt{\frac{1}{3}}$ Change both radicals to simplest radical form.

$$= 2 \cdot \sqrt{4} \cdot \sqrt{3} - \frac{1 \cdot \sqrt{3}}{\sqrt{3} \cdot \sqrt{3}}$$

$$= 4\sqrt{3} - \frac{\sqrt{3}}{3}$$

$$= (4 - \tfrac{1}{3})\sqrt{3}$$

$$= \tfrac{11}{3}\sqrt{3} \quad \text{or} \quad \frac{11\sqrt{3}}{3}$$

(C) $\sqrt[3]{81} - \sqrt[3]{\frac{1}{9}}$ Change both radicals to simplest radical form.

$$= \sqrt[3]{3^3 \cdot 3} - \sqrt[3]{\frac{3}{3^3}}$$

$$= 3\sqrt[3]{3} - \tfrac{1}{3}\sqrt[3]{3}$$

$$= (3 - \tfrac{1}{3})\sqrt[3]{3} = \tfrac{8}{3}\sqrt[3]{3}$$

Matched Problem 2   Express terms in simplest radical form and combine where possible:

(A) $\sqrt{12} - \sqrt{48}$ **(B)** $3\sqrt{8} - \sqrt{\frac{1}{2}}$ **(C)** $\sqrt[3]{\frac{1}{4}} - \sqrt[3]{16}$

♦ ## PRODUCTS

We now consider several types of products and quotients that involve radicals.

Example 3
Multiplying Radical Expressions

Multiply and simplify:

(A) $\sqrt{2}(\sqrt{10} - 3)$ **(B)** $(\sqrt{2} - 3)(\sqrt{2} + 5)$
(C) $(\sqrt{x} - 3)(\sqrt{x} + 5)$ **(D)** $(\sqrt[3]{m} + \sqrt[3]{n^2})(\sqrt[3]{m^2} - \sqrt[3]{n})$

Solution   **(A)** $\sqrt{2}(\sqrt{10} - 3)$ Use the distributive property.

$$= \sqrt{2}\sqrt{10} - \sqrt{2} \cdot 3$$

$$= \sqrt{20} - 3\sqrt{2} \qquad \text{Change the first radical to simplest radical form.}$$

$$= 2\sqrt{5} - 3\sqrt{2}$$

(B) $(\sqrt{2} - 3)(\sqrt{2} + 5)$ We can multiply using FOIL, just as we would if variables were involved.

$$= \sqrt{2}\sqrt{2} - 5\sqrt{2} + 3\sqrt{2} - 15$$

$$= 2 + 2\sqrt{2} - 15$$

$$= 2\sqrt{2} - 13$$

(C) $(\sqrt{x} - 3)(\sqrt{x} + 5) = \sqrt{x}\sqrt{x} - 5\sqrt{x} + 3\sqrt{x} - 15$

$$= x + 2\sqrt{x} - 15$$

(D) $(\sqrt[3]{m} + \sqrt[3]{n^2})(\sqrt[3]{m^2} - \sqrt[3]{n}) = \sqrt[3]{m^3} - \sqrt[3]{mn} + \sqrt[3]{m^2 n^2} - \sqrt[3]{n^3}$

$$= m - \sqrt[3]{mn} + \sqrt[3]{m^2 n^2} - n$$

Matched Problem 3   Multiply and simplify:

(A) $\sqrt{3}(\sqrt{6} - 4)$ **(B)** $(\sqrt{3} - 2)(\sqrt{3} + 4)$
(C) $(\sqrt{y} - 2)(\sqrt{y} + 4)$ **(D)** $(\sqrt[3]{x^2} - \sqrt[3]{y^2})(\sqrt[3]{x} + \sqrt[3]{y})$

Example 4
Evaluating with
Radical Expressions

Show that $2 - \sqrt{3}$ is a solution of the equation $x^2 - 4x + 1 = 0$.

Solution Substitute $2 - \sqrt{3}$ into $x^2 - 4x + 1$ and simplify. We want to show that the resulting expression is equal to 0:

$$(2 - \sqrt{3})^2 - 4(2 - \sqrt{3}) + 1 = (4 - 4\sqrt{3} + 3) - 4(2 - \sqrt{3}) + 1$$
$$= 7 - 4\sqrt{3} - 8 + 4\sqrt{3} + 1 = 0$$

Matched Problem 4 Show that $2 + \sqrt{3}$ is a solution of $x^2 - 4x + 1 = 0$.

♦ **RATIONALIZING DENOMINATORS IN QUOTIENTS**

Recall that to express $\sqrt{2}/\sqrt{3}$ in simplest radical form, we multiplied the numerator and denominator by $\sqrt{3}$ to rationalize the denominator, that is, to replace a denominator involving radicals by a rational form:

$$\frac{\sqrt{2}}{\sqrt{3}} = \frac{\sqrt{2} \cdot \sqrt{3}}{\sqrt{3} \cdot \sqrt{3}} = \frac{\sqrt{6}}{3}$$

How can we rationalize the binomial denominator in the expression below?

$$\frac{1}{\sqrt{3} - \sqrt{2}}$$

Multiplying the numerator and denominator by $\sqrt{3}$ or $\sqrt{2}$ doesn't help. Try it to see why. Instead, recall the special product

$$(a - b)(a + b) = a^2 - b^2$$

This suggests that we can multiply the numerator and denominator by the denominator, but with the middle sign changed, and effectively square both terms in the denominator. Thus.

$$\frac{1}{\sqrt{3} - \sqrt{2}} = \frac{1(\sqrt{3} + \sqrt{2})}{(\sqrt{3} - \sqrt{2})(\sqrt{3} + \sqrt{2})}$$
$$= \frac{\sqrt{3} + \sqrt{2}}{(\sqrt{3})^2 - (\sqrt{2})^2}$$
$$= \frac{\sqrt{3} + \sqrt{2}}{3 - 2} = \sqrt{3} + \sqrt{2}$$

The expressions $\sqrt{3} + \sqrt{2}$ and $\sqrt{3} - \sqrt{2}$ are often called **conjugates** of each other. To rationalize a denominator that is the sum or difference of two terms involving square roots, we multiply both the numerator and denominator by the conjugate of the denominator.

Example 5
Rationalizing the
Denominator

Rationalize denominators and simplify:

(A) $\dfrac{\sqrt{2}}{\sqrt{6}-2}$　　(B) $\dfrac{\sqrt{x}-\sqrt{y}}{\sqrt{x}+\sqrt{y}}$

Solution

(A) $\dfrac{\sqrt{2}}{\sqrt{6}-2} = \dfrac{\sqrt{2}(\sqrt{6}+2)}{(\sqrt{6}-2)(\sqrt{6}+2)} = \dfrac{\sqrt{12}+2\sqrt{2}}{6-4}$

$\quad = \dfrac{2\sqrt{3}+2\sqrt{2}}{2} = \dfrac{2(\sqrt{3}+\sqrt{2})}{2} = \sqrt{3}+\sqrt{2}$

(B) $\dfrac{\sqrt{x}-\sqrt{y}}{\sqrt{x}+\sqrt{y}} = \dfrac{(\sqrt{x}-\sqrt{y})(\sqrt{x}-\sqrt{y})}{(\sqrt{x}+\sqrt{y})(\sqrt{x}-\sqrt{y})}$

$\quad = \dfrac{x-2\sqrt{xy}+y}{x-y}$

Matched Problem 5

Rationalize denominators and simplify:

(A) $\dfrac{\sqrt{2}}{\sqrt{2}+3}$　　(B) $\dfrac{\sqrt{x}+\sqrt{y}}{\sqrt{x}-\sqrt{y}}$

Answers to
Matched Problems

1. (A) $8\sqrt{2}$　(B) $-5\sqrt[5]{2x^2y^3}$　(C) $3\sqrt[3]{mn^2}+4\sqrt{mn}$

2. (A) $-2\sqrt{3}$　(B) $\dfrac{11\sqrt{2}}{2}$　(C) $\dfrac{-3\sqrt[3]{2}}{2}$

3. (A) $3\sqrt{2}-4\sqrt{3}$　(B) $2\sqrt{3}-5$　(C) $y+2\sqrt{y}-8$
   (D) $x+\sqrt[3]{x^2y}-\sqrt[3]{xy^2}-y$

4. $(2+\sqrt{3})^2-4(2+\sqrt{3})+1 = 4+4\sqrt{3}+3-8-4\sqrt{3}+1 = 0$

5. (A) $\dfrac{2-3\sqrt{2}}{-7}$ or $\dfrac{3\sqrt{2}-2}{7}$　(B) $\dfrac{x+2\sqrt{xy}+y}{x-y}$

EXERCISE 6-7

A *Express in simplest radical form and combine where possible.*

1. $7\sqrt{3}+2\sqrt{3}$　　2. $5\sqrt{2}+3\sqrt{2}$

3. $2\sqrt{a}-7\sqrt{a}$　　4. $\sqrt{y}-4\sqrt{y}$

5. $\sqrt{n}-4\sqrt{n}-2\sqrt{n}$　　6. $2\sqrt{x}-\sqrt{x}+3\sqrt{x}$

7. $\sqrt{5}-2\sqrt{3}+3\sqrt{5}$　　8. $3\sqrt{2}-2\sqrt{3}-\sqrt{2}$

9. $\sqrt{m}-\sqrt{n}-2\sqrt{n}$　　10. $2\sqrt{x}-\sqrt{y}+3\sqrt{y}$

11. $\sqrt{18}+\sqrt{2}$　　12. $\sqrt{8}-\sqrt{2}$

13. $\sqrt{8}-2\sqrt{32}$　　14. $\sqrt{27}-3\sqrt{12}$

Multiply and simplify where possible.

15. $\sqrt{7}(\sqrt{7}-2)$　　16. $\sqrt{5}(\sqrt{5}-2)$

17. $\sqrt{2}(3-\sqrt{2})$　　18. $\sqrt{3}(2-\sqrt{3})$

19. $\sqrt{y}(\sqrt{y}-8)$　　20. $\sqrt{x}(\sqrt{x}-3)$

21. $\sqrt{n}(4-\sqrt{n})$　　22. $\sqrt{m}(3-\sqrt{m})$

23. $\sqrt{3}(\sqrt{3}+\sqrt{6})$　　24. $\sqrt{5}(\sqrt{10}+\sqrt{5})$

25. $(2-\sqrt{3})(3+\sqrt{3})$　　26. $(\sqrt{2}-1)(\sqrt{2}+3)$

27. $(\sqrt{5}+2)^2$　　28. $(\sqrt{3}-3)^2$

29. $(\sqrt{m}-3)(\sqrt{m}-4)$　　30. $(\sqrt{x}+2)(\sqrt{x}-3)$

Rationalize denominators and simplify.

31. $\dfrac{1}{\sqrt{5}+2}$　　32. $\dfrac{1}{\sqrt{11}-3}$　　33. $\dfrac{2}{\sqrt{5}+1}$

34. $\dfrac{4}{\sqrt{6}-2}$　　35. $\dfrac{\sqrt{2}}{\sqrt{10}-2}$　　36. $\dfrac{\sqrt{2}}{\sqrt{6}+2}$

37. $\dfrac{\sqrt{y}}{\sqrt{y}+3}$　　38. $\dfrac{\sqrt{x}}{\sqrt{x}-2}$

B *Express in simplest radical form and combine where possible.*

39. $\sqrt{8mn}+2\sqrt{18mn}$　　40. $\sqrt{4x}-\sqrt{9x}$

41. $\sqrt{8} - \sqrt{20} + 4\sqrt{2}$ **42.** $\sqrt{24} - \sqrt{12} + 3\sqrt{3}$

43. $\sqrt[5]{a} - 4\sqrt[5]{a} + 2\sqrt[5]{a}$ **44.** $3\sqrt[3]{u} - 2\sqrt[3]{u} - 2\sqrt[3]{u}$

45. $2\sqrt[3]{x} + 3\sqrt[3]{x} - \sqrt{x}$ **46.** $5\sqrt[5]{y} - 2\sqrt[5]{y} + 3\sqrt[4]{y}$

47. $\sqrt{\frac{1}{8}} + \sqrt{8}$ **48.** $\sqrt{\frac{2}{3}} - \sqrt{\frac{3}{2}}$

49. $\sqrt{\frac{3uv}{2}} - \sqrt{24uv}$ **50.** $\sqrt{\frac{xy}{2}} + \sqrt{8xy}$

Multiply and simplify where possible.

51. $(4\sqrt{3} - 1)(3\sqrt{3} - 2)$

52. $(2\sqrt{7} - \sqrt{3})(2\sqrt{7} + \sqrt{3})$

53. $(\sqrt{x} - \sqrt{y})(\sqrt{x} + \sqrt{y})$

54. $(2\sqrt{x} + 3)(2\sqrt{x} - 3)$

55. $(5\sqrt{m} + 2)(2\sqrt{m} - 3)$

56. $(3\sqrt{u} - 2)(2\sqrt{u} + 4)$

57. $(\sqrt[3]{4} + \sqrt[3]{9})(\sqrt[3]{2} + \sqrt[3]{3})$

58. $\sqrt[3]{4}(\sqrt[3]{2} - \sqrt[3]{16})$

59. $(\sqrt[3]{x} - 1)(\sqrt[3]{x^2} + \sqrt[3]{x} + 1)$

60. $(\sqrt[3]{x} + 1)(\sqrt[3]{x^2} - \sqrt[3]{x} + 1)$

61. $(\sqrt[3]{a} + \sqrt[3]{2})(\sqrt[3]{a^2} - \sqrt[3]{2a} + \sqrt[3]{4})$

62. $(\sqrt[3]{a} - \sqrt[3]{2})(\sqrt[3]{a^2} + \sqrt[3]{2a} + \sqrt[3]{4})$

63. $(\sqrt[6]{8} - 1)(\sqrt[6]{8} + 1)$

64. $(\sqrt[4]{9} - 1)(\sqrt[4]{9} + 1)$

Show that the given number is a solution to the given equation.

65. $x^2 - 6x + 7 = 0; \ 3 - \sqrt{2}$

66. $x^2 - 6x + 7 = 0; \ 3 + \sqrt{2}$

67. $x^2 - 2x - 4 = 0; \ 1 - \sqrt{5}$

68. $x^2 - 2x - 4 = 0; \ 1 + \sqrt{5}$

69. $x^2 - 8x + 1 = 0; \ 4 + \sqrt{15}$

70. $x^2 - 8x + 1 = 0; \ 4 - \sqrt{15}$

Rationalize denominators and simplify.

71. $\frac{\sqrt{3} + 2}{\sqrt{3} - 2}$ **72.** $\frac{\sqrt{2} - 1}{\sqrt{2} + 1}$ **73.** $\frac{\sqrt{2} + \sqrt{3}}{\sqrt{3} - \sqrt{2}}$

74. $\frac{3 - \sqrt{a}}{\sqrt{a} - 2}$ **75.** $\frac{2 + \sqrt{x}}{\sqrt{x} - 3}$ **76.** $\frac{\sqrt{5} - \sqrt{2}}{\sqrt{5} + \sqrt{2}}$

77. $\frac{3\sqrt{x}}{2\sqrt{x} - 3}$ **78.** $\frac{5\sqrt{a}}{3 - 2\sqrt{a}}$

C *Rationalize numerators and simplify.*

79. $\frac{\sqrt{3} + 2}{\sqrt{3} - 2}$ **80.** $\frac{\sqrt{2} - 1}{\sqrt{2} + 1}$ **81.** $\frac{\sqrt{2} + \sqrt{3}}{\sqrt{3} - \sqrt{2}}$

82. $\frac{3 - \sqrt{a}}{\sqrt{a} - 2}$ **83.** $\frac{2 + \sqrt{x}}{\sqrt{x} - 3}$ **84.** $\frac{\sqrt{5} - \sqrt{2}}{\sqrt{5} + \sqrt{2}}$

Express in simplest radical form and combine where possible.

85. $\frac{\sqrt{3}}{3} + 2\sqrt{\frac{1}{3}} + \sqrt{12}$ **86.** $\sqrt{\frac{1}{2}} + \frac{\sqrt{2}}{2} + \sqrt{8}$

87. $\sqrt[3]{\frac{1}{3}} + \sqrt[3]{3^5}$ **88.** $\sqrt[4]{32} - \sqrt[4]{\frac{1}{8}}$

Multiply and simplify where possible.

89. $(\sqrt[3]{x} - \sqrt[3]{y^2})(\sqrt[3]{x^2} + 2\sqrt[3]{y})$

90. $(\sqrt[5]{u^2} - \sqrt[5]{v^3})(\sqrt[5]{u^3} + \sqrt[5]{v^2})$

91. $(\sqrt[3]{x} + \sqrt[3]{y})(\sqrt[3]{x^2} - \sqrt[3]{x}\sqrt[3]{y} + \sqrt[3]{y^2})$

92. $(\sqrt[3]{x} - \sqrt[3]{y})(\sqrt[3]{x^2} + \sqrt[3]{x}\sqrt[3]{y} + \sqrt[3]{y^2})$

Rationalize denominators and simplify. (Problems 95 and 96 depend on Problems 59 and 60; Problems 97 and 98 depend on Problems 91 and 92.)

93. $\frac{2\sqrt{x} + 3\sqrt{y}}{4\sqrt{x} + 5\sqrt{y}}$ **94.** $\frac{3\sqrt{x} + 2\sqrt{y}}{2\sqrt{x} - 5\sqrt{y}}$

95. $\frac{1}{\sqrt[3]{x} - 1}$ **96.** $\frac{1}{\sqrt[3]{x} + 1}$

97. $\frac{1}{\sqrt[3]{x} + \sqrt[3]{y}}$ **98.** $\frac{1}{\sqrt[3]{x} - \sqrt[3]{y}}$

99. $\frac{1}{\sqrt{x} + \sqrt{y} - \sqrt{z}}$ **100.** $\frac{1}{\sqrt{x} - \sqrt{y} + \sqrt{z}}$

[*Hint:* Start by multiplying the numerator and denominator by $(\sqrt{x} + \sqrt{y}) + \sqrt{z}$.]

6-8 **Complex Numbers**

♦ Complex Numbers
♦ Complex Numbers and Radicals
♦ The Complex Number System

The Pythagoreans (c. 500 B.C.) found that the simple equation

$$x^2 = 2 \qquad (1)$$

had no rational number solutions. If Equation (1) were to have a solution, then a new kind of number had to be invented—the *irrational numbers*. The irrational numbers $\sqrt{2}$ and $-\sqrt{2}$ are both solutions to Equation (1). Irrational numbers were not put on a firm mathematical foundation until the last century. The rational and irrational numbers together constitute the real number system.

Is there any need to extend the real number system further? Yes, since we find that another simple equation,

$$x^2 = -1$$

has no real solutions. No real number squared is negative. Once again, we are forced to invent a new kind of number—numbers that have the possibility of being negative when they are squared. These new numbers are called the *complex numbers*. The complex numbers evolved over a long period of time, as outlined in Table 1. However, like the real numbers, it was not until the last century that the complex numbers were placed on a firm mathematical basis.

♦ **COMPLEX NUMBERS**

In order to extend the real numbers to a new system in which negative numbers have square roots, we first introduce one number whose square is -1. This number is called the **imaginary unit** and is usually denoted by i:

$$i^2 = -1 \qquad i = \sqrt{-1}$$

Table 1   **A Brief History of Complex Numbers**

Approximate Date	Person	Event
50	Heron of Alexandria	First recorded encounter of a square root of a negative number
850	Mahavira of India	Said that a negative has no square root, since it is not a square
1545	Cardano of Italy	Found that solutions to cubic equations involved square roots of negative numbers
1637	Descartes of France	Introduced the terms ''real'' and ''imaginary''
1748	Euler of Switzerland	Used i for $\sqrt{-1}$
1832	Gauss of Germany	Introduced the term ''complex number''

Our new number system also must contain numbers such as

$$2i \qquad 3 + i \qquad 4 + 7i \qquad 1 - i$$

and so forth. Our system will, in fact, consist of all numbers in the form

$$a + bi$$

where a and b are real numbers. Such a number is called a **complex number.**

In the complex number $a + bi$, a is called the **real part** of the number and bi is called the **imaginary part.** Two complex numbers are considered **equal** if they have the same real and imaginary parts.

Complex numbers can be added, subtracted, and multiplied using ordinary algebraic properties and keeping in mind that $i^2 = -1$. For example, to add $3 + 2i$ and $-4 + i$, we would add just as we would for polynomials, in essence collecting like terms:

$$(3 + 2i) + (-4 + i) = (3 - 4) + (2i + i)$$
$$= -1 + 3i$$

Subtraction works similarly:

$$(3 + 2i) - (-4 + i) = (3 + 2i) + 4 - i$$
$$= 7 + i$$

We multiply as in the FOIL method:

$$\begin{array}{cccc} & \text{F} \quad \text{O} \quad \text{I} \quad \text{L} \\ (3 + 2i)(-4 + i) = & 3(-4) + 3i - 8i + 2i^2 \\ = & -12 - 5i + 2(-1) \qquad i^2 = -1 \\ = & -14 - 5i \end{array}$$

Example 1
Adding, Subtracting, and Multiplying Complex Numbers

Carry out the indicated operations and write each answer in the form $a + bi$.

(A) $(3 + 2i) + (2 - i)$ **(B)** $(3 + 2i) - (2 - i)$
(C) $(3 + 2i)(2 - i)$ **(D)** $(2 - 3i)^2 - (4i)^2$

Solution

(A) $(3 + 2i) + (2 - i) = 3 + 2i + 2 - i$ Remove parentheses and combine like
$\qquad\qquad\qquad\qquad\quad = 5 + i$ terms.

(B) $(3 + 2i) - (2 - i) = 3 + 2i - 2 + i$ Remove parentheses and combine like
$\qquad\qquad\qquad\qquad\quad = 1 + 3i$ terms.

(C) $(3 + 2i)(2 - i) = 6 + i - 2i^2$ Multiply and replace i^2 with -1.
$\qquad\qquad\qquad\quad = 6 + i - 2(-1)$
$\qquad\qquad\qquad\quad = 6 + i + 2$
$\qquad\qquad\qquad\quad = 8 + i$

(D) $(2 - 3i)^2 - (4i)^2 = 4 - 12i + 9i^2 - 16i^2$
$\qquad\qquad\qquad\qquad\quad = 4 - 12i + 9(-1) - 16(-1)$
$\qquad\qquad\qquad\qquad\quad = 4 - 12i - 9 + 16$
$\qquad\qquad\qquad\qquad\quad = 11 - 12i$

Matched Problem 1   Carry out the indicated operations and write each answer in the form $a + bi$.

(A) $(3 + 2i) + (6 - 4i)$ **(B)** $(3 - 5i) - (1 - 3i)$
(C) $(2 - 4i)(3 + 2i)$ **(D)** $(3i)^2 - (3 - 2i)^2$

Since $i = \sqrt{-1}$, we can perform division of complex numbers in a way that is analogous to how we rationalize a denominator of a fraction involving radicals. For example, to find the quotient

$$\frac{3 + 2i}{4 - i}$$

we multiply the numerator and denominator by the number $4 + i$. This number is called the *conjugate* of $4 - i$. In general, the **conjugate of $a + bi$ is $a - bi$**. Carrying out this multiplication gives

$$\frac{(3 + 2i)(\mathbf{4 + i})}{(4 - i)(\mathbf{4 + i})} = \frac{12 + 11i + 2i^2}{16 - i^2}$$
$$= \frac{12 + 11i - 2}{16 - (-1)}$$
$$= \frac{10 + 11i}{17}$$

This last result is rewritten as $\frac{10}{17} + \frac{11}{17}i$ to obtain the final answer written as a complex number in standard form.

Example 2
Dividing Complex Numbers

Carry out the indicated division and write each answer in the form $a + bi$.

(A) $\dfrac{2 + i}{3i}$ **(B)** $\dfrac{3 + 2i}{2 - i}$

Solution   **(A)** $\dfrac{2 + i}{3i}$ Multiply the numerator and denominator by $-3i$, the conjugate of $3i$.

$$= \frac{(2 + i)(\mathbf{-3i})}{(3i)(\mathbf{-3i})} = \frac{-6i - 3i^2}{-9i^2} = \frac{-6i - 3(-1)}{(-9)(-1)}$$
$$= \frac{-6i + 3}{9} = \frac{-6i}{9} + \frac{3}{9} = \frac{1}{3} - \frac{2}{3}i$$

(B) $\dfrac{3 + 2i}{2 - i}$ Multiply the numerator and denominator by $2 + i$, the conjugate of $2 - i$.

$$= \frac{(3 + 2i)(\mathbf{2 + i})}{(2 - i)(\mathbf{2 + i})} = \frac{6 + 7i + 2i^2}{4 - i^2} = \frac{6 + 7i + 2(-1)}{4 - (-1)}$$
$$= \frac{4 + 7i}{5} = \frac{4}{5} + \frac{7}{5}i$$

We can derive a general rule for the quotient of two complex numbers:

$$\frac{a + bi}{c + di} = \frac{(a + bi)(\mathbf{c - di})}{(c + di)(\mathbf{c - di})} = \frac{(ac + bd) + (-ad + bc)i}{c^2 + d^2}$$
$$= \frac{ac + bd}{c^2 + d^2} + \frac{-ad + bc}{c^2 + d^2}i$$

However, in practice, it is easier to proceed by multiplying the numerator and denominator by the conjugate of the denominator than it is to apply this rule.

Matched Problem 2   Carry out the indicated division and write each answer in the form $a + bi$.

(A) $\dfrac{3 + i}{2i}$ **(B)** $\dfrac{2 + 4i}{3 + 2i}$

Recall that subtraction and division are defined, in general, as follows:

$$A - B = C \quad \text{if and only if} \quad A = B + C$$
$$A \div B = C \quad \text{if and only if} \quad A = BC \qquad B \neq 0, \text{ and } C \text{ is unique}$$

The results obtained by the procedures illustrated in Examples 1 and 2 are consistent with these definitions. We can summarize the definitions of equality and the operations of addition, subtraction, multiplication, and division of complex numbers as follows:

Equality and Operations on Complex Numbers

Equality $a + bi = c + di$ if and only if $a = c$ and $b = d$

Addition $(a + bi) + (c + di) = (a + c) + (b + d)i$

Subtraction $(a + bi) - (c + di) = (a - c) + (b - d)i$

Multiplication $(a + bi)(c + di) = (ac - bd) + (ad + bc)i$

Division $\dfrac{a + bi}{c + di} = \dfrac{ac + bd}{c^2 + d^2} + \dfrac{-ad + bc}{c^2 + d^2}i$

◆ ## COMPLEX NUMBERS AND RADICALS

We know that if x is a positive real number, then x has two real square roots, one the negative of the other. If x is negative, then x has no real square roots. A negative number will, however, have two complex square roots, one also the negative of the other. For example, both $2i$ and $-2i$ are square roots of -4, since $(2i)^2 = 4i^2 = 4(-1) = -4$ and $(-2i)^2 = (-2)^2i^2 = 4(-1) = -4$. More generally, if we let $x = -a$, $a > 0$, then the square roots of x are given by $\sqrt{a}\,i$ and $-\sqrt{a}\,i$, since

$$(\sqrt{a}\,i)^2 = (\sqrt{a})^2i^2 = a(-1) = -a = x$$

and

$$(-\sqrt{a}\,i)^2 = (-\sqrt{a})^2i^2 = a(-1) = -a = x$$

When working with an expression such as $\sqrt{a}\,i$, we generally write it as $i\sqrt{a}$ so that i will not accidentally end up included in the radical sign. We adopt the notation

$$\sqrt{-a} = i\sqrt{a} \qquad \text{for } a > 0 \qquad \sqrt{-9} = i\sqrt{9} = 3i$$

Example 3
Rewriting Expressions Involving $\sqrt{-a}$

Write in the form $a + bi$:

(A) $\sqrt{-4}$ (B) $4 + \sqrt{-40}$ (C) $\dfrac{-3 - \sqrt{-7}}{2}$

Solution

(A) $\sqrt{-4} = i\sqrt{4} = 2i$

(B) $4 + \sqrt{-40} = 4 + i\sqrt{40} = 4 + i\sqrt{4 \cdot 10} = 4 + 2i\sqrt{10}$

(C) $\dfrac{-3 - \sqrt{-7}}{2} = \dfrac{-3 - i\sqrt{7}}{2} = -\dfrac{3}{2} - \dfrac{\sqrt{7}}{2}i$

Matched Problem 3

Write in the form $a + bi$:

(A) $\sqrt{-16}$ (B) $5 - \sqrt{-36}$ (C) $\dfrac{-5 - \sqrt{-2}}{2}$

Example 4
Simplifying Expressions Involving $\sqrt{-a}$

Convert square roots of negative numbers to complex form, perform the indicated operations, and express your answers in the form $a + bi$.

(A) $(3 + \sqrt{-4})(2 - \sqrt{-9})$ (B) $\dfrac{1}{3 - \sqrt{-4}}$

Solution

(A) $(3 + \sqrt{-4})(2 - \sqrt{-9}) = (3 + i\sqrt{4})(2 - i\sqrt{9})$
$= (3 + 2i)(2 - 3i)$
$= 6 - 5i - 6i^2$
$= 6 - 5i - 6(-1)$
$= 6 - 5i + 6$
$= 12 - 5i$

(B) $\dfrac{1}{3 - \sqrt{-4}} = \dfrac{1}{3 - i\sqrt{4}}$
$= \dfrac{1}{3 - 2i}$
$= \dfrac{1(3 + 2i)}{(3 - 2i)(3 + 2i)} = \dfrac{3 + 2i}{9 - 4i^2}$
$= \dfrac{3 + 2i}{9 - 4(-1)} = \dfrac{3 + 2i}{9 + 4}$
$= \dfrac{3 + 2i}{13} = \dfrac{3}{13} + \dfrac{2}{13}i$

In Example 4(A), we evaluated the product $\sqrt{-4}\sqrt{-9}$ and obtained

$$\sqrt{-4}\sqrt{-9} = (2i)(3i) = 6i^2 = -6$$

On the other hand,

$$\sqrt{(-4)(-9)} = \sqrt{36} = 6$$

Thus,

$$\sqrt{-4}\sqrt{-9} \neq \sqrt{(-4)(-9)}$$

That is, the square root property

$$\sqrt{ab} = \sqrt{a}\sqrt{b}$$

does not hold for $a, b < 0$.

Matched Problem 4   Convert square roots of negative numbers to complex form, perform the indicated operations, and express your answers in the form $a + bi$.

(A) $(4 - \sqrt{-25})(3 + \sqrt{-49})$ **(B)** $\dfrac{1}{2 + \sqrt{-9}}$

♦ THE COMPLEX NUMBER SYSTEM

Some additional terminology will help to clarify the relationship between the complex numbers and the other number systems studied thus far. If $b = 0$ in $a + bi$, the number is simply a, a real number. If $b \neq 0$, $a + bi$ is called an **imaginary number;** if, in addition, $a = 0$, it is called a **pure imaginary number.** See Figure 1.

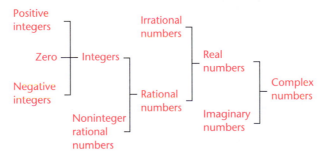

Figure 1

Complex numbers are used extensively by electrical, aeronautical, and space scientists, as well as chemists and physicists, who interpret complex numbers $a + bi$ to represent a variety of real-world quantities. Our use of the complex numbers will be in connection with solutions to second-degree equations such as $x^2 - 4x + 5 = 0$, which we will study in the next chapter.

Answers to Matched Problems

1. **(A)** $9 - 2i$ **(B)** $2 - 2i$ **(C)** $14 - 8i$ **(D)** $-14 + 12i$

2. **(A)** $\dfrac{1}{2} - \dfrac{3}{2}i$ **(B)** $\dfrac{14}{13} + \dfrac{8}{13}i$

3. **(A)** $4i$ **(B)** $5 - 6i$ **(C)** $-\dfrac{5}{2} - \dfrac{\sqrt{2}}{2}i$

4. **(A)** $47 + 13i$ **(B)** $\dfrac{2}{13} - \dfrac{3}{13}i$

EXERCISE 6-8

A *Perform the indicated operations and write each answer in the form a + bi.*

1. $(5 + 2i) + (3 + i)$
2. $(6 + i) + (2 + 3i)$
3. $(-8 + 5i) + (3 - 2i)$
4. $(2 - 3i) + (5 - 2i)$
5. $(8 + 5i) - (3 + 2i)$
6. $(9 + 7i) - (2 + 5i)$
7. $(4 + 7i) - (-2 - 6i)$
8. $(9 - 3i) - (12 - 5i)$
9. $(3 - 7i) + 5i$
10. $12 + (5 - 2i)$
11. $(5i)(3i)$
12. $(2i)(4i)$
13. $-2i(5 - 3i)$
14. $-3i(2 - 4i)$
15. $(2 - 3i)(3 + 3i)$
16. $(3 - 5i)(-2 - 3i)$
17. $(7 - 6i)(2 - 3i)$
18. $(2 - i)(3 + 2i)$
19. $(7 + 4i)(7 - 4i)$
20. $(5 - 3i)(5 + 3i)$
21. $(-1 + 3i)(2 - i)$
22. $(-2 + i)(3 - 2i)$
23. $(-3 + 4i)(4 - i)$
24. $(-5 + i)(5 - 3i)$
25. $\dfrac{1}{2 + i}$
26. $\dfrac{1}{3 - i}$
27. $\dfrac{3 + i}{2 - 3i}$
28. $\dfrac{2 - i}{3 + 2i}$
29. $\dfrac{13 + i}{2 - i}$
30. $\dfrac{15 - 3i}{2 - 3i}$
31. $\dfrac{1 + i}{1 - i}$
32. $\dfrac{3 + i}{3 - i}$
33. $\dfrac{3 - i}{3 + i}$
34. $\dfrac{1 - i}{1 + i}$

Write as a complex number a + bi.

35. $\sqrt{-16}$
36. $\sqrt{-64}$
37. $\sqrt{-72}$
38. $\sqrt{-24}$
39. $-\sqrt{-4}$
40. $-\sqrt{-25}$
41. $-\sqrt{25}$
42. $-\sqrt{4}$

B *Convert square roots of negative numbers to complex form, perform the indicated operations, and express your answers in the form a + bi.*

43. $(5 - \sqrt{-9}) + (2 - \sqrt{-4})$
44. $(-8 + \sqrt{-25}) + (3 - \sqrt{-4})$
45. $(9 - \sqrt{-9}) - (12 - \sqrt{-25})$
46. $(4 + \sqrt{-49}) - (-2 - \sqrt{-36})$
47. $(-2 + \sqrt{-49})(3 - \sqrt{-4})$
48. $(5 + \sqrt{-9})(2 - \sqrt{-1})$
49. $(-1 + \sqrt{-8})(2 - \sqrt{-10})$
50. $(-3 + \sqrt{-5})(-1 - \sqrt{-2})$
51. $(\sqrt{2} + \sqrt{-2})(\sqrt{2} + \sqrt{-2})$

52. $(\sqrt{3} + \sqrt{-3})(\sqrt{3} - \sqrt{-3})$
53. $\dfrac{5 - \sqrt{-4}}{3}$
54. $\dfrac{6 - \sqrt{-64}}{2}$
55. $\dfrac{1}{2 - \sqrt{-9}}$
56. $\dfrac{1}{3 - \sqrt{-16}}$
57. $\dfrac{1 + \sqrt{-3}}{1 - \sqrt{-3}}$
58. $\dfrac{2 - \sqrt{-5}}{2 + \sqrt{-5}}$
59. $\dfrac{1 - \sqrt{-3}}{2 + \sqrt{-5}}$
60. $\dfrac{2 - \sqrt{-5}}{1 + \sqrt{-3}}$

Perform the indicated operations and write each answer in the form a + bi.

61. $\dfrac{2}{5i}$
62. $\dfrac{1}{3i}$
63. $\dfrac{1 + 3i}{2i}$
64. $\dfrac{2 - i}{3i}$
65. $(2 - i)^2 + 3(2 - i) - 5$
66. $(2 - 3i)^2 - 2(2 - 3i) + 9$
67. $(2 + i\sqrt{3})^2 - 4(2 + i\sqrt{3}) + 7$
68. $(2 - i\sqrt{3})^2 - 4(2 - i\sqrt{3}) + 7$
69. $(1 - i\sqrt{2})^2 - 2(1 - i\sqrt{2}) + 3$
70. $(1 + i\sqrt{2})^2 - 2(1 + i\sqrt{2}) + 3$

In Problems 71–78, evaluate the expression for the given number.

71. $x^2 - 2x + 2$, for $x = 1 - i$
72. $x^2 - 2x + 2$, for $x = 1 + i$
73. $x^2 - 4x + 5$, for $x = 2 + i$
74. $x^2 - 4x + 5$, for $x = 2 - i$
75. $x^2 - 6x + 10$, for $x = 3 + i$
76. $x^2 - 6x + 10$, for $x = 3 - i$
77. $x^2 + 2x + 2$, for $x = -1 - i$
78. $x^2 + 2x + 2$, for $x = -1 + i$
79. Simplify: i^2, i^3, i^4, i^5, i^6, i^7, and i^8
80. Simplify: i^{12}, i^{13}, i^{14}, i^{15}, and i^{16}

C *Perform the indicated operations and write each answer in the form a + bi.*

81. $(a + bi) + (c + di)$
82. $(a + bi) - (c + di)$
83. $(a + bi)(a - bi)$
84. $(u - vi)(u + vi)$
85. $(a + bi)(c + di)$
86. $\dfrac{a + bi}{c + di}$
87. $(1 + i)^3$
88. $(1 - i)^3$
89. $\left(-\dfrac{1}{2} - \dfrac{\sqrt{3}}{2}i\right)^3$
90. $\left(-\dfrac{1}{2} + \dfrac{\sqrt{3}}{2}i\right)^3$

In Problems 91–94, solve the equation.

91. $y^2 = -36$ **92.** $x^2 = -25$

93. $(x - 9)^2 = -9$ **94.** $(x - 3)^2 = -4$

95. Simplify: $i^{-2}, i^{-3}, i^{-4}, i^{-5}, i^{-6}, i^{-7}$, and i^{-8}

96. Simplify: $i^{-12}, i^{-13}, i^{-14}, i^{-15}$, and i^{-16}

97. Evaluate

$$(a + bi)\left(\frac{a}{a^2 + b^2} - \frac{b}{a^2 + b^2}i\right) \qquad a, b \text{ not both } 0$$

thus showing that each nonzero complex number $a + bi$ has an inverse relative to multiplication.

98. When will

$$\frac{-b \pm \sqrt{b^2 - 4ac}}{2a}$$

represent a real number, assuming a, b, and c are all real numbers ($a \neq 0$)? When will it represent a non-real complex number?

99. Evaluate $ax^2 + bx + c$ for $x = p + qi$.

100. Evaluate $ax^2 + bx + c$ for $x = p - qi$.

CHAPTER SUMMARY

6-1   POSITIVE-INTEGER EXPONENTS

For a natural number n, the base a raised to the exponent n is $a^n = a \cdot a \cdot \;\cdots\; \cdot a$ (n factors). Natural number exponents satisfy the following **five properties of exponents** for positive integers m and n, and real numbers a and b:

1. $a^m a^n = a^{m+n}$

2. $(a^n)^m = a^{mn}$

3. $(ab)^m = a^m b^m$

4. $\left(\dfrac{a}{b}\right)^m = \dfrac{a^m}{b^m} \qquad b \neq 0$

5. $\dfrac{a^m}{a^n} = \begin{cases} a^{m-n} & \text{if } m \text{ is greater than } n \\ 1 & \text{if } m = n \\ \dfrac{1}{a^{n-m}} & \text{if } n \text{ is greater than } m \end{cases} \qquad a \neq 0$

6-2   INTEGER EXPONENTS

The concept of exponent is extended to **0** and **negative-number exponents** by

$$a^0 = 1 \qquad a \neq 0, \; 0^0 \text{ not defined}$$

$$a^{-n} = \frac{1}{a^n} \qquad \text{for } n \text{ a positive integer, } a \neq 0$$

The five basic properties of exponents continue to hold, and the fifth property can be rewritten as

$$\frac{a^m}{a^n} = a^{m-n} = \frac{1}{a^{n-m}} \qquad a \neq 0$$

6-3   SCIENTIFIC NOTATION AND APPLICATIONS

Any real number with a finite decimal representation can be written in **scientific notation**—that is, in the form $a \cdot 10^n$, where a is in the interval $[1, 10)$ and n is an integer.

6-4   RATIONAL EXPONENTS

For a natural number n, b is an **nth root** of a if $b^n = a$. For even n, a has two real nth roots when a is positive, none when a is negative; for odd n, a always has exactly one real nth root. The **nth root radical** $\sqrt[n]{a}$ represents the positive nth root, or **principal nth root,** of a if n is even and the unique nth root of a if n is odd. Here, n is called the **index** and a is the **radicand. Rational number exponents** are defined by

$$a^{1/n} = \sqrt[n]{a} \qquad a^{m/n} = (\sqrt[n]{a})^m$$

The five basic properties of exponents continue to hold for rational exponents as long as undefined roots are avoided. In particular,

$$a^{m/n} = (a^{1/n})^m = (a^m)^{1/n}$$

6-5   RADICAL FORMS AND RATIONAL EXPONENTS

In radical notation, the equations $a^{m/n} = (a^{1/n})^m = (a^m)^{1/n}$ become

$$a^{m/n} = (\sqrt[n]{a})^m = \sqrt[n]{a^m}$$

For positive real numbers a and b, and positive integers k, m, and n, the following properties of radicals hold:

1. $\sqrt[n]{a^n} = a$
2. $\sqrt[n]{ab} = \sqrt[n]{a} \cdot \sqrt[n]{b}$
3. $\sqrt[n]{\dfrac{a}{b}} = \dfrac{\sqrt[n]{a}}{\sqrt[n]{b}}$
4. $\sqrt[kn]{a^{km}} = \sqrt[n]{a^m}$

6-6   SIMPLEST RADICAL FORM

The properties of radicals can be used to convert an expression involving radicals to **simplest radical form.** In simplest radical form, the index of the radical is less than the power of any factor of the radicand, no radical occurs in a denominator, no fraction appears within a radical, and the index has no common factor with the power of the radicand. Radical Property 1 can be extended to any real number a:

$$\sqrt[n]{a^n} = \begin{cases} a & \text{if } n \text{ is odd} \\ |a| & \text{if } n \text{ is even} \end{cases}$$

6-7   BASIC OPERATIONS INVOLVING RADICALS

Expressions involving radicals often can be added, subtracted, or multiplied by making use of the distributive property for real numbers. Radicals sometimes can be removed from denominators by **rationalizing the denominator,** that is, multiplying the numerator and denominator by a factor that makes the denominator a rational number. The expressions $\sqrt{a} + \sqrt{b}$ and $\sqrt{a} - \sqrt{b}$ are **conjugates** of each other. Multiplying by the conjugate rationalizes a denominator in one of these forms.

6-8   COMPLEX NUMBERS

A **complex number** is a number of the form $a + bi$, where i is called the **imaginary unit** and a and b are real numbers. If $b = 0$, the number is real. If $b \neq 0$, $a + bi$ is called an **imaginary number;** if $a = 0$, the number is **pure imaginary.** The **conjugate of $a + bi$** is

$a - bi$. Two complex numbers $a + bi$ and $c + di$ are **equal** when $a = c$ and $b = d$. Basic operations are defined by

$$(a + bi) + (c + di) = (a + c) + (b + d)i$$
$$(a + bi) - (c + di) = (a - c) + (b - d)i$$
$$(a + bi) \cdot (c + di) = (ac - bd) + (ad + bc)i$$

Division of complex numbers makes use of rationalizing the denominator to obtain

$$\frac{a + bi}{c + di} = \frac{ac + bd}{c^2 + d^2} + \frac{-ad + bc}{c^2 + d^2}i$$

Since $i^2 = -1$, i is also denoted by $\sqrt{-1}$. Similarly, if a is positive, then $\sqrt{-a} = i\sqrt{a}$.

CHAPTER REVIEW EXERCISE

Work through all the problems in this chapter review and check answers in the back of the book. Answers to all the problems are there, and following each answer is a number in italics indicating the section in which that type of problem is discussed. Where weaknesses show up, review appropriate sections in the text.

Unless otherwise stated, all variables represent positive real numbers.

A *Simplify, using natural number exponents only.*

1. $\dfrac{x^8}{x^3}$
2. $(xy)^3$
3. $\left(\dfrac{x}{y}\right)^3$

4. $\dfrac{x^3}{x^8}$
5. $(x^3)^8$
6. $\dfrac{x^3}{x^3}$

7. $x^3 x^8$
8. $(-2x)^3$
9. $(-2x^3)(3x^8)$

Evaluate if possible, using only real numbers.

10. $\left(\dfrac{1}{3}\right)^0$
11. 3^{-2}
12. $\dfrac{1}{2^{-3}}$

13. $4^{-1/2}$
14. $(-9)^{3/2}$
15. $(-8)^{2/3}$

Write in scientific notation.

16. $4{,}280{,}000{,}000$
17. $0.000\ 031\ 8$

Write in standard notation.

18. 7.29×10^5
19. 6.03×10^{-4}

Simplify, and write answers using positive exponents only.

20. $(3x^3y^2)(2xy^5)$
21. $\dfrac{9u^8v^6}{3u^4v^8}$
22. $6(xy^3)^5$

23. $\left(\dfrac{c^2}{d^5}\right)^3$
24. $\left(\dfrac{2x^2}{3y^3}\right)^2$
25. $(x^{-3})^{-4}$

26. $\dfrac{y^{-3}}{y^{-5}}$
27. $(x^2y^{-3})^{-1}$
28. $(x^9)^{1/3}$

29. $(x^4)^{-1/2}$
30. $x^{1/3}x^{-2/3}$
31. $\dfrac{u^{5/3}}{u^{2/3}}$

Change to radical form.

32. $(3m)^{1/2}$
33. $3m^{1/2}$

Change to rational exponent form.

34. $\sqrt{2x}$
35. $\sqrt{a + b}$

Write as a complex number $a + bi$.

36. $2 - \sqrt{-9}$

Simplify, and write in simplest radical form.

37. $\sqrt[3]{375}$
38. $\sqrt{4x^2y^4}$

39. $\sqrt{\dfrac{25}{y^2}}$
40. $\sqrt{36x^4y^7}$

41. $\dfrac{1}{\sqrt{2y}}$
42. $\dfrac{6ab}{\sqrt{3a}}$

43. $\sqrt{2x^2y^5}\sqrt{18x^3y^2}$
44. $\sqrt{\dfrac{y}{2x}}$

45. $4\sqrt{x} - 7\sqrt{x}$
46. $\sqrt{7} + 2\sqrt{3} - 4\sqrt{3}$

47. $\sqrt{5}(\sqrt{5} + 2)$
48. $(\sqrt{3} - 1)(\sqrt{3} + 2)$

49. $\dfrac{\sqrt{5}}{3 - \sqrt{5}}$

Perform the indicated operations, and write the answer in the form a + bi.

50. $(-3 + 2i) + (6 - 8i)$ **51.** $(3 - 3i)(2 + 3i)$

52. $\dfrac{13 - i}{5 - 3i}$ **53.** $\dfrac{2 - i}{2i}$

B *In Problems 54–59, simplify, using natural number exponents only.*

54. $\left(\dfrac{2x^3}{y^8}\right)^2$ **55.** $(-x^2y)^2(-xy^2)^3$

56. $\dfrac{-4(x^2y)^3}{(-2x)^2}$ **57.** $(3xy^3)^2(x^2y)^3$

58. $\left(\dfrac{-2x}{y^2}\right)^3$ **59.** $\left(\dfrac{3x^3y^2}{2x^2y^3}\right)^2$

60. Convert each number to scientific notation, simplify, and write your answer in scientific notation and as a decimal fraction:

$$\frac{0.000\,052}{130(0.0002)}$$

In Problems 61–74, simplify, and write answers using positive exponents only.

61. $\dfrac{3m^4n^{-7}}{6m^2n^{-2}}$ **62.** $(x^{-3}y^2)^{-2}$

63. $\dfrac{1}{(2x^2y^{-3})^{-2}}$ **64.** $\left(-\dfrac{a^2b}{c}\right)^2\left(\dfrac{c}{b^2}\right)^3\left(\dfrac{1}{a^3}\right)^2$

65. $\left(\dfrac{8u^{-1}}{2^2u^2v^0}\right)^{-2}\left(\dfrac{u^{-5}}{u^{-3}}\right)^3$ **66.** $\left(\dfrac{9m^3n^{-3}}{3m^{-2}n^2}\right)^{-2}$

67. $(x - y)^{-2}$ **68.** $(9a^4b^{-2})^{1/2}$

69. $\left(\dfrac{27x^2y^{-3}}{8x^{-4}y^3}\right)^{1/3}$ **70.** $\dfrac{m^{-1/4}}{m^{3/4}}$

71. $(2x^{1/2})(3x^{-1/3})$ **72.** $\dfrac{3x^{-1/4}}{6x^{-1/3}}$

73. $\dfrac{5^0}{3^2} + \dfrac{3^{-2}}{2^{-2}}$ **74.** $(x^{1/2} + y^{1/2})^2$

75. If a is a square root of b, then which of the following must be true: $a^2 = b$, $b^2 = a$, $\sqrt{a} = b$, $\sqrt{b} = a$, $\sqrt{a} = |b|$, or $\sqrt{b} = |a|$?

Change to radical form.

76. $(2mn)^{2/3}$ **77.** $3x^{2/5}$

Change to rational exponent form.

78. $\sqrt[7]{x^5}$ **79.** $-3\sqrt[3]{(xy)^2}$

Simplify, and write in simplest radical form.

80. $\sqrt[3]{(2x^2y)^3}$ **81.** $3x\sqrt[3]{x^5y^4}$ **82.** $\dfrac{\sqrt{8m^3n^4}}{\sqrt{12m^2}}$

83. $\sqrt[8]{y^6}$ **84.** $-2x\sqrt[5]{36x^7y^{11}}$ **85.** $\dfrac{2x^2}{\sqrt[3]{4x}}$

86. $\sqrt[5]{\dfrac{3y^2}{8x^2}}$ **87.** $(2\sqrt{x} - 5\sqrt{y})(\sqrt{x} + \sqrt{y})$

88. $\dfrac{\sqrt{x} - 2}{\sqrt{x} + 2}$ **89.** $\dfrac{3\sqrt{x}}{2\sqrt{x} - \sqrt{y}}$

90. $\sqrt{\dfrac{2}{3}} + \sqrt{\dfrac{3}{2}}$

Perform the indicated operations, and write the answer in the form a + bi.

91. $(2 - 2\sqrt{-4}) - (3 - \sqrt{-9})$

92. $\dfrac{2 - \sqrt{-1}}{3 + \sqrt{-4}}$

93. $(3 + i)^2 - 2(3 + i) + 3$

Simplify, and write answers using positive exponents only.

94. $(x^{-1} + y^{-1})^{-1}$ **95.** $\left(\dfrac{a^{-2}}{b^{-1}} + \dfrac{b^{-2}}{a^{-1}}\right)^{-1}$

C *In Problems 96 and 97, simplify, and write in simplest radical form.*

96. $\sqrt[9]{8x^6y^{12}}$ **97.** $\sqrt[3]{3} - \dfrac{6}{\sqrt[3]{9}} + 3\sqrt[3]{\dfrac{1}{9}}$

98. Simplify $3\sqrt[3]{x^3} - 2\sqrt{x^2}$.
   (A) For x a positive number
   (B) For x a negative number

Show that the given number is a solution to the equation.

99. $x^2 + 2x + 2 = 0$; $-1 + i$

100. $x^2 - 2x - 1 = 0$; $1 + \sqrt{2}$

CHAPTER PRACTICE TEST

The following practice test is provided for you to test your knowledge of the material in this chapter. You should try to complete it in 50 minutes or less. Answers in the back of the book indicate the section in the text where the material in the question is covered. Actual tests in your class may vary from this practice test in difficulty, length, or emphasis, depending on the goals of your course or instructor.

All variables represent positive real numbers unless specified otherwise.

Rewrite in the form requested.

1. $(xy)^2 + (-x^2)^3 + \left(\dfrac{x}{y}\right)^4$ without parentheses

2. $0.000\ 034\ 5$ in scientific notation

3. 2.468×10^{10} in standard notation

4. $\sqrt{9x^3y^4z^5}$ in simplest radical form

5. $\dfrac{1}{\sqrt{3x}}$ in simplest radical form

6. $\dfrac{1}{\sqrt{x}+3}$ in simplest radical form

7. $\sqrt{125} + \sqrt[3]{49}$ in exponent notation, with integer bases as small as possible.

8. $3^{4/3} - 5^{2/5}$ in simplest radical form

9. $2^0 x^{-2} y z^{-1}$ using only positive exponents

10. $\dfrac{x^{-2}}{y} + \dfrac{x^2}{y^{-1}}$ as a single fraction using only positive exponents

In Problems 11–18, perform the indicated operations and simplify your answer.

11. $(1 + \sqrt{3})(2 - \sqrt{3})$

12. $\dfrac{1 + \sqrt{3}}{2 - \sqrt{3}}$

13. $(1 + \sqrt{3}) - (2 - \sqrt{3})$

14. $(1 + 3i) + (2 - 5i)$

15. $(1 + 3i)(2 - 5i)$

16. $\dfrac{1 + 3i}{2 - 5i}$

17. $\dfrac{1}{1 + 3i}$

18. $(1 - 2i)^2 - 2(1 - 2i) + 6$

19. Evaluate $x^2 - 2x + 5$ for $x = 1 + 2i$.

20. Evaluate $\sqrt[3]{(-2)^3} + \sqrt{(-2)^2}$.

7

Second-Degree Equations and Inequalities

The equation

$$\tfrac{1}{2}x - \tfrac{1}{3}(x + 3) = 2 - x$$

though it looks complicated, is actually a first-degree equation in one variable, since it can be transformed into the equivalent equation

$$7x - 18 = 0$$

This last equation is a special case of

$$ax + b = 0 \qquad a \neq 0$$

We have solved many equations of this type and found that each has a single solution. From a mathematical point of view we have completely answered the question of how to solve first-degree equations in one variable. We have done the same for linear inequalities in one variable.

In this chapter, we consider the next class of polynomial equations, called *second-degree equations,* or *quadratic equations.* A **quadratic equation in one variable** is any equation that can be written in the following form:

Photo reference: see Exercise 7-3, Problem 27.

> ### Quadratic Equation—Standard Form
>
> $$ax^2 + bx + c = 0 \qquad a \neq 0$$
>
> where x is a variable and a, b, and c are constants.

We will refer to this form as the **standard form** for the quadratic equation. The equations

$$2x^2 - 3x + 5 = 0 \qquad \text{and} \qquad 15 = 180t - 16t^2$$

are both quadratic equations in one variable, since they are either in the standard form or can be converted into this form. In Sections 7-1 through 7-3, we consider solution methods and applications for quadratic equations. Other equations, including certain equations involving radicals, can be converted to the form of quadratic equations and then solved. This is considered in Section 7-4.

We also consider quadratic equations in two variables, such as

$$x^2 + y^2 = 1 \qquad \text{and} \qquad y = x^2$$

A **quadratic equation in two variables** is an equation that can be written in the form

$$ax^2 + bxy + cy^2 + dx + ey + f = 0$$

We will graph elementary forms of such equations in Sections 7-5 through 7-8.

The chapter concludes with a consideration of quadratic and other nonlinear inequalities.

   7-1 ## Solving Quadratic Equations by Factoring, Square Roots, and Completing the Square

♦ Solution by Factoring
♦ Solution by Square Roots
♦ Completing the Square
♦ Solution of Quadratic Equations by Completing the Square

The technique of solving equations by factoring was introduced in Section 2-7. The technique easily solves those standard quadratic equations that can be factored. In this section we review what we know about solving by factoring and then we develop a method that is applicable to all quadratic equations. The method depends upon solving simple equations such as $x^2 = 16$ by taking square roots and upon a technique called *completing the square,* which converts equations into a readily solved form. Although this method always works, it is cumbersome. We can, however, apply the method to produce a general formula for solving quadratic equations, the subject of the next section.

♦ SOLUTION BY FACTORING

Recall the zero property for real numbers:

Zero Factor Property

For real numbers a and b:

$$a \cdot b = 0 \qquad \text{if and only if} \qquad a = 0 \text{ or } b = 0 \text{ (or both)}$$

That is, the product of real numbers is 0 if and only if one of the factors is zero. This allows us to solve equations of the form

$$(\text{Factored polynomial}) = 0$$

by setting each of the factors equal to 0. For example, to solve

$$(x - 2)(x + 1) = 0$$

we set the factors $x - 2 = 0$ and $x + 1 = 0$ to obtain

$$x = 2 \qquad \text{or} \qquad x = -1$$

as can be easily checked.

Example 1
Solving Equations by Factoring

Solve by factoring, if possible:

(A) $x^2 - 7x + 12 = 0$ **(B)** $x^2 + 6x - 2 = 0$

(C) $3 + \dfrac{5}{x} = \dfrac{2}{x^2}$

Solution

(A) $x^2 - 7x + 12 = 0$
$(x - 3)(x - 4) = 0$ The product is 0 if and only if $x - 3 = 0$ or $x - 4 = 0$.
$x - 3 = 0$ or $x - 4 = 0$
    $x = 3$ $x = 4$

(B) The polynomial cannot be factored using integer coefficients. Another method must be used. This will be done later in this section.

(C) $3 + \dfrac{5}{x} = \dfrac{2}{x^2}$ Multiply both sides by x^2, the LCM of the denominators ($x \neq 0$).
        $3x^2 + 5x = 2$ Write in standard form: $ax^2 + bx + c = 0$.
    $3x^2 + 5x - 2 = 0$ Factor the left side, if possible.
    $(3x - 1)(x + 2) = 0$
    $3x - 1 = 0$ or $x + 2 = 0$
        $3x = 1$ $x = -2$
        $x = \frac{1}{3}$

Matched Problem 1 Solve by factoring, if possible:

(A) $x^2 + 2x - 15 = 0$ (B) $x^2 + 2x + 3$

(C) $x = \dfrac{3}{2x - 5}$

♦ ## SOLUTION BY SQUARE ROOTS

The easiest type of quadratic equation to solve is the special form where the first-degree term is missing; that is, when the equation is of the form

$$ax^2 + c = 0 \qquad a \neq 0$$

We isolate x^2 and use the fact that if $x^2 = a$, then $x = \pm\sqrt{a}$. This follows from the definition of a square root. The **square root method** is illustrated in the following example.

Example 2
Solving by Square Root

Solve by the square root method:

(A) $x^2 - 8 = 0$ (B) $2x^2 - 3 = 0$
(C) $3x^2 + 27 = 0$ (D) $(x + \frac{1}{2})^2 = \frac{5}{4}$

Solution   (A) $x^2 - 8 = 0$

$x^2 = 8$ What number squared is 8?

$x = \pm\sqrt{8}$ or $\pm 2\sqrt{2}$ $\pm 2\sqrt{2}$ is a short way of writing $-2\sqrt{2}$ or $+2\sqrt{2}$.

(B) $2x^2 - 3 = 0$

$2x^2 = 3$ Don't write $2x = \pm\sqrt{3}$ next, since this would ignore the 2 when taking the square root.

$x^2 = \frac{3}{2}$ What number squared is $\frac{3}{2}$?

$x = \pm\sqrt{\frac{3}{2}}$ or $\pm\dfrac{\sqrt{6}}{2}$

(C) $3x^2 + 27 = 0$

$3x^2 = -27$ Don't write $3x = \pm\sqrt{-27}$ next. (Why?)

$x^2 = -9$ What number squared is -9?

$x = \pm\sqrt{-9} = \pm 3i$

(D) $(x + \frac{1}{2})^2 = \frac{5}{4}$ Solve for $x + \frac{1}{2}$; then solve for x.

$x + \frac{1}{2} = \pm\sqrt{\frac{5}{4}}$

$x = -\dfrac{1}{2} \pm \dfrac{\sqrt{5}}{2}$

$= \dfrac{-1 \pm \sqrt{5}}{2}$ Short for $\dfrac{-1 + \sqrt{5}}{2}$ or $\dfrac{-1 - \sqrt{5}}{2}$

Matched Problem 2   Solve by the square root method:

(A) $x^2 - 12 = 0$ (B) $3x^2 - 5 = 0$
(C) $2x^2 + 8 = 0$ (D) $(x + \frac{1}{3})^2 = \frac{2}{9}$

♦ COMPLETING THE SQUARE

The factoring and square root methods are fast and easy to use when they apply. Unfortunately, these methods cannot always be used. For example, the very simple-looking polynomial in the equation

$$x^2 + 6x - 2 = 0$$

cannot be solved by taking square roots and cannot be factored in the integers. The equation requires a new approach.

We now introduce a method, called **solution by completing the square,** that works for *all* quadratic equations. The method is based on the process of transforming the standard quadratic equation

$$ax^2 + bx + c = 0$$

into the form

$$(x + A)^2 = B$$

where A and B are constants. This last equation can be easily solved by the square root method:

$$(x + A)^2 = B$$
$$x + A = \pm\sqrt{B}$$
$$x = -A \pm \sqrt{B}$$

Before considering how the transformation above is accomplished, let's pause for a moment and consider a related problem: what number must be added to $x^2 + bx$ so that the result is the square of a linear expression? There is a rule for finding this number, based on the squares of the following binomials:

$$(x + m)^2 = x^2 + \underbrace{2mx}_{} + \overset{\uparrow}{m^2} \qquad (x - m)^2 = x^2 - \underbrace{2mx}_{} + \overset{\uparrow}{m^2}$$

m^2 is the square of $\frac{1}{2}$ the coefficient of x m^2 is the square of $\frac{1}{2}$ the coefficient of x

In either case, we see that the third term on the right of each equation is the square of $\frac{1}{2}$ the coefficient of x in the second term on the right; that is, m^2 is the square of $\frac{1}{2} \cdot 2m$. This observation leads directly to the following rule:

Completing the Square

To **complete the square** of a quadratic expression of the form

$$x^2 + bx$$

add the square of $\frac{1}{2}$ the coefficient of x, that is,

$$\left(\frac{b}{2}\right)^2 \quad \text{or} \quad \frac{b^2}{4}$$

Then factor:

$$x^2 + bx + \left(\frac{b}{2}\right)^2 = \left(x + \frac{b}{2}\right)^2$$

Example 3
Completing the Square

Complete the square and factor:

(A) $x^2 + 6x$ **(B)** $x^2 - 3x$

Solution   **(A)** To complete the square of $x^2 + 6x$, add $(\frac{6}{2})^2$, that is, 9. Thus,

$$x^2 + 6x + \mathbf{9} = (x + 3)^2$$

(B) To complete the square of $x^2 - 3x$, add $(-\frac{3}{2})^2$, that is, $\frac{9}{4}$. Thus,

$$x^2 - 3x + \tfrac{9}{4} = (x - \tfrac{3}{2})^2$$

Matched Problem 3

Complete the square and factor:

(A) $x^2 + 10x$ **(B)** $x^2 - 5x$

The rule above applies only to quadratic forms where the coefficient a of the second-degree term is 1. When solving equations where $a \neq 1$, we must divide through by the leading coefficient so that the rule may be applied. This will be illustrated in Example 6 below.

We now use the method of completing the square to solve quadratic equations. In the next section, we will use this method to develop a formula that will work for *all* quadratic equations. The process of completing the square, in addition to producing this formula, is used in many other situations in mathematics. Another use will be given in Section 7-6.

◆ **SOLUTION OF QUADRATIC EQUATIONS BY COMPLETING THE SQUARE**

The method of completing the square can be summarized as follows:

Solving Quadratic Equations by Completing the Square

1. Write the equation in standard form:

$$ax^2 + bx + c = 0$$

2. Make the coefficient of x^2 equal to 1 by dividing both sides by the existing coefficient a.

> **3.** Move the constant term to the right side.
> **4.** Complete the square on the left side, adding the same quantity to the right.
> **5.** Solve by the square root method.

The method is illustrated in the following examples.

Example 4
Solving by Completing the Square

Solve $x^2 + 6x - 2 = 0$ by the method of completing the square.

Solution $x^2 + 6x - 2 = 0$ The equation is already in standard form with coefficient of x^2 equal to 1. We thus begin with step 3. Add 2 to both sides of the equation to remove -2 from the left side.

$$x^2 + 6x = 2$$ To complete the square of the left side, add the square of $\frac{1}{2}$ the coefficient of x, that is, $(\frac{6}{2})^2$, to each side of the equation.

$$x^2 + 6x \mathbf{+ 9} = 2 \mathbf{+ 9}$$ Factor the left side.
$$(x + 3)^2 = 11$$ Solve by the square root method.
$$x + 3 = \pm\sqrt{11}$$
$$x = -3 \pm \sqrt{11}$$

Matched Problem 4   Solve $x^2 - 8x + 10 = 0$ by completing the square.

Example 5
Solving by Completing the Square

Solve $x^2 - 4x + 13 = 0$ by completing the square.

Solution $x^2 - 4x + 13 = 0$
$$x^2 - 4x = -13$$ Add 4 to each side to complete the square on the left side.

$$x^2 - 4x \mathbf{+ 4} = \mathbf{4} - 13$$
$$(x - 2)^2 = -9$$
$$x - 2 = \pm\sqrt{-9}$$
$$x - 2 = \pm 3i$$
$$x = 2 \pm 3i$$

Matched Problem 5   Solve $x^2 - 2x + 3 = 0$ by completing the square.

Example 6 **Solving by Completing the Square**	Solve $2x^2 - 4x - 3 = 0$ by completing the square.

Solution $2x^2 - 4x - 3 = 0$ Note that the coefficient of x^2 is not 1. Divide
through by the leading coefficient and proceed as in
Example 5.

$$x^2 - 2x - \tfrac{3}{2} = 0$$
$$x^2 - 2x = \tfrac{3}{2}$$
$$x^2 - 2x + 1 = \tfrac{3}{2} + 1$$
$$(x - 1)^2 = \tfrac{5}{2}$$
$$x - 1 = \pm\sqrt{\tfrac{5}{2}}$$
$$x = 1 \pm \frac{\sqrt{10}}{2} = \frac{2 \pm \sqrt{10}}{2}$$

Matched Problem 6 Solve $2x^2 + 8x + 3 = 0$ by completing the square.

**Answers to
Matched Problems**

1. **(A)** $x = -5, 3$ **(B)** Cannot be solved by factoring.   **(C)** $x = -\tfrac{1}{2}, 3$

2. **(A)** $x = \pm 2\sqrt{3}$ **(B)** $x = \pm\sqrt{\tfrac{5}{3}}$ or $\pm\dfrac{\sqrt{15}}{3}$

   (C) $x = \pm 2i$ **(D)** $x = \dfrac{-1 \pm \sqrt{2}}{3}$

3. **(A)** $x^2 + 10x + 25 = (x + 5)^2$ **(B)** $x^2 - 5x + \tfrac{25}{4} = (x - \tfrac{5}{2})^2$

4. $x = 4 \pm \sqrt{6}$

5. $x = 1 \pm i\sqrt{2}$ 6. $x = -2 \pm \sqrt{\tfrac{5}{2}}$ or $\dfrac{-4 \pm \sqrt{10}}{2}$

EXERCISE 7-1

A *Solve by factoring, if possible.*

1. $x^2 + 5x - 6 = 0$ 2. $x^2 - 4x + 5 = 0$
3. $x^2 - 7x + 10 = 0$ 4. $x^2 + 4x - 21 = 0$
5. $2x^2 + 7x - 4 = 0$ 6. $2x^2 - 7x - 15 = 0$

Solve by the square root method.

7. $x^2 - 16 = 0$ 8. $x^2 - 25 = 0$
9. $x^2 + 16 = 0$ 10. $x^2 + 25 = 0$
11. $y^2 - 45 = 0$ 12. $m^2 - 12 = 0$
13. $4x^2 - 9 = 0$ 14. $9y^2 - 16 = 0$
15. $16y^2 = 9$ 16. $9x^2 = 4$

Complete the square and factor.

17. $x^2 + 4x$ 18. $x^2 + 8x$

19. $x^2 - 6x$ 20. $x^2 - 10x$
21. $x^2 + 12x$ 22. $x^2 + 2x$

Solve by completing the square.

23. $x^2 + 4x + 2 = 0$ 24. $x^2 + 8x + 3 = 0$
25. $x^2 - 6x - 3 = 0$ 26. $x^2 - 10x - 3 = 0$

B *Write the equations in standard form and solve by factoring, if possible.*

27. $\dfrac{t}{2} = \dfrac{2}{t}$ 28. $y = \dfrac{9}{y}$

29. $\dfrac{m}{4}(m + 1) = 3$ 30. $\dfrac{A^2}{2} = A + 4$

31. $2y = \dfrac{2}{y} + 3$ 32. $L = \dfrac{15}{L - 2}$

33. $2 + \dfrac{2}{x^2} = \dfrac{5}{x}$ 34. $1 - \dfrac{3}{x} = \dfrac{10}{x^2}$

35. $\dfrac{x}{6} = \dfrac{1}{x+1}$ **36.** $x + 1 = \dfrac{2}{x}$

Solve by the square root method.

37. $y^2 = 2$ **38.** $x^2 = 3$

39. $16a^2 + 9 = 0$ **40.** $4x^2 + 25 = 0$

41. $9x^2 - 7 = 0$ **42.** $4t^2 - 3 = 0$

43. $(m - 3)^2 = 25$ **44.** $(n + 5)^2 = 9$

45. $(t + 1)^2 = -9$ **46.** $(d - 3)^2 = -4$

47. $(x - \frac{1}{3})^2 = \frac{4}{9}$ **48.** $(x - \frac{1}{2})^2 = \frac{9}{4}$

Complete the square and factor.

49. $x^2 + 3x$ **50.** $x^2 + x$

51. $u^2 - 5u$ **52.** $m^2 - 7m$

Solve by completing the square.

53. $x^2 + x - 1 = 0$ **54.** $x^2 + 3x - 1 = 0$

55. $u^2 - 5u + 2 = 0$ **56.** $n^2 - 3n - 1 = 0$

57. $m^2 - 4m + 8 = 0$ **58.** $x^2 - 2x + 3 = 0$

59. $2y^2 - 4y + 1 = 0$ **60.** $2x^2 - 6x + 3 = 0$

61. $2u^2 + 3u - 1 = 0$ **62.** $3x^2 + x - 1 = 0$

63. $2x^2 + 3x - 2 = 0$ **64.** $2x^2 + x - 3 = 0$

65. $2x^2 - 2x - 1 = 0$ **66.** $2x^2 - 3x - 1 = 0$

67. $3x^2 - 2x - 1 = 0$ **68.** $3x^2 + 3x + 1 = 0$

69. $4x^2 - 8x + 5 = 0$ **70.** $4x^2 + 5x + 6 = 0$

71. $2u^2 - 3u + 2 = 0$ **72.** $3x^2 - 5x + 3 = 0$

73. $x^2 + x + 1 = 0$ **74.** $2x^2 - 3x + 4 = 0$

Solve by the square root method.

75. $(y + \frac{5}{2})^2 = \frac{5}{2}$ **76.** $(x - \frac{3}{2})^2 = \frac{3}{2}$

77. $(x - 2)^2 = -1$ **78.** $(x + \frac{1}{2})^2 = -\frac{3}{4}$

C *In Problems 79–88, write the equations in standard form and solve.*

79. $x = \dfrac{1 + 3x}{x + 3}$ **80.** $\dfrac{1}{x} - \dfrac{1}{x^2} = \dfrac{1}{x+1}$

81. $\dfrac{x+2}{x-1} - \dfrac{6x}{x^2 - 1} = \dfrac{2x-1}{x+1}$

82. $\dfrac{2(x - 1)}{x - 2} = \dfrac{1}{x} + \dfrac{1}{x - 2}$

83. $x = \dfrac{3}{x - 2}$ **84.** $x = 2 + \dfrac{9}{x - 2}$

85. $x^2 + 2\sqrt{2}\,x - 2 = 0$ **86.** $x^2 - 2\sqrt{5}\,x + 5 = 0$

87. $x^2 - 4\sqrt{3}\,x + 13 = 0$ **88.** $x^2 + 2\sqrt{2}\,x + 3 = 0$

The method of completing the square works even if the coefficients are complex numbers. Use it for Problems 89–94.

89. $x^2 - 2ix - 4 = 0$ **90.** $x^2 + 2ix + 2 = 0$

91. $x^2 - 6ix - 9 = 0$ **92.** $x^2 - 8ix + 8 = 0$

93. $x^2 - 2ix + 2 = 0$ **94.** $x^2 - 4ix - 4 = 0$

95. Solve for x: $x^2 + mx + n = 0$

96. Solve for x: $ax^2 + bx + c = 0,\ a \neq 0$

In Problems 97 and 98, solve for the indicated variable in terms of the other letters. Use positive square roots only.

97. $a^2 + b^2 = c^2$, for a **98.** $s = \frac{1}{2}gt^2$, for t

APPLICATIONS

99. *Business* In a given city on a given day, the demand equation for gasoline is $d = 900/p$ and the supply equation is $s = p - 80$, where d and s denote the number of gallons demanded and supplied (in thousands), respectively, at a price of p cents per gallon. Find the price at which supply is equal to demand.

100. *Physics* To find the critical velocity at the top of the loop necessary to keep a steel ball on the track (see the figure), the centripetal force mv^2/r is equated to the force due to gravity mg. The mass m cancels out of the equation, and we are left with $v^2 = gr$. For a loop of radius 0.25 foot, find the critical velocity (in feet per second) at the top of the loop that is required to keep the ball on the track. Use $g = 32$ and compute your answer to two decimal places using a calculator.

7-2 The Quadratic Formula

♦ Quadratic Formula
♦ The Discriminant
♦ Which Method?

The method of completing the square can be used to solve any quadratic equation, but the process is often tedious. A more efficient method is desirable, and in this section we provide one. We will use the method of completing the square to solve the general equation

$$ax^2 + bx + c = 0 \qquad a \neq 0$$

for x in terms of coefficients a, b, and c. We thus obtain a formula that can be used to find the solution of *any* quadratic equation.

♦ **QUADRATIC FORMULA**

To apply the method of completing the square to the general equation

$$ax^2 + bx + c = 0 \qquad a \neq 0$$

we follow the steps outlined in Section 7-1.

1. The equation is given in standard form.
2. We make the leading coefficient 1 by dividing both sides of the equation by a to obtain

$$x^2 + \frac{b}{a}x + \frac{c}{a} = 0$$

3. Add $-c/a$ to both sides to clear c/a from the left side:

$$x^2 + \frac{b}{a}x = -\frac{c}{a}$$

4. Complete the square on the left side by adding the square of $\frac{1}{2}$ the coefficients of x, that is, $(b/2a)^2$, to each side:

$$x^2 + \frac{b}{a}x + \frac{b^2}{4a^2} = \frac{b^2}{4a^2} - \frac{c}{a} = \frac{b^2}{4a^2} - \frac{4ac}{4a^2} = \frac{b^2 - 4ac}{4a^2}$$

5. We now factor the left side and solve by the square root method:

$$\left(x + \frac{b}{2a}\right)^2 = \frac{b^2 - 4ac}{4a^2}$$

$$x + \frac{b}{2a} = \pm\sqrt{\frac{b^2 - 4ac}{4a^2}}$$

$$x = \frac{-b}{2a} \pm \sqrt{\frac{b^2 - 4ac}{4a^2}}$$

$$= \frac{-b}{2a} \pm \frac{\sqrt{b^2 - 4ac}}{\sqrt{4a^2}}$$

If $a > 0$, $\sqrt{4a^2} = 2a$ and

$$x = \frac{-b}{2a} \pm \frac{\sqrt{b^2 - 4ac}}{2a} = \frac{-b \pm \sqrt{b^2 - 4ac}}{2a}$$

If $a < 0$, $\sqrt{4a^2} = -2a$ and

$$x = \frac{-b}{2a} \pm \frac{\sqrt{b^2 - 4ac}}{-2a} = \frac{-b \pm \sqrt{b^2 - 4ac}}{2a}$$

In either case, we obtain the following formula for the solutions, or roots, of the equation $ax^2 + bx + c = 0$:

Quadratic Formula

$$x = \frac{-b \pm \sqrt{b^2 - 4ac}}{2a} \qquad a \neq 0$$

The **quadratic formula** should be learned and used to solve quadratic equations when simpler methods fail.

Example 1
Using the Quadratic Formula

Solve $2x^2 - 4x - 3 = 0$ using the quadratic formula.

Solution
$$2x^2 - 4x - 3 = 0$$

$$x = \frac{-b \pm \sqrt{b^2 - 4ac}}{2a} \qquad \text{Write down the quadratic formula and identify } a, b, \text{ and } c. \text{ Here, } a = 2, b = -4, c = -3.$$

$$= \frac{-(-4) \pm \sqrt{(-4)^2 - 4(2)(-3)}}{2(2)} \qquad \text{Substitute into the formula and simplify. Be careful of sign errors.}$$

$$= \frac{4 \pm \sqrt{40}}{4} = \frac{4 \pm 2\sqrt{10}}{4}$$

$$= \frac{2 \pm \sqrt{10}}{2} \qquad \text{The solutions are } \frac{2 + \sqrt{10}}{2} \text{ and } \frac{2 - \sqrt{10}}{2}.$$

Matched Problem 1 Solve $x^2 - 2x - 1 = 0$ using the quadratic formula.

Example 2
Using the
Quadratic Formula

Solve $x^2 + 11 = 6x$ using the quadratic formula.

Solution

$$x^2 + 11 = 6x \qquad \text{Write in standard form.}$$
$$x^2 - 6x + 11 = 0$$

$$x = \frac{-b \pm \sqrt{b^2 - 4ac}}{2a} \qquad a = 1, \ b = -6, \ c = 11$$

$$= \frac{-(-6) \pm \sqrt{(-6)^2 - 4(1)(11)}}{2(1)} \qquad \text{Be careful of sign errors here.}$$

$$= \frac{6 \pm \sqrt{-8}}{2}$$

$$= \frac{6 \pm 2i\sqrt{2}}{2} = 3 \pm i\sqrt{2} \qquad \text{The solutions are } 3 + i\sqrt{2} \text{ and } 3 - i\sqrt{2}.$$

Matched Problem 2   Solve $2x^2 + 3 = 4x$ using the quadratic formula.

♦ **THE DISCRIMINANT**

The expression $b^2 - 4ac$ that occurs under the radical in the quadratic formula provides useful information about the number and nature of the solutions to the equation $ax^2 + bx + c = 0$. If the expression $b^2 - 4ac$ is positive, then the radical represents a real number, and the $\pm$ sign in front of it leads to two real roots for the equation. If the expression is equal to 0, then the radical represents the number 0, and the only root is a real number. If the expression is negative, then the radical represents a complex number, and there are two complex roots due to the $\pm$ sign. Thus, $b^2 - 4ac$ determines how many roots there are to the equation and whether they are real or complex. We call $b^2 - 4ac$ the **discriminant,** since it discriminates among these three cases. We can summarize this use of the discriminant as follows:

Discriminant Test

For the equation $ax^2 + bx + c = 0$, a, b, c, real numbers; $a \neq 0$:

DISCRIMINANT $b^2 - 4ac$	ROOTS
Positive	Two real roots
0	One real root
Negative	Two nonreal, complex roots

When the equation has two complex roots, the $\pm$ sign in the quadratic formula simply changes the sign on the complex part of the number. Thus, the two roots will be complex conjugates of each other.

Example 3
Applying the Discriminant Test

Apply the discriminant test to determine the number and nature of the roots:

(A) $3x^2 - 4x + 1 = 0$ (B) $9x^2 - 6x + 1 = 0$
(C) $x^2 + 5x + 7 = 0$

Solution

(A) The discriminant $b^2 - 4ac = (-4)^2 - 4 \cdot 3 \cdot 1 = 4$ is positive. The equation has two real roots. Check that they are 1 and $\frac{1}{3}$.

(B) The discriminant $b^2 - 4ac = (-6)^2 - 4 \cdot 9 \cdot 1 = 0$, so there is one real root. Check that it is $\frac{1}{3}$.

(C) The discriminant $b^2 - 4ac = 25 - 4 \cdot 1 \cdot 7 = -3$ is negative, so there are two nonreal, complex conjugate roots. Check that these roots are $-\frac{5}{2} \pm i\sqrt{3}/2$.

Matched Problem 3

Apply the discriminant test to determine the number and nature of the roots; you do not need to solve the equation, but you can check your answer by doing so.

(A) $4x^2 - 20x + 25 = 0$ (B) $2x^2 + x - 1 = 0$ (C) $x^2 - 6x + 10 = 0$

◆   **WHICH METHOD?**

In normal practice the quadratic formula is used whenever the square root method or the factoring method does not produce results easily. These latter methods are generally faster when they apply, so they should be used whenever possible.

Note that any equation of the form

$$ax^2 + c = 0$$

always can be solved by the square root method. Also, any equation of the form

$$ax^2 + bx = 0$$

always can be solved by factoring, since $ax^2 + bx = x(ax + b)$.

To summarize:

> **To Solve $ax^2 + bx + c = 0$**
>
> **1.** If $b = 0$, solve by square roots.   To solve $2x^2 - 5 = 0$: $x^2 = = \frac{5}{2}$
> $x = \pm\sqrt{\frac{5}{2}}$
>
> **2.** If $c = 0$, solve by factoring.   To solve $2x^2 - 5x = 0$: $x(2x - 5) = 0$
> $x = 0$ or $x = \frac{5}{2}$
>
> **3.** Otherwise, try:
>   **(A)** Factoring, or
>   **(B)** Completing the square, or
>   **(C)** The quadratic formula, which *always* works

It is important to realize that the quadratic formula always can be used and will produce the same results as any other method.

Example 4
Comparing Methods

Solve $\dfrac{30}{8+x} + 2 = \dfrac{30}{8-x}$ by the most efficient method.

Solution   We solve first by the factoring method. We begin by rewriting the equation in standard form:

$$\frac{30}{8+x} + 2 = \frac{30}{8-x}$$

Multiply both sides by the LCM of the denominators: $(8 + x)(8 - x)$.

$$30(8-x) + 2(8+x)(8-x) = 30(8+x)$$
$$240 - 30x + 128 - 2x^2 = 240 + 30x$$
$$-2x^2 - 60x + 128 = 0$$

Divide both sides by -2.

$$x^2 + 30x - 64 = 0$$

Factor the left side, if possible.

$$(x+32)(x-2) = 0$$
$$x + 32 = 0 \qquad \text{or} \qquad x - 2 = 0$$
$$x = -32 \qquad\qquad\qquad x = 2$$

We also could have solved $x^2 + 30x - 64 = 0$ by using the quadratic formula:

$$x^2 + 30x - 64 = 0$$

$$x = \frac{-b \pm \sqrt{b^2 - 4ac}}{2a} \qquad a = 1,\ b = 30,\ c = -64$$

$$= \frac{-30 \pm \sqrt{30^2 - 4(1)(-64)}}{2(1)}$$

$$= \frac{-30 \pm \sqrt{1{,}156}}{2}$$

$$= \frac{-30 \pm 34}{2}$$

Thus, $x = -32$ or $x = 2$.

It is not clear which method was easier in Example 4, since ''easy'' depends on the person judging. Nevertheless, we got the same result by both methods, as expected.

Matched Problem 4   Solve $\dfrac{6}{x-2} + 2 = \dfrac{4}{x}$ by the most efficient method.

Answers to
Matched Problems

1. $x = 1 \pm \sqrt{2}$ **2.** $x = 1 \pm \dfrac{i\sqrt{2}}{2}$ or $\dfrac{2 \pm i\sqrt{2}}{2}$

3. (A) One real root   **(B)** Two real roots
(C) Two nonreal, complex conjugate roots

4. $x = \dfrac{1}{2} \pm \dfrac{\sqrt{15}}{2}i$

EXERCISE 7-2

A *Specify the constants a, b, and c for each quadratic equation when written in the standard form $ax^2 + bx + c = 0$.*

1. $2x^2 - 5x + 3 = 0$ 2. $3x^2 - 2x + 1 = 0$

3. $m = 1 - 3m^2$ 4. $2u^2 = 1 - 3u$

5. $3y^2 - 5 = 0$ 6. $2x^2 - 5x = 0$

Solve by use of the quadratic formula.

7. $x^2 + 8x + 3 = 0$ 8. $x^2 + 4x + 2 = 0$

9. $y^2 - 10y - 3 = 0$ 10. $y^2 - 6y - 3 = 0$

11. $x^2 + 3x + 5 = 0$ 12. $x^2 + 2x + 3 = 0$

13. $x^2 + 4x + 5 = 0$ 14. $x^2 + 5x + 7 = 0$

15. $x^2 - 2x + 5 = 0$ 16. $x^2 - 4x + 6 = 0$

17. $-x^2 + 3x - 3 = 0$ 18. $-x^2 + 2x - 4 = 0$

19. $-x^2 + 5x + 5 = 0$ 20. $-x^2 + x + 6 = 0$

B 21. $u^2 = 1 - 3u$ 22. $t^2 = 1 - t$

23. $y^2 + 3 = 2y$ 24. $x^2 + 8 = 4x$

25. $2m^2 + 3 = 6m$ 26. $2x^2 + 1 = 4x$

27. $p = 1 - 3p^2$ 28. $3q + 2q^2 = 1$

Apply the discriminant test to determine the number and nature of the roots.

29. $4x^2 + 5x - 6 = 0$ 30. $3x^2 + 2x + 1 = 0$

31. $9x^2 - 24x + 16 = 0$ 32. $25x^2 + 10x + 1 = 0$

33. $x^2 - 8x + 17 = 0$ 34. $x^2 + 11x + 30 = 0$

35. $10x^2 + 15x + 5 = 0$ 36. $12x^2 + 20x + 7 = 0$

37. $20x^2 - 25x + 8 = 0$ 38. $4x^2 - 30x + 53 = 0$

Solve each equation.

39. $(x - 5)^2 = 7$ 40. $(y + 4)^2 = 11$

41. $x^2 + 2x = 2$ 42. $x^2 - 1 = 3x$

43. $2u^2 + 3u = 0$ 44. $2n^2 = 4n$

45. $x^2 - 2x + 9 = 2x - 4$

46. $x^2 + 15 = 2 - 6x$

47. $y^2 = 10y + 3$ 48. $3(2x + 1) = x^2$

49. $2d^2 + 1 = 4d$ 50. $2y(3 - y) = 3$

51. $\dfrac{2}{u} = \dfrac{3}{u^2} + 1$ 52. $1 + \dfrac{8}{x^2} = \dfrac{4}{x}$

53. $\dfrac{1.2}{y - 1} + \dfrac{1.2}{y} = 1$ 54. $\dfrac{24}{10 + m} + 1 = \dfrac{24}{10 - m}$

55. $x^2 + 3x + 8 = 9 - 4x + x^2$

56. $3 - 4x^2 = 6 - 5x - 4x^2$

57. $\dfrac{x^2}{x - 1} - \dfrac{x}{x - 1} = 4$ 58. $\dfrac{x^2}{x - 4} - \dfrac{x}{x - 4} = 1$

59. $x - 2 = \dfrac{8}{x - 1} - 3$ 60. $x - 4 = \dfrac{5}{x - 2}$

Solve for the indicated variable in terms of the other letters.

61. $d = \frac{1}{2}gt^2$, for t (positive)

62. $a^2 + b^2 = c^2$, for a (positive)

63. $A = P(1 + r)^2$, for r (positive)

64. $P = EI - RI^2$, for I

C *Solve each equation.*

65. $x^2 - \sqrt{7}x + 2 = 0$

66. $3x^2 - 2\sqrt{15}x + 5 = 0$

67. $\sqrt{3}x^2 + 4x + \sqrt{3} = 0$

68. $\sqrt{2}x^2 + 2\sqrt{3}x + \sqrt{2} = 0$

69. $2x^2 + 3ix + 2 = 0$ 70. $x^2 - ix + 6 = 0$

71. $x^2 + ix - 1 = 0$ 72. $3x^2 - 5ix + 2 = 0$

73. $x^2 - \frac{1}{2}x - \frac{1}{4} = 0$ 74. $x^2 - \frac{1}{2}x - \frac{1}{3} = 0$

75. $x^2 + x - \frac{1}{2} = 0$ 76. $x^2 + x + \frac{1}{4} = 0$

Solve to two decimal places using a calculator.

77. $-3.14x^2 + x + 1.07 = 0$

78. $x^2 + 2.13x - 5.89 = 0$

79. $2.07x^2 - 3.79x + 1.34 = 0$

80. $0.61x^2 - 4.28x + 2.93 = 0$

81. $4.83x^2 + 2.04x - 3.18 = 0$

82. $5.13x^2 + 7.27x - 4.32 = 0$

Solve for x in terms of y.

83. $y^2 + xy - x^2 = 0$ 84. $x^2 + y^2 = x + y$

85. $\dfrac{x + y}{x - y} = \dfrac{x}{y}$

86. $x^2 + 3xy + y^2 - 2x + y + 1 = 0$

87. $\dfrac{1}{x^2} + \dfrac{1}{y^2} = 1$ 88. $\dfrac{1}{x^2} - \dfrac{1}{y^2} = 1$

Use the discriminant to determine which equations have real solutions.

89. $0.0134x^2 + 0.0414x + 0.0304 = 0$

90. $0.543x^2 - 0.182x + 0.00312 = 0$

91. $0.0134x^2 + 0.0214x + 0.0304 = 0$

92. $0.543x^2 - 0.182x + 0.0312 = 0$

In Problems 93–98, find the value(s) of the unspecified coefficient for which the equation will have exactly the solutions specified.

93. $2x^2 - 3x + c = 0$; one solution

94. $ax^2 + 6x + 5 = 0$; two real solutions

95. $ax^2 + 4x + 5 = 0$; two real solutions

96. $ax^2 + 2x - 9 = 0$; one real solution

97. $3x^2 + 8x + c = 0$; one real solution

98. $2x^2 - 7x + c = 0$; two real solutions

99. Show that if r_1 and r_2 are the two roots of $ax^2 + bx + c = 0$, then $r_1 r_2 = c/a$.

100. For r_1 and r_2 in Problem 99, show that $r_1 + r_2 = -b/a$.

7-3 Applications

We now consider several applications of quadratic equations. Since the equations often have two solutions, it is important to check both solutions in the original problem to see if one or the other must be rejected. (A review of the strategy for solving word problems in Sections 1-8 and 4-5 should prove helpful.)

Example 1
A Number Problem

The sum of a number and its reciprocal is $\frac{5}{2}$. Find the number.

Solution Let $x =$ The number. Its reciprocal is $1/x$. Then:

$$x + \frac{1}{x} = \frac{5}{2} \qquad \text{Clear fractions.}$$

$$2x^2 + 2 = 5x \qquad \text{Write in standard form.}$$

$$2x^2 - 5x + 2 = 0 \qquad \text{Solve by factoring.}$$

$$(2x - 1)(x - 2) = 0$$

$$x = \tfrac{1}{2} \quad \text{or} \quad 2 \qquad \text{Both answers are solutions to the problem, as can be easily checked.}$$

Matched Problem 1

If the reciprocal of a number is subtracted from the original number, the difference is $\frac{8}{3}$. Find the number.

Example 2
A Rate-Time Problem

A tank can be filled in 4 hours through two pipes when both are used. How many hours are required for each pipe to fill the tank alone if the smaller pipe requires 3 hours more than the larger one?

Solution Let

$$4 = \text{Time for both pipes to fill the tank together}$$
$$x = \text{Time for the larger pipe to fill the tank alone}$$
$$x + 3 = \text{Time for the smaller pipe to fill the tank alone}$$

Use the rate-time formula to find the rates at which the tank is filled:

$$\left(\begin{array}{c}\text{Amount of tank}\\ \text{filled per hour}\end{array}\right) \times \left(\begin{array}{c}\text{Number of}\\ \text{hours}\end{array}\right) = \left(\begin{array}{c}\text{Amount of}\\ \text{tank filled}\end{array}\right)$$

Thus, the rate at which the tank is filled when both pipes are used can be found:

$$\text{Rate} \cdot 4 \text{ hours} = 1 \text{ tank filled}$$
$$\text{Rate} = \tfrac{1}{4} \text{ tank per hour}$$

Similarly,

$$\frac{1}{x} = \text{Rate for larger pipe} \qquad \frac{1}{x} \text{ tank per hour}$$

$$\frac{1}{x+3} = \text{Rate for smaller pipe} \qquad \frac{1}{x+3} \text{ tank per hour}$$

Also,

$$\text{Sum of individual rates} = \text{Rate together}$$

$$\frac{1}{x} + \frac{1}{x+3} = \frac{1}{4} \qquad \text{Clear fractions.}$$

$$\mathbf{4x(x+3)} \cdot \frac{1}{x} + \mathbf{4x(x+3)} \cdot \frac{1}{x+3} = \mathbf{4x(x+3)} \cdot \frac{1}{4}$$

$$4(x+3) + 4x = x(x+3)$$

$$4x + 12 + 4x = x^2 + 3x$$

$$x^2 - 5x + 12 = 0 \qquad \text{Use the quadratic formula.}$$

$$x = \frac{5 \pm \sqrt{73}}{2} \qquad \begin{array}{l}\text{Since the time must be}\\ \text{positive, we discard the}\\ \text{negative root.}\end{array}$$

$$x = \frac{5 + \sqrt{73}}{2}$$

$$\approx 6.77 \text{ hours} \qquad \text{Larger pipe}$$

$$x + 3 \approx 9.77 \text{ hours} \qquad \text{Smaller pipe}$$

Matched Problem 2 Two pipes can fill a tank in 3 hours when used together. Alone, one can fill the tank 2 hours faster than the other. How long will it take each pipe to fill the tank alone? Compute the answers to two decimal places using a calculator.

Example 3
A Physical Science Problem

For a car traveling at a speed of v miles per hour, the shortest distance d (in number of feet) necessary to stop a car, under the best possible conditions, including reaction time, is given approximately by the formula $d = 0.044v^2 + 1.1v$. Estimate the speed of a car requiring 200 feet to stop after danger is realized. Compute the answer to two decimal places.

Solution

$$0.044v^2 + 1.1v = 200$$ Write in standard form.

$$0.044v^2 + 1.1v - 200 = 0$$ Use the quadratic formula.

$$v = \frac{-b \pm \sqrt{b^2 - 4ac}}{2a}$$ $a = 0.044$, $b = 1.1$, $c = -200$

$$= \frac{-1.1 \pm \sqrt{1.1^2 - 4(0.044)(-200)}}{2(0.044)}$$

$$= \frac{-1.1 \pm \sqrt{36.41}}{0.088}$$ Disregard the negative root, since we are interested only in positive v.

$$= \frac{-1.1 + 6.03}{0.088} = 56.02 \text{ miles per hour}$$ Computed to two decimal places.

Example 3 is typical of most significant real-world problems in that decimal quantities rather than convenient small numbers are involved.

Matched Problem 3   Repeat Example 3 for a car requiring 300 feet to stop after danger is realized.

Answers to Matched Problems   **1.** $-\frac{1}{3}$, 3   **2.** 5.16 hours and 7.16 hours   **3.** 71.01 miles per hour

---

EXERCISE 7-3

These problems are not grouped from easy (A) to difficult or theoretical (C). They are grouped somewhat according to type. The most difficult problems are marked with two stars (★★), those of moderate difficulty are marked with one star (★), and the easier problems are not marked.

NUMBER PROBLEMS

1. Find two consecutive positive even integers whose product is 168.

2. Find two positive numbers having a sum of 21 and a product of 104.

3. Find all numbers with the property that when the number is added to itself the sum is the same as when the number is multiplied by itself.

4. The sum of a number and its reciprocal is $\frac{10}{3}$. Find the number.

★ 5. The denominator of a fraction is 2 greater than its numerator. If both the numerator and denominator are increased by 3, the resulting fraction is equal to twice the original fraction. Find the original fraction.

★ 6. The denominator of a fraction is 1 greater than its numerator. If both the numerator and denominator are decreased by 3, the resulting fraction is equal to 4 times the original fraction. Find the original fraction.

7. The sum of a number and twice its reciprocal is equal to 3 times the product of the number and its reciprocal. Find the number.

8. Two more than a number is 1 less than 4 times the reciprocal of the number. Find the number.

9. The sum of the first n positive integers, 1, 2, 3, . . . , n, is given by the formula

$$S = \frac{n(n + 1)}{2}$$

Find n if $S = 66$.

10. The sum of the first n even integers, 1, 2, 3, . . . , n, is given by the formula $S = n(n + 1)$. Find n if $S = 156$.

GEOMETRY

The Pythagorean theorem may be needed in some of the following problems: A triangle is a right triangle if and only if the square of the longest side is equal to the sum of *the squares of the two shorter sides. That is, the triangle below is a right triangle if and only if $c^2 = a^2 + b^2$.*

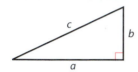

★ **11.** Approximately how far would the horizon be from an airplane 2 miles high? Assume that the radius of the earth is 4,000 miles and use a calculator to estimate the answer to the nearest mile (see the figure).

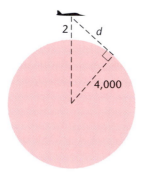

12. Repeat Problem 11 for an altitude of 3 miles.

13. Find the base and height of a triangle with area 3 square meters if its base is 3 meters longer than its height ($A = \frac{1}{2}bh$).

14. Repeat Problem 13 for an area of 2 square meters.

★ **15.** If the length and width of a 4- by 2-centimeter rectangle are each increased by the same amount, the area of the new rectangle will be twice the old. What are the dimensions (to two decimal places) of the new rectangle?

★ **16.** Repeat Problem 15 for a 3- by 5-centimeter rectangle.

17. The width of a rectangle is 2 meters less than its length. Find its dimensions (to two decimal places) if its area is 24 square meters.

18. Repeat Problem 17 for an area of 12 square meters.

19. A flag has a white cross of uniform width on a color background. Find the width of the cross so that it takes up exactly one-half the total area of a 4- by 3-foot flag.

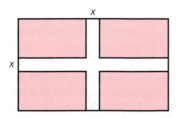

20. Repeat Problem 19 for a 5- by 3.75-foot flag.

21. A regular polygon with n sides has $n(n - 3)/2$ diagonals. Find n if the polygon has 20 diagonals.

22. Repeat Problem 21 for a polygon with 44 diagonals.

★ **23.** A rectangle has length 3 meters more than its width. If both dimensions are increased by 3 meters, the area becomes 4 times that of the original rectangle. Find the dimensions of the original rectangle.

★ **24.** A triangle has height 1 foot greater than its base. If both dimensions are increased by 2 feet, the area of the resulting triangle is 2.5 times that of the original triangle. Find the original dimensions.

PHYSICS AND ENGINEERING

25. The pressure p (in pounds per square foot) of wind blowing at v miles per hour is given by $p = 0.003v^2$. If a pressure gauge on a bridge registers a wind pressure of 7.5 pounds per square foot, what is the velocity of the wind?

26. Repeat Problem 25 for a pressure of 14.7 pounds per square foot.

27. One method of measuring the velocity of water in a stream or river is to use an L-shaped tube, as indicated in the figure. In physics, Torricelli's law tells us that the height (in feet) that the water is pushed up into the tube above the surface is related to the water's velocity (in feet per second) by the formula $v^2 = 2gh$, where g is approximately 32 feet per second per second. [*Note:* The device also can be used as a simple speedometer for a boat.] How fast is a stream flowing if $h = 0.5$ foot? Find the answer to two decimal places.

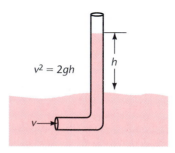

28. Repeat Problem 27 for $h = 1$ foot.

29. At 20 miles per hour a car collides with a stationary object with the same force it would have if it had been dropped $13\frac{1}{2}$ feet—that is, if it had been pushed off the roof of an average one-story house. In general, a car moving at r miles per hour hits a stationary object with a force of impact equivalent to the force with which it would hit the ground when falling from a height h given by the formula $h = 0.0336r^2$. Approximately how fast would a car have to be moving if it crashed as hard as if it had been pushed off the top of a 14-story building 142 feet high?

30. Repeat Problem 29 for a 12-story building 121 feet high.

31. For a car traveling at a speed of v miles per hour, the shortest distance d (in number of feet) necessary to stop a car under the best possible conditions (including reaction time) is given approximately by the formula $d = 0.044v^2 + 1.1v$. Estimate the speed of a car requiring 165 feet to stop after danger is realized. (See Example 3.)

32. Repeat Problem 31 for a distance of 292 feet.

33. An arrow shot vertically into the air with an initial velocity of v feet per second will reach a maximum height, if air resistance can be ignored, of $v^2/32$ feet. What is the initial velocity if the maximum height is 968 feet?

34. Repeat Problem 33 for a maximum height of 512 feet.

★ **35.** If an arrow is shot vertically into the air (from the ground) with an initial velocity of 128 feet per second, its distance y above the ground t seconds after it is released (neglecting air resistance) is given by $y = 128t - 16t^2$.
  (A) Find the time after the arrow is released when y is 0, and interpret physically.
  (B) Find the times when the arrow is 16 feet off the ground. Compute answers to two decimal places.

★ **36.** Repeat Problem 35 for an initial velocity of 128 feet per second.

★ **37.** A barrel 2 feet in diameter and 4 feet in height has a 1-inch-diameter drainpipe in the bottom. It can be shown that the height h of the surface of the water above the bottom of the barrel at time t minutes after the drain has been opened is given by the formula $h = (\sqrt{h_0} - \frac{5}{12}t)^2$, where h_0 is the water level above the drain at time $t = 0$. If the barrel is full and the drain opened, how long will it take to empty one-half of the contents?

★ **38.** How long will it take to empty three-quarters of the contents of the barrel in Problem 37?

RATE-TIME PROBLEMS

★★ **39.** A new printing press can do a job in 1 hour less than an older press. Together they can do the same job in 1.2 hours. How long would it take each press alone to do the job?

★★ **40.** One pipe can fill a tank in 5 hours less than another; together they fill the tank in 5 hours. How long would it take each pipe alone to fill the tank? Compute the answer to two decimal places.

41. Two boats travel at right angles to each other after leaving the same dock at the same time; 1 hour later they are 25 kilometers apart. If one boat travels 5 kilometers per hour faster than the other, what is the rate of each? [*Hint:* Use the Pythagorean theorem.]

42. Repeat Problem 41 if the boats are 13 kilometers apart after 1 hour and one travels 7 kilometers per hour faster than the other.

★ **43.** A speedboat takes 1 hour longer to go 24 kilometers up a river than to return. If the boat cruises at 10 kilometers per hour in still water, what is the rate of the current?

★ **44.** Repeat Problem 43 for the situation where it takes 2.5 hours longer to go upriver than to return

ECONOMICS AND BUSINESS

★ **45.** If a principal of P dollars is invested at an annual rate r (expressed as a decimal), at the end of 2 years it will be worth an amount A, with $A = P(1 + r)^2$. At what rate will $100 grow to $110.25 in 2 years?

46. Refer to Problem 45. At what rate will $200 grow to $216.32 in 2 years?

★ **47.** In a certain city the demand equation for popular records is $d = 3,000/p$, where d would be the quantity of records demanded on a given day if the selling price were p dollars per record. (Notice that as the price goes up, the number of records the people are willing to buy goes down, and vice versa.) On the other hand, the supply equation is $s = 200p - 700$, where s is the quantity of records a supplier is willing to supply at p dollars per record. (Notice that as the price goes up, the number of records a supplier is willing to sell goes up, and vice versa.) At what price will supply equal demand; that is, at what price will $d = s$? In economic theory the price at which supply equals demand is called the **equilibrium point,** the point at which the price ceases to change.

★ **48.** Repeat Problem 47 for $d = 3,000/p$ and $s = 200p - 400$.

★ **49.** A warehouse is to be built in a rectangular shape twice as long as it is wide. The property on which it is to be built is also a rectangle with sides parallel to those of the warehouse. The warehouse will sit 20 feet from the sides of the property and 40 feet from the front and back. If the property has 72,200 square feet, what are the dimensions of the warehouse?

★ **50.** Repeat Problem 49 for a property having 57,800 square feet.

★★ **51.** The total yield from a crop is the number of plants times the yield per plant. A fruit grower can achieve a yield of 40 bushels per tree when 120 trees are planted in an orchard. If the number of trees is increased, the yield per tree drops due to crowding. For each additional tree planted, the yield is expected to decrease by $\frac{1}{2}$ bushel per tree. Let x represent the number of trees planted. Write an expression for the total yield using x as the variable. What number of trees will result in a total yield of 4,200 bushels?

★★ **52.** Repeat Problem 51, but now let x represent the number of trees planted in addition to 120 existing trees. Write an expression for the total yield using x as the variable. How many additional trees will result in a total yield of 3,750 bushels?

★★ **53.** The revenue from sales of a product is equal to its selling price times the number of units sold. A toy manufacturer can sell 5,000 of a particular toy if the selling price is $8 per unit. For each $0.25 that the seller increases the price, the sales are expected to decrease by 100 units. Let x represent the number of $0.25 increases in the price. Write an expression for the resulting revenue using x as the variable. How many $0.25 increases to the price will yield a total revenue of $41,800?

★★ **54.** Repeat Problem 53, but now let x represent the selling price after the price is increased by some number of $0.25 increments. Find the price that yields a total revenue of $42,000.

★★ **55.** A resort hotel with 180 units can expect to have all units occupied at a rate of $420 per week and to lose one rental for each $10 increase in the weekly rate. What rate will yield a total weekly revenue of $87,210?

★★ **56.** Repeat Problem 55 for a total weekly revenue of $91,850.

★★ **57.** A theater expects an average daily attendance of 300 when ticket prices are $7. Average attendance will decrease by 20 for each $1 increase in price. What price will yield an average daily revenue of $2,375?

★★ **58.** Repeat Problem 57 for an average daily revenue of $2,415.

7-4   Radical Equations and Other Equations Reducible to Quadratic Form

♦ Radical Equations
♦ Other Equations Reducible to Quadratic Form

Consider the equation

$$x - 1 = \sqrt{x + 11}$$

Such an equation is called a **radical equation** since it contains a variable in the radicand. This particular equation can be rewritten as a quadratic equation and solved. In this section we solve such equations and other equations that can be rewritten in linear or quadratic form.

♦ **RADICAL EQUATIONS**

To solve the equations

$$\sqrt{x - 1} = 4 \quad \text{and} \quad x - 1 = \sqrt{x + 11}$$

we must eliminate the radicals. We use the following result:

> If $a = b$, then $a^2 = b^2$.

That is, if we square both sides of an equation, the resulting equation is still true. Moreover, any solution of the original equation is also a solution of the equation that results from squaring. Thus, let us square both sides of the equations above. For the first equation, we obtain

$$\sqrt{x - 1} = 4$$
$$x - 1 = 16$$
$$x = 17$$

and you can check that 17 is a solution. However, the second equation leads to a complication:

$$x - 1 = \sqrt{x + 11}$$
$$(x - 1)^2 = (\sqrt{x + 11})^2$$
$$x^2 - 2x + 1 = x + 11$$
$$x^2 - 3x - 10 = 0$$
$$(x - 5)(x + 2) = 0$$
$$x = 5, -2$$

Now check these two "solutions." For $x = 5$,

$$x - 1 = 5 - 1 = 4 \quad \text{and} \quad \sqrt{x + 11} = \sqrt{5 + 11} = 4$$

so 5 is a solution. However, for $x = -2$,

$$x - 1 = -2 - 1 = -3 \quad \text{but} \quad \sqrt{x + 11} = \sqrt{-2 + 11} = \sqrt{9} = 3$$

so -2 is *not* a solution. The process of squaring introduced an *extraneous solution*. In general, one can prove the following important result:

Squaring and Extraneous Solutions

If both members of an equation are raised to a natural number power, then the solution set of the original equation is a subset of the solution set of the new equation. The new equation *may* have solutions, called **extraneous solutions,** that are not solutions of the original equation.

Recall that two equations are equivalent if they have exactly the same solutions. Thus, squaring both sides of an equation preserves the equality but does not necessarily produce an equivalent equation. In Section 4-1, we saw that the same thing was true of multiplying both sides of an equation by an expression that includes a variable. For example, if we multiply $x = 3$ by $x - 1$, we obtain $x(x - 1) = 3(x - 1)$. Solving this equation yields two solutions, $x = 3$ and $x = 1$:

$$x(x - 1) = 3(x - 1)$$
$$x^2 - x = 3x - 3$$
$$x^2 - 4x + 3 = 0$$
$$(x - 1)(x - 3) = 0$$
$$x = 1, 3$$

The solution $x = 1$ is extraneous to the original equation. This is why the multiplication property for equality (given in Section 1-7) included the condition that the multiplier was not 0:

If $a = b$ and $c \neq 0$, than $ac = bc$ is an equivalent equation.

Squaring both sides of an equation is a form of multiplying both sides by the same thing, for if $a = b$, then

$$aa = ab \qquad \text{We have multiplied both sides by } a.$$
$$ab = bb \qquad \text{We have multiplied both sides of } a = b \text{ by } b.$$
$$aa = bb \qquad \text{Use the transitive property of equality:}$$
$$\qquad\qquad \text{if } r = s \text{ and } s = t, \text{ then } r = t.$$
$$a^2 = b^2$$

This is why, if variables are involved, the resulting equation may not be equivalent to the original.

In summary, when a new equation is obtained by squaring both sides of an equation, any solution of the original equation must be among those of the new equation. We need to check all the solutions at the end of the process to eliminate the extraneous ones.

Example 1
Solving a Radical Equation

Solve:

$$x + \sqrt{x - 4} = 4$$

Solution

$$x + \sqrt{x - 4} = 4 \qquad\qquad \text{Isolate the radical on one side.}$$
$$\sqrt{x - 4} = 4 - x \qquad\qquad \text{Square both sides.}$$
$$x - 4 = (4 - x)^2 \qquad\qquad \text{Multiply out.}$$
$$x - 4 = 16 - 8x + x^2 \qquad \text{Write in standard form.}$$
$$x^2 - 9x + 20 = 0$$
$$(x - 5)(x - 4) = 0$$
$$x = 4, 5$$

Check Substitute in the *original* equation.

$$x = 4: \quad 4 + \sqrt{4 - 4} = 4, \text{ so } 4 \text{ is a solution}$$
$$x = 5: \quad 5 + \sqrt{5 - 4} = 5 + 1 \neq 4, \text{ so } 5 \text{ is extraneous}$$

Matched Problem 1 Solve: $x = 5 + \sqrt{x - 3}$

Radical equations involving two radicals are usually easiest to solve if the radicals are on opposite sides of the equation.

Example 2
Solving a Radical Equation with Two Radicals

Solve:

$$\sqrt{2x+3} - \sqrt{x-2} = 2$$

Solution

$$\sqrt{2x+3} - \sqrt{x-2} = 2$$

The equation will be easier to solve with the radicals on opposite sides of the equation.

$$\sqrt{2x+3} = \sqrt{x-2} + 2$$

Square both sides.

$$2x+3 = x - 2 + 4\sqrt{x-2} + 4$$

Isolate the radical on one side.

$$x + 1 = 4\sqrt{x-2}$$

Square both sides again.

$$x^2 + 2x + 1 = 16(x-2)$$
$$x^2 - 14x + 33 = 0$$
$$(x-3)(x-11) = 0$$
$$x = 3, 11$$

Check

$x = 3$:
$$\sqrt{2 \cdot 3 + 3} - \sqrt{3-2} = \sqrt{9} - \sqrt{1}$$
$$= 3 - 1 = 2, \text{ so } 3 \text{ is a solution}$$

$x = 11$:
$$\sqrt{2 \cdot 11 + 3} - \sqrt{11-2} = \sqrt{25} - \sqrt{9}$$
$$= 5 - 3 = 2, \text{ so } 11 \text{ is a solution}$$

Matched Problem 2 Solve: $\sqrt{2x+7} + \sqrt{x+3} = 1$

♦ **OTHER EQUATIONS REDUCIBLE TO QUADRATIC FORM**

In Section 2-6, we were able to factor some higher-degree polynomials such as $x^4 - 6x^2 + 8$ by recognizing that they could be considered as quadratic in some power of x. For example,

$$x^4 - 6x^2 + 8 = (x^2)^2 - 6(x^2) + 8$$
$$= (x^2 - 2)(x^2 - 4)$$

We can apply the same technique to solve many equations that are not immediately recognizable as quadratic but that can be transformed into a quadratic form.

Example 3
Solving an Equation by Reducing to Quadratic Form

Solve:

$$x^4 - x^2 - 12 = 0$$

Solution

If you recognize that the equation is quadratic in x^2, you can solve for x^2 first, and then solve for x. It may be helpful to make the substitution $u = x^2$ and then solve the equation

$$u^2 - u - 12 = 0$$
$$(u-4)(u+3) = 0$$
$$u = 4, -3$$

Replacing u with x^2, we obtain

$$x^2 = 4 \qquad\qquad x^2 = -3$$
$$x = \pm 2 \qquad\qquad x = \pm i\sqrt{3}$$

You can check that each of these four values is a solution of the original equation. Remember, however, that the only operations that might introduce extraneous roots are multiplying by variable expressions or raising to powers. Since this solution did not involve either of these operations, extraneous roots cannot occur, and you will be checking only the accuracy of the solutions.

Matched Problem 3   Solve $x^6 + 6x^3 - 16 = 0$ for real solutions only.

In general, if an equation that is not quadratic can be transformed into the form

$$au^2 + bu + c = 0$$

where u is an expression in some other variable, then the equation is said to be in **quadratic form.** Once recognized as a quadratic form, an equation often can be solved using one of the quadratic methods.

Example 4

Solving an Equation in Quadratic Form

Solve:

$$x^{2/3} - x^{1/3} - 6 = 0$$

Solution   Let $u = x^{1/3}$, then $u^2 = x^{2/3}$. After substitution, the original equation becomes

$$u^2 - u - 6 = 0 \qquad\qquad \text{Solve by factoring.}$$
$$(u - 3)(u + 2) = 0$$
$$u = 3, -2$$

Replacing u with $x^{1/3}$, we obtain

$$x^{1/3} = 3 \qquad\qquad x^{1/3} = -2 \qquad \text{Raise both sides to the third power.}$$
$$(x^{1/3})^3 = 3^3 \qquad\qquad (x^{1/3})^3 = (-2)^3 \qquad (x^{1/3})^3 = x$$
$$x = 27 \qquad\qquad x = -8$$

Check $27^{2/3} - 27^{1/3} - 6 = 3^2 - 3 - 6 = 0$

$(-8)^{2/3} - (-8)^{1/3} - 6 = (-2)^2 - (-2) - 6 = 4 + 2 - 6 = 0$

Matched Problem 4   Solve: $x^{2/3} - x^{1/3} - 12 = 0$

Answers to Matched Problems   **1.** 7   **2.** -3 **3.** $-2, \sqrt[3]{2}$ **4.** 64, -27

EXERCISE 7-4

A *Solve.*

1. $\sqrt{x+2} = 7$ **2.** $\sqrt{3x+1} = 5$

3. $\sqrt{5x+4} = 8$ **4.** $\sqrt{x-9} = 5$

5. $x - 2 = \sqrt{x}$ **6.** $\sqrt{x} = x - 6$

7. $\sqrt{x} = x - 12$ **8.** $x - 20 = \sqrt{x}$

9. $m - 13 = \sqrt{m+7}$ **10.** $\sqrt{5n+9} = n - 1$

11. $\sqrt{x+3} = x - 3$ **12.** $x - 1 = \sqrt{x+1}$

13. $\sqrt{x+10} = x - 10$ **14.** $x - 6 = \sqrt{x+6}$

15. $x^4 - 10x^2 + 9 = 0$ **16.** $x^4 - 13x^2 + 36 = 0$

17. $x^4 - 7x^2 - 18 = 0$ **18.** $y^4 - 2y^2 - 8 = 0$

19. $\sqrt{x^2-3x} = 2$ **20.** $\sqrt{x^2+8x} = 3$

21. $\sqrt{x^2+6x} = 4$ **22.** $\sqrt{x^2+5x} = 6$

23. $x + 3 = \sqrt{x+5}$ **24.** $\sqrt{x+3} = x - 1$

25. $x - 16 = \sqrt{x+4}$ **26.** $\sqrt{x+3} = x - 9$

27. $\sqrt{3x-2} = x$ **28.** $x = \sqrt{4x-3}$

29. $x + 1 = \sqrt{4x+1}$ **30.** $\sqrt{3x+1} = x + 1$

B 31. $m - 7\sqrt{m} + 12 = 0$ **32.** $t - 11\sqrt{t} + 18 = 0$

33. $1 + \sqrt{x+5} = x$ **34.** $x - \sqrt{x+10} = 2$

35. $\sqrt{3x+1} = \sqrt{x} - 1$ **36.** $\sqrt{3x+4} = 2 + \sqrt{x}$

37. $\sqrt{3t+4} + \sqrt{t} = -3$ **38.** $\sqrt{3w-2} - \sqrt{w} = 2$

39. $\sqrt{u-2} = 2 + \sqrt{2u+3}$

40. $\sqrt{3y-2} = 3 - \sqrt{3y+1}$

41. $\sqrt{2x-1} - \sqrt{x-4} = 2$

42. $\sqrt{y-2} - \sqrt{5y+1} = -3$

43. $\sqrt{x^2-x} = \sqrt{2x-2}$ **44.** $\sqrt{x^2-2x} = \sqrt{x-2}$

45. $\sqrt{x^2-3x} = \sqrt{x-3}$ **46.** $\sqrt{x^2-2x} = \sqrt{2x-3}$

47. $\sqrt{3x+1} = \sqrt{x+4} + 1$

48. $\sqrt{2x+1} = \sqrt{x-2} + 1$

49. $\sqrt{5x-4} = \sqrt{x} + 2$

50. $\sqrt{7x+2} = \sqrt{x-2} + 4$

51. $\sqrt{3x-2} = \sqrt{x} + 2$

52. $\sqrt{4x+1} = \sqrt{x-2} + 4$

Find only real solutions.

53. $x^6 - 7x^3 - 8 = 0$ **54.** $x^6 + 3x^3 - 10 = 0$

55. $x^6 - 6x^3 + 8 = 0$ **56.** $x^6 - 26x^3 - 27 = 0$

Find all solutions.

57. $y^8 - 17y^4 + 16 = 0$ **58.** $3m^4 - 4m^2 - 7 = 0$

59. $x^4 - 13x^2 + 36 = 0$ **60.** $x^4 - 10x^2 + 9 = 0$

61. $x^4 - 5x^2 - 36 = 0$ **62.** $x^4 - 8x^2 - 9 = 0$

63. $2x^4 - 3x^2 - 5 = 0$ **64.** $3x^4 + 4x^2 - 7 = 0$

65. $x^{2/3} - 3x^{1/3} - 10 = 0$

66. $2x^{2/3} + 3x^{1/3} - 2 = 0$

67. $y^{1/2} - 3y^{1/4} + 2 = 0$ **68.** $y^{1/2} - 5y^{1/4} + 6 = 0$

69. $\dfrac{2}{x^2} + \dfrac{3}{x} + 1 = 0$ **70.** $\dfrac{3}{x^2} - \dfrac{4}{x} + 1 = 0$

71. $x^{-1} - 4x^{-2} + 3x^{-3} = 0$

72. $x^{-1} + 3x^{-2} + 2x^{-3} = 0$

73. $6x^{-2} - 5x^{-1} - 6 = 0$ **74.** $3n^{-2} - 11n^{-1} - 20 = 0$

75. $x^{-2} - 3x^{-3} - 10x^{-4} = 0$

76. $x^{-2} + 7x^{-3} + 12x^{-4} = 0$

C 77. $1 - 2x^{-2} + x^{-4} = 0$ **78.** $1 - 8x^{-2} + 16x^{-4} = 0$

79. $4x^{-4} - 17x^{-2} + 4 = 0$

80. $9y^{-4} - 10y^{-2} + 1 = 0$

81. $(m^2 - m)^2 - 4(m^2 - m) = 12$

82. $(x^2 + 2x)^2 - (x^2 + 2x) = 6$

83. $(x - 3)^4 + 3(x - 3)^2 = 4$

84. $(m - 5)^4 + 36 = 13(m - 5)^2$

85. $\sqrt{3x+6} - \sqrt{x+4} = \sqrt{2}$

86. $\sqrt{7x-2} - \sqrt{x+1} = \sqrt{3}$

87. $\dfrac{1}{\sqrt{x-2}} + \dfrac{2}{3} = 1$ **88.** $\dfrac{1}{\sqrt{x+5}} = \dfrac{\sqrt{x}}{6}$

89. $\dfrac{x}{3} + \dfrac{2}{x} = \dfrac{6x+1}{3x}$ **90.** $x + \dfrac{1}{x} = \dfrac{x+3}{2}$

91. $\dfrac{1}{x-1} + \dfrac{1}{x-2} = \dfrac{5}{6}$ **92.** $\sqrt{x} = 3 - \dfrac{2}{\sqrt{x}}$

7-5 Graphing Quadratic Polynomials

♦ Graphing $y = ax^2 + bx + c$
♦ Intercepts of Graphs and Solutions to Equations

We have seen that the graph of the linear equation $y = mx + b$, with y equal to a linear expression in x, is a straight line. In this section, we will consider the equation $y = ax^2 + bx + c$, with y equal to a quadratic expression in x. The resulting graph is called a *parabola*.

♦ GRAPHING $y = ax^2 + bx + c$

The graph of a linear equation $y = ax + b$ can be obtained by plotting a sufficient number of points on the graph and connecting them in a smooth curve. The task is made much easier by knowing beforehand that the graph is a straight line. Thus, we only need to plot two points. We can approach graphing the quadratic equation in much the same way: we start by simply plotting points, but quickly learn to take advantage of properties of the curves that will result.

We begin by graphing the simplest quadratic equation, $y = x^2$. To do so, we plot several points and join them in a smooth curve, as shown in Figure 1.

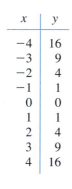

x	y
-4	16
-3	9
-2	4
-1	1
0	0
1	1
2	4
3	9
4	16

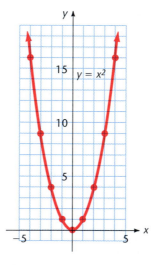

Figure 1

We do the same for $y = -x^2$ in Figure 2.

The two graphs just obtained are examples of curves called **parabolas.** Every quadratic equation of the form $y = ax^2 + bx + c$ will have a graph that is similar to one of these two examples. The graph can be obtained more efficiently if we

x	y
-4	-16
-3	-9
-2	-4
-1	-1
0	0
1	-1
2	-4
3	-9
4	-16

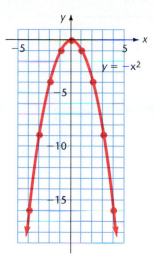

$y = -x^2$

Figure 2

know which of the two shapes is involved and where the highest or lowest point on the graph is located. The highest or lowest point on a parabola is called its **vertex.**

Parabolas

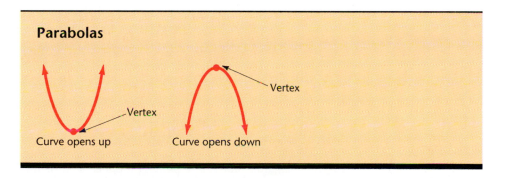

Vertex

Vertex

Curve opens up   Curve opens down

The following result relates the coefficients in $ax^2 + bx + c$ to the graph of the curve $y = ax^2 + bx + c$:

Graph of $y = ax^2 + bx + c$

1. The graph is a parabola.
2. The graph opens up if $a > 0$ or opens down if $a < 0$.
3. The vertex is located at $x = -\dfrac{b}{2a}$, $y = -\dfrac{b^2 - 4ac}{4a}$.

The y coordinate of the vertex can be obtained by substituting $x = -b/2a$ into $y = ax^2 + bx + c$. Notice that the numerator of the y coordinate in this form is the discriminant of the equation $ax^2 + bx + c = 0$.

Example 1

Graphing $y = ax^2 + bx + c$

Graph:

(A) $y = x^2 + 2$ **(B)** $y = -x^2 + 4x - 4$ **(C)** $y = x^2 - 4x + 2$

Solution   **(A)** The parabola opens up, since $a = 1$ is positive. The vertex is located at

$$x = -\frac{b}{2a} = -\frac{0}{2 \cdot 1} = 0 \qquad \text{and} \qquad y = (\mathbf{0})^2 + 2 = 2$$

We plot a few values on either side of the vertex. That is, we choose x values to the left and right of $x = 0$:

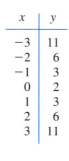

x	y
-3	11
-2	6
-1	3
0	2
1	3
2	6
3	11

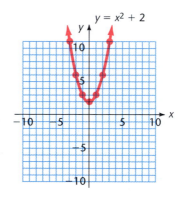

(B) The parabola opens down, since $a < 0$. The vertex is located at

$$x = -\frac{4}{2(-1)} = 2 \qquad y = -(\mathbf{2})^2 + 4(\mathbf{2}) - 4 = 0$$

Again, plot a few points on either side of the vertex:

x	y
-1	-9
0	-4
1	-1
2	0
3	-1
4	-4
5	-9

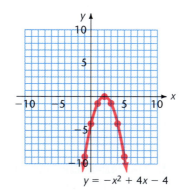

(C) The parabola opens up, since $a > 0$. The vertex is located at

$$x = -\frac{-4}{2 \cdot 1} = 2 \qquad y = (\mathbf{2})^2 - 4(\mathbf{2}) + 2 = -2$$

x	y
-1	7
0	2
1	-1
2	-2
3	-1
4	2
5	7

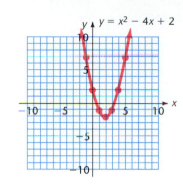

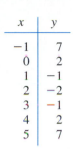

Matched Problem 1   Graph:

(A) $y = x^2 - 3$ **(B)** $y = -x^2 + 6x + 9$ **(C)** $y = -x^2 + 2x + 2$

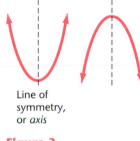

Line of
symmetry,
or *axis*

Figure 3

There are several things to notice about the graphs we have obtained thus far. First, as indicated in Figure 4, each appears symmetric about a vertical line through its vertex.

If you look closely at the tables of values plotted, you can see this symmetry. In Example 1(B), for instance, the y value is the same for both $x = 1$ and $x = 3$, the values 1 unit on either side of $x = 2$, the vertex. Similarly, the y values for $x = 0$ and $x = 4$ are the same, and so forth. This can simplify finding values to plot.

Second, the graphs appear to be exactly the same as the graphs of $y = x^2$ or $y = -x^2$, but moved away from (0, 0). For instance, we can plot the parabola from Example 1(A) and $y = x^2$ on the same axes, as shown in Figure 4, and see that it appears that one graph is the other, simply raised 2 units. All graphs of $y = ax^2 + bx + c$ can, in fact, be obtained from the graph of $y = x^2$. We explore this further in Section 7-6.

Third, there is a relationship between the points where the graph of $y = ax^2 + bx + c$ crosses the x axis and the solutions to the equation $ax^2 + bx + c = 0$. For instance, in Example 1(C), the curve crosses the x axis once between $x = 0$ and $x = 1$, and another time between $x = 3$ and $x = 4$. If we solve the equation

$$x^2 - 4x + 2 = 0$$

we obtain

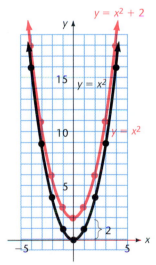

Figure 4

$$x = \frac{4 \pm \sqrt{(-4)^2 - 4(1)(2)}}{2(1)} = \frac{4 \pm \sqrt{8}}{2} = \frac{4 \pm 2\sqrt{2}}{2}$$
$$= 2 \pm \sqrt{2} \approx 3.41, \ 0.59$$

We now explore this relationship.

♦ **INTERCEPTS OF GRAPHS AND SOLUTIONS OF EQUATIONS**

The **y intercepts** of a graph are the points where the curve crosses the y axis. These points correspond to $x = 0$. The **x intercepts** of a graph are the points where the curve crosses the x axis. These points correspond to $y = 0$ (Figure 5, on page 324).

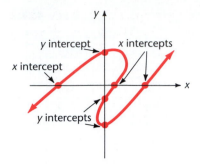

Figure 5

If the graph is described by an equation of the form

$$y = (\text{Expression in } x) \tag{1}$$

then the x intercepts correspond to the points where $y = 0$, that is, where

$$(\text{Expression in } x) = 0 \tag{2}$$

The x values that are the solutions to Equation (2) are exactly the x values of the x intercepts for Equation (1). Since we know all about solutions to

$$ax^2 + bx + c = 0 \tag{3}$$

we also know about the x intercepts of

$$y = ax^2 + bx + c \tag{4}$$

Information about the solutions to Equation (3) and the x intercepts of Equation (4) is closely related to the discriminant.

Suppose the discriminant $b^2 - 4ac = 0$. Then there is one real solution to Equation (3). On the graph, the vertex has y coordinate $-(b^2 - 4ac)/4a = 0$. That is, the vertex lies on the x axis, so there is one intercept, as illustrated:

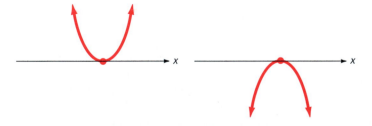

Suppose $b^2 - 4ac > 0$. Then there are two real solutions to $ax^2 + bx + c = 0$. If a is positive, then the parabola opens up. But $-(b^2 - 4ac)/4a$ is negative, so the vertex of the graph is below the x axis and there are two x intercepts:

If a is negative, the curve opens down. In this case, $-(b^2 - 4ac)/4a$ is positive and the vertex lies above the x axis. Again, there are two intercepts:

Suppose $b^2 - 4ac < 0$. Then the equation $ax^2 + bx + c = 0$ has no real solution. If a is positive, then the curve opens up. Since $-(b^2 - 4ac)/4a$ is positive, the vertex of the graph of $y = ax^2 + bx + c$ lies above the x axis and there are no intercepts:

If a is negative, the curve opens down. In this case, $-(b^2 - 4ac)/4a$ is negative, so the vertex is below the x axis and again there are no intercepts:

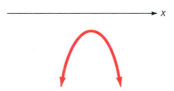

Example 2
Relating Intercepts and Solutions

Graph $y = x^2 - x - 6$. Find the x intercepts exactly by solving the equation $x^2 - x - 6 = 0$.

Solution The graph opens up. The vertex is at

$$x = -\frac{b}{2a} = \frac{1}{2} \qquad y = -\frac{b^2 - 4ac}{4a} = -\frac{1 + 24}{4} = -\frac{25}{4}$$

x	y
-1	-4
0	-6
$\frac{1}{2}$	$-\frac{25}{4}$
2	-4

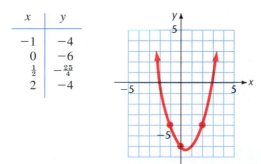

The x intercepts appear to be near $x = -2$ and $x = 3$. To find the exact intercepts we solve $x^2 - x - 6 = 0$:

$$x^2 - x - 6 = 0$$
$$(x + 2)(x - 3) = 0$$
$$x = -2 \quad \text{or} \quad x = 3$$

Thus, the x intercepts and the solutions are the same.

Matched Problem 2   Graph $y = -x^2 + 4x + 5$. Find the x intercepts exactly by solving the equation $-x^2 + 4x + 5 = 0$.

Answers to Matched Problems

1. (A)

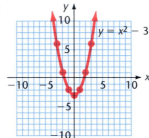

(B)

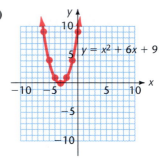

(C)

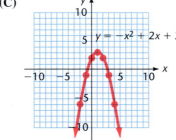

2. x intercepts $5, -1$

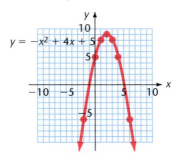

EXERCISE 7-5

In Problems 1–54, graph each equation.

A

1. $y = x^2 + 4$
2. $y = -x^2 + 2$
3. $y = -x^2 - 1$
4. $y = x^2 - 3$
5. $y = x^2 + 2x + 1$
6. $y = x^2 - 4x + 3$
7. $y = x^2 - 5x + 4$
8. $y = x^2 + 4x + 4$
9. $y = x^2 - 5x + 6$
10. $y = x^2 - x + 6$
11. $y = x^2 + x - 6$
12. $y = x^2 + 5x + 6$
13. $y = x^2 + 4x + 3$
14. $y = x^2 + 5x + 4$
15. $y = x^2 + 6x + 5$
16. $y = x^2 - 6x + 5$
17. $y = x^2 - 4x - 5$
18. $y = x^2 + 4x - 5$
19. $y = -x^2 + 4x - 3$
20. $y = -x^2 + 5x - 4$
21. $y = -x^2 + x - 6$
22. $y = -x^2 - x + 6$
23. $y = -x^2 + 4x + 5$
24. $y = -x^2 + 6x - 5$

B

25. $y = 3x^2$
26. $y = \frac{1}{3}x^2$
27. $y = -\frac{1}{2}x^2$
28. $y = -2x^2$
29. $y = 2x^2 + 1$
30. $y = -3x^2 + 4$
31. $y = -\frac{1}{3}x^2 - 2$
32. $y = \frac{1}{2}x^2 - 1$
33. $y = 2x^2 + 4x - 1$
34. $y = 2x^2 - 4x - 1$
35. $y = 2x^2 - 4x + 1$
36. $y = 2x^2 + 4x + 5$
37. $y = 2x^2 - 8x + 6$
38. $y = 2x^2 + 8x + 6$
39. $y = 2x^2 - 8x + 10$
40. $y = 2x^2 + 8x + 10$

41. $y = 2x^2 - 12x + 14$
42. $y = 2x^2 + 12x + 14$
43. $y = 2x^2 + 12x + 22$
44. $y = 2x^2 - 12x + 22$
45. $y = 3x^2 - 6x + 6$
46. $y = 3x^2 + 6x$
47. $y = 3x^2 - 6x$
48. $y = 3x^2 + 6x + 6$
49. $y = 4x^2 - 8x + 19$
50. $y = 4x^2 + 8x + 19$
51. $y = 4x^2 - 8x + 13$
52. $y = 4x^2 + 8x + 13$
53. $y = 5x^2 - 40x + 81$
54. $y = 5x^2 + 40x + 79$

C *Graph and find any x intercepts.*

55. $y = x^2 - 2x - 3$
56. $y = x^2 + 6x + 8$
57. $y = -x^2 + x + 2$
58. $y = -x^2 - 2x + 3$
59. $y = x^2 + x - 6$
60. $y = x^2 - 6x + 5$
61. $y = x^2 + 4x + 4$
62. $y = -x^2 + 6x - 9$
63. $y = x^2 - 3x + 3$
64. $y = x^2 + 4x + 5$
65. $y = -x^2 + 2x - 8$
66. $y = -x^2 - x + 12$
67. $y = -x^2 + 5x - 6$
68. $y = -x^2 + x - 1$
69. $y = 2x^2 - 7x + 5$
70. $y = 3x^2 + 11x - 4$
71. $y = -2x^2 - 3x + 2$
72. $y = -2x^2 - 5x + 12$

Determine the number of x intercepts.

73. $y = 8x^2 + 41x + 52$
74. $y = 9x^2 + 42x + 53$
75. $y = 10x^2 - 60x + 91$
76. $y = 11x^2 - 60x + 81$
77. $y = 23x^2 - 24x - 25$
78. $y = -23x^2 - 24x - 25$
79. $y = -53x^2 + 52x + 51$
80. $y = 53x^2 - 52x + 51$

7-6 Completing the Square and Graphing (Optional)

♦ Graphing $y = ax^2$, $y = x^2 + k$, and $y = (x - h)^2$
♦ Graphing $y = a(x - h)^2 + k$
♦ Graphing $y = ax^2 + bx + c$ by Geometric Transformations

In Section 7-5, it was claimed that the graph of $y = ax^2 + bx + c$ could be obtained from the graph of $y = x^2$ by operations such as shifting the graph up or down, left or right, or by turning it upside down. In this section, we describe the geometric relationship between the graphs of these two equations and show how the graph of $y = ax^2 + bx + c$ is obtained from this relationship.

◆ GRAPHING $y = ax^2$, $y = x^2 + k$, AND $y = (x - h)^2$

The equation $y = 2x^2$ doubles the y value of each point in the graph of $y = x^2$. The effect is to "stretch" the graph, as shown in Figure 1.

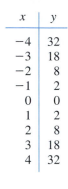

x	y
-4	32
-3	18
-2	8
-1	2
0	0
1	2
2	8
3	18
4	32

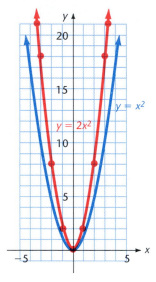

Figure 1

Multiplying by a number greater than 2 stretches it even more; multiplying by a number between 0 and 1 "flattens" the graph. Multiplying by a negative number a turns the graph over and either stretches or flattens it, depending on the size of $|a|$; see Figure 2.

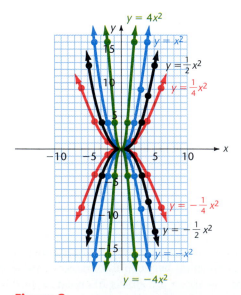

Figure 2

In Section 7-5, Example 1(A), we saw that the equation $y = x^2 + 2$ has the same graph as $y = x^2$, but raised 2 units. More generally, $y = x^2 + k$ has the same graph as $y = x^2$, but raised k units when $k > 0$ and lowered $|k|$ units when $k < 0$.

Example 1
Graphing $y = ax^2$ and $y = x^2 + k$

Graph:

(A) $y = -2x^2$ **(B)** $y = x^2 - 3$

Solution **(A)** Begin with the graph of $y = x^2$. The graph of $y = 2x^2$ is this graph stretched, as we saw in Figure 1 (repeated here as Figure 3a). The graph of $y = -2x^2$ is the graph in Figure 3a turned over the x axis, as shown in Figure 3b.

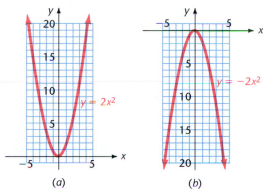

(a)   (b)

Figure 3

(B) Shift the graph of $y = x^2$ down 3 units:

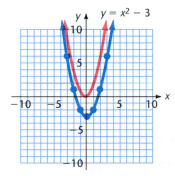

Matched Problem 1   Graph: **(A)** $y = 3x^2$ **(B)** $y = x^2 - 2$

In the form $y = (x - 2)^2$, the same y values occur as in $y = x^2$, but they occur for x values 2 units larger. For example, in $y = (x - 2)^2$, $y = 0$ when $x = 2$, whereas in $y = x^2$, $y = 0$ when $x = 0$. The effect is to shift the graph of $y = x^2$ 2 units to the right, as shown in Figure 4.

x	y
-2	16
-1	9
0	4
1	1
2	0
3	1
4	4
5	9
6	16

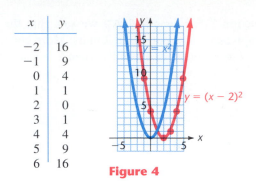

Figure 4

In general, the graph of $y = (x - h)^2$ is the same as the graph of $y = x^2$, but shifted h units to the right for $h > 0$ and $|h|$ units to the left for $h < 0$.

Example 2
Graphing $y = (x - h)^2$

Graph:

(A) $y = (x - 4)^2$ **(B)** $y = (x + 3)^2$

Solution

(A) Shift the graph of $y = x^2$ to the right 4 units as shown in Figure 5a
(B) Shift the graph of $y = x^2$ to the left 3 units as shown in Figure 5b

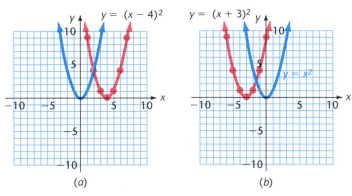

Figure 5

Matched Problem 2   Graph: **(A)** $y = (x + 5)^2$ **(B)** $y = (x - 1)^2$

Thus, the graph of $y = x^2$ leads quickly to graphs of several related forms. In summary, we have the following:

Graphs of Related Forms from the Graph of $y = x^2$

$y = ax^2$, $a > 0$
The graph of $y = x^2$
stretched ($a > 1$) or
flattened ($a < 1$)

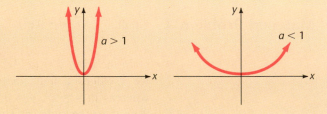

$y = x^2 + k$

The graph of $y = x^2$ raised k units if $k > 0$ or lowered $|k|$ units if $k < 0$

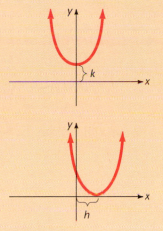

$y = (x - h)^2$

The graph of $y = x^2$ shifted h units to the right if $h > 0$ or $|h|$ units to the left if $h < 0$

♦ **GRAPHING** $y = a(x - h)^2 + k$

The related forms $y = ax^2$, $y = x^2 + k$, and $y = (x - h)^2$ may occur in combination, such as $y = a(x - h)^2 + k$. We can see the effect of these several changes in sequence:

1. Start with the graph of $y = x^2$.
2. Stretch or flatten the graph using $|a|$. If a is negative, also turn the graph upside down.
3. Shift the graph right or left $|h|$ units. If h is negative, say -2, as in $(x + 2)^2 = [x - (-2)]^2$, this means move left $|h| = 2$ units.
4. Raise or lower the graph $|k|$ units. If k is negative, say -2, as in $x^2 - 2 = x^2 + (-2)$, this means lower it $|k| = 2$ units.

Example 3
Graphing
$y = a(x - h)^2 + k$

Graph:

$$y = \tfrac{1}{2}(x - 2)^2 + 3$$

Solution   The graph is that of $y = x^2$, except that it is flattened (using $a = \tfrac{1}{2}$), shifted right 2 units ($h = 2$), and raised 3 units ($k = 3$). Plotting a few points will now locate the graph. We choose points near 2 since $y = x^2$ is shifted right 2 units:

x	y
0	5
2	3
4	5

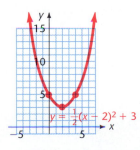

Matched Problem 3   Graph: $y = 2(x + 3)^2 - 1$

♦ **GRAPHING $y = ax^2 + bx + c$ BY GEOMETRIC TRANSFORMATIONS**

To graph

$$y = ax^2 + bx + c$$

we will rewrite the equation in the form

$$y = a(x - h)^2 + k$$

To do this we will use the technique of completing the square introduced in Section 7-1.

Example 4
Rewriting
$y = ax^2 + bx + c$ **as**
$y = a(x - h)^2 + k$

Rewrite $y = 3x^2 - 2x - 1$ in the form $y = a(x - h)^2 + k$.

Solution

$$y = 3x^2 - 2x - 1 \qquad \text{Add 1 to each side.}$$

$$y + 1 = 3x^2 - 2x \qquad \text{Divide all terms by 3 to make the coefficient of } x^2 \text{ equal to 1.}$$

$$\frac{y}{3} + \frac{1}{3} = x^2 - \frac{2}{3}x \qquad \text{Add } \tfrac{1}{9} \text{ to each side. This completes the square on the right side.}$$

$$\frac{y}{3} + \frac{1}{3} + \frac{1}{9} = x^2 - \frac{2}{3}x + \frac{1}{9}$$

$$\frac{y}{3} + \frac{4}{9} = \left(x - \frac{1}{3}\right)^2 \qquad \text{Subtract } \tfrac{4}{9} \text{ from each side.}$$

$$\frac{y}{3} = \left(x - \frac{1}{3}\right)^2 - \frac{4}{9} \qquad \text{Multiply by 3.}$$

$$y = 3\left(x - \frac{1}{3}\right)^2 - \frac{4}{3} \qquad \text{This is the desired form; } h = \tfrac{1}{3}, \; k = -\tfrac{4}{3}.$$

The steps where the left side is altered can be omitted if we are careful to maintain the equality when completing the square. When a quantity is added to the right side, it also must be subtracted from the right side to keep the two sides equal. The steps would then look like this:

$$y = 3(x^2 - \tfrac{2}{3}x) - 1$$
$$y = 3(x^2 - \tfrac{2}{3}x + \tfrac{1}{9}) - 1 - 3(\tfrac{1}{9}) \qquad \text{Add } \tfrac{1}{9} \text{ within the parentheses to complete the square. This adds } 3(\tfrac{1}{9}) \text{ to the right side, so } 3(\tfrac{1}{9}) \text{ also must be subtracted from the right side.}$$

$$y = 3(x - \tfrac{1}{3})^2 - \tfrac{4}{3}$$

While this latter approach is shorter, it is also more prone to errors.

Matched Problem 4   Rewrite $y = -2x^2 + 3x + 1$ in the form $y = a(x - h)^2 + k$.

Example 5
Graphing
$y = ax^2 + bx + c$

Graph:

$$y = 3x^2 - 2x - 1$$

Solution   Rewrite $y = 3x^2 - 2x - 1$ as $y = 3(x - \frac{1}{3})^2 - \frac{4}{3}$, the form found in Example 4. From this form we see that the graph is the graph of $y = x^2$, except that it is stretched (since $a = 3$), shifted right by $\frac{1}{3}$ unit (since $h = \frac{1}{3}$), and lowered by $|-\frac{4}{3}| = \frac{4}{3}$ units (since $k = -\frac{4}{3}$). The graph therefore opens up, and the lowest point is $(\frac{1}{3}, -\frac{4}{3})$. A few points plotted will locate the graph:

x	y
0	-1
$\frac{1}{3}$	$-\frac{4}{3}$
2	7

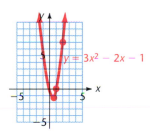

$y = 3x^2 - 2x - 1$

Matched Problem 5   Graph: $y = -2x^2 + 3x + 1$

The process of rewriting $y = ax^2 + bx + c$ in the form $y = a(x - h)^2 + k$ can be applied to the general equation:

$$y = ax^2 + bx + c \qquad \text{Subtract } c \text{ from both sides.}$$

$$y - c = ax^2 + bx \qquad \text{Divide by } a.$$

$$\frac{y}{a} - \frac{c}{a} = x^2 + \frac{b}{a}x \qquad \text{Add } (b/2a)^2 \text{ to complete the square on the right side.}$$

$$\frac{y}{a} - \frac{c}{a} + \left(\frac{b^2}{4a^2}\right) = x^2 + \frac{b}{a}x + \frac{b^2}{4a^2}$$

$$\frac{y}{a} + \frac{b^2 - 4ac}{4a^2} = \left(x + \frac{b}{2a}\right)^2 \qquad \text{Subtract } \frac{b^2 - 4ac}{4a^2} \text{ from each side.}$$

$$\frac{y}{a} = \left(x + \frac{b}{2a}\right)^2 - \frac{b^2 - 4ac}{4a^2} \qquad \text{Multiply by } a.$$

$$y = a\left(x + \frac{b}{2a}\right)^2 - \frac{b^2 - 4ac}{4a} \qquad \text{Write the square term in the form } (x - h)^2 \text{ and the constant term in the form } +k.$$

$$y = a\left(x - \frac{-b}{2a}\right)^2 + \left(-\frac{b^2 - 4ac}{4a}\right) \qquad h = -\frac{b}{2a},\ k = -\frac{b^2 - 4ac}{4a}$$

Thus, the graph of $y = ax^2 + bx + c$ is the graph of $y = x^2$ changed in the ways:

1. Stretched or flattened using a; turned over the x axis if $a < 0$
2. Shifted right or left using h, and up or down using k. The vertex $(0, 0)$ of $y = x^2$
   is shifted to the point $(h, k) = \left(-\dfrac{b}{2a}, \, -\dfrac{b^2 - 4ac}{4a}\right)$

Answers to Matched Problems

1. **(A)**

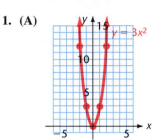

(B)

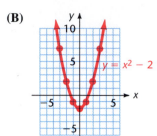

2. **(A)**

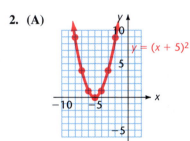

(B)

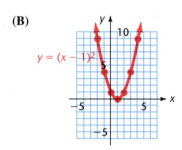

3.

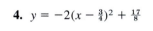

4. $y = -2(x - \frac{3}{4})^2 + \frac{17}{8}$

5.

EXERCISE 7-6

A *Graph.*

1. $y = 4x^2$	2. $y = 5x^2$
3. $y = \frac{1}{2}x^2$	4. $y = \frac{1}{3}x^2$

5. $y = -5x^2$	6. $y = -4x^2$
7. $y = -\frac{1}{3}x^2$	8. $y = -\frac{1}{2}x^2$
9. $y = x^2 - 4$	10. $y = x^2 + 4$
11. $y = x^2 + 6$	12. $y = x^2 - 6$
13. $y = x^2 + 5$	14. $y = x^2 - 5$

15. $y = x^2 - 7$

16. $y = x^2 + 7$

17. $y = (x - 3)^2$

18. $y = (x - 1)^2$

19. $y = (x + 2)^2$

20. $y = (x + 3)^2$

21. $y = (x - 4)^2$

22. $y = (x + 4)^2$

23. $y = (x + 5)^2$

24. $y = (x - 5)^2$

B 25. $y = (x - 2)^2 + 1$

26. $y = (x + 1)^2 - 3$

27. $y = (x + 2)^2 - 4$

28. $y = (x - 3)^2 + 2$

29. $y = (x + 1)^2 - 4$

30. $y = (x - 2)^2 + 3$

31. $y = (x + 5)^2 + 2$

32. $y = (x - 5)^2 - 1$

33. $y = -(x + 3)^2 - 5$

34. $y = -(x - 3)^2 - 2$

35. $y = -(x - 2)^2 - 3$

36. $y = -(x + 2)^2 - 1$

37. $y = -(x + 4)^2 + 2$

38. $y = -(x + 3)^2 - 2$

39. $y = -(x - 1)^2 - 1$

40. $y = -(x - 4)^2 + 1$

41. $y = 2(x - 1)^2 + 1$

42. $y = 3(x + 1)^2 - 2$

43. $y = \frac{1}{2}(x + 1)^2 + 3$

44. $y = -\frac{1}{3}(x + 2)^2 - 2$

45. $y = -2(x + 2)^2 - 2$

46. $y = 3(x - 1)^2 + 2$

47. $y = 2(x - 5)^2 + 3$

48. $y = 3(x + 2)^2 - 1$

49. $y = \frac{1}{2}(x - 2)^2 + 6$

50. $y = \frac{1}{2}(x + 2)^2 - 1$

51. $y = -5(x - 1)^2 - 1$

52. $y = -3(x + 3)^2 + 3$

53. $y = -\frac{1}{4}(x + 6)^2 + 2$

54. $y = -\frac{1}{4}(x + 4)^2 - 4$

C *Rewrite in the form $y = a(x - h)^2 + k$. Describe how the graph is related to the graph of $y = x^2$. Compare this description to the graphs found in Problems 55–72, Section 7-5.*

55. $y = x^2 - 2x - 3$

56. $y = x^2 + 6x + 8$

57. $y = -x^2 + x + 2$

58. $y = -x^2 - 2x + 3$

59. $y = x^2 + x - 6$

60. $y = x^2 - 6x + 5$

61. $y = x^2 + 4x + 4$

62. $y = -x^2 + 6x - 9$

63. $y = x^2 - 3x + 3$

64. $y = x^2 + 4x + 5$

65. $y = -x^2 + 2x - 8$

66. $y = -x^2 - x + 12$

67. $y = -x^2 + 5x - 6$

68. $y = -x^2 + x - 1$

69. $y = 2x^2 - 7x + 5$

70. $y = 3x^2 + 11x - 4$

71. $y = -2x^2 - 3x + 2$

72. $y = -2x^2 - 5x + 12$

Use information about the shape of the graph and the y coordinate of the vertex to determine the number of x intercepts for the graph of each equation.

73. $y = 8x^2 + 41x + 52$

74. $y = 9x^2 + 42x + 53$

75. $y = 10x^2 - 60x + 91$

76. $y = 11x^2 - 60x + 81$

77. $y = 23x^2 - 24x - 25$

78. $y = -23x^2 - 24x - 25$

79. $y = -53x^2 + 52x + 51$

80. $y = 53x^2 - 52x + 51$

  ## 7-7   Conic Sections: Circles and Parabolas

- ♦ Formula for the Distance Between Two Points
- ♦ Circles
- ♦ Parabolas

In Chapter 5 we discussed linear equations, that is, equations such as

$$2x - 3y = 5 \quad \text{and} \quad y = -\tfrac{1}{3}x + 2$$

and their straight-line graphs. These are first-degree equations in two variables. If we increase the degree of the equations by 1, what kind of graphs will we get? That is, what kind of graphs will second-degree equations such as

$$y = x^2 \qquad x^2 + y^2 = 25 \qquad \frac{x^2}{4} - \frac{y^2}{16} = 1 \qquad x^2 + 4y^2 - 3x + 7y = 4$$

or the general second-degree equation in two variables,

$$ax^2 + bxy + cy^2 + dx + ey + f = 0 \qquad (1)$$

produce? The first of these, $y = x^2$, has as its graph a parabola, as we saw in Sections 7-5 and 7-6. It can be shown that for any equation in the form (1), the graph will be one of the plane curves we would get by intersecting a plane and a general cone. For this reason, such graphs are called **conic sections.** The principal conic sections are *circles, parabolas, ellipses,* and *hyperbolas,* as shown in Figure 1.

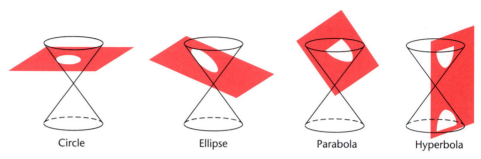

Circle   Ellipse   Parabola   Hyperbola

Figure 1

In this section and the next, we will consider some of the simpler forms of the general second-degree equation (1) and the resulting conic section graphs. Conic sections are treated more completely in a course in analytic geometry.

♦ **FORMULA FOR THE DISTANCE BETWEEN TWO POINTS**

A basic tool in studying conic sections is the **distance-between-two-points formula.** The derivation of the formula makes direct use of the Pythagorean theorem:

In a right triangle, the square of the hypotenuse is the sum of the squares of the two sides.

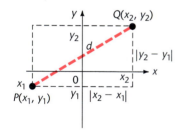

Figure 2

Let P and Q have coordinates as indicated in Figure 2. Using the Pythagorean theorem, we can write

$$d^2 = |x_2 - x_1|^2 + |y_2 - y_1|^2$$

or, since

$$|x_2 - x_1|^2 + |y_2 - y_1|^2 = (x_2 - x_1)^2 + (y_2 - y_1)^2$$

and $d > 0$, the equation can be written in the following form:

Distance Formula

$$d = \sqrt{(x_2 - x_1)^2 + (y_2 - y_1)^2}$$

Note that it doesn't matter which point you call P or Q, since $(a - b)^2 = (b - a)^2$.

<table>
<tr><td>

Example 1

Applying the Distance Formula
</td><td>

Find the distance between $(-3, 6)$ and $(4, -2)$.
</td></tr>
</table>

Solution   Let $P = (4, -2)$ and $Q = (-3, 6)$; then

$$d = \sqrt{(x_2 - x_1)^2 + (y_2 - y_1)^2}$$
$$= \sqrt{[(-3) - 4]^2 + [6 - (-2)]^2}$$
$$= \sqrt{(-7)^2 + (8)^2}$$
$$= \sqrt{113}$$

Or let $P = (-3, 6)$ and $Q = (4, -2)$; then

$$d = \sqrt{[4 - (-3)]^2 + [(-2) - 6]^2}$$
$$= \sqrt{(7)^2 + (-8)^2}$$
$$= \sqrt{113}$$

Matched Problem 1   Find the distance between $(4, -2)$ and $(3, 1)$.

♦ **CIRCLES**

We will start with the familiar definition of a circle and then use the distance formula to find the standard equation of a circle in a plane.

> **Definition of a Circle**
>
> A **circle** is the set of points equidistant from a fixed point. The fixed distance is called the **radius,** and the fixed point is called the **center.**

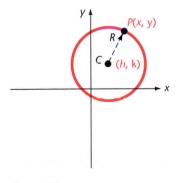

Let a circle have radius R and center at (h, k). Referring to Figure 3, we see that an arbitrary point $P(x, y)$ is on the circle if and only if

$$R = \sqrt{(x - h)^2 + (y - k)^2} \qquad \text{Distance from } C \text{ to } P \text{ is the constant } R.$$

or, equivalently,

$$R^2 = (x - h)^2 + (y - k)^2$$

which is usually written

$$(x - h)^2 + (y - k)^2 = R^2$$

Figure 3   Thus, we can state the following:

Equations of a Circle

1. Radius R and center (h, k):

$$(x - h)^2 + (y - k)^2 = R^2$$

2. Radius R and center at the origin:

$$x^2 + y^2 = R^2 \qquad \text{since} \qquad (h, k) = (0, 0)$$

Example 2
Graphing Circles

Graph:

(A) $x^2 + y^2 = 25$ **(B)** $(x - 3)^2 + (y + 2)^2 = 9$

Solution   **(A)** This is a circle with radius $\sqrt{25} = 5$ and center at the origin as shown in Figure 4a

(B) We need the equation rewritten in the form $(x - h)^2 + (y - k)^2 = R^2$:

$$(x - 3)^2 + (y + 2)^2 = 9$$

is the same as

$$(x - 3)^2 + [y - (-2)]^2 = 3^2$$

This is the equation of a circle with radius 3 and center at $(h, k) = (3, -2)$ as shown in Figure 4b

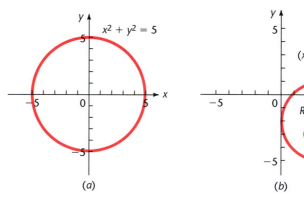

(a)   (b)

Figure 4

Matched Problem 2   Graph: **(A)** $x^2 + y^2 = 4$ **(B)** $(x + 2)^2 + (y - 3)^2 = 16$

Example 3
Finding the Equation of a Circle

What is the equation of a circle with radius 8 and center at the origin? Center at $(-5, 4)$?

Solution   If the center is at the origin, then $(h, k) = (0, 0)$; thus, the equation is

$$x^2 + y^2 = 8^2 \qquad \text{or} \qquad x^2 + y^2 = 64$$

If the center is at $(-5, 4)$, then $h = -5$ and $k = 4$, and the equation is

$$[x - (-5)]^2 + (y - 4)^2 = 8^2$$

or

$$(x + 5)^2 + (y - 4)^2 = 64$$

Matched Problem 3   What is the equation of a circle with radius $\sqrt{7}$ and center at the origin? Center at $(6, -4)$?

Example 4
**Changing an Equation
to Standard Form
and Graphing**

Graph: $x^2 + y^2 - 6x + 8y + 9 = 0$.

Solution   We transform the equation into the form

$$(x - h)^2 + (y - k)^2 = R^2$$

by completing the square relative to x and relative to y, as in Section 7-1:

$$x^2 - 6x + ? + y^2 + 8y + ? = -9 + ? + ?$$
$$x^2 - 6x + \mathbf{9} + y^2 + 8y + \mathbf{16} = -9 + \mathbf{9} + \mathbf{16}$$
$$(x - 3)^2 + (y + 4)^2 = 16$$

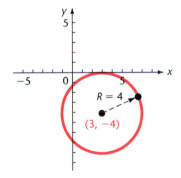

or

$$(x - 3)^2 + [y - (-4)]^2 = 16$$

This is the equation of a circle with radius 4 and center at $(h, k) = (3, -4)$ as shown in Figure 5

Figure 5

Matched Problem 4   Graph: $x^2 + y^2 + 10x - 4y + 20 = 0$

♦ **PARABOLAS**

In Section 7-5, the graph of the equation $y = ax^2 + bx + c$ was identified as a parabola. A parabola is a curve resulting from the intersection of a cone and a plane parallel to the edge of the cone. Properties of this intersection allow us to define a parabola geometrically:

Definition of a Parabola

A **parabola** is the set of all points equidistant from a fixed point and a fixed line. The fixed point is called the **focus,** and the fixed line is the **directrix.**

We will begin by finding an equation for the parabola with focus $(0, a)$, $a > 0$, and directrix $y = -a$ (see Figure 6).

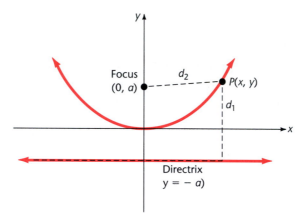

Figure 6

The two distances d_1 and d_2 are equal. Therefore,

$$d_1 = d_2$$

$$y + a = \sqrt{(x - 0)^2 + (y - a)^2}$$

$$(y + a)^2 = (x - 0)^2 + (y - a)^2$$

$$y^2 + 2ay + a^2 = x^2 + y^2 - 2ay + a^2$$

$$4ay = x^2$$

$$y = \frac{1}{4a} \cdot x^2$$

$$= kx^2 \qquad \text{where } k = \frac{1}{4a}$$

Thus, the graph of $y = x^2$ and its related forms studied in Sections 7-5 and 7-6 are, in fact, parabolas.

Proceeding in the same way, we can obtain similar equations for parabolas opening downward, to the right, and to the left. We summarize the four cases in the box.

Equations of a Parabola (Standard Forms)

$y = kx^2, \quad k > 0$
Opens up

$y = kx^2, \quad k < 0$
Opens down

$x = ky^2, \quad k > 0$
Opens right

$x = ky^2, \quad k < 0$
Opens left

The line that passes through the focus and is perpendicular to the directrix is the **axis** of the parabola. The point on the axis midway between the focus and the directrix is called the **vertex.** In each of the four cases in the box—the four **standard forms**—the vertex is at the origin $(0, 0)$. For $y = ax^2 + bx + c$, we found in Section 7-5 that the vertex is at

$$\left(-\frac{b}{2a}, -\frac{b^2 - 4ac}{4a} \right)$$

Parabolas in standard form can be sketched rather quickly. We already know how to graph $y = kx^2$. We can graph $x = ky^2$ in the same way, since we know the graph is a parabola, opening left when $k < 0$ and right when $k > 0$.

Example 5
Graphing $x = ky^2$

Graph:

$$x = -\tfrac{1}{8}y^2$$

Solution   Since $k < 0$, the x values must be negative and the parabola opens left. We look for values of y, in addition to $y = 0$, that make x easy to calculate, and then plot the graph:

x	y
0	0
-2	4
-2	-4

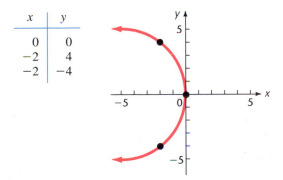

Matched Problem 5 Graph: $x = \frac{3}{5}y^2$

Parabolas are encountered frequently in the physical world. Suspension bridges, arch bridges, reflecting telescopes, radiotelescopes, radar equipment, solar furnaces, and searchlights all utilize parabolic forms in their design.

Answers to Matched Problems

1. $\sqrt{10}$ **2. (A)**

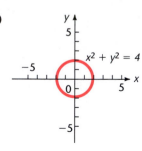

(B)

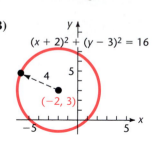

3. $x^2 + y^2 = 7$; $(x - 6)^2 + (y + 4)^2 = 7$

4. $(x + 5)^2 + (y - 2)^5 = 9$

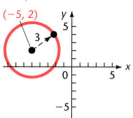

5.

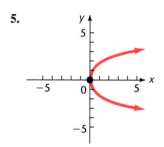

EXERCISE 7-7

A *Find the distance between each pair of points. Leave your answer in exact radical form.*

1. $(2, 4)$, $(4, 5)$ **2.** $(7, 3)$, $(8, 6)$

3. $(3, -7)$, $(-2, 1)$ **4.** $(-5, 3)$, $(-1, -2)$

5. $(-8, -2)$, $(-5, 1)$ **6.** $(2, -7)$, $(-4, -3)$

7. $(-3, 0)$, $(\frac{1}{2}, \frac{9}{2})$ **8.** $(4, 0)$, $(-\frac{5}{2}, \frac{5}{2})$

9. $(1.3, 2.5)$, $(3.8, -5.4)$ **10.** $(-2.7, 9.1)$, $(8.2, -6.3)$

Graph each circle. Identify the radius.

11. $x^2 + y^2 = 16$ **12.** $x^2 + y^2 = 36$

13. $x^2 + y^2 = 6$ **14.** $x^2 + y^2 = 10$

15. $x^2 + y^2 = \frac{25}{4}$ **16.** $x^2 + y^2 = \frac{16}{9}$

17. $x^2 + y^2 = \frac{64}{25}$ **18.** $x^2 + y^2 = \frac{49}{4}$

Write the equation of a circle with center at the origin and radius as given.

19. 7   **20.** 8   **21.** $\sqrt{5}$ **22.** $\sqrt{10}$

23. $\frac{7}{2}$ **24.** $\frac{9}{2}$ **25.** 6.2   **26.** 8.3

Graph each parabola.

27. $y^2 = 4x$ **28.** $y^2 = x$ **29.** $y^2 = -12x$

30. $y^2 = -16x$ **31.** $x^2 = y$ **32.** $x^2 = 12y$

33. $x^2 = -16y$ **34.** $x^2 = -4y$ **35.** $x = -2y^2$

36. $x = 2y^2$ **37.** $\frac{1}{3}x = y^2$ **38.** $-\frac{1}{3}x = y^2$

B *Graph each circle. Identify the radius and center.*

39. $(x - 3)^2 + (y - 4)^2 = 16$

40. $(x - 4)^2 + (y - 2)^2 = 9$

41. $(x - 4)^2 + (y + 3)^2 = 9$

42. $(x + 4)^2 + (y - 2)^2 = 25$

43. $(x + 3)^2 + (y + 3)^2 = 16$

44. $(x + 2)^2 + (y + 4)^2 = 25$

45. $(x + 2)^2 + (y + 3)^2 = 4$

46. $(x + 3)^2 + (y + 4)^2 = 5$

47. $(x - 4)^2 + (y - 3)^2 = 9$

48. $(x - 3)^2 + (y - 2)^2 = 1$

49. $(x - \frac{1}{2})^2 + (y - \frac{1}{3})^2 = \frac{7}{4}$

50. $(x + \frac{1}{4})^2 + (y + \frac{1}{2})^2 = \frac{5}{4}$

Write the equation of a circle in the form

$$(x - h)^2 + (y - k)^2 = R^2$$

with radius and center as given.

51. 7; $(3, 5)$ **52.** 2; $(4, 1)$

53. 8; $(-3, 3)$ **54.** 6; $(5, -2)$

55. $\sqrt{3}$; $(-4, -1)$ **56.** $\sqrt{14}$; $(-7, -5)$

57. $\frac{9}{4}$; $(1, \frac{1}{2})$ **58.** $\frac{16}{9}$; $(\frac{3}{4}, 2)$

59. $\sqrt[4]{2}$; $(1.1, 2)$ **60.** $\sqrt[4]{3}$; $(2, 3.1)$

Graph each parabola after first rewriting the equation in standard form.

61. $4y^2 - 8x = 0$ **62.** $3x^2 + 9y = 0$

63. $2y - 3x^2 = 0$ **64.** $5x - y^2 = 0$

65. $7x - 2y^2 = 0$ **66.** $3y - 10x^2 = 0$

67. Is the triangle with vertices $(-1, 2)$, $(2, -1)$, $(3, 3)$ an isosceles triangle, that is, are two of the sides the same length?

68. Repeat Problem 67 with vertices $(0, 3)$, $(3, 0)$, and $(-1, -1)$.

69. Is the triangle in Problem 67 equilateral, that is, are all three sides the same length?

70. Is the triangle in Problem 68 equilateral?

C *Use the method of completing the square to graph each circle. Indicate the center and radius.*

71. $x^2 + y^2 - 4x - 6y + 4 = 0$

72. $x^2 + y^2 - 6x - 4y + 4 = 0$

73. $x^2 + y^2 - 6x + 6y + 2 = 0$

74. $x^2 + y^2 + 6x - 4y - 3 = 0$

75. $x^2 + y^2 + 6x + 4y + 4 = 0$

76. $x^2 + y^2 + 4x + 4y - 8 = 0$

77. $x^2 + y^2 - 3x - 5y - \frac{15}{4} = 0$

78. $x^2 + y^2 - 5x - y - \frac{10}{4} = 0$

79. $x^2 + y^2 - x + 3y + \frac{1}{4} = 0$

80. $x^2 + y^2 + 5x + 3y + \frac{9}{4} = 0$

In Problems 81–84, find the focus and directrix of each parabola. For example, the focus of $y = kx^2$ is at $\left(0, \dfrac{1}{4k}\right)$ and the directrix is $y = -\dfrac{1}{4k}$.

81. $y^2 = 4x$ **82.** $y^2 = -12x$

83. $y = -8x^2$ **84.** $y = 16x^2$

85. Find x so that $(x, 8)$ is 13 units from $(2, -4)$.

86. Find y so that $(3, y)$ is 5 units from $(6, 9)$.

87. Find an equation of the set of points equidistant from $(-1, -1)$ and $(0, 1)$.

88. Find an equation of the set of points equidistant from $(3, 3)$ and $(6, 0)$.

89. Find the equation of the circle centered at the origin and passing through the point $(4, -3)$.

90. Find the equation of the circle centered at $(4, -3)$ and passing through the origin.

Use the definition of a parabola and the distance formula to find the equation of a parabola with directrix and focus as given.

91. $y = 4$; $(2, 2)$ **92.** $x = 2$; $(6, 4)$

93. $x = 0$; $(2, 2)$ **94.** $y = -8$; $(0, 0)$

APPLICATIONS

95. An ancient stone bridge in the form of a circular arc has a span of 80 feet (see the figure). If the height of the arch above its ends is 20 feet, find an equation of the circle containing the arch if its center is at the origin as indicated. [*Hint:* $(40, R - 20)$ must satisfy $x^2 + y^2 = R^2$.]

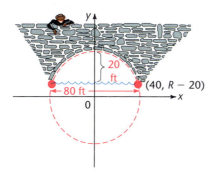

96. A cylindrical oil drum is cut to make a watering trough (see the figure). The end of the resulting trough is 20 centimeters high and 80 centimeters wide at the top. What was the radius R of the original drum?

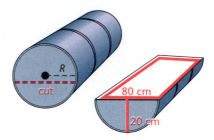

97. A parabolic concrete bridge is to span 100 meters. If the arch rises 25 meters above its ends, find the equation of the parabola, assuming it passes through the origin of a coordinate system and has its focus on the y axis [*Hint:* $(50, -25)$ must satisfy $x^2 = -4ay$.]

98. A radar bowl, in the form of a rotated parabola, is 20 meters in diameter and 5 meters deep. Find the equation of the parabola, assuming it passes through the origin of a coordinate system and has its focus on the positive y axis. (See Problem 97.)

7-8   Conic Sections: Ellipses and Hyperbolas

♦ Ellipses
♦ Hyperbolas

Here, we provide a very brief discussion of the other two conic sections shown in Figure 1 of Section 7-7. This introduction to *ellipses* and *hyperbolas* should help to increase your understanding of a more detailed development in a future course. In addition, it will provide you with some concrete experience with graphs other than straight lines, circles, and parabolas.

♦ ELLIPSES

You may be aware of many uses or occurrences of elliptical forms: orbits of satellites, orbits of planets and comets, gears and cams, and domes in buildings are just a few examples. Formally, we define an ellipse as follows:

Definition of an Ellipse

An **ellipse** is the set of all points such that the sum of the distances from each point to two fixed points is constant. The fixed points are called **foci,** and each separately is a **focus.**

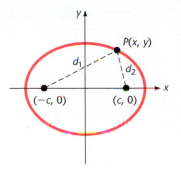

Figure 1

An ellipse can be drawn as follows. Place two pins in a piece of cardboard; these become the foci. Tie a piece of loose string, representing the constant sum, to the pins. Then move a pencil within the string, keeping it taut.

We will limit ourselves to the cases in which the foci are symmetrically located on either coordinate axis. Thus, if $(-c, 0)$ and $(c, 0)$, $c > 0$, are the foci and $2a$ is the constant sum of the distances, we obtain Figure 1.

From the triangle within the ellipse, we see that $d_1 + d_2 = 2a > 2c$, so that we must require $a > c$. Using the distance formula to rewrite $d_1 + d_2$, we get

$$d_1 + d_2 = 2a$$
$$\sqrt{(x + c)^2 + y^2} + \sqrt{(x - c)^2 + y^2} = 2a$$

After eliminating radicals and simplifying—a good exercise for the reader—we eventually obtain

$$(a^2 - c^2)(x^2) + a^2 y^2 = a^2(a^2 - c^2)$$

or

$$\frac{x^2}{a^2} + \frac{y^2}{a^2 - c^2} = 1$$

Since $a > c$, we also have $a^2 - c^2 > 0$. To simplify the equation further, we choose to let $b^2 = a^2 - c^2$, $b > 0$. Thus,

$$\frac{x^2}{a^2} + \frac{y^2}{b^2} = 1$$

Since $b^2 < a^2$ and both a and b are positive, $b < a$.

Proceeding similarly with the foci on the vertical axis, we would arrive at

$$\frac{x^2}{b^2} + \frac{y^2}{a^2} = 1$$

Combining these results, we can write

$$\frac{x^2}{m^2} + \frac{y^2}{n^2} = 1$$

as a standard form for an equation of an ellipse with foci symmetrically located on the x or y axis. We shift over to m and n to simplify our approach and because a and b have special significance in a more advanced treatment of the subject. In this form, the foci are located on the axis corresponding to the variable with the larger square in the denominator. The two cases are summarized in the box:

Equations of Ellipses (Standard Form)

Case 1. $\dfrac{x^2}{m^2} + \dfrac{y^2}{n^2} = 1, \quad m > n > 0$

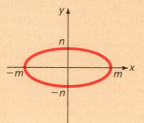

Case 2. $\dfrac{x^2}{m^2} + \dfrac{y^2}{n^2} = 1, \quad n > m > 0$

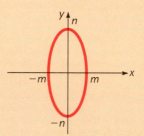

We now show how ellipses with equation in standard form can be sketched rather quickly.

Rapid Sketching of Ellipses

To graph $\dfrac{x^2}{m^2} + \dfrac{y^2}{n^2} = 1$:

Step 1. Find the x intercepts by letting $y = 0$ and solving for x.
Step 2. Find the y intercepts by letting $x = 0$ and solving for y.
Step 3. Sketch an ellipse passing through these intercepts.

Example 1
Graphing an Ellipse

Graph:

$$\frac{x^2}{16} + \frac{y^2}{9} = 1$$

Solution   *Step 1.* Find x intercepts:

$$\frac{x^2}{16} + \frac{0}{9} = 1$$
$$x^2 = 16$$
$$x = \pm\sqrt{16} = \pm 4$$

Step 2. Find y intercepts:

$$\frac{0}{16} + \frac{y^2}{9} = 1$$
$$y^2 = 9$$
$$y = \pm\sqrt{9} = \pm 3$$

Step 3. Plot the intercepts and sketch in the ellipse:

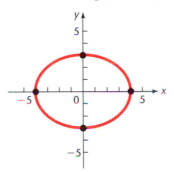

Matched Problem 1   Graph: $\dfrac{x^2}{9} + \dfrac{y^2}{16} = 1$

◆ **HYPERBOLAS**

A hyperbola is defined as follows:

> **Definition of a Hyperbola**
>
> A **hyperbola** is the set of all points such that the absolute value of the difference of the distances from each point to two fixed points is constant. The two fixed points are called **foci.**

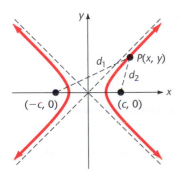

Figure 2

As with the ellipse, we will limit our investigation to cases in which the foci are symmetrically located on either coordinate axis. Thus, if $(-c, 0)$ and $(c, 0)$ are the foci and $2a$ is the constant difference, we obtain Figure 2.

Again, applying the distance formula, we get

$$|d_1 - d_2| = 2a$$
$$\left|\sqrt{(x + c)^2 + y^2} - \sqrt{(x - c)^2 + y^2}\right| = 2a$$

After eliminating radicals and absolute-value signs by appropriate use of squaring and simplifying—another good exercise for the reader—we eventually obtain

$$\left|\frac{x^2}{a^2} + \frac{y^2}{a^2 - c^2}\right| = 1$$

which looks like the equation we obtained for the ellipse. However, from Figure 2 we see that $d_1 < 2c + d_2$ so $d_1 - d_2 < 2c$; that is, $2a < 2c$. Hence, $a^2 - c^2 < 0$. To simplify the equation further, we let $-b^2 = a^2 - c^2$. Thus,

$$\frac{x^2}{a^2} - \frac{y^2}{b^2} = 1$$

Proceeding similarly with the foci on the vertical axis, we would obtain

$$\frac{y^2}{a^2} - \frac{x^2}{b^2} = 1$$

Combining these results, we can write

$$\frac{x^2}{m^2} - \frac{y^2}{n^2} = 1 \qquad \frac{y^2}{n^2} - \frac{x^2}{m^2} = 1$$

as standard forms for equations of hyperbolas with foci symmetrically located on a coordinate axis. Again, we shift over to m and n to simplify our approach and because a and b have special significance in a more advanced treatment of the subject. In summary:

Equations of Hyperbolas (Standard Forms)

$$\frac{x^2}{m^2} - \frac{y^2}{n^2} = 1$$

Opens left and right

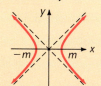

$$\frac{y^2}{n^2} - \frac{x^2}{m^2} = 1$$

Opens up and down

The equations for hyperbolas in standard form are similar to those for ellipses, except that the terms x^2/m^2 and y^2/n^2 are subtracted rather than added. In both equations we have kept the denominator m^2 associated with the variable x and n^2 associated with the variable y. The hyperbola opens in the direction corresponding to the variable with a positive coefficient when the equation is in standard form.

As an aid to graphing equations of hyperbolas in standard form, we solve each equation for y in terms of x. From the first equation, we obtain

$$y = \pm \frac{nx}{m} \sqrt{1 - \frac{m^2}{x^2}}$$

and from the second equation, we obtain

$$y = \pm \frac{nx}{m} \sqrt{1 + \frac{m^2}{x^2}}$$

For very large x, the radicals represent numbers very near to 1, since m^2/x^2 will be nearly 0. Hence, for very large x, the equations behave very much like the straight lines

$$y = \pm \frac{n}{m}x$$

These lines are called the **asymptotes** for the hyperbola and are helpful in obtaining its graph.

Asymptotes for a Hyperbola

For either $\dfrac{x^2}{m^2} - \dfrac{y^2}{n^2} = 1$ or $\dfrac{y^2}{n^2} - \dfrac{x^2}{m^2} = 1$, the lines $y = \pm \dfrac{n}{m}x$

are **asymptotes.**

The asymptotes are guidelines for the graph. The hyperbola will approach these guidelines, but never touch them, as the graph moves farther and farther away from the origin. The asymptotes can be drawn quickly by drawing a rectangle with x intercepts $\pm m$ and y intercepts $\pm n$, and then drawing the extended diagonals. You should check that the diagonals have equations $y = \pm(n/m)x$. Quick sketches of hyperbolas then can be made by following these steps:

Rapid Sketching of Hyperbolas

To graph $\dfrac{x^2}{m^2} - \dfrac{y^2}{n^2} = 1$ or $\dfrac{y^2}{n^2} - \dfrac{x^2}{m^2} = 1$:

Step 1. Draw a dashed rectangle with intercepts $x = \pm m$ and $y = \pm n$.
Step 2. Draw dashed diagonals of the rectangle and extend them to form the asymptotes.
Step 3. Determine the intercepts of the hyperbola,
Step 4. Sketch in the hyperbola, including both branches.

Be particularly careful in determining the intercepts in step 3. An error here can result in the hyperbola opening the wrong way.

<table>
<tr><td>**Example 2**
Graphing Hyperbolas</td><td>Graph:</td></tr>
</table>

Example 2
Graphing Hyperbolas

Graph:

(A) $\dfrac{x^2}{25} - \dfrac{y^2}{16} = 1$ (B) $\dfrac{y^2}{16} - \dfrac{x^2}{25} = 1$

Solution   (A) *Step 1.*   Draw a dashed rectangle with intercepts $x = \pm\sqrt{25} = \pm 5$ and $y = \pm\sqrt{16} = \pm 4$:

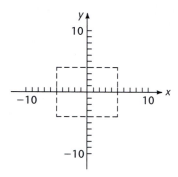

Step 2.   Draw in asymptotes, that is, the extended diagonals of the rectangle:

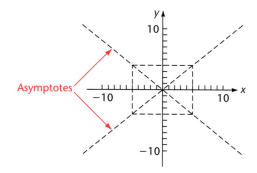

Step 3.   Determine the intercepts for the hyperbola: Let $y = 0$; then

$$\frac{x^2}{25} - \frac{0}{16} = 1$$
$$x^2 = 25$$
$$x = \pm\sqrt{25} = \pm 5$$

If we let $x = 0$, then

$$\frac{0}{25} - \frac{y^2}{16} = 1$$
$$y^2 = -16$$
$$y = \pm\sqrt{-16} = \pm 4i$$

These are complex numbers and do not represent intercepts. We conclude that the only real intercepts are $x = \pm 5$, so the hyperbola crosses the x axis and opens left and right.

Step 4. Sketch the curve as in Figure 3*a*

(B) See Figure 3*b*

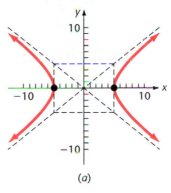
(a)

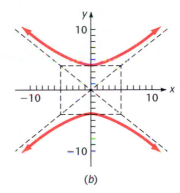
(b)

Figure 3

Matched Problem 2   Graph: **(A)** $\dfrac{x^2}{4} - \dfrac{y^2}{9} = 1$ **(B)** $\dfrac{y^2}{9} - \dfrac{x^2}{4} = 1$

Hyperbolic forms are encountered in the study of comets, the loran system of navigation for ships and aircraft, some modern architectural structures, and optics, to name just a few examples among many.

Answers to Matched Problems

1.

2. (A)
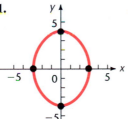

(B)
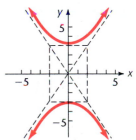

EXERCISE 7-8

A *Graph each ellipse.*

1. $\dfrac{x^2}{25} + \dfrac{y^2}{4} = 1$

2. $\dfrac{x^2}{9} + \dfrac{y^2}{4} = 1$

3. $\dfrac{x^2}{4} + \dfrac{y^2}{25} = 1$

4. $\dfrac{x^2}{4} + \dfrac{y^2}{9} = 1$

5. $\dfrac{x^2}{16} + \dfrac{y^2}{25} = 1$

6. $\dfrac{x^2}{9} + \dfrac{y^2}{16} = 1$

7. $\dfrac{x^2}{4} + \dfrac{y^2}{16} = 1$

8. $\dfrac{x^2}{25} + \dfrac{y^2}{16} = 1$

Graph each hyperbola.

9. $\dfrac{x^2}{4} - \dfrac{y^2}{25} = 1$

10. $\dfrac{x^2}{16} - \dfrac{y^2}{9} = 1$

11. $\dfrac{y^2}{25} - \dfrac{x^2}{4} = 1$

12. $\dfrac{y^2}{9} - \dfrac{x^2}{16} = 1$

13. $\dfrac{x^2}{4} - \dfrac{y^2}{9} = 1$

14. $\dfrac{x^2}{9} - \dfrac{y^2}{25} = 1$

15. $\dfrac{y^2}{9} - \dfrac{x^2}{4} = 1$

16. $\dfrac{y^2}{25} - \dfrac{x^2}{4} = 1$

17. $y^2 - \dfrac{x^2}{25} = 1$

18. $x^2 - \dfrac{y^2}{25} = 1$

19. $x^2 - \dfrac{y^2}{36} = 1$

20. $y^2 - \dfrac{x^2}{36} = 1$

Graph.

21. $\dfrac{x^2}{4} - y^2 = 1$

22. $\dfrac{x^2}{4} + y^2 = 1$

23. $x^2 + \dfrac{y^2}{9} = 1$

24. $x^2 - \dfrac{y^2}{9} = 1$

25. $\dfrac{x^2}{4} + \dfrac{y^2}{25} = 1$

26. $\dfrac{x^2}{4} + \dfrac{y^2}{4} = 1$

27. $-\dfrac{x^2}{4} + \dfrac{y^2}{25} = 1$

28. $\dfrac{x^2}{4} - \dfrac{y^2}{25} = 1$

29. $\dfrac{x^2}{25} + \dfrac{y^2}{25} = 1$

30. $-x^2 + y^2 = 1$

B *Graph each equation after first rewriting it in one of the standard forms discussed in this section. For example,*

$$9x^2 + 4y^2 = 36$$

can be rewritten in the form

$$\dfrac{x^2}{4} + \dfrac{y^2}{9} = 1$$

by dividing through by 36.

31. $4x^2 + 9y^2 = 36$

32. $4x^2 + 25y^2 = 100$

33. $4x^2 - 9y^2 = 36$

34. $25y^2 - 4x^2 = 100$

35. $4x^2 + 16y^2 = 16$

36. $4x^2 - 16y^2 = 16$

37. $16x^2 - 4y^2 = 16$

38. $16x^2 - 16y^2 = 16$

39. $-16x^2 + 4y^2 = 16$

40. $-4x^2 + 16y^2 = 16$

41. $4x^2 + 4y^2 = 16$

42. $-4x^2 + 4y^2 = 16$

43. $4x^2 - 4y^2 = 16$

44. $16x^2 + 16y^2 = 16$

Graph each equation after first rewriting it in one of the standard forms discussed in this section. For example,

$$4x^2 + y^2 = 1$$

can be rewritten in the form

$$\dfrac{x^2}{\frac{1}{4}} + y^2 = 1$$

by writing 4 as $\dfrac{1}{\frac{1}{4}}$.

45. $x^2 - 9y^2 = 1$

46. $x^2 + 9y^2 = 1$

47. $9x^2 + y^2 = 1$

48. $9x^2 - 9y^2 = 1$

49. $4x^2 - y^2 = 1$

50. $-4x^2 + y^2 = 1$

51. $x^2 + 4y^2 = 1$

52. $x^2 - 4y^2 = 1$

53. $4x^2 + 9y^2 = 1$

54. $4x^2 - 9y^2 = 1$

In Problems 55–70, graph the equation and identify the conic section.

55. $y^2 - x^2 = 1$

56. $x^2 + y = 0$

57. $x^2 + y^2 = 4$

58. $16x^2 + y^2 = 1$

59. $4x^2 + 16y^2 = 1$

60. $4y^2 - \dfrac{x^2}{4} = 1$

61. $y^2 - x = 0$

62. $x^2 + y^2 = 8$

63. $x^2 - y^2 = 16$

64. $x^2 - 4y = 0$

65. $x^2 + \dfrac{y^2}{4} = 1$

66. $x^2 + y^2 = 16$

67. $4x^2 + y^2 = 0$

68. $x^2 + 8y^2 = 2$

69. $x^2 + y^2 = 2$

70. $y^2 + x^2 = 0$

C 71. Find the coordinates of the foci for the ellipse in Problem 3.

72. Find the coordinates of the foci for the ellipse in Problem 2.

73. Find the coordinates of the foci for the hyperbola in Problem 10.

74. Find the coordinates of the foci for the hyperbola in Problem 11.

For $m \neq n$, the equation

$$\dfrac{(x - h)^2}{m^2} + \dfrac{(y - k)^2}{n^2} = 1$$

represents an ellipse centered about (h, k) rather than $(0, 0)$. Similarly,

$$\dfrac{(x - h)^2}{m^2} - \dfrac{(y - k)^2}{n^2} = 1 \text{ and } \dfrac{(y - h)^2}{n^2} - \dfrac{(x - k)^2}{m^2} = 1$$

represent hyperbolas centered about (h, k) rather than $(0, 0)$. Rewrite each equation by completing the square in x and y. Identify the conic section the equation represents.

75. $x^2 + 4y^2 - 2x - 16y + 13 = 0$

76. $x^2 + 4y^2 + 2x + 16y + 13 = 0$

77. $x^2 + y^2 + 2x + 4y + 1 = 0$

78. $x^2 + y^2 + 6x - 2y + 6 = 0$

79. $4x^2 + y^2 + 24x - 2y + 33 = 0$

80. $4x^2 + y^2 - 24x + 2y + 33 = 0$

81. $4x^2 - y^2 - 24x - 2y + 31 = 0$

82. $-4x^2 + y^2 + 24x + 2y - 27 = 0$

83. $-4x^2 + y^2 - 24x - 2y - 27 = 0$

84. $4x^2 - y^2 + 24x + 2y + 31 = 0$

7-9 Nonlinear Inequalities

♦ Quadratic Inequalities
♦ Other Inequalities
♦ Describing Solution Sets
♦ Graphing Quadratic Inequalities

We know how to solve linear inequalities in one variable, that is, inequalities such as

$$x + 2 < 8$$

where the variable occurs only to the first power. If we allow the variable to be squared, we obtain inequalities such as

$$x^2 + 2x < 8$$

This is called a **quadratic inequality.** In this section, we introduce a method for solving quadratic inequalities. The same method also can be applied to certain other inequalities that are not quadratic. In both cases, the solutions will involve combinations of intervals on a real number line, which can be described compactly using interval notation. To end the section, we relate the solutions of quadratic inequalities to the graphs of corresponding quadratic polynomials.

♦ QUADRATIC INEQUALITIES

How do we solve an inequality such as $x^2 + 2x < 8$? If we move all terms to the left and factor, then we will be able to observe something that will lead to a solution:

$$x^2 + 2x - 8 < 0$$
$$(x + 4)(x - 2) < 0$$

We are looking for values of x that will make the left side less than 0, that is, negative. What will the signs of $(x + 4)$ and $(x - 2)$ have to be so that their product is negative? They must have opposite signs!

Let us see where each of the factors is positive, negative, and 0. The point at which either factor is 0 is called a **critical point.** We will see why shortly.

Sign Analysis for $(x + 4)$:

Critical point	$(x + 4)$ is positive when	$(x + 4)$ is negative when
$x + 4 = 0$ $x = -4$	$x + 4 > 0$ $x > -4$	$x + 4 < 0$ $x < -4$

Geometrically:

Thus, $(x + 4)$ is negative for values of x to the left of the critical point and is positive for values of x to the right of the critical point.

Sign Analysis for $(x - 2)$:

Critical point	$(x - 2)$ is positive when	$(x - 2)$ is negative when
$x - 2 = 0$	$x - 2 > 0$	$x - 2 < 0$
$x = 2$	$x > 2$	$x < 2$

Geometrically:

Thus, $(x - 2)$ is negative for values of x to the left of the critical point and is positive for values of x to the right of the critical point.

Combining these results in a single geometric representation leads to the solution of the original problem:

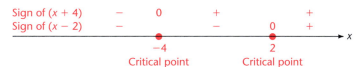

Now we can see that the factors have opposite signs for x between -4 and 2. Thus, the solution of $x^2 + 2x < 8$ is

$$-4 < x < 2$$

This discussion leads to the general result behind the sign-analysis method of solving quadratic inequalities.

> The value of x at which the linear expression $(ax + b)$ is 0 is called a **critical point.** On the real number line, $(ax + b)$ has one sign to the left of the critical point and the opposite sign to the right of the critical point.

This allows us to summarize the **sign-analysis method** for solving quadratic inequalities as follows:

Sign-Analysis Method for Quadratic Inequalities

To solve a quadratic inequality of the form

$$ax^2 + bx + c \text{ [compared to] } 0$$

The notation [compared to] represents the four possible inequalities $<$, $>$, $\leq$, and $\geq$.

1. Factor $ax^2 + bx + c$ into linear factors.
2. Determine the critical point for each factor.
3. Analyze the sign of each factor on each interval determined by the critical points.
4. Use the signs of the factors to determine the sign of the product $ax^2 + bx + c$. If the comparison is $\leq$ or $\geq$, include the critical points in the solution.

If the factoring in step 1 cannot be done, then another method (such as graphing) must be used.

Example 1
Solving by Sign Analysis

Solve and graph the solution on a real number line:

$$x^2 \geq x + 6$$

Solution

$$x^2 \geq x + 6 \qquad \text{Rewrite in the form } ax^2 + bx + c \geq 0.$$
$$x^2 - x - 6 \geq 0 \qquad \text{Factor.}$$
$$(x - 3)(x + 2) \geq 0$$

Critical points: -2 and 3

Locate these points on a real number line and indicate the sign of each factor to the left and to the right of each critical point:

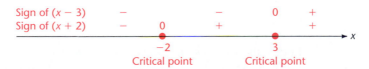

The inequality statement is satisfied when both factors have the same sign or when one or the other factor is 0. The factors have the same sign when x is to the left of -2 or to the right of 3; one or the other factor is 0 at the critical points. Thus,

Solution: $x \leq -2$ or $x \geq 3$

Graph:

Matched Problem 1   Solve and graph the solution on a real number line:

         (A) $x^2 < x + 12$ **(B)** $x^2 \geq x + 12$

It is important to recognize that in the sign-analysis method we need to obtain an expression that we are to compare to 0. The method will work for an inequality such as $(x - 3)(x + 2) \geq 0$ but will not work for an inequality such as $(x - 3)(x + 2) \geq 5$.

◆   **OTHER INEQUALITIES**

Sign analysis also can be used on other inequalities where linear terms are combined by multiplication and division and the result compared to 0.

Example 2
Sign Analysis for a
Quotient

Solve and graph the solution on a real number line:

$$\frac{x^2 - x + 1}{2 - x} \geq 1$$

Solution

$$\frac{x^2 - x + 1}{2 - x} \geq 1$$ We don't know the sign of $2 - x$. Thus, we can't multiply both sides by $2 - x$, since we don't know whether to reverse the inequality. Instead, we subtract 1 from each side to obtain an expression compared to 0.

$$\frac{x^2 - x + 1}{2 - x} - 1 \geq 0$$ Combine terms on the left side into a single fraction.

$$\frac{x^2 - x + 1}{2 - x} - \frac{2 - x}{2 - x} \geq 0$$

$$\frac{x^2 - 1}{2 - x} \geq 0$$ Factor the numerator.

$$\frac{(x - 1)(x + 1)}{2 - x} \geq 0$$ Proceed as in Example 1 to determine which values make the linear components of the expression positive or 0.

Critical points: $-1, 1, 2$

Locate these points on a real number line and indicate the sign of each first-degree form to the left and to the right of each critical point:

Sign of $(x - 1)$	$-$		$-$	0	$+$		$+$
Sign of $(x + 1)$	$-$	0	$+$		$+$		$+$
Sign of $(2 - x)$	$+$		$+$		$+$	0	$-$

$$\xrightarrow{\hspace{2cm} \bullet \hspace{2cm} \bullet \hspace{0.3cm} \bullet \hspace{2cm}} x$$

            -1 1 2

          Critical   Critical   Critical
           point   point   point

The rational expression is greater than or equal to 0 when $(x - 1)$, $(x + 1)$, and $(2 - x)$ are all positive, when two are negative and one is positive, or when the numerator is 0. Two factors are negative to the left of -1 and all are positive between 1 and 2. The equality part of the inequality holds when x is 1 or -1, but not when $x = 2$. Thus,

Solution: $x \le -1$ or $1 \le x < 2$

Graph:

Matched Problem 2 Solve and graph the solution on a real number line: $\dfrac{3}{2-x} \le \dfrac{1}{x+4}$

♦ **DESCRIBING SOLUTION SETS**

There is a convenient way to describe solution sets that result from solving quadratic and other nonlinear inequalities. These solution sets are combinations of intervals. If A and B are such intervals, then we will use $A \cup B$ to denote the set of all points formed by combining all the elements of A and all the elements of B into one set. The set $A \cup B$ is called the **union** of A and B. Thus, $A \cup B$ consists of all points that belong to one interval or the other. The word "or" is used in the way it is generally used in mathematics; that is, x may belong to set A or set B or both.

The **intersection** of sets A and B, denoted by $A \cap B$, is the set of points in A that are also in B. That is, $A \cap B$ consists of all the points that are in both the intervals A and B. If there are no points common to A and B, then we say that the sets are **disjoint,** or that the intersection is **empty,** or that the intersection is the **empty set.** In this case, we write $A \cap B = \varnothing$. Set union, intersection, and the empty set are described more fully in Appendix A.

Example 3
Unions and Intersections of Intervals

If $P = [-4, 2)$, $Q = (-1, 6]$, and $R = [3, \infty)$, graph:

(A) $P \cup Q$ and $P \cap Q$ (B) $P \cup R$ and $P \cap R$

Solution (A)

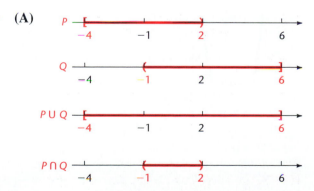

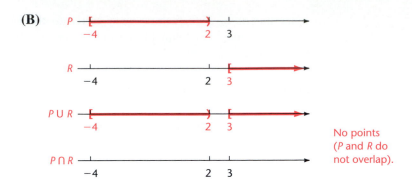

(B)

No points
(*P* and *R* do
not overlap).

Matched Problem 3   If $A = [0, 3]$, $B = (-2, 2)$, and $C = (-1, \infty)$, graph:

(A) $A \cup B$ **(B)** $A \cap B$ **(C)** $B \cup C$ **(D)** $B \cap C$

Example 4
Unions and
Intersections of
Intervals

Express the following sets using interval notation, unions, and intersections:

(A) $x \leq -2$ or $x \geq 3$ (see Example 1)
(B) $x \leq -1$ or $1 \leq x < 2$ (see Example 2)
(C) $x \geq 2$ and $x < 4$

Solution   **(A)** $(-\infty, -2] \cup [3, \infty)$ **(B)** $(-\infty, -1] \cup [1, 2)$
        (C) $[2, \infty) \cap (-\infty, 4)$ or $[2, 4)$

Matched Problem 4   Express the following sets using interval notation, unions, and intersections:

(A) $x < 4$ or $x > -3$ [see Matched Problem 1(A)]
(B) $x \leq -3$ or $x \geq 4$ [see Matched Problem 1(B)]
(C) $-4 < x \leq -\frac{5}{2}$ or $x > 2$ (see Matched Problem 2)

◆   **GRAPHING QUADRATIC INEQUALITIES**

At the start of this section, we solved the inequality $x^2 + 2x - 8 < 0$ and found the solution to be the interval $-4 < x < 2$. We can see this solution graphically by graphing $y = x^2 + 2x - 8$ and asking where on the graph $y < 0$. The graph in Figure 1 shows that $y < 0$ for $-4 < x < 2$.

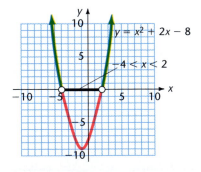

Figure 1

Example 5
Solving Graphically

Solve by graphing:

(A) $x^2 - x - 6 \geq 0$ **(B)** $x^2 + 3x + 4 < 0$

Solution

(A) Graph $y = x^2 - x - 6$:

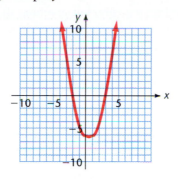

Find the x intercepts: $x^2 - x - 6 = (x - 3)(x + 2) = 0$ when $x = 3$ or -2. With this information, we can see the solution on the graph:

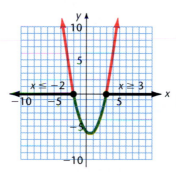

Compare this solution to Example 1.

(B) Graph $y = x^2 + 3x + 4$:

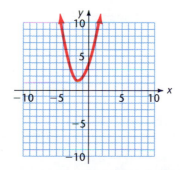

Observe that the entire graph lies above the x axis. That is, $y = x^2 + 3x + 4$ is always greater than 0. Thus, $x^2 + 3x + 4 < 0$ has no solution.

Matched Problem 5

Solve by graphing:

(A) $x^2 - x - 12 < 0$ [compared to Matched Problem 1(A)]
(B) $x^2 - x - 12 \geq 0$ [compared to Matched Problem 1(B)]
(C) $x^2 + 2x + 3 > 0$

An inequality such as

$$\frac{x^2 - x + 1}{2 - x} - 1 \geq 0 \qquad \text{See Example 2.}$$

also can be solved graphically, but graphing the expression on the left-hand side requires more sophisticated graphing techniques. However, with the aid of a computer or graphing calculator we could obtain the graph shown in Figure 2 for

$$y = \frac{x^2 - x + 1}{2 - x} - 1$$

The portion of the graph where $y \geq 0$ corresponds to the solution of the inequality. We can see that this is $x \leq -1$ or $1 \leq x < 2$, just as we found in Example 2.

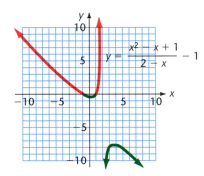

Figure 2

1. **(A)** $-3 < x < 4$ ←——————→ x
   $\qquad\qquad\qquad\qquad\quad -3 \qquad\qquad 4$

   (B) $x \leq -3$ or $x \geq 4$ ←—┤   ├—→ x
   $\qquad\qquad\qquad\qquad\qquad -3 \qquad\quad 4$

2. $-4 < x \leq -\frac{5}{2}$ or $x > 2$ ←——┤ (——→ x
   $\qquad\qquad\qquad\qquad\quad -4 \quad -\frac{5}{2} \quad\; 2$

3. **(A)** ——(——————]——→ x **(B)** ————————[——)——→ x
   $\qquad\qquad -2 \qquad\quad 3 \qquad\qquad\qquad\qquad\qquad\qquad 0 \quad\; 2$

   (C) ——(————————————→ x **(D)** ————————(——)——→ x
   $\qquad\qquad -2 \qquad\qquad\qquad\qquad\qquad\qquad\qquad\qquad -1 \quad\; 2$

4. **(A)** $(-3, 4)$ **(B)** $(-\infty, -3] \cup [4, \infty)$ **(C)** $(-4, -\frac{5}{2}] \cup (2, \infty)$
5. **(A)** $(-3, 4)$ **(B)** $(-\infty, -3] \cup [4, \infty)$

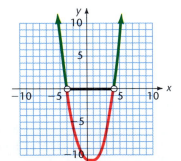
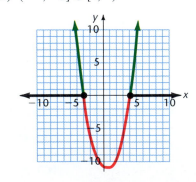

(C) All real numbers

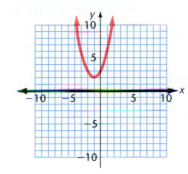

EXERCISE 7-9

A *Solve and graph each solution on a real number line.*

1. $(x - 3)(x + 4) < 0$
2. $(x + 2)(x - 4) < 0$
3. $(x - 3)(x + 4) \geq 0$
4. $(x + 2)(x - 4) > 0$
5. $x^2 + x < 12$
6. $x^2 < 10 - 3x$
7. $x^2 + 21 > 10x$
8. $x^2 + 7x + 10 > 0$
9. $x^2 + 6x + 5 \leq 0$
10. $x^2 + 6x + 5 > 0$
11. $x^2 - 9x + 8 < 0$
12. $x^2 - 9x + 8 \geq 0$
13. $x^2 - 3x - 10 > 0$
14. $x^2 - 3x - 10 < 0$
15. $x^2 + 4x - 12 \leq 0$
16. $x^2 + 4x - 12 \geq 0$
17. $x(x + 6) \geq 0$
18. $x(x - 8) \leq 0$
19. $x^2 \geq 9$
20. $x^2 > 4$

Rewrite as a single interval, if possible.

21. $[-1, 3] \cap [1, 5]$
22. $[-1, 3] \cup [1, 5]$
23. $(0, 5) \cup [-3, 2)$
24. $(0, 5] \cap [-3, 2)$
25. $(-\infty, 1) \cap (-1, \infty)$
26. $(-\infty, 1) \cup (-1, \infty)$
27. $(-\infty, 5] \cup [0, 10)$
28. $(-\infty, 5] \cap [0, 10)$
29. $(-3, 0] \cap [-2, 1)$
30. $(-3, 4) \cup [0, 8]$
31. $(-1, 1) \cup (0, \infty)$
32. $(-\infty, -1) \cap [-3, 2]$
33. $(-\infty, 3) \cup (2, 5]$
34. $[0, 2) \cup [2, 5)$
35. $[-4, -1) \cap (1, 4)$
36. $[-4, -1) \cup (1, 4)$
37. $(0, 10] \cup [-10, 0)$
38. $(0, 10] \cap [-10, 0)$

B *Solve and graph each solution on a real number line.*

39. $(x - 3)(x + 4)(x - 5) < 0$
40. $(x - 3)(x + 4)(x - 5) \geq 0$
41. $x(x - 3)(x + 4) \geq 0$
42. $x(x - 3)(x + 4) < 0$

43. $x^3 + x^2 < 12x$
44. $x^3 < 10x - 3x^2$
45. $x^3 + 21x > 10x^2$
46. $x^3 + 7x^2 + 10x > 0$
47. $\dfrac{x - 5}{x + 2} \leq 0$
48. $\dfrac{x + 2}{x - 3} < 0$
49. $\dfrac{x - 5}{x + 2} > 0$
50. $\dfrac{x + 2}{x - 3} \geq 0$
51. $\dfrac{x - 4}{x(x + 2)} \leq 0$
52. $\dfrac{x(x + 5)}{x - 3} \geq 0$
53. $\dfrac{x(x - 1)}{x + 2} \geq 0$
54. $\dfrac{x(x + 2)}{x - 4} > 0$
55. $\dfrac{x}{(x - 1)(x + 3)} < 0$
56. $\dfrac{x}{(x + 2)(x - 2)} \leq 0$
57. $\dfrac{4}{x} \geq 3$
58. $\dfrac{3}{x} \leq 2$
59. $\dfrac{1}{x} < 4$
60. $\dfrac{5}{x} > 3$
61. $x^2 + 4 \geq 4x$
62. $6x \leq x^2 + 9$
63. $x^2 + 9 < 6x$
64. $x^2 + 4 < 4x$
65. $x^2 \geq 3$
66. $x^2 < 2$
67. $\dfrac{2}{x - 3} \leq -2$
68. $\dfrac{2x}{x + 3} \geq 1$

Solve graphically.

69. $x^2 - 4 < 0$
70. $x^2 - 9 > 0$
71. $x^2 + 4x + 3 > 0$
72. $x^2 + 6x + 5 \leq 0$
73. $x^2 + x \leq 6$
74. $x^2 - 2x \geq 8$

C *Solve and graph the solution on a real number line.*

75. $\dfrac{2}{x - 3} \leq \dfrac{2}{x + 2}$
76. $\dfrac{2}{x + 1} \geq \dfrac{1}{x - 2}$
77. $\dfrac{(x - 1)(x + 3)}{x} > 0$
78. $\dfrac{x(x - 3)}{x + 2} < 0$

79. $\dfrac{1}{x+1} + \dfrac{1}{x-2} \geq 0$

80. $\dfrac{x^2 - 6x + 8}{x+2} > 0$

81. $\dfrac{(x+1)^2}{x^2 + 2x - 3} \leq 0$

82. $\dfrac{(x-1)(x+3)}{x(x+2)} > 0$

Find all values of x that make the expression a real number.

83. $\sqrt{x^2 - 3x + 2}$

84. $\sqrt{\dfrac{x-3}{x+5}}$

85. $\sqrt{\dfrac{x+1}{x-1}}$

86. $\sqrt{x^2 + 4x + 3}$

87. $\sqrt{x^2 + x + 1}$

88. $\sqrt{x^2 + 4x + 5}$

89. $\sqrt{-1 + 2x - 3x^2}$

90. $\sqrt{-2 + 5x - 7x^2}$

91. If an object is shot straight up from the ground with an initial velocity of 160 feet per second, its distance d in feet above the ground at the end of t seconds (neglecting air resistance) is given by $d = 160t - 16t^2$. Find the duration of time for which $d \geq 256$.

92. Repeat Problem 91 for $d \geq 0$.

CHAPTER SUMMARY

7-1 SOLVING QUADRATIC EQUATIONS BY SQUARE ROOTS AND BY COMPLETING THE SQUARE

The **standard form** of a **quadratic equation** is $ax^2 + bx + c = 0$, with $a \neq 0$. When $b = 0$, a quadratic equation can be solved by taking square roots. When $ax^2 + bx + c$ can be factored into two first-degree factors, the equation can be solved by setting each factor equal to 0.

Any quadratic equation can be solved by **completing the square.** To complete the square $x^2 + bx$, add the square of $\frac{1}{2}$ the coefficient of x to obtain a perfect square:

$$x^2 + bx + \left(\frac{b}{2}\right)^2 = \left(x + \frac{b}{2}\right)^2$$

To solve $ax^2 + bx + c = 0$ by completing the square, complete the square on the left side and then solve by taking square roots.

7-2 THE QUADRATIC FORMULA

The method of completing the square leads to the **quadratic formula,** a general solution of $ax^2 + bx + c = 0$:

$$x = \frac{-b \pm \sqrt{b^2 - 4ac}}{2a} \qquad a \neq 0$$

The expression $b^2 - 4ac$ is called the **discriminant** and provides information about the roots: if positive, there are two real solutions; if negative, there are two nonreal, complex conjugate solutions; if 0, there is one real solution.

7-3 APPLICATIONS

Many applications lead to quadratic equations. If an equation has two solutions, it is necessary to check both of them in the original problem.

7-4 RADICAL EQUATIONS AND OTHER EQUATIONS REDUCIBLE TO QUADRATIC FORM

Radicals can be removed from **radical equations** by raising both sides to a power, but this procedure may introduce **extraneous solutions.** Other equations may sometimes by reduced to **quadratic form** and solved.

7-5 GRAPHING QUADRATIC POLYNOMIALS

The graphs of $y = x^2$ and $y = -x^2$ are examples of **parabolas,** the first opening up, the second opening down. The origin is the **vertex** of either of these parabolas. The graph of $y = ax^2 + bx + c$ is also a parabola. The parabola opens up if $a > 0$, down if $a < 0$, and has vertex at

$$x = \frac{-b}{2a} \qquad y = -\frac{b^2 - 4ac}{4a}$$

The x intercepts of the parabola are the solutions of the equation $ax^2 + bx + c = 0$. The parabola is symmetric about its **axis,** the line $x = -b/2a$.

7-6 COMPLETING THE SQUARE AND GRAPHING (OPTIONAL)

The equation $y = ax^2 + bx + c$ can be rewritten in the form $y = a(x - h)^2 + k$ by completing the square. This yields

$$h = -\frac{b}{2a} \qquad \text{and} \qquad k = -\frac{b^2 - 4ac}{4a}$$

The graph then can be obtained from the graph of $y = x^2$. It is stretched if $|a| > 1$, or flattened if $|a| < 1$, and turned over the x axis if $a < 0$. The graph is shifted $|h|$ units right if $h > 0$, $|h|$ units left if $h < 0$, $|k|$ units up if $k > 0$, and $|k|$ units down if $k < 0$.

7-7 CONIC SECTIONS: CIRCLES AND PARABOLAS

Circles, parabolas, ellipses, and hyperbolas are the principal **conic sections**—geometric figures obtained by intersecting a plane and a cone.

The **distance between two points** (x_1, y_1) and (x_2, y_2) in the plane is given by

$$d = \sqrt{(x_2 - x_1)^2 + (y_2 - y_1)^2}$$

A **circle** is the set of all points a fixed distance **(radius)** from a fixed point **(center).** The equation of a circle of radius R with center (h, k) is $(x - h)^2 + (y - k)^2 = R^2$. If the center is the origin, this becomes $x^2 + y^2 = R^2$. A **parabola** is the set of all points equidistant from a fixed point **(focus)** and a fixed line **(directrix).** The straight line through the focus and perpendicular to the directrix is the axis of the parabola. A parabola in standard form has equation

$$x = ky^2 \qquad \text{Opens horizontally}$$

or

$$y = kx^2 \qquad \text{Opens vertically}$$

7-8 CONIC SECTIONS: ELLIPSES AND HYPERBOLAS

An **ellipse** is the set of all points such that the sum of the distance from each point to two fixed points **(foci;** each separately is a **focus)** is constant. An ellipse in standard form has equation

$$\frac{x^2}{m^2} + \frac{y^2}{n^2} = 1$$

A **hyperbola** is the set of all points such that the absolute value of the difference of the distances from each point to two fixed points **(foci)** is constant. A hyperbola in standard form as equation

$$\frac{x^2}{m^2} - \frac{y^2}{n^2} = 1 \qquad \text{Opens horizontally}$$

or

$$\frac{y^2}{n^2} - \frac{x^2}{m^2} = 1 \qquad \text{Opens vertically}$$

The lines $y = \pm \dfrac{n}{m}x$ are **asymptotes** for the hyperbola.

7-9 NONLINEAR INEQUALITIES

A **quadratic inequality** comparing $ax^2 + bx + c$ to 0 may be solved by factoring the quadratic expression and checking the signs of the factors. The points where the factors are 0 are called **critical points. Sign analysis** also can be applied to other inequalities that compare a product or quotient of linear factors to 0. Solutions of such inequalities can be written compactly in interval notation using set unions and intersections. The **union $A \cup B$** of two intervals A and B is the set of all points in A or B (or both); the **intersection $A \cap B$** is the set of all points common to both A and B.

CHAPTER REVIEW EXERCISE

Work through all the problems in this chapter review and check answers in the back of the book. Answers to all the problems are there, and following each answer is a number in italics indicating the section in which that type of problem is discussed. Where weaknesses show up, review appropriate sections in the text.

A **1.** Write $4x = 2 - 3x^2$ in standard form, $ax^2 + bx + c = 0$, and identify a, b, and c.

2. Write down the quadratic formula associated with the standard form $ax^2 + bx + c = 0$.

Find all solutions by factoring or square root methods.

3. $x^2 - 3x = 0$

4. $x^2 = 25$

5. $x^2 - 5x + 6 = 0$

6. $x^2 - 2x - 15 = 0$

7. $x^2 - 7 = 0$

8. $x^2 - 6x + 8 = 0$

9. $x^2 - 1 = 0$

10. $x^2 - x - 2 = 0$

In Problems 11–24, solve the equation or inequality.

11. $x^2 + 3x + 1 = 0$

12. $x^2 + 3x - 28 = 0$

13. $x^2 - 5x - 6 = 0$

14. $3x^2 + 13x + 14 = 0$

15. $2x^2 - 9x + 10 = 0$

16. $x^2 + 9x + 20 = 0$

17. $x^2 + 3x - 28 < 0$

18. $x^2 - 5x + 6 > 0$

19. $x^2 + x < 20$

20. $x^2 + x \geq 20$

21. $\sqrt{x + 3} = 8$

22. $\sqrt{x + 5} = x - 1$

23. $x = \sqrt{3x + 4}$

24. $x = \sqrt{2 - x}$

25. Find the discriminant of $7x^2 - 11x + 5 = 0$ and use it to determine the nature and number of roots.

26. What is the distance between the two points $(1, 3)$ and $(3, 7)$?

27. What is an equation of a circle with radius 5 and center at the origin?

28. Write an equation of a circle in the form $(x - h)^2 + (y - k)^2 = R^2$ if its center is at $(-3, 4)$ and it has a radius of 7.

29. Find the center and radius of the circle with equation $x^2 + y^2 - 2x = 0$.

30. Write as a single interval:
(A) $[-1, 5) \cup (0, 10)$ (B) $[-1, 5) \cap (0, 10)$

In Problems 31–43, graph the equation.

31. $y = x^2 + 2$

32. $y = x^2 + 2x + 2$

33. $y = -x^2 + 3x - 2$

34. $x^2 + y^2 = 9$

35. $x^2 - y^2 = 9$

36. $x^2 - 9y^2 = 9$

37. $x^2 + 9y^2 = 9$

38. $y^2 = 1 - x^2$

39. $x^2 + y^2 = 49$

40. $(x - 2)^2 + (y + 3)^2 = 16$

41. $y^2 = -2x$

42. $\dfrac{x^2}{9} + \dfrac{y^2}{16} = 1$

43. $\dfrac{y^2}{16} - \dfrac{x^2}{9} = 1$

44. Find two positive numbers whose product is 27 if one is 6 more than the other.

45. Three more than a certain number is equal to the sum of twice the number and the reciprocal of the number. Find the number.

46. A rectangle with length 3 greater than its width has area 108 square feet. Find its dimensions.

47. Two ships are traveling on courses at a right angle to each other after passing. One is steaming at a rate 5 miles per hour faster than the other. Two hours after passing, the ships are 50 miles apart. What are their speeds?

B *Find all solutions.*

48. $10x^2 = 20x$

49. $3x^2 = 36$

50. $3x^2 + 27 = 0$

51. $(x - 2)^2 = 16$

52. $3t^2 - 8t - 3 = 0$

53. $2x = \dfrac{3}{x} - 5$

54. $2x^2 - 3x + 6 = 0$

55. $x^3 - 4x = 0$

56. $x^3 - 4x^2 + 4x = 0$

57. $8x^2 - 9x - 10 = 0$

58. $x^4 - 3x^2 = 0$

59. $3x^2 = 2(x + 1)$

60. $2x(x - 1) = 3$

61. $2x^2 - 2x = 40$

62. $\dfrac{8m^2 + 15}{2m} = 13$

63. $m^2 + m - 1 = 0$

64. $u + \dfrac{3}{u} = 2$

65. $\sqrt{5x - 6} - x = 0$

66. $8\sqrt{x} = x + 15$

67. $m^4 + 5m^2 - 36 = 0$

68. $2x^{2/3} - 5x^{1/3} - 12 = 0$ **69.** $\sqrt{x} + \sqrt{x + 5} = 5$

70. $\sqrt{x - 1} + \sqrt{x - 4} = 3$ **71.** $2 + \sqrt{x - 3} = 1$

72. $\dfrac{3}{\sqrt{x - 4}} = \sqrt{x - 12}$ **73.** $x^2 \geq 4x + 21$

74. $\dfrac{1}{x} < 2$

75. $10x > x^2 + 25$

76. $x^2 + 16x \geq 8x$

In Problems 77–80, graph the equation.

77. $y = (x - 2)^2 + 3$ **78.** $y = -(x + 1)^2 - 2$

79. $4x^2 + 25y^2 = 100$ **80.** $4x^2 - 25y^2 = 100$

81. Rewrite $x^2 - 6x - 3 = 0$ in the form $(x - h)^2 = k$.

82. Find an equation of a circle with center at the origin that passes through $(12, -5)$.

83. Graph $y = 3x^2 + 5x - 2$ and find all values where the graph crosses the x axis.

84. Transform the equation

$$x^2 + y^2 + 6x - 8y = 0$$

into the form

$$(x - h)^2 + (y - k)^2 = R^2$$

Since the graph is a circle, what is its radius and what are the coordinates of its center?

85. Find the equations of the asymptotes for the hyperbola $3x^2 - 9y^2 = 36$.

86. The perimeter of a rectangle is 22 inches. If its area is 30 square inches, find the length of each side.

87. A boy can paint a picket fence in an hour less than it would take his friend to do the job. Together they can paint the fence in 3 hours and 44 minutes. How long would it take each separately to do the job?

C *Solve using any method.*

88. $(t - \tfrac{3}{2})^2 = -\tfrac{3}{2}$ **89.** $3x - 1 = \dfrac{2(x + 1)}{x + 2}$

90. $y^8 - 17y^4 + 16 = 0$

91. $\sqrt{y - 2} - \sqrt{5y + 1} = -3$

92. $\sqrt{x - 3} + \sqrt{x + 2} = \sqrt{3x + 4}$

93. $x^4 - 1 = 0$

94. $x^{4/3} - 1 = 0$

In Problems 95–97, graph the solution.

95. $\dfrac{3}{x - 4} \leq \dfrac{2}{x - 3}$ **96.** $\dfrac{x^2 + 1}{x - 1} > 0$

97. $\dfrac{x^2 - 1}{x^2 + 2x} < 0$

98. If the hypotenuse of a right triangle is 15 centimeters and its area is 54 square centimeters, what are the lengths of the two sides? [*Hint:*   If x represents one side, use the Pythagorean theorem to express the other side in terms of x; then use the formula for the area of a triangle, $A = \tfrac{1}{2}bh$.]

99. Cost equations for manufacturing companies are often quadratic in nature. (At very high or very low outputs the costs are more per unit because of inefficiency of plant operation at these extremes.) If the cost equation for producing paint is $C = x^2 - 10x + 31$, where C is the cost of producing x gallons per week (both in thousands), find:
(A) The output for a $15,000 weekly cost
(B) The output for a $6,000 weekly cost

100. A farmer selling sweet corn at a roadside stand can sell 50 dozen ears of corn at a price of $1.50 per dozen. For each $0.25 the price is raised, the farmer expects sales to decrease by 4 dozen ears of corn. What price will yield a revenue of $82.50?

CHAPTER PRACTICE TEST

The following practice test is provided for you to test your knowledge of the material in this chapter. You should try to complete it in 50 minutes or less. Answers in the back of the book indicate the section in the text where the material in the question is covered. Actual tests in your class may vary from this practice test in difficulty, length, or emphasis, depending on the goals of your course or instructor.

In Problems 1–7, solve the equation or inequality.

1. $x^2 - 5x = 0$

2. $x^2 - 9 = 0$

3. $x^2 - 5x - 9 = 0$

4. $\sqrt{x - 5} = x - 7$

5. $x^2 - 5x < 0$

6. $x^4 - 5x^2 - 6 = 0$

7. $x(x + 1)(x - 4) \geq 0$

8. Rewrite $x^2 + 4x + 7 = 0$ in the form

$$(x - h)^2 + k = 0$$

9. Find the center and radius of the circle having the equation

$$x^2 + y^2 - 4x + 6y + 3 = 0$$

10. Find the distance between the points $(2, 3)$ and $(-4, 5)$.

Graph.

11. $y = x^2 + 4$

12. $y = 2x^2 + 3x + 4$

13. $x^2 + 4y^2 = 16$

14. $x^2 + y^2 = 4$

15. $x^2 - y^2 = 4$

16. $-4x^2 + y^2 = 16$

17. $x + 4y^2 = 0$

18. $y + 4x^2 = 9$

19. Find all numbers such that the sum of the number and its reciprocal is equal to twice the number.

20. A rectangle with area 80 square centimeters has length equal to 1 centimeter more than 3 times its width. Find the dimensions.

CUMULATIVE REVIEW EXERCISES, CHAPTERS 1–7

This set of exercises reviews the major concepts and techniques of Chapters 1–7. Work through all the problems and check answers in the back of the book. Answers to all the problems are there, and following each answer is a number in italics indicating the section in which that type of problem is discussed. Where weaknesses show up, review appropriate sections in the text.

A *Translate each statement into an algebraic equation or inequality statement using x as the only variable. Do not solve the equation or inequality.*

1. The square of the quantity that is 1 more than a given number is 6 less than 5 times the number.

2. Three less than the distance between 5 and a given number is less than 2 less than the number.

3. The sum of 3 times a number and 4 times the reciprocal of the number is 5 times the quantity that is 6 more than the number.

Find, evaluate, or calculate the quantity requested.

4. $(-3)(-4) - 5(3 - 4)$

5. $-3 \cdot 4^2 - 5(3 - 4)^2$

6. $1 + 3 \times 12 \div 9 - 6$

7. The degree of the polynomial $3xy^4 + 4x^3y^3 + 5x^5y^2$

8. The LCM of 8, 12, and 18

9. The LCM of x^2, $x^2 - 3x$, and $x^2 - 9$

10. The quotient and remainder when $x^4 + x^2 + 2x + 2$ is divided by $x + 3$

11. The distance between the points -8 and 21 on the real number line

12. The distance between the points $(-8, 0)$ and $(0, 21)$ in a rectangular coordinate plane

13. $(\frac{1}{4})^{1/2} + (\frac{1}{5})^0 + (\frac{1}{6})^{-1}$

14. $\sqrt[3]{64}$

15. $64^{-2/3}$

16. $(5 - 4i)(3 - 2i)$

17. $(1 + i) + (3 - 2i) + (4 - i)$

18. $\dfrac{2 + i}{1 - i}$

19. The coordinates of points A, B, and C in the figure

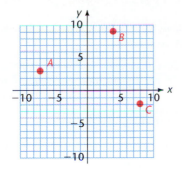

20. The slope and y intercept of the line having equation

$$y - 2 = 3(x - 4)$$

21. The center and radius of the circle having equation

$$(x - 2)^2 + (y - 3)^2 = 4$$

22. The slope of the line through the points $(-1, 4)$ and $(7, -10)$

23. The vertex of the parabola having equation

$$y = 3(x - 4)^2 + 5$$

Rewrite in the form indicated.

24. $x - [1 - 2(x - 3)]$ without grouping symbols

25. $x - 2y - 3z$ in the form $x - (?)$

26. $2xy + 3yz + 4xz + 5zx + 6yx + 7zy$ with like terms combined

27. $(x - 7)(x - 2)$ as a polynomial

28. $(x + 11)^2$ as a polynomial

29. $(x^2 - 3x + 2) - (x^3 - x^2 - 2x - 3)$ as a polynomial

30. $(x + 1)(x^2 + 2x + 3)$ as a polynomial

31. $x^3 x^{-9}$ as a power of x

32. $(xy)^3 \left(\dfrac{x}{y}\right)^4$ as a power of x times a power of y

33. $(x^2)^3 (x^{-3})^4 (x^{-5})^{-5}$ as a power of x

34. $6x + 3y + 9z$ in factored form

35. $x^2 + 8x + 16$ in factored form

36. $x^2 + 4x - 5$ in factored form

37. $x^2 + 2x - 15$ in factored form

38. $2x^2 + 5x - 3$ in factored form

39. $x^2 - 36$ in factored form

40. $x^3 + 1$ in factored form

41. $x^3 + 6x^2 + 8x$ in factored form

42. $x^2 + 2xy + 3x + 6y$ in factored form

43. $\dfrac{3ab^2}{5cd^2}$ as a fraction with denominator $10abc^2d^2$

44. $\dfrac{12a^2bc^3}{15ab^2c^2}$ reduced to lowest terms

45. $\dfrac{1}{x} + \dfrac{x^2}{3} \cdot \dfrac{x + 1}{x^3}$ as a fraction

46. $\dfrac{1}{x} + \dfrac{2}{x^2} + \dfrac{3}{x^3}$ as a fraction

47. $\dfrac{1 + \dfrac{1}{x}}{1 - \dfrac{1}{x}}$ as a simple fraction

48. $\dfrac{x^2 + 2x + 1}{x^2 - 1} \cdot \dfrac{x^2 - 2x + 1}{x^2 - x - 2}$ as a fraction in lowest terms

49. $0.000\,000\,032$ in scientific notation

50. 1.234×10^{-8} in standard decimal notation

Solve for x.

51. $3x - 5 = 7 - 9(x - 11)$ **52.** $3x - 5y = 7$

53. $(3x - 5)(x - 7) = 0$

54. $\dfrac{1}{x + 2} + \dfrac{x}{x + 1} = \dfrac{1}{2}$

55. $5x - 3 < 1 - 2(x - 3)$ **56.** $|2x + 3| \le 4$

57. $x^2 - 7x = 0$ **58.** $x^2 - 7 = 0$

59. $x^2 - 7x + 12 = 0$ **60.** $\sqrt{x + 5} = x - 7$

61. $x^4 - 2x^2 + 1 = 0$ **62.** $x^2 - 6x + 5 > 0$

Graph.

63. $-3 \le x < 2$ (on a real number line)

64. $y = 3x - 5$ **65.** $4x + 3y = 12$

66. $x + 5y \le 10$ **67.** $x + y \ge 1$

68. $y = x^2 + 3x - 4$ **69.** $x^2 + y^2 = 9$

70. $\dfrac{x^2}{9} - y^2 = 1$ **71.** $x^2 - y = 1$

72. $x^2 + \dfrac{y^2}{9} = 1$

Set up an appropriate equation or inequality and solve.

73. The product of two consecutive positive integers is 6 greater than the product of 6 times their sum. Find the integers.

74. A rectangle has length equal to 3 more than twice its width. If the rectangle has a perimeter of 42 feet, find its dimensions.

75. A law school accepts 60% of its applicants, and 80% of those accepted attend the school. If a first-year class has 336 students enrolled, how many applicants were there?

76. A bus leaves a city at 1 P.M., traveling at an average speed of 56 miles per hour. An hour later a charter bus leaves, traveling along the same route. If the charter bus averages 64 miles per hour, how long will it take to overtake the first bus?

77. A volunteer in a fundraising telethon has been successful in 25 out of the first 60 phone calls. How many consecutive successful calls must the volunteer make to raise the success rate to at least 60%?

B *Rewrite in the form indicated.*

78. $3x^2 - 13x + 14$ in factored form

79. $\dfrac{1}{x} + \dfrac{2}{x+1} + \dfrac{2}{x+3}$ as a fraction

80. $(-3 + \sqrt{5})^2 + 6(-3 + \sqrt{5}) + 4$ in simplified form

81. $\dfrac{(x+1)^2}{x^2 + 3x + 2} \cdot \dfrac{x^2 + 5x + 6}{x^2 + 4x + 3}$ as a fraction reduced to lowest terms

82. $\dfrac{x^4 - 8x}{x^2 - 2x}$ as a fraction reduced to lowest terms

Solve for x.

83. $3x^2 + 4x - 5 = 0$

84. $|3x + 4| \geq 5$

85. $\sqrt{3x + 4} + \sqrt{4x - 3} = 10$

86. $\dfrac{3x + 4}{(4x - 3)(x - 5)} \geq 0$

87. $\dfrac{1}{x} + \dfrac{1}{y} = 1$

88. $2x^2 - 9x + 10 = 0$

Graph.

89. The line passing through the point $(2, 3)$ and perpendicular to the line passing through the points $(4, 5)$ and $(-6, 7)$

90. $y = 2(x - 3)^2 + 4$

91. $x + 3y \leq 9$
  $x \geq 0$
  $y \geq 0$

92. $0 \leq x \leq 4$
  $-4 \leq y \leq 2$

In Problems 93–95, set up an appropriate equation or inequality and solve.

93. A developer sells lots with 20,000 square feet for $1.20 per square foot. For smaller lots, the price is $0.05 per square foot lower for each 1,000 square feet less than 20,000 square feet. If a lot sells for $14,250, how large is it?

94. A canoeist takes twice as long to paddle 15 miles upstream as to return the same distance downstream. If the canoeist can paddle 5 miles per hour in still water, how fast is the current?

95. If a rectangle that is 4 feet longer than it is wide has both the length and width increased by 2 feet, the resulting rectangle has area 1.6 times as large as the original rectangle. Find the original dimensions.

C 96. Solve: $4x^3 - 17x^2 + 18x = 0$

97. Rewrite as a simple fraction:

$$1 + \cfrac{1}{x + \cfrac{1}{x^{-1} + 1}}$$

98. Show that $1 + i$ is a solution to $x^3 - 3x^2 + 4x - 2 = 0$.

99. Evaluate: $(27^{-2/3} + 3i)[(-8)^{4/3} - i]^{-1}$

100. The roof shown in the figure has "pitch" equal to the slope of the line segment representing the roof line and passing through the origin of the superimposed coordinate system. The house itself is 39 feet deep and the roof overhangs the house by 1.5 feet. The attic of the house is 7 feet high. What is the pitch of the roof?

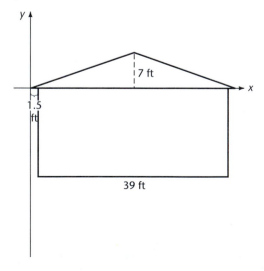

8

Systems of Equations and Inequalities

Up to this point we have dealt primarily with individual equations or inequalities. In this chapter, we will turn our attention to *systems* of equations or inequalities. For example,

$$\begin{array}{lll} (1) \quad x + 3y = 7 & (2) \quad x + 3y \le 7 & (3) \quad x^2 + y^2 = 25 \\ \quad\;\; 5x - y = 3 & \quad\;\; 5x - y < 3 & \quad\;\;\; x + 2y = 1 \end{array}$$

To solve such a system means to find all pairs of real numbers (x, y) that satisfy each equation or inequality in the system.

Systems like (1) are studied in Sections 8-1 through 8-5. Those like (2) are considered in Section 8-6, and those like (3) in Section 8-7.

Photo reference: see Section 8-6, Example 4.

8-1   Systems of Linear Equations in Two Variables

♦ Solution by Graphing
♦ Solution by Substitution
♦ Solution by Elimination Using Addition

Many practical problems can be solved conveniently using methods for two equations in two variables. For example, if a 12-foot-board is cut in two pieces so that one piece is 2 feet longer than the other piece, how long is each piece? We could solve this problem by using one equation with one variable, as discussed in Chapter 4. However, we can also proceed as follows, using two variables. Let

$$x = \text{Length of the longer piece}$$
$$y = \text{Length of the shorter piece}$$

Then
$$x + y = 12$$
$$x - y = 2$$

To **solve** this system is to find all the ordered pairs of real numbers that satisfy both equations at the same time. In general, we are interested in solving linear systems consisting of two linear equations in two variables, such as

$$
\begin{array}{ll}
3x + 4y = 12 & \\
6x - \ \ y = 18 & \text{or}
\end{array}
\qquad
\begin{array}{l}
3x = 2y - 1 \\
\ \ y = 3x + 7
\end{array}
$$

If we write all the variables on the left and the constant terms on the right, we obtain the standard form for such a system:

Linear System (Standard Form)

$$ax + by = m$$
$$cx + dy = n$$

a, b, c, d, m, and n are constants, with a, b, c, d not all 0; x and y are variables.

There are several methods for solving systems of this type. We will consider three that are widely used: solution by graphing, solution by substitution, and solution by elimination using addition.

♦ SOLUTION BY GRAPHING

We graph both equations on the same coordinate system. The coordinates of any points that the graphs have in common must be solutions to the system since they must satisfy both equations.

Example 1
Solving by Graphing

Solve by graphing:

$$x + y = 12$$
$$x - y = 2$$

Solution   Graph each equation and find coordinates of points of intersection, if they exist:

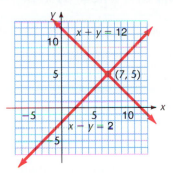

The solution appears to be $x = 7$, $y = 5$, or $(x, y) = (7, 5)$.

Check

$$7 + 5 = 12$$
$$7 - 5 = 2$$

Matched Problem 1

Solve by graphing: $x + y = 10$
$ x - y = 6$

It is clear that the system in Example 1 has exactly one solution since the lines have exactly one point of intersection. In general, two lines in the same rectangular coordinate system must be related to each other in one of three ways: (1) they intersect at one and only one point, (2) they are parallel, or (3) they coincide. Example 2 illustrates all three possibilities.

Example 2
Three Kinds
of Solutions

Solve each of the following systems by graphing:

(A) $2x - 3y = 2$ **(B)** $4x + 6y = 12$ **(C)** $2x - 3y = -6$
$x + 2y = 8$ $2x + 3y = -6$ $-x + \frac{3}{2}y = 3$

Solution

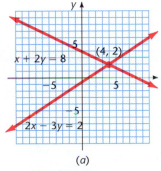

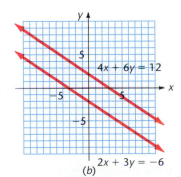

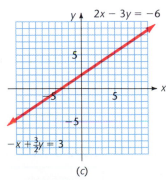

(a)

Lines intersect at one
point only:
Exactly one solution
$x = 4$, $y = 2$

(b)

Lines are parallel:
No solution
Such a system is
called **inconsistent**.

(c)

Lines coincide:
Infinite number of solutions
Such a system is
called **dependent**.

Now we know exactly what to expect when solving a system of two linear equations in two variables:

Possible Solutions to a Linear System

1. Exactly one pair of numbers
2. No solution
3. Infinitely many solutions

Matched Problem 2   Solve each of the following systems by graphing:

(A) $2x + 3y = 12$ **(B)** $x - 3y = -3$ **(C)** $2x - 3y = 12$
$\quad\quad x - 3y = -3$ $-2x + 6y = 12$ $-x + \frac{3}{2}y = -6$

Generally, graphing methods give us only rough approximations of solutions. The methods of substitution and elimination using addition, which we consider next, will yield results to any accuracy desired, when solutions exist.

◆ **SOLUTION BY SUBSTITUTION**

This method proceeds as follows:

Solving a System by Substitution

1. Choose one of the two equations in the system and solve for one variable in terms of the other. It will generally be easier to choose an equation that avoids getting involved with fractions, if possible.
2. Substitute the result into the other equation and solve the resulting linear equation in one variable.
3. Substitute this result from step 2 back into either of the original equations to find the second variable.

An example should make the process clear.

Example 3
Solving by Substitution

Solve by substitution:

$$2x - 3y = 7 \quad\quad (1)$$
$$-3x + y = -7 \quad\quad (2)$$

Solution

$$-3x + y = -7$$

Solve Equation (2), the simplest choice, for y in terms of x. Then substitute the result in the other equation.

$$y = \underbrace{3x - 7}$$

$$2x - 3\overset{\downarrow}{y} = 7$$
$$2x - 3(\mathbf{3x - 7}) = 7$$
$$2x - 9x + 21 = 7$$
$$-7x = -14$$
$$x = 2$$

Substitute $x = 2$ into one of the original equations or, more easily, into Equation (2) rewritten as $y = 3x - 7$ and solve for y.

$$y = 3 \cdot \mathbf{2} - 7$$
$$\mathbf{y = -1}$$

Thus, $(2, -1)$ is a solution to the original system, as we can readily check.

Check $(2, -1)$ must satisfy *both* equations:

Equation (1): $2(\mathbf{2}) - 3(\mathbf{-1}) = 4 + 3 = 7$
Equation (2): $-3(\mathbf{2}) + (\mathbf{-1}) = -6 - 1 = -7$

Matched Problem 3   Solve by substitution and check: $3x - 4y = 18$
$2x + y = 1$

◆   **SOLUTION BY ELIMINATION USING ADDITION**

Now we turn to another method, called **elimination using addition.** This is probably the most important method of solution, since it is readily generalized to higher-order systems. We will see this in Section 8-3. The method involves the replacement of systems of equations with simpler *equivalent systems,* by performing appropriate operations, until we obtain a system with an obvious solution. **Equivalent systems** of equations are, as you would expect, systems that have exactly the same solution set. The following result lists three operations that produce equivalent systems.

Equivalent Systems

Equivalent systems of equations result if:

(A) Two equations are interchanged.
(B) An equation is multiplied by a nonzero constant.
(C) A constant multiple of another equation is added to a given equation.

The method of solving systems of equations by elimination using addition is illustrated by the following examples.

<table>
<tr>
<td>

Example 4
Solving by Elimination

</td>
<td>

Solve by elimination using addition:

$$3x - 2y = 8 \qquad (3)$$
$$2x + 5y = -1 \qquad (4)$$

</td>
</tr>
</table>

Solution   We use the operations listed in the box above to eliminate one of the variables and thus obtain a system with an obvious solution.

$$
\begin{array}{rl}
15x - 10y = & 40 \\
\underline{4x + 10y = -2} \\
19x \quad\quad = & 38 \\
x = & 2
\end{array}
$$

If we multiply Equation (3) by 5 and Equation (4) by 2, we make the coefficients of y opposites. If we then add, we can eliminate y.

Now substitute $x = 2$ back into either of the original equations, say Equation (4), and solve for y:

$$
\begin{array}{r}
2(2) + 5y = -1 \\
5y = -5 \\
y = -1
\end{array}
$$

Thus, $(2, -1)$ is a solution to the original system. The check is left to the reader.

Matched Problem 4   Solve by elimination using addition and check your answer. Eliminate the variable x first.

$$2x + 5y = -9$$
$$3x - 4y = -2$$

<table>
<tr>
<td>

Example 5
Solving by Elimination

</td>
<td>

Solve by elimination using addition:

$$x + 3y = 2 \qquad (5)$$
$$2x + 6y = -3 \qquad (6)$$

</td>
</tr>
</table>

Solution
$$
\begin{array}{r}
-2x - 6y = -4 \\
\underline{2x + 6y = -3} \\
0 = -7
\end{array}
$$

Multiply Equation (5) by −2 and add.

A contradiction!

Our assumption that there are values for x and y that satisfy Equations (5) and (6) simultaneously must be false. Otherwise, we have proved that $0 = -7$! If you check the slopes of the two lines represented by Equations (5) and (6), you will find that they are both $-\frac{1}{3}$. However, the y intercepts of the two lines are different. Hence, the lines are parallel and conditions have been placed on the variables x and y that are impossible to meet simultaneously. The system is, therefore, said to be **inconsistent.**

Matched Problem 5   Solve by elimination using addition:
$$3x - 4y = -2$$
$$-6x + 8y = 1$$

Example 6
Solving by Elimination

Solve by elimination using addition:

$$-2x + y = -8 \qquad (7)$$
$$x - \tfrac{1}{2}y = 4 \qquad (8)$$

Solution $-2x + y = -8$ Multiply Equation (8) by 2 and add.
$$\underline{2x - y = 8}$$
$$0 = 0$$

Both sides have been eliminated. Actually, if we had multiplied Equation (8) by -2, we would have obtained Equation (7). When one equation is a constant multiple of the other, the system is said to be **dependent,** and their graphs coincide. The two equations represent the same line, $y = 2x - 8$. Thus, there are infinitely many solutions to the system—any solution of one equation will be a solution of the other. One way of expressing all solutions is to simply give the equation of the line that is the solution set, $y = 2x - 8$. Another is to look at the set of all pairs that result: $(x, 2x - 8)$. If $x = 1$, for example, then $y = 2x - 8 = 2 - 8 = -6$ and $(1, -6)$ is a solution. We also could have chosen to solve for x in terms of y. The solution would then be described as the set of all pairs $(\tfrac{1}{2}y + 4, y)$.

Matched Problem 6   Solve by elimination using addition: $6x - 3y = -2$
$$-2x + y = \tfrac{2}{3}$$

Answers to Matched Problems

1. $x = 8$, $y = 2$
2. **(A)** Graphs cross at $x = 3$ and $y = 2$.
   (B) Graphs do not cross; no solution.
   (C) Graphs coincide; infinitely many solutions.
3. $(2, -3)$ 4. $(-2, -1)$ 5. No solution
6. $(x, 2x + \tfrac{2}{3})$ is a solution for x any real number.

EXERCISE 8-1

A *Solve by graphing.*

1. $3x - 2y = 12$
$7x + 2y = 8$

2. $x + 5y = -10$
$-5x + y = 24$

3. $3x + 5y = 15$
$6x + 10y = -5$

4. $3x - 5y = 15$
$x - \tfrac{5}{3}y = 5$

5. $2x + 3y = 11$
$3x + 2y = 9$

6. $4x + y = 5$
$6x - 2y = 8$

7. $x - 5y = 15$
$2x + y = 8$

8. $6x - y = 7$
$x + 2y = 12$

9. $-2x + y = 10$
$x + 2y = 5$

10. $3x - y = 10$
$5x + y = -2$

11. $2x - 3y = 8$
$-5x + y = 6$

12. $4x + 3y = -2$
$x - 3y = 7$

Solve by substitution.

13. $2x + y = 6$
$-x + y = 3$

14. $m - 2n = 0$
$-3m + 6n = 8$

15. $3x - y = -3$
$5x + 3y = -19$

16. $2m - 3n = 9$
$m + 2n = -13$

17. $2x - 3y = 9$
$x + 4y = 10$

18. $x + 7y = 5$
$2x + 3y = 7$

19. $2x + y = -1$
$x + 2y = 4$

20. $-x + y = 5$
$-5x + 2y = -2$

21. $x - y = 9$
$x + y = 1$

22. $-x + y = 7$
$x + y = -1$

23. $8x + y = 0$
$7x - y = 15$

24. $3x + 2y = 12$
$x + 4y = 14$

Solve by elimination using addition.

25. $3p + 8q = 4$
$15p + 10q = -10$

26. $3x - y = -17$
$5x + 3y = -19$

27. $6x - 2y = 18$
$-3x + y = -9$

28. $4m + 6n = 2$
$6m - 9n = 15$

29. $3x - 2y = 5$
$-5x + 2y = -3$

30. $2x + 5y = 3$
$7x - 2y = -9$

31. $-3x + 2y = 16$
$2x + 3y = 11$

32. $7x + 4y = 16$
$2x - 3y = -12$

33. $2x + 5y = -1$
$-x + y = -10$

34. $3x + 7y = 10$
$4x + 5y = 2$

35. $3x - 5y = -2$
$2x - 7y = 6$

36. $3x - 3y = 9$
$2x + 4y = -6$

B *Solve each system by all three methods: graphing, substitution, and elimination using addition.*

37. $x - 3y = -11$
$2x + 5y = 11$

38. $5x + y = 4$
$x - 2y = 3$

39. $11x + 2y = 1$
$9x - 3y = 24$

40. $2x + y = 0$
$3x + y = 2$

Use any of the methods discussed in this section to solve each system.

41. $y = 3x - 3$
$6x = 8 + 3y$

42. $3m = 2n$
$n = -7 - 2m$

43. $\frac{1}{2}x - y = -3$
$-x + 2y = 6$

44. $y = 2x - 1$
$6x - 3y = -1$

45. $2x + 3y = 2y - 2$
$3x + 2y = 2x + 2$

46. $2u - 3v = 1 - 3u$
$4v = 7u - 2$

47. $x - 4y = -1$
$3x + 8y = 7$

48. $5x + 7y = 1$
$-10x + 3y = -2$

49. $3x + 2y = 3$
$-6x + 3y = 8$

50. $x + y = 1$
$2x + 6y = 2$

51. $2x + 3y = 2$
$4x - 9y = -1$

52. $3x - 12y = -2$
$x + y = 1$

53. $4x + 3y = 1$
$-8x + 18y = 14$

54. $2x + 2y = -1$
$3x + y = 1$

C **55.** $0.2x - 0.5y = 0.07$
$0.8x - 0.3y = 0.79$

56. $0.5m + 0.2n = 0.54$
$0.3m - 0.6n = 0.18$

57. $\frac{1}{4}x - \frac{2}{3}y = -2$
$\frac{1}{2}x - y = -2$

58. $\frac{2}{3}a + \frac{1}{2}b = 2$
$\frac{1}{2}a + \frac{1}{3}b = 1$

Estimate the solution by graphing. Then solve exactly by substitution or elimination using addition.

59. $x + y = 2$
$100x - 100y = 2$

60. $100x + 100y = 2$
$x + y = 0$

61. $y - x = -2$
$100x - 100y = 202$

62. $100x + 99y = 300$
$x + y = 0$

63. $2x - 3y = 5$
$17x - 22.5y = 43$

64. $3x + 5y = 30$
$y = -0.6x + 5.8$

In Problems 65–68, let L_1 be a line with slope m_1, y intercept b_1, and equation $A_1x + B_1y = C_1$. Let L_2 be a line with slope m_2, y intercept b_2, and equation $A_2x + B_2y = C_2$. Determine the number of solutions to the system of equations

$$A_1x + B_1y = C_1$$
$$A_2x + B_2y = C_2$$

for the given information.

65. $m_1 \neq m_2, b_1 = b_2$

66. $m_1 = m_2, b_1 \neq b_2$

67. $m_1 = m_2, b_1 = b_2$

68. $m_1 \neq m_2, b_1 \neq b_2$

Use slopes and y intercepts to determine the number of solutions to each system. Do not solve the system.

69. $1,020x + 3,210y = 1,000$
$80x + 94y = 100$

70. $689x + 654y = 864$
$357x + 7,654y = 1,098$

71. $312x - 1,560y = 3,275$
$-45x + 225y = -625$

72. $119x - 61y = 4,291$
$-218x + 1,526y = 6,543$

73. $97x - 61y = 429$
$-679x + 427y = -3,003$

74. $83x + 115y = 7,531$
$747x + 1,035y = 67,779$

  ## 8-2 Applications: Mixture Problems

Many of the applications already considered using one-equation–one-variable methods also can be set up as systems of equations using two variables. This is particularly true of the type of application known as *mixture problems*. The examples in this section will illustrate both the one-equation–one-variable method and the approach involving systems of equations. Some problems lend themselves more naturally to one approach than the other. Experience will help you recognize this and will enable you to decide which approach you prefer when both methods work equally well.

Example 1
A Money Problem

A change machine changes dollar bills into quarters and nickels. If you received 12 coins after inserting a $1 bill, how many of each type of coin did you receive?

Solution   Let

$$x = \text{Number of quarters}$$
$$y = \text{Number of nickels}$$

Then

$$
\begin{array}{ll}
x + \ y = 12 & \color{blue}\text{Number of coins} \\
25x + 5y = 100 & \color{blue}\text{Value of coins in cents}
\end{array}
$$

$$
\begin{array}{l}
-5x - 5y = -60 \\
\underline{25x + 5y = 100} \\
\ \ 20x \ \ \ \ \ \ = 40 \\
\ \ \ \ \ \ \ \ \ \ x = 2 \qquad \color{blue}\text{Quarters}
\end{array}
$$

$$
\begin{array}{l}
x + \ y = 12 \\
2 + \ y = 12 \\
\ \ \ \ \ \ y = 10 \qquad \color{blue}\text{Nickels}
\end{array}
$$

Check
$$2 + 10 = 12 \text{ coins in all}$$
$$25 \cdot 2 + 5 \cdot 10 = 50 + 50 = 100 \text{ cents, or } \$1$$

We also could have set this up using only one variable: x for the number of quarters, $12 - x$ for the number of nickels. The equation to solve would then be

$$25x + 5(12 - x) = 100$$

Matched Problem 1

Repeat Example 1 with 8 coins received in exchange for a $1 bill.

Example 2
A Money Problem

A concert brought in $27,200 on the sale of 4,000 tickets. If tickets sold for $5 and $8, how many of each were sold?

Solution   We use a one-variable approach. Let

$$x = \text{Number of \$5 tickets sold}$$

Then

$$4,000 - x = \text{Number of \$8 tickets sold}$$

We now form an equation using the value of the tickets before and after combining the two kinds of tickets:

Value before combining = Value after combining

$$\left(\begin{array}{c}\text{Value of}\\ \text{\$5 tickets}\\ \text{sold}\end{array}\right) + \left(\begin{array}{c}\text{Value of}\\ \text{\$8 tickets}\\ \text{sold}\end{array}\right) = \left(\begin{array}{c}\text{Total value}\\ \text{of all}\\ \text{tickets sold}\end{array}\right)$$

$$5x \quad + 8(4{,}000 - x) = \quad 27{,}200$$
$$5x + 32{,}000 - 8x = 27{,}200$$
$$-3x = -4{,}800$$
$$x = 1{,}600 \qquad \text{\$5 tickets}$$
$$4{,}000 - x = 2{,}400 \qquad \text{\$8 tickets}$$

Check $(\$5)(1{,}600) + (\$8)(2{,}400) = \$27{,}200$

If we had chosen a two-variable approach, we could have set

$$x = \text{Number of \$5 tickets sold}$$
$$y = \text{Number of \$8 tickets sold}$$

and obtained the system of equations

$$x + y = 4{,}000$$
$$5x + 8y = 27{,}200$$

You should try this approach.

Matched Problem 2   Suppose you receive 40 coins (nickels and quarters) worth $4. How many of each type of coin do you have?

We next consider some mixture problems involving percent. Recall that 23% in decimal form is 0.23, 6.5% is 0.065, and so on.

Example 3
A Diet Mix Problem

A zoologist wishes to prepare a special diet that contains, among other things, 120 grams of protein and 17 grams of fat. Two available food mixes specify the following percentages of protein and fat:

Mix	Protein (%)	Fat (%)
A	30	1
B	20	5

How many grams of each mix should be used to prepare the diet mix?

Solution   Let

$$x = \text{Number of grams of mix } A \text{ used}$$
$$y = \text{Number of grams of mix } B \text{ used}$$

Set up one equation for the protein requirements and one equation for the fat requirements:

$$0.3x + 0.2y = 120 \qquad \text{Protein requirements}$$
$$0.01x + 0.05y = 17 \qquad \text{Fat requirements}$$

Multiply the top equation by 10 and the bottom equation by 100 to clear decimals. This is not necessary, but is helpful.

$$3x + 2y = 1{,}200 \qquad \text{Multiply bottom equation by } -3;$$
$$x + 5y = 1{,}700 \qquad \text{then add to eliminate } x.$$

$$\begin{aligned} 3x + 2y &= 1{,}200 \\ -3x - 15y &= -5{,}100 \\ \hline -13y &= -3{,}900 \\ y &= 300 \text{ grams} \qquad \text{Mix } B \end{aligned}$$

$$\begin{aligned} x + 5y &= 1{,}700 \\ x + 5(300) &= 1{,}700 \\ x &= 200 \text{ grams} \qquad \text{Mix } A \end{aligned}$$

The zoologist should use 200 grams of mix A and 300 grams of mix B to meet the diet requirements.

Check   Protein requirement: Protein from mix A + Protein from mix B = 30% of 200 grams + 20% of 300 grams = 60 + 60 = 120 grams, as required.

Fat requirement: Fat from mix A + Fat from mix B = 1% of 200 grams + 5% of 300 grams = 2 + 15 = 17 grams, as required.

Matched Problem 3   Repeat Example 3 for a diet mixture that is to contain 110 grams of protein and 8 grams of fat.

Example 4
An Alloy Problem

A jeweler has two bars of gold alloy in stock, one 12-carat and the other 18-carat. Carats measure the purity of gold: 24-carat is pure gold, 12-carat gold is $\frac{12}{24}$ pure, 18-carat gold is $\frac{18}{24}$ pure, and so on. How many grams of each alloy must be mixed to obtain 10 grams of 14-carat gold?

Solution   Let

$$x = \text{Number of grams of 12-carat gold used}$$
$$y = \text{Number of grams of 18-carat gold used}$$

Then

$$x + y = 10 \qquad \text{Amount of new alloy}$$
$$\tfrac{12}{24}x + \tfrac{18}{24}y = \tfrac{14}{24}(10) \qquad \text{Pure gold present before mixing equals pure gold present after mixing.}$$

$$x + y = 10 \qquad \text{The second equation has been multiplied by}$$
$$6x + 9y = 70 \qquad \tfrac{24}{2} \text{ to simplify; now solve using the methods described in Section 8-1. We use elimination here.}$$

$$\begin{aligned} -6x - 6y &= -60 \\ 6x + 9y &= 70 \\ \hline 3y &= 10 \\ y &= 3\tfrac{1}{3} \text{ grams} \qquad \text{18-carat alloy} \\ x + 3\tfrac{1}{3} &= 10 \\ x &= 6\tfrac{2}{3} \text{ grams} \qquad \text{12-carat alloy} \end{aligned}$$

The checking of the solution is left to you.

Matched Problem 4   Repeat Example 4, but suppose that the jeweler has only 10-carat and pure gold in stock.

**Example 5
A Liquid
Mixture Problem**

How many milliliters of distilled water must be added to 60 milliliters of 70% acid solution to obtain a 60% solution? A 70% acid solution is 70% pure acid and 30% distilled water.

Solution   Only one variable is apparent in this problem, namely, the amount of distilled water added. Let x = Number of milliliters of distilled water added. We illustrate the situation before and after mixing, keeping in mind that the amount of acid present before mixing must equal the amount of acid present after mixing.

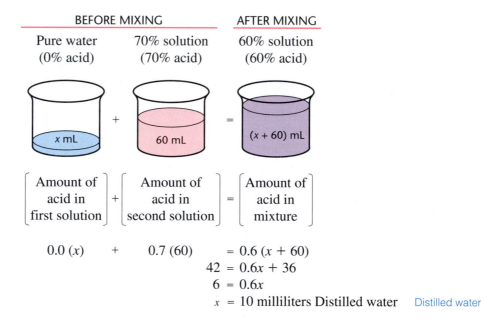

BEFORE MIXING   AFTER MIXING

Pure water   70% solution   60% solution
(0% acid)   (70% acid)   (60% acid)

x mL   60 mL $(x + 60)$ mL

$$\begin{bmatrix} \text{Amount of} \\ \text{acid in} \\ \text{first solution} \end{bmatrix} + \begin{bmatrix} \text{Amount of} \\ \text{acid in} \\ \text{second solution} \end{bmatrix} = \begin{bmatrix} \text{Amount of} \\ \text{acid in} \\ \text{mixture} \end{bmatrix}$$

$$\begin{aligned}
0.0\,(x) \quad + \quad 0.7\,(60) \quad &= 0.6\,(x + 60) \\
42 &= 0.6x + 36 \\
6 &= 0.6x \\
x &= 10 \text{ milliliters Distilled water} \qquad \text{Distilled water}
\end{aligned}$$

Check   The amount of acid in the mix is $0.0(10) + 0.7(60) = 42$ milliliters; a 60% solution should have $0.6(10 + 60) = 42$ milliliters of acid.

The information contained in the diagram in this problem also can be presented in a table:

	Amount (mL)	Acid (%)	Amount of Acid
Pure Water	x	0	0
70% Solution	60	70	0.7(60)
Mixture	$x + 60$	60	$0.6(x + 60) = 0.7(60)$

Since the mixture is to be a 60% solution, the amount of acid $0.7(60)$ must be 60% of the amount $x + 60$ in the mixture. That is, $0.7(60) = 0.6(x + 60)$.

Matched Problem 5   How many centiliters of pure alcohol must be added to 35 centiliters of a 20% solution to obtain a 30% solution? Pure alcohol is a 100% alcohol solution.

Example 6
A Liquid
Mixture Problem

A chemical stockroom has a 20% alcohol solution and a 50% alcohol solution. How many centiliters must be taken from each to obtain 24 centiliters of a 30% solution?

Solution   Here, we take a two-variable approach. Let

$$x = \text{Amount of 20\% solution used}$$
$$y = \text{Amount of 50\% solution used}$$

The information about the mixture is summarized in tabular form as follows:

	Amount (cL)	Alcohol (%)	Amount of Alcohol
20% Solution	x	20	$0.2x$
50% Solution	y	50	$0.5y$
Mixture	24	30	$0.3(24) = 0.2x + 0.5y$

The two amounts we add together must equal what is in the mixture. Thus, we get the system

$$x + y = 24 \qquad \text{Amount of mixture}$$
$$0.2x + 0.5y = 7.2 \qquad \text{Amount of alcohol}$$

Substitution yields the solution:

$$y = 24 - x$$
$$0.2x + 0.5(24 - x) = 7.2$$
$$0.2x + 12 - 0.5x = 7.2$$
$$-0.3x = -4.8$$
$$x = 16 \text{ centiliters} \qquad \text{20\% solution}$$
$$24 - x = 8 \text{ centiliters} \qquad \text{50\% solution}$$

Check $0.2(16) + 0.5(8) = 7.2$ centiliters   Amount of alcohol
             $0.3(24) = 7.2$ centiliters   30% of the mixture

The information in the table also can be shown in a diagram, as in Example 5:

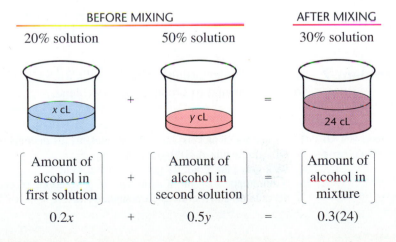

BEFORE MIXING		AFTER MIXING
20% solution 50% solution		30% solution

$$\begin{bmatrix} \text{Amount of} \\ \text{alcohol in} \\ \text{first solution} \end{bmatrix} + \begin{bmatrix} \text{Amount of} \\ \text{alcohol in} \\ \text{second solution} \end{bmatrix} = \begin{bmatrix} \text{Amount of} \\ \text{alcohol in} \\ \text{mixture} \end{bmatrix}$$

$$0.2x \qquad + \qquad 0.5y \qquad = \qquad 0.3(24)$$

Matched Problem 6   Repeat Example 6 using 10% and 40% stockroom solutions.

Example 7
A Blending Problem

A coffee shop manager wishes to blend coffee that sells for $3 per pound with coffee that sells for $4.25 per pound to produce a blend that will sell for $3.50 per pound. How much of each should be used to produce 50 pounds of the new blend?

Solution   This problem is mathematically very close to Example 6. We solve it by the one-variable method for comparison. Let

$$x = \text{Amount of \$3-per-pound coffee used}$$

Then
$$50 - x = \text{Amount of \$4.25-per-pound coffee used}$$

Value before blending = Value after blending

$$\begin{pmatrix} \text{Value of} \\ \text{\$3-per-pound} \\ \text{coffee used} \end{pmatrix} + \begin{pmatrix} \text{Value of} \\ \text{\$4.25-per-pound} \\ \text{coffee used} \end{pmatrix} = \begin{pmatrix} \text{Total value} \\ \text{of 50 pounds} \\ \text{of the blend} \end{pmatrix}$$

$$3x \quad + \quad 4.25(50 - x) \quad = \quad 3.50(50)$$
$$3x + 212.5 - 4.25x = 175$$
$$-1.25x = -37.5$$
$$x = 30 \text{ pounds} \qquad \text{\$3 coffee}$$
$$50 - x = 20 \text{ pounds} \qquad \text{\$4.25 coffee}$$

Check   ($3)(30) + ($4.25)(20) = $175   The value of the components
($3.50)(50) = $175   The value of the blend

A two-variable approach to this problem with

$$x = \text{Amount of \$3.00-per-pound coffee used}$$
$$y = \text{Amount of \$4.25-per-pound coffee used}$$

leads to the system

$$x + \quad y = 50$$
$$3x + 4.25y = 3.5(50)$$

Matched Problem 7   Repeat Example 7 using coffee that sells for $2.75 per pound and coffee that sells for $4 per pound to produce the new blend.

Answers to
Matched Problems

1. 3 quarters, 5 nickels   **2.** 30 nickels, 10 quarters
3. 300 grams mix *A*, 100 grams mix *B*
4. $2\frac{6}{7}$ grams pure gold, $7\frac{1}{7}$ grams 10-carat gold
5. 5 centiliters   **6.** 8 centiliters 10% solution, 16 centiliters 40% solution
7. 20 pounds $2.75 coffee, 30 pounds $4 coffee

EXERCISE 8-2

This set of exercises contains a variety of mixture problems classified according to subject area. The most difficult problems are marked with two stars (★★), those of moderate difficulty are marked with one star (★), and the easier problems are not marked.

Solve using either a one- or two-variable approach.

1. **Puzzle** A vending machine takes only dimes and quarters. If it contains 100 coins with a total value of $14.50, how many of each type of coin are in the machine?

2. **Puzzle** A parking meter contains only nickels and dimes. If it contains 50 coins at a total value of $3.50, how many of each type of coin are in the meter?

3. **Business** A school musical production brought in $12,600 on the sale of 3,500 tickets. If the tickets sold for $2 and $4, how many of each type were sold?

4. **Business** A concert brought in $60,000 on the sale of 8,000 tickets. If tickets sold for $6 and $10, how many of each type of ticket were sold?

5. **Chemistry** How many deciliters of alcohol must be added to 100 deciliters of a 40% alcohol solution to obtain a 50% solution?

6. **Chemistry** How many milliliters of hydrochloric acid must be added to 12 milliliters of a 30% solution to obtain a 40% solution?

7. **Puzzle** A bank gave you $1.50 in change consisting of only nickels and dimes. If there were 22 coins in all, how many of each type of coin did you receive?

8. **Puzzle** Your friend came out of a post office having spent $1.32 on thirty 4-cent and 5-cent stamps. How many of each type did he buy?

9. **Geometry** If the sum of two angles in a right triangle is 90° and their difference is 14°, find the two angles.

10. **Geometry** Find the dimensions of a rectangle with perimeter 72 inches if its length is 25% greater than its width.

11. **Chemistry** How many liters of distilled water must be added to 140 liters of an 80% alcohol solution to obtain a 70% solution?

12. **Chemistry** How many centiliters of distilled water must be added to 500 centiliters of a 60% acid solution to obtain a 50% solution?

13. **Chemistry** A chemical stockroom has a 20% alcohol solution and a 50% solution. How many deciliters of each should be used to obtain 90 deciliters of a 30% solution?

14. **Chemistry** A chemical supply company has a 30% sulfuric acid solution and a 70% sulfuric acid solution. How many liters of each should be used to obtain 100 liters of a 40% solution?

15. **Business** A tea shop manager wishes to blend tea that sells for $5 per kilogram with tea that sells for $6.50 per kilogram to produce a blend that will sell for $6 per kilogram. How much of each should be used to obtain 75 kilograms of the new blend?

16. **Business** A gourmet food store manager wishes to blend coffee that sells for $7 per kilogram with coffee that sells for $9.50 per kilogram to produce a blend that will sell for $8 per kilogram. How much of each should be used to obtain 100 kilograms of the new blend?

17. **Finance** You have inherited $20,000 and wish to invest it. If part is invested in a low-risk investment at 10% and the rest in a higher-risk investment at 15%, how much should you invest at each rate to produce the same yield as if all the money had been invested at 13%?

18. **Finance** An investor has $10,000 to invest. If part is invested in a low-risk investment at 11% and the rest in a higher-risk investment at 16%, how much should be invested at each rate to produce the same yield as if all the money had been invested at 12%?

★ 19. **Business** A packing carton contains 144 small packages, some weighing $\frac{1}{4}$ pound each and the others $\frac{1}{2}$ pound each. How many of each type are in the carton if the total contents of the carton weigh 51 pounds?

★ 20. **Biology** For a nutrition experiment, a biologist wants to prepare a special diet for experimental animals. The experiment requires a food mixture that contains, among other things, 20 ounces of protein and 6 ounces of fat. The biologist is able to purchase food mixes of the following composition:

Mix	Protein (%)	Fat (%)
A	20	2
B	10	6

How many ounces of each mix should be used to prepare the diet mix? Solve graphically and algebraically.

★ 21. **Chemistry** A chemist has two concentrations of hydrochloric acid in stock, a 50% solution and an 80% solution. How much of each should be mixed to obtain 100 milliliters of a 68% solution?

★ 22. *Business*   A newspaper printing plant has two folding machines for the final assembly of the evening newspaper, which has a circulation of 29,000 copies. The slower machine can fold papers at the rate of 6,000 per hour, and the faster machine at the rate of 10,000 per hour. If the use of the slower machine is delayed $\frac{1}{2}$ hour because of a minor breakdown, how much total time is required to fold all the papers? How much time does each machine spend on the job?

★ 23. *Earth science*   A ship using sound-sensing devices above and below water recorded a surface explosion 6 seconds sooner by is underwater device than its abovewater device. Sound travels in air at about 1,100 feet per second and in seawater at about 5,000 feet per second.
   (A) How long did it take each sound wave to reach the ship?
   (B) How far was the explosion from the ship?

★ 24. *Earth science*   An earthquake emits a primary wave and a secondary wave. Near the surface of the earth the primary wave travels at about 5 miles per second, and the secondary wave at about 3 miles per second. From the time lag between the two waves arriving at a given station, it is possible to estimate the distance to the quake. (The epicenter cen be located by obtaining distance bearings at three or more stations.) Suppose a station measured a time difference of 16 seconds between the arrival of the two waves. How long did each wave travel, and how far was the earthquake from the station?

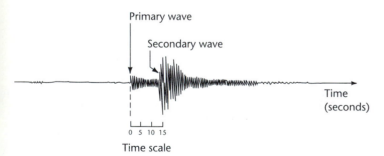

Time scale

★★ 25. *Domestic*   A 9-liter radiator contains a 50% solution of antifreeze in distilled water. How much should be drained and replaced with pure antifreeze to obtain a 70% solution?

★★ 26. *Domestic*   A 12-liter radiator contains a 60% solution of antifreeze and distilled water. How much should be drained and replaced with pure antifreeze to obtain an 80% solution?

★★ 27. *Business*   Two companies have offered you a sales position. Both jobs are essentially the same, but one company pays a straight 8% commission and the other pays $51 per week plus a 5% commission. The best sales representatives with either company rarely

have sales greater than $4,000 in any one week. Before accepting either offer, it would be helpful to know at what point both companies pay the same, and which of the companies pays more on either side of this point. Solve graphically and algebraically.

★★ 28. *Business*   Solve Problem 27 with the straight-commission company paying 7% commission and the salary-plus-commission company paying $75 per week plus 4% commission.

29. *Economics*   In a particular city, the weekly supply s and demand d for popular tape cassette recordings at price p is given by

$$d = 2,100 - 100p \qquad \$6 \le p \le \$12$$
$$s = -2,400 + 400p$$

Where is the equilibrium point? That is, at what price does the supply equal the demand?
   (A) Solve graphically.   (B) Solve algebraically.

30. *Economics*   In a particular city, the weekly supply s and demand d for popular compact disc recordings at price p is given by

$$d = 3,500 - 200p \qquad \$10 \le p \le \$15$$
$$s = -3,000 + 300p$$

Where is the equilibrium point? That is, at what price does the supply equal the demand?
   (A) Solve graphically.   (B) Solve algebraically.

31. *Business*   A company packs 110 packages, some weighing $\frac{1}{2}$ pound and some $\frac{1}{4}$ pound, in a carton for shipping. The total weight of the packages is 38 pounds. How many packages of each size are in the carton?

32. *Business*   Repeat Problem 31 for 104 packages totaling 38 pounds.

33. *Business*   Tickets to a concert cost $15 in advance and $18 at the gate. If attendance was 4,300 and total ticket receipts were $69,000, how many tickets were bought in advance? At the gate?

34. *Business*   Tickets to a minor league baseball game cost $4 for general admission and $6 for reserved seats. If attendance at a game was 2,200 and total ticket receipts were $9,400, how many tickets of each type were sold?

Solve using a two-variable approach. These are problems from Chapter 4 (Sections 4-3 and 4-5), where they were solved with a one-variable approach.

35. *Business*   If $100,000 is invested, part at an annual interest rate of 4.5% and the rest at a rate of 7%, how much should be in each part to yield $6,000 total for the year?

36. Business If $100,000 is invested, part at an annual interest rate of 4% and the rest at a rate of 8.5%, how much should be in each part to yield $6,000 total for the year?

37. Rate-Time The average speed of an express bus is 1.5 times the average speed of the local bus. If the express travels 60 miles in 40 minutes less time than the local, what are the two speeds?

38. Rate-Time The average speed of an express bus is $1\frac{1}{3}$ times the average speed of the local bus. If the express travels 40 miles in 25 minutes less time than the local, what are the two speeds?

39. Rate-Time Two clerical workers process tax forms. One can process 8 forms per hour more than the other. If the faster worker can process 286 forms in the same time the slower worker can process 260, what are their rates in forms per hour?

40. Rate-Time Two long-distance runners would be separated by $\frac{1}{2}$ mile if both ran for 1 hour at their individual average speeds. The faster runner can run $2\frac{2}{3}$ miles in the same time it takes the other to run $2\frac{1}{2}$ miles. What are their individual average speeds?

8-3 Systems of Linear Equations in Three Variables

♦ Solving Systems of Three Equations in Three Variables
♦ Application

We now proceed to systems of linear equations with three variables, that is, systems of the form

$$3x - 2y + 4z = 6$$
$$2x + 3y - 5z = -8 \qquad (1)$$
$$5x - 4y + 3z = 7$$

The triplet of numbers $x = 0$, $y = -1$, $z = 1$ is a **solution** of system (1), since each equation is satisfied by this triplet. We can also write this solution as the **ordered triplet** $(0, -1, 1)$. The set of all such ordered triplets of numbers is called the **solution set** of the system. Two systems are said to be **equivalent** if they have the same solution set.

♦ ## SOLVING SYSTEMS OF THREE EQUATIONS IN THREE VARIABLES

We will use an extension of the method of elimination by addition discussed in Section 8-1 to solve systems in the form of (1). We rely on the definition of equivalent systems and the operations that produce them given in Section 8-1.

Equivalent Systems

Equivalent systems of equations result if:

(A) Two equations are interchanged.
(B) An equation is multiplied by a nonzero constant.
(C) A constant multiple of another equation is added to a given equation.

Example 1
Solving a System in Three Variables

Solve:

$$3x - 2y + 4z = 6 \qquad (2)$$
$$2x + 3y - 5z = -8 \qquad (3)$$
$$5x - 4y + 3z = 7 \qquad (4)$$

Solution   *Step 1.*   We look at the coefficients of the variables and choose to eliminate y from Equations (2) and (4) because of the convenient coefficients -2 and -4. Multiply Equation (2) by -2 and add to Equation (4):

$$
\begin{array}{ll}
-6x + 4y - 8z = -12 & -2[\text{Equation (2)}] \\
\underline{5x - 4y + 3z = 7} & \text{Equation (4)} \\
-x \quad\quad - 5z = -5 &
\end{array}
$$

Step 2.   Now let us eliminate the same variable y from Equations (2) and (3). Multiply Equation (2) by 3 and Equation (3) by 2 and add:

$$
\begin{array}{ll}
9x - 6y + 12z = 18 & 3[\text{Equation (2)}] \\
\underline{4x + 6y - 10z = -16} & 2[\text{Equation (3)}] \\
13x \quad\quad + 2z = 2 &
\end{array}
$$

Step 3.   From steps 1 and 2 we obtain the system

$$-x - 5z = -5 \qquad (5)$$
$$13x + 2z = 2 \qquad (6)$$

We solve this system as in Section 8-1. Multiply Equation (5) by 13 and add to Equation (6) to eliminate x:

$$
\begin{array}{ll}
-13x - 65z = -65 & 13[\text{Equation (5)}] \\
\underline{13x + 2z = 2} & \\
-63z = -63 & \\
z = 1 &
\end{array}
$$

Substitute $z = 1$ back into either Equation (5) or (6)—we choose Equation (5)—to find x:

$$-x - 5z = -5$$
$$-x - 5 \cdot 1 = -5$$
$$-x = 0$$
$$x = 0$$

Step 4.   Substitute $x = 0$ and $z = 1$ back into any of the three original equations—we choose Equation (2)—to find y:

$$3x - 2y + 4z = 6$$
$$3 \cdot 0 - 2y + 4 \cdot 1 = 6$$
$$-2y + 4 = 6$$
$$-2y = 2$$
$$y = -1$$

Thus, the solution to the original system is $(0, -1, 1)$, or $x = 0$, $y = -1$, $z = 1$.

Check   To check the solution, we must check *each* equation in the original system:

$$3x - 2y + 4z = 6 \qquad\qquad 2x + 3y - 5z = -8$$
$$3 \cdot 0 - 2(-1) + 4 \cdot 1 = 6 \qquad 2 \cdot 0 + 3(-1) - 5 \cdot 1 = -8$$

$$5x - 4y + 3z = 7$$
$$5 \cdot 0 - 4(-1) + 3 \cdot 1 = 7$$

Steps in Solving Systems of Three Equations in Three Variables

Step 1. Choose two equations from the system and eliminate one of the three variables, using elimination by addition or subtraction. The result is generally one equation in two variables.

Step 2. Now eliminate the same variable from the unused equation and one of those used in step 1. We generally obtain another equation in two variables.

Step 3. The two equations from steps 1 and 2 form a system of two equations in two variables. Solve as in Section 8-1.

Step 4. Substitute the solution from step 3 into any of the three original equations and solve for the third variable to complete the solution of the original system.

Step 5. Check the solution in *each* of the three equations.

Matched Problem 1   Solve: $2x + 3y - 5z = -12$
$3x - 2y + 2z = 1$
$4x - 5y - 4z = -12$

In the process described above, if we encounter an equation that states a contradiction, such as $0 = -2$, then we must conclude that the system has no solution. We say the system is **inconsistent.** If, on the other hand, one of the equations turns out to be an identity such as $0 = 0$, the system either has infinitely many solutions or it has none. We must proceed further to determine which. Notice how this last result differs from the case of two equations in two variables. There, when we obtained $0 = 0$, we *knew* that there were infinitely many solutions. If a system has infinitely many solutions, then it is said to be **dependent.**

It is possible to interpret linear equations in three variables geometrically. The triple (x, y, z) can be used to represent a point in three-dimensional space, just as (x, y) represents a point in the plane. When this is done, a linear equation in three variables represents a plane. A system of three linear equations in three variables then represents three planes. The three planes may intersect at a point, as do the front wall, side wall, and ceiling of a room. They may intersect in a line, as do three walls that meet in a Y. Or they may not intersect at all, just as the floor, ceiling, and one wall of a room all do not meet. Thus, there is either one solution, an infinite number of solutions, or no solution. We will not pursue this geometric interpretation further in this book, but you may encounter it in later courses.

Let us look at a system that turns out to be dependent to see how the solution set can be represented. Consider the system

$$x + y - z = 2 \tag{7}$$
$$3x + 2y - z = 5 \tag{8}$$
$$5x + 2y + z = 7 \tag{9}$$

We choose to eliminate z from two equations by adding Equation (9) to Equation (7) and by adding Equation (9) to Equation (8). Doing this, we obtain the system

$$6x + 3y = 9$$
$$8x + 4y = 12$$

By multiplying the top equation by $\frac{1}{3}$ and the bottom equation by $\frac{1}{4}$, we obtain the simpler system

$$2x + y = 3 \tag{10}$$
$$2x + y = 3 \tag{11}$$

Since these two equations are the same, the original system must be dependent. In fact, if we multiply either Equation (10) or (11) by -1 and add the result to the other, we will obtain $0 = 0$. To represent the solution set of the original system, we proceed as follows. Solve Equation (10) for y in terms of x:

$$y = 3 - 2x \tag{12}$$

Now replace y by $3 - 2x$ in any of the original equations and solve for z. We use Equation (9):

$$5x + 2y + z = 7$$
$$5x + 2(3 - 2x) + z = 7$$
$$5x + 6 - 4x + z = 7$$
$$z = 1 - x$$

Thus, for *any* real number x, the ordered triplet (x, y, z) becomes

$$(x, 3 - 2x, 1 - x)$$

and is a solution of the original system. For example:

If $x = 1$, then $(\mathbf{1}, 3 - 2 \cdot \mathbf{1}, 1 - \mathbf{1}) = (1, 1, 0)$ is a solution.
If $x = -3$, then $(\mathbf{-3}, 3 - 2(\mathbf{-3}), 1 - (\mathbf{-3})) = (-3, 9, 4)$ is a solution.
And so on.

Other approaches to solving systems of three equations in three unknowns are given in the following two (optional) sections. These approaches are more readily generalized to higher-order systems.

♦ **APPLICATION**

We now consider a real-world problem that leads to a system of three equations in three variables.

Example 2
Production Scheduling

A small manufacturing plant makes three types of inflatable boats: one-person, two-person, and four-person models. Each boat requires the services of three departments, as listed in the table. The cutting, assembly, and packaging departments have available a maximum of 380, 330, and 120 workhours per week, respectively. How many boats of each type must be produced each week for the plant to operate at full capacity?

	One-Person Boat	Two-Person Boat	Four-Person Boat
Cutting Department	0.6 hour	1.0 hour	1.5 hours
Assembly Department	0.6 hour	0.9 hour	1.2 hours
Packaging Department	0.2 hour	0.3 hour	0.5 hour

Solution   Let

$$x = \text{Number of one-person boats produced per week}$$
$$y = \text{Number of two-person boats produced per week}$$
$$z = \text{Number of four-person boats produced per week}$$

The 380 workhours available in the cutting department will be used up by $0.6x$ workhour on the x one-person boats, $1.0y$ workhour on the y two-person boats, and $1.5z$ workhours on the z four-person boats, so that $380 = 0.6x + 1.0y + 1.5z$. Similar equations are obtained for the other departments:

$$0.6x + 1.0y + 1.5z = 380 \quad \text{Cutting department}$$
$$0.6x + 0.9y + 1.2z = 330 \quad \text{Assembly department}$$
$$0.2x + 0.3y + 0.5z = 120 \quad \text{Packaging department}$$

We can clear the system of decimals, if desired, by multiplying each side of each equation by 10:

$$6x + 10y + 15z = 3{,}800 \qquad (13)$$
$$6x + 9y + 12z = 3{,}300 \qquad (14)$$
$$2x + 3y + 5z = 1{,}200 \qquad (15)$$

Let us start by eliminating x from Equations (13) and (14):

$$\text{Add} \begin{cases} 6x + 10y + 15z = 3{,}800 & \text{Equation (13)} \\ -6x - 9y - 12z = -3{,}300 & -1[\text{Equation (14)}] \\ \hline y + 3z = 500 \end{cases}$$

Now we eliminate x from Equations (13) and (15):

$$\text{Add} \begin{cases} 6x + 10y + 15z = 3{,}800 \quad &\text{Equation (13)} \\ -6x - 9y - 15z = -3{,}600 \quad &-3[\text{Equation (15)}] \\ \hline y = 200 \end{cases}$$

Substituting $y = 200$ into $y + 3z = 500$ from above, we can solve for z:

$$200 + 3z = 500$$
$$3z = 300$$
$$z = 100$$

Now use Equation (13), (14), or (15) to find x. We choose Equation (15):

$$2x + 3y + 5z = 1{,}200$$
$$2x + 3(200) + 5(100) = 1{,}200$$
$$2x = 100$$
$$x = 50$$

Thus, each week the company should produce 50 one-person boats, 200 two-person boats, and 100 four-person boats to operate at full capacity. The check of the solution is left to you.

Matched Problem 2 Repeat Example 2 assuming the cutting, assembly, and packaging departments have available a maximum of 260, 234, and 82 workhours per week, respectively.

Answers to Matched Problems
1. $(-1, 0, 2)$
2. 100 one-person boats, 140 two-person boats, and 40 four-person boats

EXERCISE 8-3

A *Solve and check each system.*

1.
$x - y - 2z = 1$
$y + 4z = 7$
$z = 3$

2.
$x - 2y - 3z = 6$
$y - 2z = 8$
$z = -3$

3.
$x + y + 4z = 3$
$y - 3z = -2$
$2z = 1$

4.
$x + 2y - 2z = 2$
$2y + z = 0$
$2z = -1$

5.
$x + 2y + 9z = 4$
$3y - 6z = -1$
$3z = 2$

6.
$2x + 2y + z = 4$
$3y - z = 1$
$3z = 6$

7.
$-2x = 2$
$x - 3y = 2$
$-x + 2y + 3z = -7$

8.
$2y + z = -4$
$x - 3y + 2z = 9$
$-y = 3$

9.
$4y - z = -13$
$3y + 2z = 4$
$6x - 5y - 2z = 0$

10.
$2x + z = -5$
$x - 3z = -6$
$4x + 2y - z = -9$

B **11.**
$2x + y - z = 5$
$x - 2y - 2z = 4$
$3x + 4y + 3z = 3$

12.
$x - 3y + z = 4$
$-x + 4y - 4z = 1$
$2x - y + 5z = -3$

13.
$2a + 4b + 3c = 6$
$a - 3b + 2c = -7$
$-a + 2b - c = 5$

14.
$3u - 2v + 3w = 11$
$2u + 3v - 2w = -5$
$u + 4v - w = -5$

15.
$2x - 3y + 3z = -15$
$3x + 2y - 5z = 19$
$5x - 4y - 2z = -2$

16.
$3x - 2y - 4z = -8$
$4x + 3y - 5z = -5$
$6x - 5y + 2z = -17$

17.
$5x - 3y + 2z = 13$
$2x + 4y - 3z = -9$
$4x - 2y + 5z = 13$

18.
$4x - 2y + 3z = 0$
$3x - 5y - 2z = -12$
$2x + 4y - 3z = -4$

19.
$2x - 3y + 4z = 10$
$3x - y - z = 8$
$4x + 5y - 6z = 6$

20.
$3x + 2y + z = 5$
$4x + 5y - 6z = 4$
$5x - 4y + 3z = -5$

21.
$-x + 2y + 3z = 5$
$3x + 4y + 5z = 5$
$6x + 5y + z = 4$

22.
$2x + 3y - 4z = 2$
$4x + 5y + 6z = -10$
$3x - 5y - 7z = 4$

23.
$$x - 8y + 2z = -1$$
$$x - 3y + z = 1$$
$$2x - 11y + 3z = 2$$

24.
$$-x + 2y - z = -4$$
$$4x + y - 2z = 1$$
$$x + y - z = -1$$

25.
$$x + 2y + 3z = 4$$
$$2x + 3y + 4z = -5$$
$$3x + 5y + 7z = 6$$

26.
$$x - 4y + 7z = 1$$
$$5x - 4y - 3z = 2$$
$$6x + 5y + 4z = 3$$

C *To solve a system of four equations in four variables, use elimination to obtain a system of three equations in three variables. Then solve the resulting system by elimination.*

27.
$$w - x + y + z = 0$$
$$x - y - z = 1$$
$$y + z = 1$$
$$z = 4$$

28.
$$w + 2x + 3y + 4z = -3$$
$$x + 2y + 3z = -4$$
$$y + 2z = -4$$
$$z = -3$$

29.
$$w + 2x + 3y + 4z = 5$$
$$x + 2y + 3z = 4$$
$$y + 2z = 3$$
$$z = 2$$

30.
$$w - 2x + 3y - 4z = 5$$
$$x - 2y + 3z = 4$$
$$y - 2z = 3$$
$$z = 2$$

31.
$$w - 2x + 3y - 4z = 5$$
$$w - x + y - z = 9$$
$$w - 2x + 4y - 5z = 10$$
$$x - y + 2z = 9$$

32.
$$w + 2x + 3y + 4z = 5$$
$$w + x + y + z = 1$$
$$x + 3y + 5z = 7$$
$$x + 2y + 4z = 6$$

33.
$$4w - x = 5$$
$$-3w + 2x - y = -5$$
$$2w - 5x + 4y + 3z = 13$$
$$2w + 2x - 2y - z = -2$$

34.
$$2r - s + 2t - u = 5$$
$$r - 2s + t + u = 1$$
$$-r + s - 3t - u = -1$$
$$-r - 2s + t + 2u = -4$$

The following systems have an infinite number of solutions. Describe them as ordered triples involving only the variable z.

35.
$$x + 2y + 3z = 4$$
$$2x - 3y + 4z = -5$$
$$4x + y + 10z = 3$$

36.
$$x - y + z = -1$$
$$2x + y - 3z = -2$$
$$x + 5y - 6z = 1$$

37.
$$x + y + z = 4$$
$$-x + y - z = 2$$
$$x + 5y + z = 16$$

38.
$$x - y = 1$$
$$x + z = 10$$
$$y + z = 9$$

APPLICATIONS

39. *Geometry* A circle in a rectangular coordinate system can be written in the form $x^2 + y^2 + Dx + Ey + F = 0$. Find D, E, and F so that the circle passes through $(-2, -1)$, $(-1, -2)$, and $(6, -1)$.

40. *Geometry* Repeat Problem 39 with the circle passing through $(6, -8)$, $(6, 0)$, and $(0, -8)$.

41. *Geometry* The points $(1, 2)$, $(3, 6)$, and $(-2, 11)$ all lie on the graph of the parabola $y = ax^2 + bx + c$. Find a, b, and c.

42. *Geometry* Repeat Problem 41 for the points $(-1, 0)$, $(2, 3)$, and $(4, 11)$.

★ **43.** *Production scheduling* A garment company manufactures three shirt styles. Each style of shirt requires the services of three departments, as listed in the table. The cutting, sewing, and packaging departments have available a maximum of 1,160, 1,560, and 480 workhours per week, respectively. How many of each style shirt must be produced each week for the plant to operate at full capacity?

Department	Style A	Style B	Style C
Cutting	0.2 h	0.4 h	0.3 h
Sewing	0.3 h	0.5 h	0.4 h
Packaging	0.1 h	0.2 h	0.1 h

★ **44.** *Production scheduling* Repeat Problem 43 with the cutting, sewing, and packaging departments having available a maximum of 1,180, 1,560, and 510 workhours per week, respectively.

★ **45.** *Diet* For an experiment involving guinea pigs, a zoologist needs a food mix that contains, among other things, 23 grams of protein, 6.2 grams of fat, and 16 grams of moisture. Mixes are available with the compositions shown in the table. How many grams of each mix should be used to get the desired diet mix?

Mix	Protein (%)	Fat (%)	Moisture (%)
A	20	2	15
B	10	6	10
C	15	5	5

★ **46.** *Diet* Repeat Problem 45 assuming the diet mix is to contain 18.5 grams of protein, 4.9 grams of fat, and 13 grams of moisture.

47. *Business* An appliance company has two warehouses and two retail stores in a metropolitan area.

For a holiday sale, it wants to ship 200 refrigerators to the two stores, 80 to store A and 120 to store B. There are 140 refrigerators at warehouse I and 60 at warehouse II. The company decides to ship w refrigerators from warehouse I to store A, x from I to B, y from II to A, and z from II to B. Write a system of equations that gives the shipping requirements. Do not solve.

48. *Business* Repeat Problem 47 for the case where there are 100 refrigerators at each warehouse and the demands at stores A and B are 130 and 70, respectively.

★★ **49. *Business*** A newspaper firm uses three printing presses of different ages and capacities to print the

evening paper. With all three presses running, the paper can be printed in 2 hours. If the newest press breaks down, the older two presses can print the paper in 4 hours; if the middle press breaks down, the newest and oldest together can print the paper in 3 hours. How long would it take each press alone to print the paper? [*Hint:*   Use $(2/x) + (2/y) + (2/z) = 1$ as one of the equations.]

★★ **50. *Business*** Repeat Problem 49 for the following data: all three presses can do the job in 1 hour; the older two presses can do it in 2 hours; and the newer two presses can do it in 1 hour, 12 minutes.

  ## 8-4 **Systems of Equations and Matrices (Optional)**

♦ Matrices and Augmented Matrices
♦ Solving Linear Systems Using Augmented Matrix Methods

In solving systems of equations by elimination in the preceding sections, the coefficients of the variables and constant terms played a central role. The process can be made more efficient for generalization and computer work by the introduction of a mathematical form called a *matrix,* or in the plural, *matrices.*

♦ ### MATRICES AND AUGMENTED MATRICES

A **matrix** is a rectangular array of numbers written within brackets. Some examples are

$$\begin{bmatrix} 3 & 5 \\ 0 & -2 \end{bmatrix} \qquad \begin{bmatrix} 2 \\ -3 \\ 0 \end{bmatrix} \qquad \begin{bmatrix} 1 & -1 & 0 & 5 \end{bmatrix}$$

$$\begin{bmatrix} -1 & 2 & -5 & 0 \\ 0 & 3 & 2 & 1 \end{bmatrix} \qquad \begin{bmatrix} 1 & 0 & 0 \\ 0 & 1 & 0 \\ 0 & 0 & 1 \end{bmatrix} \qquad \begin{bmatrix} a & b & m \\ c & d & n \end{bmatrix}$$

Each number in a matrix is called an **element** of the matrix. The size of a matrix is determined by the number of rows and columns. The matrix

$$\begin{bmatrix} -1 & 2 & -5 & 0 \\ 0 & 3 & 2 & 1 \end{bmatrix}$$

has two rows,

$$-1 \quad 2 \quad -5 \quad 0 \qquad \text{and} \qquad 0 \quad 3 \quad 2 \quad 1$$

and four columns,

$$-1 \quad 2 \quad -5 \quad \text{and} \quad 0$$
$$0 \quad 3 \quad 2 \quad \text{and} \quad 1$$

We say the size of the matrix is **2 by 4,** or **2 × 4.**

Each linear system of the form

$$ax + by = m \tag{1}$$
$$cx + dy = n$$

where x and y are variables, is associated with a matrix called the **augmented matrix** of the system:

$$\begin{bmatrix} a & b & | & m \\ c & d & | & n \end{bmatrix} \tag{2}$$

This matrix contains the essential parts of system (1). The vertical bar is included only to separate the coefficients of the variables from the constant terms. Our objective is to learn how to manipulate augmented matrices in such a way that a solution to system (1) will result, if a solution exists. The manipulative process is a direct outgrowth of the elimination process discussed in Sections 8-1 and 8-3.

Recall that two linear systems are said to be **equivalent** if they have exactly the same solution set. How did we transform linear systems into equivalent linear systems? We used three operations, which we restate here for convenient reference:

Producing Equivalent Systems

A system of linear equations is transformed into an equivalent system when:

1. Two equations are interchanged.
2. An equation is multiplied by a nonzero constant.
3. A constant multiple of another equation is added to a given equation.

Paralleling the previous discussion, we say that two augmented matrices are **row-equivalent,** denoted by the symbol $\sim$ between the two matrices, if they are augmented matrices of equivalent systems of equations. Think about this. How do we transform augmented matrices into row-equivalent matrices? We use the following *row operations* which follow from the boxed material above:

> **Producing Row-Equivalent Matrices**
>
> An augmented matrix is transformed into a row-equivalent matrix if:
>
> **1.** Two rows are interchanged ($R_i \leftrightarrow R_j$).
> **2.** A row is multiplied by a nonzero constant ($kR_i \rightarrow R_i$).
> **3.** A constant multiple of another row is added to a given row
> ($R_i + kR_j \rightarrow R_i$).
>
> Here, the arrow $\rightarrow$ means "replaces." These three operations are called **row operations.**

◆ **SOLVING LINEAR SYSTEMS USING AUGMENTED MATRIX METHODS**

The use of row-equivalent matrices in solving systems in the form of system (1) is illustrated by the examples that follow.

Example 1
Solving by Matrix Methods

Solve using augmented matrix methods:

$$3x + 4y = 1 \\ x - 2y = 7 \tag{3}$$

Solution   We start by writing the augmented matrix corresponding to system (3):

$$\begin{bmatrix} 3 & 4 & | & 1 \\ 1 & -2 & | & 7 \end{bmatrix} \tag{4}$$

Our objective is to use row operations to try to transform system (4) into the form

$$\begin{bmatrix} 1 & k & | & m \\ 0 & 1 & | & n \end{bmatrix} \tag{5}$$

where k, m, and n are real numbers. The solution to system (3) can then be found, since matrix (5) will be the augmented matrix of the following system, which can be solved by substitution:

$$x + ky = m \\ y = n$$

We now proceed to use row operations to transform (4) into form (5).
Step 1.   To get a 1 in the upper left corner, we interchange rows 1 and 2 using row operation 1:

$$\begin{bmatrix} 3 & 4 & | & 1 \\ 1 & -2 & | & 7 \end{bmatrix} \underset{\sim}{\overset{R_1 \leftrightarrow R_2}{}} \begin{bmatrix} 1 & -2 & | & 7 \\ 3 & 4 & | & 1 \end{bmatrix}$$

Step 2. To get a 0 in the lower left corner, we multiply R_1 by -3 and add the result to R_2 using row operation 3. The operation here is to replace R_2 by $R_2 + (-3)R_1$ so that this changes R_2 but not R_1. Some people find it useful to write $(-3)R_1$ outside the matrix to help reduce errors in arithmetic, as shown:

$$\begin{matrix} -3 & 6 & -21 \end{matrix}$$
$$\begin{bmatrix} 1 & -2 & | & 7 \\ 3 & 4 & | & 1 \end{bmatrix} \quad R_2 + (-3)R_1 \to R_2 \quad \begin{bmatrix} 1 & -2 & | & 7 \\ 0 & 10 & | & -20 \end{bmatrix}$$

Step 3. To get a 1 in the second row, second column, we use row operation 2, multiplying R_2 by $\tfrac{1}{10}$:

$$\begin{bmatrix} 1 & -2 & | & 7 \\ 0 & 10 & | & -20 \end{bmatrix} \quad \tfrac{1}{10}R_2 \to R_2 \quad \begin{bmatrix} 1 & -2 & | & 7 \\ 0 & 1 & | & -2 \end{bmatrix}$$

We have accomplished our objective! The last matrix is the augmented matrix for the system

$$\begin{aligned} x - 2y &= 7 \\ y &= -2 \end{aligned} \tag{6}$$

which can be solved by substitution. Substituting $y = -2$ into the first equation, we obtain $x + 4 = 7$, or $x = 3$. Since system (6) is equivalent to our starting system (3), we have solved (3); that is, $x = 3$ and $y = -2$.

Check
$$3(\mathbf{3}) + 4(\mathbf{-2}) = 1 \qquad \mathbf{3} - 2(\mathbf{-2}) = 7$$

The solution process is written more compactly as follows:

Step 1:
Need a 1 here
$$\begin{bmatrix} 3 & 4 & | & 1 \\ 1 & -2 & | & 7 \end{bmatrix} \quad R_1 \leftrightarrow R_2$$

Step 2:
Need a 0 here
$$\sim \begin{bmatrix} 1 & -2 & | & 7 \\ 3 & 4 & | & 1 \end{bmatrix} \quad R_2 + (-3)R_1 \to R_2$$
$$\begin{matrix} -3 & 6 & -21 \end{matrix}$$

Step 3:
Need a 1 here
$$\sim \begin{bmatrix} 1 & -2 & | & 7 \\ 0 & 10 & | & -20 \end{bmatrix} \quad \tfrac{1}{10}R_2 \to R_2$$

$$\sim \begin{bmatrix} 1 & -2 & | & 7 \\ 0 & 1 & | & -2 \end{bmatrix}$$

Matched Problem 1 Solve using augmented matrix methods:
$$\begin{aligned} 2x - y &= -7 \\ x + 2y &= 4 \end{aligned}$$

Example 2
Solving by
Matrix Methods

Solve using augmented matrix methods:

$$\begin{aligned} 2x - 3y &= 7 \\ 3x + 4y &= 2 \end{aligned}$$

Solution

Step 1:
Need a 1 here

$$\begin{bmatrix} 2 & -3 & | & 7 \\ 3 & 4 & | & 2 \end{bmatrix} \quad \tfrac{1}{2}R_1 \to R_1$$

Step 2:
Need a 0 here

$$\sim \begin{bmatrix} 1 & -\tfrac{3}{2} & | & \tfrac{7}{2} \\ 3 & 4 & | & 2 \end{bmatrix} \quad R_2 + (-3)R_1 \to R_2$$

$$-3 \quad \tfrac{9}{2} \quad -\tfrac{21}{2}$$

Step 3:
Need a 1 here

$$\sim \begin{bmatrix} 1 & -\tfrac{3}{2} & | & \tfrac{7}{2} \\ 0 & \tfrac{17}{2} & | & -\tfrac{17}{2} \end{bmatrix} \quad \tfrac{2}{17}R_2 \to R_2$$

$$\sim \begin{bmatrix} 1 & -\tfrac{3}{2} & | & \tfrac{7}{2} \\ 0 & 1 & | & -1 \end{bmatrix}$$

The original system is therefore equivalent to

$$x - \tfrac{3}{2}y = \tfrac{7}{2}$$
$$y = -1$$

Thus, $y = -1$ and $x + \tfrac{3}{2} = \tfrac{7}{2}$, so $x = 2$. You should check the solution $(2, -1)$.

Matched Problem 2   Solve using augmented matrix methods: $5x - 2y = 12$
$2x + 3y = 1$

Example 3
Solving by
Matrix Methods

Solve using augmented matrix methods:

$$2x - y = 4$$
$$-6x + 3y = -12$$

Solution

$$\begin{bmatrix} 2 & -1 & | & 4 \\ -6 & 3 & | & -12 \end{bmatrix} \quad \begin{matrix} \tfrac{1}{2}R_1 \to R_1 \text{ (This produces a 1 in the upper left corner.)} \\ \tfrac{1}{3}R_2 \to R_2 \text{ (This simplifies } R_2 \text{.)} \end{matrix}$$

$$\sim \begin{bmatrix} 1 & -\tfrac{1}{2} & | & 2 \\ -2 & 1 & | & -4 \end{bmatrix} \quad R_2 + 2R_1 \to R_2 \text{ (This produces a 0 in the lower left corner.)}$$

$$2 \quad -1 \quad 4$$

$$\sim \begin{bmatrix} 1 & -\tfrac{1}{2} & | & 2 \\ 0 & 0 & | & 0 \end{bmatrix}$$

The last matrix corresponds to the system

$$x - \tfrac{1}{2}y = 2$$
$$0x + 0y = 0$$

Thus, $x = \tfrac{1}{2}y + 2$. Hence, for any real number y, $(\tfrac{1}{2}y + 2, y)$ is a solution. If $y = 6$, for example, then $(\tfrac{1}{2} \cdot \mathbf{6} + 2, \mathbf{6}) = (5, 6)$ is a solution; if $y = -2$, then $(1, -2)$ is a solution; and so on. Geometrically, the graphs of the two original equations are the same line and there are infinitely many solutions. In general, if we end up with a row of 0's in an augmented matrix for a two-equation–two-variable system, the system is dependent and there are infinitely many solutions.

Matched Problem 3   Solve using augmented matrix methods: $-2x + 6y = 6$
$3x - 9y = -9$

Example 4
Solving by
Matrix Methods

Solve using augmented matrix methods: $2x + 6y = -3$
$\qquad\qquad\qquad\qquad\qquad\qquad\qquad\qquad x + 3y = 2$

Solution

$$\begin{bmatrix} 2 & 6 & | & -3 \\ 1 & 3 & | & 2 \end{bmatrix} \qquad R_1 \leftrightarrow R_2$$

$$\sim \begin{bmatrix} 1 & 3 & | & 2 \\ 2 & 6 & | & -3 \end{bmatrix} \qquad R_2 + (-2)R_1 \rightarrow R_2$$

$$\sim \begin{bmatrix} \overset{-2}{1} & \overset{-6}{3} & | & \overset{-4}{2} \\ 0 & 0 & | & -7 \end{bmatrix} \qquad R_2 \text{ implies the contradiction } 0 = -7.$$

The system, seen previously in Section 8-1, Example 5, is inconsistent and has no solution—otherwise, we have proved that $0 = -7$! Thus, if in a row of an augmented matrix we obtain all 0's to the left of the vertical bar and a nonzero number to the right of the bar, the system is inconsistent and there are no solutions.

Matched Problem 4

Solve using augmented matrix methods: $\quad -2x + 6y = 6$
$\qquad\qquad\qquad\qquad\qquad\qquad\qquad\qquad\qquad\qquad 3x - 9y = 9$

Final Matrix Forms for Two Equations in Two Variables

FORM 1:
A UNIQUE
SOLUTION

FORM 2:
INFINITELY
MANY SOLUTIONS
(DEPENDENT)

FORM 3:
NO SOLUTION
(INCONSISTENT)

$$\begin{bmatrix} 1 & k & | & m \\ 0 & 1 & | & n \end{bmatrix} \qquad \begin{bmatrix} 1 & m & | & n \\ 0 & 0 & | & 0 \end{bmatrix} \qquad \begin{bmatrix} 1 & m & | & n \\ 0 & 0 & | & p \end{bmatrix}$$

where k, m, n, and p are real numbers; $p \neq 0$.

The augmented matrix method is readily applied to systems with three or more variables.

Example 5
Solving by
Matrix Methods

Solve using augmented matrix methods: $2x - 2y + z = 3$
$\qquad\qquad\qquad\qquad\qquad\qquad\qquad\qquad 3x + y - z = 7$
$\qquad\qquad\qquad\qquad\qquad\qquad\qquad\qquad x - 3y + 2z = 0$

Solution

We change to equivalent matrices, aiming at a matrix of the form

$$\begin{bmatrix} 1 & m & n & | & p \\ 0 & 1 & q & | & r \\ 0 & 0 & 1 & | & s \end{bmatrix}$$

as our final result.

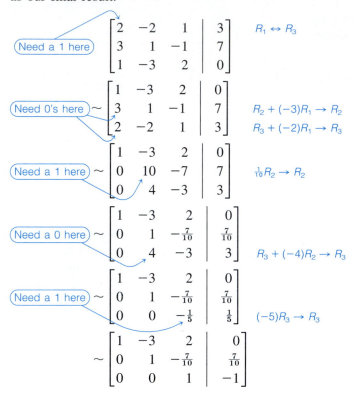

The resulting equivalent system,

$$\begin{aligned} x - 3y + 2z &= 0 \\ y - \tfrac{7}{10}z &= \tfrac{7}{10} \\ z &= -1 \end{aligned}$$

is solved by substitution:

$$\begin{aligned} y - \tfrac{7}{10}z = y + \tfrac{7}{10} = \tfrac{7}{10} \qquad &\text{so} \qquad y = 0 \\ x - 3y + 2z = x - 0 - 2 = 0 \qquad &\text{so} \qquad x = 2 \end{aligned}$$

You should check the solution $(2, 0, -1)$.

Matched Problem 5   Solve using augmented matrix methods:
$$\begin{aligned} 3x + y - 2z &= 2 \\ x - 2y + z &= 3 \\ 2x - y - 3z &= 3 \end{aligned}$$

The process of solving systems of equations described in this section is known as **gaussian elimination.** It is a powerful method that can be used to solve large-scale systems on a computer. The concept of a matrix is also a powerful mathematical tool with applications far beyond solving systems of equations. Matrices are discussed further in subsequent courses. Another aspect of matrices that leads to a different solution method is discussed in the next section.

Answers to
Matched Problems
1. $x = -2$, $y = 3$ 2. $x = 2$, $y = -1$
3. The system is dependent. For y any real number, $(3y - 3, y)$ is a solution.
4. Inconsistent—no solution.   5. $x = 1$, $y = -1$, $z = 0$

EXERCISE 8-4

A *Perform each of the indicated row operations on the following matrix:*

$$\begin{bmatrix} 1 & -3 & \bigm| & 2 \\ 4 & -6 & \bigm| & -8 \end{bmatrix}$$

1. $R_1 \leftrightarrow R_2$

2. $\frac{1}{2}R_2 \to R_2$

3. $-4R_1 \to R_1$

4. $-2R_1 \to R_1$

5. $2R_2 \to R_2$

6. $-1R_2 \to R_2$

7. $R_2 + (-4)R_1 \to R_2$

8. $R_1 + (-\frac{1}{2})R_2 \to R_1$

9. $R_2 + (-2)R_1 \to R_2$

10. $R_2 + (-3)R_1 \to R_2$

11. $R_2 + (-1)R_1 \to R_2$

12. $R_2 + (1)R_1 \to R_2$

Perform each of the indicated row operations on the following matrix:

$$\begin{bmatrix} 1 & 0 & 2 & \bigm| & 3 \\ 2 & -1 & 0 & \bigm| & 4 \\ 3 & 2 & -1 & \bigm| & 5 \end{bmatrix}$$

13. $R_1 \leftrightarrow R_3$

14. $R_2 \leftrightarrow R_3$

15. $-2R_2 \to R_2$

16. $-3R_3 \to R_3$

17. $R_1 + R_2 \to R_1$

18. $R_2 + (-1)R_3 \to R_2$

19. $R_2 + \frac{1}{2}R_3 \to R_2$

20. $R_1 + \frac{1}{2}R_2 \to R_1$

Solve using augmented matrix methods.

21. $x + y = 5$
$x - y = 1$

22. $x - y = 2$
$x + y = 6$

23. $x - 2y = 1$
$2x - y = 5$

24. $x + 3y = 1$
$3x - 2y = 14$

25. $x - 4y = -2$
$-2x + y = -3$

26. $x - 3y = -5$
$-3x - y = 5$

27. $3x - y = 2$
$x + 2y = 10$

28. $2x + y = 0$
$x - 2y = -5$

Solve the system of equations corresponding to each augmented matrix. Use x and y as the variables.

29. $\begin{bmatrix} 1 & 2 & \bigm| & 2 \\ 2 & 3 & \bigm| & -1 \end{bmatrix}$

30. $\begin{bmatrix} 2 & 1 & \bigm| & -1 \\ 5 & 3 & \bigm| & 3 \end{bmatrix}$

31. $\begin{bmatrix} 3 & 4 & \bigm| & -2 \\ 2 & 3 & \bigm| & 1 \end{bmatrix}$

32. $\begin{bmatrix} 5 & 2 & \bigm| & 4 \\ 7 & 3 & \bigm| & 0 \end{bmatrix}$

33. $\begin{bmatrix} 4 & 5 & \bigm| & 0 \\ 3 & 4 & \bigm| & 4 \end{bmatrix}$

34. $\begin{bmatrix} 8 & 3 & \bigm| & -1 \\ 3 & 1 & \bigm| & 2 \end{bmatrix}$

B *Solve using augmented matrix methods.*

35. $x + 2y = 4$
$2x + 4y = -8$

36. $2x - 3y = -2$
$-4x + 6y = 7$

37. $2x + y = 6$
$x - y = -3$

38. $3x - y = -5$
$x + 3y = 5$

39. $3x - 6y = -9$
$-2x + 4y = 6$

40. $2x - 4y = -2$
$-3x + 6y = 3$

41. $4x - 2y = 2$
$-6x + 3y = -3$

42. $-6x + 2y = 4$
$3x - y = -2$

43. $3x - y = 7$
$2x + 3y = 1$

44. $2x - 3y = -8$
$5x + 3y = 1$

45. $3x + 2y = 4$
$2x - y = 5$

46. $4x + 3y = 26$
$3x - 11y = -7$

Solve the system of equations corresponding to each augmented matrix. Use x and y as the variables.

47. $\begin{bmatrix} 3 & 6 & \bigm| & 4 \\ 1 & 2 & \bigm| & 2 \end{bmatrix}$

48. $\begin{bmatrix} 2 & 3 & \bigm| & 1 \\ 4 & 6 & \bigm| & 2 \end{bmatrix}$

49. $\begin{bmatrix} 2 & -5 & \bigm| & -4 \\ -4 & 10 & \bigm| & -8 \end{bmatrix}$

50. $\begin{bmatrix} 6 & -8 & \bigm| & -6 \\ -3 & 4 & \bigm| & -3 \end{bmatrix}$

C *Solve using augmented matrix methods.*

51. $0.2x - 0.5y = 0.07$
$0.8x - 0.3y = 0.79$

52. $0.3x - 0.6y = 0.18$
$0.5x - 0.2y = 0.54$

53. $2x + 4y - 10z = -2$
$3x + 9y - 21z = 0$
$x + 5y - 12z = 1$

54. $3x + 5y - z = -7$
$x + y + z = -1$
$2x + y + 11z = 7$

55. $3x + 8y - z = -18$
$2x + y + 5z = 8$
$2x + 4y + 2z = -4$

56. $2x + 7y + 15z = -12$
$4x + 7y + 13z = -10$
$3x + 6y + 12z = -9$

Solve the system of equations corresponding to each augmented matrix. Use x, y, and z as the variables.

57. $\begin{bmatrix} 3 & 1 & -2 & \bigm| & 4 \\ 0 & 2 & 5 & \bigm| & 1 \\ 1 & 0 & 3 & \bigm| & -2 \end{bmatrix}$

58. $\begin{bmatrix} 1 & 4 & 2 & \bigm| & 3 \\ 2 & -3 & 4 & \bigm| & -5 \\ 3 & 2 & 1 & \bigm| & 0 \end{bmatrix}$

59. $\begin{bmatrix} 1 & 2 & 3 & \bigm| & -1 \\ 1 & 1 & 2 & \bigm| & 3 \\ 2 & 4 & 7 & \bigm| & 5 \end{bmatrix}$

60. $\begin{bmatrix} 1 & 0 & 1 & \bigm| & 2 \\ 3 & 2 & -1 & \bigm| & 4 \\ 1 & 2 & -3 & \bigm| & 0 \end{bmatrix}$

61. $\begin{bmatrix} 1 & 1 & 0 & \bigm| & 2 \\ 0 & 1 & 1 & \bigm| & 3 \\ 1 & 0 & 1 & \bigm| & 4 \end{bmatrix}$

62. $\begin{bmatrix} 1 & 0 & -1 & \bigm| & 1 \\ 0 & 1 & -1 & \bigm| & 2 \\ 1 & -1 & 0 & \bigm| & 3 \end{bmatrix}$

Augmented matrix methods extend naturally to any number of variables. Solve using augmented matrix methods.

63. $x - 2y + w = 1$
$-2x + 5y + 3z - w = -5$
$3x - 6y + z + 7w = 2$
$x - y + 4z + 7w = -3$

64.
$$\begin{aligned} x - y + z - w &= 8 \\ 2x - y + 4z + w &= 20 \\ x + 4z + 8w &= 20 \\ -2x + 3y + z + 12w &= -3 \end{aligned}$$

65.
$$\begin{aligned} w + 4x + 7y + 10z &= 13 \\ x + 2y + 3z &= 4 \\ w - 3x - 4y - 4z &= -4 \\ x + y + z &= 1 \end{aligned}$$

66.
$$\begin{aligned} w + 5x + 10y + 17z &= 24 \\ 2w + 7x + 11y + 13z &= 15 \\ x + 3y + 6z &= 9 \\ 2x - 2y - 5z &= -10 \end{aligned}$$

8-5 Determinants and Cramer's Rule (Optional)

♦ Second-Order Determinants
♦ Third-Order Determinants
♦ Cramer's Rule

A matrix with the same number of rows and columns is called a **square matrix.** The matrices of coefficients for the systems of equations considered in Section 8-4 are examples. There is a number, called the *determinant of the matrix,* associated with each square matrix. We will define the determinant in this section. Determinants arise quite naturally in many areas in mathematics. We will consider their use in solving systems of equations. Other uses are suggested in the exercises.

♦ SECOND-ORDER DETERMINANTS

The square matrix

$$\begin{bmatrix} 2 & -3 \\ 5 & 1 \end{bmatrix}$$

has associated with it a number called its *determinant* and denoted by

$$\begin{vmatrix} 2 & -3 \\ 5 & 1 \end{vmatrix}$$

Because the matrix is 2×2, the determinant is referred to as a 2×2, or *second-order, determinant.*

In general, we can symbolize a **second-order determinant** as follows:

$$\begin{vmatrix} a_{11} & a_{12} \\ a_{21} & a_{22} \end{vmatrix}$$

where we use a single letter with a **double subscript** to facilitate generalization to higher-order determinants. The first subscript number indicates the row in which the element lies, and the second subscript number indicates the column. Thus, a_{21} is the element in the second row and first column and a_{12} is the element in the first row and second column. Each second-order determinant represents a real number given by the following formula:

Second-Order Determinant

$$\begin{vmatrix} a_{11} & a_{12} \\ a_{21} & a_{22} \end{vmatrix} = a_{11}a_{22} - a_{21}a_{12}$$

Example 1
Calculating a Second-Order Determinant

Calculate:

$$\begin{vmatrix} -1 & 2 \\ -3 & 4 \end{vmatrix}$$

Solution

$$\begin{vmatrix} -1 & 2 \\ -3 & 4 \end{vmatrix} = (-1)(4) - (-3)(2) = -4 - (-6) = 2$$

Matched Problem 1 Calculate: $\begin{vmatrix} 3 & -5 \\ 4 & -2 \end{vmatrix}$

◆ **THIRD-ORDER DETERMINANTS**

A **third-order determinant** is the determinant of a 3×3 matrix and represents a real number given by the following formula:

Third-Order Determinant

$$\begin{vmatrix} a_{11} & a_{12} & a_{13} \\ a_{21} & a_{22} & a_{23} \\ a_{31} & a_{32} & a_{33} \end{vmatrix} = \begin{array}{l} a_{11}a_{22}a_{33} - a_{11}a_{32}a_{23} + a_{21}a_{32}a_{13} \\ \quad - a_{21}a_{12}a_{33} + a_{31}a_{12}a_{23} - a_{31}a_{22}a_{13} \end{array} \qquad (1)$$

Note that each term in the expansion on the right of Formula (1) contains exactly one element from each row and each column. You do not need to memorize this definition since we will now proceed to find an easier way to evaluate this determinant.

The **minor of an element** in a third-order determinant is a second-order determinant obtained by deleting the row and column that contains the element. For example, in the determinant in Formula (1),

$$\text{Minor of } a_{23} = \begin{vmatrix} a_{11} & a_{12} & a_{13} \\ a_{21} & a_{22} & a_{23} \\ a_{31} & a_{32} & a_{33} \end{vmatrix} = \begin{vmatrix} a_{11} & a_{12} \\ a_{31} & a_{32} \end{vmatrix}$$

$$\text{Minor of } a_{32} = \begin{vmatrix} a_{11} & a_{12} & a_{13} \\ a_{21} & a_{22} & a_{23} \\ a_{31} & a_{32} & a_{33} \end{vmatrix} = \begin{vmatrix} a_{11} & a_{13} \\ a_{21} & a_{23} \end{vmatrix}$$

A quantity closely associated with the minor of an element is the *cofactor of an element*. The **cofactor of the element** a_{ij} in the ith row and jth column is the product of the minor of a_{ij} and $(-1)^{i+j}$. That is:

Cofactor

$$\text{Cofactor of } a_{ij} = (-1)^{i+j}(\text{Minor of } a_{ij})$$

Thus, a cofactor of an element is nothing more than a signed minor. The sign is determined by raising -1 to a power that is the sum of the numbers indicating the row and column in which the element lies. Note that $(-1)^{i+j}$ is -1 if $i + j$ is odd and 1 if $i + j$ is even. Referring again to the determinant in Formula (1),

$$\text{Cofactor of } a_{23} = (-1)^{2+3}\begin{vmatrix} a_{11} & a_{12} \\ a_{31} & a_{32} \end{vmatrix} = -\begin{vmatrix} a_{11} & a_{12} \\ a_{31} & a_{32} \end{vmatrix}$$

$$\text{Cofactor of } a_{11} = (-1)^{1+1}\begin{vmatrix} a_{22} & a_{23} \\ a_{32} & a_{33} \end{vmatrix} = \begin{vmatrix} a_{22} & a_{23} \\ a_{32} & a_{33} \end{vmatrix}$$

Example 2
Calculating Cofactors

Calculate the cofactors of -2 and 5 in the determinant:

$$\begin{vmatrix} -2 & 0 & 3 \\ 1 & -6 & 5 \\ -1 & 2 & 0 \end{vmatrix}$$

Solution

$$\text{Cofactor of } -2 = (-1)^{1+1}\begin{vmatrix} -6 & 5 \\ 2 & 0 \end{vmatrix} = \begin{vmatrix} -6 & 5 \\ 2 & 0 \end{vmatrix}$$
$$= (-6)(0) - (2)(5) = -10$$

$$\text{Cofactor of } 5 = (-1)^{2+3}\begin{vmatrix} -2 & 0 \\ -1 & 2 \end{vmatrix} = -\begin{vmatrix} -2 & 0 \\ -1 & 2 \end{vmatrix}$$
$$= -[(-2)(2) - (-1)(0)] = 4$$

Matched Problem 2

Calculate the cofactors of 2 and 3 in the determinant in Example 2.

The sign in front of the minor, $(-1)^{i+j}$, can be determined mechanically by using a checkerboard pattern of plus and minus signs over the determinant, starting with $+$ in the upper left-hand corner:

$$\begin{matrix} + & - & + \\ - & + & - \\ + & - & + \end{matrix}$$

Use either the checkerboard pattern or the exponent method, whichever is easier for you, to determine the sign in front of the minor.

Now we are ready for the result that will enable us to calculate 3×3 determinants more readily than by using the definition. The result also generalizes completely to include determinants of arbitrary order.

Expansion by Cofactors

The value of a determinant of order 3 is the sum of three products obtained by multiplying each element of any one row, or each element of any one column, by its cofactor.

To prove this result we must show that the expansions indicated by the result for any row or any column—six cases in all—produce the expression on the right of Formula (1). Proofs of some cases of this result are included in Exercise 8-5.

Example 3
Expanding by Cofactors

Evaluate

$$\begin{vmatrix} 2 & -2 & 0 \\ -3 & 1 & 2 \\ 1 & -3 & -1 \end{vmatrix}$$

(A) By expanding by the cofactors of the first row
(B) By expanding by the cofactors of the second column

Solution **(A)** $\begin{vmatrix} 2 & -2 & 0 \\ -3 & 1 & 2 \\ 1 & -3 & -1 \end{vmatrix}$ Expand by the first row.

$$= a_{11}\left(\begin{array}{c}\text{Cofactor}\\\text{of } a_{11}\end{array}\right) + a_{12}\left(\begin{array}{c}\text{Cofactor}\\\text{of } a_{12}\end{array}\right) + a_{13}\left(\begin{array}{c}\text{Cofactor}\\\text{of } a_{13}\end{array}\right)$$

$$= 2\left((-1)^{1+1}\begin{vmatrix} 1 & 2 \\ -3 & -1 \end{vmatrix}\right) + (-2)\left((-1)^{1+2}\begin{vmatrix} -3 & 2 \\ 1 & -1 \end{vmatrix}\right) + 0$$

$$= (2)(1)[(1)(-1) - (-3)(2)] + (-2)(-1)[(-3)(-1) - (1)(2)]$$

$$= (2)(5) + (2)(1) = 12$$

(B) $\begin{vmatrix} 2 & -2 & 0 \\ -3 & 1 & 2 \\ 1 & -3 & -1 \end{vmatrix}$ Expand by the second column.

$$= a_{12}\left(\begin{array}{c}\text{Cofactor}\\\text{of } a_{12}\end{array}\right) + a_{22}\left(\begin{array}{c}\text{Cofactor}\\\text{of } a_{22}\end{array}\right) + a_{32}\left(\begin{array}{c}\text{Cofactor}\\\text{of } a_{32}\end{array}\right)$$

$$= (-2)\left((-1)^{1+2}\begin{vmatrix} -3 & 2 \\ 1 & -1 \end{vmatrix}\right) + (1)\left((-1)^{2+2}\begin{vmatrix} 2 & 0 \\ 1 & -1 \end{vmatrix}\right) + (-3)\left((-1)^{3+2}\begin{vmatrix} 2 & 0 \\ -3 & 2 \end{vmatrix}\right)$$

$$= (-2)(-1)[(-3)(-1) - (1)(2)] + (1)(1)[(2)(-1) - (1)(0)]$$
$$\quad + (-3)(-1)[(2)(2) - (-3)(0)]$$

$$= (2)(1) + (1)(-2) + (3)(4) = 12$$

Matched Problem 3 Evaluate

$$\begin{vmatrix} 2 & 1 & -1 \\ -1 & -3 & 0 \\ -1 & 2 & 1 \end{vmatrix}$$

(A) By expanding by the cofactors of the first row
(B) By expanding by the cofactors of the third column

It should be clear that we can greatly reduce the work involved in evaluating a determinant by choosing to expand by a row or column with the greatest number of 0's.

Where are determinants used? Many equations and formulas have particularly simple and compact representations in determinant form that are easily remembered. *Cramer's rule* for solving a system of linear equations is one example.

♦ **CRAMER'S RULE**

Determinants arise rather naturally in the process of solving systems of linear equations. We will start by investigating two equations and two variables and then extend any results to three equations and three variables.

Instead of thinking of each system of linear equations in two variables as a different problem, let us see what happens when we attempt to solve the general system

$$a_{11}x + a_{12}y = k_1 \qquad (2A)$$
$$a_{21}x + a_{22}y = k_2 \qquad (2B)$$

once and for all in terms of the unspecified real constants a_{11}, a_{12}, a_{21}, a_{22}, k_1, and k_2.

We proceed by multiplying Equations (2A) and (2B) by suitable constants so that when the resulting equations are added, left side to left side and right side to right side, one of the variables drops out. Suppose we choose to eliminate y. What constants should we use to make the coefficients of y the same except for the signs? Multiply Equation (2A) by a_{22} and Equation (2B) by $-a_{12}$. Then add:

$$a_{11}a_{22}x + a_{12}a_{22}y = k_1a_{22} \qquad \text{\color{blue}a_{22}(2A)}$$
$$-a_{21}a_{12}x - a_{12}a_{22}y = -k_2a_{12} \qquad \text{\color{blue}$-a_{12}$(2B)}$$

$$\overline{}$$

$$a_{11}a_{22}x - a_{21}a_{12}x + 0y = k_1a_{22} - k_2a_{12}$$
$$(a_{11}a_{22} - a_{21}a_{12})x = k_1a_{22} - k_2a_{12}$$
$$x = \frac{k_1a_{22} - k_2a_{12}}{a_{11}a_{22} - a_{21}a_{12}} \qquad \text{\color{blue}We assume $a_{11}a_{22} - a_{21}a_{12} \neq 0$.}$$

That is,

$$x = \frac{\begin{vmatrix} k_1 & a_{12} \\ k_2 & a_{22} \end{vmatrix}}{\begin{vmatrix} a_{11} & a_{12} \\ a_{21} & a_{22} \end{vmatrix}}$$

Similarly, starting with the same system and eliminating x, we obtain

$$y = \frac{\begin{vmatrix} a_{11} & k_1 \\ a_{21} & k_2 \end{vmatrix}}{\begin{vmatrix} a_{11} & a_{12} \\ a_{21} & a_{22} \end{vmatrix}}$$

These results are summarized in the following rule, which is named after the Swiss mathematician, G. Cramer (1704–1752):

Cramer's Rule for Two Equations and Two Variables

Given the system

$$a_{11}x + a_{12}y = k_1 \qquad \text{with} \qquad D = \begin{vmatrix} a_{11} & a_{12} \\ a_{21} & a_{22} \end{vmatrix} \neq 0$$
$$a_{21}x + a_{22}y = k_2$$

then

$$x = \frac{\begin{vmatrix} k_1 & a_{12} \\ k_2 & a_{22} \end{vmatrix}}{D} \qquad \text{and} \qquad y = \frac{\begin{vmatrix} a_{11} & k_1 \\ a_{21} & k_2 \end{vmatrix}}{D}$$

The determinant D is called the **determinant of the coefficient matrix.** If $D \neq 0$, then the system has exactly one solution, which is given by Cramer's rule. If, on the other hand, $D = 0$, then it can be shown that the system is either inconsistent or dependent; that is, the system either has no solutions or has an infinite number of solutions.

Example 4
Applying Cramer's Rule

Solve using Cramer's rule:
$$2x - 3y = 7$$
$$-3x + y = -7$$

Solution

$$D = \begin{vmatrix} 2 & -3 \\ -3 & 1 \end{vmatrix} = -7$$

$$x = \frac{\begin{vmatrix} 7 & -3 \\ -7 & 1 \end{vmatrix}}{-7} = \frac{-14}{-7} = 2 \qquad y = \frac{\begin{vmatrix} 2 & 7 \\ -3 & -7 \end{vmatrix}}{-7} = \frac{7}{-7} = -1$$

Matched Problem 4   Solve using Cramer's rule: $3x + 2y = -3$
$$-4x + 3y = -13$$

Cramer's rule generalizes completely for any size of linear system that has the same number of variables as equations. We state the rule for three equations and three variables.

Cramer's Rule for Three Equations and Three Variables

Given the system

$$\begin{aligned} a_{11}x + a_{12}y + a_{13}z &= k_1 \\ a_{21}x + a_{22}y + a_{23}z &= k_2 \\ a_{31}x + a_{32}y + a_{33}z &= k_3 \end{aligned} \quad \text{with} \quad D = \begin{vmatrix} a_{11} & a_{12} & a_{13} \\ a_{21} & a_{22} & a_{23} \\ a_{31} & a_{32} & a_{33} \end{vmatrix} \neq 0$$

then

$$x = \frac{\begin{vmatrix} k_1 & a_{12} & a_{13} \\ k_2 & a_{22} & a_{23} \\ k_3 & a_{32} & a_{33} \end{vmatrix}}{D} \qquad y = \frac{\begin{vmatrix} a_{11} & k_1 & a_{13} \\ a_{21} & k_2 & a_{23} \\ a_{31} & k_3 & a_{33} \end{vmatrix}}{D} \qquad z = \frac{\begin{vmatrix} a_{11} & a_{12} & k_1 \\ a_{21} & a_{22} & k_2 \\ a_{31} & a_{32} & k_3 \end{vmatrix}}{D}$$

To help remember these determinant formulas for x, y, and z, observe the following:

1. Determinant D is formed from the coefficients of x, y, and z, keeping the same relative position in the determinant as found in the system.
2. Determinant D appears in the denominators for x, y, and z.
3. The numerator for x can be obtained from D by replacing the coefficients of x—that is, a_{11}, a_{21}, and a_{31}—with the constants k_1, k_2, and k_3, respectively. Similar statements can be made for the numerators for y and z.

Example 5
Applying Cramer's Rule

Use Cramer's rule to solve:

$$\begin{aligned} x + y &= 1 \\ 3y - z &= -4 \\ x + z &= 3 \end{aligned}$$

Solution   Write the system

$$\begin{aligned} x + y &= 1 \\ 3y - z &= -4 \\ x + z &= 3 \end{aligned}$$

Missing variables have 0 coefficients.

as

$$\begin{aligned} x + y + 0z &= 1 \\ 0x + 3y - z &= -4 \\ x + 0y + z &= 3 \end{aligned}$$

$$D = \begin{vmatrix} 1 & 1 & 0 \\ 0 & 3 & -1 \\ 1 & 0 & 1 \end{vmatrix} = 2$$

$$x = \frac{\begin{vmatrix} 1 & 1 & 0 \\ -4 & 3 & -1 \\ 3 & 0 & 1 \end{vmatrix}}{2} = \frac{4}{2} = 2 \qquad y = \frac{\begin{vmatrix} 1 & 1 & 0 \\ 1 & -4 & -1 \\ 1 & 3 & 1 \end{vmatrix}}{2} = \frac{-2}{2} = -1$$

$$z = \frac{\begin{vmatrix} 1 & 1 & 1 \\ 0 & 3 & -4 \\ 1 & 0 & 3 \end{vmatrix}}{2} = \frac{2}{2} = 1$$

Matched Problem 5   Use Cramer's rule to solve: $3x - z = 5$
$x - y + z = 0$
$x + y = 0$

In practice, Cramer's rule is rarely used to solve systems of order higher than 2 or 3; more efficient methods are available. Cramer's rule is, however, a valuable tool in theoretical mathematics.

The results of all the methods we have discussed for solving two equations with two variables are summarized in Table 1.

Table 1   Solving Second-Order Systems of Linear Equations

		Method	
Solutions	Graphing	Elimination or Substitution	Cramer's Rule
Exactly one	Lines intersect in exactly one point	One unique pair of numbers	$D \neq 0$
None	Lines are parallel	Contradiction occurs, such as $0 = 5$	$D = 0$
Infinite number	Lines coincide	The equation $0 = 0$ occurs	$D = 0$

Answers to   **1.** 14   **2.** 13, -4 **3. (A)** 3   **(B)** 3
Matched Problems   **4.** $x = 1, y = -3$ **5.** $x = 1, y = -1, z = -2$

EXERCISE 8-5

A *Evaluate each second-order determinant.*

1. $\begin{vmatrix} 3 & 2 \\ 4 & 3 \end{vmatrix}$ **2.** $\begin{vmatrix} 4 & 2 \\ 6 & 3 \end{vmatrix}$ **3.** $\begin{vmatrix} 2 & 8 \\ 1 & 4 \end{vmatrix}$

4. $\begin{vmatrix} 4 & 3 \\ 5 & 4 \end{vmatrix}$ **5.** $\begin{vmatrix} 2 & 4 \\ 3 & -1 \end{vmatrix}$ **6.** $\begin{vmatrix} 2 & 2 \\ -3 & 1 \end{vmatrix}$

7. $\begin{vmatrix} 5 & -4 \\ -2 & 2 \end{vmatrix}$ **8.** $\begin{vmatrix} 6 & -2 \\ -1 & -3 \end{vmatrix}$ **9.** $\begin{vmatrix} 3 & -3.1 \\ -2 & 1.2 \end{vmatrix}$

10. $\begin{vmatrix} -1.4 & 3 \\ -0.5 & -2 \end{vmatrix}$ **11.** $\begin{vmatrix} \frac{1}{2} & 2 \\ \frac{1}{4} & 3 \end{vmatrix}$ **12.** $\begin{vmatrix} \frac{1}{2} & \frac{2}{3} \\ 6 & 9 \end{vmatrix}$

13. $\begin{vmatrix} \frac{2}{3} & \frac{1}{2} \\ \frac{1}{6} & \frac{1}{4} \end{vmatrix}$ **14.** $\begin{vmatrix} \frac{1}{4} & \frac{3}{4} \\ \frac{3}{2} & 4 \end{vmatrix}$

In Problems 15–22, use the determinant

$$\begin{vmatrix} a_{11} & a_{12} & a_{13} \\ a_{21} & a_{22} & a_{23} \\ a_{31} & a_{32} & a_{33} \end{vmatrix}$$

Write the minor of each element:

15. a_{11} **16.** a_{33} **17.** a_{23} **18.** a_{22}

Write the cofactor of each element.

19. a_{11} **20.** a_{33} **21.** a_{23} **22.** a_{22}

In Problems 23–30, use the determinant

$$\begin{vmatrix} -2 & 3 & 0 \\ 5 & 1 & -2 \\ 7 & -4 & 8 \end{vmatrix}$$

Write the minor of each element. Leave your answer in determinant form.

23. a_{11} **24.** a_{22} **25.** a_{32} **26.** a_{21}

Write the cofactor of each element and evaluate.

27. a_{11} **28.** a_{22} **29.** a_{32} **30.** a_{21}

Solve using Cramer's rule.

31. $x + 2y = 1$ **32.** $x + 2y = 3$
     $x + 3y = -1$ $x + 3y = 5$

33. $2x + y = 1$ **34.** $x + 3y = 1$
     $5x + 3y = 2$ $2x + 8y = 0$

35. $2x - y = -3$ **36.** $2x + y = 1$
     $-x + 3y = 4$ $5x + 3y = 2$

B *Evaluate each determinant using cofactors.*

37. $\begin{vmatrix} 1 & 0 & 0 \\ -2 & 4 & 3 \\ 5 & -2 & 1 \end{vmatrix}$ **38.** $\begin{vmatrix} 2 & -3 & 5 \\ 0 & -3 & 1 \\ 0 & 6 & 2 \end{vmatrix}$

39. $\begin{vmatrix} 0 & 1 & 5 \\ 3 & -7 & 6 \\ 0 & -2 & -3 \end{vmatrix}$ **40.** $\begin{vmatrix} 4 & -2 & 0 \\ 9 & 5 & 4 \\ 1 & 2 & 0 \end{vmatrix}$

41. $\begin{vmatrix} 4 & -4 & 6 \\ 2 & 8 & -3 \\ 0 & -5 & 0 \end{vmatrix}$ **42.** $\begin{vmatrix} 3 & -2 & -8 \\ -2 & 0 & -3 \\ 1 & 0 & -4 \end{vmatrix}$

43. $\begin{vmatrix} -1 & 2 & -3 \\ -2 & 0 & -6 \\ 4 & -3 & 2 \end{vmatrix}$ **44.** $\begin{vmatrix} 0 & 2 & -1 \\ -6 & 3 & 1 \\ 7 & -9 & -2 \end{vmatrix}$

45. $\begin{vmatrix} 1 & 4 & 1 \\ 1 & 1 & -2 \\ 2 & 1 & -1 \end{vmatrix}$ **46.** $\begin{vmatrix} 3 & 2 & 1 \\ -1 & 5 & 1 \\ 2 & 3 & 1 \end{vmatrix}$

47. $\begin{vmatrix} 1 & 4 & 3 \\ 2 & 1 & 6 \\ 3 & -2 & 9 \end{vmatrix}$ **48.** $\begin{vmatrix} 4 & -6 & 3 \\ -1 & 4 & 1 \\ 5 & -6 & 3 \end{vmatrix}$

Solve using Cramer's rule.

49. $x + y = 0$ **50.** $x + y = -4$
     $2y + z = -5$ $2y + z = 0$
     $-x + z = -3$ $-x + z = 5$

51. $x + y = 1$ **52.** $x + y = -4$
     $2y + z = 0$ $2y + z = 3$
     $-x + z = 0$ $-x + z = 7$

53. $y + z = -4$ **54.** $x - z = 2$
     $x + 2z = 0$ $2x - y = 8$
     $x - y = 5$ $x + y + z = 2$

55. $2y - z = -4$ **56.** $2x + y = 2$
     $x - y - z = 0$ $x - y + z = -1$
     $x - y + 2z = 6$ $x + y + z = -1$

57. $2a + 4b + 3c = 6$ **58.** $3u - 2v + 3w = 11$
     $a - 3b + 2c = -7$ $2u + 3v - 2w = -5$
     $-a + 2b - c = 5$ $u + 4v - w = -5$

C *Assuming that expansion by cofactors applies to determinants of arbitrary order, use it to evaluate the fourth- and fifth-order determinants in Problems 59–62.*

59. $\begin{vmatrix} 0 & 1 & 0 & 1 \\ 2 & 4 & 7 & 6 \\ 0 & 3 & 0 & 1 \\ 0 & 6 & 2 & 5 \end{vmatrix}$ **60.** $\begin{vmatrix} 2 & 6 & 1 & 7 \\ 0 & 3 & 0 & 0 \\ 3 & 4 & 2 & 5 \\ 0 & 9 & 0 & 2 \end{vmatrix}$

61. $\begin{vmatrix} 2 & 0 & 0 & 0 & 0 \\ 0 & 3 & 0 & 0 & 0 \\ 0 & 0 & 2 & 0 & 0 \\ 0 & 0 & 0 & 1 & 0 \\ 0 & 0 & 0 & 0 & 4 \end{vmatrix}$ **62.** $\begin{vmatrix} -2 & 0 & 0 & 0 & 0 \\ 9 & -1 & 0 & 0 & 0 \\ 2 & 1 & 3 & 0 & 0 \\ -1 & 4 & 2 & 2 & 0 \\ 7 & -2 & 3 & 5 & 5 \end{vmatrix}$

63. Show that

$$\begin{vmatrix} x & y & 1 \\ 2 & 3 & 1 \\ -1 & 2 & 1 \end{vmatrix} = 0$$

is the equation of a line that passes through $(2, 3)$ and $(-1, 2)$.

64. Show that

$$\begin{vmatrix} x & y & 1 \\ x_1 & y_1 & 1 \\ x_2 & y_2 & 1 \end{vmatrix} = 0$$

is the equation of a line that passes through (x_1, y_1) and (x_2, y_2).

65. In analytic geometry it is shown that the area of a triangle with vertices (x_1, y_1), (x_2, y_2), and (x_3, y_3) is the absolute value of

$$\frac{1}{2} \begin{vmatrix} x_1 & y_1 & 1 \\ x_2 & y_2 & 1 \\ x_3 & y_3 & 1 \end{vmatrix}$$

Use this result to find the area of a triangle with given vertices $(-1, 4)$, $(4, 8)$, $(1, 1)$.

66. Find the area of a triangle with given vertices $(-1, 2)$, $(2, 5)$, and $(6, -3)$. (See Problem 65.)

67. Justify one case of the result stated for expansion by cofactors by expanding the left side of Formula (1) using the cofactors of the first row. Simplify to obtain the right side of the formula.

68. Justify one case of the result stated for expansion by cofactors by expanding the left side of Formula (1) using the cofactors of the first column. Simplify to obtain the right side of the formula.

It is clear that $x = 0$, $y = 0$, $z = 0$ is a solution to each of the following systems. Use Cramer's rule to determine whether this solution is unique. [Hint: If $D \neq 0$, what can you conclude? If $D = 0$, what can you conclude?]

69.
$$\begin{aligned} x - 4y + 9z &= 0 \\ 4x - y + 6z &= 0 \\ x - y + 3z &= 0 \end{aligned}$$

70.
$$\begin{aligned} 3x - y + 3z &= 0 \\ 5x + 5y - 9z &= 0 \\ -2x + y - 3z &= 0 \end{aligned}$$

8-6 Systems of Linear Inequalities

♦ Systems of Inequalities
♦ Application

We say that the ordered pair of numbers (x, y) is a **solution of a system of linear inequalities** in two variables if the ordered pair satisfies each inequality in the system, just as in systems of linear equations in two variables. Thus, the **graph of a system of linear inequalities** is the intersection of the graphs of each inequality in the system—that is, the set of all points that satisfy each inequality in the system. We will limit our investigation of solutions of systems of inequalities to graphical methods.

♦ **SYSTEMS OF INEQUALITIES**

The examples that follow will illustrate the process for solving a system of linear inequalities. First, recall from Section 5-5 that a linear equation divides the plane into two half-planes. The solution of a linear inequality is one of the half-planes, and a test point will indicate which. For example, the solution of $x + 3y \leq 6$ is the half-plane shaded in Figure 1.

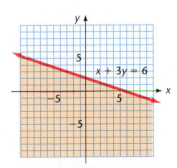

Figure 1

Example 1
Solving a System of Linear Inequalities

Solve the following linear system graphically:

$$0 \leq x \leq 8$$
$$0 \leq y \leq 4$$

Solution This system is equivalent to the system

$\left. \begin{array}{l} x \geq 0 \\ x \leq 8 \\ y \geq 0 \\ y \leq 4 \end{array} \right\}$ The solution to the system is the intersection of all four solution sets.

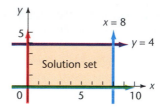

Matched Problem 1 Solve graphically: $2 \leq x \leq 6$
$ 1 \leq y \leq 3$

Example 2
Solving a System of Linear Inequalities

Solve graphically:

$$3x + 5y \leq 60$$
$$4x + 2y \leq 40$$
$$x \geq 0$$
$$y \geq 0$$

Solution

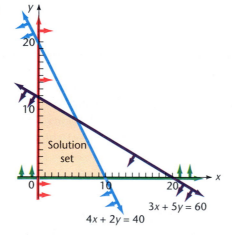

Matched Problem 2 Solve graphically: $x + 2y \geq 12$
$ 3x + 2y \geq 24$
$ x \geq 0$
$ y \geq 0$

Example 3
Solving a System of Linear Inequalities

Solve graphically:

$$x + y \leq 1$$
$$y - x \leq 1$$
$$y \geq -1$$

Solution

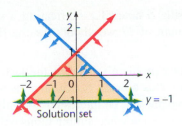

Matched Problem 3   Solve graphically: $x + 3y \leq 6$
$x - 2y \leq 2$
$x \geq -2$

♦ **APPLICATION**

Example 4
Production Scheduling

A manufacturer of sailboards makes a standard model and a competition model. The relevant manufacturing data are shown in the table. What combinations of boards can be produced each week so as not to exceed the number of workhours available in each department per week?

	Standard Model (Workhours per Board)	Competition Model (Workhours per Board)	Maximum Workhours Available per Week
Fabricating	6	8	120
Finishing	1	3	30

Solution   Let x and y be the respective number of standard and competition boards produced per week. These variables are restricted as follows:

$$6x + 8y \leq 120 \quad \text{Fabricating}$$
$$x + 3y \leq 30 \quad \text{Finishing}$$
$$x \geq 0$$
$$y \geq 0$$

The solution set of this system of inequalities is the shaded area in the figure below and is referred to as the **feasible region.** Any point within the shaded area would represent a possible production schedule. Any point outside the shaded area would represent an impossible schedule. For example, it would be possible to produce 10 standard boards and 5 competitive boards per week, but it would not be possible to produce 13 standards boards and 10 competition boards per week.

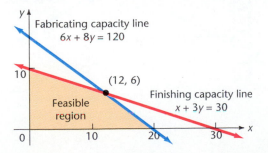

Matched Problem 4   Repeat Example 4 using 5 hours for fabricating a standard board (in place of 6 hours) and a maximum of 27 workhours for the finishing department.

There is an important problem related to the last example. Suppose that each standard model board yields a profit of $50 and each competition model yields a profit of $80. The company will be interested not only in what production combinations are possible, but also in what combinations will provide the largest possible profit. The problem of finding the most profitable combination leads to an area of mathematics called **linear programming,** which you may encounter in a subsequent course. A major result in linear programming states that the best combination will occur at one of the corners of the feasible region we graphed in Example 4. We can therefore find the most profitable solution by comparing the profits at each of the corner points:

Corner	Profit $= 50x + 80y$	
$(0, 0)$	$50 \cdot 0 + 80 \cdot 0 = 0$	
$(20, 0)$	$50 \cdot 20 + 80 \cdot 0 = 1{,}000$	
$(12, 6)$	$50 \cdot 12 + 80 \cdot 6 = 1{,}080$	Most profitable combination
$(0, 10)$	$50 \cdot 0 + 80 \cdot 10 = 800$	

Problems 49 and 50 in Exercise 8-6 illustrate linear programming problems in which the desired best combination involves the smallest possible cost, rather than the largest profit.

Answers to Matched Problems

1.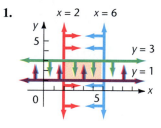

2.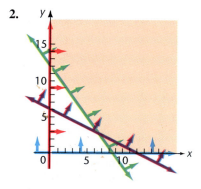

3.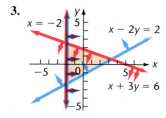

4.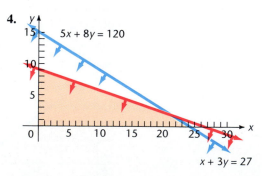

EXERCISE 8-6

A *Find the solution set of each system graphically.*

1. $-2 \leq x < 2$
$-1 < y \leq 6$

2. $-4 \leq x < -1$
$-2 < y \leq 5$

3. $-1 \leq y \leq 2$
$x \geq -3$
$x + y \leq 1$

4. $1 \leq x \leq 3$
$x + y \leq 5$
$y \geq 0$

5. $0 \leq y \leq 3$
$x - y \geq -3$
$x - y \leq 1$

6. $-2 \leq x \leq 1$
$y - x \leq 1$
$y - x \geq -2$

7. $2x + y \leq 8$
$0 \leq x \leq 3$
$0 \leq y \leq 5$

8. $x + 3y \leq 12$
$0 \leq x \leq 8$
$0 \leq y \leq 3$

9. $2x + y \leq 8$
$x + 3y \leq 12$
$x \geq 0$
$y \geq 0$

10. $x + 2y \leq 10$
$3x + y \leq 15$
$x \geq 0$
$y \geq 0$

11. $6x + 3y \leq 24$
$3x + 6y \leq 30$
$x \geq 0$
$y \geq 0$

12. $2x + y \leq 10$
$x + 2y \leq 8$
$x \geq 0$
$y \geq 0$

13. $3x + 4y \geq 8$
$4x + 3y \geq 24$
$x \geq 0$
$y \geq 0$

14. $x + 2y \geq 8$
$2x + y \geq 10$
$x \geq 0$
$y \geq 0$

15. $3x + 4y \leq 12$
$2x + 5y \leq 10$
$x \geq 0$
$y \geq 0$

16. $2x + y \leq 6$
$3x + 5y \leq 15$
$x \geq 0$
$y \geq 0$

17. $2x + 3y \leq 6$
$4x + 3y \leq 12$
$x \geq 0$
$y \geq 0$

18. $7x + 2y \leq 14$
$x + y \leq 6$
$x \geq 0$
$y \geq 0$

B *Solve graphically. Find the coordinates of all corner points in the solution set.*

19. $2x + y \leq 13$
$x + 2y \leq 14$
$x \geq 0$
$y \geq 0$

20. $2x + y \leq 10$
$x + 3y \leq 25$
$x \geq 0$
$y \geq 0$

21. $4x + y \leq 15$
$x + 2y \leq 16$
$x \geq 0$
$y \geq 0$

22. $x + 3y \leq 15$
$2x + y \leq 15$
$x \geq 0$
$y \geq 0$

23. $x + 3y \leq 11$
$2x + y \leq 12$
$x \geq 0$
$y \geq 0$

24. $x + y \leq 10$
$x + 3y \leq 14$
$x \geq 0$
$y \geq 0$

25. $2x + y \leq 14$
$x + 2y \leq 16$
$x \geq 0$
$y \geq 0$

26. $x + y \leq 11$
$x + 2y \leq 20$
$x \geq 0$
$y \geq 0$

27. $2x + 3y \leq 2$
$4x + 9y \leq 5$
$x \geq 0$
$y \geq 0$

28. $x + 2y \leq 2$
$6x + 4y \leq 6$
$x \geq 0$
$y \geq 0$

Solve graphically.

29. $x + y \leq 7$
$2x + 3y \leq 18$
$3x + 2y \leq 18$
$x \geq 0$
$y \geq 0$

30. $x + y \leq 8$
$5x + 2y \leq 30$
$2x + 5y \leq 30$
$x \geq 0$
$y \geq 0$

31. $x + y \leq 5$
$3x + y \leq 12$
$x + 3y \leq 12$
$x \geq 0$
$y \geq 0$

32. $x + y \leq 6$
$x + 2y \leq 10$
$2x + y \leq 10$
$x \geq 0$
$y \geq 0$

33. $2x + y \leq 10$
$x + 2y \leq 8$
$x + 6y \leq 18$
$x \geq 0$
$y \geq 0$

34. $x + 2y \leq 10$
$3x + y \leq 18$
$x + 5y \leq 20$
$x \geq 0$
$y \geq 0$

35. $2y - x \geq 0$
$2y + x \leq 4$
$x \geq -2$

36. $y - 2x \leq 4$
$2y + x \leq 1$
$y \geq -1$

C 37. $3x + 5y \geq 60$
$4x + 2y \geq 40$
$2 \leq x \leq 14$
$6 \leq y \leq 18$

38. $2x + y \geq 8$
$x + 2y \geq 10$
$1 \leq x \leq 7$
$3 \leq y \leq 9$

39. $2x - 3y \geq 0$
$4x + 9y \leq 36$
$x \geq 0$
$0 \leq y \leq 1$

40. $3x - 5y \geq 0$
$2x + 3y \leq 18$
$x \geq 0$
$0 \leq y \leq 2$

41. $x + y \leq 10$
$20x + 7y \leq 140$
$8x + 15y \leq 120$
$x \geq 0$
$y \geq 0$

42. $x + y \leq 8$
$x + 6y \leq 30$
$5x + y \leq 35$
$x \geq 0$
$y \geq 0$

APPLICATIONS

43. *Manufacturing—resource allocation* A manufacturing company makes two types of water skis: a trick ski and a slalom ski. The trick ski requires 6 workhours for fabricating and 1 workhour for finishing. The slalom ski requires 4 workhours for fabricating and 1 workhour for finishing. The maximum workhours available per day for fabricating and finishing are 108 and 24, respectively. If x is the number of trick skis and y is the number of slalom skis produced per day, write a system of inequalities that indicates appropriate restraints on x and y. Find the set of feasible solutions graphically for the number of each type of ski that can be produced.

44. Repeat Problem 43 for the situation where the number of workhours available per day for fabricating and finishing are 160 and 30, respectively.

45. Refer to Problem 43. Find the most profitable combination of skis to produce if the profit is $15 on each trick ski and $12 on each slalom ski.

46. Refer to Problem 44. Find the most profitable combination of skis to produce if the profit is $15 on each trick ski and $12 on each slalom ski.

47. *Nutrition*   A dietitian in a hospital is to arrange a special diet using two foods. Each ounce of food M contains 30 units of calcium, 10 units of iron, and 10 units of vitamin A. Each ounce of food N contains 10

units of calcium, 10 units of iron, and 30 units of vitamin A. The minimum requirements in the diet are 360 units of calcium, 160 units of iron, and 240 units of vitamin A. If x is the number of ounces of food M used and y is the number of ounces of food N used, write a system of linear inequalities that reflects the conditions indicated. Find the set of feasible solutions graphically for the amount of each kind of food that can be used.

48. Repeat Problem 47 for the situation where the minimum requirements in the diet are 480 units of calcium, 180 units of iron, and 300 units of vitamin A.

49. Refer to Problem 47. Find the least costly combination of foods M and N to use if each ounce of M costs 8 cents and each ounce of N costs 10 cents.

50. Refer to Problem 48. Find the least costly combination of foods M and N to use if each ounce of M costs 8 cents and each ounce of N costs 10 cents.

8-7 Systems Involving Second-Degree Equations

In this section we consider some systems that involve at least one second-degree equation in two variables. The methods used to solve these systems are illustrated by the examples.

Example 1
A System with One Quadratic Equation

Solve the system:

$$4x^2 + y^2 = 25$$
$$2x + 3y = 5$$

Solution   In this type of problem, involving one linear equation, the substitution principle is effective. Solve the linear equation for one variable in terms of the other, and then substitute into the nonlinear equation to obtain a quadratic equation in one variable.

$$4x^2 + y^2 = 25 \qquad \text{Solve } 2x + 3y = 5 \text{ for } y \text{ in terms of } x.$$

$$2x + 3y = 5$$

$$3y = 5 - 2x$$

$$y = \frac{5 - 2x}{3} \qquad \text{Substitute into } 4x^2 + y^2 = 25.$$

$$4x^2 + y^2 = 25$$

$$4x^2 + \left(\frac{5 - 2x}{3}\right)^2 = 25 \qquad \text{Simplify, and write in standard quadratic form, } ax^2 + bx + c = 0.$$

$$4x^2 + \frac{25 - 20x + 4x^2}{9} = 25 \qquad \text{Clear fractions.}$$

$$36x^2 + 25 - 20x + 4x^2 = 225$$

$$40x^2 - 20x - 200 = 0 \qquad \text{Multiply both sides by } \tfrac{1}{20}.$$

$$2x^2 - x - 10 = 0 \qquad \text{Solve—we use the factoring method.}$$

$$(2x - 5)(x + 2) = 0$$

$$2x - 5 = 0 \qquad \text{or} \qquad x + 2 = 0$$
$$x = \tfrac{5}{2} \qquad\qquad\qquad x = -2$$

These values are substituted back into the linear equation

$$y = \frac{5 - 2x}{3}$$

to find the corresponding values for y. If we substitute these values back into the second-degree equation, we may obtain extraneous roots. You should try it and see why. Remember: a solution of a system is an ordered pair of numbers that satisfies *both* equations.

$$\text{For } x = \tfrac{5}{2}: \quad y = \frac{5 - 2(\tfrac{5}{2})}{3} \qquad \text{For } x = -2: \quad y = \frac{5 - 2(-2)}{3}$$

$$= 0 \qquad\qquad\qquad\qquad = 3$$

Thus, $(\tfrac{5}{2}, 0)$ and $(-2, 3)$ are the solutions to the system, as can easily be checked.

The equations in this example can be graphed. The first equation is an ellipse, and the second is a straight line. The intersection points of the ellipse and the line are the solutions of the system:

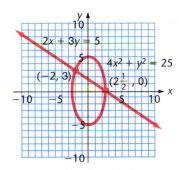

Graphing the equations provides another check on the solutions. In situations where the system cannot be solved algebraically, graphing allows us to estimate the real solutions.

Matched Problem 1   Solve the system: $2x^2 - y^2 = 1$
$$3x + y = 2$$

Example 2
**A System with Two
Quadratic Equations**

Solve the system:

$$x^2 - y^2 = 5$$
$$x^2 + 2y^2 = 17$$

Solution For this system, where both equations are second-degree, we use elimination.

$$
\begin{aligned}
x^2 - \ y^2 &= 5 \\
x^2 + 2y^2 &= 17 \\
\hline
-3y^2 &= -12 \\
y^2 &= 4 \\
y &= \pm 2
\end{aligned}
$$
    Proceed as with linear equations—
    subtract to eliminate x.

For $y = 2$:
$$
\begin{aligned}
x^2 - (2)^2 &= 5 \\
x^2 &= 9 \\
x &= \pm 3
\end{aligned}
$$
    Substitute into either original
    equation; then solve for x.

For $y = -2$:
$$
\begin{aligned}
x^2 - (-2)^2 &= 5 \\
x^2 &= 9 \\
x &= \pm 3
\end{aligned}
$$
    Substitute into either original
    equation; then solve for x.

Thus, $(3, -2)$, $(3, 2)$, $(-3, -2)$, and $(-3, 2)$ are the four solutions to the system. Checking the solutions is left to you. The graph of the system shows the reasonableness of the solutions:

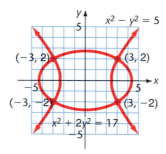

Matched Problem 2 Solve the system:
$$2x^2 - 3y^2 = 5$$
$$3x^2 + 4y^2 = 16$$

Example 3
**A System with
Complex Solutions**

Solve the system:

$$x^2 + 4y - 2y^2 = 5 \qquad (1)$$
$$xy - y^2 = 0 \qquad (2)$$

Solution The presence of four unlike terms in the equations means that elimination will not work here. Instead, we solve the second equation for one variable in terms of the other, and then use substitution.

$$
\begin{aligned}
y(x - y) &= 0 \\
y = 0 \qquad \text{or} \qquad x - y &= 0 \\
y &= x
\end{aligned}
$$
    Factor Equation (2).
    Substitute each of these in turn into
    Equation (1) and proceed as before.

For $y = 0$, replace y with 0 in Equation (1) and solve for x:

$$x^2 + 4(0) - 2(0)^2 = 5$$
$$x^2 = 5$$
$$x = \pm\sqrt{5} \qquad (\sqrt{5}, 0) \text{ and } (-\sqrt{5}, 0) \text{ are solutions.}$$

For $y = x$, replace y with x in Equation (1) and solve for x:

$$x^2 + 4x - 2x^2 = 5$$
$$-x^2 + 4x - 5 = 0$$
$$x = \frac{4 \pm \sqrt{16 - 20}}{2} = 2 \pm i$$

Substitute these values back into $y = x$ to find corresponding values of y. For $x = 2 + i$, $y = 2 + i$; for $x = 2 - i$, $y = 2 - i$. Thus, $(2 + i, 2 + i)$ and $(2 - i, 2 - i)$ are solutions.

Thus, $(\sqrt{5}, 0)$, $(-\sqrt{5}, 0)$, $(2 + i, 2 + i)$, and $(2 - i, 2 - i)$ are the four solutions to the system. Checking the solutions is left to you. These equations require more advanced techniques to graph efficiently and their graphs are therefore omitted.

CAUTION

We could have started the solution to Example 3 by solving $xy - y^2 = 0$ for x as follows:

$$xy - y^2 = 0$$
$$xy = y^2$$
$$x = y$$

However, we would have lost the two solutions $(\sqrt{5}, 0)$ and $(-\sqrt{5}, 0)$. Dividing both sides of $xy = y^2$ by the variable y results in this loss.

Do not divide both sides of an equation by an expression involving a variable for which you are solving.

The operation may result in the loss of solutions.

Matched Problem 3   Solve the system: $x^2 + xy - y^2 = 4$
$$2x^2 - xy = 0$$

Example 3 is somewhat specialized. However, it suggests how substitution may be effective for some problems involving an xy term.

To get an idea of how many real solutions you might expect from a system involving second-degree equations in two variables, you have only to look at the number of ways two conics can intersect at points. In general, a system of one linear and one quadratic equation can have at most two real solutions, and a system of two quadratic equations can have at most four real solutions. Some of the solutions may be complex, and in some cases there may be an infinite number of solutions when the two equations represent the same graph. The graphs of Examples 1 and 2 illustrate this result. Figure 1 provides additional examples.

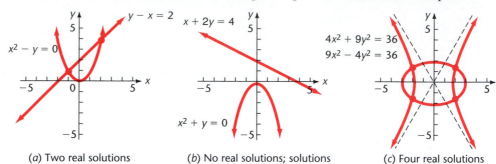

(a) Two real solutions   (b) No real solutions; solutions   (c) Four real solutions
                                     are nonreal complex

Figure 1

Answers to   **1.** $(1, -1)$, $(\frac{5}{7}, -\frac{1}{7})$ **2.** $(2, 1)$, $(2, -1)$, $(-2, 1)$, $(-2, -1)$
Matched Problems   **3.** $(0, 2i)$, $(0, -2i)$, $(2i, 4i)$, $(-2i, -4i)$

EXERCISE 8-7

A *Solve each system.*

1. $x^2 + y^2 = 25$
   $y = -4$

2. $x^2 + y^2 = 169$
   $x = -12$

3. $x^2 + y^2 = 36$
   $x = 4$

4. $x^2 + y^2 = 1$
   $y = \frac{1}{2}$

5. $y^2 = 2x$
   $x = y - \frac{1}{2}$

6. $x + y = 24$
   $y = 3x^2$

7. $x^2 - y^2 = 9$
   $x - y = 3$

8. $8x^2 - y^2 = 16$
   $y = 2x$

9. $x^2 + 4y^2 = 32$
   $x + 2y = 0$

10. $x^2 + \dfrac{y^2}{4} = 1$
    $y = x - 1$

11. $\dfrac{x^2}{9} + y^2 = 1$
    $x + y = 1$

12. $2x^2 - 3y^2 = 25$
    $x + y = 0$

13. $x^2 = 2y$
    $3x = y + 5$

14. $y^2 = -x$
    $x - 2y = 5$

15. $x^2 - y^2 = 3$
    $x^2 + y^2 = 5$

16. $2x^2 + y^2 = 24$
    $x^2 - y^2 = -12$

17. $x^2 - 2y^2 = 1$
    $x^2 + 4y^2 = 25$

18. $x^2 + y^2 = 10$
    $16x^2 + y^2 = 25$

B **19.–28.** Graph each system in Problems 1–10. Check the solution you found by comparing it to the graph.

Solve each system for real solutions only. Check your solution by graphing.

29. $y = x^2$
   $x - y = 3$

30. $x + y = 1$
   $y = -x^2$

31. $x^2 + \dfrac{y^2}{25} = 1$
    $x - y = 8$

32. $\dfrac{x^2}{4} + y^2 = 1$
    $x + y = 3$

33. $x^2 + \dfrac{y^2}{4} = 1$

   $\dfrac{x^2}{4} - y^2 = 1$

34. $\dfrac{x^2}{25} + y^2 = 1$

    $\dfrac{x^2}{36} - y^2 = 1$

35. $x^2 + y^2 = 36$

   $\dfrac{x^2}{25} + \dfrac{y^2}{16} = 1$

36. $\dfrac{x^2}{25} + \dfrac{y^2}{16} = 1$

    $x^2 + y^2 = 9$

In Problems 37–46, solve the system.

37. $xy - 6 = 0$
   $x - y = 4$

38. $xy = -4$
   $y - x = 2$

39. $x^2 + xy - y^2 = -5$
   $y - x = 3$

40. $x^2 - 2xy + y^2 = 1$
   $x - 2y = 2$

41. $2x^2 - 3y^2 = 10$
   $x^2 + 4y^2 = -17$

42. $2x^2 + 3y^2 = -4$
   $4x^2 + 2y^2 = 8$

43. $x^2 + y^2 = 20$
   $x^2 = y$

44. $x^2 - y^2 = 2$
   $y^2 = x$

45. $x^2 + y^2 = 16$
   $y^2 = 4 - x$

46. $x^2 + y^2 = 5$
   $x^2 = 4(2 - y)$

47. Find the dimensions of a rectangle with area 32 square feet and perimeter 36 feet.

48. Find two numbers such that their sum is 1 and their product is 1.

51. $x^2 + 2xy + y^2 = 36$
$x^2 - xy = 0$

52. $2x^2 - xy + y^2 = 8$
$(x - y)(x + y) = 0$

53. $x^2 - 2xy + 2y^2 = 16$
$x^2 - y^2 = 0$

54. $x^2 + xy - 3y^2 = 3$
$x^2 + 4xy + 3y^2 = 0$

C *Solve each system.*

49. $2x^2 + y^2 = 18$
$xy = 4$

50. $x^2 - y^2 = 3$
$xy = 2$

CHAPTER SUMMARY

8-1 SYSTEMS OF LINEAR EQUATIONS IN TWO VARIABLES

The standard form of a system of linear equations in two variables is

$$ax + by = m$$
$$cx + dy = n$$

The **solutions of the system** are all ordered pairs (x, y) that satisfy both equations. Solutions can be estimated by **solving graphically**—that is, by graphing each equation and estimating the intersection point, if any, from the graph. Solutions also can be found by **substitution**—that is, by solving one equation for one variable in terms of the other, substituting the result in the other equation, and then solving.

Equivalent systems are those with exactly the same solution set. The following operations produce equivalent systems:

1. Two equations are interchanged.
2. An equation is multiplied by a nonzero constant.
3. A constant multiple of another equation is added to a given equation.

These operations can be used to solve the system by **elimination**—that is, by eliminating one variable.

A system that has no solution is called **inconsistent;** one that has an infinite number of solutions is called **dependent.**

8-2 APPLICATION: MIXTURE PROBLEMS

Applied problems often can be solved using systems of equations. It remains important to check solutions in the original problem.

8-3 SYSTEMS OF LINEAR EQUATIONS IN THREE VARIABLES

A system of three equations in three variables can be solved by elimination. **Solutions, equivalent systems,** and **dependent systems** have the same meaning as for the two-equation–two-variable case.

8-4 SYSTEMS OF EQUATIONS AND MATRICES (OPTIONAL)

A **matrix** is a rectangular array of numbers—the **elements** of the matrix—written within brackets. The **augmented matrix** of the standard system of two equations in two variables is

$$\begin{bmatrix} a & b & | & m \\ c & d & | & n \end{bmatrix}$$

The operations that are performed on the system only need to be done on the coefficients and constants in the augmented matrix to obtain a **row-equivalent matrix**—that is, an augmented matrix of an equivalent system. These **row operations** are:

1. Two rows are interchanged.
2. A row is multiplied by a nonzero constant.
3. A constant multiple of another row is added to a given row.

The **augmented matrix method** is a simple, concise way to solve systems and can be extended to large-scale systems.

8-5   DETERMINANTS AND CRAMER'S RULE (OPTIONAL)

The **determinant** of the 2×2 matrix

$$\begin{bmatrix} a_{11} & a_{12} \\ a_{21} & a_{22} \end{bmatrix}$$

is denoted by

$$\begin{vmatrix} a_{11} & a_{12} \\ a_{21} & a_{22} \end{vmatrix}$$

and is defined to be $a_{11}a_{22} - a_{21}a_{12}$. The determinant of the 3×3 matrix

$$\begin{bmatrix} a_{11} & a_{12} & a_{13} \\ a_{21} & a_{22} & a_{23} \\ a_{31} & a_{32} & a_{33} \end{bmatrix}$$

is denoted by

$$\begin{vmatrix} a_{11} & a_{12} & a_{13} \\ a_{21} & a_{22} & a_{23} \\ a_{31} & a_{32} & a_{33} \end{vmatrix}$$

and is defined to be

$$a_{11}a_{22}a_{33} - a_{11}a_{32}a_{23} + a_{21}a_{32}a_{13} - a_{21}a_{12}a_{33} + a_{31}a_{12}a_{23} - a_{31}a_{22}a_{13}$$

The **minor of an element** in a third-order determinant is a second-order determinant obtained by deleting the row and column that contains the element. The **cofactor of the element** a_{ij} in the ith row and jth column is the product of the minor of a_{ij} and $(-1)^{i+j}$. Determinants can be evaluated using **expansion by cofactors:** the value of a determinant of order 3 is the sum of three products obtained by multiplying each element of any one row, or each element of any one column, by its cofactor. This generalizes to determinants of any size. **Cramer's rule** for two equations and two variables gives the solution to the system in terms of determinants. Given the system

$$\begin{aligned} a_{11}x + a_{12}y &= k_1 \\ a_{21}x + a_{22}y &= k_2 \end{aligned} \qquad \text{with} \qquad D = \begin{vmatrix} a_{11} & a_{12} \\ a_{21} & a_{22} \end{vmatrix} \neq 0$$

then

$$x = \frac{\begin{vmatrix} k_1 & a_{12} \\ k_2 & a_{22} \end{vmatrix}}{D} \qquad \text{and} \qquad y = \frac{\begin{vmatrix} a_{11} & k_1 \\ a_{21} & k_2 \end{vmatrix}}{D}$$

Cramer's rule generalizes completely for any size of linear system that has the same number of variables as equations.

8-6 SYSTEMS OF LINEAR INEQUALITIES

The **graph of a system of linear inequalities** is the intersection of half-planes representing the graph of each inequality.

8-7 SYSTEMS INVOLVING SECOND-DEGREE EQUATIONS

Systems involving second-degree equations sometimes can be solved by substitution or elimination. Graphing the equations can confirm the reasonableness of the solution.

CHAPTER REVIEW EXERCISE

Work through all the problems in this chapter review and check answers in the back of the book. Answers to all the problems are there, and following each answer is a number in italics indicating the section in which that type of problem is discussed. Where weaknesses show up, review appropriate sections in the text. (Problems 33–44, 67–70, and 72–80 depend on optional Sections 8-4 and 8-5.)

A *Solve by graphing. The solutions have integer values.*

1. $x - y = 5$
$x + y = 7$

2. $2x + 3y = 12$
$x + 2y = 7$

3. $x - y = 5$
$2x - 5y = 1$

Solve graphically.

4. $2x + y \le 8$
$2x + 3y \le 12$
$x \ge 0$
$y \ge 0$

5. $x + y \le 10$
$2x + y \le 8$
$x \ge 0$
$y \ge 0$

6. $x - y \le 8$
$y - x \le 6$
$x \ge 0$
$y \ge 0$

Solve by substitution.

7. $2x + 3y = 7$
$3x - y = 5$

8. $x - 3y = 14$
$2x + y = 7$

Solve by elimination using addition.

9. $2x + 3y = 7$
$3x - y = 5$

10. $5x + 2y = -1$
$2x - 3y = -27$

11. $5x + 3y = 26$
$8x - 2y = -6$

12. $3x - 5y = 9$
$-2x + 3y = -5$

Solve.

13. $y + 2z = 4$
$x - z = -2$
$x + y = 1$

14. $x^2 + y^2 = 2$
$2x - y = 3$

15. $x^2 - y^2 = 7$
$x^2 + y^2 = 25$

16. $x + 2y + z = 3$
$2x + 3y + 4z = 3$
$x + 2y + 3z = 1$

17. $x + y + z = 8$
$2x - y + z = 3$
$y - 2z = -5$

18. $x + y + z = 8$
$x - 2y + 3z = 4$
$2x + 5z = 12$

19. $x - 3y = 8$
$-4x + 12y = 32$

20. $x^2 + y^2 = 25$
$4x - 3y = 0$

21. $x^2 + y = 2$
$x^2 + y^2 = 4$

22. $x + 2y + z = 4$
$x - z = 0$
$x + y = 2$

23. $x + 2y + z = 4$
$x + z = 0$
$x + y = 2$

24. $x + 2y = 6$
$3x + 4z = 0$
$-x + z = -4$

25. $x^2 - 9y^2 = 36$
$x + 9y^2 = 6$

26. $x^2 - 9y^2 = 36$
$x - 9y^2 = 16$

27. $x^2 - 9y^2 = 36$
$3x - y = 0$

28. $x + 3y + 5z = 10$
$2x - 2y + z = 9$
$5x + z = 8$

29. $2x + 5y = 10$
$3x + 8y = 5$

30. $2x + 3y = 6$
$5x + 7y = -2$

31. $x^2 + y^2 = 9$
$x - y = 0$

32. $y = 4x^2$
$y = 11x + 3$

Evaluate each determinant.

33. $\begin{vmatrix} 1 & 2 \\ 3 & 4 \end{vmatrix}$

34. $\begin{vmatrix} -3 & 5 \\ 9 & -15 \end{vmatrix}$

35. $\begin{vmatrix} 9 & 7 \\ 6 & 8 \end{vmatrix}$

36. $\begin{vmatrix} \frac{1}{2} & 2 \\ \frac{1}{12} & \frac{2}{3} \end{vmatrix}$

Perform the indicated row operations on the following matrix:

$$\begin{bmatrix} 1 & 2 & 3 \\ 4 & 5 & 6 \end{bmatrix}$$

37. $R_1 \leftrightarrow R_2$

38. $R_2 + (-1)R_1 \to R_2$

39. $\frac{1}{4}R_2 \to R_2$

40. $R_1 + \frac{1}{2}R_2 \to R_1$

In Problems 41–44, solve using Cramer's rule.

41. $3x + 8y = 9$
$2x + 5y = -3$

42. $4x + 3y = 5$
$3x + 4y = 4$

43. $x + y = 3$
$x - y = 1$

44. $7x + 18y = -2$
$2x + 5y = 2$

45. If you have 30 nickels and dimes worth $2.30 in your pocket, how many of each do you have?

46. If $6,000 is to be invested, part at 10% and the rest at 6%, how much should be invested at each rate so that the total annual return from both investments is $440?

47. A mixture of two kinds of candy, one worth $2.40 per pound and the other $3.00 per pound, results in 10 pounds of candy with a value of $28. How much of each kind is included in the mix?

48. A concert charges $8 for tickets purchased in advance and $10 for tickets purchased at the door. Attendance of 3,700 yielded gate receipts of $35,000. How many advance tickets were sold? How many tickets were sold at the door?

B *Solve graphically. Find the coordinates of all corner points.*

49. $2x + y \geq 8$
$x + 3y \geq 12$
$x \geq 0$
$y \geq 0$

50. $x + y \leq 9$
$2x + 3y \leq 22$
$x \geq 0$
$y \geq 0$

51. $x + 2y \geq 17$
$3x + y \geq 16$
$x \geq 0$
$y \geq 0$

52. $x + y \leq 10$
$x + 3y \leq 26$
$3x + y \leq 24$
$x \geq 0$
$y \geq 0$

53. $x + y \geq 12$
$5x + y \geq 20$
$2x + 5y \geq 30$
$x \geq 0$
$y \geq 0$

Solve.

54. $3x - 2y = -1$
$-6x + 4y = 3$

55. $3x - 2y - 7z = -6$
$-x + 3y + 2z = -1$
$x + 5y + 3z = 3$

56. $3x^2 - y^2 = -6$
$2x^2 + 3y^2 = 29$

57. $x^2 = y$
$y = 2x - 2$

58. $2x + 3y - 4z = -25$
$5x - 4y - 3z = 16$
$-x + y + 2z = 0$

59. $x + 3y + 2z = 2$
$x - 3y - 2z = -1$
$2x + 9y - 8z = 2$

60. $x^2 + y^2 = 25$
$y + 2x^2 = 22$

61. $x^2 + y^2 = 25$
$x^2 + 4y^2 = 16$

62. $x^2 - 4y = 0$
$-4x + y^2 = 0$

63. $x^2 + \dfrac{y^2}{9} = 1$
$x - y^2 = 1$

64. $x^2 + y^2 = 9$
$x^2 + \dfrac{y^2}{9} = 1$

65. $\dfrac{x^2}{9} - \dfrac{y^2}{16} = 1$
$4x + 3y = 0$

66. $9x^2 - y = 0$
$y = 4x + 13$

Solve the system represented by each augmented matrix. The variables are x, y, and, if needed, z.

67. $\begin{bmatrix} 3 & 2 & | & 3 \\ 1 & 3 & | & 8 \end{bmatrix}$

68. $\begin{bmatrix} 3 & -1 & | & 0 \\ -2 & 1 & | & -1 \end{bmatrix}$

69. $\begin{bmatrix} 1 & 2 & | & 5 \\ 0 & 3 & | & 6 \end{bmatrix}$

70. $\begin{bmatrix} 4 & 3 & | & 10 \\ 1 & 2 & | & 5 \end{bmatrix}$

71. $\begin{bmatrix} 1 & 2 & 3 & | & 4 \\ 1 & 0 & 1 & | & 6 \\ 0 & -1 & 1 & | & 5 \end{bmatrix}$

72. $\begin{bmatrix} 1 & -1 & 1 & | & 0 \\ 0 & 1 & -1 & | & -1 \\ 2 & -1 & -1 & | & -3 \end{bmatrix}$

73. $\begin{bmatrix} 1 & -1 & 1 & | & 3 \\ 1 & 1 & -1 & | & 4 \\ 0 & -2 & -2 & | & -1 \end{bmatrix}$

74. $\begin{bmatrix} 1 & 2 & 3 & | & 4 \\ 1 & 0 & 1 & | & 6 \\ 2 & 6 & 8 & | & 10 \end{bmatrix}$

Evaluate each determinant.

75. $\begin{vmatrix} 1 & 1 & 1 \\ 0 & 1 & 2 \\ 3 & 2 & 1 \end{vmatrix}$

76. $\begin{vmatrix} 1 & 2 & 1 \\ 2 & 1 & 1 \\ 3 & 0 & 1 \end{vmatrix}$

77. $\begin{vmatrix} 13 & 14 & 15 \\ 0 & 16 & 17 \\ 0 & 0 & 18 \end{vmatrix}$

78. $\begin{vmatrix} 0 & 2 & 4 \\ 1 & 3 & 5 \\ 2 & 4 & 6 \end{vmatrix}$

In Problems 79 and 80, solve using Cramer's Rule.

79. $x + y + z = 15$
$x + z = 10$
$6y - 5z = 0$

80. $x + y + z = 12$
$2x - z = 1$
$2y - z = 3$

81. The perimeter of a rectangle is 22 centimeters. If its area is 30 square centimeters, find the length of each side.

82. A chemist has one 40% and one 70% solution of acid in stock. How much of each should be used to get 100 grams of a 49% solution?

83. Find the equation of the circle, in the form $x^2 + y^2 + Dx + EY + F = 0$, if the points $(5, 7)$, $(-1, -1)$, and $(6, 0)$ lie on the circle.

84. A homeowner who wishes to pack books prior to a move buys two sizes of boxes, 1 cubic foot and 1.5 cubic feet. The 30 boxes purchased have a capacity of 39 cubic feet. How many of each size were purchased?

C *In Problems 85–94, solve the system.*

85. $\frac{1}{4}x - \frac{3}{4}y = -\frac{3}{8}$
$-\frac{2}{3}x + 2y = -1$

86. $x^2 + 2xy - y^2 = -4$
$x^2 - xy = 0$

87. $\dfrac{x^2}{9} - \dfrac{y^2}{4} = 1$
$x^2 + y^2 = 25$

88. $\dfrac{x^2}{9} - \dfrac{y^2}{4} = 1$
$\dfrac{x^2}{16} - \dfrac{y^2}{9} = 1$

89. $x + y = 10$
$xy = 1$

90. $x^2 + y^2 = 4$
$xy = 1$

91. $x^2 + y^2 = 9$
$x^2 - y^2 = 0$

92. $x^2 - y = 0$
$y - x^2 = -1$

93. $\frac{1}{3}x + \frac{1}{3}y + \frac{1}{3}z = 1$
$\frac{1}{4}y + \frac{1}{4}z = 1$
$\frac{1}{12}x + \frac{1}{6}y + \frac{1}{4}z = 1$

94. $0.2x + 0.4y + 0.4z = 1.4$
$0.5x + 1.5y + 2.0z = 8.0$
$0.4x + 0.4y + 0.8z = 3.2$

95. Find two numbers whose product and sum are both 5.

96. Find the equation of the parabola, in the form $y = ax^2 + bx + c$, that passes through the points $(0, \frac{1}{4})$, $(1, \frac{7}{4})$, and $(2, \frac{29}{4})$.

97. A container holds 120 packages. Some of the packages weigh $\frac{1}{2}$ pound each, and the rest weigh $\frac{1}{3}$ pound each. If the total contents of the container weigh 48 pounds, how many are there of each type of package?

98. A lab assistant wishes to obtain a food mix that contains, among other things, 27 grams of protein, 5.4 grams of fat, and 19 grams of moisture. Mixes with the compositions listed in the table are available. How many grams of each mix should be used to get the desired diet mix? Set up a system of equations and solve using augmented matrix methods.

Mix	Protein (%)	Fat (%)	Moisture (%)
A	30	3	10
B	20	5	20
C	10	4	10

99. Refer to Problem 98. Suppose that mix *C* is unavailable and that the other two mixes must be combined to yield a diet that has at least 27 grams of protein, 5.4 grams of fat, and 19 grams of moisture. Set up a system of inequalities to represent these restrictions and solve graphically.

100. Refer to Problem 99. If mix *A* and mix *B* each cost 3 cents per gram, what is the cheapest combination of the two products that will meet the requirements?

CHAPTER PRACTICE TEST

The following practice test is provided for you to test your knowledge of the material in this chapter. You should try to complete it in 50 minutes or less. Answers in the back of the book indicate the section in the text where the material in the question is covered. Actual tests in your class may vary from this practice test in difficulty, length, or emphasis, depending on the goals of your course or instructor. (Problems 6–9, 16, and 17 depend on optional sections 8-4 and 8-5.)

Solve.

1. $x + 3y = 5$
$3x + 2y = 1$

2. $3x + 4y = 1$
$2x - 3y = 12$

3. $x + y + z = 6$
$2x + y = 8$
$3y + 4z = 10$

4. $x + 2y + 3z = 4$
$2x - y + z = 5$
$3x + y + 4z = 6$

5. $x + 2y + 3z = 4$
$-x + y + 3z = 5$
$2x + y = -1$

6. The system with variables x and y, and augmented matrix

$$\begin{bmatrix} 1 & 2 & | & 3 \\ 3 & 2 & | & 1 \end{bmatrix}$$

7. The system with variables x, y, and z, and augmented matrix

$$\begin{bmatrix} 1 & 2 & 2 & | & 4 \\ 1 & 2 & 1 & | & 5 \\ 2 & 1 & 1 & | & 6 \end{bmatrix}$$

8. The system with variables x, y, and z, and augmented matrix

$$\begin{bmatrix} 1 & 2 & 3 & | & 6 \\ 2 & 3 & 1 & | & 6 \\ 3 & 1 & 2 & | & 6 \end{bmatrix}$$

9. The system

$$2x + y + 3z = 1$$
$$3x + 2y + 7z = 0$$
$$4x + y + 6z = 0$$

using the fact that

$$\begin{vmatrix} 2 & 1 & 3 \\ 3 & 2 & 7 \\ 4 & 1 & 6 \end{vmatrix} = 5$$

10. $x^2 + y^2 = 16$
$x^2 - y = 4$

11. $x^2 + y^2 = 16$
$x - y = 0$

12. $x^2 + y^2 = 16$
$$\dfrac{x^2}{25} + \dfrac{y^2}{36} = 1$$

In Problems 13–15, solve the system graphically.

13. $4x + 3y \le 12$
$x + 5y \le 10$
$x \ge 0$
$y \ge 0$

14. $3x - 2y \le 6$
$2x + 3y \ge -6$
$3x - 2y \ge -6$

15. $2x + 3y \ge 6$
$3x + 2y \ge 6$
$x \ge 0$
$y \ge 0$

16. Find the final result of the following row operations, in order, on the matrix

$$\begin{bmatrix} 1 & 2 & | & 3 \\ 4 & 5 & | & 6 \end{bmatrix}$$

(A) $R_1 \leftrightarrow R_2$

(B) $3R_2 \to R_2$

(C) $R_1 + (-1)R_2 \to R_1$

17. Evaluate: $\begin{vmatrix} 2 & 3 \\ 4 & 5 \end{vmatrix} - \begin{vmatrix} 1 & 2 & 3 \\ 0 & 4 & 5 \\ 6 & 7 & 8 \end{vmatrix}$

18. Find two numbers whose product and sum are both equal to $-\frac{1}{2}$.

19. General admission tickets to a minor league baseball game cost $1.50 and reserved seats cost $2.50. A crowd of 760 resulted in gate receipts of $1,280. How many of each type of ticket were sold?

20. Fertilizer mix *A* has 18% nitrogen content and 12% potash. Mix *B* has 12% nitrogen and 15% potash. How many pounds of each mix are needed to supply exactly 26.4 pounds of nitrogen and 24.6 pounds of potash?

Functions

The relationships

$$y = mx + b \qquad \text{See Chapter 5.}$$

$$y = ax^2 + bx + c \qquad \text{See Chapter 7.}$$

share the important characteristic that for each value of x just one value of y is determined. These are examples of a relationship, or correspondence, between variables that is called a *function*. The function concept is among the most important in mathematics and it will be studied extensively in this and subsequent courses. The concept is introduced in this chapter and related to several algebraic topics studied in previous chapters.

Photo reference: see Exercise 9-3, Problem 77

9-1   Functions

♦ Functions
♦ Common Ways of Specifying Functions
♦ Vertical-Line Test for Graphs
♦ Functions Specified by Equations
♦ A Brief History of the Function Concept

One of the most important mathematical aspects of the sciences is establishing relationships between various phenomena. Once a relation is known, predictions can be made. An engineer can use a formula to predict pressures on a bridge for various wind speeds, an economist can predict unemployment rates given various levels of government spending, a chemist can use a formula to predict the pressure of an enclosed gas given its temperature, and so on. Establishing and working with such relationships is so fundamental to both pure and applied science that it is desirable to describe them in the precise language of mathematics.

We build a basis for a mathematical description by looking at the variables connected with phenomena such as those mentioned above. In each case, the values of one variable provide information about the related second variable:

VARIABLE	RELATED VARIABLE
Wind speed	Pressure on bridge
Government spending	Unemployment rate
Temperature	Pressure of gas

The key feature of these relationships is that for any value of the given variable, the related variable should be completely determined. For example, once the wind speed is known, the pressure on the bridge should be completely determined. That is, the pressure on the bridge is a unique number corresponding to the wind speed. The rule of correspondence that determines the pressure from the wind speed is called a *function*, and we say that *pressure is a function of wind speed*.

In this section, we define what a function is and consider ways in which a function can be specified.

♦ **FUNCTIONS**

In addition to the scientific relationships mentioned above, we also can look for functions in more commonplace settings. For example:

To each item on the shelf in a grocery store there corresponds a price.

To each square there corresponds an area.

To each student there corresponds a grade point average (GPA).

To each letter on a telephone dial there corresponds a number on the dial.

Once again, for each value of the first variable, there is a unique corresponding value of the related variable:

VARIABLE	RELATED VARIABLE
Grocery item	Price
Square	Area
Student	GPA
Letter on phone dial	Number on phone dial

If we denote the first variable by x and the related variable by y, we say that y *is a function of x* if for each value of x there is a unique corresponding value for y. The actual *rule of correspondence* is called a *function*. Here, we allow the word "value" to represent an object—for example, a grocery item, student, or letter—as well as a number. We thus have the following informal definition of a function:

Informal Definition of a Function

A **function** is a rule that produces a correspondence between a first variable x and a second variable y such that to each value of x there corresponds *one and only one* value for y. In this case, we say that **y is a function of x.**

If a variable y is a function of a variable x, we call the variable denoted by x the **independent variable** and the variable denoted by y the **dependent variable.** The value of the dependent variable is determined by, or dependent on, the value of the independent variable. The set of allowable values for the independent variable is called the **domain** of the function. The set of values for the dependent variable is called the **range** of the function. (See Appendix A for a review of set terminology.)

For most of the functions you will encounter in beginning college mathematics courses, the variables being related will represent numbers. That is, both the domain and the range will be sets of numbers. The examples of the phone dial and the student-to-GPA correspondence indicate that this is not necessarily the case. The function concept is more general than just relationships between numerical variables. This leads to a more formal, general definition of a function:

Formal Definition of Function

A **function** is a rule that produces a correspondence between a first set of elements, called the **domain,** and a second set, called the **range,** such that to each element in the domain there corresponds *one and only one* element in the range.

Example 1
Identifying Functions

The correspondence rules in this example are given by arrows. For example, read $1 \rightarrow 5$ as ''1 corresponds to 5.'' Indicate which rules are functions.

(A) DOMAIN   RANGE   **(B)** DOMAIN   RANGE   **(C)** DOMAIN   RANGE

        1 ———→ 5   −2 ———→ −1   3 ———→ 1
        2 ———→ 7   0 ———→ 0   ↘ 3
        3 ———→ 9   2 ↗   7 ———→ 8
                               4 ———→ 1   9 ———→ 9

Solution **(A)** Function; exactly one range value corresponds to each domain value.
(B) Function; exactly one range value corresponds to each domain value.
(C) Not a function; two range values correspond to the domain value 3.

Matched Problem 1   Indicate which correspondence rules are functions.

(A) DOMAIN   RANGE   **(B)** DOMAIN   RANGE   **(C)** DOMAIN   RANGE

       −5 ———→ 6   1 ———→ 5   −1 ↘
       −3 ↗   2 ———→ 6   0 ↘
        3 ———→ 0   ↗ 7   1 ———→ 5
                               3 ———→ 8   3 ↗

For the remainder of this text we will restrict ourselves to functions where both variables represent numbers. For our purposes, the informal definition of a function will be sufficient.

◆   **COMMON WAYS OF SPECIFYING FUNCTIONS**

The arrow method of specifying functions illustrated in Example 1 is convenient only when the domain has a small number of elements. Given the domain consisting of the numbers 1, 2, and 3, we can also specify the function from Example 1(A) in several other ways:

1. By an equation such as $y = 2x + 3$

2. By a table:

x	y
1	5
2	7
3	9

3. By a graph:

4. By a set of ordered pairs: $(1, 5), (2, 7), (3, 9)$

If a correspondence rule is specified by a set of ordered pairs of elements, then the set of first components forms the domain and the set of second components forms the range. Tables and ordered pairs are useful for specifying functions when the number of elements in the domain is small or for describing part of a function rule. Equations and graphs can be used effectively when the domain and range are sets of real numbers.

<table>
<tr><td valign="top">

Example 2

Identifying Functions

</td><td valign="top">

Given the set of ordered pairs: $(0, 0)$, $(1, -1)$, $(1, 1)$, $(4, -2)$, $(4, 2)$

(A) Write the correspondence rule using arrows as in Example 1. Indicate the domain and range.

(B) Graph the set in a rectangular coordinate system.

(C) Is the rule a function? Explain.

</td></tr>
</table>

Solution   **(A)** Correspondence rule:

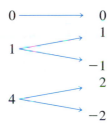

(B)

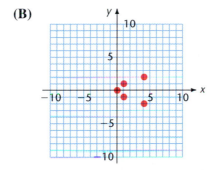

(C) The rule is not a function, since some domain values are associated with more than one range value. In particular, 1 is associated with both -1 and 1, and 4 is associated with both -2 and 2.

Matched Problem 2   Given the set of ordered pairs: $(-2, 4)$, $(-1, 1)$, $(0, 0)$, $(1, 1)$, $(2, 4)$

(A) Write the correspondence rule using arrows as in Example 1. Indicate the domain and range.

(B) Graph the set in a rectangular coordinate system.

(C) Is the rule a function? Explain.

♦   **VERTICAL-LINE TEST FOR GRAPHS**

For a rule described by ordered pairs to be a function, no two ordered pairs can have the same first coordinate and different second coordinates. If we graph the ordered pairs, no two pairs will be on the same vertical line. We can apply the **vertical-line test** to determine whether a rule specified by a graph or by an equation is a function.

Vertical-Line Test

If y is a function of x, then the graph of all ordered pairs (x, y) belonging to the function has the property that no vertical line crosses the graph more than once. Conversely, if no vertical line crosses a graph more than once, the graph is the graph of a function.

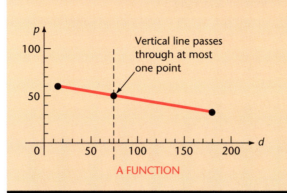

Vertical line passes through at most one point

A FUNCTION

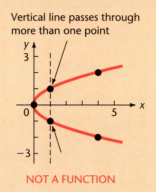

Vertical line passes through more than one point

NOT A FUNCTION

Example 3

Applying the Vertical-Line Test

Apply the vertical-line test to determine whether the ordered pairs, graph, or equation determine y as a function of x:

(A) $(0, 0)$, $(1, -1)$, $(1, 1)$, $(4, -2)$, $(4, 2)$

(B)

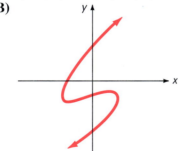

(C) $y = x^2$

Solution **(A)** The graph of the ordered pairs fails the vertical-line test, so the set of pairs does not describe a function.

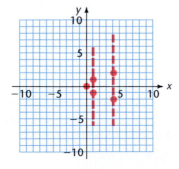

(B) The graph fails the vertical-line test, so it does not determine y as a function of x.

(C) The graph of the equation passes the vertical-line test, so the equation does determine a function.

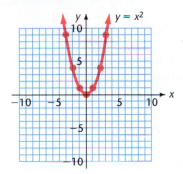

Matched Problem 3   Apply the vertical-line test to determine whether the ordered pairs, graph, or equation determine y as a function of x:

(A) $(-2, 4)$, $(-1, 1)$, $(0, 0)$, $(1, 1)$, $(2, 4)$

(B)

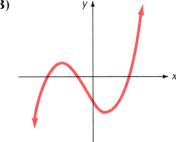

(C) $x = y^2$

♦   **FUNCTIONS SPECIFIED BY EQUATIONS**

The functions of most interest to us will be specified by equations in two variables. All the following equations give y as a function of x:

$$y = 2x + 3 \qquad y = \sqrt{x}$$

$$y = x^2 \qquad y = \frac{1}{x - 1}$$

$$y = |x|$$

In each case, for any allowable value for x we obtain exactly one value for y. For example, using the function given by $y = 2x + 3$:

If $x = 2$, then $y = 2(2) + 3 = 7$.

If $x = -1$, then $y = 2(-1) + 3 = 1$.

Using the function given by $y = \sqrt{x}$:

If $x = 9$, then $y = \sqrt{9} = 3$.   *Recall that the symbol $\sqrt{x}$ represents only the positive square root of x, so that $\sqrt{x}$ gives only one number corresponding to x.*

If $x = 36$, then $y = \sqrt{36} = 6$.

The domains of functions are often restricted due to physical constraints. For instance, a measurement of a length or weight must be at least 0. The allowable values of x are the values in the domain.

If a domain is not specified, and unless stated to the contrary, the domain will be assumed to consist of all real values of x for which the equation provides a real value of y.

In the first three equations in this section (bottom of page 431), any real number can be substituted for x, so the domain in each case is the set of all real numbers. In the case of $y = \sqrt{x}$, the domain is all $x \geq 0$. For $y = 1/(x - 1)$, the domain is all $x \neq 1$.

Example 4
Finding the Domain

Assuming x is the independent variable, find the domain of the function specified by the indicated equation.

(A) $y = 2x - 1$ **(B)** $y = \sqrt{x - 2}$ **(C)** $y = \sqrt{x^2 - 4}$ **(D)** $y = \dfrac{x + 1}{x - 1}$

Solution   **(A)** For each real x, y is defined and is real. Thus,

$$\text{Domain:}\quad R \qquad \text{\small The set of real numbers}$$

(B) For y to be real, $x - 2$ cannot be negative; that is,

$$x - 2 \geq 0$$
$$x \geq 2$$

Thus,

$$\text{Domain:}\quad x \geq 2 \quad \text{or} \quad [2, \infty)$$

(C) For y to be real, $x^2 - 4$ cannot be negative; that is,

$$x^2 - 4 \geq 0 \qquad \text{\small Solve the inequality by sign analysis.}$$
$$(x - 2)(x + 2) \geq 0$$

Sign of $(x - 2)$: $-$ $-$ 0 $+$
Sign of $(x + 2)$: $-$ 0 $+$ $+$
Sign of product: $+$ 0 $-$ 0 $+$

$$\xrightarrow{\hspace{3cm}} x$$
$$-2 \qquad 2$$

Thus,

$$\text{Domain:}\quad x \leq -2 \quad \text{or} \quad x \geq 2, \quad \text{or} \quad (-\infty, -2] \cup [2, \infty)$$

(D) For y to be defined, $x - 1$ cannot be 0; that is, $x \neq 1$. Thus,

$$\text{Domain:} \quad x \neq 1 \quad \text{or} \quad (-\infty, 1) \cup (1, \infty)$$

Matched Problem 4 Assuming x is the independent variable, find the domain of the function specified by the indicated equation.

(A) $y = \dfrac{1}{x^2 - 4}$ **(B)** $y = \sqrt{2 - x}$ **(C)** $y = \sqrt{4 - x^2}$ **(D)** $y = x^2 - 4$

All equations in two variables specify correspondence rules, but when does such an equation specify a function? Suppose x is to be the independent variable and y the dependent variable. If we can solve the equation uniquely for y in terms of x, then the equation does specify y as a function of x. On the other hand, if we can find a value for x that gives two or more corresponding values for y, then the equation does not specify y as a function of x. This will happen, for example, when solving $y^2 = x$ for y: $y = \pm\sqrt{x}$.

Example 5
Specifying Functions by Equations

Given the equation, with independent variable x and dependent variable y, decide whether the equation determines y as a function of x.

(A) $x^2 + y^2 = 4$ **(B)** $x^2 + y = 4$

Solution **(A)** Solve the equation for y:

$$x^2 + y^2 = 4$$
$$y^2 = 4 - x^2$$
$$y = \pm\sqrt{4 - x^2}$$

Recall that the solution to $y^2 = a$ is $y = \pm\sqrt{a}$.

Is there a value of x that will produce more than one value of y? Yes. For example, if $x = 0$, then

$$y = \pm\sqrt{4} = \pm 2$$

Thus, two values of y result from one value of x. Therefore, this rule is not a function.

(B) Solve the equation for y:

$$x^2 + y = 4$$
$$y = 4 - x^2$$

Thus, since we can solve the equation for y uniquely, it does specify y as a function of x. For example, if we let $x = 1$, then $y = 3$, and no other value.

Matched Problem 5 Given the equation, with independent variable x and dependent variable y, decide whether the equation determines y as a function of x.

(A) $3 = x^2 - y$ **(B)** $y^2 = x - 3$

In Example 5, we also could have tested whether the equations defined y as a function of x by considering the graphs of the equations, as shown in Figure 1, and applying the vertical-line test.

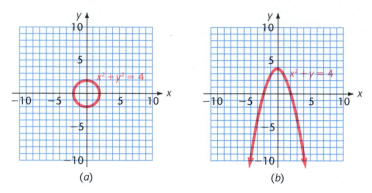

(a)   (b)

Figure 1

Thus, we have several ways to test whether an equation specifies y as a function of x:

1. Graph the equation and apply the vertical-line test.
2. Solve the equation for y in terms of x and check whether each x produces a unique y.
3. Look for a particular x that has two or more corresponding y values; this determines that the equation does *not* specify a function.

◆ A BRIEF HISTORY OF THE FUNCTION CONCEPT

The history of the function concept shows the tendency of mathematicians to extend and generalize an idea. The word ''function'' appears to have been first used by Leibniz (1646–1716) in 1694 to stand for any quantity associated with a curve. By 1718, Johann Bernoulli (1667–1748) considered a function any expression made up of constants and a single variable. Later in the same century, Euler (1707–1783) came to regard a function as any equation made up of constants and variables. Euler made extensive use of the extremely important notation $f(x)$, which we will consider in the next section, although its origin is generally attributed to Clairaut (1713–1765).

The form of the definition of a function that was used until well into this century was the one we called the informal definition. It was formulated by Dirichlet (1805–1859). He also introduced the terms independent variable, dependent variable, domain, and range. The formal definition of a function given in this section followed the development of the theory of sets by Georg Cantor (1845–1918) in the early part of this century.

The function concept is one of the most important in mathematics, and as such it plays a central role as a guide for the selection and development of material in many mathematics courses. In the remainder of this chapter, we will consider functions related to algebraic topics already studied. A different kind of function will be introduced in Chapter 10, and Chapter 11 will focus on certain functions where the domain consists of positive integers.

Answers to Matched Problems

1. **(A)** Function   **(B)** Not a function   **(C)** Function

2. **(A)**
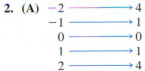
   (B)
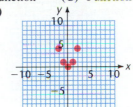

(C) Yes, the rule is a function. Each number in the domain corresponds to exactly one number in the range.

3. **(A)** Function   **(B)** Function   **(C)** Not a function

4. **(A)** All real numbers except ± 2 **(B)** $x \leq 2$ or $[-\infty, 2]$
   (C) $-2 \leq x \leq 2$ or $[-2, 2]$ **(D)** All real numbers

5. **(A)** A function   **(B)** Not a function

EXERCISE 9-1

A *Indicate whether each rule is or is not a function.*

1. DOMAIN   RANGE
   3 ⟶ 0
   5 ⟶ 1
   7 ⟶ 2

2. DOMAIN   RANGE
   −1 ⟶ 5
   −2 ⟶ 7
   −3 ⟶ 9

3. DOMAIN   RANGE
   3 ⟶ 5
      ⟶ 6
   4 ⟶ 7
   5 ⟶ 8

4. DOMAIN   RANGE
   8 ⟶ 0
   9 ⟶ 1
      ⟶ 2
   10 ⟶ 3

5. DOMAIN   RANGE
   3
      ⟶ 5
   6
   9
      ⟶ 6
   12

6. DOMAIN   RANGE
   2
   −1
      ⟶ 6
   0
   1

7. DOMAIN   RANGE
      ⟶ 1
   1 ⟶ 2
      ⟶ 3

8. DOMAIN   RANGE
   1 ⟶ 1
   2 ⟶ 2
   3 ⟶ 3

9. DOMAIN   RANGE
   1
   2 ⟶ 0
   3

10. DOMAIN   RANGE
       ⟶ −1
    0 ⟶ 0
       ⟶ 1

11. DOMAIN   RANGE
    1 ⟶ −1
    2 ⟶ −2
    3 ⟶ −3

12. DOMAIN   RANGE
    1 ⟶ 2
    2 ⟶ 3
    3 ⟶ 1

For each graph, determine whether it is the graph of a function with x the independent variable.

13.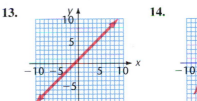

14.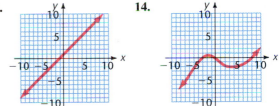

15.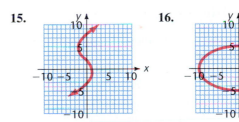

16.

17.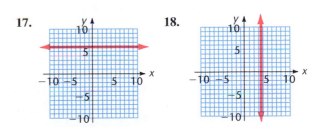

18.

19.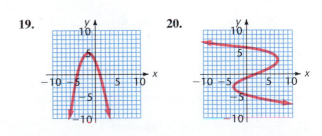

20.

21.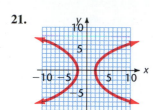

22.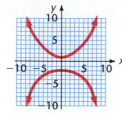

23.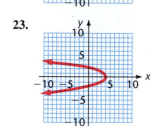

24.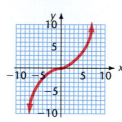

Graph each set of ordered pairs. Determine whether the rule represented by each set is a function.

25. $(1, 1), (2, 1), (3, 2), (3, 3)$

26. $(2, 4), (4, 2), (2, 0), (4, -2)$

27. $(-1, -2), (0, -1), (1, 0), (2, 1), (3, 2), (4, 1)$

28. $(-2, 0), (0, 2), (2, 0)$

29. $(1, 2), (2, 3), (3, 4), (4, 3), (3, 2), (2, 1)$

30. $(2, 4), (1, 3), (3, 1), (4, 2)$

31. $(1, 1), (2, 2), (3, 3), (1, -1), (2, -2), (3, -3)$

32. $(1, 1), (2, 2), (3, 3), (-1, 1), (-2, 2), (-3, 3)$

B *Determine whether each equation specifies y as a function of x by solving for y.*

33. $\dfrac{y + 1}{3} = x$

34. $2y + 1 = x$

35. $y - 1 = x^2 - 3x$

36. $x^3 - y = 0$

37. $y^2 = x$

38. $x^2 + y^2 = 25$

39. $x = y^2 - y$

40. $x = (y - 1)(y + 2)$

41. $x - 3 = \dfrac{y}{x}$

42. $2x - 3y = 5$

43. $xy - x - y - 1 = 0$

44. $0 = x^2 + xy - y$

45. $\dfrac{y - 1}{x} = x + 2$

46. $y^2 = x^2 + 1$

47. $x^2 = y^2 + 1$

48. $\dfrac{y - 4}{x} = x - 4$

49. $9x^2 = 36 - 4y^2$

50. $y^2 - 4 = x^2$

Graph each equation and apply the vertical-line test to determine whether the equation specifies a function.

51. $\dfrac{y + 1}{3} = x$

52. $2y + 1 = x$

53. $y - 1 = x^2 - 3x$

54. $x^3 - y = 0$

55. $y^2 = x$

56. $x^2 + y^2 = 25$

57. $\dfrac{y - 1}{x} = x + 2$

58. $y^2 = x^2 + 1$

59. $x^2 = y^2 + 1$

60. $\dfrac{y - 4}{x} = x - 4$

61. $9x^2 = 36 - 4y^2$

62. $y^2 - 4 = x^2$

Graph all pairs (x, y) satisfying the equation for the given domain. Determine whether the rule represented by each set is a function.

63. $y = 6 - 2x$
Domain: $\{0, 1, 2, 3, 4\}$

64. $y = \dfrac{x}{2} - 4$
Domain: $\{0, 1, 2, 3, 4\}$

65. $y^2 = x$
Domain: $\{0, 1, 4\}$

66. $y = x^2$
Domain: $\{-2, 0, 2\}$

67. $x^2 + y^2 = 4$
Domain: $\{-2, 0, 2\}$

68. $x^2 + y^2 = 9$
Domain: $\{-3, 0, 3\}$

69. $y = |x|$
Domain: $\{-2, 0, 2\}$

70. $|y| = x$
Domain: $\{0, 1, 4\}$

The equations in Problems 71–90 each specify a function. Determine the domain.

71. $y = 5 - x$

72. $y = 5x + 2$

73. $y = 3x^2 - 2x + 1$

74. $y = (5 - x)^2$

75. $y = \dfrac{1}{x}$

76. $y = \dfrac{1}{x - 1}$

77. $y = \dfrac{x - 1}{(x + 2)(x - 3)}$

78. $y = \dfrac{(x + 5)}{(x - 4)(x + 3)}$

79. $y = \dfrac{3x}{x^2 + x - 12}$

80. $y = \dfrac{x + 1}{x^2 - 7x + 12}$

81. $y = \sqrt{4 - x}$

82. $y = \sqrt{x - 5}$

83. $y = \sqrt{\dfrac{x - 1}{x + 3}}$

84. $y = \sqrt{x^2 + 3x - 10}$

85. $y = \sqrt{x^2 - x - 6}$

86. $y = \sqrt{\dfrac{x + 1}{x - 2}}$

87. $y = x^{3/4}$

88. $y = x^{5/6}$

89. $y = x^{6/5}$

90. $y = x^{4/3}$

C *Graph each set of ordered pairs in Problems 91–96. Determine whether the rule represented by each set is a function.*

91. The set of all pairs (x, y), where $y = x/2$ and x is an integer between -10 and 10, inclusive

92. The set of all pairs (x, y), where $y = x + 3$ and x is an integer between -10 and 10, inclusive

93. The set of all pairs (x, y), where $0 \leq y \leq x$, $0 \leq x \leq 3$, and x and y are integers

94. The set of all pairs (x, y), where $0 \leq y < |x|$, $-2 \leq x \leq 2$, and x and y are integers

95. The set of all pairs (x, y), where $y = |x|/x$, $0 < |x| \leq 3$, and x is an integer

96. The set of all pairs (x, y), where $y = (-1)^x$ and x is an integer between -2 and 2, inclusive

Problems 97 and 98 refer to the standard touchtone telephone keys, as shown here. Let f be the rule that takes each letter on the keys and associates it with the number corresponding to it. Let g be the rule that takes each number on the keys and associates it with the letters corresponding to it.

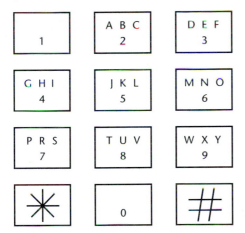

97. Is rule f a function? If so, find the value corresponding to X, the domain, and range of the function.

98. Is rule g a function? If so, find the value corresponding to 7, the domain, and range of the function.

APPLICATIONS

99. **Physics** If an arrow is shot straight upward from the ground with an initial velocity of 160 feet per second, its distance d (in feet) above the ground at the end of t seconds (neglecting air resistance) is given by

$$d = 160t - 16t^2 \qquad 0 \leq t \leq 10$$

(A) Graph this rule (t is the independent variable).
(B) What are its domain and range?
(C) Is it a function?

100. **Physics** The distance s (in feet) that an object falls (neglecting air resistance) in t seconds is given by

$$s = 16t^2 \qquad t \geq 0$$

(A) Graph this rule (t is the independent variable).
(B) What are its domain and range?
(C) Is it a function?

Behavioral Science Problems 101 and 102 refer to the following situation. A laboratory experiment may yield the table and graph shown below:

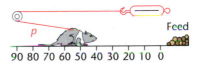

Distance d (in centimeters)	Pull Toward Food p (in grams)
30	64
50	60
70	56
90	52
110	48
130	44
150	40
170	36

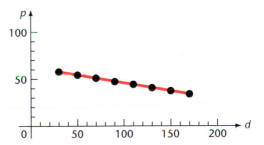

Both the table and the graph establish the same function f with domain values d and range values p. Assume the graph is actually a straight line when all possible values are plotted for $30 \leq d \leq 170$.

101. Find the equation for the graph and use it to find the function value corresponding to $d = 75$.

102. Suppose the same equation remains valid not just for $30 \leq d \leq 170$, but for all values of d and p that are physically possible. Graph the resulting function. What are the domain and range of the resulting function?

9-2 Function Notation and the Algebra of Functions

♦ The Function Symbol $f(x)$
♦ Sums, Differences, Products, and Quotients
♦ Composition of Functions

In this section we introduce notation that allows us to name functions and describe function rules in compact form. The notation also makes it easy to build more complicated functions out of components, using addition, subtraction, multiplication, division, and a new operation called *composition*.

♦ **THE FUNCTION SYMBOL $f(x)$**

In algebra we use different letters to denote names for numbers. In essentially the same way, we will now use different letters to denote names for functions. For example, f and g may be used to name the following two functions specified by equations:

$$f: \quad y = 2x + 1$$
$$g: \quad y = x^2 + 2x - 3$$

If x represents an element in the domain of a function f, we will use the symbol

$$f(x)$$

in place of y to designate the number in the range of the function f that corresponds to x (see Figure 1).

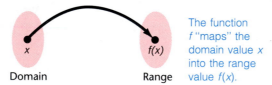

The function f "maps" the domain value x into the range value $f(x)$.

Figure 1

This new function symbol is *not* the product of f and x. The symbol $f(x)$ is read "f of x" or "the value of f at x." The variable x is the independent variable; y and $f(x)$ both name the dependent variable.

This new function notation is extremely useful. For example, in place of the more formal representation of the functions f and g above, we can now write

$$f(x) = 2x + 1 \qquad \text{and} \qquad g(x) = x^2 + 2x - 3$$

The function symbols $f(x)$ and $g(x)$ also allow us to describe the correspondence for particular values of x. For example, if we write $f(3)$ and $g(5)$, then each symbol indicates in a concise way that these are range values of particular functions associated with particular domain values. Let us find $f(3)$ and $g(5)$. To find $f(3)$, we replace x by 3 wherever x occurs in $f(x) = 2x + 1$, and evaluate the right side:

$$\begin{aligned} f(3) &= 2 \cdot 3 + 1 \\ &= 6 + 1 \\ &= 7 \end{aligned}$$

Thus,

$$f(3) = 7 \qquad \text{The function } f \text{ assigns the range value 7 to the domain value 3; the ordered pair } (3, 7) \text{ belongs to } f.$$

To find $g(5)$, we replace x by 5 wherever x occurs in $g(x) = x^2 + 2x - 3$, and evaluate the right side:

$$\begin{aligned} g(5) &= 5^2 + 2 \cdot 5 - 3 \\ &= 25 + 10 - 3 \\ &= 32 \end{aligned}$$

Thus,

$$g(5) = 32 \qquad \text{The function } g \text{ assigns the range value 32 to the domain value 5; the ordered pair } (5, 32) \text{ belongs to } g.$$

It is important to understand and remember the definition of $f(x)$:

The Function Symbol $f(x)$

For any element x in the domain of the function f, the function symbol $f(x)$ represents the element in the range of f corresponding to that x.

We may think of x as an **input** value into the function rule. Then $f(x)$ is the corresponding **output** value. The input–output way of looking at a function rule suggests viewing a function as a "machine" that converts inputs x into outputs y. Figure 2 illustrates this idea. The function notation then can be thought of as playing the role of the "function machine": $f(\ )$ converts x into $f(x)$.

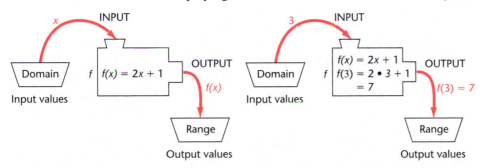

Figure 2 "Function machine": exactly one output for each input

For the function $f(x) = 2x + 1$, the rule or machine takes each domain value, multiplies it by 2 and then adds 1 to the result to produce the range value. Different rules inside the machine result in different functions.

Example 1
Using the *f(x)* Notation

Let $f(x) = \dfrac{x}{2} + 1$ and $g(x) = 1 - x^2$. Find:

(A) $f(6)$ **(B)** $g(-2)$ **(C)** $f(4) + g(0)$

Solution   **(A)** $f(\mathbf{6}) = \dfrac{\mathbf{6}}{2} + 1 = 3 + 1 = 4$

(B) $g(\mathbf{-2}) = 1 - (\mathbf{-2})^2 = 1 - 4 = -3$

(C) $f(\mathbf{4}) + g(\mathbf{0}) = \overbrace{\left(\dfrac{\mathbf{4}}{2} + 1\right)}^{f(4)} + \overbrace{(1 - \mathbf{0}^2)}^{g(0)} = 3 + 1 = 4$

Matched Problem 1   Let $f(x) = \dfrac{x}{3} - 2$ and $g(x) = 4 - x^2$. Find:

(A) $f(9)$ **(B)** $g(-2)$ **(C)** $f(0) + g(2)$

Example 2
Using the *f(x)* Notation

For $f(x) = 5x + 3$, find:

(A) $f(p)$ **(B)** $f(p + 1)$ **(C)** $f(2q)$

Solution   **(A)** $f(p) = 5p + 3$ Replace x in $f(x) = 5x + 3$ with p.
(B) $f(p + 1) = 5(p + 1) + 3$ Replace x in $f(x) = 5x + 3$ with $p + 1$ and simplify.
$\qquad\qquad\quad = 5p + 5 + 3$
$\qquad\qquad\quad = 5p + 8$
(C) $f(2q) = 5(2q) + 3$ Replace x in $f(x) = 5x + 3$ with $2q$ and simplify.
$\qquad\qquad = 10q + 3$

Matched Problem 2   For $f(x) = 2 - 3x$, find:

(A) $f(m)$ **(B)** $f(m - 1)$ **(C)** $f(5p)$

Example 3
Using the *f(x)* Notation

Let $f(x) = \dfrac{x}{2} + 1$ and $g(x) = 1 - x^2$. Find:

(A) $f(g(3))$ **(B)** $g(f(-4))$

Solution   **(A)** $f(g(3)) = f(1 - 3^2)$ Evaluate $g(3) = 1 - 3^2 = -8$ first; then evaluate f for this value.

$\qquad\qquad = f(-8) = \dfrac{-8}{2} + 1 = -3$

(B) $g(f(-4)) = g\left(\dfrac{-4}{2} + 1\right)$

$\qquad\qquad = g(-1) = 1 - (-1)^2 = 0$

Matched Problem 3   Let $f(x) = \dfrac{x}{3} - 2$ and $g(x) = 3 - x^2$. Find:

(A) $f(g(3))$ (B) $g(f(-3))$

◆ **SUMS, DIFFERENCES, PRODUCTS, AND QUOTIENTS**

Given functions $f(x)$ and $g(x)$, we can define several related functions using the ordinary operations of addition, subtraction, multiplication, and division.

Sums, Differences, Products, and Quotients

For functions $f(x)$ and $g(x)$,

$$(f + g)(x) = f(x) + g(x)$$
$$(f - g)(x) = f(x) - g(x)$$
$$(f \cdot g)(x) = f(x) \cdot g(x)$$
$$\left(\frac{f}{g}\right)(x) = \frac{f(x)}{g(x)}$$

The new functions $f + g$, $f - g$, $f \cdot g$, and f/g are called the **sum, difference, product,** and **quotient** of f and g, respectively. The domain of $f + g$, $f - g$, and $f \cdot g$ is the set of numbers for which *both f and g are defined*. The domain of f/g is the set of numbers for which *both f and g are defined* and $g(x)$ *is not 0*.

Example 4
Finding the Sum, Difference, Product, and Quotient

For $f(x) = x^2 - 1$ and $g(x) = x + 1$, find $f + g$, $f - g$, $f \cdot g$, and f/g. Give the domain of each new function.

Solution $(f + g)(x) = f(x) + g(x) = (x^2 - 1) + (x + 1) = x^2 + x$

$(f - g)(x) = f(x) - g(x) = (x^2 - 1) - (x + 1) = x^2 - x - 2$

$(f \cdot g)(x) = f(x) \cdot g(x) = (x^2 - 1)(x + 1) = x^3 + x^2 - x - 1$

$\left(\dfrac{f}{g}\right)(x) = \dfrac{f(x)}{g(x)} = \dfrac{x^2 - 1}{x + 1}$

The domain of $f + g$, $f - g$, and $f \cdot g$ is the set of all real numbers. The domain of f/g is the set of all real numbers excluding -1; that is, $x \neq -1$. As long as we keep this restricted domain in mind, we can simplify

$$\left(\frac{f}{g}\right)(x) = \frac{f(x)}{g(x)} = \frac{x^2 - 1}{x + 1} = x - 1 \qquad x \neq -1$$

Matched Problem 4   For $f(x) = x^2 - x - 2$ and $g(x) = x - 1$, find $f + g$, $f - g$, $f \cdot g$, and f/g. Give the domain of each new function.

♦ COMPOSITION OF FUNCTIONS

Since functions are rules, we can combine them to create new functions by applying the rules in sequence. That is, if we start with a value x and then apply functions f and g in succession, the combination of the rules is another function called the *composition* of f and g. More precisely, since it makes a difference which function is applied first, we call $g(f(x))$ the **composition of f by g.** See Figure 3. The domain of $g(f(x))$ is the set of all numbers in the domain of f for which $f(x)$ is in the domain of g.

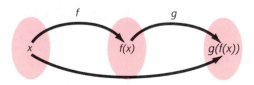

Figure 3   Composition of f by g

Example 5

Composition of Functions

For $f(x) = x^2 - 1$ and $g(x) = x + 1$, find:

(A) $g(f(x))$ **(B)** $f(g(x))$

Solution   **(A)** Beginning with x and applying f, we obtain $f(x) = x^2 - 1$. Now applying g to this result, we get

$$g(f(x)) = g(\overbrace{x^2 - 1}^{f(x)}) = (x^2 - 1) + 1$$

The rule for g is to take the input and add 1. Apply this rule to $f(x) = x^2 - 1$.

$$= x^2$$

(B) $f(g(x)) = f(\overbrace{x + 1}^{g(x)}) = (x + 1)^2 - 1$

The rule for f is to square the input and subtract 1. Apply this rule to $g(x) = x + 1$.

$$= x^2 + 2x + 1 - 1$$
$$= x^2 + 2x$$

Matched Problem 5   For $f(x) = x^2 - x - 2$ and $g(x) = x - 1$, find:

(A) $g(f(x))$ **(B)** $f(g(x))$

Answers to Matched Problems

1. (A) 1   **(B)** 0   **(C)** -2 **2. (A)** $2 - 3m$ **(B)** $5 - 3m$ **(C)** $2 - 15p$
3. (A) -4 **(B)** -6 **4.** $(f + g)(x) = x^2 - 3$; domain R
  $(f - g)(x) = x^2 - 2x - 1$; domain R
  $(f \cdot g)(x) = x^3 - 2x^2 - x + 2$; domain R
  $\left(\dfrac{f}{g}\right)(x) = \dfrac{x^2 - x - 2}{x - 1}$; domain $x \neq 1$
5. (A) $x^2 - x - 3$ **(B)** $x^2 - 3x - 1$

EXERCISE 9-2

A *If $f(x) = 3x - 2$, $g(x) = x - x^2$, and $h(x) = x(x - 1)$, find the following:*

1. $f(2)$ **2.** $f(1)$ **3.** $f(-2)$

4. $f(-1)$ **5.** $f(0)$ **6.** $f(4)$

7. $g(2)$ **8.** $g(1)$ **9.** $g(4)$

10. $g(5)$ **11.** $g(-2)$ **12.** $g(-1)$

13. $h(2)$ **14.** $h(-2)$ **15.** $h(0)$

16. $h(1)$ **17.** $h(-6)$ **18.** $h(7)$

19. $f(x + 1)$ **20.** $g(x + 1)$ **21.** $h(x + 1)$

22. $f(x - 1)$ **23.** $g(x - 1)$ **24.** $h(x - 1)$

Let $f(x) = 10x - 7$, $g(x) = 6 - 2x$, $F(x) = 3x^2$, and $G(x) = x - x^2$. Find the following:

25. $f(-2)$ **26.** $F(-1)$ **27.** $g(2)$

28. $G(-3)$ **29.** $g(0)$ **30.** $G(0)$

B   **31.** $f(3) + g(2)$ **32.** $F(2) + G(3)$

33. $2g(-1) - 3G(-1)$ **34.** $4G(-2) - g(-3)$

35. $\dfrac{f(2) \cdot g(-4)}{G(-1)}$ **36.** $\dfrac{F(-1) \cdot G(2)}{g(-1)}$

37. $g(u - 2)$ **38.** $f(v + 1)$ **39.** $G(3a)$

40. $F(2c)$ **41.** $g(2 + h)$ **42.** $F(2 + h)$

43. $\dfrac{g(2 + h) - g(2)}{h}$ **44.** $\dfrac{F(2 + h) - F(2)}{h}$

45. $\dfrac{f(3 + h) - f(3)}{h}$ **46.** $\dfrac{G(2 + h) - G(2)}{h}$

47. $F(g(1))$ **48.** $G(F(1))$ **49.** $g(f(1))$

50. $g(G(0))$ **51.** $f(G(1))$ **52.** $G(g(2))$

If $f(x) = 3x - 2$, $g(x) = x - x^2$, and $h(x) = x(x - 1)$, find the indicated function and simplify. Give its domain.

53. $(f + g)(x)$ **54.** $(f + h)(x)$ **55.** $(g + h)(x)$

56. $(f - g)(x)$ **57.** $(g - h)(x)$ **58.** $(h - f)(x)$

59. $(g - f)(x)$ **60.** $(h - g)(x)$ **61.** $(f - h)(x)$

62. $(f \cdot g)(x)$ **63.** $(f \cdot h)(x)$ **64.** $(g \cdot h)(x)$

65. $(f/g)(x)$ **66.** $(g/h)(x)$ **67.** $(h/f)(x)$

68. $(g/f)(x)$ **69.** $(h/g)(x)$ **70.** $(f/h)(x)$

Let $f(x) = 10x - 7$, $g(x) = 6 - 2x$, $F(x) = 3x^2$, and $G(x) = x - x^2$. Find the following:

71. $f(g(x))$ **72.** $f(F(x))$ **73.** $f(G(x))$

74. $g(G(x))$ **75.** $g(F(x))$ **76.** $g(f(x))$

77. $F(f(x))$ **78.** $F(g(x))$ **79.** $F(G(x))$

80. $G(f(x))$ **81.** $G(g(x))$ **82.** $G(F(x))$

83. $f(x + h)$ **84.** $g(x + h)$ **85.** $F(x + h)$

86. $G(x + h)$

C **87.** $\dfrac{f(x + h) - f(x)}{h}$ **88.** $\dfrac{g(x + h) - g(x)}{h}$

89. $\dfrac{F(x + h) - F(x)}{h}$ **90.** $\dfrac{G(x + h) - G(x)}{h}$

91. Let $A(w) = \dfrac{w - 3}{w + 5}$. Find $A(5)$, $A(0)$, $A(-5)$, and $A(x - 5)$.

92. Let $h(s) = \dfrac{s}{s - 2}$. Find $h(3)$, $h(0)$, $h(2)$, and $h(x + 2)$.

APPLICATIONS

93. *Business*   If $a(x)$ represents the total sales of automobile model A during the first x days of the year and $b(x)$ represents the total sales of model B, what does $(a + b)(x)$ represent?

94. *Business*   If $R(x)$ represents the revenue from sales of x items and $C(x)$ represents the cost of producing the x items, what does $(R - C)(x)$ represent?

★ **95.** *Business*   If $f(x) = x$ and $g(x)$ represents the number of sales of some item at a price of x dollars, what does $(f \cdot g)(x)$ represent?

★ **96.** *Business*   If $f(x) = x$ and $C(x)$ represents the cost of producing x items, what does $(C/f)(x)$ represent?

Each of the statements in Problems 97–100 can be described by a function. Write an equation that specifies the function.

97. *Cost function*   The cost $C(x)$ of x records at \$8.60 per record. (The cost depends on the number of records purchased.)

98. *Cost function*   The daily cost $C(x)$ of manufacturing x pairs of skis if fixed costs are \$800 per day and the variable costs are $60 per pair of skis. (The cost per day depends on the number of skis manufactured per day.)

99. Temperature conversion The temperature in Celsius degress $C(F)$ can be found from the temperature in Fahrenheit degrees F by subtracting 32 from the Fahrenheit temperature and multiplying the difference by $\frac{5}{9}$.

100. Earth science The pressure $P(d)$ in the ocean (in pounds per square inch) depends on the depth d (in feet) . To find the pressure, divide the depth by 33, add 1 to the quotient, then multiply the result by 15.

101. Distance-rate-time Let the distance in miles that a car travels at 30 miles per hour in t hours be given by $d(t) = 30t$. Find:

(A) $d(1)$ and $d(10)$ **(B)** $\dfrac{d(2 + h) - d(2)}{h}$

★ **102. Physics** The distance in feet that an object falls in t seconds in a vacuum is given by $s(t) = 16t^2$. Find:

(A) $s(0)$, $s(1)$, $s(2)$, and $s(3)$

(B) $\dfrac{s(2 + h) - s(2)}{h}$

What happens as h tends to 0? Interpret physically.

9-3 Graphing Functions

- ◆ The Graph of a Function
- ◆ Linear and Quadratic Functions
- ◆ Higher-Degree Polynomial Functions
- ◆ Application

The **graph of a function defined by an equation** is the graph of the equation $y = f(x)$. More generally, the **graph of a function** is the graph of all the ordered pairs of numbers $(x, f(x))$ that constitute the function. The linear and quadratic equations

$$y = ax + b \qquad \text{and} \qquad y = ax^2 + bx + c$$

studied in Chapters 5 and 7 define **linear** and **quadratic functions,** respectively. We thus already know a great deal about the graphs of these functions. Linear and quadratic functions are important special cases of a larger class of functions called **polynomial functions,** in which $f(x)$ is specified by a polynomial in x. In this section we will look at graphing functions in general and the graphs of linear and quadratic functions in particular. We also consider the difficulties connected with graphing higher-degree polynomial functions.

◆ THE GRAPH OF A FUNCTION

We graph functions, in general, by plotting enough points to see the shape of the function for the domain values of interest to us.

Example 1
Graphing a Function

Graph: $f(x) = x + \dfrac{1}{x}$

Solution   We set up a table of values, just as we have been doing to graph equations. Note first that the domain of the function is $x \neq 0$. Therefore, the graph will be in two parts, one to the left of the y axis, the other to the right. We graph the right half first:

x	$f(x) = x + \dfrac{1}{x}$
1	2
2	2.5
3	3.33 . . .
4	4.25
5	5.2
10	10.1

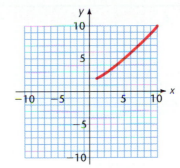

To see the behavior near the y axis, we use x values near 0:

x	$f(x) = x + \dfrac{1}{x}$
0.5	2.5
0.1	10.1
0.01	100.01

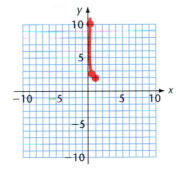

This gives a clear view of the right half of the function. Proceeding similarly with the negative x values gives us the complete graph shown:

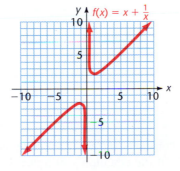

Matched Problem 1 Graph: $f(x) = x - \dfrac{1}{x}$

With enough time and patience any function can be graphed in this way, but the process is obviously tedious. Technology offers us considerable assistance. Graphing calculators and software graphing packages for personal computers can provide graphs such as this almost instantaneously.

◆ LINEAR AND QUADRATIC FUNCTIONS

Any function defined by an equation of the form

$$f(x) = ax + b \qquad \text{Linear function}$$

where a and b are constants and x is a variable, is called a **linear function.** The graph of this function is the graph of the equation $y = ax + b$, and we know from Section 5-2 that this graph is a nonvertical straight line with slope a and y intercept b.

Example 2
Graphing a Linear Function

Graph the linear function defined by

$$f(x) = \frac{x}{3} + 1$$

and indicate its slope and y intercept.

Solution

x	$f(x)$
-3	0
0	1
3	2

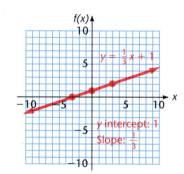

Matched Problem 2 Graph the linear function defined by

$$f(x) = -\frac{x}{2} + 3$$

and indicate its slope and y intercept.

Any function defined by an equation of the form

$$f(x) = ax^2 + bx + c \qquad a \neq 0 \qquad \text{Quadratic function}$$

where a, b, and c are constants and x is a variable, is called a **quadratic function.** Its graph is the graph of the equation $y = ax^2 + bx + c$. We know from Section 7-5 that this graph is a parabola with vertex at $x = -b/2a$ and with y intercept c. The x intercepts, if any, are given by

$$x = \frac{-b \pm \sqrt{b^2 - 4ac}}{2a}$$

The parabola opens up if $a > 0$ and down if $a < 0$.

Example 3
Graphing a Quadratic Function

Graph $f(x) = -2x^2 + 10x + 5$. Find the coordinates of the vertex and any intercepts.

Solution   Since $a = -2$, the parabola opens down. The vertex is at

$$x = -\frac{b}{2a} = \frac{-10}{-4} = \frac{5}{2}$$

At this value of x,

$$f(x) = -2\left(\frac{5}{2}\right)^2 + 10\left(\frac{5}{2}\right) + 5 = \frac{35}{2}$$

The y intercept is 5. The x intercepts are at

$$x = \frac{-10 \pm \sqrt{100 + 40}}{-4} = \frac{-10 \pm \sqrt{140}}{-4} \approx -0.46,\ 5.46$$

With this information we can sketch a graph of the parabola:

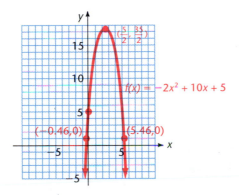

Matched Problem 3   Graph $f(x) = 3x^2 - 9x + 8$. Find the coordinates of the vertex and any intercepts.

♦ **HIGHER-DEGREE POLYNOMIAL FUNCTIONS**

Linear and quadratic functions are special cases of a general class of functions called **polynomial functions,** that is, functions whose values are determined by use of a polynomial. In general, a function f defined by an equation of the form

$$f(x) = a_n x^n + a_{n-1} x^{n-1} + \cdots + a_1 x + a_0 \qquad a_n \neq 0$$

where the coefficients a_i are constants and n is a nonnegative integer, is called an **nth-degree polynomial function.** The following equations define polynomial functions of various degrees:

$$f(x) = 3x - 3 \qquad \text{First-degree (linear)}$$
$$g(x) = 2x^2 - 3x + 2 \qquad \text{Second-degree (quadratic)}$$
$$P(x) = x^3 - 2x^2 + x - 1 \qquad \text{Third-degree (cubic)}$$
$$Q(x) = x^4 - 5 \qquad \text{Fourth-degree (quartic)}$$

A lot of information about polynomial functions can be obtained from techniques studied in more advanced courses. For instance, it can be shown that graphs of polynomial functions have no holes or breaks in them—the graphs are continuous, smooth curves. Also, the number of "turns" in the graph can be at most one less than the degree of the polynomial. We will graph third- and fourth-degree polynomial functions by plotting points. It is helpful to know the possible shapes so that we can be fairly confident that enough points have been plotted. Third-degree polynomials will have one of these four shapes:

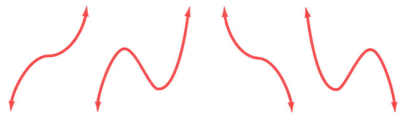

The first two shapes occur when the coefficient of x^3 is positive, the second two when it is negative. Fourth-degree polynomials will have one of these shapes:

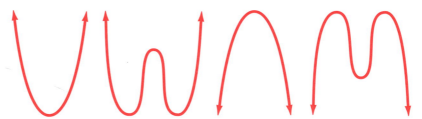

Again, the first two shapes occur when the coefficient of x^4 is positive, the second two when it is negative.

Example 4
Graphing a Cubic Function

Graph $P(x) = x^3 + 3x^2 - x - 3$.

Solution

x	$P(x)$
-4	-15
-3	0
-2	3
-1	0
0	-3
1	0
2	15

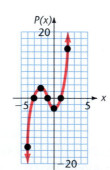

The shape obtained is one of the expected ones, so the graph is essentially complete.

Matched Problem 4   Graph $P(x) = x^3 - 4x^2 - 4x + 16$. Start with values of x between -3 and 5.

♦ **APPLICATION**

Example 5
A Geometric
Application

A rectangular dog pen is to be made with 160 feet of fencing.

(A) If x represents the width of the pen, express its area $A(x)$ in terms of x.
(B) What is the domain of the function A as determined by the physical restrictions?
(C) Graph $A(x)$.
(D) What dimensions will enclose the largest possible area?

Solution   **(A)** Draw a figure and label the sides:

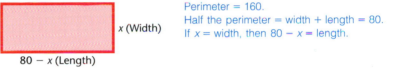

Perimeter = 160.
Half the perimeter = width + length = 80.
If x = width, then $80 - x$ = length.

x (Width)

$80 - x$ (Length)

$$A(x) = (\text{Length})(\text{Width}) = (80 - x)x \qquad \text{Area depends on width } x.$$

(B) The area cannot be negative. Hence, we need

$$(80 - x)x \geq 0$$

Solving this by sign analysis yields

Domain: $0 \leq x \leq 80$ Inequality notation
          $[0, 80]$ Interval notation

(C) Since

$$A(x) = (80 - x)x = -x^2 + 80x \qquad \text{A quadratic function}$$

the graph is a parabola that opens down, has vertex at $x = 40$, $y = 1{,}600$, and has x intercepts at 0 and 80 as shown in the figure.

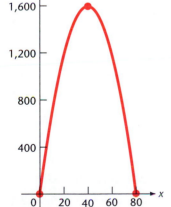

$A(x) = -x^2 + 80x$

(D) From the graph, the largest value of $A(x)$ occurs at the vertex, that is, when $x = 40$. Therefore, a 40- by 40-foot pen will enclose the largest possible area, 1,600 square feet.

Matched Problem 5   Work Example 5 with the added assumption that a large barn is to be used as one side of the pen, as shown:

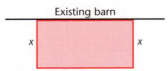

Existing barn

x x

Answers to Matched Problems

1.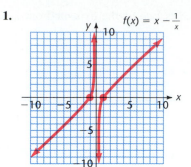
$f(x) = x - \frac{1}{x}$

2.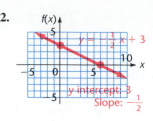
$y = -\frac{1}{2}x + 3$
y intercept: 3
Slope: $-\frac{1}{2}$

3.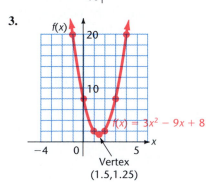
$f(x) = 3x^2 - 9x + 8$
Vertex (1.5, 1.25)

4.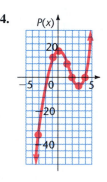

5. (A) $A(x) = (160 - 2x)x$ **(B)** Domain: $0 \le x \le 80$ or $[0, 80]$
(C)   **(D)** 40 by 80 feet

(graph showing y values 3,200; 2,400; 1,600; 800 with x values 40, 80)

EXERCISE 9-3

A *Graph each of the following linear functions. Indicate the slope and y intercept for each.*

1. $f(x) = 2x - 4$

2. $g(x) = \dfrac{x}{2}$

3. $h(x) = 4 - 2x$

4. $f(x) = -\dfrac{x}{2} + 3$

5. $g(x) = -\frac{2}{3}x + 4$

6. $f(x) = 3$

7. $f(x) = 120 + 10x$

8. $g(x) = 100 + 25x$

9. $h(x) = 300x - 1,500$

10. $h(x) = 600x - 10,000$

11. $g(x) = 80,000 - 10,000x$

12. $f(x) = 120,000 - 8,000x$

Graph each of the following quadratic functions. Find the vertex and any intercepts.

13. $f(x) = x^2 + 8x + 16$ **14.** $h(x) = x^2 - 2x - 3$

15. $f(x) = x^2 - 2x + 4$ **16.** $f(x) = x^2 - 10x + 25$

17. $h(x) = 2 + 4x - x^2$ **18.** $g(x) = -x^2 - 6x - 4$

19. $f(x) = 6x - x^2$ **20.** $G(x) = 16x - 2x^2$

21. $F(x) = x^2 - 4$ **22.** $g(x) = x^2 + 4$

23. $F(x) = 4 - x^2$ **24.** $G(x) = 9 - x^2$

25. $f(x) = x^2 - 7x + 10$ **26.** $g(x) = x^2 - 5x + 2$

27. $g(x) = 4 + 3x - x^2$ **28.** $h(x) = 2 - 5x - x^2$

29. $f(x) = \frac{1}{2}x^2 + 2x$

30. $f(x) = 2x^2 - 12x + 14$

31. $f(x) = -2x^2 - 8x - 2$

32. $f(x) = -\frac{1}{2}x^2 + 4x - 4$

B *Graph each polynomial function.*

33. $P(x) = x^3 - 5x^2 + 2x + 8;\quad -2 \le x \le 5$

34. $P(x) = x^3 + 2x^2 - 5x - 6;\quad -4 \le x \le 3$

35. $P(x) = x^3 + 4x^2 - x - 4;\quad -5 \le x \le 2$

36. $P(x) = x^3 - 2x^2 - 5x + 6;\quad -3 \le x \le 4$

37. $P(x) = x^3 + 2$ **38.** $P(x) = x^3 - 1$

39. $P(x) = 3 - x^3$ **40.** $P(x) = 5 - x^3$

41. $P(x) = (x - 1)^3$ **42.** $P(x) = (x + 1)^3$

43. $P(x) = (x + 2)^3$ **44.** $P(x) = (x - 2)^3$

45. $P(x) = (x + 1)^3 - 2$

46. $P(x) = (x - 1)^3 + 2$

C **47.** $P(x) = x^4 + 4x^3 - 7x^2 - 22x + 24;\quad -4 \le x \le 2$

48. $P(x) = x^4 - 4x^3 - 7x^2 + 22x + 24;\quad -2 \le x \le 4$

49. $P(x) = x^4 - 2x^2 + 1;\quad -5 \le x \le 5$

50. $P(x) = x^4 - 8x^2 + 16;\quad -3 \le x \le 3$

51. $P(x) = x^4 - 4x^2$ **52.** $P(x) = x^4 - x^2$

53. $P(x) = (x + 1)^4$ **54.** $P(x) = (x - 1)^4$

55. $P(x) = x^4 - 2x^3 - 2x^2 + 8x - 8$

56. $P(x) = x^4 - 2x^2 + 16x - 15$

57. $P(x) = x^4 + 4x^3 - x^2 - 20x - 20$

58. $P(x) = x^4 - 4x^2 - 4x - 1$

Graph each function for $-10 \le x \le 10$:

59. $f(x) = \dfrac{1}{x}$ **60.** $f(x) = 1 - \dfrac{1}{x}$

61. $f(x) = 1 - \sqrt{x}$ **62.** $f(x) = \sqrt{x} + 1$

63. $f(x) = |x|$ **64.** $f(x) = |x + 1|$

APPLICATIONS

65. *Cost equation* The cost equation for a particular company to produce stereos is found to be

$$C = g(n) = 96,000 + 80n$$

where \$96,000 is fixed costs (tooling and overhead) and \$80 is the variable cost per unit (material, labor, and so on). Graph this function for $0 \le n \le 1,000$.

66. *Cost equation* Repeat Problem 65 for $C = h(n) = 40,000 + 140n$.

67. *Demand equation* After extensive surveys the research department of a stereo equipment manufacturer determined that the demand equation for a particular model is

$$n = f(p) = 7,500 - 100p \qquad 25 \le p \le 75$$

where n is the number of units that retailers are likely to purchase each week if the price is set at p dollars per unit.

(A) Graph the function $f(p)$ for the domain indicated.

(B) The revenue from sales of n units will be np. Express the revenue as a function of p, $R(p)$.

(C) Graph $R(p)$ and determine from the graph the price that yields the largest possible weekly revenue.

68. *Demand equation* Repeat Problem 67 for the demand equation

$$n = f(p) = 8,000 - 40p \qquad 100 \le p \le 200$$

69. *Construction* A rectangular feeding pen for cattle is to be made with 100 meters of fencing.

(A) If x represents the width of the pen, express its area $A(x)$ in terms of x.

(B) What is the domain of the function A (determined by the physical restrictions)?

(C) Graph the function for this domain.

(D) What dimensions for the pen will produce the largest area? What is the largest area?

70. *Construction* Work Problem 69 with the added assumption that a large straight river is to be used as one side of the pen.

71. *Demand equation* A company can manufacture a particular product for \$10 each. The demand equation for the product is

$$n = f(p) = 80 - p \qquad 10 \le p \le 80$$

where n units are the expected weekly sales when the price is set at p dollars per unit.

(A) Graph the function $f(p)$ for the domain indicated.

(B) The profit from sales of n units will be $n(p - 10)$. Express the profit as a function of p, $P(p)$.

(C) Graph $P(p)$ and determine from the graph the price that yields the largest possible weekly profit.

72. *Demand equation* Repeat Problem 71 for a unit cost of \$15 and the demand equation

$$n = f(p) = 120 - p \qquad 15 \le p \le 120$$

★ **73.** *Packaging* A candy box is to be made out of a rectangular piece of cardboard that measures 8 by 12 inches. Equal-sized squares (x by x inches) will be cut out of each corner, and then the ends and sides will be folded up to form a rectangular box.

(A) Write the volume of the box $V(x)$ in terms of x.

(B) Considering the physical limitations, what is the domain of the function V?

(C) Graph the function for this domain.

(D) From the graph, estimate to the nearest half-inch the size of the square that must be cut from each corner to yield a box with the largest volume. What is the largest volume?

★★ 74. *Packaging* A parcel delivery service will deliver only packages with length plus girth (distance around) not exceeding 108 inches. A packaging company wishes to design a box with a square base (x by x inches) that will have a maximum volume but meet the delivery service's restrictions.

(A) Write the volume of the box $V(x)$ in terms of x.

(B) Considering the physical limitations imposed by the delivery service, what is the domain of the function V?

(C) Graph the function for this domain.

(D) From the graph, estimate to the nearest inch the dimensions of the box with the largest volume. What is the largest volume?

75. *Packaging* Repeat Problem 74 for a box with a base that is twice as long as the box is wide. Let x be the width of the box so that the base is x by $2x$.

76. *Packaging* Repeat Problem 73 for a piece of cardboard measuring 12 by 20 inches.

★ 77. *Agriculture* The total yield from a crop is the number of plants times the yield per plant. A fruit grower can achieve a yield of 40 bushels per tree when 120 trees are planted in an orchard. If the number of trees is increased, the yield per tree drops due to crowding. For each additional tree planted, the yield is expected to decrease by $\frac{1}{4}$ bushel per tree. Let x represent the number of additional trees planted.

(A) Express the total yield as a function of x, $Y(x)$.

(B) Graph the function $Y(x)$.

(C) Use the graph to determine the number of additional trees to plant in order to obtain the largest possible total yield.

★ 78. *Agriculture* Repeat Problem 77 for an orchard of 100 trees that yield 30 bushels per tree.

★ 79. *Business* The revenue from sales of a product is its selling price times the number of units sold. A toy manufacturer can sell 5,000 of a particular toy if the

selling price is $8 each. For each quarter that the seller increases the price, the sales are expected to decrease by 100 units. Let x represent the number of quarters by which the price is increased.

(A) Express the resulting revenue as a function of x, $R(x)$.

(B) Graph the function $R(x)$.

(C) Use the graph to determine how much the price should be increased to yield the largest possible revenue.

★ 80. *Business* Repeat Problem 79 for a toy that will have sales of 4,000 at a price of $12 and for which sales will decrease by 150 units for every half-dollar that the price is increased. Let x represent the number of half-dollars by which the price is increased.

★ 81. *Business* A resort hotel with 180 units can expect to have all units occupied at a rate of $420 per week and to lose one rental for each $10 increase in the weekly rate. Let x be the number of $10 increases added to the price.

(A) Express the total revenue as a function of x, $R(x)$.

(B) Graph the function $R(x)$.

(C) Use the graph to determine how much the price should be increased to yield the largest possible revenue.

★ 82. *Business* Repeat Problem 81 for a period during which the hotel can expect to have all 180 units rented at $600 per week and to lose one rental for each $15 increase in the weekly rate. Let x be the number of $15 increases added to the price.

★ 83. *Business* A theater expects an average daily attendance of 300 when ticket prices are $7 each. Average attendance will decrease by 20 for each $1 increase in price. Let x be the number of $1 increases added to the price.

(A) Express the total revenue as a function of x, $R(x)$.

(B) Graph the function $R(x)$.

(C) Use the graph to determine how much the price should be increased to yield the largest possible revenue.

★ 84. *Business* Repeat Problem 83 for a theater that expects an average daily attendance of 400 when ticket prices are $8 each and for which average attendance will decrease by 20 for each half-dollar increase in the price. Let x be the number of half-dollar increases added to the price.

9-4   Inverse Functions

♦ Inverse Functions
♦ Graphs of Inverse Functions
♦ One-to-One Correspondence and Inverses

In this section we discuss another method for obtaining new functions from old ones. In particular, we will try to "reverse" the input–output rule for a function to obtain a new function called the *inverse function*. This will not always be possible but can be done for those functions that meet the additional condition that each output comes from a unique input. Such functions are called *one-to-one*. We will need this method in Chapter 10 to obtain logarithmic functions from exponential functions.

♦ **INVERSE FUNCTIONS**

A function f associates a unique output value $f(x)$ with each input value x. Can this process be reversed? For example, if

$$y = f(x) = 2x - 1$$

what input x yields output 7? That is, if $f(x) = 7$, what is x? This amounts to solving $7 = 2x - 1$ for x:

$$7 = 2x - 1$$
$$8 = 2x$$
$$x = 4$$

Thus, for the given range value 7 we have found the corresponding domain value 4. Proceeding in the same way, we could find the corresponding domain value for each range value for the function f.

The process just described leads to a new function, the **inverse of f,** which is denoted by f^{-1}. More precisely:

> **Inverse Function**
>
> If y is a function f of x and each value of y is obtained from a unique value of x, then we say that f has an **inverse function f^{-1}.** The inverse function is the rule that reverses f. That is, if $b = f(a)$, then $a = f^{-1}(b)$.

In the notation for the inverse function, -1 is *not* an exponent:

$$f^{-1}(x) \qquad \text{does } not \text{ mean} \qquad \frac{1}{f}$$

The range for f becomes the domain for f^{-1}, and the domain for f becomes the range for f^{-1}. In terms of ordered pairs, if (a, b) belongs to f, then (b, a) belongs to f^{-1}. From $b = f(a)$ and $a = f^{-1}(b)$, we obtain

$$a = f^{-1}(b) = f^{-1}(f(a))$$
$$b = f(a) = f(f^{-1}(b))$$

That is:

$$f^{-1}(f(x)) = x \qquad \textbf{and} \qquad f(f^{-1}(x)) = x$$

When f is specified by an equation, we may be able to find an equation that specifies f^{-1} as we did above for the particular function $f(x) = 2x - 1$ and a particular value of y. We solve for x in terms of y and then interchange the variables:

$$f: \quad y = 2x - 1 \qquad \text{Solve for } x \text{ in terms of } y.$$
$$y + 1 = 2x$$
$$x = \frac{y + 1}{2} \qquad \text{Inverse rule.}$$

Thus,

$$x = f^{-1}(y) = \frac{y + 1}{2} \qquad \text{Interchange variables to form an inverse rule}$$
$$\text{with } x \text{ as the name of the independent variable.}$$
$$y = f^{-1}(x) = \frac{x + 1}{2}$$

Note that $(4, 7)$ is an element of function f and $(7, 4)$ is an element of the inverse function f^{-1}; that is, $f(4) = 7$ and $f^{-1}(7) = 4$. Thus, to find the inverse for $y = f(x)$, where $f(x)$ is specified by an algebraic expression in x:

1. Solve for x in terms of y, if possible, to obtain $x = f^{-1}(y)$.
2. Interchange the variables to obtain the function in the usual notation $y = f^{-1}(x)$.

Step 1 may not be possible. This means either that no inverse function exists or that the rule cannot be found this way. In practice, we can do either of the two steps first. That is, it doesn't matter when we reverse the names of the variables. We can interchange the variables first and then solve for y in terms of x if we choose. Thus:

Finding the Inverse Rule

To find an inverse rule for $y = f(x)$, where $f(x)$ is specified by an algebraic expression in x:

1. Interchange the variables, writing $x = f(y)$.
2. Solve for y in terms of x, if possible, to obtain $y = f^{-1}(x)$.

<table>
<tr><td>

Example 1
Finding the Inverse Rule

</td><td>

The function $f(x) = \sqrt{x}$ has an inverse. Find the inverse rule and its domain and range.

</td></tr>
</table>

Solution For our given function f we have

$$
\begin{aligned}
\text{Rule:} & \quad y = f(x) = \sqrt{x} \\
\text{Domain:} & \quad x \geq 0 \\
\text{Range:} & \quad y \geq 0
\end{aligned}
$$

For the inverse rule, the domain and range will be reversed; that is, the inputs for the inverse will be the outputs from the original function. We find the inverse rule by interchanging variables and solving for y in terms of x:

$$y = f(x) = \sqrt{x} \qquad \text{Interchange the variables.}$$
$$x = f(y) = \sqrt{y} \qquad \text{Solve for } y.$$
$$x^2 = y \qquad \text{This is the inverse rule.}$$

Thus, we have for f^{-1}.

$$
\begin{aligned}
\text{Rule:} & \quad y = f^{-1}(x) = x^2 \\
\text{Domain:} & \quad x \geq 0 \\
\text{Range:} & \quad y \geq 0
\end{aligned}
$$

Matched Problem 1 The function $f(x) = x^3$ has an inverse. Find the inverse rule and its domain and range.

<table>
<tr><td>

Example 2
Finding and Using the Inverse Rule

</td><td>

The function $f(x) = 3x + 2$ has an inverse. Find:

(A) $f^{-1}(x)$ **(B)** $f^{-1}(5)$ **(C)** $f^{-1}[f(5)]$ **(D)** $f^{-1}[f(x)]$

</td></tr>
</table>

Solution **(A)**

$$
\begin{aligned}
f: & \quad y = 3x + 2 \\
f^{-1}: & \quad x = 3y + 2 \qquad \text{Interchange variables } x \text{ and } y. \\
& \qquad\qquad\qquad\quad \text{Solve for } y \text{ in terms of } x. \\
& \quad x - 2 = 3y \\
& \quad y = \frac{x - 2}{3}
\end{aligned}
$$

Thus,

$$f^{-1}(x) = \frac{x - 2}{3}$$

(B) $f^{-1}(5) = \dfrac{5 - 2}{3} = \dfrac{3}{3} = 1$

(C) $f^{-1}[f(5)] = \dfrac{f(5) - 2}{3}$ We are just verifying the basic result $f^{-1}(f(x)) = x$.

$\qquad = \dfrac{17 - 2}{3} = \dfrac{15}{3} = 5$

(D) $f^{-1}[f(x)] = \dfrac{f(x) - 2}{3}$ See comment in part (C).

$\qquad = \dfrac{(3x + 2) - 2}{3} = x$

Matched Problem 2   The function $g(x) = \dfrac{x}{3} - 2$ has an inverse. Find:

(A) $g^{-1}(x)$ **(B)** $g^{-1}(-2)$ **(C)** $g^{-1}(g(3))$ **(D)** $g^{-1}(g(x))$

◆ GRAPHS OF INVERSE FUNCTIONS

If a function $f(x)$ has an inverse, there is a relationship between the graphs of $f(x)$ and $f^{-1}(x)$. A point (a, b) is on one graph if and only if (b, a) is on the other. The points (a, b) and (b, a) are symmetric about the 45° line through $(0, 0)$, that is, about the line $y = x$, as shown in Figure 1. The graph of f^{-1} is therefore the graph of f reflected about the 45° line as a mirror image.

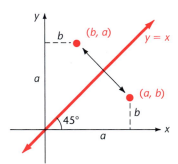

Figure 1

Example 3
Graphing an Inverse Function

Graph the function and its inverse.

(A) $f(x) = 2x - 1$ **(B)** $f(x) = \sqrt{x}$

Solution   **(A)** We already know that

$$f^{-1}(x) = \frac{x + 1}{2}$$

We can rewrite $(x + 1)/2$ as $\frac{1}{2}x + \frac{1}{2}$ to see that the graph is a straight line. Thus, we get the graphs shown in Figure 2*a*.

(B) We already know that $f^{-1}(x) = x^2$ for $x \geq 0$. The two graphs are as shown in Figure 2b.

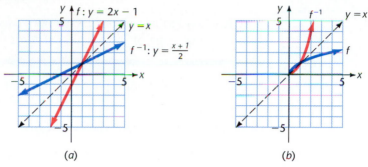

(a) (b)

Figure 2

Matched Problem 3   Graph the function $f(x) = x^3$ and its inverse.

Figure 3 illustrates some functions that have inverse functions, and Figure 4 illustrates some that do not. In Figure 3, each domain value corresponds to exactly one domain value. In Figure 4, each domain value corresponds to exactly one range value, but some range values correspond to more than one domain value: in Figure 4a, y_1 corresponds to x_1 and x_2; in Figure 4b, y_1 corresponds to x_1, x_2, and x_3.

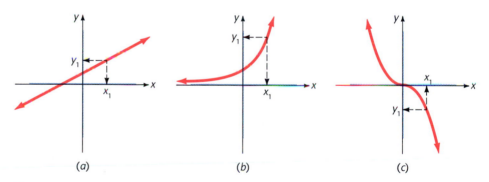

(a) (b) (c)

Figure 3   Functions that have an inverse function

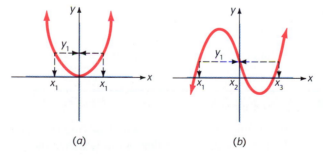

(a) (b)

Figure 4   Functions that do not have an inverse function

Given the graph of a function, we can see whether it has an inverse that is a function:

Horizontal-Line Test for an Inverse Function

A function has an inverse if each horizontal line in the coordinate system crosses the graph of the function at most once.

Notice how this test corresponds to the results in Figures 3 and 4.

Example 4
Applying the Horizontal-Line Test

Determine from its graph whether the given function has an inverse:

(A) $g(x) = x^2 - 4$ **(B)** $f(x) = 3x + 2$

Solution   **(A)** From the graph, $g(x)$ has no inverse.

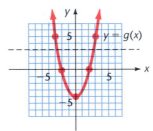

Horizontal line crosses $g(x)$ twice.

(B) From the graph, $f(x)$ has an inverse.

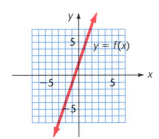

No horizontal line will cross $f(x)$ twice.

Matched Problem 4   Determine from its graph whether the given function has an inverse:

(A) $g(x) = 3 - 4x$ **(B)** $f(x) = 3 - x^2$

♦ ONE-TO-ONE CORRESPONDENCE AND INVERSES

Not all functions have inverses. Applying the horizontal-line test to the graph of a function is one way to determine whether the function has an inverse function. There is also an algebraic criterion. We know that for a function, each x value corresponds to a unique y value. We need to know whether each y value also corresponds to a unique x value. If so, there will then be a *one-to-one correspondence* between the x and y values. A **one-to-one correspondence** exists between two sets if each element in the first set corresponds to exactly one element in the second set, and each element in the second set corresponds to exactly one element in the first set. Consider the following two functions f and g and their inverse rules:

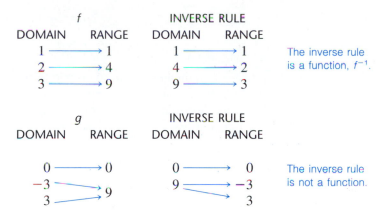

Function f is a one-to-one correspondence between domain and range values, and f^{-1} is also a function. Function g is not a one-to-one correspondence between domain and range values. In this case, the inverse rule is not a function. More generally, we have the following result:

Inverses and One-to-One Functions

A function f has an inverse function if and only if f defines a one-to-one correspondence between domain and range values of f. In this case, we say that f is a **one-to-one function** and note that

$$f[f^{-1}(y)] = y \quad \text{and} \quad f^{-1}[f(x)] = x$$

This result is interpreted schematically in Figure 5.

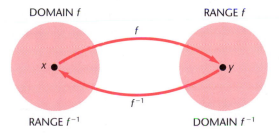

Figure 5

For a function to be one-to-one, the only way two output values can be the same is if their original input values are actually the same. Algebraically, this can be stated as:

f is one-to-one if and only if $f(p) = f(q)$ implies $p = q$.

For example, $f(x) = 2x + 1$ has a graph that is a nonvertical straight line, so it passes the horizontal-line test and has an inverse function. To see that f is one-to-one, suppose $f(p) = f(q)$. Then

$$f(p) = f(q)$$
$$2p + 1 = 2q + 1$$
$$2p = 2q$$
$$p = q$$

That is, $f(p) = f(q)$ implies $p = q$, so f is one-to-one. The C-level exercises for this section include other examples like this. For the remainder of the text, however, we will rely on the horizontal-line test or solving for x in terms of y to determine whether a function has an inverse function.

Answers to   **1.** $f^{-1}(x) = \sqrt[3]{x}$; Domain: all real numbers,   Range: all real numbers
Matched Problems   **2. (A)** $3x + 6$ **(B)** 0   **(C)** 3   **(D)** x
3.

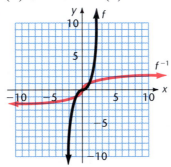

4. (A) Yes   **(B)** No

EXERCISE 9-4

A *Determine whether the function defined by each set of ordered pairs has an inverse.*

1. $(-2, 0), (-1, 1), (0, 2), (1, 3), (2, 4)$

2. $(-2, -2), (-1, -1), (0, 0), (1, 1), (2, 2)$

3. $(-2, 0), (-1, 1), (0, 2), (1, 1), (2, 0)$

4. $(-2, 2), (-1, 1), (0, 0), (1, 1), (2, 2)$

Find the inverse rule f^{-1} and its domain for the function f defined by each set of ordered pairs.

5. $(-2, 1), (-1, 2), (0, 3), (1, 4), (2, 5)$

6. $(-2, -2), (-1, -4), (0, -6), (1, -8), (2, -10)$

7. $(-2, \frac{1}{5}), (-1, \frac{1}{4}), (0, \frac{1}{3}), (1, \frac{1}{2}), (2, 1)$

8. $(-2, \frac{1}{9}), (-1, \frac{1}{3}), (0, 1), (1, 3), (2, 9)$

Graph the function, its inverse, and $y = x$ on the same coordinate system.

9. f in Problem 5

10. f in Problem 6

11. f in Problem 7

12. f in Problem 8

Find the inverse rule, $f^{-1}(x)$. Evaluate $f^{-1}(-1)$, $f^{-1}[f(-1)]$, and $f[f^{-1}(-1)]$.

13. $f(x) = 3x - 2$ **14.** $f(x) = 2x + 3$

15. $f(x) = \dfrac{x}{3} - 2$ **16.** $f(x) = \dfrac{x}{2} + 5$

In Problems 17–22, determine whether the function graphed has an inverse function.

17.
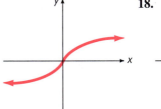

18.
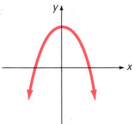

19.   20.
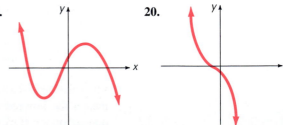

21.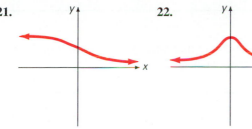

22.

B 23. For $f(x) = 3x - 2$, find:
   (A) $f^{-1}(x)$ (B) $f^{-1}(2)$ (C) $f[f^{-1}(3)]$

24. For $g(x) = 2x + 3$, find:
   (A) $g^{-1}(x)$ (B) $g^{-1}(5)$ (C) $g[g^{-1}(4)]$

25. For $F(x) = \dfrac{x}{3} - 2$; find:

   (A) $F^{-1}(x)$ (B) $F^{-1}(-1)$ (C) $F^{-1}[F(4)]$

26. For $G(x) = \dfrac{x}{2} + 5$, find:

   (A) $G^{-1}(x)$ (B) $G^{-1}(8)$ (C) $G^{-1}[G(-4)]$

For each function find the inverse function in the form of an equation.

27. $f(x) = \sqrt{x} + 1$ **28.** $f(x) = \sqrt{x} - 3$

29. $f(x) = \sqrt{x - 3}$ **30.** $f(x) = \sqrt{x + 1}$

31. $f(x) = \sqrt{x + 1} - 3$ **32.** $f(x) = \sqrt{x - 3} + 1$

33. $f(x) = \dfrac{1}{x + 1}$ **34.** $f(x) = \dfrac{1}{x - 1}$

35. $f(x) = \sqrt[3]{x}$ **36.** $f(x) = \sqrt[5]{x}$

37. $f(x) = \sqrt[3]{x - 1}$ **38.** $f(x) = \sqrt[3]{x + 1}$

39. $f(x) = x^5 - 2$ **40.** $f(x) = x^5 + 3$

41. $f(x) = 1 - \dfrac{1}{x}$ **42.** $f(x) = 1 + \dfrac{1}{x}$

43. $f(x) = \dfrac{x + 3}{x - 1}$ **44.** $f(x) = \dfrac{x - 1}{x + 3}$

45. $f(x) = -x^2$, for $x \geq 0$

46. $f(x) = -x^4$, for $x \geq 0$

47–66. For each function f in Problems 27–46, show that $f^{-1}[f(x)] = x$.

67–74. For each function f in Problems 13–16 and 27–30, graph f and f^{-1} in the same coordinate system, along with $y = x$.

C *Show that the function f is one-to-one by showing that $f(p) = f(q)$ implies $p = q$.*

75. $f(x) = 3x - 5$ **76.** $f(x) = 5x + 3$

77. $f(x) = \sqrt{x + 1}$ **78.** $f(x) = \sqrt{x - 1}$

79. $f(x) = \dfrac{1}{x + 1}$ **80.** $f(x) = \dfrac{1}{x - 1}$

81–96. For each function f in Problems 31–46, graph f and f^{-1} in the same coordinate system, along with $y = x$.

 ## 9-5 **Variation**

    ♦   Direct Variation
    ♦   Inverse Variation
    ♦   Joint Variation
    ♦   Combined Variation

In reading scientific material, one is likely to come across statements such as "the pressure of an enclosed gas varies directly as the absolute temperature," or "the frequency of vibration of air in an organ pipe varies inversely as the length of the pipe." These statements have precise mathematical meaning in that they represent particular types of functions. The purpose of this section is to investigate these special functions and also more complicated statements such as "the force of attraction between two bodies varies jointly as their masses and inversely as the square of the distance between the two bodies."

♦   **DIRECT VARIATION**

The statement **y varies directly as x,** or alternatively, **y is proportional to x,** means that the quotient y/x is a constant. That is:

Direct Variation

y varies directly as x, or y is proportional to x, means

$$y = kx \qquad k \neq 0 \tag{1}$$

where k is a constant called the **constant of variation.**

Similarly, the statement "y varies directly as the square of x" means

$$y = kx^2 \qquad k \neq 0 \tag{2}$$

and so on. Equation (1) defines a linear function $f(x) = kx$, and Equation (2) a quadratic function $g(x) = kx^2$.

Direct variation is illustrated by the familiar formulas

$$C = \pi D \qquad \text{and} \qquad A = \pi r^2$$

where the first formula asserts that the circumference C of a circle varies directly as the diameter D, and the second formula states that the area A of a circle varies directly as the square of the radius r. In both cases, π is the constant of variation.

Example 1
Finding the Constant in Direct Variation

Translate each statement into an appropriate equation, and find the constant of variation if $y = 16$ when $x = 4$.

(A) y varies directly as x.   **(B)** y varies directly as the cube of x.

Solution   **(A)** y varies directly as x means

$$y = kx \qquad \text{Do not forget } k.$$

To find the constant of variation k, substitute $x = 4$ and $y = 16$ and solve for k:

$$16 = k \cdot 4$$

$$k = \frac{16}{4} = 4 \qquad \text{Constant of variation}$$

Thus, $k = 4$, and the equation is

$$y = 4x$$

(B) y varies directly as the cube of x means

$$y = kx^3 \qquad \text{Do not forget } k.$$

To find k, substitute $x = 4$ and $y = 16$:

$$16 = k \cdot 4^3$$

$$k = \frac{16}{64} = \frac{1}{4} \qquad \text{Constant of variation}$$

Thus, the equation is

$$y = \frac{1}{4}x^3$$

Matched Problem 1   If $y = 4$ when $x = 8$, find the constant of variation and equation for each statement:

(A) y varies directly as x.   **(B)** y varies directly as the cube root of x.

♦   **INVERSE VARIATION**

The statement **y varies inversely as x,** or alternatively, **y is inversely proportional to x,** means that the product yx is a constant. Equivalently:

Inverse Variation

y varies inversely as x, or y is inversely proportional to x, means

$$y = \frac{k}{x} \qquad k \neq 0$$

where k is again a constant called the **constant of variation.**

As in the case of direct variation, we also may discuss y varying inversely as the square of x, and so on.

An illustration of inverse variation is given in the distance-rate-time formula $d = rt$ in the form $t = d/r$ for a fixed distance d. In driving a fixed distance, say $d = 400$ miles, time varies inversely as the rate; that is,

$$t = \frac{400}{r}$$

where 400 is the constant of variation. As the rate increases, the time decreases, and vice versa.

Example 2
Finding the Constant in Inverse Variation

Translate each statement into an appropriate equation, and find the constant of variation if $y = 16$ when $x = 4$.

(A) y varies inversely as x.
(B) y varies inversely as the square root of x.

Solution **(A)** y varies inversely as x means

$$y = \frac{k}{x} \qquad \text{Do not forget } k.$$

To find k, substitute $x = 4$ and $y = 16$:

$$16 = \frac{k}{4}$$

$$k = 64 \qquad \text{Constant of variation}$$

Thus, the equation is

$$y = \frac{64}{x}$$

(B) y varies inversely as the square root of x means

$$y = \frac{k}{\sqrt{x}}$$

To find k, substitute $x = 4$ and $y = 16$:

$$16 = \frac{k}{\sqrt{4}}$$

$$k = 32 \qquad \text{Constant of variation}$$

Thus, the equation is

$$y = \frac{32}{\sqrt{x}}$$

Matched Problem 2 If $y = 4$ when $x = 8$, find the constant of variation and equation for each statement:

(A) y varies inversely as x.   **(B)** y varies inversely as the square of x.

◆ **JOINT VARIATION**

The statement **w varies jointly as x and y** means that w varies directly as the product of x and y. That is:

> ## Joint Variation
>
> w varies jointly as x and y means
>
> $$w = kxy \qquad k \neq 0$$
>
> where k is the constant of variation.

Similarly, if

$$w = kxyz^2 \qquad k \neq 0$$

we would say that "w varies jointly as x, y, and the square of z," and so on. For example, the area of a rectangle varies jointly as its length and width (recall $A = lw$), and the volume of a right circular cylinder varies jointly as the square of its radius and its height (recall $V = \pi r^2 h$). What is the constant of variation in each case?

In the joint variation $w = kxy$, w is a function of either x or y if the other is held fixed. It is also true that w is a function of *both* x and y, but we will not consider this kind of function in this text.

◆ COMBINED VARIATION

The basic types of variation introduced above are often combined. For example, the statement "w varies jointly as x and y and inversely as the square of z" means

$$w = k \cdot \frac{xy}{z^2} \qquad k \neq 0 \qquad$$ We do not write $w = \dfrac{kxy}{kz^2}$. This is wrong, because it eliminates the proportionality constant k.

Thus, the statement "the force of attraction F between two bodies varies jointly as their masses m_1 and m_2 and inversely as the square of the distance d between the two bodies" means

$$F = k \cdot \frac{m_1 m_2}{d^2} \qquad k \neq 0$$

Assuming k is positive, if either of the two masses is increased, the force of attraction increases; on the other hand, if the distance is increased, the force of attraction decreases.

Example 3
Combined Variation in Boyle's Law

The pressure P of an enclosed gas varies directly as the absolute temperature T and inversely as the volume V (Boyle's law). If 500 cubic feet of gas yields a pressure of 10 pounds per square foot at an absolute temperature of 300 K, what will be the pressure of the same gas if the volume is decreased to 300 cubic feet and the temperature increased to 360 K? [The unit of measure for absolute temperature, the kelvin, is the same size as a Celsius degree, but 0 on the Kelvin scale is the same as $-273°C$. This is the temperature at which molecular action is assumed to stop and is called *absolute 0*.]

Solution   *Method 1.*   Write the equation of variation $P = k(T/V)$, and find k using the first set of values:

$$10 = k \cdot \frac{300}{500}$$

$$k = \frac{50}{3}$$

Hence, the equation of variation for this particular gas is

$$P = \frac{50}{3} \cdot \frac{T}{V}$$

Now find the new pressure P using the second set of values:

$$P = \frac{50}{3} \cdot \frac{360}{300} = 20 \text{ pounds per square foot}$$

Method 2.   This method is generally faster than method 1. Write the equation of variation $P = k(T/V)$; then convert to the equivalent form:

$$\frac{PV}{T} = k$$

If P_1, V_1, and T_1 are the first set of values for the gas and P_2, V_2, and T_2 are the second set, then

$$\frac{P_1 V_1}{T_1} = k \qquad \text{and} \qquad \frac{P_2 V_2}{T_2} = k$$

Hence,

$$\frac{P_1 V_1}{T_1} = \frac{P_2 V_2}{T_2}$$

Since all values are known except P_2, substitute and solve. Thus,

$$\frac{(10)(500)}{300} = \frac{P_2(300)}{360}$$

$$P_2 = 20 \text{ pounds per square foot}$$

Matched Problem 3   The length L of skid marks of a car's tires after brakes are applied varies directly as the square of the speed v of the car. If skid marks of 20 feet are produced at 30 miles per hour, how fast would the same car be going if it produced skid marks of 80 feet? Solve in two ways as in Example 3.

Example 4	The frequency of pitch f of a given musical string varies directly as the square root
Combined Variation in Music	of the tension T and inversely as the length L. What is the effect on the frequency if the tension is increased by a factor of 4 and the length is cut in half?

Solution Write the equation of variation:

$$f = \frac{k\sqrt{T}}{L} \qquad \text{or equivalently} \qquad \frac{f_2 L_2}{\sqrt{T_2}} = \frac{f_1 L_1}{\sqrt{T_1}}$$

We are given that $T_2 = 4T_1$ and $L_2 = 0.5L_1$. Substituting in the second equation, we have

$$\frac{f_2[0.5L_1]}{\sqrt{4T_1}} = \frac{f_1 L_1}{\sqrt{T_1}} \qquad \text{\textit{Solve for} } f_2.$$

$$\frac{f_2[0.5L_1]}{2\sqrt{T_1}} = \frac{f_1 L_1}{\sqrt{T_1}}$$

$$f_2 = \frac{2\sqrt{T_1}f_1 L_1}{0.5L_1\sqrt{T_1}} = 4f_1$$

Thus, the frequency of pitch is increased by a factor of 4.

Matched Problem 4 The weight w of an object on or above the surface of the earth varies inversely as the square of the distance d between the object and the center of the earth. If an object on the surface of the earth is moved into space so as to double its distance from the earth's center, what effect will this move have on its weight?

Answers to Matched Problems
1. **(A)** $y = \frac{1}{2}x$ **(B)** $y = 2\sqrt[3]{x}$ **2. (A)** $y = 32/x$ **(B)** $y = 256/x^2$
3. $v = 60$ miles per hour 4. It will be one-fourth as heavy.

EXERCISE 9-5

A *In Problems 1–12, translate each statement into an equation using k as the constant of variation.*

1. F varies directly as the square of v.

2. u varies directly as v.

3. The pitch or frequency f of a guitar string of a given length varies directly as the square root of the tension T of the string.

4. Geologists have found in studies of earth erosion that the erosive force (sediment-carrying power) P of a swiftly flowing stream varies directly as the sixth power of the velocity v of the water.

5. y varies inversely as the square root of x.

6. I varies inversely as t.

7. The biologist Reaumur suggested in 1735 that the length of time t that it takes fruit to ripen during the growing season varies inversely as the sum T of the average daily temperatures during the growing season.

8. In a study on urban concentration, F. Auerbach discovered an interesting law. After arranging all the cities of a given country according to their population size, starting with the largest, he found that the population P of a city varied inversely as the number n indicating its position in the ordering.

9. R varies jointly as S, T, and V.

10. g varies jointly as x and the square of y.

11. The volume of a cone V varies jointly as its height h and the square of the radius r of its base.

12. The amount of heat h put out by an electrical appliance (in calories) varies jointly as time t, resistance R in the circuit, and the square of the current I.

13. If y varies directly as x, and $y = 7.5$ when $x = 3$, find y when $x = 8$.

14. If y varies inversely as x, and $y = 15$ when $x = 4$, find y when $x = 10$.

15. If y varies inversely as x, and $y = 9$ when $x = 8$, find y when $x = 6$.

16. If y varies directly as x, and $y = 22$ when $x = 7$, find y when $x = 21$.

Solve using either of the two methods illustrated in Example 3.

17. u varies directly as the square root of v. If $u = 2$ when $v = 2$, find u when $v = 8$.

18. y varies directly as the square of x. If $y = 20$ when $x = 2$, find y when $x = 5$.

19. L varies inversely as the square root of M. If $L = 9$ when $M = 9$, find L when $M = 3$.

20. I varies inversely as the cube of t. If $I = 4$ when $t = 2$, find I when $t = 4$.

B *Translate each statement into an equation using k as the constant of variation.*

21. U varies jointly as a and b and inversely as the cube of c.

22. w varies directly as the square of x and inversely as the square root of y.

23. The maximum safe load L for a horizontal beam varies jointly as its width w and the square of its height h and varies inversely as its length l.

24. Joseph Cavanaugh, a sociologist, found that the number of long-distance phone calls n between two cities in a given time period varied (approximately) jointly as the populations P_1 and P_2 of the two cities and inversely as the distance d between the two cities.

Solve using either of the two methods illustrated in Example 3.

25. Q varies jointly as m and the square of n and inversely as P. If $Q = -4$ when $m = 6$, $n = 2$, and $P = 12$, find Q when $m = 4$, $n = 3$, and $P = 6$.

26. w varies jointly as x, y, and z and inversely as the square of t. If $w = 2$ when $x = 2$, $y = 3$, $z = 6$, and $t = 3$, find w when $x = 3$, $y = 4$, $z = 2$, and $t = 2$.

27. The weight w of an object on or above the surface of the earth varies inversely as the square of the distance d between the object and the center of the earth. If a girl weighs 100 pounds on the surface of the earth, how much would she weigh (to the nearest pound) 400 miles above the earth's surface? (Assume that the radius of the earth is 4,000 miles.)

28. Repeat Problem 27 for a satellite weighing 20,000 pounds on earth now located at an altitude of 180 miles.

29. A car was struck by a truck at an intersection. The truck driver had slammed on her brakes and left skid marks 250 feet long. The driver told the police she had been driving at 55 miles per hour. The police know that the length L of skid marks when brakes are applied varies directly as the square of the speed v of the vehicle. At 30 miles per hour under the same driving conditions, the truck leaves skid marks 40 feet long. How fast was the truck driver actually going when she applied her brakes?

30. Repeat Problem 29 for a driver who claims to have been traveling at 30 miles per hour but leaves skid marks 160 feet long.

31. Ohm's law states that the current I in a wire varies directly as the electromotive force E and inversely as the resistance R. If $I = 22$ amperes when $E = 110$ volts and $R = 5$ ohms, find I if $E = 220$ volts and $R = 11$ ohms.

32. Repeat Problem 31 if $I = 1.5$ amperes when $E = 6$ volts and $R = 2$ ohms.

33. See Problem 23. If a beam that is 2 inches wide, 4 inches high, and 8 feet long has a maximum safe load L of 320 pounds, what is the maximum safe load of a beam of similar material that is 2 inches wide by 6 inches high by 12 feet long?

34. See Problem 23. If a beam that is 4 inches wide, 4 inches high, and 16 feet long has a maximum safe load L of 320 pounds, what is the maximum safe load of a beam of similar material that is 2 inches wide by 12 inches high by 18 feet long?

35. The unit *worker-hours* is used to measure the number of workers times the number of hours each has worked. If a project requires a fixed number of worker-hours to complete, then the number of hours required of each worker varies inversely with the number of workers. If 6 persons working 30 hours each can complete such a project, how many hours each must 9 workers work to complete the project?

36. Repeat Problem 35 for a project that can be completed by 10 workers putting in 40 hours each, when we want to know how much each of 16 workers would have to work to complete the project.

37. See Problem 24. Suppose two cities located 900 miles apart and having populations of 3.2 million and 5.4 million generate 100,000 phone calls between the two cities during a certain period. How many calls would be generated between two cities with populations of 400,000 and 1.5 million if the cities are 100 miles apart?

38. Anthropologists, in their study of race and human genetic groupings, often use an index called the *cephalic index*. The cephalic index C varies directly as the width w of the head and inversely as the length l of the head (both when viewed from the top). If an Indian in Baja California (Mexico) has measurements of $C = 75$, $w = 6$ inches, and $l = 8$ inches, what is C for an Indian in northern California with $w = 8.1$ inches and $l = 9$ inches?

C 39. If the horsepower P required to drive a speedboat through water varies directly as the cube of the speed v of the boat, what change in horsepower is required to double the speed of the boat?

40. The intensity of illumination E on a surface varies inversely as the square of its distance d from a light source. What is the effect on the total illumination on a book if the distance between the light source and the book is doubled?

41. The frequency of vibration f of a musical string varies directly as the square root of the tension T and inversely as the length L of the string. If the tension of the string is increased by a factor of 4 and the length of the string is doubled, what is the effect on the frequency?

42. In an automobile accident the destructive force F of a car varies (approximately) jointly as the weight w of the car and the square of the speed v of the car. (This is why accidents at high speed are generally so serious.) What would be the effect on the destructive force of a car if its weight were doubled and its speed were doubled?

ADDITIONAL APPLICATIONS

The following problems include applications from many different areas and are arranged according to subject area. The more difficult problems are marked with two stars (★★), the moderately difficult problems are marked with one star (★), and the easier problems are not marked.

43. *Astronomy* The square of the time t required for a planet to make one orbit around the sun varies directly as the cube of its mean (average) distance d from the sun. Write the equation of variation using k as the constant of variation.

★ 44. *Astronomy* The centripetal force F of a body moving in a circular path at constant speed varies inversely as the radius r of the path. What happens to F if r is double?

45. *Astronomy* The length of time t a satellite takes to complete a circular orbit of the earth varies directly as the radius r of the orbit and inversely as the orbital velocity v of the satellite. If $t = 1.42$ hours when $r = 4,050$ miles and $v = 18,000$ miles per hour (*Sputnik I*), find t for $r = 4,300$ miles and $v = 18,500$ miles per hour.

46. *Life science* The number N of gene mutations resulting from x-ray exposure varies directly as the size of the x-ray dose r. What is the effect on N if r is quadrupled?

47. *Life science* In biology there is an approximate rule, called the *bioclimatic rule* for temperature climates, which states that the difference d in time for fruit to ripen (or insects to appear) varies directly as the change in altitude h. If $d = 4$ days when $h = 500$ feet, find d when $h = 2,500$ feet.

48. *Physics and engineering* Over a fixed distance d, speed r varies inversely as time t. Police use this relationship to set up speed traps. (The graph of the resulting function is a hyperbola.) If in a given speed trap $r = 30$ miles per hour when $t = 6$ seconds, what would be the speed of a car if $t = 4$ seconds?

★ 49. *Physics and engineering* The length L of skid marks of a car's tires (when the brakes are applied) varies directly as the square of the speed v of the car. How is the length of skid marks affected by doubling the speed?

50. *Physics and engineering* The time t required for an elevator to lift a weight varies jointly as the weight w and the distance d through which it is lifted and inversely as the power P of the motor. Write the equation of variation using k as the constant of variation.

51. *Physics and engineering* The total pressure P of the wind on a wall varies jointly as the area of the wall A and the square of the velocity of the wind v. If $P = 120$ pounds when $A = 100$ square feet and $v = 20$ miles per hour, find P if $A = 200$ square feet and $v = 30$ miles per hour.

★★ 52. *Physics and engineering* The thrust T of a given type of propeller varies jointly as the fourth power of its diameter d and the square of the number of revolutions per minute n it is turning. What happens to the thrust if the diameter is doubled and the number of revolutions per minute is cut in half?

53. *Psychology* In early psychological studies on sensory perception (hearing, seeing, feeling, and so on), the question was asked: "Given a certain level of

stimulation S, what is the minimum amount of added stimulation ΔS that can be detected?'' A German physiologist, E. H. Weber (1795–1878) formulated, after many experiments, the famous law that now bears his name: ''The amount of change ΔS that will be just noticed varies directly as the magnitude S of the stimulus.''

(A) Write the law as an equation of variation.

(B) If a person lifting weights can just notice a difference of 1 ounce at the 50-ounce level, what will be the least difference she will be able to notice at the 500-ounce level?

(C) Determine the just noticeable difference in illumination a person is able to perceive at 480 candlepowers if he is just able to perceive a difference of 1 candlepower at the 60-candlepower level.

54. *Psychology*   In their study of intelligence, psychologists often use an index called *intelligence quotient* (IQ). IQ varies directly as mental age (MA) and inversely as chronological age (CA) up to the age of 15. If a 12-year-old boy with a mental age of 14.4 has an IQ of 120, what will be the IQ of an 11-year-old girl with a mental age of 15.4?

55. *Music*   The frequency of vibration of air in an open organ pipe varies inversely as the length of the pipe. If the air column in an open 32-foot pipe vibrates 16 times per second (low C), how fast would the air vibrate in a 16-foot pipe?

56. *Music*   The frequency of pitch f of a musical string varies directly as the square root of the tension T and inversely as the length l and the diameter d. Write the equation of variation using k as the constant of variation. (It is interesting to note that if pitch depended on only length, then pianos would have to have strings varying from 3 inches to 38 feet.)

57. *Photography*   The f-stop numbers N on a camera, known as *focal ratios,* vary directly as the focal length F of the lens and inversely as the diameter d of the diaphragm opening (effective lens opening). Write the equation of variation using k as the constant of variation.

★ 58. *Photography*   In taking pictures using flashbulbs, the lens opening (f-stop number) N varies inversely as the distance d from the object being photographed. What adjustment should you make on the f-stop number if the distance between the camera and the object is doubled?

★ 59. *Chemistry*   Atoms and molecules that make up the air constantly fly about like microscopic missiles. The velocity v of a particular particle at a fixed temperature varies inversely as the square root of its molecular weight w. If an oxygen molecule in air at room temperature has an average velocity of 0.3 mile per second, what will be the average velocity of a hydrogen molecule, given that the hydrogen molecule is one-sixteenth as heavy as the oxygen molecule?

60. *Chemistry*   The Maxwell–Boltzmann equation states that the average velocity v of a molecule varies directly as the square root of the absolute temperature T and inversely as the square root of its molecular weight w. Write the equation of variation using k as the constant of variation.

61. *Business*   The amount of work A completed varies jointly as the number of workers W used and the time t they spend. If 10 workers can finish a job in 8 days, how long will it take 4 workers to do the same job?

62. *Business*   The simple interest I earned in a given time varies jointly as the principal p and the interest rate r. If \$100 at 4% interest earns \$8, how much will \$150 at 3% interest earn in the same period?

★ 63. *Geometry*   The volume of a sphere varies directly as the cube of its radius r. What happens to the volume if the radius is doubled?

★ 64. *Geometry*   The surface area S of a sphere varies directly as the square of its radius r. What happens to the area if the radius is cut in half?

CHAPTER SUMMARY

9-1   FUNCTIONS

Informally, we say that **y is a function of x** if for every value of the variable x there is exactly one corresponding value of the variable y. The variable x is called the **independent variable,** and the set of allowable values of x is called the **domain** of the function. The variable y is called the **dependent variable,** and the set of values of y corresponding to values of x in the domain is called the **range** of the function. More formally, a **function** is a rule that produces a correspondence between a first set of elements (the **domain**) and a second set (the **range**) such that to each element in the domain there corresponds exactly one element in the range. A function also can be thought of as a set of ordered pairs in which different pairs cannot have the same first component. Functions may be specified by diagrams, tables, equations, sets of ordered pairs, or graphs. A function with domain and range sets of real numbers can be graphed by graphing all pairs (x, y), where x represents the domain values and y the corresponding range values. A graph represents a function if no vertical line crosses it more than once; this is the **vertical-line test.**

9-2 FUNCTION NOTATION AND THE ALGEBRA OF FUNCTIONS

If a function is denoted by f, the notation $f(x)$ denotes the value in the range corresponding to x in the domain. The values in the domain can be thought of as **input** values, and those in the range as **output** values. If f and g are functions, the **sum of f and g** is defined to be the function $(f + g)(x) = f(x) + g(x)$. The **difference $f - g$** is the function $(f - g)(x) = f(x) - g(x)$, the **product of f and g** is the function $(f \cdot g)(x) = f(x) \cdot g(x)$, and the **quotient f/g** is the function $(f/g)(x) = f(x)/g(x)$. The **composition of f by g** is the function that results from applying f first and then applying g to the result: $g(f(x))$.

9-3 GRAPHING FUNCTIONS

A function $f(x) = ax + b$ is called a **linear function;** $f(x) = ax^2 + bx + c$ is a **quadratic function** for $a \neq 0$. The graph of a linear function is a nonvertical straight line; that of a quadratic function is a parabola. If $f(x)$ is specified by a polynomial in x, the function is called a **polynomial function.** Polynomial functions are graphed by plotting a sufficient number of points and joining these points with a smooth curve.

9-4 INVERSE FUNCTIONS

A function f is **one-to-one** if every range element corresponds to exactly one domain element. The "reverse" rule for such a function f is called the **inverse of f** and is denoted f^{-1}. The inverse of a function specified by an equation $y = f(x)$ may, in some cases, be found by interchanging variables, obtaining $x = f(y)$, and solving for y. The graph of the inverse function f^{-1} is the graph of f reflected about the line $y = x$. A function has an inverse if its graph passes the **horizontal-line test:** no horizontal line crosses its graph more than once.

9-5 VARIATION

Direct, inverse, joint, and combined variations are relations that occur often in applications:

Direct:	y varies directly as x	$y = kx$
Inverse:	y varies inversely as x	$y = \dfrac{k}{x}$
Joint:	w varies jointly as x and y	$w = kxy$
Combined:	w varies directly as x and inversely as y	$w = \dfrac{kx}{y}$

In each case, k is called the **constant of variation.**

CHAPTER REVIEW EXERCISE

Work through all the problems in this chapter review and check answers in the back of the book. Answers to all the problems are there, and following each answer is a number in italics indicating the section in which that type of problem is discussed. Where weaknesses show up, review appropriate sections in the text.

A *Which of the rules in Problems 1–12 are functions with x as the independent variable?*

1.

2.

3.
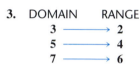

4.
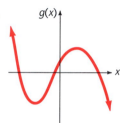

5.
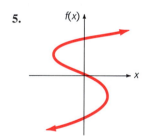

6.
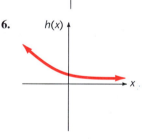

7. $y = x^3 - 2x$

8. $y^2 = x$

9. $x^2 + y^2 = 25$

10. $(1, 2), (1, -2), (0, 3)$

11. $(-1, 2), (1, 3), (2, 4)$

12. $(-2, 3), (0, 3), (2, 3)$

Find the domain and range of each function.

13. $(1, 2), (3, 4), (5, 6)$

14. $(-1, 0), (1, 0), (3, 0), (5, 0)$

15. $f(x) = \dfrac{1}{x}$

16. $f(x) = \dfrac{1}{\sqrt{x}}$

17. $f(x) = \dfrac{1}{\sqrt{x-1}}$

18. $f(x) = \sqrt{x-1}$

19. $f(x) = x - 1$

20. $f(x) = \dfrac{1}{x-1}$

In Problems 21–32, let $f(x) = 6 - x$ and $G(x) = x - 2x^2$. Find:

21. $f(6), f(0), f(-3)$

22. $G(2), G(0), G(-1)$

23. $f(m), f(x + h)$

24. $G(c), G(x + h)$

25. $(f + G)(x)$

26. $(f - G)(x)$

27. $(G - f)(x)$

28. $(f \cdot G)(x)$

29. $(f/G)(x)$

30. $(G/f)(x)$

31. $f(G(x))$

32. $G(f(x))$

33. Graph $f(x) = 2x - 4$. Indicate its slope and y intercept.

34. Graph $g(x) = \dfrac{x^2}{2}$.

35. Graph the function M defined by the ordered pairs, $(0, 5), (2, 7), (4, 9)$, its inverse M^{-1}, and $y = x$, all on the same coordinate system.

36. What are the domain and range of M^{-1} in Problem 35?

Find f^{-1}.

37. $f(x) = 3x + 5$

38. $f(x) = \dfrac{3}{x + 5}$

39. $f(x) = \sqrt{3x + 5}$

40. $f(x) = 3\sqrt{x} + 5$

41. $f(x) = \dfrac{3}{x} + 5$

In Problems 42–44, translate each statement into an equation using k as the constant of variation.

42. m varies directly as the square of n.

43. P varies inversely as the cube of Q.

44. A varies jointly as a and b.

45. If m varies directly as n and $m = 27$ when $n = 6$, find m when $n = 10$.

46. If P varies inversely as Q and $P = 27$ when $Q = 6$, find P when $Q = 10$.

47. If A varies jointly as a and b, and $A = 162$ when $a = 6$ and $b = 9$, find A when $a = 8$ and $b = 7$.

48. The cost $C(x)$ for renting a business copying machine is \$200 for 1 month plus 5 cents a copy for x copies. Express this functional relationship in terms of an equation, and graph it for $0 \le x \le 3,000$.

49. The value $V(t)$ of a piece of machinery is a linear function of the time t the machinery is owned. If the piece of machinery is purchased for \$80,000 and is worth \$5,000 after 10 years, find $V(t)$. Graph the function for $0 \le t \le 10$.

B **50.** Does every function have an inverse? Explain.

Let $f(x) = 4 - x^2$ and $g(x) = x - 2$. Find:

51. $(g - f)(x)$

52. $(f - g)(x)$

53. $(f \cdot g + f)(x)$

54. $(f \cdot g - g)(x)$

55. $(f/g)(x)$

56. $(g/f)(x)$

57. $f(g(-2))$

58. $g(f(3))$

59. $f(g(x))$

60. $g(f(x))$

61. $f(f(x))$

62. $g(g(x))$

63. If $f(x) = 2x - 3$, find:

(A) $f(3 + h)$ (B) $\dfrac{f(3 + h) - f(3)}{h}$

64. Graph $g(t) = -\frac{3}{2}t + 6$ and indicate its slope and y intercept.

65. Graph $f(x) = x^2 - 4x + 5$. Find its vertex and intercepts.

66. Graph $P(x) = x^3 - 2x^2 - 5x + 6$ for $-3 \le x \le 4$.

In Problems 67–69, indicate which functions are one-to-one in the problems indicated. In Problems 1–9, you identified which were functions.

67. Problems 1–3

68. Problems 4–6

69. Problems 7–9

70. Graph $g(t) = 96t - 16t^2$. Indicate its vertex and intercepts.

Determine whether each function has an inverse. If it does, give the rule or graph for the inverse. If it does not, explain why not.

71.

72.

73.
```
1 ──────→  1
2 ──────→ -1
3 ──────→ -3
```

74.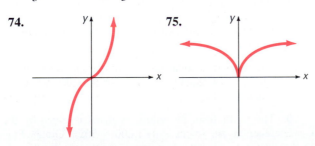

75.

76.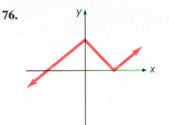

77. $(1, 3), (5, 7), (9, 11)$

78. $(1, 3), (3, 1), (5, 7), (7, 5)$

79. $(1, 3), (3, 5), (5, 3), (7, 5)$

80. $y = x^3 + 5$ **81.** $y = x^2 - 4$ **82.** $y = 3x + 2$

In Problems 83–85, determine the domain of the function.

83. $f(x) = \sqrt{\dfrac{x + 2}{x - 5}}$ **84.** $f(x) = (x + 2)(x - 5)$

85. $f(x) = \dfrac{x + 2}{x - 5}$

86. Let $M(x) = \dfrac{x + 3}{2}$. Find $M^{-1}(x)$ and verify that $M[M^{-1}(x)] = x$.

In Problems 87–89, translate each statement into an equation using k as the constant of variation.

87. y varies directly as x and inversely as z.

88. y varies directly as the cube of x and inversely as the square root of z.

89. A varies directly as a and the cube of b and inversely as the square of c.

90. If y varies directly as x and inversely as z, and $y = 4$ when $x = 6$ and $z = 2$, find y when $x = 4$ and $z = 4$.

91. If y varies directly as the cube of x and inversely as the square root of z, and $y = 12$ when $x = 2$ and $z = 4$, find y when $x = 1$ and $z = 9$.

92. The revenue function for a company producing stereo radios (for a particular model) is

$$R = f(p) = 6{,}000p - 30p^2 \qquad 0 \le p \le 200$$

where p is the price per unit.
(A) Graph f.
(B) At what price p will the revenue be largest? What is the largest revenue?

93. A publisher of paperback books expects sales of 4,000 per month when the price is $5 per book. For each $1 increase in price, sales are expected to drop by 400. Thus, if the publisher increases the price by x dollars per book, the total monthly revenue will be

$$R(x) = (5 + x)(4{,}000 - 400x)$$

(A) Graph R.
(B) At what price will the monthly revenue be largest? What is the largest monthly revenue?

94. If $g(t) = 1 - t^2$, find:

(A) $g(2 + h)$ **(B)** $\dfrac{g(2 + h) - g(2)}{h}$

95. If $f(x) = 1/x$, $x \neq 0$, find:

(A) $f(2 + h)$ **(B)** $\dfrac{f(2 + h) - f(2)}{h}$

96. Let $f(x) = x^2$, $x \geq 0$.
(A) Find $f^{-1}(x)$.
(B) Graph f and f^{-1} on the same coordinate system, along with $y = x$.
(C) Find $f^{-1}(9)$ and $f^{-1}[f(x)]$.

97. Repeat Problem 96 for $f(x) = 1/x$, $x \neq 0$

98. Let $E(x) = 2^x$, $x = -2, -1, 0, 1, 2$. Graph E, E^{-1}, and $y = x$ on the same coordinate system.

99. The time t required for an elevator to lift a weight varies jointly as the weight w and the distance d through which it is lifted and inversely as the power P of the motor. Write the equation of variation using k as the constant of variation. If it takes a 400-horsepower motor 4 seconds to lift an 800-kilogram elevator 8 meters, how long will it take the same motor to lift a 1,600-kilogram elevator 24 meters?

★ **100.** The total force F of a wind on a wall varies jointly as the area of the wall A and the square of the velocity of the wind v. How is the total force on a wall affected if the area is cut in half and the wind velocity is doubled?

CHAPTER PRACTICE TEST

The following practice test is provided for you to test your knowledge of the material in this chapter. You should try to complete it in 50 minutes or less. Answers in the back of the book indicate the section in the text where the material in the question is covered. Actual tests in your class may vary from this practice test in difficulty, length, or emphasis, depending on the goals of your course or instructor.

Determine whether each rule defines a function with x as the independent variable.

1. $(1, 0)$, $(1, 1)$, $(0, 1)$, $(0, 0)$

2.

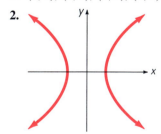

3. $y - x^2 = 4$ **4.** $y^2 - x^2 = 4$

Let $f(x) = x^2 - 2$ and $g(x) = x + 3$. Find:

5. $f(3)$ **6.** $f(g(x))$ **7.** $g(x + h)$

Graph:

8. $f(x) = 3x + 4$ **9.** $f(x) = x^2 + 2$

Find the domain of each function.

10. $f(x) = \sqrt{x^2 - 4}$ **11.** $g(x) = \dfrac{1}{x^2 - 4}$

Determine whether each function has an inverse function and explain your reasoning.

12. $(-1, 0)$, $(0, 1)$, $(1, 2)$

13.

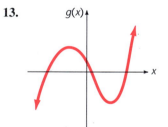

14. $h(x) = x^2 - 1$

Each of the functions in Problems 15–17 has an inverse function. Find the rule for the inverse, and give the domain of the inverse.

15. $(1, 3)$, $(2, 4)$, $(3, 5)$ **16.** $y = \sqrt{x} + 2$

17. $f(x) = \dfrac{1}{x - 1}$

18. If y varies directly as x and $y = 18$ when $x = 8$, find y when $x = 20$.

19. If z varies directly as the square of x and inversely as y, and $z = 4$ when $x = 2$ and $y = 6$, find z when $x = 3$ and $y = 15$.

20. The profit from the sale of x items is given by the function $P(x) = (x - 4)(1{,}000 - 50x)$. Graph $P(x)$ and determine the largest possible profit.

10

Exponential and Logarithmic Functions

For a fixed $b > 0$, $b \neq 1$, the function

$$f(x) = b^x$$

is called an *exponential function*. This function has an inverse, called a *logarithm* or *logarithmic function*. In this chapter, we explore these two new classes of functions.

Photo reference: see Exercise 10-5, Problem 57.

 # 10-1 Exponential Functions

♦ Exponential Functions
♦ Graphing an Exponential Function
♦ Compound Interest and Base e
♦ Basic Exponential Properties

In this section and the next, we will consider two new kinds of functions that use variable exponents in their definitions. To start, note that

$$f(x) = 2^x \qquad \text{and} \qquad g(x) = x^2$$

are not the same function. The function g has a variable base x with a fixed exponent 2; for the function f, the base is fixed and the exponent varies. The function g is a quadratic function, which we have already discussed; the function f is a new function called an *exponential function*, which we introduce in this section.

♦ **EXPONENTIAL FUNCTIONS**

An **exponential function** is a function defined by an equation of the following form:

Exponential Function

$$f(x) = b^x \qquad b > 0, b \neq 1$$

Here, b is a constant, called the **base**, and the exponent x is a variable. The replacement set for the exponent, the **domain of f,** is the set of real numbers R. We know what b^r means for a rational exponent r, but we must determine what b^x means when the exponent x is irrational. We will see that the **range of f** is the set of positive real numbers. We require b to be positive to avoid complex numbers such as $(-2)^{1/2}$, and we impose the condition that $b \neq 1$ since $1^x = 1$ would simply be a constant function.

♦ **GRAPHING AN EXPONENTIAL FUNCTION**

We begin to graph an exponential function such as $f(x) = 2^x$ by preparing a table of x values, and corresponding y values, plotting the resulting points, and then joining these points with a smooth curve (Figure 1).

x	$f(x)$
-3	$\frac{1}{8}$
-2	$\frac{1}{4}$
-1	$\frac{1}{2}$
0	1
1	2
2	4
3	8

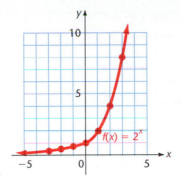

Figure 1

However, 2^x has not yet been defined for all real numbers. The values of x in our table are all integers. We also could have computed powers of 2 for noninteger, rational exponents, such as $2^{2/3}$, $2^{-3/5}$, $2^{1.4}$, and $2^{-3.15}$. The question of exactly what an expression such as $2^{\sqrt{2}}$, with 2 raised to an irrational power, should mean is not easy to answer. A precise definition of $2^{\sqrt{2}}$ must wait for more advanced courses, where it can be shown that b^x names a real number for any positive real number b and any real number x. It also can be shown that the graph of $f(x) = 2^x$ is as indicated in Figure 1.

It is useful to compare the graphs of $y = 2^x$ and $y = (\frac{1}{2})^x = (2^{-1})^x = 2^{-x}$ by plotting both on the same coordinate system (Figure 2a).

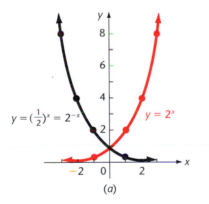
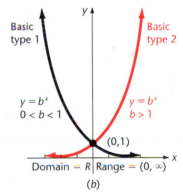

Figure 2

The graphs in Figure 2a are typical of the graphs of $f(x) = b^x$. See Figure 2b. The graph of $f(x) = b^x$ for $b > 1$ will be shaped like the graph of $y = 2^x$, and the graph of $f(x) = b^x$ for $0 < b < 1$ will be shaped like the graph of $y = (\frac{1}{2})^x$. Note in both cases that the x axis is a horizontal asymptote and the graphs will never touch it.

We can see from the graphs that an exponential function is either increasing or decreasing. That is, the y values are increasing or decreasing as the x values increase. Thus, an exponential function is one-to-one and has an inverse that is a function. This fact will be important in the next section when we define a logarithmic function as an inverse of an exponential function.

Example 1
Graphing an Exponential Function

Graph $y = \frac{1}{2}(4^x)$ for $-3 \le x \le 3$.

Solution

x	y
-3	0.01
-2	0.03
-1	0.13
0	0.50
1	2.00
2	8.00
3	32.00

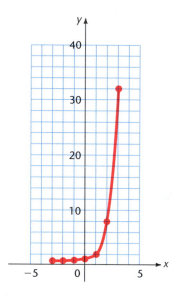

Matched Problem 1   Graph $y = \frac{1}{2}(4^{-x})$ for $-3 \le x \le 3$.

Exponential functions are often referred to as *growth functions* because of their widespread use in describing different kinds of growth. These functions are used to describe the population growth of people, wildlife, and bacteria; radioactive decay (negative growth); the increasing concentration of a new chemical substance in a chemical reaction; the increase or decline in the temperature of a substance being heated or cooled; light absorption (negative growth) as it passes through air, water, or glass; the decline of atmospheric pressure as altitude is increased; and the growth in learning a skill such as swimming or typing relative to practice. Exponential functions are also used to calculate the growth of money at compound interest.

♦   **COMPOUND INTEREST AND BASE *e***

When a sum of money earns interest at a rate r for a certain period of time, at the end of the period the interest earned is the rate times the original sum. For example, if $1,000 is invested for 1 year at an annual rate of 5%, the interest earned is $(0.05)1,000 = \$50$. The original $1,000 becomes $1,050 when the interest is added. More generally, for one interest period,

Amount at end   Original sum   Interest rate
$$A = P + Pr$$
$$= P(1 + r)$$

Here, A is called the **amount,** P the **principal,** and r the **rate.** If the interest remains invested, that is, if it is *compounded,* then after two periods the amount becomes

$$A = P + \overset{\substack{\text{Interest}\\ \text{on } P \text{ for}\\ \text{first period}}}{\widehat{Pr}} + \overset{\substack{\text{Interest on } P + Pr\\ \text{for second}\\ \text{period}}}{\overbrace{(P + Pr)r}} \qquad \text{Factor out } P \text{ from } P + Pr$$

$$= P(1 + r) + P(1 + r)r \qquad P(1 + r) \text{ is a common factor.}$$

$$= P(1 + r)(1 + r)$$

$$= P(1 + r)^2$$

If the interest is compounded for n periods, the amount becomes:

Compound Interest

$$A = P(1 + r)^n$$

where A = amount, P = principal, r = interest rate per period, and n = number of periods.

Interest is frequently stated as an annual rate but compounded more often. An annual rate of 6% compounded twice a year, or *semiannually,* means a rate of $6\%/2 = 3\%$ for each 6-month period. Similarly, 6% compounded four times a year, or *quarterly,* means a rate of $6\%/4 = 1.5\%$ for each 3-month period. More generally, an annual rate a compounded n times per year means a rate $r = a/n$ for each period.

Example 2
Compound Interest

Calculate the amount A if a principal of $1,000 is invested for 1 year using a calculator:

(A) At an annual rate of 4.5% compounded semiannually
(B) At an annual rate of 4.5% compounded quarterly
(C) At an annual rate of 4.5% compounded monthly

Solution **(A)** The number of periods is 2 and the rate per period is $0.045/2 = 0.0225$. The compound interest formula yields

$$A = 1,000(1 + 0.0225)^2 = \$1,045.51$$

(B) $n = 4$ and $r = 0.045/4 = 0.011\ 25$, so

$$A = 1,000(1.011\ 25)^4 = \$1,045.77$$

(C) $n = 12$ and $r = 0.045/12 = 0.003\ 75$, so

$$A = 1,000(1.003\ 75)^{12} = \$1,045.94$$

Matched Problem 2   Calculate the amount A if a principal of \$1,000 is invested for 1 year using a calculator:

(A) At an annual rate of 3% compounded semiannually
(B) At an annual rate of 3% compounded quarterly
(C) At an annual rate of 3% compounded monthly

For an annual rate a compounded n times per year, the compound interest formula yields an amount

$$A = P\left(1 + \frac{a}{n}\right)^n$$

after 1 year. We can rewrite this formula using $r = a/n$ and, equivalently, $n = a/r$:

$$A = P\left(1 + \frac{a}{n}\right)^n = P(1 + r)^{a/r} = P[(1 + r)^{1/r}]^a$$

As n increases, that is, as we compound more often, r gets smaller and smaller. With the aid of a calculator, we can compute values of $(1 + r)^{1/r}$ for ever decreasing positive values of r:

r	$(1 + r)^{1/r}$
1	2
0.1	2.593 74 . . .
0.01	2.704 81 . . .
0.001	2.716 92 . . .
0.000 1	2.718 14 . . .
0.000 01	2.718 26 . . .
0.000 001	2.718 28 . . .

The values of $(1 + r)^{1/r}$ are approaching an irrational number denoted by e. To eight decimal places,

$$e \approx 2.718\ 281\ 83$$

By taking n extremely large, that is, by compounding almost continuously, $(1 + r)^{1/r} \approx e$, and the formula thus becomes $A = Pe^a$. Since the amount after 1 year is obtained by multiplying the principal by e^a, the amount after t years is given by the following formula:

Continuously Compounded Interest

$$A = P(e^a)^t = Pe^{at}$$

where A = amount, P = principal, a = annual interest rate, t = number of years, and $e \approx 2.718\ 281\ 83$.

Because of the importance of e^x and e^{-x}, tables for their evaluation are readily available (see Table I following Appendix C). Scientific calculators also can be used to evaluate these functions directly. We will rely on calculators, rather than tables, to obtain values for e^x.

Example 3
Continuously
Compounded Interest

If $100 is invested at 12% compounded continuously:

(A) Find the amount after 5 years.
(B) Graph the amount in the account relative to time for a period of 10 years.

Solution

(A) The amount $A = 100e^{0.12(5)} = \$182.21$.
(B) We need to graph

$$A = 100e^{0.12t} \qquad 0 \leq t \leq 10$$

We set up a table of values using a calculator (values of A have been rounded to the nearest integer), graph the points from the table, and then join them with a smooth curve, as shown:

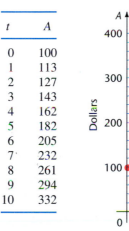

t	A
0	100
1	113
2	127
3	143
4	162
5	182
6	205
7	232
8	261
9	294
10	332

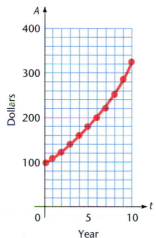

Matched Problem 3

If $1,000 is invested at 9% compounded continuously:

(A) Find the amount after 5 years.
(B) Graph the amount in the account relative to time for a period of 10 years.

The use of e as a base extends far beyond its use in the continuously compounded interest formula. For both theoretical and practical purposes, it is by far the most frequently used exponential base, and $f(x) = e^x$ is often referred to as *the* exponential function because of its widespread use.

Example 4
Graphing an
Exponential Function

Graph $y = 10e^{-0.5x}$, $-3 \leq x \leq 3$, using a calculator to obtain function values.

Solution

x	y
−3	44.82
−2	27.18
−1	16.49
0	10.00
1	6.07
2	3.68
3	2.23

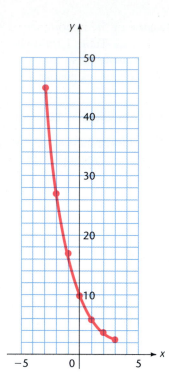

Matched Problem 4   Graph $y = 10e^{0.5x}$, $-3 \leq x \leq 3$, using a calculator to obtain function values.

♦ ## BASIC EXPONENTIAL PROPERTIES

In earlier sections we discussed five properties of exponents for rational exponents. It can be shown that these same properties hold for irrational exponents as well. Thus, we now assume that all five properties of exponents hold for any real exponents as long as the bases involved are positive.

Exponent Properties

For real numbers s and t, and positive real numbers a and b:

1. $b^s b^t = b^{s+t}$

2. $(b^s)^t = b^{st}$

3. $(ab)^s = a^s b^s$

4. $\left(\dfrac{a}{b}\right)^s = \dfrac{a^s}{b^s}$ $b \neq 0$

5. $\dfrac{b^s}{b^t} = b^{s-t}$ $b \neq 0$

As a consequence of an exponential function being either increasing or decreasing and thus one-to-one, we have the following:

For $a, b > 0$, $a, b \neq 1$,

$$a^m = a^n \qquad \text{if and only if} \qquad m = n$$

and for $m \neq 0$,

$$a^m = b^m \qquad \text{if and only if} \qquad a = b$$

Thus, if $2^{15} = 2^{3x}$, then $3x = 15$ and $x = 5$. Also, if $x^{15} = 2^{15}$, then $x = 2$.

Answers to Matched Problems

1.

x	y
-3	32.00
-2	8.00
-1	2.00
0	0.50
1	0.13
2	0.03
3	0.01

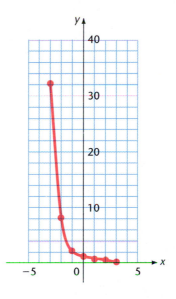

2. **(A)** \$1,030.22 **(B)** \$1,030.34 **(C)** \$1,030.42
3. **(A)** \$1,568.31
   (B) $A = 1,000e^{0.09t}$

t	A
0	1,000
1	1,094
2	1,197
3	1,300
4	1,433
5	1,568
6	1,716
7	1,877
8	2,054
9	2,248
10	2,460

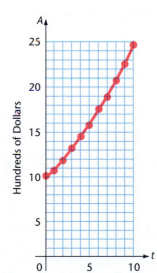

4. $y = 10e^{0.5x}$

x	y
-3	2.23
-2	3.68
-1	6.07
0	10.00
1	16.49
2	27.18
3	44.82

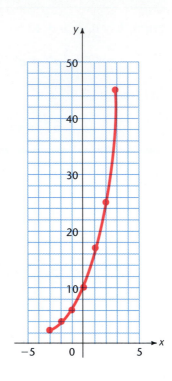

EXERCISE 10-1

A *Evaluate each of the following using a calculator.*

1. $3^{\sqrt{2}}$ **2.** $2^{\sqrt{3}}$ **3.** $5^{-\sqrt{2}}$

4. $2^{-\sqrt{5}}$ **5.** e^2 **6.** e^3

7. e^{-3} **8.** e^{-2} **9.** $10^{\sqrt{10}}$

10. $10^{-\sqrt{10}}$ **11.** $e^{\sqrt{2}}$ **12.** $e^{\sqrt{3}}$

Find the amount resulting from a principal of $1,000 earning interest at an annual rate a, compounded n times per year for t years.

13. $a = 6\%$, $n = 6$, $t = 5$

14. $a = 3\%$, $n = 6$, $t = 4$

15. $a = 7.5\%$, $n = 4$, $t = 8$

16. $a = 9\%$, $n = 4$, $t = 10$

17. $a = 1.5\%$, $n = 2$, $t = 20$

18. $a = 10.5\%$, $n = 2$, $t = 15$

19. $a = 18\%$, $n = 12$, $t = 1$

20. $a = 19.5\%$, $n = 12$, $t = 1$

Find the amount resulting from a principal of $1,000 earning interest at an annual rate a, compounded continuously for t years.

21. $a = 5\%$, $t = 30$ **22.** $a = 4\%$, $t = 36$

23. $a = 8\%$, $t = 10$ **24.** $a = 2\%$, $t = 12$

Graph each exponential function for $-3 \le x \le 3$ using a calculator to obtain function values.

25. $y = 3^x$ **26.** $y = 2^x$

27. $y = (\frac{1}{3})^x = 3^{-x}$ **28.** $y(\frac{1}{2})^x = 2^{-x}$

29. $y = 4 \cdot 3^x = 4(3^x)$ **30.** $y = 5 \cdot 2^x$

31. $y = 2^{x+3}$ **32.** $y = 3^{x+1}$

33. $y = 7(\frac{1}{2})^{2x} = 7 \cdot 2^{-2x}$ **34.** $y = 11 \cdot 2^{-2x}$

35. $y = e^x$ **36.** $y = e^{-x}$

B 37. $y = 2^x - 3$ **38.** $y = 3^x - 2$

39. $y = (\frac{1}{2})^x + 1$ **40.** $y = (\frac{1}{2})^x - 1$

41. $y = e^x + 4$ **42.** $y = e^x - 3$

43. $y = e^{-x} - 2$ **44.** $y = e^{-x} + 1$

45. $y = 3^{x+1}$ **46.** $y = 2^{x-1}$

47. $y = e^{x-2}$ **48.** $y = e^{x+2}$

49. $y = (\frac{1}{3})^{x-2}$ **50.** $y = (\frac{1}{2})^{x+3}$

51. $y = e^{-x+1}$ **52.** $y = e^{-x-2}$

C 53. $y = 2 \cdot 3^x + 4$ **54.** $y = 3 \cdot 2^x - 1$

55. $y = 3(\frac{1}{2})^x + 1$ **56.** $y = 2(\frac{1}{3})^x - 4$

57. $y = 2e^x + 1$ **58.** $y = 3e^x - 2$

59. $y = 4e^{-x} - 3$

60. $y = 3e^{-x} + 4$

61. $y = 10e^{-0.12x}$

62. $y = 100e^{0.25x}$

63. $y = 2e^{3x-4}$

64. $y = 4e^{3x-2}$

Graph each exponential function for $-2 \leq x \leq 2$.

65. $y = 10e^{-x^2}$

66. $y = -10e^{-x^2}$

For the function $f(x) = ab^x$, a is $f(0)$; that is, a is the value of the function at 0. Graph each function for $-3 \leq x \leq 3$ using a vertical scale expressed in terms of a.

67. $f(x) = a(2^x)$

68. $f(x) = a(3^{-x})$

69. $f(x) = a(\frac{1}{2})^x$

70. $f(x) = a(\frac{1}{3})^x$

Graph $f(x)$ and $f^{-1}(x)$ on the same coordinate system.

71. $f(x) = 2^x$

72. $f(x) = 10^x$

73. $f(x) = 10^{-x}$

74. $f(x) = 2^{-x}$

75. $f(x) = e^x$

76. $f(x) = e^{-x}$

APPLICATIONS

★ **77.** **Earth science**   The atmospheric pressure P, in pounds per square inch, can be calculated approximately using the formula

$$P = 14.7e^{-0.21x}$$

where x is altitude relative to sea level in miles. Graph the equation for $-1 \leq x \leq 5$.

★ **78.** **Bacterial growth**   If bacteria in a certain culture double every hour, write a formula that gives the number of bacteria N in the culture after n hours, assuming the culture has N_0 bacteria to start with.

79. **Radioactive decay**   Radioactive strontium-90 has a half-life of 28 years; that is, in 28 years one-half of any amount of strontium-90 will change to another substance because of radioactive decay. If we place a bar containing 100 milligrams of strontium-90 in a nuclear reactor, the amount of strontium-90 that will be left after t years is given by $A = 100(\frac{1}{2})^{t/28}$. Graph this exponential function for $t = 0$, 28, $2(28)$, $3(28)$, $4(28)$, $5(28)$, and $6(28)$, and join these points with a smooth curve.

80. **Radioactive decay**   Radioactive argon-39 has half-life of 4 minutes; that is, in 4 minutes one-half of any amount of argon-39 will change to another substance because of radioactive decay. If we start with A_0 milligrams of argon-39, the amount left after t minutes is given by $A = A_0(\frac{1}{2})^{t/4}$. Graph this exponential function for $A_0 = 100$ and $t = 0$, 4, 8, 12, 16, and 20, and join these points with a smooth curve.

★ **81.** **Sociology—small-group analysis**   Two sociologists, Stephan and Mischler, found that when the members of a discussion group of 10 were ranked according to the number of times each participated, the number of times $N(i)$ the ith-ranked person participated was given approximately by the exponential function

$$N(i) = N_1 e^{-0.11(i-1)} \qquad 1 \leq i \leq 10$$

where N_1 is the number of times the top-ranked person participated in the discussion. Graph the exponential function, using $N_1 = 100$.

82. **Defining 0^0**   The expression 0^0 is not defined. Graph the two equations $y = 0^x$ and $y = x^0$ for values of $x \neq 0$. If 0^0 were defined, it would appear on both curves as the value corresponding to $x = 0$. This should give you some idea of why we can't define 0^0.

  # 10-2 **Logarithmic Functions**

♦ Logarithmic Functions
♦ From Logarithmic to Exponential, and Vice Versa
♦ Solving $y = \log_b x$ for b or x
♦ Logarithmic-Exponential Identities

Since an exponential function is one-to-one, it has an inverse. The inverse is called a **logarithmic function.** Now you will see why we placed special emphasis on the general concept of inverse functions in Section 9-4. If you know about a particular function f, then by applying your knowledge of inverses in general, you will automatically know a lot about f^{-1}. For example, the graph of f^{-1} is the graph of f reflected across the line $y = x$, and the domain and range of f^{-1} are, respectively, the range and domain of f.

◆ LOGARITHMIC FUNCTIONS

If we start with the exponential function

$$f: \quad y = 2^x$$

and interchange the variables x and y, we obtain the inverse of f:

$$f^{-1}: \quad x = 2^y$$

The graphs of f, f^{-1}, and $y = x$ are shown in Figure 1.

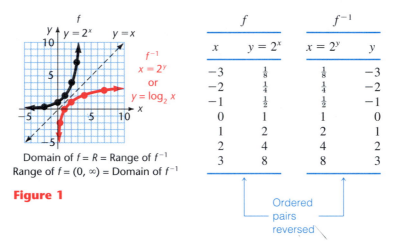

Domain of $f = R$ = Range of f^{-1}
Range of $f = (0, \infty)$ = Domain of f^{-1}

Figure 1

Ordered pairs reversed

The new function f^{-1} is given the name **logarithmic function with base 2.** Since we can't "algebraically" solve $x = 2^y$ for y, this new function is symbolized as follows:

$$y = \log_2 x$$

This is read "y equals the log of x to the base 2," or "y equals log to the base 2 of x."

Thus,

$$y = \log_2 x \quad \text{is equivalent to} \quad x = 2^y$$

That is, $\log_2 x$ is the power to which 2 must be raised to obtain x. Symbolically, $x = 2^y = 2^{\log_2 x}$. For example, $\log_2 32 = 5$ since $2^5 = 32$.

In general, we define the **logarithmic function with base b** to be the inverse of the exponential function with base b, $b > 0$ and $b \neq 1$.

Definition of Logarithmic Function

For $b > 0$ and $b \neq 1$:

$$y = \log_b x \quad \text{is equivalent to} \quad x = b^y$$

That is, log to the base b of x is the power to which b must be raised to obtain x.

$$y = \log_{10} x \quad \text{is equivalent to} \quad x = 10^y$$
$$y = \log_e x \quad \text{is equivalent to} \quad x = e^y$$

Logarithms with the two common bases, 10 and e, are given simpler notations:

$$\log_{10} x = \log x \qquad \log_e x = \ln x$$

Values of these two functions are available in tables (Tables II and III), and most scientific calculators can calculate both $\log x$ and $\ln x$ directly. Look for keys labeled "log" and "ln" on your calculator. We will depend on calculators for values when necessary.

Remember that $y = \log_b x$ and $x = b^y$ define the same function, and as such can be used interchangeably. In particular, $y = \log x$ and $x = 10^y$ mean the same thing. Similarly, $y = \ln x$ and $x = e^y$ also mean the same thing.

Since the domain of an exponential function includes all real numbers and its range is the set of positive real numbers, the **domain** of a logarithmic function is the set of all positive real numbers and its **range** is the set of all real numbers. Thus, $\log_2 3$ is defined, but $\log_2 0$ and $\log_2(-5)$ are not defined.

Example 1 **Calculating Logarithms**	Calculate the following logarithms without the aid of a calculator: **(A)** $\log_2 8$ **(B)** $\log_2(\tfrac{1}{8})$ **(C)** $\log_4 2$ **(D)** $\log_4 8$
Solution	**(A)** $\log_2 8$ is the power to which 2 must be raised to obtain 8. What power of 2 yields 8? Since $2^3 = 8$, $\log_2 8 = 3$. **(B)** What power of 2 yields $\tfrac{1}{8}$? Since $\dfrac{1}{8} = \dfrac{1}{2^3} = 2^{-3}$ we get $\log_2(\tfrac{1}{8}) = -3$. **(C)** What power of 4 yields 2? Since $\sqrt{4} = 4^{1/2} = 2$, $\log_4 2 = \tfrac{1}{2}$. **(D)** What power of 4 yields 8? Since $8 = 2^3 = (2^2)^{3/2} = 4^{3/2}$, $\log_4 8 = \tfrac{3}{2}$.
Matched Problem 1	Calculate the following logarithms without the aid of a calculator: **(A)** $\log_8 2$ **(B)** $\log_8(\tfrac{1}{8})$ **(C)** $\log_8 32$ **(D)** $\log_8 4$

Typical graphs of logarithmic functions are shown in Figure 2.

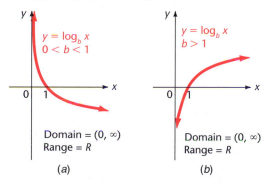

Figure 2 Typical logarithmic graphs

♦ **FROM LOGARITHMIC TO EXPONENTIAL, AND VICE VERSA**

It will often be helpful to be able to convert logarithmic forms to equivalent exponential forms, and vice versa. We rely on the definition of a logarithm to do so.

Example 2
From Logarithmic to
Exponential Form

Change to an equivalent exponential form:

(A) $\log_2 8 = 3$ (B) $\log_{25} 5 = \frac{1}{2}$ (C) $\log_2(\frac{1}{4}) = -2$

Solution
(A) $\log_2 8 = 3$ is equivalent to $8 = 2^3$.
(B) $\log_{25} 5 = \frac{1}{2}$ is equivalent to $5 = 25^{1/2}$.
(C) $\log_2(\frac{1}{4}) = -2$ is equivalent to $\frac{1}{4} = 2^{-2}$.

Matched Problem 2

Change to an equivalent exponential form:

(A) $\log_3 27 = 3$ (B) $\log_{36} 6 = \frac{1}{2}$ (C) $\log_3(\frac{1}{9}) = -2$

Example 3
From Exponential to
Logarithmic Form

Change to an equivalent logarithmic form:

(A) $49 = 7^2$ (B) $3 = \sqrt{9}$ (C) $\frac{1}{5} = 5^{-1}$

Solution
(A) $49 = 7^2$ is equivalent to $\log_7 49 = 2$.
(B) $3 = \sqrt{9}$ is equivalent to $\log_9 3 = \frac{1}{2}$. $\sqrt{9} = 9^{1/2}$
(C) $\frac{1}{5} = 5^{-1}$ is equivalent to $\log_5(\frac{1}{5}) = -1$.

Matched Problem 3

Change to an equivalent logarithmic form:

(A) $64 = 4^3$ (B) $2 = \sqrt[3]{8}$ (C) $\frac{1}{16} = 4^{-2}$

♦ **SOLVING** $y = \log_b x$ **FOR** b **OR** x

Converting from logarithmic to exponential form allows us to solve $y = \log_b x$ for b or x if we are given the other value and the value of $\log_b x$. All values in the examples that follow are chosen so that the problems can be solved without tables or a calculator.

Example 4
Solving for b or x

Solve for b or x as indicated.

(A) Find x if $\log_3 x = -2$.   (B) Find b if $\log_b 1,000 = 3$.

Solution
(A) Write $\log_3 x = -2$ in equivalent exponential form:

$$x = 3^{-2}$$

$$= \frac{1}{3^2} = \frac{1}{9}$$

Thus, $\log_3(\frac{1}{9}) = -2$.

(B) Write $\log_b 1{,}000 = 3$ in equivalent exponential form:

$$1{,}000 = b^3 \qquad \text{Write 1,000 as a third power.}$$
$$10^3 = b^3 \qquad \text{Recall that for } a,\ b > 0,$$
$$\qquad\qquad\qquad a^m = b^m \text{ if and only if } a = b.$$
$$b = 10$$

Thus, $\log_{10} 1{,}000 = 3$.

Matched Problem 4 Solve for b or x as indicated.

(A) Find x if $\log_2 x = -3$. **(B)** Find b if $\log_b 100 = 2$.

♦ **LOGARITHMIC-EXPONENTIAL IDENTITIES**

The general properties of inverse functions,

$$f^{-1}[f(x)] = x \qquad \text{and} \qquad f[f^{-1}(x)] = x$$

yield useful identities when applied to the exponential and logarithmic functions. For $f(x) = b^x$ and $f^{-1}(x) = \log_b x$, we get

$$f^{-1}[f(x)] = x \qquad\qquad f[f^{-1}(x)] = x$$
$$\log_b[f(x)] = x \qquad\qquad b^{f(x)} = x$$
$$\log_b b^x = x \qquad\qquad b^{\log_b x} = x$$

Thus, we have these logarithmic-exponential identities:

Logarithmic-Exponential Identities

For $b > 0$, $b \neq 1$:

1. $\log_b b^x = x$ That is, the power to which b must be raised to obtain b^x is x.

2. $b^{\log_b x} = x,\ x > 0$ That is, the power to which b must be raised to obtain x is $\log_b x$.

In particular, for $b = 10$, $\log 10^x = x$ and $10^{\log x} = x$. For $b = e$, $\ln e^x = x$ and $e^{\ln x} = x$.

Example 5
Applying the Logarithmic-Exponential Identities

Find each of the following:

(A) $\log 10^5$ **(B)** $\log 0.01$ **(C)** $\ln e^{2x+1}$
(D) $\log_4 1$ **(E)** $10^{\log 7}$ **(F)** $e^{\ln x}$

Solution **(A)** $\log 10^5 = \log_{10} 10^5 = 5$
(B) $\log 0.01 = \log_{10} 0.01 = \log_{10} 10^{-2} = -2$
(C) $\ln e^{2x+1} = \log_e e^{2x+1} = 2x + 1$ **(D)** $\log_4 1 = \log_4 4^0 = 0$
(E) $10^{\log 7} = 10^{\log_{10} 7} = 7$ **(F)** $e^{\ln x} = e^{\log_e x} = x$

Matched Problem 5 Find each of the following:

(A) $\log_{10} 10^{-5}$ (B) $\log_5 25$ (C) $\log_{10} 1$

(D) $\log_e e^{m+n}$ (E) $10^{\log 4}$ (F) $e^{\log_e(x+1)}$

Answers to Matched Problems

1. (A) $\frac{1}{3}$ (B) -1 (C) $\frac{5}{3}$ (D) $\frac{2}{3}$
2. (A) $27 = 3^3$ (B) $6 = 36^{1/2}$ (C) $\frac{1}{9} = 3^{-2}$
3. (A) $\log_4 64 = 3$ (B) $\log_8 2 = \frac{1}{3}$ (C) $\log_4(\frac{1}{16}) = -2$
4. (A) $x = \frac{1}{8}$ (B) $b = 10$
5. (A) -5 (B) 2 (C) 0 (D) $m + n$ (E) 4 (F) $x + 1$

EXERCISE 10-2

A *Calculate the following logarithms without the aid of a calculator:*

1. $\log_3 9$
2. $\log_3 27$
3. $\log_3(\frac{1}{27})$
4. $\log_3(\frac{1}{9})$
5. $\log_3 3$
6. $\log_3 \sqrt{3}$
7. $\log_3(1/\sqrt{3})$
8. $\log_3(\frac{1}{3})$
9. $\log_9 81$
10. $\log_9 9$
11. $\log_9 27$
12. $\log_9(\frac{1}{27})$
13. $\log_9(\frac{1}{9})$
14. $\log_9 3$
15. $\log_9(\frac{1}{3})$
16. $\log_9(\frac{1}{81})$
17. $\log_9 1$
18. $\log_3 1$
19. $\log_{10} 10^5$
20. $\log_5 5^3$
21. $\log_2 2^{-4}$
22. $\log_{10} 10^{-7}$
23. $\log_6 36$
24. $\log_3 9$
25. $\log_{10} 1{,}000$
26. $\log_{10} 0.001$

Rewrite in exponential form.

27. $\log_3 9 = 2$
28. $\log_2 4 = 2$
29. $\log_3 81 = 4$
30. $\log_5 125 = 3$
31. $\log_{10} 1{,}000 = 3$
32. $\log_{10} 100 = 2$
33. $\log_e 1 = 0$
34. $\log_8 1 = 0$

Rewrite in logarithmic form.

35. $64 = 8^2$
36. $25 = 5^2$
37. $10{,}000 = 10^4$
38. $1{,}000 = 10^3$
39. $u = v^x$
40. $a = b^c$
41. $9 = 27^{2/3}$
42. $8 = 4^{3/2}$

Solve for b or x.

43. $\log_2 x = 2$
44. $\log_3 x = 2$
45. $\log_b 16 = 2$
46. $\log_b 10^{-3} = -3$
47. $\log_b 3 = \frac{1}{2}$
48. $\log_b 4 = \frac{1}{3}$
49. $\log_5 x = 3$
50. $\log_5 x = \frac{3}{2}$
51. $\log_5 x = \frac{2}{3}$
52. $\log_5 x = -2$

B *Rewrite in exponential form.*

53. $\log_{10} 0.001 = -3$
54. $\log_{10} 0.01 = -2$
55. $\log_{81} 3 = \frac{1}{4}$
56. $\log_4 2 = \frac{1}{2}$
57. $\log_{1/2} 16 = -4$
58. $\log_{1/3} 27 = -3$
59. $\log_a N = e$
60. $\log_k u = v$

Rewrite in logarithmic form.

61. $0.01 = 10^{-2}$
62. $0.001 = 10^{-3}$
63. $1 = e^0$
64. $1 = (\frac{1}{2})^0$
65. $\frac{1}{8} = 2^{-3}$
66. $\frac{1}{8} = (\frac{1}{2})^3$
67. $\frac{1}{3} = 81^{-1/4}$
68. $\frac{1}{2} = 32^{-1/5}$
69. $7 = \sqrt{49}$
70. $11 = \sqrt{121}$

Find each of the following:

71. $\log_b b^u$
72. $\log_b b^{uv}$
73. $\log_e e^{1/2}$
74. $\log_e e^{-3}$
75. $\log_{23} 1$
76. $\log_{17} 1$
77. $\ln e^3$
78. $\ln e^{3/5}$
79. $\ln e^{\sqrt{2}}$
80. $\ln e^e$

Solve for b or x.

81. $\log_4 x = \frac{1}{2}$
82. $\log_{25} x = \frac{1}{2}$
83. $\log_b 1{,}000 = \frac{3}{2}$
84. $\log_b 4 = \frac{2}{3}$
85. $\log_b 1 = 0$
86. $\log_b b = 1$

C *In Problems 87–90:*

(A) Graph $f(x)$ and $f^{-1}(x)$ in the same coordinate system.
(B) Determine the domain and range of f and f^{-1}.
(C) Give the standard name for f^{-1}.

87. $f(x) = 10^x$
88. $f(x) = e^x$
89. $f(x) = (\frac{1}{2})^x$
90. $f(x) = (\frac{1}{10})^x$

91. For $f(x) = 1^x$ discuss the domain and range for f. Does f have an inverse?

92. Why is 1 not a suitable logarithmic base? [*Hint:* Try to find $\log_1 5.1$.]

93. If $\log_b x = 3$, find $\log_b(1/x)$.

94. If $\ln x = 2$, find $\ln(1/x)$.

95. Prove that $\ln(1/x) = -\ln x$.

96. Prove that $\log_b(1/x) = -\log_b x$.

  # 10-3 Properties of Logarithmic Functions

- Basic Logarithmic Properties
- Use of the Logarithmic Properties

Since logarithms are exponents—remember, $\log_b x$ is the power to which b must be raised to obtain x—they have properties related to the laws of exponents. These properties enable us to convert multiplication problems into addition problems, division problems into subtraction problems, and power and root problems into multiplication problems. Moreover, they enable us to solve exponential equations such as $2 = 10^x$.

♦ BASIC LOGARITHMIC PROPERTIES

Properties of Logarithmic Functions

If b, M, and N are positive real numbers, $b \neq 1$, and p is a real number, then:

1. $\log_b b^u = u$ Logarithmic-exponential identity
2. $\log_b MN = \log_b M + \log_b N$
3. $\log_b \dfrac{M}{N} = \log_b M - \log_b N$
4. $\log_b M^p = p \log_b M$
5. $\log_b 1 = 0$

The first property, the logarithmic-exponential identity from Section 10-2, follows from the definition of the logarithmic function. The last property is a restatement of $b^0 = 1$. The other three properties are logarithmic versions of the following laws of exponents:

2. $b^m b^n = b^{m+n}$

3. $\dfrac{b^m}{b^n} = b^{m-n}$

4. $(b^m)^p = b^{pm}$

For example, consider Property 2. If we let

$$\log_b M = m \qquad \log_b N = n$$

and convert to exponential form,

$$b^m = M \qquad\qquad b^n = N$$

then

$$
\begin{aligned}
\log_b MN &= \log_b b^m b^n \\
&= \log_b b^{m+n} \\
&= m + n \qquad \text{Property 1} \\
&= \log_b M + \log_b N
\end{aligned}
$$

The other two properties can be established in a similar manner.

◆ USE OF THE LOGARITHMIC PROPERTIES

We will now see how logarithmic properties can be used to convert multiplication, division, and power and root problems into alternate forms.

Example 1
Rewriting Using Logarithmic Properties

Rewrite using the properties of logarithms:

(A) $\log_b 3x$ **(B)** $\log_b \dfrac{x}{5}$ **(C)** $\log_b x^7$

(D) $\log_b \dfrac{mn}{pq}$ **(E)** $\log_b (mn)^{2/3}$ **(F)** $\log_b \dfrac{x^8}{y^{1/5}}$

Solution

(A) $\log_b 3x = \log_b 3 + \log_b x$ Property 2: $\log_b MN = \log_b M + \log_b N$
Notice that the logarithm of a product has been converted into a sum of logarithms.

(B) $\log_b \dfrac{x}{5} = \log_b x - \log_b 5$ Property 3: $\log_b \dfrac{M}{N} = \log_b M - \log_b N$

Here, the logarithm of a quotient has been converted into a difference of logarithms.

(C) $\log_b x^7 = 7 \log_b x$ Property 4: $\log_b M^p = p \log_b M$
The logarithm of a power has been converted into the power times a logarithm.

(D) $\log_b \dfrac{mn}{pq} = \log_b mn - \log_b pq$ Property 3
$\qquad\qquad = \log_b m + \log_b n - (\log_b p + \log_b q)$ Property 2
$\qquad\qquad = \log_b m + \log_b n - \log_b p - \log_b q$

(E) $\log_b (mn)^{2/3} = \tfrac{2}{3} \log_b mn$ Property 4
$\qquad\qquad\quad = \tfrac{2}{3}(\log_b m + \log_b n)$ Property 2

(F) $\log_b \dfrac{x^8}{y^{1/5}} = \log_b x^8 - \log_b y^{1/5}$ Property 3

$\qquad\qquad = 8 \log_b x - \tfrac{1}{5} \log_b y$ Property 4

Matched Problem 1

Rewrite using the properties of logarithms:

(A) $\log_b \dfrac{r}{uv}$ **(B)** $\log_b \left(\dfrac{m}{n} \right)^{3/5}$ **(C)** $\log_b \dfrac{u^{1/3}}{v^5}$

Example 2 **Evaluating Using** **Logarithmic Properties**	If $\ln 3 = 1.10$ and $\ln 7 = 1.95$, find: **(A)** $\ln \frac{7}{3}$ **(B)** $\ln \sqrt[3]{21}$

Solution **(A)** $\ln \frac{7}{3} = \ln 7 - \ln 3 = 1.95 - 1.10 = 0.85$
(B) $\ln \sqrt[3]{21} = \ln(21)^{1/3} = \frac{1}{3}\ln(3 \cdot 7)$
$\qquad\qquad = \frac{1}{3}(\ln 3 + \ln 7)$
$\qquad\qquad \approx \frac{1}{3}(1.10 + 1.95) \approx 1.02$

Matched Problem 2 If $\ln 5 = 1.609$ and $\ln 8 = 2.079$, find:

(A) $\ln \dfrac{5^{10}}{8}$ **(B)** $\ln \sqrt[4]{\dfrac{8}{5}}$

Finally, we note that since logarithmic functions are one-to-one:

$$\log_b m = \log_b n \qquad \text{if and only if} \qquad m = n$$

Thus, if $\log x = \log 32.15$, then $x = 32.15$.

Example 3 **Applying Logarithmic** **Properties**	Solve for x: $$\log_b x = \frac{2}{3}\log_b 27 + 2\log_b 2 - \log_b 3$$

Solution $\log_b x = \frac{2}{3}\log_b 27 + 2\log_b 2 - \log_b 3$ Express the right side in terms of a single logarithm.

$\qquad = \log_b(27^{2/3}) + \log_b(2^2) - \log_b 3$ Property 4: $p\log_b M = \log_b M^p$
$\qquad = \log_b 9 + \log_b 4 - \log_b 3$ $27^{2/3} = 9;\ 2^2 = 4$

$\qquad = \log_b \dfrac{9 \cdot 4}{3} = \log_b 12$ Properties 2 and 3:
$\log_b M + \log_b N = \log_b M \cdot N$

and $\log_b M - \log_b N = \log_b \dfrac{M}{N}$

Thus,

$$\log_b x = \log_b 12$$

so,

$$x = 12$$

Matched Problem 3 Solve for x: $\log_b x = \frac{2}{3}\log_b 8 + \frac{1}{2}\log_b 9 - \log_b 6$

Answers to
Matched Problems
1. (A) $\log_b r - \log_b u - \log_b v$ **(B)** $\frac{3}{5}(\log_b m - \log_b n)$
    (C) $\frac{1}{3}\log_b u - 5\log_b v$
2. (A) 14.01 (to four significant digits)*   **(B)** 0.1175 (to four significant digits)
3. $x = 2$

*Significant digits are discussed in Appendix B.

EXERCISE 10-3

A *Rewrite using properties of logarithms.*

1. $\log_b uv$
2. $\log_b rt$
3. $\log_b(A/B)$

4. $\log_b(p/q)$
5. $\log_b u^5$
6. $\log_b w^{25}$

7. $\log_b N^{3/5}$
8. $\log_b u^{-2/3}$
9. $\log_b \sqrt{Q}$

10. $\log_b \sqrt[5]{M}$
11. $\log_b uvw$
12. $\log_b(u/vw)$

Write each expression in terms of a single logarithm with a coefficient of 1. For example: $\log_b u^2 - \log_b v = \log_b(u^2/v)$

13. $\log_b A + \log_b B$

14. $\log_b P + \log_b Q + \log_b R$

15. $\log_b X - \log_b Y$

16. $\log_b x^2 - \log_b y^3$

17. $\log_b w + \log_b x - \log_b y$

18. $\log_b w - \log_b x - \log_b y$

If $\log_b 2 = 0.69$, $\log_b 3 = 1.10$, *and* $\log_b 5 = 1.61$, *find the logarithm to the base b of each of the following numbers to three decimal places:*

19. 30
20. 6
21. $\frac{2}{5}$

22. $\frac{5}{3}$
23. 27
24. 16

If $\log_b 2 = 0.431$ *and* $\log_b 3 = 0.683$, *find the logarithm to the base b of each of the following numbers to three decimal places:*

25. $\frac{1}{3}$
26. $\frac{1}{2}$
27. $\frac{2}{3}$

28. $\frac{3}{2}$
29. 6
30. 8

31. 9
32. 12
33. $\sqrt{2}$

34. $\sqrt{3}$

B *Rewrite using properties of logarithms.*

35. $\log_b u^2 v^7$
36. $\log_b u^{1/2} v^{1/3}$

37. $\log_b \dfrac{1}{a}$
38. $\log_b \dfrac{1}{M^3}$

39. $\log_b \dfrac{\sqrt[3]{N}}{p^2 q^3}$
40. $\log_b \dfrac{m^5 n^3}{\sqrt{p}}$

41. $\log_b \sqrt[4]{\dfrac{x^2 y^3}{\sqrt{z}}}$
42. $\log_b \sqrt[5]{\left(\dfrac{x}{y^4 z^9}\right)^3}$

Write each expression in terms of a single logarithm with a coefficient of 1.

43. $2 \log_b x - \log_b y$

44. $\log_b m - \frac{1}{2} \log_b n$

45. $3 \log_b x + 2 \log_b y - 4 \log_b z$

46. $\frac{1}{3} \log_b w - 3 \log_b x - 5 \log_b y$

47. $\frac{1}{5}(2 \log_b x + 3 \log_b y)$

48. $\frac{1}{3}(\log_b x - \log_b y)$

49. $\log_b 3 + 2 \log_b x + \log_b y$

50. $\log_b x + 2 \log_b y - 3 \log_b z$

51. $\frac{1}{2}(\log_b x - \log_b y)$

52. $\frac{1}{3}(\log_b x + 2 \log_b y - 3 \log_b z)$

53. $\log_b 2 + 3 \log_b x + 4 \log_b y$

54. $\log_b 2 - 3 \log_b x - 4 \log_b y$

If $\log_b 2 = 0.69$, $\log_b 3 = 1.10$, *and* $\log_b 5 = 1.61$, *find the logarithm to the base b of the following numbers to three decimal places:*

55. 7.5
56. 1.5
57. $\sqrt[3]{2}$

58. $\sqrt{3}$
59. $\sqrt{0.9}$
60. $\sqrt[3]{\frac{3}{2}}$

If $\log_b 2 = 0.431$, $\log_b 3 = 0.683$, *and* $\log_b 7 = 1.209$, *find the logarithm to the base b of each of the following numbers to three decimal places:*

61. $\frac{6}{7}$
62. $\frac{7}{6}$
63. $\frac{14}{3}$

64. 10.5
65. $\sqrt{42}$
66. $1/\sqrt{14}$

C *In Problems 67–72, solve for x.*

67. $\frac{3}{2} \log_b 4 - \frac{2}{3} \log_b 8 + 2 \log_b 2 = \log_b x$

68. $3 \log_b 2 + \frac{1}{2} \log_b 25 - \log_b 20 = \log_b x$

69. $\log_b 6 + \frac{1}{2} \log_b 16 - \log_b 216 = \log_b x$

70. $\frac{1}{3} \log_b 8 - \frac{1}{2} \log_b 16 + \log_b 24 = \log_b x$

71. $2 \log_b 2 + 3 \log_b 3 - 4 \log_b 4 = \log_b x$

72. $5 \log_b 2 - 4 \log_b 3 + 2 \log_b 4 = \log_b x$

73. Write $\log_b y - \log_b c + kt = 0$ in exponential form free of logarithms.

74. Write $\log_e x - \log_e 100 = -0.08t$ in exponential form free of logarithms.

75. Prove that $\log_b(M/N) = \log_b M - \log_b N$ for b, M, $N > 0$ following the proof of Property 2 in this section.

76. Prove that $\log_b M^p = p \log_b M$ for b, $M > 0$.

77. Prove that $\log_b MN = \log_b M + \log_b N$ by starting with $M = b^{\log_b M}$ and $N = b^{\log_b N}$.

78. Prove that $\log_b(M/N) = \log_b M - \log_b N$ by starting with $M = b^{\log_b M}$ and $N = b^{\log_b N}$.

10-4   Logarithms to Various Bases

♦ Common and Natural Logarithms—Calculator Evaluation
♦ Change-of-Base Formula
♦ Graphing Logarithmic Functions

John Napier (1550–1617) is credited with the invention of logarithms. They evolved out of an interest in simplifying the computations connected with investigations in astronomy. This new computational tool was immediately accepted by the scientific world. Now, with the availability of calculators, logarithms have lost most of their importance as a computational device. However, the logarithmic concept has been greatly generalized since its conception, and logarithmic functions are used widely in both theoretical and applied sciences. We need logarithmic functions to solve exponential equations (such as $2 = 1.08^x$) in population growth studies and the mathematics of finance.

Of all possible logarithmic bases, only two are used extensively: the base e and the base 10. Before we can use logarithms in certain practical problems, we need to be able to find the logarithm of any positive number to either base 10 or base e. And conversely, if we are given the logarithm of a number to base 10 or base e, we need to be able to find the number. In this context, "find" means to approximate to a sufficient number of decimal places, just as we "find" $\sqrt{2}$. Historically, tables were used for this purpose, but now readily available and inexpensive scientific calculators are faster and far more accurate than almost any table you might use.

♦ ## COMMON AND NATURAL LOGARITHMS—CALCULATOR EVALUATION

Common logarithms (also called **Briggsian logarithms**) are logarithms with base 10. **Natural logarithms** (also called **Napierian logarithms**) are logarithms with base e. As mentioned in Section 10-2, most scientific calculators have a key labeled "log" (or "LOG") and a key labeled "ln" (or "LN"). The former represents a common, or base 10, logarithm and the latter a natural, or base e, logarithm. In summary:

> **Logarithmic Notation**
>
> **Common logarithm:** $\log x = \log_{10} x$
> **Natural logarithm:** $\ln x = \log_e x$

Example 1
Calculating Logarithms

Use a scientific calculator to find each to six decimal places:

(A) $\log 3{,}184$ **(B)** $\ln 0.000\,349$ **(C)** $\log(-3.24)$

Solution   The entry of $\log x$ or $\ln x$ in your calculator will depend on the particular calculator being used. In some calculators, the number is entered, followed by the "log" or "ln" key. In others, the order is reversed. Your calculator should display the following answers:

(A) 3.502 973   (B) $-7.960\,439$ (C) "ERROR"

The error message in part (C) also will vary from calculator to calculator, but you should get some indication that $\log(-3.24)$ is not possible, since the domain of any logarithmic function is positive numbers only.

Matched Problem 1   Use a scientific calculator to find each to six decimal places:

(A) $\log 0.013\,529$ (B) $\ln 28.693\,28$ (C) $\ln(-0.438)$

Example 2
Calculating Logarithms

Use a scientific calculator to evaluate each to three decimal places:

(A) $n = \dfrac{\log 2}{\log 1.1}$ (B) $n = \dfrac{\ln 3}{\ln 1.08}$

Solution   First, note that

$$\frac{\log 2}{\log 1.1} \neq \log 2 - \log 1.1$$

The fraction $\dfrac{\log 2}{\log 1.1}$ is not the same as the fraction $\log \dfrac{2}{1.1}$ that occurs in the logarithmic property $\log_b M/N = \log_b M - \log_b N$.

(A) $n = \dfrac{\log 2}{\log 1.1} = \dfrac{0.301\,03\ldots}{0.041\,39\ldots} = 7.272\,540\,9\ldots \approx 7.273$

(B) $n = \dfrac{\ln 3}{\ln 1.08} = \dfrac{1.098\,612\,3\ldots}{0.076\,961\ldots} = 14.274\,915\ldots \approx 14.275$

Matched Problem 2   Use a scientific calculator to evaluate each to two decimal places:

(A) $n = \dfrac{\ln 2}{\ln 1.1}$ (B) $n = \dfrac{\log 3}{\log 1.08}$

We now turn to the second problem: given the logarithm of a number, find the number. We make direct use of the logarithmic-exponential relationships. Recall:

Logarithmic-Exponential Relationships

$$\log x = y \quad \text{is equivalent to} \quad x = 10^y$$
$$\ln x = y \quad \text{is equivalent to} \quad x = e^y$$

Example 3
Solving for x

Find x to three significant digits, given the indicated logarithms.

(A) $\log x = -9.315$ **(B)** $\ln x = 2.386$

Solution **(A)** $\log x = -9.315$ Change to equivalent exponential form.
$$x = 10^{-9.315}$$
$$= 4.84 \times 10^{-10}$$ Notice that the answer is displayed in scientific notation in the calculator. The form of the display will vary, depending on the calculator used.

(B) $\ln x = 2.386$ Change to equivalent exponential form.
$$x = e^{2.386}$$
$$= 10.9$$

Matched Problem 3 Find x to four significant digits, given the indicated logarithms.

(A) $\ln x = -5.062$ **(B)** $\log x = 12.0821$

♦ **CHANGE-OF-BASE FORMULA**

If we can find the logarithm of a number to one base, then by means of the following **change-of-base formula** we can find the logarithm of the number to any other base.

Change-of-Base Formula

$$\log_b N = \frac{\log_a N}{\log_a b} = \frac{\log N}{\log b} = \frac{\ln N}{\ln b}$$

The change-of-base formula may be derived as follows:

$$y = \log_b N$$
$$N = b^y$$ Definition of logarithm. If two positive quantities are equal, their logarithms are equal. Apply $\log_a$ to each side.

$$\log_a N = \log_a b^y$$ Apply Property 4.
$$\log_a N = y \log_a b$$ Divide

$$y = \frac{\log_a N}{\log_a b}$$

$$\log_b N = \frac{\log_a N}{\log_a b}$$ Since $y = \log_b N$

Example 4
Evaluating $\log_b x$

Find $\log_5 14$ to three decimal places using common logarithms.

Solution
$$\log_5 14 = \frac{\log 14}{\log 5} = \frac{1.146\ 128\ \ldots}{0.698\ 97\ \ldots} = 1.639\ 738\ 5\ \ldots \approx 1.640$$

Matched Problem 4 Find $\log_7 729$ to four decimal places.

♦ GRAPHING LOGARITHMIC FUNCTIONS

With the aid of a scientific calculator to find values, logarithmic functions can be graphed by plotting points.

Example 5
Graphing Logarithmic Functions

Graph:

(A) $y = 4 \log x$ **(B)** $y = 3 \ln 2x$ **(C)** $y = \log_{0.5} x$

Solution   **(A)** ──────────

x	y
0.01	−8
0.1	−4
1	0
2	1.20
5	2.80
10	4

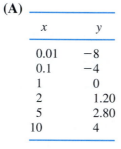

(B) ──────────

x	y
0.1	−4.8
0.5	0
1	2.08
2	4.16
3	5.37
4	6.23
6	7.45
8	8.32

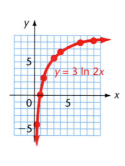

(C) To find values for y, use $\log_{0.5} x = \dfrac{\ln x}{\ln 0.5}$.

x	y
0.1	3.32
0.5	1
1	0
2	−1
3	−1.58
4	−2
6	−2.58

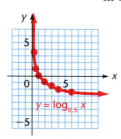

Matched Problem 5   Graph: **(A)** $y = 2 \log 10x$ **(B)** $y = \frac{1}{2} \ln 6x$ **(C)** $y = 2 \log_{1/4} x$

Answers to
Matched Problems

1. **(A)** $-1.868\,734$ **(B)** $3.356\,663$ **(C)** Not possible
2. **(A)** 7.27 **(B)** 14.27 3. **(A)** $0.006\,333$ **(B)** 1.208×10^{12}
4. 3.3875
5. **(A)**

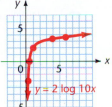

(B)

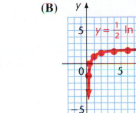

(C)

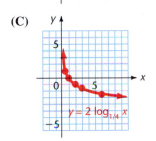

EXERCISE 10-4

A *Use a calculator to find each to four decimal places.*

1. $\log 82{,}734$
2. $\log 843{,}250$
3. $\log 0.001\,439$
4. $\log 0.035\,604$
5. $\ln 43.046$
6. $\ln 2{,}843{,}100$
7. $\ln 0.081\,043$
8. $\ln 0.000\,032\,4$
9. $\ln(-2.345)$
10. $\log(-12.987)$
11. $\log 0.005$
12. $\ln 0.0006$
13. $\log 0.000\,007$
14. $\ln 12{,}345{,}000$
15. $\log 1.000\,001$
16. $\ln 1.000\,000\,01$

Use a calculator to find x to four significant digits.

17. $\log x = 5.3027$
18. $\log x = 1.9168$
19. $\log x = -3.1773$
20. $\log x = -2.0411$
21. $\ln x = 3.8655$
22. $\ln x = 5.0884$
23. $\ln x = -0.3916$
24. $\ln x = -4.1083$
25. $\log x = 2.4680$
26. $\log x = -3.5678$
27. $\log x = -5.4321$
28. $\log x = 1.3579$
29. $\ln x = -9.7531$
30. $\ln x = 10.8642$
31. $\ln x = 1.2345$
32. $\ln x = -2.3344$

B *Evaluate each of the following to three decimal places using a calculator:*

33. $n = \dfrac{\log 2}{\log 1.15}$
34. $n = \dfrac{\log 2}{\log 1.12}$
35. $n = \dfrac{\ln 3}{\ln 1.15}$
36. $n = \dfrac{\ln 4}{\ln 1.2}$
37. $x = \dfrac{\ln 0.5}{-0.21}$
38. $t = \dfrac{\log 200}{2 \log 2}$

Use the change-of-base formula and a calculator with either log or ln to find each to four decimal places.

39. $\log_5 372$
40. $\log_4 23$
41. $\log_8 0.0352$
42. $\log_2 0.005\,439$
43. $\log_3 0.1483$
44. $\log_{12} 435.62$
45. $\log_2 4.32$
46. $\log_4 5.678$
47. $\log_3 0.5$
48. $\log_8 5.12$
49. $\log_{12} 150$
50. $\log_{1/2} 10$
51. $\log_{1/2} 4$
52. $\log_6 216$

Graph.

53. $y = \frac{1}{2} \ln 5x$
54. $y = 3 \ln \dfrac{x}{4}$
55. $y = 2 \log x^2$
56. $y = 4 \log(\frac{1}{2}x)$
57. $y = 4 \log_5 x$
58. $y = 10 \log_{0.10} x$

C *Find x to five significant digits using a calculator.*

59. $x = \log(5.3147 \times 10^{12})$

60. $x = \log(2.0991 \times 10^{17})$

61. $x = \ln(6.7917 \times 10^{-12})$

62. $x = \ln(4.0304 \times 10^{-8})$

63. $\log x = 32.068\ 523$

64. $\log x = -12.731\ 64$

65. $\ln x = -14.667\ 13$

66. $\ln x = 18.891\ 143$

Graph.

67. $y = \log(x + 2)$

68. $y = \ln(x + 4)$

69. $y = (\ln x) + 4$

70. $y = (\log x) + 2$

71. $y = 2(\ln x) - 3$

72. $y = 4(\log x) - 3$

73. $y = 2[\log(x + 3)] - 4$

74. $y = 4[\ln(x + 3)] - 2$

  # 10-5 Exponential and Logarithmic Equations

♦ Exponential Equations
♦ Logarithmic Equations

Equations involving exponential and logarithmic functions, such as

$$2^{3x-2} = 5 \qquad \text{and} \qquad \log(x + 3) + \log x = 1$$

are called **exponential** and **logarithmic equations,** respectively. Logarithmic properties play a central role in their solution.

♦ **EXPONENTIAL EQUATIONS**

The following examples illustrate the use of logarithmic properties in solving exponential equations.

Example 1
Solving an Exponential Equation

Solve $2^{3x-2} = 5$ for x to four decimal places.

Solution

$$2^{3x-2} = 5$$

How can we get x out of the exponent? Use logarithms. If two positive quantities are equal, their logarithms are equal.

$$\log 2^{3x-2} = \log 5$$

Use $\log_b M^p = p \log_b M$ to get $3x - 2$ out of the exponent position.

$$(3x - 2)\log 2 = \log 5$$

$$3x - 2 = \frac{\log 5}{\log 2}$$

Remember: $\dfrac{\log 5}{\log 2} \neq \log 5 - \log 2$

$$x = \frac{1}{3}\left(2 + \frac{\log 5}{\log 2}\right)$$

$$= 1.4406$$

To four decimal places

Matched Problem 1   Solve $35^{1-2x} = 7$ for x to four decimal places.

Example 2
Doubling Time in an Interest Problem

Recall that if a principal P is invested at an annual interest rate r compounded annually, then the amount of money A in the account after n years, assuming no withdrawals, is given by

$$A = P(1 + r)^n$$

How long will it take the money to double if it is invested at 6% compounded annually?

Solution   To double, the principal P must grow to an amount $2P$. Thus, we need to solve the following equation for n:

$$2P = P(1.06)^n \qquad \text{Divide both sides by } P.$$

$$2 = 1.06^n \qquad \text{Take the common or natural logarithm of both sides.}$$

$$\log 2 = \log 1.06^n$$

$$\log 2 = n \log 1.06 \qquad \text{Note how logarithmic properties are used to get } n \text{ out of the exponent position.}$$

$$n = \frac{\log 2}{\log 1.06}$$

$$= \frac{0.301\ 03}{0.025\ 30} \approx 11.89$$

$$= 12 \text{ years} \qquad \text{To the next whole year}$$

Matched Problem 2   How long will it take the money in Example 2 to double if the interest rate is changed to 9% compounded annually?

Example 3
An Exponential Equation for Atmospheric Pressure

The atmospheric pressure P (in pounds per square inch) at x miles above sea level is given approximately by

$$P = 14.7e^{-0.21x}$$

At what altitude will the atmospheric pressure be half the pressure at sea level? Compute the answer to two significant digits.

Solution   Sea-level pressure is the pressure at $x = 0$. Thus,

$$P = 14.7e^0 = 14.7$$

Half the pressure at sea level is therefore $14.7/2 = 7.35$. Now our problem is to find x so that $P = 7.35$. We solve $7.35 = 14.7e^{-0.21x}$ for x:

$$7.35 = 14.7e^{-0.21x}$$ Divide both sides by 14.7 to simplify.

$$0.5 = e^{-0.21x}$$ Take the natural logarithm of both sides.

$$\ln 0.5 = \ln e^{-0.21x}$$ Why use natural logarithms? Try the common logarithm to see why.

$$\ln 0.5 = -0.21x$$

$$x = \frac{\ln 0.5}{-0.21}$$ Use a calculator.

$$= 3.3 \text{ miles}$$ To two significant digits

Matched Problem 3   Using the formula in Example 3, find the altitude (in miles, to two significant digits) where the atmospheric pressure will be one-eighth that at sea level.

♦ LOGARITHMIC EQUATIONS

The next two examples illustrate approaches to solving some types of logarithmic equations.

Example 4
Solving a Logarithmic Equation

Solve $\log(x + 3) + \log x = 1$, and check.

Solution

$$\log(x + 3) + \log x = 1$$ Combine the left side using $\log M + \log N = \log MN$.

$$\log[x(x + 3)] = 1$$ Change to the equivalent exponential form.

$$x(x + 3) = 10^1$$ Write in standard form: $ax^2 + bx + c = 0$.

$$x^2 + 3x - 10 = 0$$ Solve.

$$(x + 5)(x - 2) = 0$$

$$x = -5, 2$$

Check $x = -5$: $\log(-5 + 3) + \log(-5)$ is not defined   Why?
Thus, -5 must be discarded.
$x = 2$: $\log(2 + 3) + \log 2 = \log 5 + \log 2$
$= \log(5 \cdot 2) = \log 10 = 1$
Thus, 2 is the only solution.

Remember, answers should be checked in the original equation to determine whether any should be discarded. An extraneous solution can be introduced in shifting from logarithmic to exponential form because the logarithm has a restricted domain.

Matched Problem 4   Solve $\log(x - 15) = 2 - \log x$, and check.

Example 5
Solving a Logarithmic Equation

Solve $(\ln x)^2 = \ln x^2$.

Solution

$$(\ln x)^2 = \ln x^2$$
$$(\ln x)^2 = 2 \ln x$$
$$(\ln x)^2 - 2 \ln x = 0$$
$$(\ln x)(\ln x - 2) = 0$$

This is a quadratic equation in $\ln x$. Move all nonzero terms to the left and factor.

$\ln x = 0$ or $\ln x - 2 = 0$
$x = e^0$ $\ln x = 2$
$ = 1$ $x = e^2$ Check these results.

Matched Problem 5

Solve $\log x^2 = (\log x)^2$, and check.

Answers to Matched Problems
1. 0.2263   2. More than double in 9 years, but not quite double in 8 years
3. 9.9 miles   4. $x = 20$ 5. $x = 100, 1$

EXERCISE 10-5

A *Solve to three significant digits.*

1. $10^{-x} = 0.0347$
2. $10^x = 14.3$
3. $10^{3x+1} = 92$
4. $10^{5x-2} = 348$
5. $e^x = 3.65$
6. $e^{-x} = 0.0142$
7. $e^{2x-1} = 405$
8. $e^{3x+5} = 23.8$
9. $5^x = 18$
10. $3^x = 4$
11. $2^{-x} = 0.238$
12. $3^{-x} = 0.074$
13. $5^x = 10$
14. $5^{-x} = 10$
15. $10^{-3x} = 144$
16. $10^{-4x} = 16$
17. $e^{2x} = 8$
18. $e^{3x} = 18$

Solve exactly.

19. $\log 5 + \log x = 2$
20. $\log x - \log 8 = 1$
21. $\log x + \log(x - 3) = 1$
22. $\log(x - 9) + \log 100x = 3$
23. $\log 4 + \frac{1}{2} \log x = \log 3$
24. $\log 2 - \frac{1}{2} \log x = \frac{1}{4}$
25. $\log x = 1 - \log x$
26. $2 \log x = 3 - \log x$

B *Solve to three significant digits.*

27. $2 = 1.05^x$
28. $3 = 1.06^x$
29. $e^{-1.4x} = 13$
30. $e^{0.32x} = 632$
31. $123 = 500e^{-0.12x}$
32. $438 = 200e^{0.25x}$

Solve exactly.

33. $\log x - \log 5 = \log 2 - \log(x - 3)$
34. $\log(6x + 5) - \log 3 = \log 2 - \log x$
35. $(\ln x)^3 = \ln x^4$
36. $(\log x)^3 = \log x^4$
37. $\ln(\ln x) = 1$
38. $\log(\log x) = 1$
39. $x^{\log x} = 100x$
40. $3^{\log x} = 3x$

C *In Problems 41–46, solve for the indicated letter in terms of all others using common or natural logarithms, whichever produces the simplest results.*

41. $I = I_0 e^{-kx}$, for x *X-ray absorption*
42. $A = P(1 + i)^n$, for n *Compound interest*
43. $N = 10 \log \dfrac{I}{I_0}$, for I *Sound intensity— decibels*
44. $t = \dfrac{-1}{k}(\ln A - \ln A_0)$, for A *Radioactive decay*
45. $I = \dfrac{E}{R}(1 - e^{-Rt/L})$, for t *Electric circuits*
46. $S = R\dfrac{(1 + i)^n - 1}{i}$, for n *Future value of an annuity*

47. Find the fallacy:
$$3 > 2$$
$$(\log \tfrac{1}{2})3 > (\log \tfrac{1}{2})2$$
$$3 \log \tfrac{1}{2} > 2 \log \tfrac{1}{2}$$
$$\log(\tfrac{1}{2})^3 > \log(\tfrac{1}{2})^2$$
$$(\tfrac{1}{2})^3 > (\tfrac{1}{2})^2$$
$$\tfrac{1}{8} > \tfrac{1}{4}$$

48. Find the fallacy:
$$-2 < -1$$
$$\ln e^{-2} < \ln e^{-1}$$
$$2 \ln e^{-1} < \ln e^{-1}$$
$$2 < 1$$

APPLICATIONS

49. *Interest doubling time* Find the doubling time for money invested at a rate of 12% compounded annually.

50. *Interest doubling time* Find the doubling time for money invested at a rate of 8% compounded annually.

51. *Interest tripling time* Find the tripling time for money invested at 4% compounded annually.

52. *Interest quadrupling time* Find the time for money invested at 4% compounded annually to quadruple.

★ 53. *Interest—Rule of 72* The doubling time for money invested at a rate of p percent compounded annually is approximately $72/p$ years. Test the plausibility of this result by solving

$$2 = \left(1 + \frac{p}{100}\right)^n$$

for n with various values of p, and compare your answers to $72/p$. Note that the rate as a decimal is $p/100$.

★ 54. *Interest—Rule of 72* The doubling time for money invested at a rate of p percent compounded continuously is approximately $72/p$ years. Verify this by solving

$$2P = Pe^{(p/100)n}$$

for n. Your answer should be a fraction with denominator p and numerator approximately 72. The rule uses 72 rather than the exact numerator to make the division easier.

55. *Bacterial growth* A single cholera bacterium divides every $\frac{1}{2}$ hour to produce two complete cholera bacteria. If we start with a colony of 5,000 bacteria, then after t hours we will have

$$A = 5{,}000 \cdot 2^{2t}$$

bacteria. How long will it take for A to equal 1,000,000?

56. *Bacterial growth* Refer to Problem 55. How long will it take for A to equal 175,000?

57. *Astronomy* An optical instrument is required to observe stars beyond the sixth magnitude, the limit of ordinary vision. However, even optical instruments have their limitations. The limiting magnitude L of an optical telescope with lens diameter D (in inches) is given by

$$L = 8.8 + 5.1 \log D$$

Find the limiting magnitude for a homemade 6-inch reflecting telescope.

58. *Astronomy* For an optical telescope as described in Problem 57, find the diameter of a lens that would have a limiting magnitude of 20.6.

★ 59. *World population* A mathematical model for world population growth over short periods of time is given by

$$P = P_0 e^{rt}$$

where $P_0 =$ Population at $t = 0$, $r =$ Rate compounded continuously, $t =$ Time in years, and $P =$ Population at time t. How long will it take the earth's population to double if it continues to grow at its current rate of 2% per year (compounded continuously)? [*Hint:* Given $r = 0.02$, find t so that $P = 2P_0$.]

★★ 60. *World population* If the world population is now 4 billion people and if it continues to grow at 2% per year (compounded continuously), how long will it be before there is only 1 square yard of land per person? Use the formula in Problem 59 and the fact that there are 1.7×10^{14} square yards of land on earth.

★ 61. *Nuclear reactors—strontium-90* Radioactive strontium-90 is used in nuclear reactors and decays according to

$$A = Pe^{-0.0248t}$$

where P is the amount present at $t = 0$ and A is the amount remaining after t years. Find the half-life of strontium-90; that is, find t so that $A = 0.5P$.

62. *Nuclear reactors—strontium-90* Refer to Problem 61. Find the time required for strontium-90 to decay to one-tenth of the quantity present now.

★ 63. *Archaeology—carbon-14 dating* Cosmic-ray bombardment of the atmosphere produces neutrons, which in turn react with nitrogen to produce radioactive carbon-14. Radioactive carbon-14 enters all living tissues through carbon dioxide, which is first absorbed by plants. As long as a plant or animal is alive, carbon-14 is maintained in a constant amount in its tissues. Once dead, however, it ceases taking in carbon and, to the slow beat of time, the carbon-14 diminishes by radioactive decay according to the equation

$$A = A_0 e^{-0.000124t}$$

where t is time in years. Estimate the age A of a skull uncovered in an archaeological site if 6% of the original amount A_0 of carbon-14 is still present. [*Hint:* Find t such that $A = 0.06A_0$.]

64. *Archaeology—carbon-14 dating* Refer to Problem 63. Find the age of a bone uncovered in an archaeological site if 10% of the original amount of carbon-14 is still present.

★ **65.** *Sound intensity—decibels* Because of the extraordinary range of sensitivity of the human ear (a range of over 1,000 million million to 1), it is helpful to use a logarithmic scale to measure sound intensity over this range rather than an absolute scale. The unit of measure is called the *decibel,* after the inventor of the telephone, Alexander Graham Bell. If we let N be the number of decibels, I the power of the sound in question (in watts per cubic centimeter), and I_0 the power of sound just below the threshold of hearing (approximately 10^{-16} watt per square centimeter), then

$$I = I_0 10^{N/10}$$

Show that this formula can be written in the form

$$N = 10 \log \frac{I}{I_0}$$

66. *Sound intensity—decibels* Use the formula in Problem 65 (with $I_0 = 10^{-16}$ watt per square centimeter) to find the decibel ratings of the following sounds:
(A) Whisper (10^{-13} watt per square centimeter)
(B) Normal conversation (3.16×10^{-10} watt per square centimeter)
(C) Heavy traffic (10^{-8} watt per square centimeter)
(D) Jet plane with afterburner (10^{-1} watt per square centimeter)

★ **67.** *Earth science* For relatively clear bodies of fresh or salt water, light intensity is reduced according to the exponential function

$$I = I_0 e^{-kd}$$

where I is the intensity at d feet below the surface and I_0 is the intensity at the surface; k is called the coefficient of extinction. Two of the clearest bodies of water in the world are the freshwater Crystal Lake in Wisconsin ($k = 0.0485$) and the saltwater Sargasso Sea off the West Indies ($k = 0.009\ 42$). Find the depths (to the nearest foot) in these two bodies of water at which the light is reduced to 1% of that at the surface.

68. *Earth science* Refer to Problem 67. Find the depths at which light is reduced to 50% of that at the surface.

69. *Psychology—learning* In learning a particular task, such as typing or swimming, one progresses faster at the beginning and then levels off. If we plot the level of performance against time, we obtain a curve of the type shown in the figure. This is called a learning curve and can be very closely approximated by an exponential equation of the form $y = a(1 - e^{-cx})$, where a and c are positive constants. Curves of this type have applications in psychology, education, and industry. Suppose a particular person's history of learning to type is given by the exponential equation $N = 80(1 - e^{-0.08n})$, where N is the number of words per minute typed after n weeks of instruction. Approximately how many weeks did it take the person to learn to type 60 words per minute?

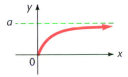

70. *Psychology—learning* Refer to Problem 69. Approximately how many weeks did it take the person to learn to type 75 words per minute?

CHAPTER SUMMARY

10-1 EXPONENTIAL FUNCTIONS

An **exponential function** is a function defined by $f(x) = b^x$, where the **base** $b > 0$ and $b \neq 1$. The **domain** of an exponential function is R; the **range** is the set of positive real numbers. The **amount** A of money accumulated from an original **principal** P compounded at a rate r per period for n periods is given by $A = P(1 + r)^n$. If the compounding is continuous at an annual rate a for t years, the formula becomes $A = Pe^{at}$. The number $e \approx 2.718\ 28\ldots$ is approached by taking smaller and smaller values of r in $(1+r)^{1/r}$ and is used for the exponential function $f(x) = e^x$. The basic properties of exponents continue to hold for b^x. In addition, $b^m = b^n$ if and only if $m = n(b > 0, b \neq 1)$, and $a^m = b^m$ if and only if $a = b$ ($a, b > 0$, $a, b \neq 1$, $m \neq 0$).

10-2 LOGARITHMIC FUNCTIONS

The inverse of an exponential function b^x is a **logarithmic function,** $y = \log_b x$; thus, $y = \log_b x$ means $b^y = x$. The **domain** of a logarithmic function is the set of positive real numbers; the **range** is the set of all real numbers. Logarithms to the **base 10** and **base e** are denoted **log x** and **ln x**, respectively. The **logarithmic-exponential identities** are

$$\log_b b^x = x \qquad \text{and} \qquad b^{\log_b x} = x$$

10-3 PROPERTIES OF LOGARITHMIC FUNCTIONS

Logarithms satisfy these basic properties:

1. $\log_b b^M = M$
2. $\log_b MN = \log_b M + \log_b N$
3. $\log_b \dfrac{M}{N} = \log_b M - \log_b N$
4. $\log_b M^p = p \log_b M$
5. $\log_b 1 = 0$

Also, $\log_b M = \log_b N$ if and only if $M = N$.

10-4 LOGARITHMS TO VARIOUS BASES

Logarithms using **base 10** are called **common logarithms;** those with **base e** are **natural logarithms.** These two types are denoted by $\log_{10} x = \log x$ and $\log_e x = \ln x$. The **change-of-base formula,**

$$\log_b N = \frac{\log_a N}{\log_a b}$$

converts logarithms using base a to logarithms using base b.

10-5 EXPONENTIAL AND LOGARITHMIC EQUATIONS

Equations involving an exponential function are called **exponential equations;** those involving logarithms are called **logarithmic equations.**

CHAPTER REVIEW EXERCISE

Work through all the problems in this chapter review and check answers in the back of the book. Answers to all the problems are there, and following each answer is a number in italics indicating the section in which that type of problem is discussed. Where weaknesses show up, review appropriate sections in the text.

You will find a scientific calculator useful in many of the problems in this exercise.

A **1.** Write $m = 10^n$ in logarithmic form with base 10.

2. Write $\log x = y$ in exponential form.

Evaluate without using a calculator or table.

3. $\log_3 27$ **4.** $\log_4 2$ **5.** $\log_4 32$

6. $\log_4(\frac{1}{4})$ **7.** $\log_3 \sqrt{3}$ **8.** $\log_3(\frac{1}{9})$

Solve for x exactly.

9. $\log_2 x = 3$ **10.** $\log_x 25 = 2$

11. $\log_{10} x = 2$ **12.** $\log_8 x = -1$

13. $\log_x 9 = 1$ **14.** $\log_x(\frac{1}{8}) = 3$

Solve for x to three significant digits.

15. $10^x = 17.5$ **16.** $e^x = 143{,}000$

17. $2^x = 10$ **18.** $3^x = \frac{1}{10}$

In Problems 19–22, solve for x exactly.

19. $\log x - 2 \log 3 = 2$ **20.** $\log x + \log(x - 3) = 1$

21. $\log 3 + \log x = \log 300$

22. $\frac{1}{2} \log x = \log 7$

B **23.** Write $\ln y = x$ in exponential form.

24. Write $x = e^y$ in logarithmic form with base e.

Evaluate without using a calculator or table.

25. $\log_{1/4} 16$

26. $\log_{1/2} \left(\frac{1}{8}\right)$

27. $\log_{1/2} 32$

28. $\log_{1/4} 2$

Solve for x exactly.

29. $\log_x 9 = -2$

30. $\log_{16} x = \frac{3}{2}$

31. $\log_x e^5 = 5$

32. $10^{\log_{10} x} = 33$

33. $\ln x = 0$

34. $\log_x(\frac{1}{2}) = 2$

35. $\ln e^x = x$

36. $\log x = 0$

Solve for x to three significant digits.

37. $25 = 5(2)^x$

38. $4,000 = 2,500e^{0.12x}$

39. $0.01 = e^{-0.05x}$

40. $500 = 5^x$

41. $e^{-3x} = e^6$

42. $10^{-x} = 500$

Solve for x exactly.

43. $\log 3x^2 - \log 9x = 2$

44. $\log x - \log 3 = \log 4 - \log(x + 4)$

45. $(\log x)^3 = \log x^9$

46. $\ln(\log x) = 1$

Evaluate. Express your answer to three decimal places.

47. $\log_5 23$

48. $\log_2 6$

49. $\log_{12} 10$

50. $\log_{121} 1$

Graph.

51. $y = 3e^{-2x}$

52. $y = 10^{x/2}$

53. $y = 2 \log 3x$

54. $y = 2 \ln \frac{x}{3}$

55. $y = 1.08^x$

C *Solve for y.*

56. $\ln y = -5t + \ln C$

57. $\ln y = 3x + 4$

58. $e^y = 3x + 4$

Graph both f and f^{-1} using the same coordinate system.

59. $f(x) = \ln x$

60. $f(x) = \log x$

61. $f(x) = \log_{1/2} x$

62. $f(x) = \log_2 x$

In Problems 63–66, solve for x exactly.

63. $\log(\log x) = 0$

64. $\ln x = e$

65. $-3 \log x = \log\frac{1}{x^3}$

66. $\log(-x) = \log x$

67. Explain why 1 cannot be used as a logarithmic base.

68. Prove that $\log_b (M/N) = \log_b M - \log_b N$.

APPLICATIONS

69. *Interest* Find the amount if $1,000 is invested at an annual interest rate of 6% compounded quarterly for 3 years.

70. *Interest* Find the amount if $1,000 is invested at an annual interest rate of 4% compounded continuously for 10 years.

71. *Population growth* Many countries in the world have a population growth rate of 3% (or more) per year. At this rate how long will it take a population to double? Use the population growth model

$$P = P_0(1.03)^t$$

which assumes annual compounding. Compute the answer to three significant digits.

72. *Population growth* Repeat Problem 71 using the continuous population growth model

$$P = P_0 e^{0.03t}$$

which assumes continuous compounding. Compute the answer to three significant digits.

73. *Carbon-14 dating* Refer to Problem 63, Exercise 10-5. How long (to three significant digits) will it take for the amount A of carbon-14 to diminish to 1% of the original amount A_0 after the death of a plant or animal?

$$A = A_0 e^{-0.000124t} \qquad \text{where } t \text{ is time in years}$$

74. *X-ray absorption* Solve $x = -(1/k)\ln(I/I_0)$ for I in terms of the other letters.

75. *Amortization—time payments* Solve $r = P\{i/[1 - (1 + i)^{-n}]\}$ for n in terms of the other letters.

CHAPTER PRACTICE TEST

The following practice test is provided for you to test your knowledge of the material in this chapter. You should try to complete it in 50 minutes or less. Answers in the back of the book indicate the section in the text where the material in the question is covered. Actual tests in your class may vary from this practice test in difficulty, length, or emphasis, depending on the goals of your course or instructor.

In Problems 1 and 2, evaluate without using a calculator or tables.

1. $\log_3 9 - \log_4 2$ **2.** $\log_{1/2}(\tfrac{1}{4}) - \log_3 \sqrt{3}$

3. Evaluate $\log_b 12$ given that $\log_b 2 = 0.356$ and $\log 3 = 0.565$.

Rewrite as a single logarithm with coefficient 1.

4. $\log_2 x - 3\log_2(x + 1)$ **5.** $\log x - 2\log y + 3\log z$

Solve for x exactly.

6. $\log_2 x = 5$ **7.** $\log_x 5 = 2$

8. $\ln e^2 = \log x$ **9.** $\log x + \log(x - 3) = 1$

10. $\log x^3 = 3$ **11.** $\ln \dfrac{1}{x} = -1$

Solve for x.

12. $2^x = 10$ **13.** $e^x = 10$

14. $1,000 = 50e^{0.008x}$ **15.** $10^x = 0.004$

In Problems 16 and 17, graph the equation.

16. $y = \log_2 x$ **17.** $y = (\tfrac{1}{2})^x$

18. Graph both $f(x) = e^x$ and $f^{-1}(x)$ using the same coordinate system. What are the domain and range for each function? Name f^{-1}.

19. Find the amount if $100 is invested at an annual interest rate of 9% compounded semiannually for 10 years.

20. *Newton's law of cooling* expresses the difference in temperature between an object and its surrounding medium as a function of time:

$$T - M = C_0 e^{kt}$$

Solve this equation for *t*.

CUMULATIVE REVIEW EXERCISE, CHAPTERS 1–10

This set of exercises reviews the major concepts and techniques of Chapters 1–10. Work through all the problems and check answers in the back of the book. Answers to all the problems are there, and following each answer is a number in italics indicating the section in which that type of problem is discussed. Where weaknesses show up, review appropriate sections in the text.

A *In Problems 1–15, find, evaluate, or calculate the quantity requested.*

1. $(-3)[7 - (-4)] - |-3|$

2. $[(-3)(5)]^2 - 4 \cdot 5^2$

3. The LCM of $3x^2$, $4xy$, $6y^2$

4. The quotient and remainder when $x^4 - 2x^2 + 3$ is divided by $x + 2$

5. $x^2 y^0 z^{-1}$ for $x = 2$, $y = 3$, and $z = 4$

6. The slope of the line through the points $(-5, 4)$ and $(3, 2)$

7. The equation of the line with slope 2 passing through the point $(3, -4)$

8. $8^{2/3} - 4^{3/2}$

9. The discriminant of $3x^2 - 4x - 5$ and the number of solutions to $3x^2 - 4x - 5 = 0$

10. $\begin{vmatrix} 3 & 4 \\ 5 & 6 \end{vmatrix}$

11. $f(3)$ for $f(x) = x^2 - \dfrac{1}{x} + \sqrt{x + 1}$

12. $(2 + 3i) + (1 + i)(2 - i)$

13. $(1 + i) \div (1 - i)$

14. The coordinates of the vertex of the parabola

$$y = 2x^2 + 4x - 7$$

15. $x^2 - x + 1$ for $x = \dfrac{1}{2} + \dfrac{\sqrt{3}}{2}i$

16. Translate the following statement into an algebraic equation or inequality statement using *x* as the only variable: 3 times the sum of a number and its reciprocal is greater than the quantity 4 less than the number. Do not solve the equation or inequality.

17. When the price of a commodity is x dollars per ton, the demand will be $3,000 - 12x$ tons. The revenue is the price times the demand. Express the revenue as a function $R(x)$.

18. The value of a piece of machinery is a linear function $f(t)$ of its age t. The machinery costs \$60,000 and its value decreases at the rate of \$8,000 per year. Determine $f(t)$.

Rewrite in the form indicated.

19. $1 - [x - (y - z)]$ without grouping symbols

20. The sum of $3x^2 - 2x + 1$ and $x^3 - 2x^2 + 3x - 4$ as a polynomial with like terms combined

21. The product of $3x + 2$ and $x + 4$ as a polynomial with like terms combined

22. The square of $2x - 3$ as a polynomial with like terms combined

23. $(x^2)^3 x^{-3}$ as a power of x

24. $2x^2y^3 + 6x^3y$ as a product with all common factors factored out

25. $x^2 + 8x + 16$ in factored form

26. $4x^2 + 17x + 4$ in factored form

27. $x^2 - 64$ in factored form

28. $\dfrac{2xy}{3z^2}$ as a fraction with numerator $4xy^2z^3$

29. $\dfrac{x^2 + 3x + 2}{x^2 + 2x + 1}$ as a fraction in lowest terms

30. $(x^2y)^{-1}\dfrac{z^{-2}}{(xy^2)^{-3}}$ as a fraction in lowest terms using only positive exponents

31. 3.456×10^5 in standard notation

32. $0.003\ 456$ in scientific notation

33. $\sqrt{50} + \sqrt{72} - \sqrt{98}$ in simplified form

34. $\dfrac{2x}{\sqrt{3x}}$ in simplest radical form for $x > 0$

35. $\sqrt{\dfrac{x^5}{9}}$ in simplest radical form for $x > 0$

36. $3x + 4y = 5$ in slope-intercept form

37. The system $\begin{aligned} 2x + \ \ y - \ \ z &= 4 \\ x + 3y + 2z &= 5 \\ -x + 2y + 3z &= 6 \end{aligned}$ in augmented matrix form

Solve.

38. $3x - 4(x - 5) = 5x - 3(x - 4)$

39. $2x^2 - 3x - 4 = 0$

40. $x^2 - 8x - 9 = 0$

41. $\begin{aligned} x + 3y &= -1 \\ 2x + \ \ y &= 8 \end{aligned}$

42. $(x - 1)(2x - 1)(x + 3) = 0$

43. $-3x + 4 < 2x - 6$

44. $|x - 2| \le 5$

45. $\dfrac{1}{x} + \dfrac{2}{x^2} = 1$

46. $\sqrt{x + 1} = x - 5$

47. $\begin{aligned} x + y &= 2 \\ x^2 + y^2 &= 4 \end{aligned}$

48. $y = 3x + 5$; solve for x

49. $\begin{aligned} x + y + z &= 3 \\ y + z &= -2 \\ z &= 1 \end{aligned}$

50. $x^2 - 9 = 0$

Graph.

51. $y = -2x + 1$

52. $y = x^2 + 7$

53. $f(x) = 3x - 5$

54. $3x + y \ge 6$

55. $\begin{aligned} x + 3y &\le 6 \\ 2x + \ \ y &\le 4 \\ x &\ge 0 \\ y &\ge 0 \end{aligned}$

56. $x^2 + y^2 = 4$

57. $\dfrac{x^2}{4} + y^2 = 1$

58. $x^2 - \dfrac{y^2}{4} = 1$

59. $x = y^2$

Set up an appropriate equation or inequality and solve.

60. Find two numbers with sum 24 and product 135.

61. How much of a 70% alcohol solution must be added to 6 gallons of a 40% alcohol solution to obtain a 60% solution?

62. The enrollment at a selective admissions college varies directly with the number of applicants. If 3,200 applicants yield a freshman class of 600, how many applicants are necessary to achieve a class of 750?

63. What is the amount after 10 years if \$10,000 is invested at an interest rate of 4.5% compounded annually?

64. What is the amount after 10 years if \$10,000 is invested at an interest rate of 4.5% compounded continuously?

B *Find, evaluate, or calculate the quantity requested.*

65. $f(g(2))$ for $f(x) = 1/x$ and $g(x) = \sqrt{x + 7}$

66. The LCM of $x^2 + 2x + 1$, $x^2 - 1$, and $x^2 - 2x + 1$

67. The center and radius of the circle

$$x^2 + 2x + y^2 - 4y = 4$$

68. The equation of the line parallel to $3x + 3y = 10$ and passing through the point $(1, 1)$

69. The inverse function for $f(x) = \dfrac{1}{x - 1}$

70. $x^3 - 64$ in factored form

71. $x^4 - 64$ in factored form

72. $ax + 3bx + 2ay + 6by$ in factored form

73. $\dfrac{3}{x} + \dfrac{4}{x + 1}$ as a fraction in lowest terms

74. $\dfrac{(x + 1)^2}{(x + 2)(x + 3)} \div \dfrac{x + 1}{(x + 2)^2}$ as a fraction in lowest terms

75. $\dfrac{\dfrac{1}{x} + 2}{2 - \dfrac{1}{x}}$ as a simple fraction in lowest terms

76. $\begin{vmatrix} 3 & 2 & 1 \\ 1 & 0 & -1 \\ 2 & 1 & 0 \end{vmatrix}$

Solve.

77. $3x^2 + 5x + 7 = 0$

78. $\sqrt{x - 1} = 5 - \sqrt{x + 4}$

79. $\dfrac{x + 3}{x - 2} = x + \dfrac{2x + 1}{x - 2}$

80. $|2x + 3| \geq 4$

81. $x^3 - 5x^2 + 6x = 0$

82. $4x + 5y = -8$
     $6x + 2y = 10$

83. $5 = e^{3x}$

84. $\log x = 1.7$

85. $\dfrac{1}{x} + \dfrac{2}{y} = 3$; solve for y

86. $\log_b 32 = -5$

Graph.

87. $x + y \leq 3.5$
     $2x + y \leq 6$
     $x + 2y \leq 6$
     $x \geq 0$
     $y \geq 0$

88. $f(x)$ and $f^{-1}(x)$, on the same coordinate system, where $f(x) = x^2$ and $x \geq 0$.

89. $y = 2(x - 1)^2 + 3$

90. $y = \ln(x - 2)$

91. Find the dimensions of a rectangle with perimeter 26 feet and area 40 square feet.

92. A rectangle of length x is surrounded by a fence that is 1,000 meters long. Express the area enclosed as a function of x, and find the length x that results in the largest possible area.

93. One copying machine can finish a duplicating job in 20 minutes. A second machine can do the same job in 30 minutes. How long will the job take if both machines are used?

94. A small furniture company produces quality hardwood desks and chairs. Each desk requires 2 hours of labor in the carpentry shop and 2 hours in the finishing shop. Each chair requires 1 hour in the carpentry shop and 1.5 hours in the finishing shop. The carpentry shop has 120 hours of time available and the finishing shop has 150 hours. How many chairs and desks can be produced within the available time? Set up a system of inequalities and graph the solution.

95. The time required to complete a research and development project varies directly with the contracted cost of the project and inversely with the size of the staff assigned to it. If a $2 million project can be completed by a staff of 45 people in 60 days, how long will it take a staff of 25 to complete a $1 million project?

C *Find, evaluate, or calculate the quantity requested.*

96. $6x^2 + 11x - 72$ in factored form

97. $x^6 - 1$ in factored form

98. $\dfrac{2 + \dfrac{3}{x}}{1 + \dfrac{2}{x}} - \dfrac{3}{x + 2}$ as a simple fraction in lowest terms

Set up an appropriate equation or inequality and solve.

99. An automobile radiator has a capacity of 3 gallons. The coolant mixture in the radiator is currently 40% antifreeze. How much should be drained and replaced with antifreeze to raise the percentage of antifreeze to 75%?

100. A truck driver completes a round trip of 600 total miles. For the return portion of the trip, the trucker drives an average of 5 miles per hour faster than on the outward portion, and takes 40 minutes less time. At what average speeds did the trucker drive? How long did the total trip take?

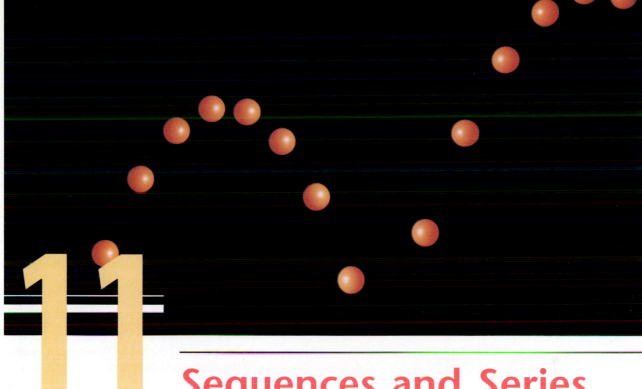

Sequences and Series

Most of the functions we have considered thus far have had domain R, the set of real numbers, or intervals in this set. In this chapter, we consider functions where the domain is the set of natural numbers, N, or particular subsets of N. These functions are called *sequences*. When all the values of a sequence are added, we get a sum called a *series*.

Photo reference: see Exercise 11-3, Problem 67

11-1   **Sequences and Series**

♦ Sequences
♦ Series

Intuitively, a *sequence* of numbers is simply a list presented in order, such as

$$1, 2, 4, 8, 16, 32, \ldots$$

In this example, 4 is in the third place, 16 is in the fifth place, etc. The rule by which place corresponds to the number in that place is a function. This observation allows a careful consideration of sequences in mathematical terms.

♦ SEQUENCES

An **infinite sequence** is a function a whose domain is the set of all natural numbers $N = \{1, 2, 3, \ldots, n, \ldots\}$. The range of the function is $a(1), a(2), a(3), \ldots, a(n), \ldots$, which is usually written in the following form:

Infinite Sequence

$$a_1, a_2, a_3, \ldots, a_n, \ldots \qquad \text{where } a_n = a(n)$$

The elements in the range are called the **terms of the sequence;** a_1 is the first term, a_2 the second term, and a_n the nth term. For example, if

$$a_n = \frac{n-1}{n} \tag{1}$$

the function a is a sequence with terms

$$0, \frac{1}{2}, \frac{2}{3}, \frac{3}{4}, \ldots, \frac{n-1}{n}, \ldots \qquad \text{Replace } n \text{ in Equation (1) with } 1, 2, 3, \text{ and so on.}$$

If the domain of a function a is the set of natural numbers $\{1, 2, 3, \ldots, n\}$ for some fixed n, then a is called a **finite sequence.** Thus, if

$$a_1 = 5$$

and

$$a_n = a_{n-1} + 2 \qquad n = 2, 3, 4$$

the function a is a finite sequence with terms

$$5, 7, 9, 11 \qquad a_2 = a_{2-1} + 2 = a_1 + 2 = 5 + 2 = 7, \text{ and so on.}$$

The two examples presented above illustrate two common ways in which sequences are specified:

1. The nth term a_n is expressed as a function of n.
2. One or more terms are given, and the nth term is expressed by means of preceding terms. In this case, the sequence is said to be defined **recursively.**

Example 1
Terms Defined as a
Function of n

Find the first four terms of a sequence whose nth term is

$$a_n = \frac{1}{2^n}$$

Solution $a_1 = \dfrac{1}{2^1} = \dfrac{1}{2}$ $\qquad a_2 = \dfrac{1}{2^2} = \dfrac{1}{4}$ $\qquad a_3 = \dfrac{1}{2^3} = \dfrac{1}{8}$ $\qquad a_4 = \dfrac{1}{2^4} = \dfrac{1}{16}$

Matched Problem 1 Find the first four terms of a sequence whose nth term is

$$a_n = \frac{n}{n^2 + 1}$$

Example 2
Terms Defined
Recursively

Find the first five terms of a sequence specified by

$$a_1 = 5 \qquad a_n = \tfrac{1}{2}a_{n-1} \quad \text{for } n \geq 2$$

Solution $a_1 = 5$
$a_2 = \tfrac{1}{2}a_{2-1} = \tfrac{1}{2}a_1 = \tfrac{1}{2}(5) = \tfrac{5}{2}$
$a_3 = \tfrac{1}{2}a_{3-1} = \tfrac{1}{2}a_2 = \tfrac{1}{2}(\tfrac{5}{2}) = \tfrac{5}{4}$
$a_4 = \tfrac{1}{2}a_{4-1} = \tfrac{1}{2}a_3 = \tfrac{1}{2}(\tfrac{5}{4}) = \tfrac{5}{8}$
$a_5 = \tfrac{1}{2}a_{5-1} = \tfrac{1}{2}a_4 = \tfrac{1}{2}(\tfrac{5}{8}) = \tfrac{5}{16}$

Matched Problem 2 Find the first five terms of a sequence specified by

$$a_1 = 3 \qquad a_n = a_{n-1} + 4 \quad \text{for } n \geq 2$$

Now let us look at the problem in reverse. That is, given the first few terms of a sequence and assuming the sequence continues in the indicated pattern, find a_n in terms of n.

Example 3
Finding a Rule for a_n

Find a rule for a_n in terms of n for the given sequences.

(A) 5, 6, 7, 8, . . . $\qquad$ **(B)** 2, -4, 8, -16, . . .

Solution **(A)** $a_n = n + 4$
(B) $a_n = (-1)^{n+1}2^n$ $\qquad$ Note how $(-1)^{n+1}$ functions as a sign alternator.

When a sequence is "specified" by its first few terms, the rule is assumed to be obvious. Actually, there will be many rules that fit the first few terms, so there is not a unique answer to the problem. See Problems 93–96 in Exercise 11-1.

Matched Problem 3   Find a_n in terms of n for the given sequences.

(A) $3, 5, 7, 9, \ldots$ (B) $1, -\frac{1}{2}, \frac{1}{4}, -\frac{1}{8}, \ldots$

◆   **SERIES**

The sum of the terms of a sequence is called a **series**. Thus, if $a_1, a_2, a_3, \ldots, a_n$, are the terms of a finite sequence, then

$$S_n = a_1 + a_2 + a_3 + \cdots + a_n$$

is the **finite series** associated with that sequence. If the sequence is infinite, the corresponding series is called an **infinite series.** We will restrict our attention to finite series in this section.

A series is often represented in a compact form using **summation notation,** as follows:

Summation Notation

$$S_n = \sum_{k=1}^{n} a_k = a_1 + a_2 + a_3 + \cdots + a_n$$

The terms of the series on the right are obtained from the summation expression in the middle by successively replacing the subscript k in a_k with integers, starting with 1 and ending with n. Thus, if a sequence is given by

$$\frac{1}{2}, \frac{1}{4}, \frac{1}{8}, \ldots, \frac{1}{2^n}$$

the corresponding series is given by

$$S_n = \frac{1}{2} + \frac{1}{4} + \frac{1}{8} + \cdots + \frac{1}{2^n} \qquad \text{or} \qquad S_n = \sum_{k=1}^{n} \frac{1}{2^k}$$

Example 4

Interpreting $\sum$ Notation

Write the following series without summation notation:

$$S_5 = \sum_{k=1}^{5} \frac{k-1}{k}$$

Solution $S_5 = \displaystyle\sum_{k=1}^{5} \frac{k-1}{k}$ Replace k in $\frac{k-1}{k}$ successively with 1, 2, 3, 4, and 5; then add.

$$= \frac{1-1}{1} + \frac{2-1}{2} + \frac{3-1}{3} + \frac{4-1}{4} + \frac{5-1}{5}$$

$$= 0 + \tfrac{1}{2} + \tfrac{2}{3} + \tfrac{3}{4} + \tfrac{4}{5}$$

Matched Problem 4   Write the following series without summation notation:

$$S_6 = \sum_{k=1}^{6} \frac{(-1)^{k+1}}{2k-1}$$

Example 5

Using $\sum$ Notation

Write the following series using summation notation:

$$S_6 = 1 - \tfrac{1}{2} + \tfrac{1}{3} - \tfrac{1}{4} + \tfrac{1}{5} - \tfrac{1}{6}$$

Solution   We first note that the nth term of the series is given by

$$a_n = (-1)^{n+1}\frac{1}{n}$$ Note that $(-1)^{n+1}$ gives the sequence 1, −1, 1, −1, 1, −1, ...

Hence,

$$S_6 = \sum_{k=1}^{6} (-1)^{k+1}\frac{1}{k}$$

Matched Problem 5   Write the following series using summation notation:

$$S_5 = 1 - \tfrac{2}{3} + \tfrac{4}{9} - \tfrac{8}{27} + \tfrac{16}{81}$$

Answers to Matched Problems

1. $a_1 = \tfrac{1}{2}, a_2 = \tfrac{2}{5}, a_3 = \tfrac{3}{10}, a_4 = \tfrac{4}{17}$ 2. $a_1 = 3, a_2 = 7, a_3 = 11, a_4 = 15, a_5 = 19$
3. (A) $a_n = 2n + 1$ (one choice of many)
   (B) $a_n = (-1)^{n+1}/2^{n-1}$ or $a_n = (-\tfrac{1}{2})^{n-1}$ (two choices of many)
4. $S_6 = 1 - \tfrac{1}{3} + \tfrac{1}{5} - \tfrac{1}{7} + \tfrac{1}{9} - \tfrac{1}{11}$
5. $S_5 = \displaystyle\sum_{k=1}^{5} (-\tfrac{2}{3})^{k-1}$ or $S_5 = \displaystyle\sum_{k=1}^{5} (-1)^{k+1}(\tfrac{2}{3})^{k-1}$ (two choices of many)

EXERCISE 11-1

A *Write the first four terms for each sequence.*

1. $a_n = n - 2$

2. $a_n = n + 3$

3. $a_n = 2n + 3$

4. $a_n = 3n + 2$

5. $a_n = 2^n$

6. $a_n = 3^n$

7. $a_n = \dfrac{n-1}{n+1}$

8. $a_n = \dfrac{n-2}{n+2}$

9. $a_n = \left(1 - \dfrac{1}{n}\right)^n$

10. $a_n = \left(1 + \dfrac{1}{n}\right)^n$

11. $a_n = (-2)^{n+1}$ **12.** $a_n = (-3)^{n+1}$

13. $a_n = \dfrac{(-1)^n}{n}$ **14.** $a_n = \dfrac{(-1)^{n+1}}{n^2}$

Write the requested term of the sequence.

15. The 8th term of the sequence in Problem 1.

16. The 10th term of the sequence in Problem 2.

17. The 18th term of the sequence in Problem 3.

18. The 28th term of the sequence in Problem 4.

19. The 6th term of the sequence in Problem 5.

20. The 6th term of the sequence in Problem 6.

21. The 100th term of the sequence in Problem 7.

22. The 99th term of the sequence in Problem 8.

23. The 8th term of the sequence in Problem 9.

24. The 8th term of the sequence in Problem 10.

25. The 7th term of the sequence in Problem 11.

26. The 6th term of the sequence in Problem 12.

27. The 20th term of the sequence in Problem 13.

28. The 10th term of the sequence in Problem 14.

Write each series in expanded form without summation notation.

29. $S_6 = \displaystyle\sum_{k=1}^{6} \dfrac{1}{k}$ **30.** $S_5 = \displaystyle\sum_{k=1}^{5} \dfrac{1}{k+1}$

31. $S_5 = \displaystyle\sum_{k=1}^{5} k$ **32.** $S_7 = \displaystyle\sum_{k=1}^{7} (k+1)$

33. $S_4 = \displaystyle\sum_{k=1}^{4} (k-1)^2$ **34.** $S_4 = \displaystyle\sum_{k=1}^{4} k^2$

35. $S_3 = \displaystyle\sum_{k=1}^{3} \dfrac{1}{10^k}$ **36.** $S_5 = \displaystyle\sum_{k=1}^{5} \dfrac{1}{2^k}$

37. $S_6 = \displaystyle\sum_{k=1}^{6} (\tfrac{1}{5})^k$ **38.** $S_5 = \displaystyle\sum_{k=1}^{5} (\tfrac{1}{3})^k$

39. $S_4 = \displaystyle\sum_{k=1}^{4} (-1)^k$ **40.** $S_6 = \displaystyle\sum_{k=1}^{6} (-1)^{k+1}$

41. $S_5 = \displaystyle\sum_{k=1}^{5} (-1)^k k$ **42.** $S_6 = \displaystyle\sum_{k=1}^{6} (-1)^{k+1} k$

B *Write the first five terms of each sequence.*

43. $a_n = (-1)^{n+1} n^2$ **44.** $a_n = (-1)^{n+1} \dfrac{1}{2^n}$

45. $a_n = \dfrac{1}{3}\left(1 - \dfrac{1}{10^n}\right)$ **46.** $a_n = \dfrac{1}{4}\left(1 - \dfrac{1}{5^n}\right)$

47. $a_n = (n+1)[1 - (-1)^{n+1}]$ **48.** $a_n = n[1 - (-1)^n]$

49. $a_1 = 1;\ a_n = 2a_{n-1}$ for $n \ge 2$

50. $a_1 = 1;\ a_n = 3a_{n-1}$ for $n \ge 2$

51. $a_1 = 7;\ a_n = a_{n-1} - 4$ for $n \ge 2$

52. $a_1 = 3;\ a_n = a_{n-1} + 2$ for $n \ge 2$

53. $a_1 = 4;\ a_n = \tfrac{1}{4}a_{n-1}$ for $n \ge 2$

54. $a_1 = 2;\ a_n = 2a_{n-1}$ for $n \ge 2$

55. $a_1 = 1;\ a_n = na_{n-1}$ for $n \ge 2$

56. $a_1 = 1;\ a_n = -na_{n-1}$ for $n \ge 2$

57. $a_1 = 1;\ a_n = a_{n-1} - n$ for $n \ge 2$

58. $a_1 = 1;\ a_n = a_{n-1} + n$ for $n \ge 2$

59. $a_1 = x;\ a_n = a_{n-1} \cdot \dfrac{x}{n}$ for $n \ge 2$

60. $a_1 = \dfrac{x^2}{2};\ a_n = a_{n-1} \cdot \dfrac{x^2}{(2n)(2n-1)}$ for $n \ge 2$

61. $a_1 = a_2 = 1;\ a_n = a_{n-1} + a_{n-2}$ for $n \ge 3$ *(Fibonacci sequence)*

62. $a_1 = a_2 = 1;\ a_n = a_{n-1} - a_{n-2}$ for $n \ge 3$

Find a rule for a_n in terms of n.

63. $4, 5, 6, 7, \ldots$ **64.** $-2, -1, 0, 1, \ldots$

65. $3, 6, 9, 12, \ldots$ **66.** $-2, -4, -6, -8, \ldots$

67. $\tfrac{1}{2}, \tfrac{2}{3}, \tfrac{3}{4}, \tfrac{4}{5}, \ldots$ **68.** $\tfrac{1}{2}, \tfrac{3}{4}, \tfrac{5}{6}, \tfrac{7}{8}, \ldots$

69. $1, -1, 1, -1, \ldots$ **70.** $1, -2, 3, -4, \ldots$

71. $-2, 4, -8, 16, \ldots$ **72.** $1, -3, 5, -7, \ldots$

73. $x, \dfrac{x^2}{2}, \dfrac{x^3}{3}, \dfrac{x^4}{4}, \ldots$ **74.** $x, -x^3, x^5, -x^7, \ldots$

Write each series in expanded form without summation notation.

75. $\displaystyle\sum_{k=1}^{5} \dfrac{(-1)^k}{k}$ **76.** $\displaystyle\sum_{k=1}^{5} \dfrac{(-1)^{k+1}}{k+1}$

77. $\displaystyle\sum_{k=1}^{4} \dfrac{(-2)^{k+1}}{k}$ **78.** $\displaystyle\sum_{k=1}^{5} (-1)^{k+1}(2k-1)^2$

79. $\sum_{k=1}^{3} \dfrac{1}{k} x^{k+1}$

80. $\sum_{k=1}^{5} x^{k-1}$

81. $\sum_{k=1}^{5} \dfrac{(-1)^{k+1}}{k} x^k$

82. $\sum_{k=0}^{4} \dfrac{(-1)^k x^{2k+1}}{2k+1}$

Write each series using summation notation.

83. $S_4 = 1^3 + 2^3 + 3^3 + 4^3$

84. $S_6 = 0 + 1 + 2 + 3 + 4 + 5$

85. $S_4 = 1^2 + 2^2 + 3^2 + 4^2$

86. $S_5 = 2 + 3 + 4 + 5 + 6$

87. $S_5 = \dfrac{1}{2} + \dfrac{1}{2^2} + \dfrac{1}{2^3} + \dfrac{1}{2^4} + \dfrac{1}{2^5}$

88. $S_4 = 1 - \frac{1}{2} + \frac{1}{3} - \frac{1}{4}$

89. $S_n = 1 + \dfrac{1}{2^2} + \dfrac{1}{3^2} + \cdots + \dfrac{1}{n^2}$

90. $S_n = 2 + \dfrac{3}{2} + \dfrac{4}{3} + \cdots + \dfrac{n+1}{n}$

91. $S_n = 1 - 4 + 9 + \cdots + (-1)^{n+1} n^2$

92. $S_n = \dfrac{1}{2} - \dfrac{1}{4} + \dfrac{1}{8} + \cdots + \dfrac{(-1)^{n+1}}{2^n}$

C *In Problems 93–96, find the first four terms of the sequence.*

93. $a_n = -9n^3 + 63n^2 - 132n + 80$; compare to Example 3(B)

94. $a_n = (\frac{1}{16})(-9n^3 + 72n^2 - 177n + 130)$; compare to Matched Problem 3(B)

95. $a_n = \dfrac{2^n}{48}(-n^3 + 15n^2 - 86n + 192)$; compare to Example 3(A)

96. $a_n = \dfrac{1}{2^{n+5}}(n^3 - 8n^2 + 9n + 46)$; compare to Matched Problem 3(A)

The sequence with a_1 chosen arbitrarily from the positive real numbers and

$$a_n = \dfrac{(a_{n-1})^2 + p}{2a_{n-1}} \quad \text{for } n \geq 2, \, p \text{ a positive number}$$

can be used to find $\sqrt{p}$ to any decimal accuracy desired. This is illustrated in Problems 97 and 98.

97. **(A)** Find the first four terms of the sequence to three decimal places:

$$a_1 = 3 \qquad a_n = \dfrac{a_{n-1}^2 + 2}{2a_{n-1}} \quad \text{for } n \geq 2$$

(A small calculator will be useful, but not necessary.)

(B) Compare the terms found in part (A) with the decimal approximation of $\sqrt{2}$ from a table or calculator.

(C) Repeat parts (A) and (B), letting a_1 be any other positive number, say 1.

98. **(A)** Find the first four terms of the sequence to three decimal places:

$$a_1 = 2 \qquad a_n = \dfrac{a_{n-1}^2 + 5}{2a_{n-1}} \quad \text{for } n \geq 2$$

(B) Find $\sqrt{5}$ in a table or by a calculator and compare with part (A).

(C) Repeat parts (A) and (B), letting a_1 be any other positive number, say 3.

99. Find the first three terms of the sequence

$$a_n = \dfrac{1}{\sqrt{5}}\left(\dfrac{1+\sqrt{5}}{2}\right)^n - \dfrac{1}{\sqrt{5}}\left(\dfrac{1-\sqrt{5}}{2}\right)^n$$

Compare these terms to the first three terms of the sequence in Problem 61. This rule will, in fact, give the nth term of the Fibonacci sequence. See also Problem 100.

100. Fibonacci, also known as Leonardo of Pisa, was a thirteenth century Italian mathematician who played a major role in introducing Arabic mathematics, including the familiar Hindu–Arabic numerals we now use, to Europe. His book *Liber abaci* (1202) included a problem that can be paraphrased as follows:

A single pair of newborn rabbits is put into an enclosure at the beginning of January. A pair of rabbits breeds a new pair at the beginning of the second month following birth and another pair at the beginning of each month after that. How many pairs of rabbits are there at the beginning of each month, beginning with January?

Find the sequence of numbers

$a_n =$ Number of rabbits at the beginning of the nth month

Compare to Problem 61.

11-2 Arithmetic Sequences and Series

♦ Arithmetic Sequences
♦ Arithmetic Series

The sequence

$$5, 9, 13, 17, \ldots$$

has the property that each term after the first can be obtained from the preceding one by adding 4 to it. This is an example of an *arithmetic sequence.*

♦ **ARITHMETIC SEQUENCES**

In general, a sequence

$$a_1, a_2, a_3, \ldots, a_n, \ldots$$

is called an **arithmetic sequence** (or **arithmetic progression**) if there exists a constant d, called the **common difference,** such that each term is d more than the preceding term. That is,

> **Arithmetic Sequence**
>
> $$a_n = a_{n-1} + d \qquad \text{for every } n > 1$$

Example 1
Identifying Arithmetic Sequences

Determine whether the sequence is an arithmetic sequence. If it is, find its common difference.

(A) $1, 2, 3, 5, \ldots$ **(B)** $3, 5, 7, 9, \ldots$

Solution

(A) The second and third terms are each 1 more than the preceding term. However, the fourth term, 5, is 2 more than the preceding term. This is not an arithmetic sequence.

(B) This is an arithmetic sequence with $d = 2$.

Matched Problem 1

Determine whether the sequence is an arithmetic sequence. If it is, find its common difference.

(A) $-4, -1, 2, 5, \ldots$ **(B)** $2, 4, 8, 16, \ldots$

Arithmetic sequences have several convenient properties. For example, we can derive formulas for the nth term in terms of n and the sum of any number of consecutive terms. To obtain an nth-term formula, we note that if a is an arithmetic sequence, then

$$a_2 = a_1 + d$$

$$a_3 = a_2 + d = a_1 + 2d$$

$$a_4 = a_3 + d = a_1 + 3d$$

This suggests the following:

The nth-Term Formula for an Arithmetic Sequence

$$a_n = a_1 + (n - 1)d \qquad \text{for every } n > 1$$

Example 2
Applying the nth-Term Formula

If the 1st and 10th terms of an arithmetic sequence are 3 and 30, respectively, find the 50th term of the sequence.

Solution First find d:

$$a_n = a_1 + (n - 1)d$$
$$a_{10} = a_1 + (10 - 1)d$$
$$30 = 3 + 9d$$
$$d = 3$$

Now find a_{50}:

$$a_{50} = 3 + (50 - 1)3 \qquad \text{Use } a_n = a_1 + (n - 1)d.$$
$$= 3 + 49 \cdot 3$$
$$= 150$$

Matched Problem 2   If the 1st and 15th terms of an arithmetic sequence are -5 and 23, respectively, find the 73d term of the sequence.

♦   **ARITHMETIC SERIES**

The sum of the terms of an arithmetic sequence is called an **arithmetic series**. We can derive two useful formulas for finding the **sum of the first n terms of an arithmetic series**.

Consider first the sum

$$S = 1 + 2 + \cdots + (n - 1) + n$$

Rewrite the terms in reverse order without affecting S:

$$S = n + (n - 1) + \cdots + 2 + 1$$

Now add the corresponding sides of these two equations:

$$
\begin{aligned}
S &= \quad 1 \quad + \quad 2 \quad + \cdots + (n - 1) + \quad n \\
S &= \quad n \quad + (n - 1) + \cdots + \quad 2 \quad + \quad 1 \\
\hline
2S &= (n + 1) + (n + 1) + \cdots + (n + 1) + (n + 1)
\end{aligned}
$$

Thus,

$$2S = n(n + 1) \qquad \text{There are } n \text{ terms on the right.}$$

$$S = \frac{n(n + 1)}{2}$$

The same technique can be used to sum an arbitrary arithmetic series. Let

$$S_n = a_1 + (a_1 + d) + \cdots + [a_1 + (n - 2)d] + [a_1 + (n - 1)d]$$

Reversing the order of the sum, we obtain

$$S_n = [a_1 + (n - 1)d] + [a_1 + (n - 2)d] + \cdots + (a_1 + d) + a_1$$

Adding left-hand sides and corresponding elements of the right-hand sides of the two equations, we see that

$$2S_n = [2a_1 + (n - 1)d] + [2a_1 + (n - 1)d] + \cdots + [2a_1 + (n - 1)d]$$

$$= n[2a_1 + (n - 1)d]$$

Dividing by 2 gives us a formula for S_n:

Sum Formula for an Arithmetic Series—Form 1

$$S_n = \frac{n}{2}[2a_1 + (n - 1)d] \qquad \text{Use this form when given the number of terms } n\text{, the first term } a_1\text{, and the common difference } d.$$

By replacing $a_1 + (n - 1)d$ with a_n, we obtain a second useful formula for the sum:

Sum Formula for an Arithmetic Series—Form 2

$$S_n = \frac{n}{2}(a_1 + a_n) \qquad \text{Use this form when given the number of terms } n\text{, the first term } a_1\text{, and the last term } a_n.$$

Example 3
Applying the Sum Formula—Form 1

Find the sum of the first 26 terms of an arithmetic series if the first term is -7 and $d = 3$.

Solution

$$S_n = \frac{n}{2}[2a_1 + (n - 1)d] \qquad \text{Use sum formula—form 1.}$$

$$S_{26} = \tfrac{26}{2}[2(-7) + (26 - 1)3]$$

$$= 793$$

Matched Problem 3 Find the sum of the first 52 terms of an arithmetic series if the first term is 23 and $d = -2$.

Example 4
Applying the Sum Formula—Form 2

Find the sum of all the odd numbers from 51 to 99, inclusive.

Solution First find n:

$$a_n = a_1 + (n - 1)d$$
$$99 = 51 + (n - 1)2$$
$$n = 25$$

Now find S_{25}:

$$S_n = \frac{n}{2}(a_1 + a_n) \qquad \text{Use sum formula—form 2.}$$
$$S_{25} = \tfrac{25}{2}(51 + 99)$$
$$= 1{,}875$$

Matched Problem 4 Find the sum of all the even numbers from -22 to 52, inclusive.

Answers to Matched Problems
1. **(A)** Arithmetic sequence with $d = 3$ **(B)** Not an arithmetic sequence
2. $a_{73} = 139$ 3. $S_{52} = -1{,}456$ 4. 570

EXERCISE 11-2

A *Determine whether the sequence is an arithmetic sequence. If it is, find its common difference.*

1. $2, 4, 8, \ldots$

2. $7, 6.5, 6, \ldots$

3. $-11, -16, -21, \ldots$

4. $\tfrac{1}{2}, \tfrac{1}{6}, \tfrac{1}{18}, \ldots$

5. $5, -1, -7, \ldots$

6. $12, 4, \tfrac{4}{3}, \ldots$

7. $\tfrac{1}{2}, \tfrac{2}{3}, \tfrac{3}{4}, \ldots$

8. $16, 48, 80, \ldots$

Let $a_1, a_2, a_3, \ldots, a_n, \ldots$ be an arithmetic sequence. In Problems 9–26, find the indicated quantities.

9. $a_1 = -5, d = 4; a_2 = ?, a_3 = ?, a_4 = ?$

10. $a_1 = -18, d = 3; a_2 = ?, a_3 = ?, a_4 = ?$

11. $a_1 = -3, d = 5; a_{15} = ?, S_{11} = ?$

12. $a_1 = 3, d = 4; a_{22} = ?, S_{21} = ?$

13. $a_1 = 1, a_2 = 5; S_{21} = ?$

14. $a_1 = 5, a_2 = 11; S_{11} = ?$

B 15. $a_1 = 7, a_2 = 5; a_{15} = ?$

16. $a_1 = -3, d = -4; a_{10} = ?$

17. $a_1 = 3, a_{20} = 117; d = ?, a_{101}?$

18. $a_1 = 7, a_8 = 28; d = ?, a_{25} = ?$

19. $a_1 = -12, a_{40} = 22; S_{40} = ?$

20. $a_1 = 24, a_{24} = -28; S_{24} = ?$

21. $a_1 = \tfrac{1}{3}, a_2 = \tfrac{1}{2}; a_{11} = ?, S_{11} = ?$

22. $a_1 = \tfrac{1}{6}, a_2 = \tfrac{1}{4}, a_{19} = ?, S_{19} = ?$

23. $a_3 = 13, a_{10} = 55; a_1 = ?$

24. $a_9 = -12, a_{13} = 3; a_1 = ?$

25. $S_{51} = \displaystyle\sum_{k=1}^{51}(3k + 3) = ?$

26. $S_{40} = \displaystyle\sum_{k=1}^{40}(2k - 3) = ?$

In Problems 27–38, sum the arithmetic series.

27. $1 + 2 + 3 + 4 + \cdots + 100$

28. $1 + 2 + 3 + 4 + \cdots + 150$

29. $2 + 4 + 6 + 8 + \cdots + 160$

30. $2 + 4 + 6 + 8 + \cdots + 100$

31. $1 + 3 + 5 + 7 + \cdots + 151$

32. $1 + 3 + 5 + 7 + \cdots + 101$

33. $3 + 6 + 9 + 12 + \cdots + 180$

34. $3 + 6 + 9 + 12 + \cdots + 210$

C **35.** $22 + 24 + 26 + 28 + \cdots + 134$

36. $101 + 103 + 105 + 107 + \cdots + 499$

37. $81 + 84 + 87 + 90 + \cdots + 360$

38. $80 + 84 + 88 + 92 + \cdots + 360$

39. Find $g(1) + g(2) + g(3) + \cdots + g(51)$ if $g(t) = 5 - t$.

40. Find $f(1) + f(2) + f(3) + \cdots + f(20)$ if $f(x) = 2x - 5$.

41. Find $g(1) + g(2) + g(3) + \cdots + g(100)$ if $g(x) = 4 - 3x$.

42. Find $f(1) + f(2) + f(3) + \cdots + f(100)$ if $f(x) = 5 + 3x$

In Problems 43–46, for the given a and b, find x so that a, x, b is an arithmetic sequence.

43. $a = 10$, $b = 60$ **44.** $a = 19$, $b = 85$

45. $a = 11$, $b = 58$ **46.** $a = 10$, $b = 55$

47. Find the sum of all the even integers between 21 and 135.

48. Find the sum of all the odd integers between 100 and 500.

49. Find the sum of all multiples of 3 between 80 and 362.

50. Find the sum of all multiples of 4 between 79 and 361.

51. Show that the sum of the first n odd natural numbers is n^2, using appropriate formulas from this section.

52. Show that the sum of the first n even natural numbers is $n + n^2$, using appropriate formulas from this section.

53. Show that the sum of the first n multiples of 3 is $\frac{3}{2}[n(n + 1)]$, using appropriate formulas from this section.

54. Show that the sum of the first n multiples of 4 is $2n(n + 1)$, using appropriate formulas from this section.

APPLICATIONS

55. *Physics*   An object falling from rest in a vacuum near the surface of the earth falls 16 feet during the 1st second, 48 feet during the 2d second, 80 feet during the 3d second, and so on.
  (A) How far will the object fall during the 11th second?
  (B) How far will the object fall in 11 seconds?
  (C) How far will the object fall in t seconds?

56. *Business*   In investigating different job opportunities, you find that firm A will start you at $10,000 per year and guarantee you a raise of $500 each year, while firm B will start you at $13,000 per year but will only guarantee you a raise of $200 each year. Over a 15-year period, which firm will pay the greatest total amount?

57. *Graphing*   Find the first 5 terms of the sequence

$$a_n = n^2 - (n + 1)^2$$

How are these terms related to the differences between successive y values in the following table?

x	$y = x^2$
0	0
1	1
2	4
3	9
4	16
5	25

This problem suggests that in graphing $y = x^2$, we can plot the vertex $(0, 0)$ and then move up successively 1, 3, 5, 7, and 9 units as we move 1 unit after another to the right. We can then plot the left half of the parabola by symmetry. Use this technique to graph $y = x^2$. How many units up will you move in going from $x = 5$ to $x = 6$?

58. *Graphing*   The technique outlined in Problem 57 can be applied to any quadratic equation $y = x^2 + bx + c$. Plot the vertex and then move up successively 1, 3, 5, 7, . . . units for each unit moved to the right; then plot the left half of the parabola by symmetry. Use this technique to graph $y = x^2 + 4x + 6$.

11-3 Geometric Sequences and Series

♦ Geometric Sequences
♦ Geometric Series
♦ Infinite Geometric Series

In the sequence

$$2, -4, 8, -16, \ldots$$

each term after the first can be obtained from the preceding term by multiplying it by -2. This is an example of a *geometric sequence.*

♦ **GEOMETRIC SEQUENCES**

In general, a sequence

$$a_1, a_2, a_3, \ldots, a_n, \ldots$$

is called a **geometric sequence** (or **geometric progression**) if there exists a non-zero constant r, called the **common ratio,** such that each term is r times the preceding term. That is:

> **Geometric Sequence**
>
> $$a_n = ra_{n-1} \qquad \text{for every } n > 1$$

Example 1
Identifying Geometric Sequences

Determine whether the sequence is a geometric sequence. If it is, find the common ratio r.

(A) $2, 6, 8, 10, \ldots$ **(B)** $-1, 3, -9, 27, \ldots$

Solution **(A)** The ratios between successive terms are $\frac{6}{2} = 3$, $\frac{8}{6} = \frac{4}{3}, \ldots$. Since the ratio is not constant, this is not a geometric sequence.
(B) This is a geometric sequence with $r = -3$.

Matched Problem 1

Determine whether the sequence is a geometric sequence. If it is, find the common ratio r.

(A) $\frac{1}{4}, \frac{1}{2}, 1, 2, \ldots$ **(B)** $\frac{1}{2}, \frac{1}{4}, \frac{1}{16}, \frac{1}{256}, \ldots$

Just as with arithmetic sequences, geometric sequences have several convenient properties. There are formulas for the nth term in terms of n and the sum of any number of consecutive terms. To obtain an nth-term formula, we note that if a is a geometric sequence, then

$$a_2 = ra_1$$

$$a_3 = ra_2 = r^2 a_1$$

$$a_4 = ra_3 = r^3 a_1$$

This suggests the following:

The nth-Term Formula for a Geometric Sequence

$$a_n = a_1 r^{n-1} \qquad \text{for every } n > 1$$

Example 2
Applying the nth-Term Formula

Find the seventh term of the geometric sequence $1, \frac{1}{2}, \frac{1}{4}, \ldots$.

Solution

$$r = \tfrac{1}{2}$$
$$a_n = a_1 r^{n-1}$$
$$a_7 = 1\left(\frac{1}{2}\right)^{7-1} = \frac{1}{2^6} = \frac{1}{64}$$

Matched Problem 2   Find the eighth term of the geometric sequence $\frac{1}{64}, -\frac{1}{32}, \frac{1}{16}, \ldots$.

Example 3
Applying the nth-Term Formula

If the first and tenth terms of a geometric sequence are 1 and 2, respectively, find the common ratio r.

Solution

$$a_n = a_1 r^{n-1}$$
$$2 = 1r^{10-1}$$
$$r = 2^{1/9} \approx 1.08 \qquad \text{Calculation by calculator.}$$

Matched Problem 3   If the first and eighth terms of a geometric sequence are 2 and 16, respectively, find the common ratio r.

◆ GEOMETRIC SERIES

A **geometric series** is any series whose terms form a geometric sequence. As was the case with an arithmetic series, we can derive two very useful formulas for finding the **sum of the first n terms of a geometric series.**

For example, consider the geometric series

$$S = 1 + 3 + 9 + 27 + 81 + 243$$

Multiplying both sides by the common ratio 3, we obtain

$$3S = 3 + 9 + 27 + 81 + 243 + 729$$

Subtracting the first sum from the second,

$$2S = 729 - 1$$

$$3S - S = (3 + 9 + 27 + 81 + 243 + 729)$$
$$- (1 + 3 + 9 + 27 + 81 + 243) = 729 - 1$$

$$S = \frac{729 - 1}{2} = 364$$

The same technique works for geometric series in general. Let

$$S_n = a_1 + a_1 r + a_1 r^2 + a_1 r^3 + \cdots + a_1 r^{n-1}$$

and multiply both sides of the equation by r, to obtain

$$r S_n = a_1 r + a_1 r^2 + a_1 r^3 + \cdots + a_1 r^{n-1} + a_1 r^n$$

Now subtract the left side of the second equation from the left side of the first, and the right side of second equation from the right side of the first, to obtain

$$S_n - r S_n = a_1 - a_1 r^n \qquad \text{Note how many terms dropped out on the right side.}$$

$$S_n (1 - r) = a_1 - a_1 r^n$$

Dividing by $1 - r$ yields the following formula:

Sum Formula for a Geometric Series—Form 1

$$S_n = \frac{a_1 - a_1 r^n}{1 - r} = \frac{a_1(1 - r^n)}{1 - r} \qquad r \neq 1 \qquad \text{Use this form when given the number of terms } n, \text{ the first term } a_1, \text{ and the common ratio } r.$$

Since $a_n = a_1 r^{n-1}$, or $r a_n = a_1 r^n$, the sum formula also can be written as:

Sum Formula for a Geometric Series—Form 2

$$S_n = \frac{a_1 - r a_n}{1 - r} \qquad r \neq 1 \qquad \text{Use this form when given the common ratio } r, \text{ the first term } a_1, \text{ and the last term } a_n.$$

Example 4
Applying the Sum Formula

Find the sum (to three significant digits) of the first 20 terms of a geometric series if the first term is 1 and $r = 2$.

Solution

$$S_n = \frac{a_1 - a_1 r^n}{1 - r}$$ Use sum formula—form 1.

$$= \frac{1 - 1 \cdot 2^{20}}{1 - 2} \approx 1{,}050{,}000$$ Calculation by calculator.

Matched Problem 4 Find the sum (to four significant digits) of the first 14 terms of a geometric series if the first term is $\frac{1}{64}$ and $r = -2$.

♦ **INFINITE GEOMETRIC SERIES**

Given a geometric series, what happens to the sum S_n as n increases? To answer this question, we first write the sum formula in the more convenient form

$$S_n = \frac{a_1 - a_1 r^n}{1 - r} = \frac{a_1}{1 - r} - \frac{a_1 r^n}{1 - r}$$

It is possible to show that if $|r| < 1$, that is, if $-1 < r < 1$, then r^n will approach 0 as n increases. For example, if $r = \frac{1}{10}$, then

$$r^n = 0.\underbrace{000\ 000\ 000\ \ldots\ 01}_{n-1 \text{ zeros}}$$

which gets as close as we want to 0 by taking n large enough. Therefore, the term $a_1 r^n / (1 - r)$ will eventually be almost 0, and S_n can be made as close to $a_1 / (1 - r)$ as we wish. Thus, we have the following definition:

Sum of an Infinite Geometric Series with $-1 < r < 1$

$$S_\infty = \frac{a_1}{1 - r} \qquad |r| < 1$$

We call S_∞ the **sum of an infinite geometric series**. If $|r| \geq 1$, an infinite geometric series has no sum, since r^n can be made as large in absolute value as we like by making n sufficiently large. For example, what happens to r^n if $r = 3$ and n gets large?

Example 5
Summing an Infinite Geometric Series

Sum:

(A) $1 + \frac{1}{2} + \frac{1}{4} + \cdots$ (B) $1 - \frac{1}{2} + \frac{1}{4} - \frac{1}{8} + \cdots$

Solution **(A)** This is an infinite geometric series with $a_1 = 1$ and $r = \frac{1}{2}$. Therefore,

$$S_\infty = \frac{1}{1 - \frac{1}{2}} = \frac{1}{\frac{1}{2}} = 2$$

(B) This is an infinite geometric series with $a_1 = 1$ and $r = -\frac{1}{2}$. Therefore,

$$S_\infty = \frac{1}{1 + \frac{1}{2}} = \frac{1}{\frac{3}{2}} = \frac{2}{3}$$

Matched Problem 5 Sum: **(A)** $1 - \frac{1}{3} + \frac{1}{9} - \frac{1}{27} + \cdots$ **(B)** $1 + \frac{1}{3} + \frac{1}{9} + \frac{1}{27} + \cdots$

Example 6
Repeating Decimals as Infinite Geometric Series

Represent each repeating decimal as the quotient of two integers. Here, the bar over the last two digits indicates that these digits repeat indefinitely.

(A) $0.45\overline{45}$ **(B)** $0.3\overline{45}45$

Solution **(A)** $0.45\overline{45} = 0.45 + 0.0045 + 0.000045 + \cdots$
The right side of the equation is an infinite geometric series with $a_1 = 0.45$ and $r = 0.01$. Thus,

$$S_\infty = \frac{a_1}{1 - r} = \frac{0.45}{1 - 0.01} = \frac{0.45}{0.99} = \frac{5}{11}$$

Hence, $0.45\overline{45}$ and $\frac{5}{11}$ name the same rational number. Check the result by dividing 5 by 11.
(B) $0.3\overline{45}45 = 0.3 + 0.045 + 0.00045 + \cdots$
The terms following 0.3 on the right side of the equation represent an infinite geometric series with $a_1 = 0.045$ and $r = 0.01$. The sum of these terms is, therefore,

$$\frac{a_1}{1 - r} = \frac{0.045}{0.99} = \frac{45}{990} = \frac{5}{110}$$

Thus,

$$0.3\overline{45}45 = 0.3 + \frac{5}{110} = \frac{38}{110} = \frac{19}{55}$$

Matched Problem 6 Represent each repeating decimal as the quotient of two integers.

(A) $0.81\overline{81}81$ **(B)** $1.28\overline{18}181$

Answers to Matched Problems
1. **(A)** Geometric sequence with $r = 2$ **(B)** Not a geometric sequence
2. -2 3. $r \approx 1.346$ 4. -85.33
5. **(A)** $\frac{3}{4}$ **(B)** $\frac{3}{2}$ 6. **(A)** $\frac{9}{11}$ **(B)** $\frac{141}{110}$

EXERCISE 11-3

A *Determine whether each sequence is a geometric sequence. If it is, find the common ratio r.*

1. $2, -4, 8, \ldots$
2. $7, 6.5, 6, \ldots$
3. $-11, -16, -21, \ldots$
4. $\frac{1}{2}, \frac{1}{6}, \frac{1}{18}, \ldots$
5. $5, -1, -7, \ldots$
6. $12, 4, \frac{4}{3}, \ldots$
7. $\frac{1}{2}, \frac{2}{3}, \frac{3}{4}, \ldots$
8. $16, 48, 80, \ldots$

Let $a_1, a_2, a_3, \ldots, a_n, \ldots$ be a geometric sequence. Find each of the indicated quantities.

9. $a_1 = -6, r = -\frac{1}{2}; a_2 = ?, a_3 = ?, a_4 = ?$
10. $a_1 = 12, r = \frac{2}{3}, a_2 = ?, a_3 = ?, a_4 = ?$
11. $a_1 = 81, r = \frac{1}{3}; a_{10} = ?$
12. $a_1 = 64, r = \frac{1}{2}; a_{13} = ?$
13. $a_1 = 3, a_7 = 2{,}187, r = 3; S_7 = ?$
14. $a_1 = 1, a_7 = 729, r = -3; S_7 = ?$

B 15. $a_1 = 100, a_6 = 1; r = ?$
16. $a_1 = 10, a_{10} = 30; r = ?$
17. $a_1 = 5, r = -2; S_{10} = ?$
18. $a_1 = 3, r = 2; S_{10} = ?$
19. $a_1 = 9, a_4 = \frac{8}{3}; a_2 = ?, a_3 = ?$
20. $a_1 = 12, a_4 = -\frac{4}{9}; a_2 = ?, a_3 = ?$
21. $S_7 = \displaystyle\sum_{k=1}^{7} (-3)^{k-1} = ?$
22. $S_7 = \displaystyle\sum_{k=1}^{7} 3^k = ?$

Sum each geometric series.

23. $2 + 6 + 18 + \cdots + 162$
24. $2 + 6 + 18 + \cdots + 4{,}374$
25. $1 - 5 + 25 - \cdots + 15{,}625$
26. $1 - 5 + 25 - \cdots + 625$
27. $\frac{1}{4} + 1 + 4 + \cdots + 1{,}024$
28. $4 + \frac{1}{2} + \frac{1}{16} + \cdots + \frac{1}{1{,}024}$
29. $1 + \frac{1}{3} + \frac{1}{9} + \cdots + \frac{1}{243}$
30. $\frac{1}{5} + 2 + 20 + \cdots + 20{,}000$
31. $1 + 1.1 + 1.21 + \cdots + 2.14358881$
32. $2 + 2.1 + 0.21 + \cdots + 0.0000021$

Find the sum of each infinite geometric series that has a sum.

33. $3 + 1 + \frac{1}{3} + \cdots$
34. $16 + 4 + 1 + \cdots$
35. $2 + 4 + 8 + \cdots$
36. $4 + 6 + 9 + \cdots$
37. $2 - \frac{1}{2} + \frac{1}{8} - \cdots$
38. $21 - 3 + \frac{3}{7} - \cdots$
39. $1 + \frac{1}{4} + \frac{1}{16} + \cdots$
40. $2 + \frac{2}{5} + \frac{2}{25} + \cdots$
41. $2 + \frac{5}{2} + \frac{25}{8} + \cdots$
42. $1 + 4 + 16 + \cdots$

Find x between a and b so that a, x, b is a geometric sequence.

43. $a = 10, b = 12.1$
44. $a = 4, b = 100$
45. $a = 8, b = 12$
46. $a = 6, b = 10$
47. $a = 11, b = 12$
48. $a = 6, b = 7$

C *In Problems 49–58, represent each repeating decimal fraction as the quotient of two integers.*

49. $0.77\overline{77}$
50. $0.55\overline{55}$
51. $0.54\overline{5454}$
52. $0.27\overline{2727}$
53. $3.216\overline{216216}$
54. $5.63\overline{6363}$
55. $0.384615\overline{384615}$
56. $0.06\overline{6}$
57. $0.05\overline{5}$
58. $2.142857\overline{142857}$

59. Write $0.999\overline{9}$ as a geometric series and sum it to show why $0.999999 \ldots = 1$.

60. Write $0.4999\overline{9}$ as a geometric series and sum it to show why $0.4999999 \ldots = \frac{1}{2}$.

APPLICATIONS

★ 61. *Business* If P dollars is invested at $r\%$ compounded annually, the amount A present after n years forms a geometric sequence with a constant ratio $(1 + r)$. Write a formula for the amount present after n years. How long will it take for a sum of money P to double if it is invested at 6% interest compounded annually?

★ 62. *Business* Repeat Problem 61 for an interest rate of 8%.

★★ 63. *Business* An *annuity* is a sequence of regular payments. Suppose each payment is the same amount P and earns interest at rate r until the next payment, and for each payment period thereafter. If this continues for $n + 1$ payments, then the first payment is worth $P(1 + r)^n$ since it earns interest for n periods. The second payment is worth $P(1 + r)^{n-1}$ since it earns interest for 1 less period, etc. The last payment is worth just P since it has not yet earned any interest. The total value of the $n + 1$ payments is, therefore,

$$P(1 + r)^n + P(1 + r)^{n-1} + \cdots + P(1 + r) + P$$

Find a formula for this sum.

** **64.** ***Business*** Refer to Problem 63. On the first of each month, beginning January 1, a person puts $100 into an account that earns 0.5% interest per month. How much is in the account immediately after the December 1 deposit? How much is in the account at the end of December?

65. ***Engineering*** A rotating flywheel coming to rest rotates 300 revolutions the first minute. If in each subsequent minute it rotates two-thirds as many times as in the preceding minute, how many revolutions will the wheel make before coming to rest?

66. ***Engineering*** Refer to Problem 65. If in each subsequent minute the flywheel rotates only one-half as many times as in the preceding minute, how many revolutions will it make before coming to rest?

67. ***Physics*** A bouncing ball regains 70% of its previous height on each bounce. If the ball is dropped from a height of 5 feet, how far will the ball travel vertically before coming to rest?

68. ***Physics*** The first swing of a bob on a pendulum is 10 inches. If on each subsequent swing it travels 0.9

as far as on the preceding swing, how far will the bob travel before coming to rest?

69. ***Economics*** The government, through a subsidy program, distributes $1,000,000. If we assume that each individual or agency spends 0.8 of what is received, and 0.8 of this is spent, and so on, how much total increase in spending results from this government action? (Let $a_1 = \$800,000$.)

★ **70.** ***Zeno's paradox*** Visualize a hypothetical 440-yard oval racetrack that has tapes stretched across the track at the halfway point and at each point that marks the halfway point of each remaining distance thereafter. A runner running around the track has to break the first tape before the second, the second before the third, and so on. From this point of view it appears that the runner will never finish the race. (This famous paradox is attributed to the Greek philosopher Zeno, 495–435 B.C.) If we assume the runner runs at 440 yards per minute, the times between tape breakings form an infinite geometric sequence. What is the sum of this sequence?

11-4 **Binomial Formula**

♦ Factorial
♦ Binomial Formula

The binomial form

$$(a + b)^n$$

with n a natural number, appears in many mathematical contexts. The coefficients in its expansion play an important role in probability studies. In this section we will give an informal derivation of the famous *binomial formula,* which will enable us to expand $(a + b)^n$ directly for any natural number n, however large. First, we introduce the concept of *factorial,* which is useful in stating the formula.

♦ **FACTORIAL**

For n a natural number, **n factorial,** denoted by **$n!$**, is the product of the first n natural numbers. **Zero factorial** is defined to be 1. Symbolically:

---

n Factorial

$$0! = 1$$
$$n! = n \cdot (n - 1) \cdot \cdots \cdot 2 \cdot 1$$

$1! = 1$

$2! = 2 \cdot 1 = 2$

$3! = 3 \cdot 2 \cdot 1 = 6$

$4! = 4 \cdot 3 \cdot 2 \cdot 1 = 24$

$5! = 5 \cdot 4 \cdot 3 \cdot 2 \cdot 1 = 120$

It is also useful to note that

$$n! = n \cdot (n - 1)! \qquad 8! = 8 \cdot 7!$$

Many scientific calculators have a factorial key for evaluating the products involved.

Example 1
Evaluating *n*!

Evaluate:

(A) 4!   **(B)** 6!   **(C)** $\dfrac{7!}{6!}$ **(D)** $\dfrac{8!}{5!}$

Solution

(A) $4! = 4 \cdot 3 \cdot 2 \cdot 1 = 24$

(B) $6! = 6 \cdot 5 \cdot 4 \cdot 3 \cdot 2 \cdot 1 = 720$

(C) $\dfrac{7!}{6!} = \dfrac{7 \cdot 6!}{6!} = 7$

(D) $\dfrac{8!}{5!} = \dfrac{8 \cdot 7 \cdot 6 \cdot 5!}{5!} = 336$

Matched Problem 1

Evaluate: **(A)** 7!   **(B)** $\dfrac{8!}{7!}$ **(C)** $\dfrac{6!}{3!}$

A form involving factorials that is very useful is given by the following notation:

$$\binom{n}{r}$$

For integers r and n, with $0 \le r \le n$,

$$\binom{n}{r} = \dfrac{n!}{(n - r)!\,r!}$$

Note: In other contexts, $\dbinom{n}{r}$ is also represented by C_r^n, $_nC_r$, and $C_{n,r}$.

Example 2

Evaluating $\dbinom{n}{r}$

Find:

(A) $\dbinom{5}{2}$ (B) $\dbinom{4}{4}$

Solution   (A) $\dbinom{5}{2} = \dfrac{5!}{(5-2)!2!} = \dfrac{5!}{3!2!} = 10$

(B) $\dbinom{4}{4} = \dfrac{4!}{(4-4)!4!} = \dfrac{4!}{0!4!} = 1$

Matched Problem 2   Find: (A) $\dbinom{9}{2}$ (B) $\dbinom{5}{5}$

If we begin to compute the values of $\dbinom{n}{r}$ for each positive integer n, we obtain the following values:

$$\binom{0}{0} = 1$$

$$\binom{1}{0} = 1 \qquad \binom{1}{1} = 1$$

$$\binom{2}{0} = 1 \qquad \binom{2}{1} = 2 \qquad \binom{2}{2} = 1$$

$$\binom{3}{0} = 1 \qquad \binom{3}{1} = 3 \qquad \binom{3}{2} = 3 \qquad \binom{3}{3} = 1$$

$$\binom{4}{0} = 1 \qquad \binom{4}{1} = 4 \qquad \binom{4}{2} = 6 \qquad \binom{4}{3} = 4 \qquad \binom{4}{4} = 1$$

If we just consider the computed values, they appear as

```
1                              This triangle is often displayed as      1
1   1                                                                  1 1
1   2   1                                                             1 2 1
1   3   3   1                                                        1 3 3 1
1   4   6   4   1                                                   1 4 6 4 1
```

This arrangement is called **Pascal's triangle.** The numbers for subsequent rows can be obtained from the following rule: each row begins and ends in a 1; every other entry is the sum of the entry directly above it and the entry to the left of the one directly above it. For example, the last line in Pascal's triangle above is obtained from the line above it as follows:

```
1   3   3   1
1   4   6   4   1
```

Following this procedure, the next row in the triangle would be given by:

$$\begin{array}{ccccccccc} 1 & & 4 & & 6 & & 4 & & 1 \\ & \downarrow & & \downarrow & & \downarrow & & \downarrow & \\ 1 & & 5 & & 10 & & 10 & & 5 & & 1 \end{array}$$

In general, for $0 < r < n$, the rule is

$$\binom{n}{r} = \binom{n-1}{r-1} + \binom{n-1}{r}$$

You are asked to show that this is true in Problem 58, Exercise 11-4.

♦ ## BINOMIAL FORMULA

Now let us try to discover a formula for the expansion of $(a + b)^n$, for n a natural number. We begin by expanding $(a + b)^n$ for small values of n:

$$(a + b)^1 = a + b$$

$$(a + b)^2 = a^2 + 2ab + b^2$$

$$(a + b)^3 = a^3 + 3a^2b + 3ab^2 + b^3$$

$$(a + b)^4 = a^4 + 4a^3b + 6a^2b^2 + 4ab^3 + b^4$$

$$(a + b)^5 = a^5 + 5a^4b + 10a^3b^2 + 10a^2b^3 + 5ab^4 + b^5$$

We make the following observations:

1. The expansion of $(a + b)^n$ has $n + 1$ terms.
2. The power of a starts at n and decreases by 1 for each term until it is 0 in the last term.
3. The power of b starts at 0 in the first term and increases by 1 for each term until it is n in the last term.
4. The sum of the powers of a and b in each term is the constant n.
5. The coefficients appear to be the same numbers as found above in Pascal's triangle. The coefficient of $a^k b^{n-k}$ corresponds to $\binom{n}{k}$.

Thus, it appears that:

Binomial Formula

$$(a + b)^n = \sum_{k=0}^{n} \binom{n}{k} a^{n-k} b^k \qquad n \geq 1$$

This result is known as the **binomial formula.** Its general proof requires a method called *mathematical induction,* which is considered in more advanced courses.

Example 3
Using the Binomial Formula

Use the binomial formula to expand $(x + y)^6$.

Solution

$$(x + y)^6 = \sum_{k=0}^{6} \binom{6}{k} x^{6-k} y^k$$

$$= \binom{6}{0} x^6 + \binom{6}{1} x^5 y + \binom{6}{2} x^4 y^2 + \binom{6}{3} x^3 y^3 + \binom{6}{4} x^2 y^4 + \binom{6}{5} x y^5 + \binom{6}{6} y^6$$

$$= x^6 + 6x^5 y + 15x^4 y^2 + 20x^3 y^3 + 15x^2 y^4 + 6xy^5 + y^6$$

Alternatively, we could have found the coefficients by determining the row in Pascal's triangle corresponding to $n = 6$:

$$
\begin{array}{ccccccccccccc}
 & & & & & & 1 & & & & & & \\
 & & & & & 1 & & 1 & & & & & \\
 & & & & 1 & & 2 & & 1 & & & & \\
 & & & 1 & & 3 & & 3 & & 1 & & & \\
 & & 1 & & 4 & & 6 & & 4 & & 1 & & \\
 & 1 & & 5 & & 10 & & 10 & & 5 & & 1 & \\
1 & & 6 & & 15 & & 20 & & 15 & & 6 & & 1
\end{array}
$$

Matched Problem 3

Use the binomial formula to expand $(x + 1)^5$.

Example 4
Using the Binomial Formula

Use the binomial formula to find the fourth term in the expansion of $(x - 2)^{20}$.

Solution

$$\text{Fourth term} = \binom{20}{3} x^{20-3} (-2)^3 \qquad \text{Since the first term is given when } k = 0, \text{ the fourth term is given when } k = 3.$$

$$= \frac{20!}{(20-3)!\,3!} x^{17} (-2)^3$$

$$= \frac{20 \cdot 19 \cdot 18}{3 \cdot 2 \cdot 1} x^{17} (-8)$$

$$= -9{,}120 x^{17}$$

Note that here it is not practical to expand Pascal's triangle to include the row for $n = 20$ just to find this coefficient.

Matched Problem 4

Use the binomial formula to find the fifth term in the expansion of $(u - 1)^{18}$.

EXERCISE 11-4

A *Evaluate.*

1. $6!$ **2.** $4!$ **3.** $\dfrac{20!}{19!}$

4. $\dfrac{5!}{4!}$ **5.** $\dfrac{10!}{7!}$ **6.** $\dfrac{9!}{6!}$

7. $\dfrac{6!}{4!2!}$ **8.** $\dfrac{5!}{2!3!}$ **9.** $\dfrac{9!}{0!(9-0)!}$

10. $\dfrac{8!}{8!(8-8)!}$ **11.** $\dfrac{8!}{2!(8-2)!}$ **12.** $\dfrac{7!}{3!(7!-3)!}$

Write as the quotient of two factorials.

13. 9 **14.** 12

15. $6 \cdot 7 \cdot 8$ **16.** $9 \cdot 10 \cdot 11 \cdot 12$

17. $20 \cdot 19$ **18.** $21 \cdot 20 \cdot 19$

19. $31 \cdot 30 \cdot 29 \cdot 28$ **20.** $40 \cdot 39 \cdot 38 \cdot 37 \cdot 36$

B *Evaluate.*

21. $\dbinom{9}{5}$ **22.** $\dbinom{5}{2}$ **23.** $\dbinom{6}{5}$ **24.** $\dbinom{7}{1}$

25. $\dbinom{9}{9}$ **26.** $\dbinom{5}{0}$ **27.** $\dbinom{17}{13}$ **28.** $\dbinom{20}{16}$

29. $\dbinom{36}{2}$ **30.** $\dbinom{80}{2}$ **31.** $\dbinom{32}{3}$ **32.** $\dbinom{50}{4}$

Expand using the binomial formula.

33. $(a + b)^7$ **34.** $(a + b)^8$

35. $(a + b)^{10}$ **36.** $(a + b)^9$

37. $(u + v)^5$ **38.** $(x + y)^4$

39. $(y - 1)^4$ **40.** $(x - 2)^5$

41. $(2x - y)^5$ **42.** $(m + 2n)^6$

Find the indicated term in each expansion.

43. $(a + b)^{16}$; 3d term   **44.** $(a + b)^{24}$; 4th term

45. $(a + b)^{35}$; 4th term   **46.** $(a + b)^{30}$; 3d term

47. $(u + v)^{15}$; 7th term   **48.** $(a + b)^{12}$; 5th term

49. $(2m + n)^{12}$; 11th term   **50.** $(x + 2y)^{20}$; 3d term

51. $[(w/2) - 2]^{12}$; 7th term   **52.** $(x - 3)^{10}$; 4th term

C *Powers of numbers can be evaluated using the binomial formula. For example, to evaluate 1.01^6 substitute $a = 1$ and $b = 0.01$ into*

$$(a + b)^6 = a^6 + 6a^5b + 15a^4b^2 + 20a^3b^3 + 15a^2b^4 + 6ab^5 + b^6$$

to obtain

$$
\begin{aligned}
(1 + 0.01)^6 &= 1 + 6(0.01) + 15(0.0001) + 20(0.000\ 001) \\
&\quad + 15(0.000\ 000\ 01) \\
&\quad + 6(0.000\ 000\ 000\ 1) \\
&\quad + 0.000\ 000\ 000\ 001 \\
&= 1 + 0.06 + 0.0015 + 0.000\ 020 \\
&\quad + 0.000\ 000\ 15 + 0.000\ 000\ 000\ 6 \\
&\quad + 0.000\ 000\ 000\ 001 \\
&= 1.061\ 520\ 150\ 601
\end{aligned}
$$

In Problems 53–56, use the binomial formula to evaluate each power.

53. 1.1^5 **54.** 0.9^5

55. 1.01^{10} **56.** 0.99^6

57. Show that: $\dbinom{n}{r} = \dbinom{n}{n-r}$

58. Show that: $\dbinom{n}{r} = \dbinom{n-1}{r-1} + \dbinom{n-1}{r}$

59. Use the binomial formula to expand $(1 + i)^6$, where i is the imaginary unit in the complex numbers.

60. Use the binomial formula to expand $(1 - i)^6$, where i is the imaginary unit in the complex numbers.

CHAPTER SUMMARY

11-1 SEQUENCES AND SERIES

An **infinite sequence** is a function whose domain is the set of all natural numbers. A **finite sequence** is a function whose domain is a set $\{1, 2, 3, \ldots, n\}$. The element associated with the natural number n is called the **nth term** of the sequence and is usually denoted with a subscript as a_n. The sum of the terms of a sequence is called a **series**—an **infinite series** if the sequence is infinite. **Summation notation** is useful for representing a series:

$$S_n = \sum_{k=1}^{n} a_k = a_1 + a_2 + \cdots + a_n$$

11-2 ARITHMETIC SEQUENCES AND SERIES

An **arithmetic sequence** has nth term $a_n = a_{n-1} + d = a_1 + (n-1)d$, where the constant d is called the **common difference.** The sum of the first n terms of an arithmetic sequence is called an **arithmetic series** and is given by the formula

$$S_n = \frac{n}{2}[2a_1 + (n-1)d] = \frac{n}{2}(a_1 + a_n)$$

11-3 GEOMETRIC SEQUENCES AND SERIES

A **geometric sequence** has nth term $a_n = ra_{n-1} = a_1 r^{n-1}$, where the constant $r \neq 0$ is called the **common ratio.** A **geometric series** is the sum of terms in a geometric sequence. For n terms, the series sum is

$$S_n = \frac{a_1 - a_1 r^n}{1 - r} = \frac{a_1(1 - r^n)}{1 - r} = \frac{a_1 - ra_n}{1 - r} \qquad r \neq 1$$

For an infinite series with $|r| < 1$, the sum is

$$S_\infty = \frac{a_1}{1 - r}$$

11-4 BINOMIAL FORMULA

The **binomial formula** is

$$(a + b)^n = \sum_{k=0}^{n} \binom{n}{k} a^{n-k} b^k$$

where

$$\binom{n}{k} = \frac{n!}{(n - k)!k!}$$

and $n! = n(n - 1) \cdots \cdot 2 \cdot 1$, $1! = 1$, and $0! = 1$. The number $n!$ is called n **factorial.** The coefficients $\binom{n}{k}$ also can be obtained from **Pascal's triangle:**

```
      1
     1   1
    1   2   1
   1   3   3   1
  1   4   6   4   1
        . . .
```

In this array, each row begins and ends in a 1; every other entry is the sum of the entry directly above it and the entry to the left of the one directly above it.

CHAPTER REVIEW EXERCISE

Work through all the problems in this chapter review and check answers in the back of the book. Answers to all the problems are there, and following each answer is a number in italics indicating the section in which that type of problem is discussed. Where weaknesses show up, review appropriate sections in the text.

A *Determine whether the sequence is an arithmetic or geometric sequence, or neither. If it is an arithmetic sequence, give the common difference. If it is a geometric sequence, give the common ratio.*

1. $16, -8, 4, \ldots$

2. $5, 7, 9, \ldots$

3. $-8, -5, -2, \ldots$

4. $2, 3, 5, 8, \ldots$

5. $-1, 2, -4, \ldots$

6. $1, 2, 4, 7, \ldots$

7. $10, 7, 4, 1, \ldots$

8. $3, 4, \frac{16}{3}, \frac{64}{9}, \ldots$

9. $1, 4, 9, 16, \ldots$

10. $10, 5, 2.5, 1.25, \ldots$

Write the first four terms of each sequence.

11. $a_n = 2n + 3$

12. $a_n = 32(\frac{1}{2})^n$

13. $a_n = n^2 - 4$

14. $a_n = n^2 + n$

15. $a_n = (-1)^n(n + 1)$

16. $a_n = \dfrac{(-1)^{n+1}}{n}$

17. $a_1 = -8; a_n = a_{n-1} + 3$ for $n \geq 2$

18. $a_1 = -1; a_n = (-2)a_{n-1}$ for $n \geq 2$

Find a_{10}.

19. For the sequence in Problem 11

20. For the sequence in Problem 12

21. For the sequence in Problem 13

22. For the sequence in Problem 14

23. For the sequence in Problem 15

24. For the sequence in Problem 16

25. For the sequence in Problem 17

26. For the sequence in Problem 18

Find S_{10}.

27. For the sequence in Problem 11

28. For the sequence in Problem 12

29. For the sequence in Problem 17

30. For the sequence in Problem 18

Find S_5.

31. For the sequence in Problem 13

32. For the sequence in Problem 14

33. For the sequence in Problem 15

34. For the sequence in Problem 16

Evaluate.

35. $6!$

36. $\dfrac{22!}{19!}$

37. $\dfrac{7!}{2!(7-2)!}$

38. $\dbinom{7}{2}$

39. $\dbinom{8}{6}$

40. $\dbinom{6}{5}$

B *Find the term, difference, or ratio requested.*

41. The 5th term in the arithmetic sequence with $a_1 = 3$ and $a_{11} = 28$

42. The 1st term in the arithmetic sequence with $a_3 = 3$ and $a_{103} = 153$

43. The 25th term in the arithmetic sequence with $a_1 = -30$ and $a_{51} = 70$

44. The 7th term in the geometric sequence with $a_1 = \frac{1}{2}$, $a_5 = 40.5$ and r positive

45. The 1st term in the geometric sequence with $a_2 = \sqrt{5}$, $a_6 = 25\sqrt{5}$ and r positive

46. The common ratio in the geometric sequence with $a_1 = 2$, $a_3 = 6$ and r positive

47. The common difference in the arithmetic sequence with $a_5 = \frac{3}{2}$ and $a_{15} = 4$

48. The common difference in the arithmetic sequence with $a_{80} - a_{60} = 400$

Rewrite without summation notation and find each sum.

49. $S_4 = \displaystyle\sum_{k=1}^{4} \frac{3}{2k}$

50. $S_4 = \displaystyle\sum_{k=1}^{4} \frac{(-1)^{k+1}}{2^k}$

51. $S_{10} = \displaystyle\sum_{k=1}^{10} (2k - 8)$

52. $S_7 = \displaystyle\sum_{k=1}^{7} \frac{16}{2^k}$

Find the sum of each geometric series.

53. $S_5 = \frac{1}{2} + \frac{7}{6} + \frac{49}{18} + \frac{343}{54} + \frac{2,401}{162}$

54. $S_{12} = 40 + 20 + 10 + 5 + \cdots + 40(\frac{1}{2})^{11}$

55. $S_\infty = 40 + 20 + 10 + \cdots$

56. $S_\infty = 27 - 18 + 12 - \cdots$

57. $S_\infty = \frac{1}{3} - \frac{1}{9} + \frac{1}{27} - \cdots$

Write in summation notation.

58. $\dfrac{1}{3} - \dfrac{1}{9} + \dfrac{1}{27} - \cdots + \dfrac{(-1)^{n+1}}{3^n}$

59. $10 + 7 + 4 + 1 + \cdots + (-38)$

60. $3 + 4 + \frac{16}{3} + \frac{64}{9}$

61. $1 + 4 + 9 + 16 + \cdots + 121$

62. $10 + 5 + 2.5 + 1.25 + \cdots + 0.3125$

Evaluate.

63. $\dfrac{20!}{18!(20 - 18)!}$

64. $\dbinom{16}{12}$

65. $\dbinom{11}{11}$

66. $\dbinom{20}{2}$

67. $\dbinom{15}{3}$

C *Expand using the binomial formula.*

68. $(x - y)^5$

69. $(x + 1)^7$

Find the term requested.

70. The 10th term in the expansion of $(2x - y)^{12}$

71. The term involving $a^{12}b^8$ in the expansion of $(a + b)^{20}$

In Problems 72 and 73, write each number as the quotient of two integers.

72. $0.727\overline{272}$

73. $1.234\overline{234}$

74. Find the complete row in Pascal's triangle that begins

$$1 \quad 8 \quad 28 \quad \cdots$$

75. A free-falling body travels $g/2$ feet in the first second, $3g/2$ during the next second, $5g/2$ feet the next, and so on, where g is the gravitational constant. Find the distance fallen during the 25th second and the total distance fallen from the start to the end of the 25th second. Express answers in terms of g.

CHAPTER PRACTICE TEST

The following practice test is provided for you to test your knowledge of the material in this chapter. You should try to complete it in 50 minutes or less. Answers in the back of the book indicate the section in the text where the material in the question is covered. Actual tests in your class may vary from this practice test in difficulty, length, or emphasis, depending on the goals of your course or instructor.

Find the sixth term in each sequence.

1. $a_n = (n - 1)n$

2. $a_1 = 6$, $a_n = 2a_{n-1} + 1$ for $n > 1$

3. $a_1 = 0$, $a_2 = 2$; $a_n = a_{n-1} + a_{n-2}$ for $n > 2$

4. The arithmetic sequence with first term 5 and common difference 3

5. The geometric sequence with first term 8 and common ratio $-\frac{1}{2}$

Determine whether the sequence is arithmetic, geometric, or neither. Find the common difference or ratio if one exists.

6. $3, -5, \frac{25}{3}, -\frac{125}{9}, \quad \ldots$

7. $3, 5, 3, 5, \quad \ldots$

Find the common difference or ratio for each sequence.

8. The arithmetic sequence with $a_4 = 1$ and $a_{15} = 23$

9. The geometric sequence with nth term $\dfrac{(-1)^{n+1}}{10^n}$

10. The arithmetic sequence with first term 3 and the sum of the first 50 terms equal to 250

In Problems 11–15, find the sum of each series.

11. $5 + 8 + 11 + \cdots + 38$

12. $1 + \frac{1}{2} + \frac{1}{4} + \cdots + \frac{1}{1,024}$

13. $1 + \frac{1}{2} + \frac{1}{4} + \frac{1}{8} + \cdots$

14. $\displaystyle\sum_{k=1}^{8} a_k$, where $a_n = 7 - 2n$

15. $1 + 2 + 3 + \cdots + 200$

16. Evaluate: $6! - \dbinom{9}{3}$

17. Write in summation notation: $2 + \frac{2}{3} + \frac{2}{9} + \cdots + \frac{2}{243}$

18. Expand $(x - y)^6$ using the binomial formula.

19. Expand $(a + 3)^5$ using the binomial formula.

20. Find the 8th term in the expansion of $(a + b)^{18}$.

Appendix A
SETS

This appendix introduces basic set terminology, relationships, and operations. The mathematical use of the word *set* does not differ much from the way it is used in everyday language. Words such as ''set,'' ''collection,'' ''bunch,'' and ''flock'' all convey the same idea. Thus, we think of a set as any collection of objects with the important property that, given any object, it is either a member of the set or it is not. For example, the letter p belongs to our alphabet, but the Greek letter π does not.

◆ **SET MEMBERSHIP AND EQUALITY**

If an object a is in set A, we say that a **is an element of** or **is a member of** set A and write

$$a \in A$$

If an object **is not an element of** set A, we write

$$a \notin A$$

Sets are often specified by **listing** their elements between braces { }. For example,

$$\{2, 3, 5, 7\}$$

represents the set with elements 2, 3, 5, and 7. For this set, $3 \in \{2, 3, 5, 7\}$, and $4 \notin \{2, 3, 5, 7\}$. If two sets A and B have exactly the same elements, they are said to be **equal,** and we write

$$A = B$$

We write

$$A \neq B$$

if sets A and B are **not equal.**

The order of listing the elements in a set does not matter; thus,

$$\{3, 4, 5\} = \{4, 3, 5\} = \{5, 4, 3\}$$

Also, elements in a set are not listed more than once. For example, the set of letters in the word ''letter'' is

$$\{e, l, t, r\}$$

Example 1 **Set Membership** **and Equality**	If $A = \{2, 4, 6\}$, $B = \{3, 5, 7\}$, and $C = \{4, 6, 2\}$, replace each question mark with $\in$, $\notin$, $=$, or $\neq$, as appropriate:

(A) $4 \, ? \, A$ **(B)** $3 \, ? \, A$ **(C)** $7 \, ? \, B$
(D) $2 \, ? \, B$ **(E)** $A \, ? \, C$ **(F)** $A \, ? \, B$

Solution   **(A)** $4 \in A$ **(B)** $3 \notin A$ **(C)** $7 \in B$
(D) $2 \notin B$ **(E)** $A = C$ **(F)** $A \neq B$

Matched Problem 1   If $P = \{1, 3, 5\}$, $Q = \{2, 3, 4\}$, and $R = \{3, 4, 2\}$, replace each question mark with $\in$, $\notin$, $=$, or $\neq$, as appropriate:

(A) $1 \, ? \, P$ **(B)** $3 \, ? \, Q$ **(C)** $5 \, ? \, R$
(D) $4 \, ? \, P$ **(E)** $P \, ? \, Q$ **(F)** $Q \, ? \, R$

◆ SUBSETS

We may be interested in sets within sets, called *subsets*. We say that a set A is a **subset** of set B if every element in set A is in set B. For example, the set of all women in a mathematics class would form a subset of all students in the class. The notation

$$A \subset B$$

is used to indicate that A is a subset of B.

A set with no elements is called the **empty** or **null** set. It is symbolized by

$$\varnothing$$

For example, the set of all months of the year beginning with the letter B is an empty, or null, set and would be designated by $\varnothing$. For any set A, $\varnothing \subset A$ and $A \subset A$.

Example 2
Subsets

Which of the following are subsets of the set $\{2, 3, 5, 7\}$?

$A = \{2, 3, 5\}$ $B = \{2\}$ $C = \{2, 3, 5, 7\}$ $D = \varnothing$ $E = \{1, 2, 3\}$
$F = 3$

Solution   The sets A, B, C, and D are all subsets of $\{2, 3, 5, 7\}$. In each case every element in the subset is in $\{2, 3, 5, 7\}$. The set $\{1, 2, 3\}$ is not a subset of $\{2, 3, 5, 7\}$, since $1 \notin \{2, 3, 5, 7\}$. The object 3 is different from the set $\{3\}$. We have $3 \in \{2, 3, 5, 7\}$ and $\{3\} \subset \{2, 3, 5, 7\}$. However, 3, and therefore F, is not a subset of $\{2, 3, 5, 7\}$.

Matched Problem 2   Which of the following are subsets of the set $\{2, 4, 6, 8\}$?

$A = \{2, 4\}$ $B = \{2, 4, 6, 8\}$ $C = \{4\}$ $D = 4$ $E = \varnothing$
$F = \{2, 4, 6, 8, 10\}$

♦ **SPECIFYING SETS**

The method of specifying a set by listing the elements, as in Example 1, is clear and convenient for small sets. However, if we are interested in specifying a set with a large number of elements, say, the set of all whole numbers from 10 to 10,000, then listing these elements would be impractical. The **rule method** for specifying sets takes care of situations of this type, as well as others. Using the rule method we would write

$$\{x \mid x \text{ is a whole number from 10 to 10,000}\}$$

which is read "the set of all elements x such that x is a whole number from 10 to 10,000." The vertical bar represents "such that."

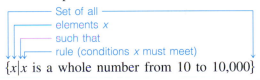

Set of all
elements x
such that
rule (conditions x must meet)
$\{x \mid x$ is a whole number from 10 to 10,000$\}$

Example 3
Describing Sets

Let $M = \{3, 4, 5, 6\}$ and $N = \{4, 5, 6, 7, 8\}$. Describe the set of all elements in M that are also in N using **(A)** the rule method and **(B)** the listing method.

Solution   **(A)** Rule method: $\{x \mid x \in M \text{ and } x \in N\}$
(B) Listing method: $\{4, 5, 6\}$

Matched Problem 3   Using the sets M and N in Example 3, describe the set of all elements that are in either M or N or both, using (A) the rule method and (B) the listing method.

♦ **UNION AND INTERSECTION**

Sets determined by "and" such as in Example 3 occur frequently in practice; the same is true for those determined by "or" as in Matched Problem 3. For this reason, we make the following definitions:

$$A \cap B = \textbf{Intersection} \text{ of the sets } A \text{ and } B$$
$$= \text{Set of all elements belonging to both sets } A \text{ and } B$$
$$= \{x | x \in A \text{ and } x \in B\}$$

$$A \cup B = \textbf{Union} \text{ of the sets } A \text{ and } B$$
$$= \text{Set of elements belonging to set } A \text{ or to set } B$$
$$= \{x | x \in A \text{ or } x \in B\}$$

The word ''or'' here is used in an inclusive sense meaning one or the other or both.

Example 4
Unions and Intersections

Let $A = \{2, 3, 5, 7\}$ and $B = \{1, 2, 3, 4\}$. Find $A \cap B$ and $A \cup B$.

Solution

$$A \cap B = \{2, 3\}$$ The set of all elements belonging to set A and to set B

$$A \cup B = \{1, 2, 3, 4, 5, 7\}$$ The set of all elements belonging to set A or to set B or to both.

Matched Problem 4   Let $A = \{2, 4, 6, 8\}$ and $B = \{1, 2, 3, 4\}$. Find $A \cap B$ and $A \cup B$.

♦ **VENN DIAGRAMS**

Set relations and operations may be visualized in diagrams called **Venn diagrams.**

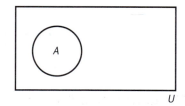

Here the rectangle represents all objects under consideration; this set is called the **universal set** and usually is denoted by U. It depends on the context in which one is working. If, for instance, you were interested in sets of students at a particular college, the universal set could be all students at that school. In the diagram, A represents a subset of U. To represent $B \subset A$, draw

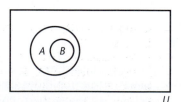

The intersection and union of two sets A and B are represented by the shaded regions in the following diagrams:

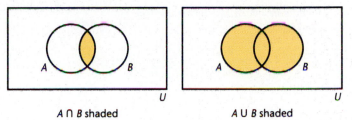

A ∩ B shaded   A ∪ B shaded

◆ COMPLEMENTS

In the Venn diagram

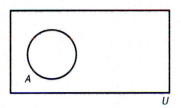

the region outside the circle represents all those elements in the universal set that are not in A, that is, the set

$$\{x \mid x \in U \text{ but } x \notin A\}$$

This set is called the **complement** of A and is usually denoted by A'.

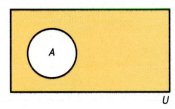

A' shaded

Example 5
Set Complements

Let $U = \{1, 2, 3, 4, 5, 6, 7, 8, 9, 10\}$ and $A = \{2, 3, 5, 7\}$. Give the complement of A by the listing method.

Solution

$$A' = \{1, 4, 6, 8, 9, 10\}$$

Matched Problem 5   Let U be as in Example 5 and $B = \{2, 4, 6, 8, 10\}$. Give B' by the listing method.

Answers to
Matched Problems

1. (A) $1 \in P$ (B) $3 \in Q$ (C) $5 \notin R$ (D) $4 \notin P$ (E) $P \neq Q$ (F) $Q = R$
2. A, B, C, E
3. Rule method: $\{x \mid x \in M \text{ or } x \in N\}$
   Listing method: $\{3, 4, 5, 6, 7, 8\}$
4. $A \cap B = \{2, 4\}$, $A \cup B = \{1, 2, 3, 4, 6, 8\}$
5. $B' = \{1, 3, 5, 7, 9\}$

EXERCISE A

A *In Problems 1–10, indicate which statements are true (T) and which are false (F):*

1. $4 \in \{2, 3, 4\}$ **2.** $7 \notin \{2, 3, 4\}$

3. $6 \notin \{2, 3, 4\}$ **4.** $7 \in \{2, 3, 4\}$

5. $\{3, 4, 5\} = \{5, 3, 4\}$

6. $\{1, 2, 3, 4\} = \{4, 3, 2, 1\}$

7. $\{3, 5, 7\} \neq \{4, 7, 5, 3\}$ **8.** $\{4, 6, 3\} \neq \{6, 3, 4\}$

9. $\{2, 3\} \subset \{2, 3, 4\}$ **10.** $\{4, 5\} \subset \{2, 3, 4\}$

Given sets

$$P = \{1, 3, 5, 7\} \qquad Q = \{2, 4, 6, 8\} \qquad R = \{5, 1, 7, 3\}$$

replace each question mark with $\in$, $\notin$, $=$, or $\neq$, as appropriate:

11. $5 \; ? \; P$ **12.** $6 \; ? \; Q$ **13.** $6 \; ? \; R$ **14.** $4 \; ? \; P$

15. $P \; ? \; R$ **16.** $Q \; ? \; R$ **17.** $P \; ? \; Q$ **18.** $R \; ? \; P$

B *Indicate the following sets by using the listing method. If the set is empty, write $\varnothing$.*

19. $\{x | x$ is a counting number between 5 and 10$\}$

20. $\{x | x$ is a counting number between 10 and 15$\}$

21. $\{x | x$ is a counting number between 7 and 8$\}$

22. $\{x | x$ is a counting number between 10 and 11$\}$

23. $\{x | x$ is a day of the week$\}$

24. $\{x | x$ is a month of the year$\}$

25. $\{x | x$ is a letter in "alababa"$\}$

26. $\{x | x$ is a letter in "millimeter"$\}$

27. $\{u | u$ is a state in the United States smaller than Rhode Island$\}$

28. $\{u | u$ is a day of the week starting with the letter $k\}$

If $U = \{1, 2, 3, 4, 5, 6, 7, 8\}$ and
$$A = \{1, 2, 3, 4\} \qquad B = \{2, 4, 6, 8\} \qquad C = \{1, 3, 5, 7\}$$

indicate each set by using the listing method:

29. $A \cap B$ **30.** $A \cap C$ **31.** $B \cap C$

32. $A \cup B$ **33.** $B \cup C$ **34.** $A \cup C$

35. A' **36.** B' **37.** C'

C 38. $\varnothing'$ **39.** $(A \cup B)'$ **40.** $(A \cap B)'$

41. $A' \cap B'$ **42.** $A' \cup B'$

43. List all the subsets of $\{1, 2\}$. (There are a total of four.)

44. List all the subsets of $\{1, 2, 3\}$. (There are a total of eight.)

A general Venn diagram for three sets A, B, and C is

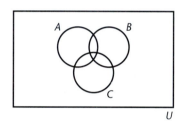

Use such a Venn diagram and shade the indicated set:

45. $(A \cup B) \cap C$ **46.** $(A \cap B) \cup C$

47. $(A \cap B) \cap C$ **48.** $(A \cap B) \cap C'$

49. $(A \cup B)' \cap C'$ **50.** $A \cap (B \cup C)'$

Appendix B
SIGNIFICANT DIGITS

Most calculations involving problems of the real world deal with figures that are only approximate. It would therefore seem reasonable to assume that a final answer could not be any more accurate than the least accurate figure used in the calculation. This is an important point, since calculators tend to give the impression that greater accuracy is achieved than is warranted.

Suppose we wish to compute the length of the diagonal of a rectangular field from measurements of its sides of 237.8 meters and 61.3 meters. Using the Pythagorean theorem and a calculator, we find

$$d = \sqrt{237.8^2 + 61.3^2}$$
$$= 245.573\ 878 \ldots$$

The calculator answer suggests an accuracy that is not justified. What accuracy is justified? To answer this question, we introduce the idea of **significant digits.**

The measurement 61.3 meters indicates that the measurement was made to the nearest tenth of a meter; that is, the actual width is between 61.25 and 61.35 meters. The number 61.3 has three significant digits. If we had written, instead, 61.30 meters as the width, then the actual width would be between 61.295 and 61.305 meters, and our measurement, 61.30 meters, would have four significant digits.

In all cases except one, the number of significant digits in a number is found by counting the digits from left to right, starting with the first nonzero digit and ending with the last digit present.

The significant digits in the following numbers are underlined:

719.37   82,395   5.600   0.000 830   0.000 08

As we said, this definition takes care of all cases except one. Consider, for example, the number 7,800. It is not clear whether the number has been rounded to the hundreds place, the tens place, or the units place. This ambiguity can be resolved by writing this type of number in scientific notation:

7.8×10^3 has two significant digits

7.80×10^3 has three significant digits

7.800×10^3 has four significant digits

All three are equal to 7,800 when written without powers of 10.

In calculations involving multiplication, division, powers, and roots, we adopt the following convention:

We will round off the answer to match the number of significant digits in the number with the least number of significant digits used in the calculation.

In computing the length of the diagonal of the field, we would write the answer to three significant digits because the width, the least accurate of the two numbers involved, has three significant digits:

$$d = 246 \text{ meters} \qquad \text{Three significant digits}$$

One final note: in rounding a number that is exactly halfway between a larger and a smaller number, we will use the convention of making the final result even.

Example 1

Round each number to three significant digits:

(A) 43.0690   **(B)** 48.05   **(C)** 48.15   **(D)** $8.017\,632 \times 10^{-3}$

Solution

(A) 43.1

(B) 48.0 ⎱ Use the convention of making the digit before the
(C) 48.2 ⎰ 5 even if it is odd or leaving it alone if it is even.

(D) 8.02×10^{-3}

Matched Problem 1

Round each number to three significant digits:

(A) 3.1495   **(B)** 0.004 135   **(C)** 32,450
(D) $4.314\,764\,09 \times 10^{12}$

Answers to Matched Problem

1. **(A)** 3.15   **(B)** 0.004 14   **(C)** 32,400   **(D)** 4.31×10^{12}

Tables

Table I   Values of e^x and e^{-x} (0.00 to 3.00)

x	e^x	e^{-x}	x	e^x	e^{-x}	x	e^x	e^{-x}
0.00	1.0000	1.000 00	0.50	1.6487	0.606 53	1.00	2.7183	0.367 88
0.01	1.0101	0.990 05	0.51	1.6653	0.600 50	1.01	2.7456	0.364 22
0.02	1.0202	0.980 20	0.52	1.6820	0.594 52	1.02	2.7732	0.360 59
0.03	1.0305	0.970 45	0.53	1.6989	0.588 60	1.03	2.8011	0.357 01
0.04	1.0408	0.960 79	0.54	1.7160	0.582 75	1.04	2.8292	0.353 45
0.05	1.0513	0.951 23	0.55	1.7333	0.576 95	1.05	2.8577	0.349 94
0.06	1.0618	0.941 76	0.56	1.7507	0.571 21	1.06	2.8864	0.346 46
0.07	1.0725	0.932 39	0.57	1.7683	0.565 53	1.07	2.9154	0.343 01
0.08	1.0833	0.923 12	0.58	1.7860	0.559 90	1.08	2.9447	0.339 60
0.09	1.0942	0.913 93	0.59	1.8040	0.554 33	1.09	2.9743	0.336 22
0.10	1.1052	0.904 84	0.60	1.8221	0.548 81	1.10	3.0042	0.332 87
0.11	1.1163	0.895 83	0.61	1.8404	0.543 35	1.11	3.0344	0.329 56
0.12	1.1275	0.886 92	0.62	1.8589	0.537 94	1.12	3.0649	0.326 28
0.13	1.1388	0.878 10	0.63	1.8776	0.532 59	1.13	3.0957	0.323 03
0.14	1.1503	0.869 36	0.64	1.8965	0.527 29	1.14	3.1268	0.319 82
0.15	1.1618	0.860 71	0.65	1.9155	0.522 05	1.15	3.1582	0.316 64
0.16	1.1735	0.852 14	0.66	1.9348	0.516 85	1.16	3.1899	0.313 49
0.17	1.1853	0.843 66	0.67	1.9542	0.511 71	1.17	3.2220	0.310 37
0.18	1.1972	0.835 27	0.68	1.9739	0.506 62	1.18	3.2544	0.307 28
0.19	1.2092	0.826 96	0.69	1.9937	0.501 58	1.19	3.2871	0.304 22
0.20	1.2214	0.818 73	0.70	2.0138	0.496 59	1.20	3.3201	0.301 19
0.21	1.2337	0.810 58	0.71	2.0340	0.491 64	1.21	3.3535	0.298 20
0.22	1.2461	0.802 52	0.72	2.0544	0.486 75	1.22	3.3872	0.295 23
0.23	1.2586	0.794 53	0.73	2.0751	0.481 91	1.23	3.4212	0.292 29
0.24	1.2712	0.786 63	0.74	2.0959	0.477 11	1.24	3.4556	0.289 38
0.25	1.2840	0.778 80	0.75	2.1170	0.472 37	1.25	3.4903	0.286 50
0.26	1.2969	0.771 05	0.76	2.1383	0.467 67	1.26	3.5254	0.283 65
0.27	1.3100	0.763 38	0.77	2.1598	0.463 01	1.27	3.5609	0.280 83
0.28	1.3231	0.755 78	0.78	2.1815	0.458 41	1.28	3.5966	0.278 04
0.29	1.3364	0.748 26	0.79	2.2034	0.453 84	1.29	3.6328	0.275 27
0.30	1.3499	0.740 82	0.80	2.2255	0.449 33	1.30	3.6693	0.272 53
0.31	1.3634	0.733 45	0.81	2.2479	0.444 86	1.31	3.7062	0.269 82
0.32	1.3771	0.726 15	0.82	2.2705	0.440 43	1.32	3.7434	0.267 14
0.33	1.3910	0.718 92	0.83	2.2933	0.436 05	1.33	3.7810	0.264 48
0.34	1.4049	0.711 77	0.84	2.3164	0.431 71	1.34	3.8190	0.261 85
0.35	1.4191	0.704 69	0.85	2.3396	0.427 41	1.35	3.8574	0.259 24
0.36	1.4333	0.697 68	0.86	2.3632	0.423 16	1.36	3.8962	0.256 66
0.37	1.4477	0.690 73	0.87	2.3869	0.418 95	1.37	3.9354	0.254 11
0.38	1.4623	0.683 86	0.88	2.4109	0.414 78	1.38	3.9749	0.251 58
0.39	1.4770	0.677 06	0.89	2.4351	0.410 66	1.39	4.0149	0.249 08
0.40	1.4918	0.670 32	0.90	2.4596	0.406 57	1.40	4.0552	0.246 60
0.41	1.5068	0.663 65	0.91	2.4843	0.402 52	1.41	4.0960	0.244 14
0.42	1.5220	0.657 05	0.92	2.5093	0.398 52	1.42	4.1371	0.241 71
0.43	1.5373	0.650 51	0.93	2.5345	0.394 55	1.43	4.1787	0.239 31
0.44	1.5527	0.644 04	0.94	2.5600	0.390 63	1.44	4.2207	0.236 93
0.45	1.5683	0.637 63	0.95	2.5857	0.386 74	1.45	4.2631	0.234 57
0.46	1.5841	0.631 28	0.96	2.6117	0.382 89	1.46	4.3060	0.232 24
0.47	1.6000	0.625 00	0.97	2.6379	0.379 08	1.47	4.3492	0.229 93
0.48	1.6161	0.618 78	0.98	2.6645	0.375 31	1.48	4.3939	0.227 64
0.49	1.6323	0.612 63	0.99	2.6912	0.371 58	1.49	4.4371	0.225 37
0.50	1.6487	0.606 53	1.00	2.7183	0.367 88	1.50	4.4817	0.223 13

Table I   *(continued)*

x	e^x	e^{-x}	x	e^x	e^{-x}	x	e^x	e^{-x}
1.50	4.4817	0.223 13	2.00	7.3891	0.135 34	2.50	12.182	0.082 085
1.51	4.5267	0.220 91	2.01	7.4633	0.133 99	2.51	12.305	0.081 268
1.52	4.5722	0.218 71	2.02	7.5383	0.132 66	2.52	12.429	0.080 460
1.53	4.6182	0.216 54	2.03	7.6141	0.131 34	2.53	12.554	0.079 659
1.54	4.6646	0.214 38	2.04	7.6906	0.130 03	2.54	12.680	0.078 866
1.55	4.7115	0.212 25	2.05	7.7679	0.128 73	2.55	12.807	0.078 082
1.56	4.7588	0.210 14	2.06	7.8460	0.127 45	2.56	12.936	0.077 305
1.57	4.8066	0.208 05	2.07	7.9248	0.126 19	2.57	13.066	0.076 536
1.58	4.8550	0.205 98	2.08	8.0045	0.124 93	2.58	13.197	0.075 774
1.59	4.9037	0.203 93	2.09	8.0849	0.123 69	2.59	13.330	0.075 020
1.60	4.9530	0.201 90	2.10	8.1662	0.122 46	2.60	13.464	0.074 274
1.61	5.0028	0.199 89	2.11	8.2482	0.121 24	2.61	13.599	0.073 535
1.62	5.0531	0.197 90	2.12	8.3311	0.120 03	2.62	13.736	0.072 803
1.63	5.1039	0.195 93	2.13	8.4149	0.118 84	2.63	13.874	0.072 078
1.64	5.1552	0.193 98	2.14	8.4994	0.117 65	2.64	14.013	0.071 361
1.65	5.2070	0.192 05	2.15	8.5849	0.116 48	2.65	14.154	0.070 651
1.66	5.2593	0.190 14	2.16	8.6711	0.115 33	2.66	14.296	0.069 948
1.67	5.3122	0.188 25	2.17	8.7583	0.114 18	2.67	14.440	0.069 252
1.68	5.3656	0.186 37	2.18	8.8463	0.113 04	2.68	14.585	0.068 563
1.69	5.4195	0.184 52	2.19	8.9352	0.111 92	2.69	14.732	0.067 881
1.70	5.4739	0.182 68	2.20	9.0250	0.110 80	2.70	14.880	0.067 206
1.71	5.5290	0.180 87	2.21	9.1157	0.109 70	2.71	15.029	0.066 537
1.72	5.5845	0.179 07	2.22	9.2073	0.108 61	2.72	15.180	0.065 875
1.73	5.6407	0.177 28	2.23	9.2999	0.107 53	2.73	15.333	0.065 219
1.74	5.6973	0.175 52	2.24	9.3933	0.106 46	2.74	15.487	0.064 570
1.75	5.7546	0.173 77	2.25	9.4877	0.105 40	2.75	15.643	0.063 928
1.76	5.8124	0.172 04	2.26	9.5831	0.104 35	2.76	15.800	0.063 292
1.77	5.8709	0.170 33	2.27	9.6794	0.103 31	2.77	15.959	0.062 662
1.78	5.9299	0.168 64	2.28	9.7767	0.102 28	2.78	16.119	0.062 039
1.79	5.9895	0.166 96	2.29	9.8749	0.101 27	2.79	16.281	0.061 421
1.80	6.0496	0.165 30	2.30	9.9742	0.100 26	2.80	16.445	0.060 810
1.81	6.1104	0.163 65	2.31	10.074	0.099 261	2.81	16.610	0.060 205
1.82	6.1719	0.162 03	2.32	10.176	0.098 274	2.82	16.777	0.059 606
1.83	6.2339	0.160 41	2.33	10.278	0.097 296	2.83	16.945	0.059 013
1.84	6.2965	0.158 82	2.34	10.381	0.096 328	2.84	17.116	0.058 426
1.85	6.3598	0.157 24	2.35	10.486	0.095 369	2.85	17.288	0.057 844
1.86	6.4237	0.155 67	2.36	10.591	0.094 420	2.86	17.462	0.057 269
1.87	6.4883	0.154 12	2.37	10.697	0.093 481	2.87	17.637	0.056 699
1.88	6.5535	0.152 59	2.38	10.805	0.092 551	2.88	17.814	0.056 135
1.89	6.6194	0.151 07	2.39	10.913	0.091 630	2.89	17.993	0.055 576
1.90	6.6859	0.149 57	2.40	11.023	0.090 718	2.90	18.174	0.055 023
1.91	6.7531	0.148 08	2.41	11.134	0.089 815	2.91	18.357	0.054 476
1.92	6.8210	0.146 61	2.42	11.246	0.088 922	2.92	18.541	0.053 934
1.93	6.8895	0.145 15	2.43	11.359	0.088 037	2.93	18.728	0.053 397
1.94	6.9588	0.143 70	2.44	11.473	0.087 161	2.94	18.916	0.052 866
1.95	7.0287	0.142 27	2.45	11.588	0.086 294	2.95	19.106	0.052 340
1.96	7.0993	0.140 86	2.46	11.705	0.085 435	2.96	19.298	0.051 819
1.97	7.1707	0.139 46	2.47	11.822	0.084 585	2.97	19.492	0.051 303
1.98	7.2427	0.138 07	2.48	11.941	0.083 743	2.98	19.688	0.050 793
1.99	7.3155	0.136 70	2.49	12.061	0.082 910	2.99	19.886	0.050 287
2.00	7.3891	0.135 34	2.50	12.182	0.082 085	3.00	20.086	0.049 787

Table II   Common Logarithms

x	0	1	2	3	4	5	6	7	8	9
1.0	0.0000	0.004321	0.008600	0.01284	0.01703	0.02119	0.02531	0.02938	0.03342	0.03743
1.1	0.04139	0.04532	0.04922	0.05308	0.05690	0.06070	0.06446	0.06819	0.07188	0.07555
1.2	0.07918	0.08279	0.08636	0.08991	0.09342	0.09691	0.1004	0.1038	0.1072	0.1106
1.3	0.1139	0.1173	0.1206	0.1239	0.1271	0.1303	0.1335	0.1367	0.1399	0.1430
1.4	0.1461	0.1492	0.1523	0.1553	0.1584	0.1614	0.1644	0.1673	0.1703	0.1732
1.5	0.1761	0.1790	0.1818	0.1847	0.1875	0.1903	0.1931	0.1959	0.1987	0.2014
1.6	0.2041	0.2068	0.2095	0.2122	0.2148	0.2175	0.2201	0.2227	0.2253	0.2279
1.7	0.2304	0.2330	0.2355	0.2380	0.2405	0.2430	0.2455	0.2480	0.2504	0.2529
1.8	0.2553	0.2577	0.2601	0.2625	0.2648	0.2673	0.2695	0.2718	0.2742	0.2765
1.9	0.2788	0.2810	0.2833	0.2856	0.2878	0.2900	0.2923	0.2945	0.2967	0.2989
2.0	0.3010	0.3032	0.3054	0.3075	0.3096	0.3118	0.3139	0.3160	0.3181	0.3201
2.1	0.3222	0.3243	0.3263	0.3284	0.3304	0.3324	0.3345	0.3365	0.3385	0.3404
2.2	0.3424	0.3444	0.3464	0.3483	0.3502	0.3522	0.3541	0.3560	0.3579	0.3598
2.3	0.3617	0.3636	0.3655	0.3674	0.3692	0.3711	0.3729	0.3747	0.3766	0.3784
2.4	0.3802	0.3820	0.3838	0.3856	0.3874	0.3892	0.3909	0.3927	0.3945	0.3962
2.5	0.3979	0.3997	0.4014	0.4031	0.4048	0.4065	0.4082	0.4099	0.4116	0.4133
2.6	0.4150	0.4166	0.4183	0.4200	0.4216	0.4232	0.4249	0.4265	0.4281	0.4298
2.7	0.4314	0.4330	0.4346	0.4362	0.4378	0.4393	0.4409	0.4425	0.4440	0.4456
2.8	0.4472	0.4487	0.4502	0.4518	0.4533	0.4548	0.4564	0.4579	0.4594	0.4609
2.9	0.4624	0.4639	0.4654	0.4669	0.4683	0.4698	0.4713	0.4728	0.4742	0.4757
3.0	0.4771	0.4786	0.4800	0.4814	0.4829	0.4843	0.4857	0.4871	0.4886	0.4900
3.1	0.4914	0.4928	0.4942	0.4955	0.4969	0.4983	0.4997	0.5011	0.5024	0.5038
3.2	0.5051	0.5065	0.5079	0.5092	0.5105	0.5119	0.5132	0.5145	0.5159	0.5172
3.3	0.5185	0.5198	0.5211	0.5224	0.5237	0.5250	0.5263	0.5276	0.5289	0.5302
3.4	0.5315	0.5328	0.5340	0.5353	0.5366	0.5378	0.5391	0.5403	0.5416	0.5428
3.5	0.5441	0.5453	0.5465	0.5478	0.5490	0.5502	0.5514	0.5527	0.5539	0.5551
3.6	0.5563	0.5575	0.5587	0.5599	0.5611	0.5623	0.5635	0.5647	0.5658	0.5670
3.7	0.5682	0.5694	0.5705	0.5717	0.5729	0.5740	0.5752	0.5763	0.5775	0.5786
3.8	0.5798	0.5809	0.5821	0.5832	0.5843	0.5855	0.5866	0.5877	0.5888	0.5899
3.9	0.5911	0.5922	0.5933	0.5944	0.5955	0.5966	0.5977	0.5988	0.5999	0.6010
4.0	0.6021	0.6031	0.6042	0.6053	0.6064	0.6075	0.6085	0.6096	0.6107	0.6117
4.1	0.6128	0.6138	0.6149	0.6160	0.6170	0.6180	0.6191	0.6201	0.6212	0.6222
4.2	0.6232	0.6243	0.6253	0.6263	0.6274	0.6284	0.6294	0.6304	0.6314	0.6325
4.3	0.6335	0.6345	0.6355	0.6365	0.6375	0.6385	0.6395	0.6405	0.6415	0.6425
4.4	0.6435	0.6444	0.6454	0.6464	0.6474	0.6484	0.6493	0.6503	0.6513	0.6522
4.5	0.6532	0.6542	0.6551	0.6561	0.6571	0.6580	0.6590	0.6599	0.6609	0.6618
4.6	0.6628	0.6637	0.6646	0.6656	0.6665	0.6675	0.6684	0.6693	0.6702	0.6712
4.7	0.6721	0.6730	0.6739	0.6749	0.6758	0.6767	0.6776	0.6785	0.6794	0.6803
4.8	0.6812	0.6821	0.6830	0.6839	0.6848	0.6857	0.6866	0.6875	0.6884	0.6893
4.9	0.6902	0.6911	0.6920	0.6928	0.6937	0.6946	0.6955	0.6964	0.6972	0.6981
5.0	0.6990	0.6998	0.7007	0.7016	0.7024	0.7033	0.7042	0.7050	0.7059	0.7067
5.1	0.7076	0.7084	0.7093	0.7101	0.7110	0.7118	0.7126	0.7135	0.7143	0.7152
5.2	0.7160	0.7168	0.7177	0.7185	0.7193	0.7202	0.7210	0.7218	0.7226	0.7235
5.3	0.7243	0.7251	0.7259	0.7267	0.7275	0.7284	0.7292	0.7300	0.7308	0.7316
5.4	0.7324	0.7332	0.7340	0.7348	0.7356	0.7364	0.7372	0.7380	0.7388	0.7396

Table II   *(continued)*

x	0	1	2	3	4	5	6	7	8	9
5.5	0.7404	0.7412	0.7419	0.7427	0.7435	0.7443	0.7451	0.7459	0.7466	0.7474
5.6	0.7482	0.7490	0.7497	0.7505	0.7513	0.7520	0.7528	0.7536	0.7543	0.7551
5.7	0.7559	0.7566	0.7574	0.7582	0.7589	0.7597	0.7604	0.7612	0.7619	0.7627
5.8	0.7634	0.7642	0.7649	0.7657	0.7664	0.7672	0.7679	0.7686	0.7694	0.7701
5.9	0.7709	0.7716	0.7723	0.7731	0.7738	0.7745	0.7752	0.7760	0.7767	0.7774
6.0	0.7782	0.7789	0.7796	0.7803	0.7810	0.7818	0.7825	0.7832	0.7839	0.7846
6.1	0.7853	0.7860	0.7868	0.7875	0.7882	0.7889	0.7896	0.7903	0.7910	0.7917
6.2	0.7924	0.7931	0.7938	0.7945	0.7952	0.7959	0.7966	0.7973	0.7980	0.7987
6.3	0.7993	0.8000	0.8007	0.8014	0.8021	0.8028	0.8035	0.8041	0.8048	0.8055
6.4	0.8062	0.8069	0.8075	0.8082	0.8089	0.8096	0.8102	0.8109	0.8116	0.8122
6.5	0.8129	0.8136	0.8142	0.8149	0.8156	0.8162	0.8169	0.8176	0.8182	0.8189
6.6	0.8195	0.8202	0.8209	0.8215	0.8222	0.8228	0.8235	0.8241	0.8248	0.8254
6.7	0.8261	0.8267	0.8274	0.8280	0.8287	0.8293	0.8299	0.8306	0.8312	0.8319
6.8	0.8325	0.8331	0.8338	0.8344	0.8351	0.8357	0.8363	0.8370	0.8376	0.8382
6.9	0.8388	0.8395	0.8401	0.8407	0.8414	0.8420	0.8426	0.8432	0.8439	0.8445
7.0	0.8451	0.8457	0.8463	0.8470	0.8476	0.8482	0.8488	0.8494	0.8500	0.8506
7.1	0.8513	0.8519	0.8525	0.8531	0.8537	0.8543	0.8549	0.8555	0.8561	0.8567
7.2	0.8573	0.8579	0.8585	0.8591	0.8597	0.8603	0.8609	0.8615	0.8621	0.8627
7.3	0.8633	0.8639	0.8645	0.8651	0.8657	0.8663	0.8669	0.8675	0.8681	0.8686
7.4	0.8692	0.8698	0.8704	0.8710	0.8716	0.8722	0.8727	0.8733	0.8739	0.8745
7.5	0.8751	0.8756	0.8762	0.8768	0.8774	0.8779	0.8785	0.8791	0.8797	0.8802
7.6	0.8808	0.8814	0.8820	0.8825	0.8831	0.8837	0.8842	0.8848	0.8854	0.8859
7.7	0.8865	0.8871	0.8876	0.8882	0.8887	0.8893	0.8899	0.8904	0.8910	0.8915
7.8	0.8921	0.8927	0.8932	0.8938	0.8943	0.8949	0.8954	0.8960	0.8965	0.8971
7.9	0.8976	0.8982	0.8987	0.8993	0.8998	0.9004	0.9009	0.9015	0.9020	0.9025
8.0	0.9031	0.9036	0.9042	0.9047	0.9053	0.9058	0.9063	0.9069	0.9074	0.9079
8.1	0.9085	0.9090	0.9096	0.9101	0.9106	0.9112	0.9117	0.9122	0.9128	0.9133
8.2	0.9138	0.9143	0.9149	0.9154	0.9159	0.9165	0.9170	0.9175	0.9180	0.9186
8.3	0.9191	0.9196	0.9201	0.9206	0.9212	0.9217	0.9222	0.9227	0.9232	0.9238
8.4	0.9243	0.9248	0.9253	0.9258	0.9263	0.9269	0.9274	0.9279	0.9284	0.9289
8.5	0.9294	0.9299	0.9304	0.9309	0.9315	0.9320	0.9325	0.9330	0.9335	0.9340
8.6	0.9345	0.9350	0.9355	0.9360	0.9365	0.9370	0.9375	0.9380	0.9385	0.9390
8.7	0.9395	0.9400	0.9405	0.9410	0.9415	0.9420	0.9425	0.9430	0.9435	0.9440
8.8	0.9445	0.9450	0.9455	0.9460	0.9465	0.9469	0.9474	0.9479	0.9484	0.9489
8.9	0.9494	0.9499	0.9504	0.9509	0.9513	0.9518	0.9523	0.9528	0.9533	0.9538
9.0	0.9542	0.9547	0.9552	0.9557	0.9562	0.9566	0.9571	0.9576	0.9581	0.9586
9.1	0.9590	0.9595	0.9600	0.9605	0.9609	0.9614	0.9619	0.9624	0.9628	0.9633
9.2	0.9638	0.9643	0.9647	0.9652	0.9657	0.9661	0.9666	0.9671	0.9675	0.9680
9.3	0.9685	0.9689	0.9694	0.9699	0.9703	0.9708	0.9713	0.9717	0.9722	0.9727
9.4	0.9731	0.9736	0.9741	0.9745	0.9750	0.9754	0.9759	0.9763	0.9768	0.9773
9.5	0.9777	0.9782	0.9786	0.9791	0.9795	0.9800	0.9805	0.9809	0.9814	0.9818
9.6	0.9823	0.9827	0.9832	0.9836	0.9841	0.9845	0.9850	0.9854	0.9859	0.9863
9.7	0.9868	0.9872	0.9877	0.9881	0.9886	0.9890	0.9894	0.9899	0.9903	0.9908
9.8	0.9912	0.9917	0.9921	0.9926	0.9930	0.9934	0.9939	0.9943	0.9948	0.9952
9.9	0.9956	0.9961	0.9965	0.9969	0.9974	0.9978	0.9983	0.9987	0.9991	0.9996

Table III   **Natural Logarithms ($\ln x = \log_e x$)**

ln 10 = 2.3026	**5 ln 10 = 11.5130**	**9 ln 10 = 20.7233**
2 ln 10 = 4.6052	**6 ln 10 = 13.8155**	**10 ln 10 = 23.0259**
3 ln 10 = 6.9078	**7 ln 10 = 16.1181**	
4 ln 10 = 9.2103	**8 ln 10 = 18.4207**	

x	.00	.01	.02	.03	.04	.05	.06	.07	.08	.09
1.0	0.0000	0.0100	0.0198	0.0296	0.0392	0.0488	0.0583	0.0677	0.0770	0.0862
1.1	0.0953	0.1044	0.1133	0.1222	0.1310	0.1398	0.1484	0.1570	0.1655	0.1740
1.2	0.1823	0.1906	0.1989	0.2070	0.2151	0.2231	0.2311	0.2390	0.2469	0.2546
1.3	0.2624	0.2700	0.2776	0.2852	0.2927	0.3001	0.3075	0.3148	0.3221	0.3293
1.4	0.3365	0.3436	0.3507	0.3577	0.3646	0.3716	0.3784	0.3853	0.3920	0.3988
1.5	0.4055	0.4121	0.4187	0.4253	0.4318	0.4383	0.4447	0.4511	0.4574	0.4637
1.6	0.4700	0.4762	0.4824	0.4886	0.4947	0.5008	0.5068	0.5128	0.5188	0.5247
1.7	0.5306	0.5365	0.5423	0.5481	0.5539	0.5596	0.5653	0.5710	0.5766	0.5822
1.8	0.5878	0.5933	0.5988	0.6043	0.6098	0.6152	0.6206	0.6259	0.6313	0.6366
1.9	0.6419	0.6471	0.6523	0.6575	0.6627	0.6678	0.6729	0.6780	0.6831	0.6881
2.0	0.6931	0.6981	0.7031	0.7080	0.7129	0.7178	0.7227	0.7275	0.7324	0.7372
2.1	0.7419	0.7467	0.7514	0.7561	0.7608	0.7655	0.7701	0.7747	0.7793	0.7839
2.2	0.7885	0.7930	0.7975	0.8020	0.8065	0.8109	0.8154	0.8198	0.8242	0.8286
2.3	0.8329	0.8372	0.8416	0.8459	0.8502	0.8544	0.8587	0.8629	0.8671	0.8713
2.4	0.8755	0.8796	0.8838	0.8879	0.8920	0.8961	0.9002	0.9042	0.9083	0.9123
2.5	0.9163	0.9203	0.9243	0.9282	0.9322	0.9361	0.9400	0.9439	0.9478	0.9517
2.6	0.9555	0.9594	0.9632	0.9670	0.9708	0.9746	0.9783	0.9821	0.9858	0.9895
2.7	0.9933	0.9969	1.0006	1.0043	1.0080	1.0116	1.0152	1.0188	1.0225	1.0260
2.8	1.0296	1.0332	1.0367	1.0403	1.0438	1.0473	1.0508	1.0543	1.0578	1.0613
2.9	1.0647	1.0682	1.0716	1.0750	1.0784	1.0818	1.0852	1.0886	1.0919	1.0953
3.0	1.0986	1.1019	1.1053	1.1086	1.1119	1.1151	1.1184	1.1217	1.1249	1.1282
3.1	1.1314	1.1346	1.1378	1.1410	1.1442	1.1474	1.1506	1.1537	1.1569	1.1600
3.2	1.1632	1.1663	1.1694	1.1725	1.1756	1.1787	1.1817	1.1848	1.1878	1.1909
3.3	1.1939	1.1969	1.2000	1.2030	1.2060	1.2090	1.2119	1.2149	1.2179	1.2208
3.4	1.2238	1.2267	1.2296	1.2326	1.2355	1.2384	1.2413	1.2442	1.2470	1.2499
3.5	1.2528	1.2556	1.2585	1.2613	1.2641	1.2669	1.2698	1.2726	1.2754	1.2782
3.6	1.2809	1.2837	1.2865	1.2892	1.2920	1.2947	1.2975	1.3002	1.3029	1.3056
3.7	1.3083	1.3110	1.3137	1.3164	1.3191	1.3218	1.3244	1.3271	1.3297	1.3324
3.8	1.3350	1.3376	1.3403	1.3429	1.3455	1.3481	1.3507	1.3533	1.3558	1.3584
3.9	1.3610	1.3635	1.3661	1.3686	1.3712	1.3737	1.3762	1.3788	1.3813	1.3838
4.0	1.3863	1.3888	1.3913	1.3938	1.3962	1.3987	1.4012	1.4036	1.4061	1.4085
4.1	1.4110	1.4134	1.4159	1.4183	1.4207	1.4231	1.4255	1.4279	1.4303	1.4327
4.2	1.4351	1.4375	1.4398	1.4422	1.4446	1.4469	1.4493	1.4516	1.4540	1.4563
4.3	1.4586	1.4609	1.4633	1.4656	1.4679	1.4702	1.4725	1.4748	1.4770	1.4793
4.4	1.4816	1.4839	1.4861	1.4884	1.4907	1.4929	1.4951	1.4974	1.4996	1.5019
4.5	1.5041	1.5063	1.5085	1.5107	1.5129	1.5151	1.5173	1.5195	1.5217	1.5239
4.6	1.5261	1.5282	1.5304	1.5326	1.5347	1.5369	1.5390	1.5412	1.5433	1.5454
4.7	1.5476	1.5497	1.5518	1.5539	1.5560	1.5581	1.5602	1.5623	1.5644	1.5665
4.8	1.5686	1.5707	1.5728	1.5748	1.5769	1.5790	1.5810	1.5831	1.5851	1.5872
4.9	1.5892	1.5913	1.5933	1.5953	1.5974	1.5994	1.6014	1.6034	1.6054	1.6074
5.0	1.6094	1.6114	1.6134	1.6154	1.6174	1.6194	1.6214	1.6233	1.6253	1.6273
5.1	1.6292	1.6312	1.6332	1.6351	1.6371	1.6390	1.6409	1.6429	1.6448	1.6467
5.2	1.6487	1.6506	1.6525	1.6544	1.6563	1.6582	1.6601	1.6620	1.6639	1.6658
5.3	1.6677	1.6696	1.6715	1.6734	1.6752	1.6771	1.6790	1.6808	1.6827	1.6845
5.4	1.6864	1.6882	1.6901	1.6919	1.6938	1.6956	1.6974	1.6993	1.7011	1.7029

Note: ln 35,200 = ln (3.52 × 10⁴) = ln 3.52 + 4 ln 10

       ln 0.008 64 = ln (8.64 × 10⁻³) = ln 8.64 − 3 ln 10

Table III   *(continued)*

x	.00	.01	.02	.03	.04	.05	.06	.07	.08	.09
5.5	1.7047	1.7066	1.7084	1.7102	1.7120	1.7138	1.7156	1.7174	1.7192	1.7210
5.6	1.7228	1.7246	1.7263	1.7281	1.7299	1.7317	1.7334	1.7352	1.7370	1.7387
5.7	1.7405	1.7422	1.7440	1.7457	1.7475	1.7492	1.7509	1.7527	1.7544	1.7561
5.8	1.7579	1.7596	1.7613	1.7630	1.7647	1.7664	1.7681	1.7699	1.7716	1.7733
5.9	1.7750	1.7766	1.7783	1.7800	1.7817	1.7834	1.7851	1.7867	1.7884	1.7901
6.0	1.7918	1.7934	1.7951	1.7967	1.7984	1.8001	1.8017	1.8034	1.8050	1.8066
6.1	1.8083	1.8099	1.8116	1.8132	1.8148	1.8165	1.8181	1.8197	1.8213	1.8229
6.2	1.8245	1.8262	1.8278	1.8294	1.8310	1.8326	1.8342	1.8358	1.8374	1.8390
6.3	1.8405	1.8421	1.8437	1.8453	1.8469	1.8485	1.8500	1.8516	1.8532	1.8547
6.4	1.8563	1.8579	1.8594	1.8610	1.8625	1.8641	1.8656	1.8672	1.8687	1.8703
6.5	1.8718	1.8733	1.8749	1.8764	1.8779	1.8795	1.8810	1.8825	1.8840	1.8856
6.6	1.8871	1.8886	1.8901	1.8916	1.8931	1.8946	1.8961	1.8976	1.8991	1.9006
6.7	1.9021	1.9036	1.9051	1.9066	1.9081	1.9095	1.9110	1.9125	1.9140	1.9155
6.8	1.9169	1.9184	1.9199	1.9213	1.9228	1.9242	1.9257	1.9272	1.9286	1.9301
6.9	1.9315	1.9330	1.9344	1.9359	1.9373	1.9387	1.9402	1.9416	1.9430	1.9445
7.0	1.9459	1.9473	1.9488	1.9502	1.9516	1.9530	1.9544	1.9559	1.9573	1.9587
7.1	1.9601	1.9615	1.9629	1.9643	1.9657	1.9671	1.9685	1.9699	1.9713	1.9727
7.2	1.9741	1.9755	1.9769	1.9782	1.9796	1.9810	1.9824	1.9838	1.9851	1.9865
7.3	1.9879	1.9892	1.9906	1.9920	1.9933	1.9947	1.9961	1.9974	1.9988	2.0001
7.4	2.0015	2.0028	2.0042	2.0055	2.0069	2.0082	2.0096	2.0109	2.0122	2.0136
7.5	2.0149	2.0162	2.0176	2.0189	2.0202	2.0215	2.0229	2.0242	2.0255	2.0268
7.6	2.0281	2.0295	2.0308	2.0321	2.0334	2.0347	2.0360	2.0373	2.0386	2.0399
7.7	2.0412	2.0425	2.0438	2.0451	2.0464	2.0477	2.0490	2.0503	2.0516	2.0528
7.8	2.0541	2.0554	2.0567	2.0580	2.0592	2.0605	2.0618	2.0631	2.0643	2.0656
7.9	2.0669	2.0681	2.0694	2.0707	2.0719	2.0732	2.0744	2.0757	2.0769	2.0782
8.0	2.0794	2.0807	2.0819	2.0832	2.0844	2.0857	2.0869	2.0882	2.0894	2.0906
8.1	2.0919	2.0931	2.0943	2.0956	2.0968	2.0980	2.0992	2.1005	2.1017	2.1029
8.2	2.1041	2.1054	2.1066	2.1078	2.1090	2.1102	2.1114	2.1126	2.1138	2.1150
8.3	2.1163	2.1175	2.1187	2.1199	2.1211	2.1223	2.1235	2.1247	2.1258	2.1270
8.4	2.1282	2.1294	2.1306	2.1318	2.1330	2.1342	2.1353	2.1365	2.1377	2.1389
8.5	2.1401	2.1412	2.1424	2.1436	2.1448	2.1459	2.1471	2.1483	2.1494	2.1506
8.6	2.1518	2.1529	2.1541	2.1552	2.1564	2.1576	2.1587	2.1599	2.1610	2.1622
8.7	2.1633	2.1645	2.1656	2.1668	2.1679	2.1691	2.1702	2.1713	2.1725	2.1736
8.8	2.1748	2.1759	2.1770	2.1782	2.1793	2.1804	2.1815	2.1827	2.1838	2.1849
8.9	2.1861	2.1872	2.1883	2.1894	2.1905	2.1917	2.1928	2.1939	2.1950	2.1961
9.0	2.1972	2.1983	2.1994	2.2006	2.2017	2.2028	2.2039	2.2050	2.2061	2.2072
9.1	2.2083	2.2094	2.2105	2.2116	2.2127	2.2138	2.2148	2.2159	2.2170	2.2181
9.2	2.2192	2.2203	2.2214	2.2225	2.2235	2.2246	2.2257	2.2268	2.2279	2.2289
9.3	2.2300	2.2311	2.2322	2.2332	2.2343	2.2354	2.2364	2.2375	2.2386	2.2396
9.4	2.2407	2.2418	2.2428	2.2439	2.2450	2.2460	2.2471	2.2481	2.2492	2.2502
9.5	2.2513	2.2523	2.2534	2.2544	2.2555	2.2565	2.2576	2.2586	2.2597	2.2607
9.6	2.2618	2.2628	2.2638	2.2649	2.2659	2.2670	2.2680	2.2690	2.2701	2.2711
9.7	2.2721	2.2732	2.2742	2.2752	2.2762	2.2773	2.2783	2.2793	2.2803	2.2814
9.8	2.2824	2.2834	2.2844	2.2854	2.2865	2.2875	2.2885	2.2895	2.2905	2.2915
9.9	2.2925	2.2935	2.2946	2.2956	2.2966	2.2976	2.2986	2.2996	2.3006	2.3016

Answers to Selected Problems

CHAPTER 1

EXERCISE 1-1

1. F   **3.** F   **5.** T   **7.** T   **9.** F   **11.** F

13.

15.

17.

19. $-3, 0, 5$ (Infinitely many more answers are possible, except for 0.)

21. $\frac{2}{3}, -\frac{7}{8}, 2.65$ are three of infinitely many.   **23.** $\{6, 7, 8, 9\}$ **25.** $\{a, s, t, u\}$ **27.** $\varnothing$ **29.** $\{5\}$

31. $\varnothing$ **33.** $\{-2, 2\}$ **35. (A)** T **(B)** F **(C)** T   **37. (A)** $3, 4$ **(B)** $-2, -1$ **(C)** $-3, -2$

39. (A) $2, 3$ **(B)** $-5, -4$ **(C)** $6, 7$ **41. (A)** $16, 17$ **(B)** $-21, -20$ **(C)** $5, 6$

43. (A) $5, 6$ **(B)** $-4, -3$ **(C)** $11, 12$ **45.** Three terms: $3x, -4y, 5xy$ **47.** One term: $x(2y - 3z)$

49. Two terms: $xyz, w(x + y + z)$ **51.** Four factors: $2, a, b, c$ **53.** Two factors: $x + 2, x - 3$

55. Three factors: $3x, x + 3y, 3y + x$ **57.** F   **59.** T   **61.** T   **63.** F   **65.** F   **67.** F   **69.** T

71. F   **73. (A)** $\{1, 2, 3, 4, 6\}$ **(B)** $\{2, 4\}$ **75. (A)** $\{2, 4, 6, 8, 10\}$ **(B)** $\{4, 6\}$ **77.** $\frac{1}{11}$ **79.** $\frac{41}{333}$

81. $\frac{281}{495}$ **83.** $\dfrac{23,433}{99,900} = \dfrac{7,811}{33,300}$

85. (A) $0.88888888\ldots$ **(B)** $0.27272727\ldots$
(C) $2.23606797\ldots$ **(D)** $1.37500000\ldots$

87. (A) 0.4 **(B)** -0.075 **(C)** $1.732050808\ldots$
(D) $0.003\,003\,003\ldots$

89. $\{P, V\}, \{P, S\}, \{P, T\}, \{V, S\}, \{V, T\}, \{S, T\}$

91. $\{A, K, Q\}, \{A, K, J\}, \{A, K, 10\}, \{A, Q, J\}, \{A, Q, 10\}, \{A, J, 10\}, \{K, Q, J\}, \{K, Q, 10\}, \{K, J, 10\}, \{Q, J, 10\}$

EXERCISE 1-2

1. $11 - 5 = \frac{12}{2}$ **3.** $4 > -18$ **5.** $-12 < -3$ **7.** $x \geq -8$ **9.** $-2 < x < 2$ **11.** $<$ **13.** $>$

15. $<$ **17.** $=$ **19.** $<$ **21.** $>$ **23.** $<, <$ **25.** $>$ **27.** $>$ **29.** $<$ **31.** $<$ **33.** $>$

35. $>$ **37.** Greater than   **39.**

41.

43.

45.

47.

49.

51.

53. $x - 8 > 0$ **55.** $x + 4 \geq 0$

57. $80 = 2x + 3$ **59.** $-3 \leq x < 4$ **61.** $26 = x - 12$ **63.** $x < 2x - 6$ **65.** $6x = 3x + 4$

67. $x - 6 = 5(x + 7)$ **69.** $63 \leq \frac{9}{5}C + 32 \leq 72$ **71.** $x + (x + 1) + (x + 2) = 186$ **73.** $t = -5$

75. $5x + 7x = 12x$ **77.** $3 - x$ **79.** Symmetric property   **81.** Transitive property for equality

83. Trichotomy property   **85.** Substitution principle   **87.** Transitive property for equality or substitution principle

89. Substitution principle   **91.** Substitution principle

93. ''Is'' does not translate into ''equal'' in this case. (The number 8 is actually an element in the set of even numbers.) The properties of equality do not apply.

95. $90 = x(2x - 3)$ **97.** $2x + 2(3x - 10) = 210$ **99. (A)** $A = x(200 - x)$ **(B)** $0 \leq x \leq 200$

EXERCISE 1-3

1. 23 **3.** 18 **5.** 14 **7.** $3 + x$ **9.** $(5 \cdot 7)z$ **11.** mn **13.** $(9 + 11) + M$ **15.** $3x + 3$

17. $2x + x^2$ **19.** $7x$ **21.** $x + y$ **23.** Commutative property for addition

25. Associative property for addition   **27.** Commutative property for multiplication   **29.** Distributive property

554

31. Associative property for multiplication   **33.** Identity property for addition
35. Identity property for multiplication   **37.** Distributive property   **39.** $x + 9$ **41.** $20y$ **43.** $u + 15$
45. $21x$ **47.** x **49.** Commutative property for addition
51. Commutative property for multiplication   **53.** Commutative property for addition
55. Associative property for multiplication   **57.** $x + y + z + 12$ **59.** $3x + 4y + 11$ **61.** $x + y + 5$
63. $36mnp$ **65.** $x^2 + 5x$ **67.** $2(x + 9)$ **69.** $8x + 16$ **71.** $6(x + 2)$ **73.** $a(x + y)$ **75.** $y(1 + x)$
77. $a^2 + ab$ **79.** $xy + y$ **81.** (B) and (D) are false, since $12 - 4 \neq 4 - 12$ and $12 \div 4 \neq 4 \div 12$.
83. False; $8 \div (2 + 2) \neq (8 \div 2) + (8 \div 2)$.   **85.** False, $12 \times (6 \div 2) \neq (12 \times 6) \div (12 \times 2)$.   **87.** Closed
89. Closed   **91.** Closed   **93.** Not closed
95. *1.* Commutative +   *2.* Associative +   *3.* Distributive   *4.* Substitution =
97. *1.* Commutative +   *2.* Associative +   *3.* Associative +   *4.* Substitution =   *5.* Commutative +   *6.* Associative +

EXERCISE 1-4

1. -7 **3.** 6 **5.** 2 **7.** 27 **9.** 0 **11.** -10 **13.** 4 **15.** -6 **17.** 12 **19.** Sometimes
21. -3 **23.** -2 **25.** -6 **27.** -5 **29.** -5 **31.** -5 **33.** -1 **35.** -6 **37.** 28
39. -2 **41.** 2 **43.** 8 **45.** -5 **47.** 7 or -7 **49.** -5 **51.** 5 **53.** -29.191 **55.** 76.025
57. -16.179 **59.** 39.596 **61.** 16.179 **63.** -10.405 **65.** -5 **67.** -11 **69.** -3 **71.** 6
73. -5 **75.** $-\frac{1}{6}$ **77.** $-\frac{5}{6}$ **79.** $\frac{1}{12}$ **81.** $\frac{1}{4}$ **83.** $-\frac{1}{6}$ **85.** True
87. False; $(+7) - (-3) = +10$, $(-3) - (+7) = -10$ **89.** True
91. False; $|(+9) + (-3)| = +6$, $|+9| + |-3| = +12$ **93.** True   **95.** False; $5 - (3 + 2) \neq (5 - 3) + 2$
97. *1.* Commutative +   *2.* Associative +   *3.* Inverse +   *4.* Identity +
99. *1.* Definition of subtraction   *2.* Problem 98   *3.* Associative +   *4.* Definition of subtraction
101. $23.75   **103.** $+1$; that is, 1 over par

EXERCISE 1-5

1. 15 **3.** 3 **5.** -18 **7.** -3 **9.** 0 **11.** 0 **13.** Not defined   **15.** Not defined   **17.** -7
19. -1 **21.** 11 **23.** 9 **25.** Both are 8   **27.** -10 **29.** 3 **31.** -14 **33.** -5 **35.** -70
37. 0 **39.** 10 **41.** 40 **43.** 12 **45.** -18 **47.** -2 **49.** -5 **51.** 4 **53.** -1 **55.** 0
57. 6 **59.** 5 **61.** 1 **63.** Not defined   **65.** -12 **67.** Not defined   **69.** 56 **71.** 0 **73.** -6
75. 8 **77.** 0 **79.** -50 **81.** 0 **83.** Not defined (cannot divide by 0)   **85.** 100 **87.** Not defined
89. $\frac{35}{24}$ **91.** $-\frac{29}{48}$ **93.** Never   **95.** Always   **97.** When x and y are of opposite signs
99. *1.* Identity +   *2.* Distributive   *3.* Addition =   *4.* Inverse and associative +   *5.* Inverse +
    6. Identity +   *7.* Symmetric =
101. *1.* Distributive   *2.* Inverse +   *3.* Definition of multiplication   *4.* Inverse +
103. $+8$; that is, 8 over par for a score of 80
105. **(A)** 6 mph forward; that is, $+6$ **(B)** 6 mph backward; that is, -6

EXERCISE 1-6

1. 7 **3.** 3 **5.** 9 **7.** 5 **9.** 18 **11.** 5 **13.** $10x^{11}$ **15.** $12y^{10}$ **17.** 20×10^9
19. 35×10^{17} **21.** $17x$ **23.** x **25.** $8x$ **27.** $-13t$ **29.** $3x + 5y$ **31.** $4 + 4x$ **33.** $4m - 6n$
35. $9u - 4v$ **37.** $-2m - 24n$ **39.** $5u - 6v$ **41.** $4x - 3$ **43.** $x + 11$ **45.** -12 **47.** -25
49. 84 **51.** $k + m$ **53.** 1 **55.** $7m$ **57.** $3m + 4$ **59.** $2m + 5$ **61.** $6x^{2m+1}$ **63.** $10x^{7m}$
65. $-3x^2y$ **67.** $3y^3 + 4y^2 - y - 3$ **69.** $3a^2 - b^2$ **71.** $-7x + 9y$ **73.** $-5x + 3y$ **75.** $4x - 6$
77. $-8x + 12$ **79.** $10t - 18$ **81.** $x - 14$ **83.** -30 **85.** 15 **87.** $27 - a$ **89.** $b + c - a$
91. -3 **93.** $3a - 2$ **95.** $-\frac{7}{18}$ **97.** $-\frac{23}{108}$ **99.** 6, 12, 10; mn

EXERCISE 1-7

1. 9 **3.** -7 **5.** 4 **7.** 6 **9.** 4 **11.** -1 **13.** 15 **15.** -7 **17.** -2 **19.** 11 **21.** 18
23. All real numbers   **25.** 9 **27.** No solution   **29.** 10 **31.** 1 **33.** All real numbers   **35.** $\frac{9}{2}$
37. $-\frac{5}{2}$ **39.** 10 **41.** $\frac{5}{6}$ **43.** $\frac{1}{8}$ **45.** 32 **47.** 4 **49.** $\frac{33}{20}$ **51.** 4 **53.** No solution
55. All real numbers   **57.** 0 **59.** No solution   **61.** $x = 15 - \frac{1}{2}y$ **63.** $x = y - 1$ **65.** $y = 8 - x$
67. $x = 6 - 2y$ **69.** $y = \frac{4}{9}x$ **71.** $y = \frac{5}{3}x - 2$ **73.** $y = \frac{3}{2}x - 1$ **75.** Subtraction property
77. Transitive property or substitution principle   **79.** Symmetric property   **81.** Division property   **83.** (A)
85. (A)   **87.** Neither   **89.** $\frac{17}{2}$ **91.** -0.2 **93.** $\frac{8}{15}$ **95.** $\frac{7}{20}$ **97.** $y = x$ **99.** $x = \frac{2}{3}y + \frac{2}{9}$

EXERCISE 1-8

1. $2x$ **3.** $x - 3$ **5.** $2x + 2$ **7.** $3x - 4$ **9.** $-2x + 1$ **11.** $2x = x - 3;\ x = -3$

13. $2x + 2 = 3x - 4;\ x = 6$ **15.** $-2x + 1 = 3x;\ x = \frac{1}{5}$ **17.** $x + 2 = 2x - 5;\ x = 7$

19. $6x + 5 = -3x - 1;\ x = -\frac{2}{3}$ **21.** $(2x - 3) + x = 12;$ 5 ft and 7 ft   **23.** $2x + 2(x - 6) = 36;\ 12 \times 6$ ft

25. $2(2x + 3) + 2x = 66;\ 23 \times 10$ cm   **27.** $\frac{x}{6} - 2 = \frac{x}{4} + 1;\ -36$ **29.** $x + (x + 2) = (x + 4) + 5;\ 7, 9, 11$

31. $2x + 2 \cdot \frac{x}{6} = 84;\ 36 \times 6$ m   **33.** $\frac{3x}{5} - 4 = \frac{x}{3} + 8;\ 45$ **35.** $7x = 4x - 12;\ x = -4$

37. $2x + 3 = 3x - 12;\ x = 15$ **39.** 8 in.   **41.** $x + (x + 1) + (x + 2) = 96;\ 31, 32, 33$

43. $x + (x + 2) + (x + 4) = 42;\ 12, 14, 16$ **45.** $4x + x = 55;\ x = 11;$ the numbers are 44 and 11.

47. $5x + x = 48;\ x = 8;$ the numbers are 40 and 8.   **49.** $\frac{2P}{5} + 70 + \frac{P}{4} = P;\ 200$ cm

51. $\frac{D}{3} + 6 + \frac{D}{2} = D;\ 36$ km   **53.** $3x = 2(40 - x) + 10;\ x = 18$ **55.** $3x - (70 - x) = 3(70 - x) - 7x;\ x = 20$

57. $3x - 20 = w + 20;\ 20$ ft   **59.** $\frac{1}{2}(180 - x) - \frac{1}{6}x = 10;\ 120°$

CHAPTER 1 REVIEW EXERCISE

1. **(A)** 143, 12   **(B)** 143, 0, 12   **(C)** 143, -1, 0, 12   **(D)** $3.127\overline{127}$, 143, -1.43, $\frac{2}{3}$, -1, 0, 12, $-\frac{3}{7}$ **(E)** $\sqrt{5}$, π *(1-1)*

2. **(A)** **(B)** *(1-2)*   **3.** 5^{13} *(1-6)*   **4.** x^{10} *(1-6)*

5. 17 *(1-5)*   **6.** 13 *(1-5)*   **7.** -5 *(1-4)*   **8.** -13 *(1-4)*

9. 6 *(1-4)*   **10.** -3 *(1-4)*   **11.** 3 *(1-4)*   **12.** -12 *(1-4)*   **13.** 28 *(1-5)*   **14.** -18 *(1-5)*

15. -4 *(1-5)*   **16.** 6 *(1-5)*   **17.** Not defined *(1-5)*   **18.** 0 *(1-5)*   **19.** 4 *(1-5)*   **20.** -14 *(1-5)*

21 -5 *(1-5)*   **22.** -12 *(1-5)*   **23.** 8 *(1-4)*   **24.** 5 *(1-4)*   **25.** -3 *(1-4)*   **26.** -2 *(1-4)*

27. -5 *(1-4)*   **28.** -5 *(1-4)*   **29.** $x + 10$ *(1-3)*   **30.** $15x$ *(1-3)*   **31.** $8xy$ *(1-3)*

32. $x + y + z + 12$ *(1-3)*   **33.** $x + 2$ *(1-3)*   **34.** x *(1-3)*   **35.** $3x - 15$ *(1-3)*   **36.** $7a + 7b$ *(1-3)*

37. $3x + 14$ *(1-3)*   **38.** $2x + 1$ *(1-3)*   **39.** $>$ *(1-2)*   **40.** $<$ *(1-2)*   **41.** $<$ *(1-2)*   **42.** $<$ *(1-2)*

43. $>$ *(1-2)*   **44.** $<$ *(1-2)*   **45.** $6a$ *(1-6)*   **46.** $5a$ *(1-6)*   **47.** $-8x + 8y$ *(1-6)*

48. $5x - 3y + 2z - 1$ *(1-6)*   **49.** $12x^3y + x^2y^2 - 6xy^3$ *(1-6)*   **50.** $-xy + 3xy^2$ *(1-6)*   **51.** 6 *(1-5)*

52. 2 *(1-4)*   **53.** 4 *(1-4)*   **54.** 1 *(1-4)*   **55.** 4 *(1-5)*   **56.** -26 *(1-5)*   **57.** 10 *(1-5)*

58. 35 *(1-5)*   **59.** -2 *(1-5)*   **60.** Not defined *(1-5)*   **61.** -30 *(1-6)*   **62.** -180 *(1-6)*

63. -108 *(1-6)*   **64.** $\frac{1}{60}$ *(1-6)*   **65.** 6 *(1-5)*   **66.** -2 *(1-4)*   **67.** 8 *(1-4)*   **68.** -6 *(1-4)*

69. -9 *(1-5)*   **70.** -7 *(1-5)*   **71.** -1 *(1-5)*   **72.** -4 *(1-7)*   **73.** $-4y - 4$ *(1-7)*   **74.** $\frac{10}{7}$ *(1-7)*

75. All real numbers *(1-7)*   **76.** $\frac{8}{7}$ *(1-7)*   **77.** $\frac{y}{y - 1}$ *(1-7)*   **78.** $x - 1 > 0$ *(1-2)*   **79.** $2x + 3 \geq 0$ *(1-2)*

80. $50 = 2x - 10$ *(1-2)*   **81.** $x < 2x - 12$ *(1-2)*   **82.** $-5 \leq x < 5$ *(1-2)*   **83.** $x + 8 = 5(x - 6)$ *(1-2)*

84. $x(x - 10) = 1,200$ *(1-2)*   **85.** $2x + 2(x + 5) = 43$ *(1-2)*   **86.** Commutative property for addition *(1-3)*

87. Associative property for addition *(1-3)*   **88.** Commutative property for multiplication *(1-3)*

89. Associative property for multiplication *(1-3)*   **90.** Additive identity *(1-3)*   **91.** Additive inverse *(1-3)*

92. Distributive property *(1-3)*   **93.** -42 *(1-5)*   **94.** 6 *(1-5)*   **95.** $\frac{58}{33}$ *(1-5)*   **96.** 6 *(1-8)*

97. 15, 16, 17 *(1-8)*   **98.** 7, 8 *(1-8)*   **99.** 36, 18, and 12 in. *(1-8)*

100. $2x + 2\left(\frac{3x}{5} - 2\right) = 76;\ 25$ by 13 cm. *(1-8)*

CHAPTER 1 PRACTICE TEST

1. 41 *(1-4)*   **2.** 25 *(1-4)*   **3.** -324 *(1-5)*   **4.** -12 *(1-6)*   **5.** -117 *(1-6)*   **6.** x^9 *(1-6)*

7. $ab + ac - ad$ *(1-3)*   **8.** $3x(1 - 2y + 3x)$ *(1-3)*   **9.** $5x^2y - xy$ *(1-6)*

10. *(1-2)*   **11.** *(1-2)*   **12.** $3x + x(x + 2)$ *(1-2)*

13. $x^2 < (x - 4)(x + 1)$ *(1-2)*   **14.** $x + 3 = 2(x - 3)$ *(1-4)*   **15.** $x = -\frac{19}{2}$ *(1-7)*   **16.** $x = 2$ *(1-7)*

17. All real numbers *(1-7)*   **18.** $a = \frac{b + c}{bc - 1}$ *(1-7)*   **19.** $4(x - 5) = x + 1;\ x = 7$ *(1-8)*

20. $x + \frac{1}{3}x + (x - 2) = 40;\ x = 18;$ the sides are 18, 6, and 16 in. *(1-8)*

CHAPTER 2

EXERCISE 2-1

1. Binomial, 2   **3.** Trinomial, 6   **5.** Binomial, 3   **7.** Monomial, 8   **9.** Monomial, 0   **11.** -3 **13.** 3
15. 1   **17.** $9x - 3$ **19.** $-2x - 4$ **21.** $7x^2 - x - 12$ **23.** $7 + 11a + 7b$ **25.** $12x + 5y$ **27.** $-5x$
29. $-3x^2$ **31.** $9x^2$ **33.** $-x + 1$ **35.** $-y^2 - 2$ **37.** $-3x^2y$ **39.** $3y^3 + 4y^2 - y - 3$ **41.** $3a^2 - b^2$
43. $-7x + 9y$ **45.** $-5x + 3y$ **47.** $4x - 6$ **49.** $-8x + 12$ **51.** $10t - 18$ **53.** $x - 14$
55. $-m + 2n$ **57.** $-y$ **59.** y **61.** $-x - y$ **63.** $x + y$ **65.** $y + z$ **67.** $2x^4 + 3x^3 + 7x^2 - x - 8$
69. $4x^5 - 7x^4 - x^3 + 3x^2 + 1$ **71.** $10xy - 3x^2y - x^2y^2$ **73.** $-3x^3 + x^2 + 3x - 2$
75. $2xy - 3x^2y + xy^2 - 4x^2y^2$ **77.** $-4x^4 - 3x^3 + x^2 + 2x - 1$ **79.** $3xy + 6x^2y - 5xy^2 + 10x^2y^2$
81. $-4x^5 - x^4 - 5x^3 + 2x^2 + x$ **83.** $5xy + 3x^2y - 4xy^2 + 6x^2y^2$ **85.** $-t + 27$ **87.** $2x - w$ **89.** -3
91. $3x - 2$ **93.** $P = 2x + 2(x - 5) = 4x - 10$ **95.** Value in cents $= 5x + 10(x - 5) + 25(x - 3) = 40x - 125$
97. $x + 2x + 2x + 3x = 8x$ **99.** $8t + 12 \cdot 2t = 32t$

EXERCISE 2-2

1. y^5 **3.** $10y^5$ **5.** $-24x^{20}$ **7.** $6u^{16}$ **9.** c^3d^4 **11.** $15x^2y^3z^5$ **13.** $y^2 + 7y$ **15.** $10y^2 - 35y$
17. $3a^5 + 6a^4$ **19.** $2y^3 + 4y^2 - 6y$ **21.** $7m^6 - 14m^5 - 7m^4 + 28m^3$ **23.** $10u^4v^3 - 15u^2v^4$
25. $2c^3d^4 - 4c^2d^4 + 8c^4d^5$ **27.** $6y^3 + 19y^2 + y - 6$ **29.** $m^3 - 2m^2n - 9mn^2 - 2n^3$
31. $6m^4 + 2m^3 - 5m^2 + 4m - 1$ **33.** $a^3 + b^3$ **35.** $2x^4 + x^3y - 7x^2y^2 + 5xy^3 - y^4$ **37.** $x^2 + 5x + 6$
39. $a^2 + 4a - 32$ **41.** $t^2 - 16$ **43.** $m^2 - n^2$ **45.** $4t^2 - 11t + 6$ **47.** $3x^2 - 7xy - 6y^2$ **49.** $4m^2 - 49$
51. $30x^2 - 2xy - 12y^2$ **53.** $6s^2 - 11st + 3t^2$ **55.** $x^4 + x^3y + xy^3 + y^4$ **57.** $2x^3 + 6xy - x^2y^2 - 3y^3$
59. $9x^2 + 12x + 4$ **61.** $4x^2 - 20xy + 25y^2$ **63.** $36u^2 + 60uv + 25v^2$ **65.** $4m^2 - 20mn + 25n^2$
67. $x^4 - 2x^2 + 1$ **69.** $x^4 + 2x^2y^2 + y^4$ **71.** $x^3 + 6x^2y + 12xy^2 + 8y^3$ **73.** $-x^2 + 17x - 11$
75. $2x^3 - 13x^2 + 25x - 18$ **77.** $9x^3 - 9x^2 - 18x$ **79.** $12x^5 - 19x^3 + 12x^2 + 4x - 3$
81. $9x^2 + 6xy + y^2 - 12x - 4y + 4$ **83.** $x^2 + 2xy + y^2 - x - y - 2$ **85.** -7 **87.** -1 **89.** $m + n$
91. Area $= y(y - 8) = y^2 - 8y$ **93.** $(8 + 0.25x)(5{,}000 - 100x) = -25x^2 + 450x + 40{,}000$
95. $x[40 - 0.5(x - 120)]$ **97.** $x(x + 1) = (x - 1)(x + 3); x^2 + x = x^2 + 2x - 3$ **99.** $x = 3$

EXERCISE 2-3

1. $3x(z - 2)$ **3.** $2(4x + y)$ **5.** $3x(2x + 3)$ **7.** $7xy(2x - y)$ **9.** $(x - 3)(2x + z)$ **11.** $(a + b)(c - d)$
13. $(x - y)^2$ **15.** $(c + d)(ab - 1)$ **17.** $x^3(x^2 + x + 1)$ **19.** $a^2b(1 - ab - a^2)$ **21.** $(x - 5)^2$
23. $(x + 6)^2$ **25.** Not a perfect square   **27.** $(x - 10)^2$ **29.** $(x + 12)^2$ **31.** Not a perfect square
33. $c(ab + bd + de)$ **35.** $xyz(yz^2 - xy^2 + x^2z)$ **37.** $(v - 5)(v + 5)$ **39.** $(3x - 2)(3x + 2)$
41. Not factorable   **43.** $(3x - 4y)(3x + 4y)$ **45.** $(2x - 3y)(2x + 3y)$ **47.** $(5x - 8y)(5x + 8y)$
49. Not factorable   **51.** $(a - 2)(a - 3)$ **53.** $(x - 1)(x + 1)$ **55.** $6(x - 2)$ **57.** $-3(x - 1)$
59. $(x^2 + 3)(2x + 1)$ **61.** $(ab - c^2)(c - ab)$ **63.** $(3x - y)(y + 2)$ **65.** $(x^2 + 3)(2x + 1)$
67. $(ab - c^2)(c - ab)$ **69.** $(3x - y)(y + 2)$ **71.** $3(x - 3)(x + 3)$ **73.** $5(x - 5)(x + 5)$
75. $3(x - 2y)(x + 2y)$ **77.** $(xy - 1)(xy + 1)$ **79.** $(ab - cd)(ab + cd)$ **81.** $(x^2 + 3)(2x^2 - 3)$
83. $(x^2y^2 + 3)(x - y)$ **85.** $(b - a)(4a - b)$ **87.** $x(yz - 1)(y - z)$ **89.** $(2 + 3z)(x - 3y)$
91. $(a + c)(3a + b)$ **93.** $(2 - y)(2 + y)(3 - x)$ **95.** $(a + d)(b + c)$ **97.** $(z - y)(y - 2x)$

EXERCISE 2-4

1. $(x + 1)(x + 2)$ **3.** $(x + 2)(x + 5)$ **5.** $(x - 3)(x + 1)$ **7.** $(x + 4)(x - 1)$ **9.** $(x - 4)(x - 5)$
11. $(x - 1)(x - 5)$ **13.** $(x - 3)^2$ **15.** $(x + 7)^2$ **17.** $(x - 3y)^2$ **19.** $(x + 5)(x - 1)$ **21.** $(x + 3)(x - 6)$
23. $(x - 2)(x + 6)$ **25.** $(x - 1)(x - 8)$ **27.** Not factorable   **29.** $(2x + 1)(x - 1)$ **31.** $(2x + 5)(x + 1)$
33. Not factorable   **35.** Not factorable   **37.** $4(x^2 - 10)$ **39.** $2(x - 10)(x + 10)$ **41.** Not factorable
43. $(2x - 5y)(x - 2y)$ **45.** $(2x + 3)(2x - 1)$ **47.** $(4x - 1)(x + 1)$ **49.** Not factorable   **51.** $(2x + 3)^2$
53. $(x - 2)(x + 1)$ **55.** $(3x + 2)^2$ **57.** $(3x - 2)(2x - 1)$ **59.** $(3x + y)(2x - y)$ **61.** Not factorable
63. Not factorable   **65.** $(3x + y)^2$ **67.** Not factorable   **69.** Not factorable   **71.** $(4x + 3)(2x - 5)$
73. Not factorable   **75.** $3(2x - y)(x + 4y)$ **77.** $(x^2 + 1)(x^2 + 3)$ **79.** $(x^2 + 1)(x^2 - 1) = (x^2 + 1)(x + 1)(x - 1)$
81. $(x^2 + 1)^2$ **83.** $(2x^2 + 1)(2x^2 - 1)$ **85.** $(3x^2 - 2)^2$ **87.** $(x^4 + 1)(x^4 - 1) = (x^4 + 1)(x^2 + 1)(x + 1)(x - 1)$
89. $(x - 1)^2(x^2 + x + 1)^2$ **91.** Not factorable   **93.** $(x^3 + 2)(x^3 - 4)$ **95.** $(x^3 + 3)(x^3 + 2)$
97. $(2x^2 - 3)(x^2 + 1)$ **99.** $(x^3 - 2)(x^3 + 3)$

EXERCISE 2-5

1. $(4x - 1)(x + 3)$ **3.** $(2x + 3)(2x - 1)$ **5.** Not factorable   **7.** $(4x - 1)(x - 6)$ **9.** $(6x - 1)(x - 4)$
11. $(2x + 5)(2x - 3)$ **13.** Not factorable   **15.** $(2x + 3)(x - 5)$ **17.** $(3x - 1)(x + 3)$ **19.** Not factorable
21. $(7x - 2)(x + 4)$ **23.** $(3x + 5)(x + 2)$ **25.** Not factorable   **27.** $(6x - 5)(x - 4)$ **29.** $(2x + 3)(3x - 1)$
31. $(2x + 5)(3x - 4)$ **33.** $(3x + 5)(2x - 3)$ **35.** Not factorable   **37.** $(4x + 3)(2x - 1)$ **39.** Not factorable
41. Not factorable   **43.** $(4x + 1)(2x - 9)$ **45.** $(x - 2y)(4x - 5y)$ **47.** Not factorable
49. $2(3x - 2y)(x + 3y)$ **51.** $(4x - 3)(2x + 3)$ **53.** $(6x + 5)(x - 6)$ **55.** $(6x + 5)(4x + 3)$
57. $(6x + 5)(4x - 3)$ **59.** $(12x - 1)(2x + 9)$ **61.** $3(4x - 1)(2x + 3)$ **63.** $(4x + 5y)(3x + y)$
65. Not factorable   **67.** $(4x + 3y)(x - 5y)$ **69.** Not factorable   **71.** $(9x - 4)(2x + 5)$
73. $(3x + 4y)(6x - 5y)$ **75.** $-17, -7, -3, 3, 7, 17$ **77.** $0, 4, 6$ **79.** $(2x^2 + 1)(3x^2 + 2)$
81. $(x^2 + y^2)(2x^2 + 3y^2)$ **83.** $(2x^3 - 3)(x + 1)(x^2 - x + 1)$ **85.** $(4x^2 + 3)(3x^2 + 4)$ **87.** $2(6x^2 - 1)(x^2 + 6)$
89. $3(x^3 - 3)(2x + 1)(4x^2 - 2x + 1)$

EXERCISE 2-6

1. $5(v - 5)(v + 5)$ **3.** $(v - 5)(v^2 + 5v + 25)$ **5.** $21(2m - 1)(2m + 1)$ **7.** $(y + 4)(y^2 - 4y + 16)$
9. $(x + 1)(x^2 - x + 1)$ **11.** $(m - n)(m^2 + mn + n^2)$ **13.** $(2x + 3)(4x^2 - 6x + 9)$ **15.** $3uv^2(2u - v)$
17. $2(x - 2)(x + 2)$ **19.** $2x(x^2 + 4)$ **21.** $3x(2x - y)(2x + y)$ **23.** $2x(x + 1)(x^2 - x + 1)$
25. $6(x + 2)(x + 4)$ **27.** $3x(x^2 - 2x + 5)$ **29.** $(x^2 - 3)(x^2 + 3)$ **31.** $(x^2 + 3y^2)(x^4 - 3x^2y^2 + 9y^4)$
33. $(x^2 + 2)^2$ **35.** $(x^3 + 3)^2$ **37.** $(xy - 4)(xy + 4)$ **39.** $(ab + 2)(a^2b^2 - 2ab + 4)$ **41.** $2xy(2x + y)(x + 3y)$
43. $4(u + 2v)(u^2 - 2uv + 4v^2)$ **45.** $5y^2(6x + y)(2x - 7y)$ **47.** $(y + 2)(x + y)$ **49.** $(x - 5)(x + y)$
51. $(a - 2b)(x - y)$ **53.** $(3c - 4d)(5a + b)$ **55.** $(x - 2)(x + 1)(x - 1)$
57. $(y - x)[(y - x) - 1] = (y - x)(y - x - 1)$ **59.** $(xy + 2)(xy - 3)$ **61.** $(z^2 - 3)(z^2 + 2)$ **63.** $x(x + 1)$
65. $x(x - 1)(2x - 1)$ **67.** $-5(x + 4)^2(x - 1)^3$ **69.** $6(x + 2)^3(x - 4)^4$ **71.** $(x^4 + 2)(x^4 - 2)$
73. $(r^2 + s^2)(r - s)(r + s)$ **75.** $(x^2 - 4)(x^2 + 1) = (x - 2)(x + 2)(x^2 + 1)$
77. $[(x - 3) - 4y][(x - 3) + 4y] = (x - 3 - 4y)(x - 3 + 4y)$ **79.** $[(a - b) - 2(c - d)][(a - b) + 2(c - d)]$
81. $[5(2x - 3y) - 3ab][5(2x - 3y) + 3ab]$ **83.** $(x - 1)(x^2 + x + 1)(x + 1)(x^2 - x + 1)$ **85.** $(2x - 1)(x - 2)(x + 2)$
87. $[5 - (a + b)][5 + (a + b)]$ **89.** $(x^2 + 1)(x + 1)(x - 1)$ **91.** $(x - y)(x + y)(x^2 + xy + y^2)(x^2 - xy + y^2)$
93. $(x + 2)(x - 2)(x^2 - 2x + 4)(x^2 + 2x + 4)$ **95.** $(x^4 + y^4)(x^2 + y^2)(x + y)(x - y)$
97. $[4x^2 - (x - 3y)][4x^2 + (x - 3y)]$ **99.** $(x - 2)(x^2 + 3)$ **101.** $(x^4 + 1)(x - 1)$ **103.** $(3x - 1)(x^2 + 4)$
105. $(x + 2 + y)(x + 2 - y)$ **107.** There are no integers p and q with $p + q = -1$, $pq = 1$

EXERCISE 2-7

1. $-1, -2$ **3.** $-2, -5$ **5.** $3, -1$ **7.** $-4, 1$ **9.** $4, 5$ **11.** $1, 5$ **13.** 3 **15.** -7 **17.** $1, -\frac{1}{2}$
19. $-1, -\frac{5}{2}$ **21.** $0, -5$ **23.** $0, -4$ **25.** $12, -1$ **27.** $1, -5$ **29.** $-\frac{2}{3}, 4$ **31.** $-\frac{3}{2}, \frac{1}{2}$ **33.** $\frac{1}{4}, -1$
35. Not solvable by factoring   **37.** $-\frac{3}{2}$ **39.** $2, -1$ **41.** $-\frac{2}{3}$ **43.** $\frac{2}{3}, \frac{1}{2}$ **45.** $-1, 3$ **47.** $-\frac{2}{3}, 1$
49. Not solvable by factoring   **51.** $-\frac{1}{2}, 3$ **53.** $-2, 2$ **55.** $-6, 2$ **57.** $-\frac{1}{2}, 2$ **59.** $\frac{1}{2}, 2$ **61.** $2, -3$
63. Not solvable by factoring   **65.** $-\frac{3}{4}, \frac{5}{2}$ **67.** $3, -3$ **69.** $1, 2, -2$ **71.** $3, -3$ **73.** $-2, 1$
75. $-1, 1$ **77.** $-1, 1$ **79.** 10 **81.** -9 and -8 **83.** $-3, \frac{7}{2}$ **85.** Base = 6 cm, Height = 12 cm
87. 11 by 3 in.   **89.** 9 by 12 in.

CHAPTER 2 REVIEW EXERCISE

1. 2 *(2-1)*   **2.** 7 *(2-1)*   **3.** **(A)** 5 **(B)** 3 *(2-1)*   **4.** **(A)** 5 **(B)** 11 *(2-2)*   **5.** $x^2 + 2x + 1$ *(2-1)*
6. $-x^2 + 2x + 9$ *(2-1)*   **7.** $2x^3 + 5x^2 - 8x - 20$ *(2-2)*   **8.** $x^2 - 9$ *(2-2)*   **9.** -6 *(2-1)*   **10.** $2x$ *(2-1)*
11. $-x^2 + 6x - 2$ *(2-1)*   **12.** $x^3 - x^2 - 14x + 24$ *(2-2)*   **13.** $x^2 - 4x + 10$ *(2-1)*   **14.** $x^2 + 5x - 3$ *(2-1)*
15. $5x^2 - x + 5$ *(2-2)*   **16.** $6x^4 - 5x^3 + 8x^2 + 5x + 4$ *(2-2)*   **17.** $2x^2 - 2$ *(2-1)*
18. $2x^3 - 4x^2 + 12x$ *(2-1)*   **19.** $2x - 2$ *(2-1)*   **20.** $(x - 2)(x - 3)$ or $x^2 - 5x + 6$ *(2-2)*
21. $3x^3(x + 3)$ *(2-3)*   **22.** $x^2y(2y^2 + x)$ *(2-3)*   **23.** $3x(2x + 5)$ *(2-3)*   **24.** $(x - 4)(x + 4)$ *(2-3)*
25. Not factorable *(2-3)*   **26.** $14(x - 2)(x + 2)$ *(2-3)*   **27.** $3xy(2x + 3 + 4y)$ *(2-3)*   **28.** $(3xy + 4)(2x + 3)$ *(2-3)*
29. $(x - 5)(2x + 1)$ *(2-3)*   **30.** $(x + 3)(4x - 1)$ *(2-4)*   **31.** $(2x + 5)(2x - 5)$ *(2-6)*
32. $(x + y)(x + 2y)$ *(2-4)*   **33.** Not factorable *(2-6)*   **34.** $(x + 2)(x + 2)$ or $(x + 2)^2$ *(2-3)*
35. $(2x + 1)(x - 3)$ *(2-4)*   **36.** $(x - 3)(x - 3)$ or $(x - 3)^2$ *(2-3)*   **37.** $(3x + 2)(x - 1)$ *(2-3)*
38. $(a - 2)(a^2 + 2a + 4)$ *(2-6)*   **39.** $(x + 2)(x + 4)$ *(2-4)*   **40.** $(x - 2y)(2x - y)$ *(2-4)*
41. $(x + 4)(x^2 - 4x + 16)$ *(2-6)*   **42.** $(a - 3b)(a + 3b)$ *(2-6)*   **43.** $(x + 3)^2$ *(2-3)*   **44.** $(x - 4)^2$ *(2-3)*
45. $(x - 2)(x - 6)$ *(2-4)*   **46.** $(x + 3)(x + 6)$ *(2-4)*   **47.** $(x - 11)(x + 1)$ *(2-4)*   **48.** $(x + 13)(x - 2)$ *(2-4)*
49. $(x - 6)^2$ *(2-3)*   **50.** Not factorable *(2-4)*   **51.** $(2x - 5)(4x^2 + 10x + 25)$ *(2-6)*
52. $(x^2 + 1)(x + 1)(x - 1)$ *(2-6)*   **53.** Not factorable *(2-6)*   **54.** $(2x + 3)^2$ *(2-3)*   **55.** $(3x - 2)^2$ *(2-3)*

56. $(2x + 1)(x + 1)$ *(2-4)*   **57.** $(2x - 3)(x - 1)$ *(2-4)*   **58.** $(2x + 1)(3x - 1)$ *(2-4)*
59. $(x - y)(x^2 + xy + y^2)$ *(2-6)*   **60.** $(x + 2y)^2$ *(2-3)*   **61.** Not factorable *(2-4)*   **62.** Not factorable *(2-4)*
63. $-4, -3$ *(2-7)*   **64.** $-2, 2$ *(2-7)*   **65.** $2, -5$ *(2-7)*   **66.** $6, -1$ *(2-7)*   **67.** $4, -5$ *(2-7)*
68. $6, -2$ *(2-7)*   **69.** 1 *(2-7)*   **70.** $-1, 1$ *(2-7)*   **71.** Not solvable by factoring *(2-7)*
72. -2 *(2-7)*   **73.** Not solvable by factoring *(2-7)*   **74.** 1 *(2-7)*   **75.** 1 *(2-7)*
76. Not solvable by factoring *(2-7)*   **77.** $(x^2 + 1)(x^4 - x^2 + 1)$ *(2-6)*   **78.** $x^2(2x + 1)(x + 3)$ *(2-6)*
79. $a^3(x - 2)^2$ *(2-6)*   **80.** $3xy(x + y)(x - y)$ *(2-6)*   **81.** $(x^2 + 1)(x - 3)$ *(2-6)*   **82.** $x(x^4 + 1)(x + 1)$ *(2-6)*
83. $2x(x + 1)(x - 2)$ *(2-6)*   **84.** $(x + y + 1)(x + y - 1)$ *(2-6)*   **85.** $(x^2 + x + 1)(x^2 - x + 1)(x - 1)(x + 1)$ *(2-6)*
86. $(2x + 5)(4x^2 - 10x + 25)$ *(2-6)*   **87.** $(3x + 2)(x^2 - 5)$ *(2-6)*   **88.** $x^2(x + 3)(x - 1)$ *(2-6)*
89. $-x^3y(xy + 1)^2$ *(2-6)*   **90.** $2a(a + 1)(a^2 - a + 1)$ *(2-6)*   **91.** $-1, \frac{3}{2}$ *(2-7)*   **92.** $-4, 5$ *(2-7)*
93. $-\frac{3}{2}, \frac{2}{3}$ *(2-7)*   **94.** $\frac{1}{4}, -\frac{8}{3}$ *(2-7)*   **95.** 8 *(2-7)*   **96.** 5 *(2-7)*   **97.** 8 *(2-7)*
98. 6 by 30 in. *(2-7)*   **99.** Base $= 30$ cm, Height $= 10$ cm *(2-7)*   **100.** 5 by 7 ft *(2-7)*

CHAPTER 2 PRACTICE TEST

1. **(A)** 3 **(B)** -4 **(C)** 4 *(2-1)*   **2.** **(A)** 6 **(B)** 13 **(C)** $-6x^5y^8$ *(2-1)*   **3.** $x^2 + 2x + 6$ *(2-1)*
4. $4x^3 + 17x^2 + 35x + 25$ *(2-2)*   **5.** $2x^2 - 15x + 28$ *(2-2)*   **6.** $2x^3 - 3x - 1$ *(2-1)*   **7.** $-2x$ *(2-1)*
8. $2ab(b + 2ab + 3a)$ *(2-3)*   **9.** $5(x - 4)(x + 4)$ *(2-3)*   **10.** $(x + 6)^2$ *(2-3)*   **11.** $(x - 4)(x - 8)$ *(2-4)*
12. $(x - 3)(x + 15)$ *(2-4)*   **13.** $(x - 4)(x^2 + 4x + 16)$ *(2-6)*   **14.** $(x + 4y)(x - 3y)$ *(2-3)*
15. $(3x + 4)(2x - 5)$ *(2-5)*   **16.** $(x^2 - 3)(x^2 + 8)$ *(2-6)*   **17.** $2, 5$ *(2-7)*   **18.** $2, -\frac{1}{2}$ *(2-7)*
19. Not solvable by factoring *(2-7)*   **20.** 7 *(2-7)*   **21.** $1\frac{2}{3}$ by 3 ft *(2-7)*

CUMULATIVE REVIEW EXERCISES, CHAPTERS 1–2

1. $3(x - 2) = x + 4$ *(1-2)*   **2.** $(x - 2)(x + 3) = 84$ *(1-2)*   **3.** $3x + 2 < \frac{1}{2}x + 4$ *(1-2)*   **4.** -13 *(1-4)*
5. 40 *(1-5)*   **6.** -6 *(1-4)*   **7.** 2 *(1-5)*   **8.** 78 *(1-5)*   **9.** -42 *(1-5)*   **10.** -41 *(1-6)*
11. 5 *(1-4)*   **12.** -4 *(2-1)*   **13.** **(A)** -4 **(B)** -1 *(2-1)*   **14.** **(A)** 3 **(B)** 8 *(2-1)*
15. x^{16} *(1-6)*   **16.** $3x + 24$ *(1-3)*   **17.** $-2x + 2y$ *(1-6)*   **18.** $2xyz$ *(1-3, 1-6)*   **19.** $3x^2 - x + 15$ *(2-1)*
20. $-x^2 + 10x + 13$ *(2-1)*   **21.** $x^3 + 3x^2 + 3x + 1$ *(2-2)*   **22.** $3x^2 + 2x - 8$ *(2-2)*   **23.** $(x + 2)^2$ *(2-2)*
24. $(x + 2)(x + 3)$ *(2-4)*   **25.** $(x - 12)(x + 2)$ *(2-4, 2-5)*   **26.** $3x^3(x - 3)$ *(2-3)*   **27.** $x(x + 1)(x - 1)$ *(2-3)*
28. $(x - 1)(x^2 + x + 1)$ *(2-6)*   **29.** Not factorable *(2-3)*   **30.** $(x - 5)^2$ *(2-3)*   **31.** $-\frac{7}{2}$ *(1-7)*
32. $\frac{1}{4}$ *(1-7)*   **33.** $\frac{498}{35}$ *(1-7)*   **34.** $\frac{6}{5}$ *(1-7)*   **35.** $\frac{1}{2}$ *(1-7)*   **36.** $0, 1, -2$ *(2-7)*
37. $x = \dfrac{y - 3}{2}$ *(1-7)*   **38.** $x = \dfrac{3 - 4y}{6}$ *(1-7)*   **39.** $4, -1$ *(2-7)*   **40.** $(x - 2) + 3 = 5x - 4$; $x = \frac{5}{4}$ *(1-8)*
41. $x + (x + 1) + (x + 2) = 111$; $x = 36$ *(1-8)*   **42.** $x + 2(x + 2) = 3(x + 1) + 1$; all integers x *(1-8)*
43. 5 by 11 in. *(1-8)*   **44.** 3, 4, and 9 ft *(1-8)*   **45.** $(x - 2)^2 - 1 = x + 3$ *(1-2)*
46. $2x^2 = (x + 5)^2 - 14$ *(1-2)*   **47.** $|x - 5| = x - 9$ *(1-4)*   **48.** -5 *(1-3)*   **49.** -3 *(1-6)*
50. $-\frac{1}{5}$ *(1-3)*   **51.** 9 *(1-6)*   **52.** 11 *(1-4)*   **53.** 3 *(1-5)*   **54.** x^{21} *(1-6)*   **55.** $2x^3 - 3x^2 + 4x$ *(1-6)*
56. $x^4 - 2x^3 + x^2$ *(2-2)*   **57.** $3x^2y^4 + 2x^4y^6$ *(2-1)*   **58.** $x^6 + 2x^4 + 2x^2 - 15$ *(2-2)*   **59.** $(x + \frac{7}{2})^2$ *(2-3)*
60. $(x - 15)(x + 4)$ *(2-4, 2-5)*   **61.** $(x - 2)(x^2 + 2x + 4)$ *(2-6)*   **62.** $(x^2 + 1)(x + 1)$ *(2-6)*
63. $x(x^2 + 4)$ *(2-3)*   **64.** $(2x + 3)(x - 1)$ *(2-4, 2-5)*   **65.** $(2x + 3)(x - 4)$ *(2-4, 2-5)*
66. $(x - 12)(x + 5)$ *(2-4, 2-5)*   **67.** $(2x + 3)^2$ *(2-4, 2-5)*   **68.** $x(x + 3)^2$ *(2-3)*   **69.** $\frac{17}{3}$ *(1-7)*
70. $-\frac{2}{17}$ *(1-7)*   **71.** 4 *(2-7)*   **72.** $3, -8$ *(2-7)*   **73.** $-\frac{66}{7}$ *(1-7)*   **74.** $-\frac{3}{4}$ *(1-7)*
75. $\dfrac{L + 6}{2} + \dfrac{4}{5}L + L = 72$; 18, 24, and 30 cm *(1-8)*   **76.** $7w^2 = 175w$; 5 by 35 ft *(2-8)*
77. $w(w + 7) = 60$; 5 by 12 ft *(2-8)*   **78.** $(x - 2)(x + 2) = x^2 - 4$; any positive x *(2-8)*
79. $x(-x - 5) = 6$; $x = -3$; -2 *(1-8)*   **80.** $2x^2 = (x + 5)^2 - 14$; $x = 11$ *(2-8)*
81. $2x^3 - 2x^2 - x + 1$ *(2-2)*   **82.** $2x^3 - 16x^2 - 24x$ *(2-2)*   **83.** $2x^4 - x^3 + 3x^2 + x + 4$ *(2-2)*
84. $(2x + 5)(x - 6)$ *(2-4, 2-5)*   **85.** $(4x + 3)(x - 10)$ *(2-4, 2-5)*   **86.** $(2x - 15)(x + 2)$ *(2-4, 2-5)*
87. $(a + b)(x - y)$ *(2-6)*   **88.** $(x^2 + 2)^2$ *(2-6)*   **89.** $x(x^2 + 1)(x - 2)(x + 2)$ *(2-6)*
90. $(2x - ay)(4x^2 + 2axy + a^2y^2)$ *(2-6)*   **91.** $(x - 2)(x + 2)(x^2 - 2x + 4)(x^2 + 2x + 4)$ *(2-6)*
92. -4 *(2-7)*   **93.** -8 *(2-7)*   **94.** No solution *(1-7)*   **95.** $0, \frac{11}{6}$ *(2-7)*   **96.** $x = \dfrac{1 + y}{1 - y}$ *(1-7)*
97. $3w^2 = 2(\frac{1}{2} \cdot w \cdot 3w)$; The rectangle has width w, length $3w$; the triangle has base w, height $3w$; w can be any positive number. *(2-7)*
98. $3w^2 = 6(8w)$; 16 by 48 *(2-7)*

99. $\dfrac{x + (x + 2) + (x + 4) + (x + 6)}{4} = \dfrac{7}{8}(x + 6)$; $x = 18$; 18, 20, 22, and 24   *(1-8)*

100. $(3C + 10) + C = 102$; $C = 23$; $A = 2C - 1 = 45$; $A + B = 79$; $B = 34$; that is, 45 A problems, 34 B problems, 23 C problems   *(1-8)*

CHAPTER 3

EXERCISE 3-1

1. $\dfrac{2}{5}$ **3.** $\dfrac{5}{3}$ **5.** $\dfrac{1}{2x^2}$ **7.** $\dfrac{2x^2}{3y}$ **9.** $-\dfrac{x^3}{y^4}$ **11.** $-\dfrac{a^2}{b^2}$ **13.** $\dfrac{3(x - 9)}{y}$ **15.** $\dfrac{2x - 1}{3x}$

17. $\dfrac{(x + 3)^5}{x(x - 1)^2}$ **19.** $\dfrac{(x + 5)^3}{x^2}$ **21.** $\dfrac{x}{2}$ **23.** $\dfrac{1}{n}$ **25.** $\dfrac{x - 3}{x(x + 2)}$ **27.** $\dfrac{x + 2}{x}$ **29.** $12xy$ **31.** $14x^3y$

33. $3y$ **35.** $4b^2$ **37.** $\dfrac{x + 2}{3x}$ **39.** $\dfrac{x - 3}{x + 3}$ **41.** $\dfrac{2x - 3y}{2xy}$ **43.** $\dfrac{x + 1}{x + 2}$ **45.** $\dfrac{x + 2}{x - 2}$ **47.** $\dfrac{x + 1}{x + 3}$

49. $\dfrac{x + 2}{x + y}$ **51.** $\dfrac{x + 5}{2x}$ **53.** $\dfrac{x^2 + 2x + 4}{x + 2}$ **55.** $\dfrac{x + 3}{3x}$ **57.** $\dfrac{x}{2(x - 3)}$ **59.** $\dfrac{x + 5}{x + 10}$ **61.** $\dfrac{x^2 - x + 1}{x - 1}$

63. $\dfrac{x + 3}{x - 2}$ **65.** $\dfrac{2x}{y}$ **67.** $\dfrac{x^2 + 2}{x - 2}$ **69.** $\dfrac{3}{4a}$ **71.** $\dfrac{x - 1}{x + 1}$ **73.** $\dfrac{2x - 3}{2x + 3}$ **75.** $3x^2 + 3xy$ **77.** $x^2 - y^2$

79. $2x$ **81.** $x(x + 2)$ **83.** $\dfrac{x - y}{3x}$ **85.** $\dfrac{x - y}{2x + y}$ **87.** $\dfrac{x^2 + y^2}{(x + y)^2}$ **89.** $x + 1$

91. Already in lowest term   **93.** $\dfrac{x + y}{u + v}$ **95.** $3 + 3h + h^2$ **97.** $-\dfrac{2(x + 3)(x + 8)}{(x - 2)^5}$ **99.** $2x + h + 2$

EXERCISE 3-2

1. $\dfrac{8}{9}$ **3.** $\dfrac{27}{28}$ **5.** $\dfrac{6}{b}$ **7.** $\dfrac{y}{x}$ **9.** $\dfrac{2x^2}{3z^2}$ **11.** $\dfrac{3x^3y}{2}$ **13.** $\dfrac{3x^2}{z}$ **15.** $\dfrac{48}{175}$ **17.** $\dfrac{5}{2}$ **19.** $\dfrac{3}{2}$

21. $\dfrac{3c}{a}$ **23.** $\dfrac{x}{9y^2}$ **25.** $54y^2$ **27.** $\dfrac{128x}{75y^3}$ **29.** $48y^2z$ **31.** $\dfrac{1}{48y^2z}$ **33.** $\dfrac{16xy}{3}$ **35.** $\dfrac{9xy}{8c}$

37. $\dfrac{-45u^2}{16v^2}$ **39.** $\dfrac{c^3d^2}{a^6b^6}$ **41.** $6xy$ **43.** $-\dfrac{y}{4z}$ **45.** $-\dfrac{9z}{2}$ **47.** $-\dfrac{2c^3d}{a}$ **49.** $\dfrac{x}{2}$ **51.** $\dfrac{x}{x - 3}$

53. $\dfrac{3y}{x + 3}$ **55.** $\dfrac{1}{2y}$ **57.** $t(t - 4)$ **59.** $(x - 2)(x - 3) = x^2 - 5x + 6$ **61.** $(x + 2)^2 = x^2 + 4x + 4$

63. $\dfrac{(x + 1)^2}{(x + 3)^2}$ **65.** $\dfrac{(x + 1)^2}{(x + 3)^2}$ **67.** $\dfrac{1}{(x + 5)^2}$ **69.** $\dfrac{m}{(m - n)^2}$ **71.** $\dfrac{1}{m}$ **73.** $-x(x - 2)$ or $2x - x^2$

75. $\dfrac{a^2}{2}$ **77.** $\dfrac{2}{y}$ **79.** xyz **81.** $-\dfrac{3}{4}$ **83.** -1 **85.** $x^2 - 4 = (x - 2)(x + 2)$ **87.** $x + 1$

89. $\dfrac{x + 1}{x - 1}$ **91.** $\dfrac{(x - y)^2}{y^2(x + y)}$ **93.** $x \neq 1$ **95.** $x \neq -3$ **97.** $x \neq 0, 1$ **99.** $x \neq 0, 1$

EXERCISE 3-3

1. $3x$ **3.** x **5.** v^3 **7.** $12x^2$ **9.** $(x + 1)(x - 2)$ **11.** $3y(y + 3)$ **13.** $\dfrac{7x + 2}{5x^2}$ **15.** 2

17. $\dfrac{1}{y + 3}$ **19.** $\dfrac{3 - 2x}{k}$ **21.** $\dfrac{12x + y}{4y}$ **23.** $\dfrac{2 + y}{y}$ **25.** $\dfrac{u^3 + uv - v^2}{v^3}$ **27.** $\dfrac{9x^2 + 8x - 2}{12x^2}$

29. $\dfrac{5x - 1}{(x + 1)(x - 2)}$ **31.** $\dfrac{7y - 6}{3y(y + 3)}$ **33.** $\dfrac{x + y}{y^2}$ **35.** $-\dfrac{y}{(x + y)}$ **37.** $\dfrac{x^2z - y^2z}{xy^2}$ **39.** $24x^3y^2$

41. $75x^2y^2$ **43.** $18(x - 1)^2$ **45.** $24(x - 7)(x + 7)^2$ **47.** $(x - 2)(x + 2)^2$ **49.** $12x^2(x + 1)^2$

51. $\dfrac{8v - 6u^2v^2 + 3u^3}{36u^3v^3}$ **53.** $\dfrac{15t^2 + 14t - 6}{36t^3}$ **55.** $\dfrac{2}{t - 1}$ **57.** $\dfrac{5a^2 - 2a - 5}{(a + 1)(a - 1)}$ **59.** $\dfrac{5x + 55}{12(x - 5)^2(x + 5)}$

61. $\dfrac{15x - 11}{18(x - 1)^2}$ **63.** $\dfrac{-4}{(x - 1)(x + 3)}$ **65.** $\dfrac{2s^2 + s - 2}{2s(s - 2)(s + 2)}$ **67.** $\dfrac{2(x + 4)}{(x - 2)(x + 2)^2}$ **69.** $\dfrac{3}{x + 3}$

71. $\dfrac{2}{x^2 - 1}$ **73.** $-(x + 1)^2(x - 1)$ **75.** $\dfrac{1}{x + 1/2}$ **77.** $\dfrac{7}{y - 3}$ **79.** -1 **81.** $\dfrac{-17}{15(x - 1)}$

83. $\dfrac{x + 3}{(x - 2)(x + 7)}$ **85.** $\dfrac{(3x + 1)(x + 3)}{12x^2(x + 1)^2}$ **87.** $\dfrac{xy^2 - xy + y^2}{x^3 - y^3}$ **89.** $\dfrac{1}{x + 2}$ **91.** 1 **93.** $\frac{1}{2}$

95. $\dfrac{x^2 + 2x + 2}{x(x + 1)(x + 2)}$ **97.** $\dfrac{b^2 - a^2}{a^2b^2}$ **99.** $\dfrac{a^2 + b^2}{a^2 - b^2}$

EXERCISE 3-4

1. $3x + 1$ **3.** $2y^2 + y - 3$ **5.** $3x + 1, R = 3$ **7.** $4x - 1, R = 10$ **9.** $3x - 4, R = -1$ **11.** $x + 2$
13. $2x + 3$ **15.** $3x^2 + 4x$ **17.** $4x^3 - x^2$ **19.** $x^2 + 4x + 11, R = 26$ **21.** $x^2 + x + 2, R = 2$
23. $2x^2 + 5x + 16, R = 46$ **25.** $2x^2 - 3x + 4, R = -6$ **27.** $x^3 + 5x^2 + 23x + 95, R = 385$
29. $x^3 - x^2 + 5x - 7, R = 19$ **31.** $4x + 1, R = -4$ **33.** $4x + 6, R = 25$ **35.** $x - 4, R = 3$ **37.** $x^2 + x + 1$
39. $x^3 + 3x^2 + 9x + 27$ **41.** $4a + 5, R = -7$ **43.** $x^2 + 3x - 5$ **45.** $x^2 + 3x + 8, R = 27$
47. $x^2 + 2x + 2, R = 8$ **49.** $x^2 - x - 1, R = 5$ **51.** $x^3 + 4x^2 + 16x + 61, R = 249$
53. $x^3 - 2x^2 + 4x - 11, R = 27$ **55.** 170 **57.** -2 **59.** -5 **61.** 14 **63.** 538 **65.** 53
67. $3x^3 + x^2 - 2, R = -4$ **69.** $4x^2 - 2x - 1, R = -2$ **71.** $2x^3 + 6x^2 + 32x + 84, R = 186x - 170$
73. $3x^2 - 4x + 1$ **75.** $x^2 - x + 1, R = -4x - 6$ **77.** $x^3 + 2x^2 + 4x + 8$ **79.** $Q = x - 3, R = x + 2$
81. $Q = 1, R = -6x^2 + 12x$ **83.** $Q = x^2 + x - 4, R = 2x + 6$ **85.** $Q = x, R = -4x^2 + 5x + 2$
87. $-0.389\,000$ **89.** $-1.234\,625$ **91.** $-43.817\,000$ **93.** $5.297\,889$ **95.** $-7.096\,000$
97. $-5.009\,007$ **99.** $13.211\,102$

EXERCISE 3-5

1. $\frac{3}{4}$ **3.** $\frac{9}{10}$ **5.** $\frac{8}{13}$ **7.** $\frac{22}{51}$ **9.** $\dfrac{3y}{4x}$ **11.** $-\dfrac{x}{5y}$ **13.** $\dfrac{a^2}{c^2}$ **15.** $\dfrac{x}{2y}$ **17.** xy **19.** $\dfrac{3xy}{2}$

21. $-\dfrac{1}{1 + x}$ **23.** $\dfrac{1}{3x}$ **25.** $\dfrac{a}{b}$ **27.** $\dfrac{3}{b}$ **29.** $\dfrac{z}{y}$ **31.** $\frac{1}{3}$ **33.** $\dfrac{2a}{3b}$ **35.** $\dfrac{1}{x - 3}$ **37.** $\dfrac{x + y}{x}$

39. $\dfrac{1}{y - x}$ **41.** $\dfrac{x^2 - x + 1}{x}$ **43.** $\dfrac{9 + 3x + x^2}{3x}$ **45.** $-\dfrac{2 + x}{(1 + x)^2}$ **47.** $\dfrac{4(a + b)}{a^2b^2}$ **49.** $\dfrac{2a + b}{2a - b}$

51. $\dfrac{x - y}{x + y}$ **53.** $x + 2$ **55.** $\dfrac{x^2 + 4x}{x + 2}$ **57.** 1 **59.** $\dfrac{x + 3}{(x + 4)}$ **61.** $\dfrac{x(x - 2y)}{x + 2y}$ **63.** $\dfrac{x + 4}{2x(x - 1)}$

65. $-\frac{1}{2}$ **67.** $\dfrac{x^2 + 1}{x}$ **69.** $\dfrac{1}{1 - x}$ **71.** $\dfrac{-x - 3}{x - 1}$ **73.** $\dfrac{x - 1}{x^2}$ **75.** $-x$ **77.** $\dfrac{3x + 5}{2x + 3}$

79. $r = \dfrac{2r_R r_G}{r_R + r_G}$

CHAPTER 3 REVIEW EXERCISE

1. $\frac{2}{5}$ *(3-1)* **2.** a^2b *(3-1)* **3.** $\dfrac{x}{x + 5}$ *(3-1)* **4.** $\dfrac{x - 2}{x + 2}$ *(3-1)* **5.** $3y^2$ *(3-1)* **6.** $6x^2y^2z$ *(3-1)*

7. $36x^3$ *(3-3)* **8.** $6x^3y^3$ *(3-3)* **9.** $x^4 - x^2$ *(3-3)* **10.** $\dfrac{3x^2}{2(z + 3)}$ *(3-1)* **11.** $\dfrac{x + 1}{x - 1}$ *(3-1)*

12. $\dfrac{3x + 2}{3x}$ *(3-3)* **13.** $\dfrac{2x + 11}{6x}$ *(3-3)* **14.** $\dfrac{2 - 9x^2 - 8x^3}{12x^3}$ *(3-3)* **15.** $\dfrac{2xy}{ab}$ *(3-2)* **16.** $2(x + 1)$ *(3-3)*

17. $\dfrac{2}{m + 1}$ *(3-3)* **18.** $\dfrac{x + 7}{(x - 2)(x + 1)}$ *(3-3)* **19.** $\dfrac{(d - 2)^2}{d + 2}$ *(3-2)* **20.** $\dfrac{-1}{(x + 2)(x + 3)}$ *(3-3)*

21. $\dfrac{1}{x^2 + x}$ *(3-3)* **22.** $\dfrac{-x}{2x + 8}$ *(3-3)* **23.** 2 *(3-2)* **24.** $\dfrac{x^2}{x - 1}$ *(3-3)* **25.** $\dfrac{x^2}{y}$ *(3-2)*

26. $\dfrac{x^4 + 8xy^3}{6y^2}$ *(3-3)* **27.** $\frac{3}{8}$ *(3-5)* **28.** $\frac{11}{6}$ *(3-5)* **29.** $\dfrac{y - 2}{y + 1}$ *(3-5)* **30.** $\dfrac{4}{5x}$ *(3-5)*

31. $\dfrac{-2}{3(3 + x)}$ *(3-5)* **32.** $\dfrac{x}{3}$ *(3-5)* **33.** $\dfrac{y}{x - 1}$ *(3-5)* **34.** $\dfrac{x + 4}{x - 4}$ *(3-1)* **35.** $\dfrac{x + 1}{x^2 + x + 1}$ *(3-1)*

36. $\dfrac{x - 2}{x + 3}$ *(3-1)* **37.** Already reduced *(3-1)* **38.** $x^2 + 6x + 9$ *(3-1)* **39.** $x + 1$ *(3-1)*
40. $6x^3 + 24x^2 + 24x$ *(3-1)* **41.** $(x + 1)^2(x - 2)^2$ *(3-3)* **42.** $60(x^2 + x - 6)$ *(3-3)*

43. $x^4 - 18x^2 + 81$ *(3-3)*   **44.** $x^2 - 2x + 4$ *(3-4)*   **45.** $x + 3, R = 18$ *(3-4)*
46. $x^2 - 2x - 1, R = -4$ *(3-4)*   **47.** $x, R = 0$ *(3-4)*   **48.** $x^2 + 2x + 1, R = 3$ *(3-4)*
49. $x^3 - 2x^2 + 5x - 10, R = 19$ *(3-4)*   **50.** $x^3 + x^2 + x + 1, R = 0$ *(3-4)*   **51.** $x^2 - 1, R = 2x + 1$ *(3-4)*
52. $\dfrac{12a^2b^2 - 40a^2 - 5b}{30a^2b^3}$ *(3-3)*   **53.** $\dfrac{5 - 2x}{2x - 3}$ *(3-3)*   **54.** $\dfrac{2y^4}{9a^4}$ *(3-2)*   **55.** $\dfrac{5x - 12}{3(x - 4)(x + 4)}$ *(3-2)*
56. $\dfrac{x}{x + 1}$ *(3-2)*   **57.** $\dfrac{x - y}{x}$ *(3-5)*   **58.** $\dfrac{y}{x^3 - y^3}$ *(3-3)*   **59.** $\dfrac{x}{y(x + y)}$ *(3-5)*   **60.** $x + 1$ *(3-2)*
61. $\dfrac{x^2 + 24x - 9}{12x(x - 3)(x + 3)^2}$ *(3-3)*   **62.** $\dfrac{x^2 + 3x + 1}{x^2 + x}$ *(3-3)*   **63.** 1 *(3-2)*   **64.** $\dfrac{1}{x + 3}$ *(3-3)*   **65.** $\dfrac{1}{x}$ *(3-3)*
66. 0 *(3-5)*   **67.** 0 *(3-5)*   **68.** $\dfrac{-1}{s + 2}$ *(3-3)*   **69.** -1 *(3-2)*   **70.** $\dfrac{y^2}{x}$ *(3-2)*   **71.** $\dfrac{x - y}{x + y}$ *(3-5)*
72. $\dfrac{(x + 1)(x - 2)}{2x}$ *(3-5)*   **73.** $\dfrac{x + 3}{x(x + 1)(x + 2)}$ *(3-3)*   **74.** $\dfrac{-2 + 2x - 4x^2}{(x - 1)^2(x + 1)^2}$ *(3-3)*
75. Quotient $x^3 - 3x^2 + 7x - 15$, Remainder 31, so value is 31   *(3-4)*

CHAPTER 3 PRACTICE TEST

1. $\dfrac{2yz}{5x}$ *(3-1)*   **2.** $\dfrac{x - 1}{x + 3}$ *(3-1)*   **3.** $12x^2y^2z^4$ *(3-3)*   **4.** $(x + 1)^2(x + 2)$ *(3-3)*   **5.** $40yz^6$ *(3-1)*
6. $x^2 + 4x + 3$ *(3-1)*   **7.** $\dfrac{2x - 3}{x - 1}$ *(3-3)*   **8.** $\dfrac{2x^2 + 3x + 4}{x^3}$ *(3-3)*   **9.** $\dfrac{4x^2}{7y}$ *(3-2)*   **10.** $\dfrac{3x}{2yz}$ *(3-2)*
11. $\dfrac{1}{x - 1}$ *(3-2)*   **12.** $\dfrac{-5x}{x^2 + x - 6}$ *(3-3)*   **13.** $\dfrac{-1}{x^3 - x}$ *(3-3)*   **14.** $\dfrac{x - 2}{x}$ *(3-5)*   **15.** $\dfrac{x - 1}{2}$ *(3-3)*
16. $1 - x$ *(3-5)*   **17.** -1 *(3-5)*   **18.** $x^2 + 1, R = 0$ *(3-4)*   **19.** $x + 1, R = x + 3$ *(3-4)*   **20.** 580 *(3-4)*

CHAPTER 4

EXERCISE 4-1

1. 13   **3.** 8   **5.** -6 **7.** $-\frac{1}{12}$ **9.** 30   **11.** 20   **13.** 10   **15.** 3   **17.** $-\frac{7}{4}$ **19.** 3
21. 10   **23.** -9 **25.** $\frac{25}{9}$ **27.** 4   **29.** 5   **31.** 3   **33.** 3   **35.** 6   **37.** 4   **39.** No solution
41. 5   **43.** $\frac{53}{11}$ **45.** No solution   **47.** 1   **49.** $\frac{4}{7}$ **51.** No solution   **53.** 0   **55.** -3 **57.** -5
59. -11 **61.** -2 **63.** $-\frac{2}{5}$ **65.** $a - 2b$ **67.** $\dfrac{b + 1}{b - 1}$ **69.** $\dfrac{a - cb}{c - 1}$ **71.** $\frac{7}{13}$ **73.** $\frac{3}{10}$ **75.** $\frac{31}{24}$
77. -4 **79.** No solution   **81.** $\frac{2}{3}$ **83.** -2 **85.** 4, -3 **87.** No solution   **89.** $-\frac{5}{16}$ **91.** $-\frac{5}{7}$
93. $-\frac{16}{9}$ **95.** $\dfrac{abc}{a + b + c}$ **97.** $\dfrac{a^2 - b^2}{a^2 + b^2}$ **99.** $\frac{7}{4}$

EXERCISE 4-2

1. $\frac{3}{2}$ **3.** $\frac{5}{2}$ **5.** $\frac{25}{2}$ **7.** 20   **9.** 7   **11.** 1,750   **13.** 100   **15.** 60 quarters   **17.** 40 m
19. 162 km   **21.** 81   **23.** 188   **25.** $\frac{12}{38}$ **27.** $\frac{4}{36}$ **29.** $\frac{5}{31}$ **31.** $\frac{8}{28}$ **33.** $\frac{8}{36}$ **35.** 4   **37.** 5
39. $14\frac{2}{3}$ in.   **41.** $13\frac{1}{3}$ m   **43.** $\dfrac{x}{52} = \dfrac{9}{46}$; 10.17 mL   **45.** $\dfrac{x}{23} = \dfrac{10}{35}$; 6.57 in.   **47.** $\dfrac{x}{5} = \dfrac{1}{0.26}$; 19.23 liters
49. $\dfrac{x}{4} = \dfrac{1}{1.0567}$; 3.79 liters   **51.** $\dfrac{x}{1} = \dfrac{1}{0.6215}$; 1.61 km   **53.** $\dfrac{x}{100} = \dfrac{1}{2.54}$; 39.37 in.
55. $\dfrac{x}{100} = \dfrac{1}{1.54}$; 64.94 British pounds   **57.** $\dfrac{x}{65} = \dfrac{1.5}{10}$; \$9.75   **59.** $\dfrac{x}{500} = \dfrac{240}{200}$; \$600   **61.** $\dfrac{66}{x} = \dfrac{12}{1}$; \$5.50
63. *A*: 60°, *B*: 144°, *C*: 156°   **65.** *A*: 126°, *B*: 162°, *C*: 72°   **67.** *A*: 72°, *B*: 90°, *C*: 144°, *D*: 54°   **69.** 40°39′
71. 50°25′48″
73. 70°36′18″   **75.** 30.2625°   **77.** 60.8017°   **79.** 10.7050°   **81.** 15.2056°

83. 60.4125°   **85.** 72 mi   **87.** 0°20′50″   **89.** $\dfrac{x}{300} = \dfrac{250}{25}$; 3,000 trout   **91.** $\dfrac{f}{1200} = \dfrac{\pi \cdot 6^2}{\pi \cdot 120^2}$; 3 kg

93. $\dfrac{x}{5^3} = \dfrac{4,188.79}{10^3}$; 523.6 in.³   **95.** $\dfrac{x}{8} = \dfrac{234}{3}$; $624   **97.** 210 ft   **99.** 90 yd

EXERCISE 4-3

1. (A) $5.65/hr **(B)** $293.80 **(C)** 88.5 hr   **3. (A)** 69,100 elephants/yr **(B)** 1,036,500 elephants **(C)** 8.8 yr
5. 5 hr   **7.** 6 hr   **9.** 15 hr   **11.** 65 min   **13.** 50 min   **15.** 5 hr   **17.** 3.43 hr   **19.** 71.25 min
21. 7.5 days   **23.** 30 mph and 45 mph   **25.** 4 mph   **27.** 80 and 88 forms/hr   **29.** 9.6 mi   **31.** 15 mph
33. 670 m   **35.** 90 and 120 copies/min   **37.** 50 mph

EXERCISE 4-4

1. $-\frac{1}{4}$ **3.** $\frac{8}{3}$ **5.** $-\frac{3}{4}$ **7.** $\frac{16}{11}$ **9.** $\frac{9}{2}$ **11.** $\frac{22}{3}$ **13.** 1   **15.** 4   **17.** $\frac{2}{5}$ **19.** -2
21. $h = \dfrac{2A}{b}$ **23.** $a = \dfrac{2A - bh}{h}$ **25.** $x = \dfrac{y - 7}{3}$ **27.** $\theta = \dfrac{180l}{\pi r}$ **29.** $t = \dfrac{S - P}{dS}$ **31.** $a = \dfrac{3V}{4\pi b^2}$
33. $p = \dfrac{Pq}{1 - p}$ **35.** $I = \dfrac{9R}{ERA}$ **37.** $v = \dfrac{Ft + mv_0}{m}$ **39.** $V = \dfrac{QP - y - F}{Q}$ **41.** 0, 2, 4   **43.** 3, 4
45. 0, $\frac{5}{2}$ **47.** $-1, 5$ **49.** $-\frac{3}{2}, \frac{2}{3}$ **51.** $f = \dfrac{ab}{a + b}$ **53.** $D = \dfrac{Q}{C_0}\left(C - \dfrac{QC_h}{2}\right)$ **55.** $T_2 = \dfrac{T_1 P_2 V_2}{P_1 V_1}$
57. $x = \dfrac{5y - 3}{3y - 2}$

EXERCISE 4-5

1. 10, 12, 14   **3.** $\frac{5}{2}$ and $\frac{1}{4}$ **5. (A)** .345 **(B)** Approx. 3,103 **(C)** Approx. 8,696   **7.** 45°C   **9.** 4,400 m
11. 5,400 m   **13.** 180 cm   **15.** 85.8 ft   **17.** Solve $\dfrac{N}{600} = \dfrac{500}{60}$; $N = 5,000$ chipmunks.   **19.** $4\frac{10}{11} \approx 4.91$ gal
21. $238,500   **23.** 42 oz   **25.** 205 points   **27. (A)** 216 mi **(B)** 225 mi
29. (A) 15 in. **(B)** 20 in. **(C)** 22.5 in. **(D)** 24 in. **(E)** 25 in. **(F)** 18 in. **(G)** 18.75 in.   **31.** $35.80
33. 10,000   **35.** $6,000 at 7% and $4,000 at 4.5%   **37.** $9,374   **39.** $34,400
41. Solve $x - 0.2x = 160$; $x = $200.   **43.** 5,300 copies   **45.** 800   **47.** 2,000   **49.** 35   **51.** 7 by 19 in.
53. $\pi/6$ radian   **55.** 45°   **57.** $2\pi/3$ radians   **59.** 315°   **61.** π **63.** $500\pi/9$ **65.** $4\pi/3$
67. $\dfrac{25\pi}{2}$ **69.** 8 sec   **71.** 7 cm²   **73.** Solve $20x = 50(180)$; $x = 450$ kg.   **75.** 200,000 mi/sec
77. 60 mph   **79.** $31.20   **81.** $616   **83.** CA: 52 or 53; DE: 1 or 2
85. CA: 572,307 with 52 districts, 561,509 with 53; DE: 666,000 with 1 district, 333,000 with 2   **87.** 30 m.
89. 15 min, or $\frac{1}{4}$ hr

CHAPTER 4 REVIEW EXERCISE

1. 2 *(4-1)*   **2.** $\frac{12}{35}$ *(4-1)*   **3.** 1 *(4-1)*   **4.** $\frac{3}{4}$ *(4-1)*   **5.** 9 *(4-1)*   **6.** $-\frac{10}{9}$ *(4-1)*   **7.** 60 *(4-1)*
8. -12 *(4-1)*   **9.** All real numbers *(4-1)*   **10.** No solution *(4-1)*   **11.** $\frac{9}{4}$ *(4-1)*   **12.** 20 *(4-1)*
13. 11 *(4-1)*   **14.** $b = \dfrac{2A}{h}$ *(4-4)*   **15.** $y = \dfrac{15 - 3x}{5}$ *(4-4)*   **16.** 15 *(4-5)*   **17.** 100 *(4-5)*
18. 220,000 *(4-5)*   **19.** $266\frac{2}{3}$ ft² *(4-5)*   **20. (A)** $\frac{1}{470}$ gal/ft² **(B)** Approx. 8.5 gal **(C)** 1,880 ft² *(4-2)*
21. 2,200 *(4-5)*   **22.** $50,000 *(4-5)*   **23.** $300 *(4-3)*   **24.** 50 in. *(4-2)*   **25.** 25.2 mL alcohol *(4-2)*
26. 3,300 squirrels *(4-2)*   **27.** 941.18 mL *(4-2)*   **28.** 19.05 mm *(4-2)*   **29.** 28.80 CI *(4-2)*
30. 198° *(4-2)*   **31.** 666 *(4-2)*   **32.** $78,000 *(4-5)*   **33.** $244.44 *(4-5)*   **34.** 3.5 by 6 m *(4-5)*
35. 5 m *(4-5)*   **36.** 120 lb *(4-5)*   **37.** 5,625 *(4-2)*   **38.** 5 *(4-1)*   **39.** No solution *(4-1)*
40. 1, 4 *(4-1)*   **41.** $\frac{5}{2}$ *(4-1)*   **42.** 10 *(4-1)*   **43.** 1, -3 *(4-1)*   **44.** 0, $\frac{1}{2}$, $-\frac{3}{2}$ *(4-1)*   **45.** -2 *(4-1)*
46. -5 *(4-1)*   **47.** $-\frac{3}{5}$ *(4-1)*   **48.** No solution *(4-1)*   **49.** $\frac{3}{4}$ *(4-1)*   **50.** 3 *(4-1)*
51. No solution *(4-1)*   **52.** $L = \dfrac{2S}{n} - a$, or $L = \dfrac{2S - an}{n}$ *(4-4)*   **53.** $M = \dfrac{P}{1 - dt}$ *(4-4)*   **54.** 6:1 *(4-2)*
55. 3:10 *(4-2)*   **56.** 162 *(4-2)*   **57.** 14 and 19.6 in. *(4-2)*   **58.** 4.2 hr *(4-3)*   **59.** 35.5 min *(4-3)*
60. 80 *(4-3)*   **61.** 59°F *(4-3)*   **62.** $-\frac{13}{5}$ *(4-1)*   **63.** -15 *(4-1)*   **64.** No solution *(4-1)*

65. All real numbers except 0 and -1 *(4-1)*   **66.** $x = \dfrac{abc}{b - a}$ *(4-4)*   **67.** $x = \dfrac{5y + 3}{2y - 4}$ *(4-4)*

68. $f_1 = \dfrac{ff_2}{f_2 - f}$ *(4-4)*   **69.** 20 days   *(4-5)*   **70.** 24   *(4-1)*   **71.** $\frac{3}{4}$ *(4-1)*

72. Freight: 52 mph, passenger: 78 mph   *(4-3)*   **73.** 8 hr   *(4-3)*   **74.** $16\frac{2}{3}$ yr   *(4-5)*   **75.** $-40°$ *(4-4)*

CHAPTER 4 PRACTICE TEST

1. $\frac{13}{5}$ *(4-1)*   **2.** $-\frac{1}{2}$ *(4-1)*   **3.** -2 *(4-1)*   **4.** 2   *(4-1)*   **5.** $5, \frac{4}{3}$ *(4-1)*   **6.** No solution   *(4-1)*

7. $\frac{5}{2}$ *(4-2)*   **8.** $y = -\frac{3}{4}x + \frac{5}{4}$ *(4-4)*   **9.** $x = \dfrac{3y}{5y - 4}$ *(4-4)*   **10.** $y = \dfrac{3x^4 - 6}{5}$ *(4-4)*   **11.** $26\frac{2}{3}$ min   *(4-3)*

12. 700   *(4-2)*   **13.** 112   *(4-2)*   **14.** 23.8 mL   *(4-2)*   **15.** $8\frac{1}{3}$ loads   *(4-2)*   **16.** 4 hr   *(4-3)*

CHAPTER 5

EXERCISE 5-1

1. $A(-10, 10)$, $B(10, -10)$, $C(16, 14)$, $D(16, 0)$, $E(-14, -16)$, $F(0, 4)$

3. $A(5, 5)$, $B(8, 2)$, $C(-5, 5)$, $D(-3, 8)$, $E(-5, -6)$, $F(-7, -7)$

5. **7.** **9.** $A(-3\frac{1}{2}, 2)$, $B(-2, -4\frac{1}{2})$, $C(0, -2\frac{1}{2})$, $D(2\frac{1}{2}, 0)$

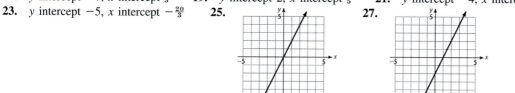

11. $A(2\frac{1}{2}, 1)$, $B(-2\frac{1}{2}, 3\frac{1}{2})$, $C(-2, -4\frac{1}{2})$, $D(3\frac{1}{2}, -3)$ **13.**   **15.**

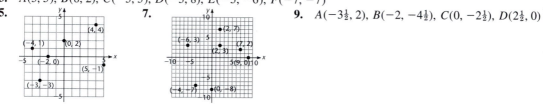

17. y intercept -7, x intercept $\frac{7}{3}$ **19.** y intercept 2, x intercept $\frac{2}{5}$ **21.** y intercept -4, x intercept 6

23. y intercept -5, x intercept $-\frac{20}{3}$ **25.**   **27.**

29.   **31.**   **33.**   **35.**

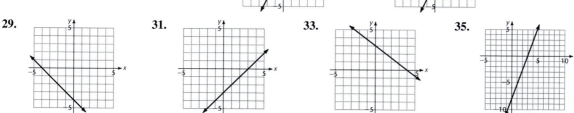

37.   **39.**   **41.**   **43.**

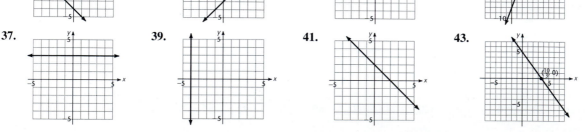

45.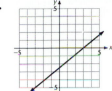

47.

49.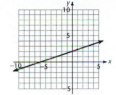

51.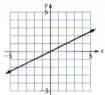

53.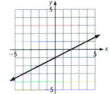

55.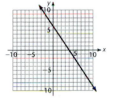

57.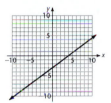

59.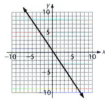

61. $y = -\frac{3}{4}x + \frac{3}{2}$

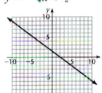

63. $y = 2x - 4$

65. $3x - y = 4$

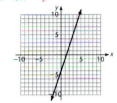

67. $x + 2y = -6$

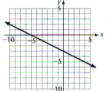

69.

71.

73.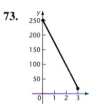

75.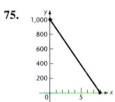

77.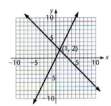

79.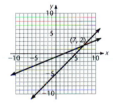

81.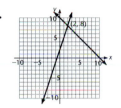

83.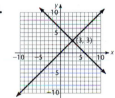

85.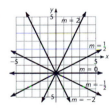

87.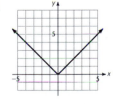

89.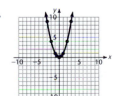

91.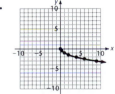

93.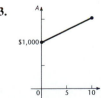

95.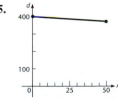

97.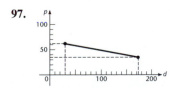

99.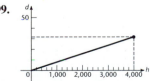

EXERCISE 5-2

1. Slope: 2
y intercept: −3

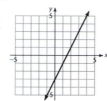

3. Slope: −1
y intercept: 2

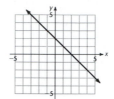

5. $y = 5x - 2$ **7.** $y = -2x + 4$ **9.** $y - 4 = 2(x - 5)$ or $y = 2x - 6$ **11.** $y - 1 = -2(x - 2)$ or $y = -2x + 5$

13. $y = x$ **15.** $y = 2x + 3$ **17.** 2   **19.** $\frac{1}{2}$ **21.** −2   **23.** $-\frac{1}{2}$

25. $y - 6 = 2(x - 5)$ or $y - 2 = 2(x - 3)$; $y = 2x - 4$ **27.** $y - 5 = \frac{1}{2}(x - 10)$ or $y - 1 = \frac{1}{2}(x - 2)$; $y = \frac{1}{2}x$

29. $y = -2x + 12$ **31.** $y = -\frac{1}{2}x + 6$

33. Slope: $-\frac{1}{3}$
y intercept: 2

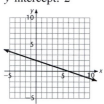

35. Slope: $-\frac{1}{2}$
y intercept: 2

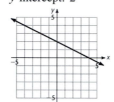

37. Slope: $-\frac{2}{3}$
y intercept: 2

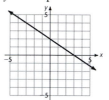

39. Slope: $\frac{2}{5}$
y intercept: −3

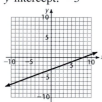

41. Slope: $\frac{1}{6}$
y intercept: 2

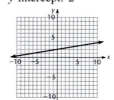

43. Slope: $-\frac{2}{3}$
y intercept: $\frac{5}{3}$

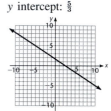

45. Slope: $\frac{1}{2}$
y intercept: $-\frac{3}{2}$

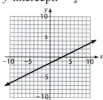

47. Slope: $\frac{3}{4}$
y intercept: $-\frac{3}{2}$

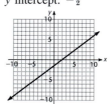

49. Slope: $-\frac{3}{2}$
y intercept: $\frac{3}{4}$

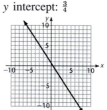

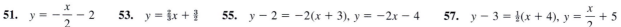

51. $y = -\dfrac{x}{2} - 2$ **53.** $y = \frac{2}{3}x + \frac{3}{2}$ **55.** $y - 2 = -2(x + 3)$, $y = -2x - 4$ **57.** $y - 3 = \frac{1}{2}(x + 4)$, $y = \dfrac{x}{2} + 5$

59. $y = -3x$ **61.** $y = -2x - 2$ **63.** $y = \frac{1}{3}x - \frac{2}{3}$ **65.** $y = -\frac{1}{2}x + \frac{9}{2}$ **67.** $y = \frac{1}{2}x + \frac{1}{2}$ **69.** $y = -\frac{3}{2}x + \frac{21}{8}$

71. $\frac{1}{3}$ **73.** $-\frac{1}{4}$ **75.** $-\frac{4}{3}$ **77.** $-\frac{5}{12}$ **79.** $y - 4 = \frac{1}{3}(x + 6)$ or $y - 7 = \frac{1}{3}(x - 3)$, $y = \dfrac{x}{3} + 6$

81. $y = -\frac{1}{4}(x + 4)$ or $y + 2 = -\frac{1}{4}(x - 4)$, $y = -\dfrac{x}{4} - 1$ **83.** $y = -\frac{4}{3}x + \frac{4}{3}$ **85.** $y = -\frac{5}{12}x - \frac{11}{24}$

87. $x = -3$, $y = 5$ **89.** $x = -1$, $y = 22$ **91.** Neither   **93.** Perpendicular

95. **(A)** $y = \frac{3}{5}x + \frac{17}{5}$ **(B)** $y = -\frac{5}{3}x + \frac{17}{3}$ **97.** **(A)** $y = -3x + 2$ **(B)** $y = \frac{1}{3}x + 2$

99. **(A)** $y = -x + 2$ **(B)** $y = x$ **101.** **(A)** $y = 5$ **(B)** $x = -2$

103. (A) $R = \frac{3}{2}C + 3$ (B)

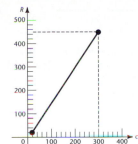

(C) $158 **105.** $F = \frac{9}{5}C + 32$ or $C = \frac{5}{9}F - \frac{160}{9}$

107. (A) $V = -1{,}800t + 20{,}000$ (B) $12,800; $5,600 (C) $-1{,}800$ (D)

EXERCISE 5-3

1. $-8 \le x \le 7$

3. $-6 \le x < 6$

5. $x \ge -6$

7. $x > -1$

9. $(-2, 6]$

11. $(-7, 8)$

13. $(-\infty, -2]$

15. $(-\infty, 0)$

17. $[-7, 2); -7 \le x < 2$

19. $(-\infty, 0]; x \le 0$ **21.** $(-\infty, -3]; x \le -3$ **23.** $(-1, 8); -1 < x < 8$ **25.** $x < 5$ or $(-\infty, 5)$

27. $x \ge 3$ or $[3, \infty)$ **29.** $1 \le x; [1, \infty);$

31. $x < 10; (-\infty, 10);$ **33.** $-5 \le x; [-5, \infty);$

35. No solution **37.** $N < -8$ or $(-\infty, -8)$ **39.** $t > 2$ or $(2, \infty)$

41. $m > 3$ or $(3, \infty)$ **43.** $x > \frac{1}{2}; (\frac{1}{2}, \infty);$

45. All real numbers; $(-\infty, \infty);$ **47.** $B \ge -4$ or $[-4, \infty)$

49. $-2 < t \le 3$ or $(-2, 3]$ **51.** $-1 \le x < 1$ or $[-1, 1)$

53. $-\frac{1}{2} \le x \le \frac{3}{4}$ or $[-\frac{1}{2}, \frac{3}{4}]$ **55.** $-8 \le x < -3$ or $[-8, -3)$

57. $2x - 3 \ge -6; x \ge -\frac{3}{2}$ **59.** $15 - 3x < 6; x > 3$ **61.** $3x + 3 < 75; x < 24$ **63.** $\frac{1}{2}x + \frac{2}{3}x \le x + 5; x \le 30$

65. $-\frac{1}{2}x - 1 > x; x > -\frac{2}{3}$ **67.** $q < -14$ or $(-\infty, -14)$ **69.** $x \ge 4.5$ or $[4.5, \infty)$

71. $-20 \le x \le 20$ or $[-20, 20]$ **73.** $-30 \le x < 18$ or $[-30, 18)$

75. $-14 < x \le 11$ or $(-14, 11]$ **77.** $\frac{1}{6} \le x < \frac{7}{6}$ or $[\frac{1}{6}, \frac{7}{6})$

79. $0 \le x < 1$ or $[0, 1)$ **81.** Positive **83.** (A) F (B) T (C) T

85. Dividing by $a - 1$ in step 4 will reverse the sense of the inequality, since $a < 1, a - 1 < 0$.

87. **1.** Definition of $<$ **2.** Product of positive numbers is positive; **3.** distributive property; **4.** inequality property 1

89. **1.** $b - a > 0$, definition of $>$ **2.** $c(b - a) < 0$, product of numbers of opposite sign is negative **3.** $cb - ca < 0$, distributive property **4.** $cb < ca$, inequality Property 1

91. $8{,}000 \le h \le 20{,}000$ or $[8{,}000, 20{,}000]$ **93.** $x > 600$ **95.** 2,154 or more **97.** 15 or more

99. If r is the worker's maximum running rate and R is the train's rate, then he will escape running toward the train if $r > R/3 = 7R/21$, and he will escape running away from the train if $r > 3R/7 = 9R/21$. Thus, his chances are better if he runs toward the train!

EXERCISE 5-4

1. $2\pi - 6$ **3.** $5 - \sqrt{24}$ **5.** $\sqrt{5}$ **7.** 4 **9.** $5 - \sqrt{5}$ **11.** $5 - \sqrt{5}$ **13.** 12 **15.** 12 **17.** 9

19. 3 **21.** 3 **23.** 4 **25.** 4 **27.** 9 **29.** 17 **31.** 14 **33.** $x = \pm 7$

35. $-7 \le x \le 7$ **37.** $x \le -7$ or $x \ge 7$ **39.** $y = 2$ or 8

41. $2 < y < 8$ **43.** $y < 2$ or $y > 8$ **45.** $u = -11$ or -5

47. $-11 \leq u \leq -5$ **49.** $u \leq -11$ or $u \geq -5$

51. $x < 0$ or $x > 2$ **53.** $1 \leq x \leq 7$ **55.** $x = -4, \frac{4}{3}$

57. $-\frac{9}{5} \leq x \leq 3$ **59.** $y < 3$ or $y > 5$ **61.** $t = -\frac{4}{5}, \frac{18}{5}$ **63.** $-\frac{5}{7} < u < \frac{23}{7}$ **65.** $x \leq -6$ or $x \geq 9$

67. $-35 < C < -\frac{5}{9}$ **69.** $|x| = 2$ **71.** $|x| > 2$ **73.** $|x| \leq 2$ **75.** $|x - 10| < 0.01$ **77.** $|x + 5| < 0.1$

79. $|x + 8| < 5$ **81.** $|x - 5| < 0.01$ **83.** $x \geq 0$ **85.** $x \geq 5$ **87.** $x \leq -8$ **89.** $x \geq -\frac{3}{4}$

91. $x \leq \frac{2}{5}$ **93.** $x = 0, y = 1$ (many answers possible)

95. *Case 1.* If $x = 0$, then both sides are 0 and the equality holds.

     Case 2. Suppose $x > 0$.

         a. If $y > 0$, then the equality is $xy = xy$.

         b. If $y < 0$, then $xy < 0$, $|xy| = -xy = x(-y) = |x| \cdot |y|$.

     Case 3. Suppose $x < 0$.

         a. If $y > 0$, then $xy < 0$, $|xy| = -xy = (-x)y = |x| \cdot |y|$.

         b. If $y < 0$, then $xy > 0$, $|xy| = xy = (-x)(-y) = |x| \cdot |y|$.

EXERCISE 5-5

1. **3.** **5.** **7.**

9. 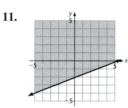 **11.** **13.** **15.**

17. 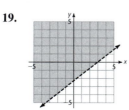 **19.** **21.** **23.**

25. 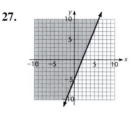 **27.** **29.** **31.**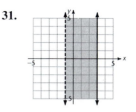

33. 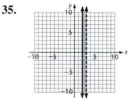 **35.** **37.** 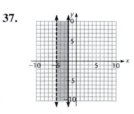 **39.**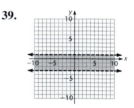

41.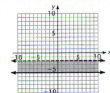

43.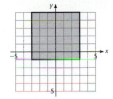

45.

47.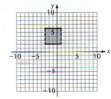

49.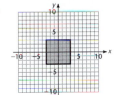

51.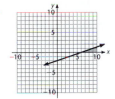

53.

55.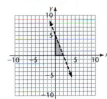

57.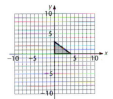

59.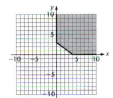

61.

63.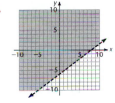

65.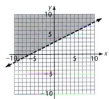

67.

69. **(A)** $5x + 6.5y$ **(B)** $5x + 6.5y \geq 133$ **(C)**

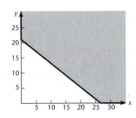

71. **(A)** $0.06x + 0.09y$ **(B)** $0.06x + 0.09y \geq 180$ **(C)**

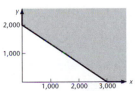

73. **(A)** $0.25x + 0.6y$ **(B)** $0.25x + 0.6y \geq 300$ **(C)**

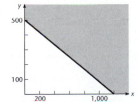

75. **(A)** $6x + 8y$ **(B)** $6x + 8y \leq 120$ **(C)**

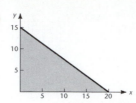

77. **(A)** $0.2x + 0.1y$ **(B)** $0.2x + 0.1y \geq 23$ **(C)**

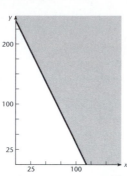

CHAPTER 5 REVIEW EXERCISE

1. $(-1, 8]$ *(5-3)*   **2.** $(2, \infty)$ *(5-3)*   **3.** $(-\infty, 10]$ *(5-3)*   **4.** $[-5, 10)$ *(5-3)*

5. $-4 \leq x < 8$ ⊢────────→ x   *(5-3)*   **6.** $x < 3$ ←────────→ x   *(5-3)*

7. $6 \leq x$ ⊢────────→ x   *(5-3)*   **8.** $2 < x \leq 5$ ⊢────────→ x   *(5-3)*

9. $3x - 5 > \dfrac{x}{2}$ *(5-3)*   **10.** $|x - 20| < 0.01$ or $19.99 < x < 20.01$ *(5-4)*

11. $25x + 10y \geq 460$ *(5-5)*   **12.** $-3 < x + 2 < 3$ *(5-4)*   **13.** $4x + 3y$ *(5-5)*

14. $|a - b| = |b - a| = |-8 - 11| = |11 - (-8)| = 19$ *(5-4)*   **15.** $A(1, 4), B(-3, 0), C(-2, -4), D(2, -3)$ *(5-1)*

16.

[graph showing points $(-3, 2)$, $(4, 0)$, $(0, -3)$, $(-3, -3)$, $(2, -4)$]

(5-1)   **17.** $x = -2$ *(5-3)*   **18.** $x = 2$ *(5-3)*   **19.** $x < -2$ *(5-4)*

20. $1 < x < 6$ *(5-3)*   **21.** $x = \pm 6$ *(5-4)*   **22.** $-6 < x < 6$ *(5-4)*   **23.** $x < -6$ or $x > 6$ *(5-4)*

24. $y = -14, -4$ *(5-4)*   **25.** $-14 < y < -4$ *(5-4)*   **26.** $y < -14$ or $y > -4$ *(5-4)*   **27.** $x \leq -12$ *(5-3)*

28. $\left[-\frac{2}{15}, \frac{8}{15}\right)$ or $-\frac{2}{15} \leq x < \frac{8}{15}$ *(5-3)*   **29.** Slope $= -2$, y intercept $= -3$ *(5-2)*   **30.** $2x + y = 8$ *(5-2)*

31. 2 *(5-2)*   **32.** $y = 2x + 1$ *(5-2)*

33.

[graph]

(5-1)   **34.**

[graph]

(5-1)   **35.**

[graph]

(5-5)

36.

[graph]

(5-5)   **37.**

[graph]

(5-1)   **38.**

[graph]

(5-1)

39. 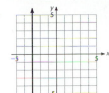 *(5-2)* **40.** 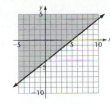 *(5-5)* **41.** 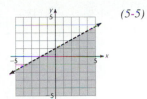 *(5-5)*

42. 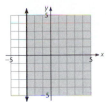 *(5-5)* **43.**  *(5-5)*

44. $-4 \leq x < 3$ 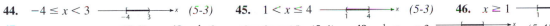 *(5-3)* **45.** $1 < x \leq 4$ *(5-3)* **46.** $x \geq 1$ *(5-3)*

47. *(5-3)* **48.** $\frac{1}{2}, 3$ *(5-4)* **49.** $\frac{1}{2} \leq x \leq 3$ *(5-4)*

50. $x < \frac{1}{2}$ or $x > 3$ *(5-4)* **51.** $x \geq -19$ *(5-3)*

52. $-6 < x \leq -1$ *(5-3)*

53. Slope $= -\frac{1}{2}$, y intercept $= -3$ *(5-2)* **54.** $y = -\frac{1}{3}x + 1$ *(5-2)* **55.** $2x + 3y = 0$ *(5-2)*

56. $y = 2x - 10$ *(5-2)* **57.** Vertical: $x = 5$; horizontal: $y = -2$ *(5-2)* **58.** $y = \frac{3}{2}x + 11$ *(5-2)*

59. 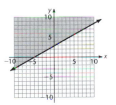 *(5-5)* **60.** 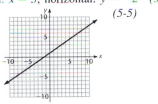 *(5-5)* **61.** 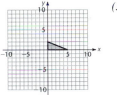 *(5-5)*

62. 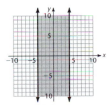 *(5-5)* **63.** 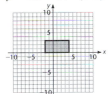 *(5-4)* **64.** 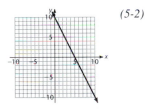 *(5-2)*

65. 14 *(5-4)* **66.** $y = -50x + 3{,}000$ *(5-3)* **67.** $-1 \leq x \leq 4$ or $[-1, 4]$ *(5-4)*

68. $x \geq \frac{25}{7}$ *(5-3)* **69.** $-3 \leq x \leq 6$ *(5-3)* **70.** No solution *(5-4)*

71. $x \geq \frac{3}{2}$ *(5-4)* **72.** $x \leq \frac{3}{2}$ *(5-4)* **73.** **(A)** Neither **(B)** Parallel *(5-2)*

74. 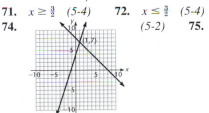 *(5-2)* **75.** 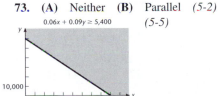 *(5-5)*

CHAPTER 5 PRACTICE TEST

1. $[\frac{1}{2}, \frac{7}{2}]$ or $\frac{1}{2} \leq x \leq \frac{7}{2}$ *(5-4)* **2.** $(-\infty, \frac{1}{2}]$ or $x \leq \frac{1}{2}$ *(5-3)* **3.** $[-2, \frac{3}{2})$ or $-2 \leq x < \frac{3}{2}$ *(5-3)*

4. $-3 \leq x < 6$ *(5-3)* **5.** $y = -\frac{2}{3}x + \frac{20}{3}$ *(5-2)* **6.** $1 - 2x > 3$ or $1 - 2x < -3$ *(5-3)* **7.** $-\frac{3}{2}$ *(5-2)*

8. $y = -4x + 26$ *(5-2)* **9.** $y = -\frac{1}{2}x - \frac{1}{2}$ *(5-2)*

10. Parallel: $x = 2$ and $x - 6 = 0$, $y = 4x + 5$ and $12x - 3y = 6$; perpendicular: $x = 2$ and $y = 3$, $x - 6 = 0$ and $y = 3$ *(5-2)*

11. *(5-1)*

12. *(5-2)*

13. *(5-1)*

14. *(5-5)*

15. *(5-1)*

16. *(5-5)*

17. *(5-5)*

18. *(5-3, 5-5)*

(a)

(b)

19. 80 or more   *(5-3)*   **20.** $14x + 12y \le 1{,}000$ *(5-5)*

CUMULATIVE REVIEW EXERCISES, CHAPTERS 1–5

1. $2x - 3 > 0$ *(1-2)*   **2.** $x + \dfrac{1}{x} \ge 1$ *(3-1)*   **3.** $\dfrac{x}{x+5} = \dfrac{3}{5}$ *(4-2)*   **4.** $x(x+1) = 812$ *(1-8)*

5. $25x + 10(x-8) = 1390$ *(1-2)*   **6.** $\frac{14}{9}$ *(1-6)*   **7.** $6\frac{3}{4}$ *(1-6)*   **8.** $x^3 - x$ *(3-3)*

9. $x^3 - 2x^2 + 4x - 4, R = 1$ *(3-4)*   **10.** 14   *(5-4)*   **11.** $\frac{1}{3}$ *(5-2)*   **12.** $y = 5x - 23$ *(5-2)*

13. $A: \frac{3}{2}$ $B: (6, -3)$ *(5-1)*   **14.** $\dfrac{3z}{2x^3y}$ *(3-1)*   **15.** $\dfrac{6ab^4c^4}{a^2b^3c^4}$ *(3-1)*   **16.** $33ab + 24bc + 6ac$ *(1-6)*

17. $x^2 + 8x + 15$ *(2-2)*   **18.** $\dfrac{3x+1}{x(x+1)}$ *(3-3)*   **19.** $\dfrac{x+1}{x^2}$ *(3-5)*   **20.** $x^3 + 3x^2 + x + 1$ *(2-1)*

21. $x^3 + x^2 + 5x + 7$ *(2-1)*   **22.** $3x - 5$ *(2-1)*   **23.** $(0, 5]$ *(5-3)*   **24.** $(x+3)(x-3)$ *(2-3)*

25. $(x+4)^2$ *(2-3)*   **26.** $(x-1)(x-2)$ *(2-4)*   **27.** $x(x^2+1)$ *(2-3)*   **28.** $(x-3)(x^2+3x+9)$ *(2-6)*

29. $\dfrac{x}{x-1}$ *(3-2)*   **30.** $\dfrac{(x+1)^2}{x^2}$ *(3-2)*   **31.** $\dfrac{x+4}{x+3}$ *(3-2)*   **32.** $\frac{1}{7}$ *(1-7)*   **33.** $-\frac{7}{3}, \frac{9}{11}$ *(2-7)*

34. $\frac{11}{25}$ *(4-1)*   **35.** $x < \frac{1}{7}$ *(5-3)*   **36.** $-\frac{16}{3} < x < \frac{2}{3}$ *(5-4)*   **37.** -2 *(2-7)*   **38.** $-\frac{55}{7}$ *(1-7)*

39. *(5-1)*   **40.** *(5-2)*   **41.** *(5-5)*

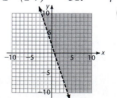
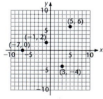
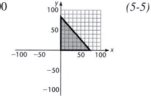

42. 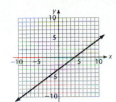 *(5-2)* **43.** *(5-5)* **44.** 20 by 35 cm *(1-8)*

45. 4 ft and 7 ft *(1-8)* **46.** 116 *(4-2)* **47.** $\frac{81}{5}$ *(1-8)* **48.** 550 *(4-5)* **49.** $0 < w \le 12$ *(5-3)*

50. 4 P.M., 600 mi *(4-3)* **51.** -1 *(5-2)* **52.** $y = -x + 1$ *(5-2)* **53.** $(2, 2)$ *(5-1)*

54. $1, R = -3$ *(3-4)* **55.** $(x - 1)(x - 3)(x + 4)$ *(3-3)* **56.** $y = -\frac{2}{3}x + \frac{4}{3}$ *(5-2)* **57.** $\frac{2x + 1}{x^2 - 1}$ *(3-3)*

58. $\frac{x^2 + 3x + 2}{x^2 + 1}$ *(3-2)* **59.** $x - x^2$ *(3-2)* **60.** $\frac{1}{x - 1}$ *(3-5)* **61.** $(x - 3)(2x + 1)$ *(2-4, 2-5)*

62. $x(x - 1)(x^2 + x + 1)$ *(2-6)* **63.** $(a + b)(x^2 - 2)$ *(2-3)* **64.** $x^4 + x^3 - x^2 - 7x - 6$ *(2-2)*

65. -1 *(4-1)* **66.** $-1, 2$ *(4-1)* **67.** 1 *(4-1)* **68.** $-\frac{17}{11}$ *(4-1)* **69.** 0 *(4-1)* **70.** $1, 5$ *(2-7)*

71. $x \le -6$ or $x \ge 1$ *(5-4)* **72.** $y = -\frac{3}{4}x + \frac{5}{4}$ *(4-3)* **73.** 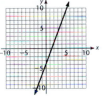 *(5-1)*

74. 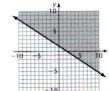 *(5-5)* **75.** 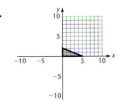 *(5-5)* **76.** Yes, time would be 1 min 59.16 sec *(4-3)*

77. 1,045,900 *(4-2)* **78.** 22 or more *(5-3)* **79.** 12 *(4-1)* **80.** 66 *(4-2)* **81.** $24\frac{4}{9} \le C \le 40\frac{5}{9}$ *(5-3)*

82. 15.2 by 26.6 cm *(4-5)* **83.** $-\frac{101}{112}$ *(1-6)* **84.** 2 *(5-2)* **85.** $y = -\frac{1}{2}x + 8$ *(5-2)*

86. $(x - 1)(x^2 + x + 1)(x + 1)(x^2 - x + 1)$ *(2-6)* **87.** $(x - 2)(x + 2)(x - 3)$ *(2-6)* **88.** $3xy(x - 2y)(x + 2y)$ *(2-6)*

89. $\frac{x^2 - x}{x^2 - 2x - 3}$ *(3-3)* **90.** $\frac{1}{x - 1}$ *(3-2)* **91.** $\frac{-b}{a}$ *(3-5)* **92.** $\frac{x^2(6 - x)}{(x + 2)^3}$ *(3-1)* **93.** $\frac{1}{(a - b)^2}$ *(3-5)*

94. $x = \frac{y^2 + yz + z^2}{y + z}$ *(4-4)* **95.** $-\frac{4}{5}$ *(4-1)* **96.** 8 *(4-1)* **97.** 2 by 6 m *(4-5)*

98. Any real number except 0 and -1 *(4-5)* **99.** 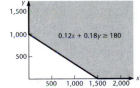 *(5-5)* **100.** 11:50 A.M. *(4-3)*

CHAPTER 6

EXERCISE 6-1

1. 10 **3.** 6 **5.** 12 **7.** 2 **9.** u^7v^7 **11.** 4 **13.** $\frac{a^8}{b^8}$ **15.** 3 **17.** 6 **19.** 2 **21.** 7

23. 12 **25.** $8x^{12}$ **27.** $3x^2$ **29.** $\frac{3}{4m^2}$ **31.** $x^{10}y^{10}$ **33.** $\frac{m^5}{n^5}$ **35.** $20y^{11}$ **37.** 48×10^{17}

39. 10^{14} **41.** x^6 **43.** $m^6 n^{15}$ **45.** $\dfrac{c^6}{d^{15}}$ **47.** $\dfrac{3u^4}{v^2}$ **49.** $2^4 s^8 t^{16}$ or $16 s^8 t^{16}$ **51.** $6x^5 y^{15}$ **53.** $\dfrac{m^4 n^{12}}{p^8 q^4}$

55. $\dfrac{u^3}{v^9}$ **57.** $9x^4$ **59.** $\dfrac{b^4}{a}$ **61.** $-\dfrac{b}{2a^2}$ **63.** 1 **65.** -1 **67.** $\dfrac{-1}{x^5}$ **69.** $-\dfrac{wy}{x^3}$ **71.** c

73. $a^2 - b^2$ **75.** $ab(b - a)$ **77.** $\dfrac{(x+1)^4(-2x-7)}{x^8}$ **79.** $\dfrac{(x-y)^2}{2(x+y)^2}$ **81.** x^8 **83.** x^n **85.** x^{2n+2}

87. $\dfrac{u^2}{v^4}$ **89.** x **91.** xy **93.** $\dfrac{1}{x^2 y^2}$ **95.** 1 **97.** $(x^n - 1)(x^n + 1)$ **99.** $(x^n + 1)(x^{2n} - x^n + 1)$

EXERCISE 6-2

1. 1 **3.** 1 **5.** $\dfrac{1}{3^3}$ **7.** $\dfrac{1}{m^7}$ **9.** 4^3 **11.** y^5 **13.** 10^2 **15.** y **17.** 1 **19.** 10^{10}

21. x^{11} **23.** $\dfrac{1}{z^5}$ **25.** $\dfrac{1}{10^7}$ **27.** 10^{12} **29.** y^8 **31.** $u^{10} v^6$ **33.** $\dfrac{x^4}{y^6}$ **35.** $\dfrac{x^2}{y^3}$ **37.** 1

39. 10^2 **41.** y **43.** 10 **45.** 3×10^{16} **47.** y^9 **49.** $3^2 m^2 n^2$ **51.** $\dfrac{2^3 m^3}{n^9}$ **53.** $\dfrac{n^{15}}{m^{12}}$ **55.** $\dfrac{3^3}{2^2}$

57. 1 **59.** $\dfrac{4y^3}{3x^5}$ **61.** $\dfrac{a^9}{8b^4}$ **63.** $\dfrac{1}{x^7}$ **65.** $\dfrac{n^8}{m^{12}}$ **67.** $\dfrac{m^3 n^3}{8}$ **69.** $\dfrac{1}{a^3 b^5}$ **71.** $\dfrac{16x^4}{9y^8}$ **73.** $\dfrac{8x^{15}}{27y^{15}}$

75. $\dfrac{t^2}{x^2 y^{10}}$ **77.** 4 **79.** $\dfrac{1}{a^2 - b^2}$ **81.** $\dfrac{1}{xy}$ **83.** $-cd$ **85.** $\dfrac{xy}{x+y}$ **87.** $\dfrac{(y-x)^2}{x^2 y^2}$ **89.** $\dfrac{y-x}{y}$

91. $b + ac$ **93.** $a + b$ **95.** $y^4 - x^4$ **97.** $a^{-4} b^{-5}(a + b^3)$ **99.** $x^{-3} y^{-4} z^{-3}(x^2 y^2 + z)$

EXERCISE 6-3

1. 7×10 **3.** 8×10^2 **5.** 8×10^4 **7.** 8×10^{-3} **9.** 8×10^{-8} **11.** 5.2×10 **13.** 6.3×10^{-1}
15. 3.4×10^2 **17.** 8.5×10^{-2} **19.** 6.3×10^3 **21.** 6.8×10^{-6} **23.** 800 **25.** 0.04 **27.** $300,000$
29. 0.0009 **31.** $56,000$ **33.** 0.0097 **35.** $430,000$ **37.** $0.000\,000\,38$ **39.** 5.46×10^9
41. 7.29×10^{-8} **43.** 1.23×10^{-11} **45.** 6.789×10^{12} **47.** $1.020\,030\,04 \times 10^{-1}$ **49.** $1.234\,000\,567 \times 10^{12}$
51. 4.5514×10^{10} **53.** 1×10^{13} **55.** 1×10^{-5} **57.** $83,500,000,000$ **59.** $0.000\,000\,000\,006\,14$
61. $2,000,010,000$ **63.** $0.000\,010\,020\,03$ **65.** $0.000\,001\,539$ **67.** $12,270,000$ **69.** $865,000$
71. $0.000\,000\,000\,000\,000\,000\,000\,001\,7$ **73.** 9×10^4 **75.** 6×10^{-4} **77.** 3×10^5 **79.** 5×10^4
81. 3×10 or 30 **83.** 3×10^{-4} or 0.0003 **85.** $6,600,000,000,000,000,000,000$ or 6.6×10^{21} tons
87. $4,000,000$ or 4×10^6; 250 **89.** 665 or 6.65×10^2 **91.** $1,827,000$ or 1.827×10^6

EXERCISE 6-4

1. 5 **3.** Not a real number   **5.** 4 **7.** -4 **9.** -4 **11.** 64 **13.** 4 **15.** x **17.** $\dfrac{1}{x^{1/5}}$

19. x^2 **21.** ab^3 **23.** $\dfrac{x^3}{y^4}$ **25.** $x^2 y^3$ **27.** $\frac{2}{5}$ **29.** $\frac{8}{125}$ **31.** $\frac{1}{4}$ **33.** $\frac{1}{6}$ **35.** $\frac{1}{125}$ **37.** 25

39. $\frac{1}{9}$ **41.** $\frac{4}{9}$ **43.** $\frac{625}{256}$ **45.** Not a real number   **47.** $\dfrac{1}{x^{1/2}}$ **49.** $n^{1/12}$ **51.** x^4 **53.** $\dfrac{2v^2}{u}$

55. $\dfrac{1}{x^2 y^3}$ **57.** $\dfrac{x^4}{y^3}$ **59.** $2x^2 y$ **61.** $2x^{2/3} y^2$ **63.** $\dfrac{2}{10^2} = \frac{1}{50} = 0.02$ **65.** $\dfrac{x}{2y^3}$ **67.** $\dfrac{y^{5/2}}{x^2 z^5}$ **69.** $\frac{5}{4} x^4 y^2$

71. $64 y^{1/3}$ **73.** $x^{\frac{8}{9}} - x^{\frac{7}{10}}$ **75.** $12m - 6m^{35/4}$ **77.** $2x + 3x^{1/2} y^{1/2} + y$ **79.** $x + 2x^{1/2} y^{1/2} + y$ **81.** $x - 1$

83. $x^3 - y^3$ **85.** $\dfrac{2}{a} + \dfrac{5}{a^{1/2} b^{1/2}} - \dfrac{3}{b}$ **87.** $a^{1/2} b^{1/3}$ **89.** x **91.** $\dfrac{1}{x^m}$ **93.** $\sqrt[n]{x}\sqrt[n]{y} = \sqrt[n]{xy}$

95. $(\sqrt[n]{x})^n = x$; $\sqrt[n]{x^n} = x$ **97.** No, false when x and y are negative and n is even.
99. No, false when x is negative and n is even.

EXERCISE 6-5

1. $\sqrt{11}$ **3.** $\sqrt[3]{5}$ **5.** $\sqrt[5]{u^3}$ **7.** $4\sqrt[7]{(y)^3}$ **9.** $\sqrt[7]{(4y)^3}$ **11.** $\sqrt[5]{(4ab^3)^2}$ **13.** $\sqrt{a + b}$ **15.** $6^{1/2}$
17. $m^{1/4}$ **19.** $y^{3/5}$ **21.** $(xy)^{3/4}$ **23.** $(x^2 - y^2)^{1/2}$ **25.** $6\sqrt{2}$ **27.** $2\sqrt{19}$ **29.** 2 **31.** 4 **33.** 2^3
35. 3^2 **37.** 3 **39.** $\sqrt{3}$ **41.** $x^2 \sqrt{x}$ **43.** $xy^2 \sqrt{xy}$ **45.** x **47.** $\dfrac{1}{x^3}$ **49.** $\dfrac{x\sqrt{x}}{2\sqrt{6}}$ or $\dfrac{x\sqrt{6x}}{12}$

51. $-5\sqrt[5]{y^2}$ **53.** $\sqrt[7]{(1 + m^2n^2)^3}$ **55.** $\dfrac{1}{\sqrt[3]{w^2}}$ **57.** $\dfrac{1}{\sqrt[5]{(3m^2n^3)^3}}$ **59.** $\sqrt{a} + \sqrt{b}$ **61.** $\sqrt[3]{(a^3 + b^3)^2}$

63. $(a + b)^{2/3}$ **65.** $-3x(a^3b)^{1/4}$ **67.** $(-2x^3y^7)^{1/9}$ **69.** $\dfrac{3}{y^{1/3}}$ or $3y^{-1/3}$ **71.** $\dfrac{-2x}{(x^2 + y^2)^{1/2}}$ or $-2x(x^2 + y^2)^{-1/2}$

73. $m^{2/3} - n^{1/2}$ **75.** $2\sqrt[3]{5}$ **77.** $2\sqrt[5]{2}$ **79.** $5\sqrt[3]{3}$ **81.** $4\sqrt[3]{4}$ **83.** 3 **85.** 25 **87.** 2 **89.** $x\sqrt[3]{x}$

91. x^2 **93.** $3xy^2$ **95.** $7x^8y^3\sqrt[4]{7x^3y^3}$ **97.** No, true only for $x \geq 0$. **99.** Yes

EXERCISE 6-6

1. y **3.** $2u$ **5.** $7x^2y$ **7.** $3\sqrt{2}$ **9.** $m\sqrt{m}$ **11.** $2x\sqrt{2x}$ **13.** $\frac{1}{3}$ **15.** $\dfrac{1}{y}$ **17.** $\dfrac{\sqrt{5}}{5}$

19. $\dfrac{\sqrt{5}}{5}$ **21.** $\dfrac{\sqrt{y}}{y}$ **23.** $\dfrac{\sqrt{y}}{y}$ **25** $3xy^2\sqrt{xy}$ **27.** $3x^4y^2\sqrt{2y}$ **29.** $\dfrac{\sqrt{2x}}{2x}$ **31.** $2x\sqrt{3x}$

33. $\dfrac{3\sqrt{2ab}}{2b}$ **35.** $\dfrac{\sqrt{42xy}}{7y}$ **37.** $\dfrac{3m^2\sqrt{2mn}}{2n}$ **39.** $2x^2y$ **41.** $2xy^2\sqrt[3]{2xy}$ **43.** $2xy^2\sqrt[3]{4x}$ **45.** $4x^2y^3\sqrt{2}$

47. $z\sqrt[5]{24x^3y^4}$ **49.** $yz\sqrt{24x^3z}$ **51.** $\sqrt{x}$ **53.** 4 **55.** $6m^3n^3$ **57.** $2\sqrt[3]{9}$ **59.** $\dfrac{2a\sqrt{3ab}}{3b}$

61. Is in the simplest radical form **63.** $\dfrac{2x}{3y^2}$ **65.** $-3m^2n^2\sqrt[5]{3m^2n}$ **67.** $\sqrt[3]{x^2(x - y)}$ **69.** $x^2y\sqrt[3]{6xy}$

71. $2x^2y$ **73.** $6xy\sqrt[3]{y}$ **75.** $\dfrac{\sqrt[3]{x}}{2}$ **77.** $2x^2y\sqrt[3]{4x^2y}$ **79.** $-\sqrt[3]{6x^2y^2}$ **81.** $\sqrt[3]{(x - y)^2}$ **83.** $\dfrac{\sqrt[4]{12x^3y^3}}{2x}$

85. $-x\sqrt{x^2 + 2}$ **87.** $4x^9y^3\sqrt[3]{2y}$ **89.** $mn\sqrt[12]{3^7m^5n^2}$ **91.** $x^n(x + y)^{n+2}$ **93.** $x^3y^{n^2+1}$

95. **(A)** $6x$ **(B)** $-2x$ **97.** **(A)** $2x$ **(B)** 0 **99.** **(A)** x **(B)** $-5x$

EXERCISE 6-7

1. $9\sqrt{3}$ **3.** $-5\sqrt{a}$ **5.** $-5\sqrt{n}$ **7.** $4\sqrt{5} - 2\sqrt{3}$ **9.** $\sqrt{m} - 3\sqrt{n}$ **11.** $4\sqrt{2}$ **13.** $-6\sqrt{2}$

15. $7 - 2\sqrt{7}$ **17.** $3\sqrt{2} - 2$ **19.** $y - 8\sqrt{y}$ **21.** $4\sqrt{n} - n$ **23.** $3 + 3\sqrt{2}$ **25.** $3 - \sqrt{3}$

27. $9 + 4\sqrt{5}$ **29.** $m - 7\sqrt{m} + 12$ **31.** $\sqrt{5} - 2$ **33.** $\dfrac{\sqrt{5} - 1}{2}$ **35.** $\dfrac{\sqrt{5} + \sqrt{2}}{3}$ **37.** $\dfrac{y - 3\sqrt{y}}{y - 9}$

39. $8\sqrt{2mn}$ **41.** $6\sqrt{2} - 2\sqrt{5}$ **43.** $-\sqrt[5]{a}$ **45.** $5\sqrt[3]{x} - \sqrt{x}$ **47.** $\dfrac{9\sqrt{2}}{4}$ **49.** $\dfrac{-3\sqrt{6uv}}{2}$

51. $38 - 11\sqrt{3}$ **53.** $x - y$ **55.** $10m - 11\sqrt{m} - 6$ **57.** $5 + \sqrt[3]{18} + \sqrt[3]{12}$ **59.** $x - 1$

61. $a + 2$ **63.** 1 **65.** $(3 - \sqrt{2})^2 - 6(3 - \sqrt{2}) + 7 = 9 - 6\sqrt{2} + 2 - 18 + 6\sqrt{2} + 7 = 0$

67. $(1 - \sqrt{5})^2 - 2(1 - \sqrt{5}) - 4 = (1 - 2\sqrt{5} + 5) - 2 + 2\sqrt{5} - 4 = 0$

69. $(4 + \sqrt{15})^2 - 8(4 + \sqrt{15}) + 1 = (16 + 8\sqrt{15} + 15) - 32 - 8\sqrt{15} + 1 = 0$ **71.** $-7 - 4\sqrt{3}$ **73.** $5 + 2\sqrt{6}$

75. $\dfrac{x + 5\sqrt{x} + 6}{x - 9}$ **77.** $\dfrac{6x + 9\sqrt{x}}{4x - 9}$ **79.** $\dfrac{-1}{7 - 4\sqrt{3}}$ **81.** $\dfrac{1}{5 - 2\sqrt{6}}$ **83.** $\dfrac{x - 4}{x - 5\sqrt{x} + 6}$ **85.** $3\sqrt{3}$

87. $\frac{10}{3}\sqrt[3]{9}$ **89.** $x + 2\sqrt[3]{xy} - \sqrt[3]{x^2y^2} - 2y$ **91.** $x + y$ **93.** $\dfrac{8x + 2\sqrt{xy} - 15y}{16x - 25y}$ **95.** $\dfrac{\sqrt[3]{x^2} + \sqrt[3]{x} + 1}{x - 1}$

97. $\dfrac{\sqrt[3]{x^2} - \sqrt[3]{x}\sqrt[3]{y} + \sqrt[3]{y^2}}{x + y}$ **99.** $\dfrac{(\sqrt{x} + \sqrt{y} + \sqrt{z})[(x + y - z) - 2\sqrt{xy}]}{(x + y - z)^2 - 4xy}$

EXERCISE 6-8

1. $8 + 3i$ **3.** $-5 + 3i$ **5.** $5 + 3i$ **7.** $6 + 13i$ **9.** $3 - 2i$ **11.** -15 or $-15 + 0i$ **13.** $-6 - 10i$

15. $15 - 3i$ **17.** $-4 - 33i$ **19.** 65 or $65 + 0i$ **21.** $1 + 7i$ **23.** $-8 + 19i$ **25.** $\frac{2}{5} - \frac{1}{5}i$

27. $\frac{3}{13} + \frac{11}{13}i$ **29.** $5 + 3i$ **31.** i **33.** $\frac{4}{5} - \frac{3}{5}i$ **35.** $4i$ **37.** $6i\sqrt{2}$ **39.** $-2i$ **41.** -5

43. $7 - 5i$ **45.** $-3 + 2i$ **47.** $8 + 25i$ **49.** $-2 + 4\sqrt{5} + (4\sqrt{2} + \sqrt{10})i$ **51.** $4i$ **53.** $\frac{5}{3} - \frac{2}{3}i$

55. $\frac{2}{13} + \frac{3}{13}i$ **57.** $\dfrac{-1}{2} + \dfrac{\sqrt{3}}{2}i$ **59.** $\dfrac{2 - \sqrt{15}}{9} - \dfrac{\sqrt{5} + 2\sqrt{3}}{9}i$ **61.** $-\frac{2}{5}i$ or $0 - \frac{2}{5}i$ **63.** $\frac{3}{2} - \frac{1}{2}i$

65. $4 - 7i$ **67.** 0 **69.** 0 **71.** 0 or $0 + 0i$ **73.** 0 **75.** 0 **77.** 0 **79.** $-1, -i, 1, i, -1, -i, 1$

81. $(a + c) + (b + d)i$ **83.** $a^2 + b^2$ or $(a^2 + b^2) + 0i$ **85.** $(ac - bd) + (ad + bc)i$ **87.** $-2 + 2i$ **89.** 1

91. $\pm 6i$ **93.** $9 \pm 3i$ **95.** $-1, i, 1, -i, -1, i, 1$ **97.** 1 **99.** $(ap^2 - aq^2 + bp + c) + (2apq + bq)i$

CHAPTER 6 REVIEW EXERCISE

1. x^5 *(6-1)*   **2.** x^3y^3 *(6-1)*   **3.** $\dfrac{x^3}{y^3}$ *(6-1)*   **4.** $\dfrac{1}{x^5}$ *(6-1)*   **5.** x^{24} *(6-1)*   **6.** 1 *(6-1)*

7. x^{11} *(6-1)*   **8.** $-8x^3$ *(6-1)*   **9.** $-6x^{11}$ *(6-1)*   **10.** 1 *(6-2)*   **11.** $\frac{1}{9}$ *(6-2)*   **12.** 8 *(6-2)*

13. $\frac{1}{2}$ *(6-4)*   **14.** Not a real number *(6-4)*   **15.** 4 *(6-4)*   **16.** 4.28×10^9 *(6-3)*   **17.** 3.18×10^{-5} *(6-3)*

18. 729,000 *(6-3)*   **19.** 0.000 603 *(6-3)*   **20.** $6x^4y^7$ *(6-2)*   **21.** $\dfrac{3u^4}{v^2}$ *(6-2)*   **22.** $6x^5y^{15}$ *(6-2)*

23. $\dfrac{c^6}{d^{15}}$ *(6-2)*   **24.** $\dfrac{4x^4}{9y^6}$ *(6-2)*   **25.** x^{12} *(6-2)*   **26.** y^2 *(6-2)*   **27.** $\dfrac{y^3}{x^2}$ *(6-2)*   **28.** x^3 *(6-4)*

29. $\dfrac{1}{x^2}$ *(6-4)*   **30.** $\dfrac{1}{x^{1/3}}$ *(6-4)*   **31.** u *(6-4)*   **32.** $\sqrt{3m}$ *(6-5)*   **33.** $3\sqrt{m}$ *(6-5)*   **34.** $(2x)^{1/2}$ *(6-5)*

35. $(a+b)^{1/2}$ *(6-5)*   **36.** $2-3i$ *(6-8)*   **37.** $5\sqrt[3]{3}$ *(6-6)*   **38.** $2xy^2$ *(6-6)*   **39.** $\dfrac{5}{y}$ *(6-6)*

40. $6x^2y^3\sqrt{y}$ *(6-6)*   **41.** $\dfrac{\sqrt{2y}}{2y}$ *(6-6)*   **42.** $2b\sqrt{3a}$ *(6-6)*   **43.** $6x^2y^3\sqrt{xy}$ *(6-6)*   **44.** $\dfrac{\sqrt{2xy}}{2x}$ *(6-6)*

45. $-3\sqrt{x}$ *(6-7)*   **46.** $\sqrt{7}-2\sqrt{3}$ *(6-7)*   **47.** $5+2\sqrt{5}$ *(6-7)*   **48.** $1+\sqrt{3}$ *(6-7)*   **49.** $\dfrac{5+3\sqrt{5}}{4}$ *(6-7)*

50. $3-6i$ *(6-8)*   **51.** $15+3i$ *(6-8)*   **52.** $2+i$ *(6-8)*   **53.** $-\frac{1}{2}-i$ *(6-8)*   **54.** $\dfrac{4x^6}{y^{16}}$ *(6-1)*

55. $-x^7y^8$ *(6-1)*   **56.** $-x^4y^3$ *(6-1)*   **57.** $9x^8y^9$ *(6-1)*   **58.** $\dfrac{-8x^3}{y^6}$ *(6-1)*   **59.** $\dfrac{9x^2}{4y^2}$ *(6-1)*

60. 2×10^{-3} or 0.002 *(6-3)*   **61.** $\dfrac{m^2}{2n^5}$ *(6-2)*   **62.** $\dfrac{x^6}{y^4}$ *(6-2)*   **63.** $\dfrac{4x^4}{y^6}$ *(6-2)*   **64.** $\dfrac{c}{a^2b^4}$ *(6-2)*

65. $\frac{1}{4}$ *(6-2)*   **66.** $\dfrac{n^{10}}{9m^{10}}$ *(6-2)*   **67.** $\dfrac{1}{(x-y)^2}$ *(6-2)*   **68.** $\dfrac{3a^2}{b}$ *(6-4)*   **69.** $\dfrac{3x^2}{2y^2}$ *(6-4)*   **70.** $\dfrac{1}{m}$ *(6-4)*

71. $6x^{1/6}$ *(6-4)*   **72.** $\dfrac{x^{1/12}}{2}$ *(6-4)*   **73.** $\frac{5}{9}$ *(6-2)*   **74.** $x+2x^{1/2}y^{1/2}+y$ *(6-4)*

75. $a^2=b$ and $\sqrt{b}=|a|$ *(6-4)*   **76.** $\sqrt[3]{4m^2n^2}$ *(6-5)*   **77.** $3\sqrt[5]{x^2}$ *(6-5)*   **78.** $x^{5/7}$ *(6-5)*

79. $-3(xy)^{2/3}$ *(6-5)*   **80.** $2x^2y$ *(6-6)*   **81.** $3x^2y\sqrt[3]{x^2y}$ *(6-6)*   **82.** $\dfrac{n^2\sqrt{6m}}{3}$ *(6-6)*   **83.** $\sqrt[4]{y^3}$ *(6-6)*

84. $-6x^2y^2\sqrt[5]{3x^2y}$ *(6-6)*   **85.** $x\sqrt[3]{2x^2}$ *(6-6)*   **86.** $\dfrac{\sqrt[5]{12x^3y^2}}{2x}$ *(6-7)*   **87.** $2x-3\sqrt{xy}-5y$ *(6-7)*

88. $\dfrac{x-4\sqrt{x}+4}{x-4}$ *(6-7)*   **89.** $\dfrac{6x+3\sqrt{xy}}{4x-y}$ *(6-7)*   **90.** $\dfrac{5\sqrt{6}}{6}$ *(6-7)*   **91.** $-1-i$ *(6-8)*

92. $\frac{4}{13}-\frac{7}{13}i$ *(6-8)*   **93.** $5+4i$ *(6-8)*   **94.** $\dfrac{xy}{x+y}$ *(6-2)*   **95.** $\dfrac{a^2b^2}{a^3+b^3}$ *(6-2)*   **96.** $y\sqrt[3]{2x^2y}$ *(6-6)*

97. 0 *(6-7)*   **98.** (A) x (B) $5x$ *(6-6)*
99. $(-1+i)^2+2(-1+i)+2=(1-2i-1)-2+2i+2=0$ *(6-8)*
100. $(1+\sqrt{2})^2-2(1+\sqrt{2})-1=(1+2\sqrt{2}+2)-2-2\sqrt{2}-1=0$ *(6-7)*

CHAPTER 6 PRACTICE TEST

1. $x^2y^2-x^6+\dfrac{x^4}{y^4}$ *(6-1)*   **2.** 3.45×10^{-5} *(6-3)*   **3.** 24,680,000,000 *(6-3)*   **4.** $3xy^2z^2\sqrt{xz}$ *(6-6)*

5. $\dfrac{\sqrt{3x}}{3x}$ *(6-6)*   **6.** $\dfrac{\sqrt{x}-3}{x-9}$ *(6-7)*   **7.** $5^{3/2}+7^{2/3}$ *(6-2)*   **8.** $3\sqrt[3]{3}-\sqrt[5]{25}$ *(6-6)*   **9.** $\dfrac{y}{x^2z}$ *(6-2)*

10. $\dfrac{1+x^4y^2}{x^2y}$ *(6-2)*   **11.** $-1+\sqrt{3}$ *(6-7)*   **12.** $5+3\sqrt{3}$ *(6-7)*   **13.** $-1+2\sqrt{3}$ *(6-7)*

14. $3-2i$ *(6-8)*   **15.** $17+i$ *(6-8)*   **16.** $-\frac{13}{29}+\frac{11}{29}i$ *(6-8)*   **17.** $\frac{1}{10}-\frac{3}{10}i$ *(6-8)*   **18.** 1 *(6-8)*
19. 0 *(6-8)*   **20.** 0 *(6-6)*

CHAPTER 7

EXERCISE 7-1

1. $-6, 1$ **3.** $2, 5$ **5.** $\frac{1}{2}, -4$ **7.** ± 4 **9.** $\pm 4i$ **11.** $\pm 3\sqrt{5}$ **13.** $\pm\frac{3}{2}$ **15.** $\pm\frac{3}{4}$
17. $x^2 + 4x + 4 = (x + 2)^2$ **19.** $x^2 - 6x + 9 = (x - 3)^2$ **21.** $x^2 + 12x + 36 = (x + 6)^2$ **23.** $-2 \pm \sqrt{2}$
25. $3 \pm 2\sqrt{3}$ **27.** $-2, 2$ **29.** $3, -4$ **31.** $-\frac{1}{2}, 2$ **33.** $\frac{1}{2}, 2$ **35.** $2, -3$ **37.** $\pm\sqrt{2}$ **39.** $\pm\frac{3}{4}i$
41. $\pm\sqrt{\frac{7}{9}}$ or $\pm\dfrac{\sqrt{7}}{3}$ **43.** $8, -2$ **45.** $-1 \pm 3i$ **47.** $-\frac{1}{3}, 1$ **49.** $x^2 + 3x + \frac{9}{4} = (x + \frac{3}{2})^2$
51. $u^2 - 5u + \frac{25}{4} = (u - \frac{5}{2})^2$ **53.** $\dfrac{-1 \pm \sqrt{5}}{2}$ **55.** $\dfrac{5 \pm \sqrt{17}}{2}$ **57.** $2 \pm 2i$ **59.** $\dfrac{2 \pm \sqrt{2}}{2}$
61. $\dfrac{-3 \pm \sqrt{17}}{4}$ **63.** $\frac{1}{2}, -2$ **65.** $\dfrac{1 \pm \sqrt{3}}{2}$ **67.** $-\frac{1}{3}, 1$ **69.** $1 \pm \frac{1}{2}i$ **71.** $\dfrac{3 \pm i\sqrt{7}}{4}$ **73.** $\dfrac{-1 \pm i\sqrt{3}}{2}$
75. $\dfrac{-5 \pm \sqrt{10}}{2}$ **77.** $2 \pm i$ **79.** $-1, 1$ **81.** No solution   **83.** $-1, 3$ **85.** $-\sqrt{2} \pm 2$ **87.** $2\sqrt{3} \pm i$
89. $i \pm \sqrt{3}$ **91.** $3i$ **93.** $(1 \pm \sqrt{3})i$ **95.** $x = \dfrac{-m \pm \sqrt{m^2 - 4n}}{2}$ **97.** $a = \sqrt{c^2 - b^2}$ **99.** 90¢/gal

EXERCISE 7-2

1. $a = 2, b = -5, c = 3$ **3.** $a = 3, b = 1, c = -1$ **5.** $a = 3, b = 0, c = -5$ **7.** $-4 \pm \sqrt{13}$ **9.** $5 \pm 2\sqrt{7}$
11. $-\dfrac{3}{2} \pm \dfrac{\sqrt{11}}{2}i$ **13.** $-2 \pm i$ **15.** $1 \pm 2i$ **17.** $\dfrac{3}{2} \pm \dfrac{\sqrt{3}}{2}i$ **19.** $\dfrac{5 \pm 3\sqrt{5}}{2}$ **21.** $\dfrac{-3 \pm \sqrt{13}}{2}$
23. $1 \pm i\sqrt{2}$ **25.** $\dfrac{3 \pm \sqrt{3}}{2}$ **27.** $\dfrac{-1 \pm \sqrt{13}}{6}$ **29.** Two real roots   **31.** One real root
33. Two nonreal, complex roots   **35.** Two real roots   **37.** Two nonreal, complex roots   **39.** $5 \pm \sqrt{7}$
41. $-1 \pm \sqrt{3}$ **43.** $0, -\frac{3}{2}$ **45.** $2 \pm 3i$ **47.** $5 \pm 2\sqrt{7}$ **49.** $\dfrac{2 \pm \sqrt{2}}{2}$ **51.** $1 \pm i\sqrt{2}$ **53.** $\frac{2}{5}, 3$
55. $\frac{1}{7}$ **57.** 4 **59.** $3, -3$ **61.** $t = \sqrt{\dfrac{2d}{g}}$ or $\dfrac{\sqrt{2dg}}{g}$ **63.** $r = -1 + \sqrt{\dfrac{A}{P}}$ or $-1 + \dfrac{\sqrt{AP}}{P}$ **65.** $\dfrac{\sqrt{7} \pm i}{2}$
67. $-\sqrt{3}, \dfrac{-\sqrt{3}}{3}$ **69.** $\frac{1}{2}i, -2i$ **71.** $\dfrac{\pm\sqrt{3} - i}{2}$ **73.** $\dfrac{1 \pm \sqrt{5}}{4}$ **75.** $\dfrac{-1 \pm \sqrt{3}}{2}$ **77.** $-0.45, 0.76$
79. $1.35, 0.48$ **81.** $-1.05, 0.63$ **83.** $x = \dfrac{y \pm \sqrt{5y^2}}{2}$ or $\dfrac{y \pm y\sqrt{5}}{2}$ **85.** $x = y \pm \sqrt{2y^2}$ or $y \pm y\sqrt{2}$
87. $x = \pm\sqrt{\dfrac{y^2}{y^2 - 1}}$ or $\dfrac{\pm y\sqrt{y^2 - 1}}{y^2 - 1}$ **89.** Has two real solutions, since discriminant is positive
91. Has no real solutions, since discriminant is negative   **93.** $c = \frac{9}{8}$ **95.** $a < \frac{4}{5}$ **97.** $c = \frac{16}{3}$
99. $\left(\dfrac{-b + \sqrt{b^2 - 4ac}}{2a}\right)\left(\dfrac{-b - \sqrt{b^2 - 4ac}}{2a}\right) = \dfrac{b^2 - (b^2 - 4ac)}{4a^2} = \dfrac{c}{a}$

EXERCISE 7-3

1. $12, 14$ **3.** $0, 2$ **5.** $\frac{1}{3}, \dfrac{-6}{-4}$ **7.** $1, 2$ **9.** 11 **11.** 127 mi
13. Height $= \dfrac{\sqrt{33}}{2} - \dfrac{3}{2}$ m; Base $\dfrac{\sqrt{33}}{2} + \dfrac{3}{2}$ m   **15.** 5.12 by 3.12 cm   **17.** 6 by 4 m   **19.** 1 ft
21. 8 **23.** 2 by 5 m   **25.** 50 mph   **27.** 5.66 ft/sec   **29.** 65 mph   **31.** 50 mph   **33.** 176 ft/sec
35. **(A)** $t = 8$; the arrow returns to the ground at time $t = 8$ sec.   **(B)** 0.13 sec, 7.87 sec
37. 1.41 min   **39.** 2 hr; 3 hr   **41.** 15 kph; 20 kph   **43.** 2 km/hr   **45.** 0.05 (or 5%)   **47.** $6
49. 150 by 300 ft   **51.** $x[40 - \frac{1}{2}(x - 120)]$; 140 trees   **53.** $(8 + 0.25x)(5{,}000 - 100x)$; 6 or 12
55. $510 or $1,710   **57.** $9.50 or $12.50

EXERCISE 7-4

1. 47 **3.** 12 **5.** 4 **7.** 16 **9.** 18 **11.** 6 **13.** 15 **15.** $\pm 1, \pm 3$ **17.** $\pm 3, \pm i\sqrt{2}$
19. $-1, 4$ **21.** $2, -8$ **23.** -1 **25.** 21 **27.** $1, 2$ **29.** $0, 2$ **31.** $9, 16$ **33.** 4

35. No solution   **37.** No solution   **39.** No solution   **41.** 5, 13   **43.** 1, 2   **45.** 1, 3   **47.** 5

49. 4   **51.** 9   **53.** $-1, 2$ **55.** $\sqrt[3]{4}, \sqrt[3]{2}$ **57.** $\pm 1, \pm 2, \pm 2i, \pm i$ **59.** $\pm 2, \pm 3$ **61.** $\pm 3, \pm 2i$

63. $\pm \dfrac{\sqrt{10}}{2}, \pm i$ **65.** $-8, 125$ **67.** 1, 16   **69.** $-1, -2$ **71.** 1, 3   **73.** $\frac{2}{3}, -\frac{3}{2}$ **75.** $-2, 5$

77. ± 1 **79.** $\pm 2, \pm \frac{1}{2}$ **81.** $-2, 3, \frac{1}{2} \pm \sqrt{7}/2 i$ **83.** $3 \pm 2i, 4, 2$ **85.** 4   **87.** 11   **89.** 1, 5

91. 4, 1.4

EXERCISE 7-5

1.

3.

5.

7.

9.

11.

13.

15.

17.

19.

21.

23.

25.

27.

29.

31.

33.

35.

37.

39.

41.

43.

45.

47.

49.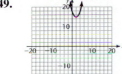

51.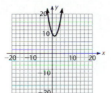

53.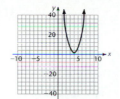

55.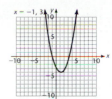

57.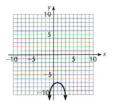

59.

61.

63.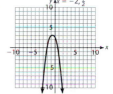

65.

67.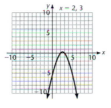

69.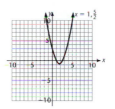

71.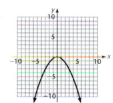

73. 2   **75.** 0   **77.** 2   **79.** 2

EXERCISE 7-6

1.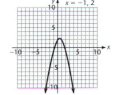

3.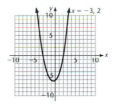

5.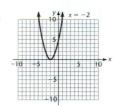

7.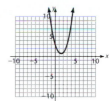

9.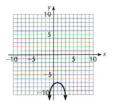

11.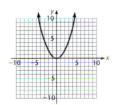

13.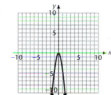

15.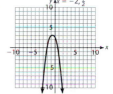

17.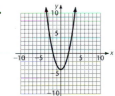

19.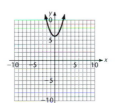

21.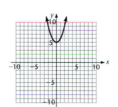

23.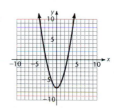

25.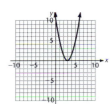

27.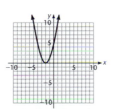

29.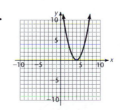

31.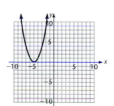

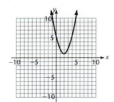
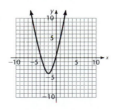
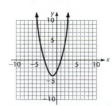
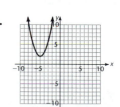

33. 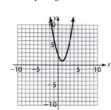 **35.** 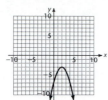 **37.** 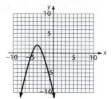 **39.**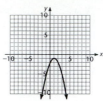

41. 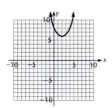 **43.** 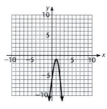 **45.** 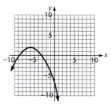 **47.**

49. 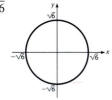 **51.** 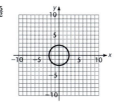 **53.**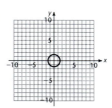

55. $(x-1)^2 - 4$; moved right 1, down 4

57. $-(x-\frac{1}{2})^2 + \frac{9}{4}$; moved right $\frac{1}{4}$, reflected about x axis, moved up $\frac{9}{4}$ **59.** $(x+\frac{1}{2})^2 - \frac{25}{4}$; moved left $\frac{1}{2}$, down $\frac{25}{4}$

61. $(x+2)^2$; moved left 2 **63.** $(x-\frac{3}{2})^2 + \frac{3}{4}$; moved right $\frac{3}{2}$, up $\frac{3}{4}$

65. $-(x-1)^2 - 7$; moved right 1, reflected about x axis, moved down 7

67. $-(x-\frac{5}{2})^2 + \frac{1}{4}$; moved right $\frac{5}{2}$, reflected about x axis, moved up $\frac{1}{4}$

69. $2(x-\frac{7}{4})^2 - \frac{9}{8}$; moved right $\frac{7}{4}$, stretched by a factor of 2, moved down $\frac{9}{8}$

71. $-2(x+\frac{3}{4})^2 + \frac{25}{8}$; moved left $\frac{3}{4}$, stretched by a factor of 2, reflected about x axis, moved up $\frac{25}{8}$

73. 2 **75.** 0 **77.** 2 **79.** 2

EXERCISE 7-7

1. $\sqrt{5}$ **3.** $\sqrt{89}$ **5.** $\sqrt{18}$ or $3\sqrt{2}$ **7.** $\sqrt{32.5}$ **9.** $\sqrt{68.66}$ **11.** $R = 4$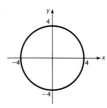

13. $R = \sqrt{6}$ 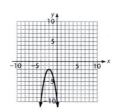 **15.** $R = \frac{5}{2}$ 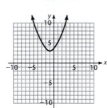 **17.** $R = \frac{8}{5}$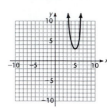

19. $x^2 + y^2 = 49$ **21.** $x^2 + y^2 = 5$ **23.** $x^2 + y^2 = \frac{49}{4}$ **25.** $x^2 + y^2 = 38.44$

27. 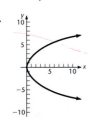 **29.** 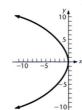 **31.** **33.**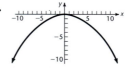

35. 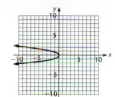   **37.** 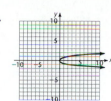   **39.** 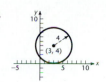   **41.**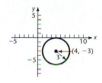

43.   **45.**    **47.** 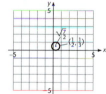   **49.**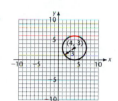

51. $(x - 3)^2 + (y - 5)^2 = 49$ **53.** $(x + 3)^2 + (y - 3)^2 = 64$

55. $(x + 4)^2 + (y + 1)^2 = 3$ **57.** $(x - 1)^2 + (y - \frac{1}{2})^2 = \frac{81}{16}$ **59.** $(x - 1.1)^2 + (y - \frac{1}{2})^2 = \sqrt{2}$

61. $x = \frac{1}{2}y^2$ 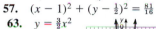   **63.** $y = \frac{3}{2}x^2$ 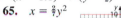   **65.** $x = \frac{2}{7}y^2$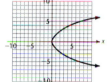

67. Yes, since two sides have length $\sqrt{17}$.   **69.** No, the lengths of the three sides are $\sqrt{17}$, $\sqrt{17}$, and $\sqrt{18}$.

71. $(x - 2)^2 + (y - 3)^2 = 9$ **73.** $(x - 3)^2 + (y + 3)^2 = 16$

75. $(x + 3)^2 + (y + 2)^2 = 9$ 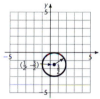   **77.** $(x - \frac{3}{2})^2 + (y - \frac{5}{2})^2 = \frac{49}{4}$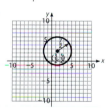

79. $(x - \frac{1}{2})^2 + (y + \frac{3}{2})^2 = \frac{9}{4}$ **81.** Focus $(1, 0)$; directrix $x = -1$

83. Focus $(0, -\frac{1}{32})$; directrix $y = \frac{1}{32}$ **85.** $x = -3, 7$ **87.** $2x + 4y + 1 = 0$ **89.** $x^2 + y^2 = 25$

91. $(x - 2)^2 = -4(y - 3)$ or $x^2 - 4x + 4y - 8 = 0$ **93.** $4(x - 1) = (y - 2)^2$ or $4x = y^2 - 4y + 8$

95. $x^2 + y^2 = 50^2$ **97.** $x^2 = -100y$

EXERCISE 7-8

1. 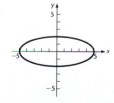   **3.**    **5.** 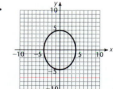   **7.**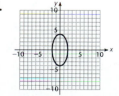

9.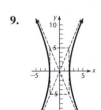

11.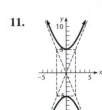

13.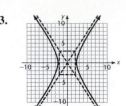

15.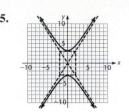

17.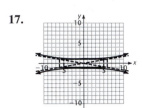

19.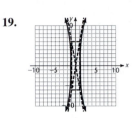

21.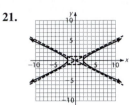

23.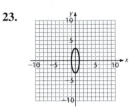

25.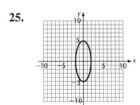

27.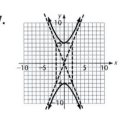

29.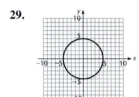

31. $\dfrac{x^2}{9} + \dfrac{y^2}{4} = 1$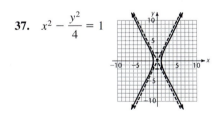

33. $\dfrac{x^2}{9} - \dfrac{y^2}{4} = 1$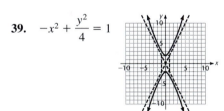

35. $\dfrac{x^2}{4} + y^2 = 1$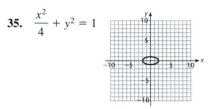

37. $x^2 - \dfrac{y^2}{4} = 1$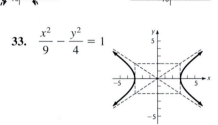

39. $-x^2 + \dfrac{y^2}{4} = 1$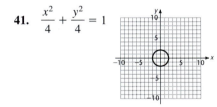

41. $\dfrac{x^2}{4} + \dfrac{y^2}{4} = 1$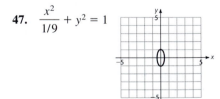

43. $\dfrac{x^2}{4} - \dfrac{y^2}{4} = 1$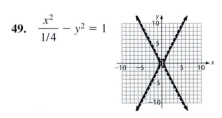

45. $x^2 - \dfrac{y^2}{1/9} = 1$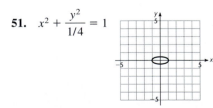

47. $\dfrac{x^2}{1/9} + y^2 = 1$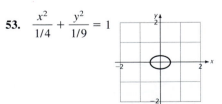

49. $\dfrac{x^2}{1/4} - y^2 = 1$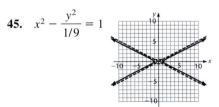

51. $x^2 + \dfrac{y^2}{1/4} = 1$

53. $\dfrac{x^2}{1/4} + \dfrac{y^2}{1/9} = 1$

55. Hyperbola 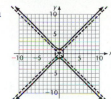   **57.** Circle    **59.** Ellipse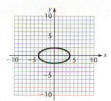

61. Parabola 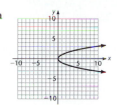   **63.** Hyperbola 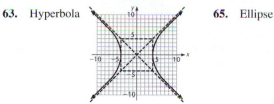   **65.** Ellipse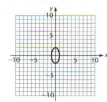

67. Point 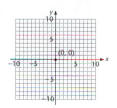   **69.** Circle 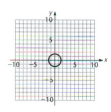   **71.** $(0, -\sqrt{21}), (0, \sqrt{21})$ **73.** $(5, 0), (-5, 0)$

75. Ellipse, $\dfrac{(x-1)^2}{4} + (y-2)^2 = 1$ **77.** Circle, $(x+1)^2 + (y+2)^2 = 4$ **79.** Ellipse, $(x+3)^2 + \dfrac{(y-1)^2}{4} = 1$

81. Hyperbola, $(x-3)^2 - \dfrac{(y+1)^2}{4} = 1$ **83.** Hyperbola, $-\dfrac{(x+3)^2}{16} + \dfrac{(y-1)^2}{64} = 1$

EXERCISE 7-9

1. $(-4, 3)$ **3.** $(-\infty, -4] \cup [3, \infty)$ **5.** $(-4, 3)$

7. $(-\infty, 3) \cup (7, \infty)$ **9.** $[-5, -1]$

11. $(1, 8)$ **13.** $(-\infty, -2) \cup (5, \infty)$

15. $[-6, -2]$ **17.** $(-\infty, -6] \cup [0, \infty)$

19. $(-\infty, -3] \cup [3, \infty)$ **21.** $[1, 3]$ **23.** $[-3, 5)$ **25.** $(-1, 1)$ **27.** $(-\infty, 10)$

29. $[-2, 0]$ **31.** $(-1, \infty)$ **33.** $(-\infty, 5]$ **35.** $\varnothing$ **37.** Not a single interval

39. $(-\infty, -4) \cup (3, 5)$

41. $[-4, 0] \cup [3, \infty)$

43. $(-\infty, -4) \cup (0, 3)$

45. $(0, 3) \cup (7, \infty)$

47. $(-2, 5]$ **49.** $(-\infty, -2) \cup (5, \infty)$ **51.** $(-\infty, -2) \cup (0, 4]$

53. $(-2, 0] \cup [1, \infty)$

55. $(-\infty, -3) \cup (0, 1)$

57. $(0, \frac{4}{3}]$ **59.** $(-\infty, 0) \cup (\frac{1}{4}, \infty)$

61. All real numbers; graph is the whole real line.   **63.** No solution   **65.** $(-\infty, -\sqrt{3}] \cup [\sqrt{3}, \infty)$

67. $[2, 3)$ **69.** $(-2, 2)$

71. $(-\infty, -3) \cup (-1, \infty)$ **73.** $[-3, 2]$

75. $(-2, 3)$ **77.** $(-3, 0) \cup (1, \infty)$ **79.** $(-1, \frac{1}{2}] \cup (2, \infty)$

81. $(-3, 1)$ **83.** $(-\infty, 1] \cup [2, \infty)$ **85.** $(-\infty, -1] \cup (1, \infty)$ **87.** All real numbers

89. No solution   **91.** $2 \le t \le 8$, or $[2, 8]$

CHAPTER 7 REVIEW EXERCISE

1. $a = 3$, $b = 4$, $c = -2$ *(7-2)*   **2.** $x = \dfrac{-b \pm \sqrt{b^2 - 4ac}}{2a}$ *(7-2)*   **3.** $0, 3$ *(7-1)*   **4.** ± 5 *(7-1)*

5. $2, 3$ *(7-1)*   **6.** $-3, 5$ *(7-1)*   **7.** $\pm\sqrt{7}$ *(7-1)*   **8.** $2, 4$ *(7-1)*   **9.** $1, -1$ *(7-1)*   **10.** $2, -1$ *(7-1)*

11. $\dfrac{-3 \pm \sqrt{5}}{2}$ *(7-2)*   **12.** $-7, 4$ *(7-1, 7-2)*   **13.** $6, -1$ *(7-1, 7-2)*   **14.** $-2, -\frac{7}{3}$ *(7-2)*   **15.** $\frac{5}{2}, 2$ *(7-2)*

16. $-5, -4$ *(7-1, 7-2)*   **17.** $(-7, 4)$ *(7-9)*   **18.** $(-\infty, 2) \cup (3, \infty)$ *(7-9)*   **19.** $(-5, 4)$ *(7-9)*

20. $(-\infty, -5] \cup [4, \infty)$ 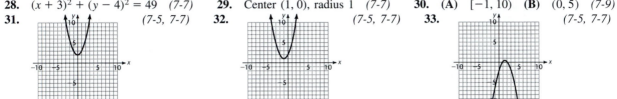 x *(7-9)*   **21.** 61 *(7-4)*   **22.** 4 *(7-4)*   **23.** 4 *(7-4)*   **24.** 1 *(7-4)*

25. -19; two nonreal, complex roots   *(7-2)*   **26.** $\sqrt{20}$ or $2\sqrt{5}$ *(7-7)*   **27.** $x^2 + y^2 = 25$ *(7-7)*

28. $(x + 3)^2 + (y - 4)^2 = 49$ *(7-7)*   **29.** Center $(1, 0)$, radius 1 *(7-7)*   **30.** **(A)** $[-1, 10]$ **(B)** $(0, 5)$ *(7-9)*

31. *(7-5, 7-7)*   **32.** *(7-5, 7-7)*   **33.** *(7-5, 7-7)*

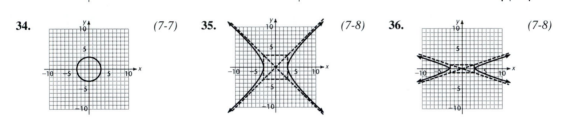

34. *(7-7)*   **35.** *(7-8)*   **36.** *(7-8)*

37. *(7-8)*   **38.** *(7-8)*   **39.** *(7-7)*

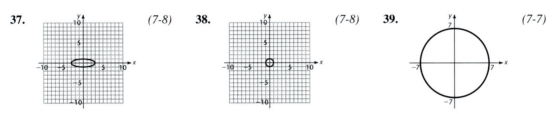

40. *(7-7)*   **41.** *(7-7)*   **42.** *(7-8)*

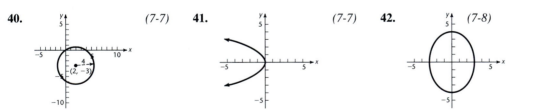

43. *(7-8)*   **44.** $3, 9$ *(7-3)*   **45.** $\dfrac{3 \pm \sqrt{5}}{2}$ *(7-3)*   **46.** 9 by 12 ft   *(7-3)*

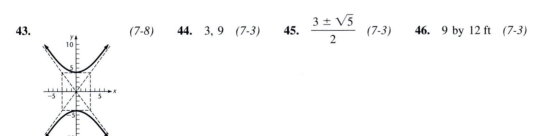

47. 15 mph, 20 mph   *(7-3)*   **48.** $0, 2$ *(7-1)*   **49.** $\pm 2\sqrt{3}$ *(7-1)*   **50.** $\pm 3i$ *(7-1)*   **51.** $-2, 6$ *(7-1)*

52. $\dfrac{-1}{3}, 3$ *(7-1)*   **53.** $\frac{1}{2}, -3$ *(7-1)*   **54.** $\dfrac{3 \pm i\sqrt{39}}{4}$ *(7-2)*   **55.** $0, 2, -2$ *(7-1)*   **56.** $0, 2$ *(7-1)*

57. $\dfrac{9 \pm \sqrt{401}}{16}$ *(7-2)*   **58.** $0, \pm\sqrt{3}$ *(7-1)*   **59.** $\dfrac{1 \pm \sqrt{7}}{3}$ *(7-2)*   **60.** $\dfrac{1 \pm \sqrt{7}}{2}$ *(7-2)*   **61.** $-4, 5$ *(7-1)*

62. $\frac{3}{4}, \frac{5}{2}$ *(7-1)*   **63.** $\frac{-1 \pm \sqrt{5}}{2}$ *(7-2)*   **64.** $1 \pm i\sqrt{2}$ *(7-2)*   **65.** $2, 3$ *(7-4)*   **66.** $9, 25$ *(7-4)*

67. $\pm 2, \pm 3i$ *(7-4)*   **68.** $64, \frac{-27}{8}$ *(7-4)*   **69.** 4 *(7-4)*   **70.** 5 *(7-4)*   **71.** No solution *(7-4)*

72. 13 *(7-4)*   **73.** $(-\infty, -3] \cup [7, \infty)$ *(7-9)*   **74.** $(-\infty, 0) \cup (\frac{1}{2}, \infty)$ *(7-9)*   **75.** No solution *(7-9)*

76. $(-\infty, -8) \cup [0, \infty)$ *(7-9)*   **77.** *(7-5)*   **78.** *(7-5)*

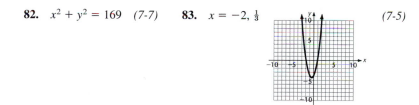

79. *(7-8)*   **80.** *(7-8)*   **81.** $(x - 3)^2 = 12$ *(7-1)*

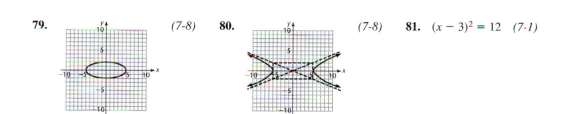

82. $x^2 + y^2 = 169$ *(7-7)*   **83.** $x = -2, \frac{1}{3}$ *(7-5)*

84. $(x + 3)^2 + (y - 4)^2 = 25$; radius = 5, center = $(-3, 4)$ *(7-7)*   **85.** $y = \pm \frac{\sqrt{3}}{3}x$ *(7-8)*   **86.** 6 by 5 in. *(7-3)*

87. 7 hr and 8 hr *(7-3)*   **88.** $\frac{3 \pm i\sqrt{6}}{2}$ or $\frac{3}{2} \pm \frac{\sqrt{6}}{2}i$ *(7-2)*   **89.** $\frac{-3 \pm \sqrt{57}}{6}$ *(7-2)*

90. $\pm 1, \pm 2, \pm 2i, \pm i$ *(7-4)*   **91.** $3, \frac{9}{4}$ *(7-4)*   **92.** 7 *(7-4)*   **93.** $\pm 1, \pm i$ *(7-1, 7-4)*   **94.** $\pm 1, \pm i$ *(7-4)*

95. $(-\infty, 1] \cup (3, 4)$ *(7-9)*   **96.** $(1, \infty)$ *(7-9)*

97. $(-2, -1) \cup (0, 1)$ *(7-9)*   **98.** 9 cm, 12 cm *(7-3)*

99. **(A)** 2,000 or 8,000 **(B)** 5,000 *(7-3)*   **100.** \$2.75 *(7-3)*

CHAPTER 7 PRACTICE TEST

1. $0, 5$ *(7-1)*   **2.** ± 3 *(7-1)*   **3.** $\frac{5 \pm \sqrt{61}}{2}$ *(7-2)*   **4.** 9 *(7-4)*   **5.** $(0, 5)$ *(7-9)*

6. $\pm\sqrt{6}, \pm i$ *(7-4)*   **7.** $[-1, 0] \cup [4, \infty)$ *(7-9)*   **8.** $[x - (-2)]^2 + 3 = 0$ *(7-1, 7-6)*

9. Center $(2, -3)$, radius $\sqrt{10}$ *(7-7)*   **10.** $2\sqrt{10}$ *(7-7)*

11. *(7-5, 7-7)*   **12.** *(7-5, 7-7)*   **13.** *(7-8)*

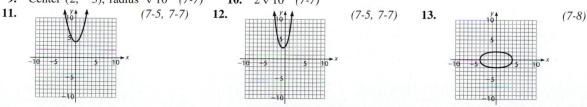

14. 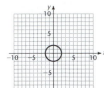 *(7-7)*   **15.** 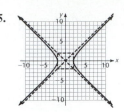 *(7-8)*   **16.** 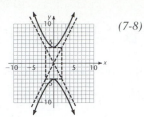 *(7-8)*

17. 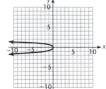 *(7-8)*   **18.** 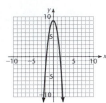 *(7-8)*   **19.** ± 1 *(7-3)*   **20.** 5 by 16 cm *(7-3)*

CUMULATIVE REVIEW EXERCISES, CHAPTERS 1–7

1. $(x + 1)^2 = 5x - 6$ *(1-2)*   **2.** $|x - 5| - 3 < x - 2$ *(1-4)*   **3.** $3x + 4 \cdot \dfrac{1}{x} = 5(x + 6)$ *(1-2, 3-2)*

4. 17 *(1-3)*   **5.** -53 *(1-6)*   **6.** -1 *(1-3)*   **7.** 7 *(2-1)*   **8.** 72 *(3-3)*   **9.** $x^2(x^2 - 9)$ *(3-3)*

10. $x^3 - 3x^2 + 10x - 28, R = 86$ *(3-4)*   **11.** 29 *(1-4)*   **12.** $\sqrt{505}$ *(7-7)*   **13.** 7.5 *(6-2)*

14. 4 *(6-4)*   **15.** $\frac{1}{16}$ *(6-4)*   **16.** $7 - 22i$ *(6-8)*   **17.** $8 - 2i$ *(6-8)*   **18.** $\frac{1}{2} + \frac{3}{2}i$ *(6-8)*

19. $A(-7, 3), B(4, 9), C(8, -2)$ *(5-1)*   **20.** Slope 3, y intercept -10 *(5-3)*   **21.** Center $(2, 3)$, radius 2 *(7-7)*

22. $-\frac{7}{4}$ *(5-2)*   **23.** $(4, 5)$ *(7-5)*   **24.** $3x - 7$ *(2-1)*   **25.** $x - (2y + 3z)$ *(2-1)*

26. $8xy + 10yz + 9xz$ *(1-6)*   **27.** $x^2 - 9x + 14$ *(2-2)*   **28.** $x^2 + 22x + 121$ *(2-2)*

29. $-x^3 + 2x^2 - x + 5$ *(2-1)*   **30.** $x^3 + 3x^2 + 5x + 3$ *(2-2)*   **31.** x^{-6} *(6-2)*   **32.** $x^7 y^{-1}$ *(6-2)*

33. x^{19} *(6-2)*   **34.** $3(2x + y + 3z)$ *(2-3)*   **35.** $(x + 4)^2$ *(2-3)*   **36.** $(x + 5)(x - 1)$ *(2-4, 2-5)*

37. $(x + 5)(x - 3)$ *(2-4, 2-5)*   **38.** $(2x - 1)(x + 3)$ *(2-4, 2-5)*   **39.** $(x + 6)(x - 6)$ *(2-3)*

40. $(x + 1)(x^2 - x + 1)$ *(2-6)*   **41.** $x(x + 2)(x + 4)$ *(2-6)*   **42.** $(x + 3)(x + 2y)$ *(2-3)*   **43.** $\dfrac{6a^2 b^3 c}{10abc^2 d^2}$ *(3-1)*

44. $\dfrac{4ac}{5b}$ *(3-1)*   **45.** $\dfrac{x + 4}{3x}$ *(3-3)*   **46.** $\dfrac{x^2 + 2x + 3}{x^3}$ *(3-3)*   **47.** $\dfrac{x + 1}{x - 1}$ *(3-5)*   **48.** $\dfrac{x - 1}{x - 2}$ *(3-2)*

49. 3.2×10^{-8} *(6-3)*   **50.** 0.000 000 012 34 *(6-3)*   **51.** $\frac{111}{12}$ *(1-7)*   **52.** $\dfrac{5y + 7}{3}$ *(1-7)*

53. $\frac{5}{3}, 7$ *(2-7)*   **54.** $-3, 0$ *(4-1)*   **55.** $x < \frac{10}{7}$ *(5-3)*   **56.** $-\frac{7}{2} \leq x \leq \frac{1}{2}$ *(5-4)*   **57.** $0, 7$ *(7-1)*

58. $\pm\sqrt{7}$ *(7-1)*   **59.** 3, 4 *(2-7, 7-1)*   **60.** 11 *(7-4)*   **61.** ± 1 *(2-7, 7-4)*   **62.** $(-\infty, 1) \cup (5, \infty)$ *(7-9)*

63. 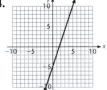 *(5-3)*   **64.** 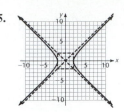 *(5-2)*

65. 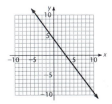 *(5-2)*   **66.** 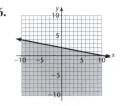 *(5-5)*

67. 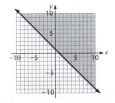 *(5-5)*   **68.** 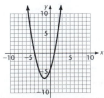 *(7-5)*

69. 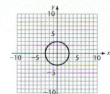   *(7-7)*   **70.** 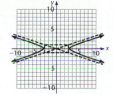   *(7-8)*

71. 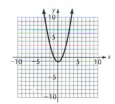   *(7-5, 7-7)*   **72.** 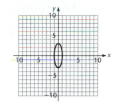   *(7-8)*

73. 12 and 13 *(7-3)*   **74.** 6 by 15 ft *(1-8)*   **75.** 700 *(4-5)*   **76.** 7 hr *(4-3)*

77. 28 or more *(5-3)*   **78.** $(3x - 7)(x - 2)$ *(2-4, 2-5)*   **79.** $\dfrac{5x^2 + 12x + 3}{x(x + 1)(x + 3)}$ *(3-3)*   **80.** 0 *(6-6)*

81. 1 *(3-2)*   **82.** $x^2 + 2x + 4$ *(3-1)*   **83.** $\dfrac{-2 \pm \sqrt{19}}{3}$ *(7-2)*   **84.** $(-\infty, -3] \cup [\tfrac{1}{3}, \infty)$ *(5-4)*

85. 7 *(7-4)*   **86.** $[-\tfrac{4}{3}, \tfrac{3}{4}) \cup (5, \infty)$ *(7-9)*   **87.** $x = \dfrac{y}{y - 1}$ *(4-3)*   **88.** $\tfrac{5}{2}, 2$ *(7-2)*

89. 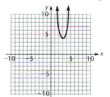   *(5-2)*   **90.** 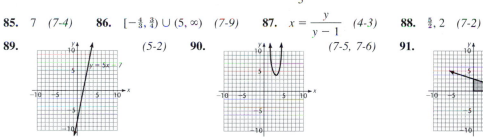   *(7-5, 7-6)*   **91.** 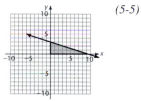   *(5-5)*

92. 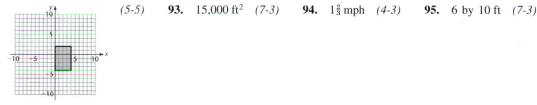   *(5-5)*   **93.** 15,000 ft² *(7-3)*   **94.** $1\tfrac{2}{3}$ mph *(4-3)*   **95.** 6 by 10 ft *(7-3)*

96. 0, 2, $\tfrac{9}{4}$ *(7-2)*   **97.** $\dfrac{x^2 + 3x + 1}{x^2 + 2x}$ *(3-5)*

98. $(1 + i)^3 - 3(1 + i)^2 + 4(1 + i) - 2 = (1 + i)(2i) - 3(2i) + 4 + 4i - 2 = 2i - 2 - 6i + 4 + 4i - 2 = 0$ *(6-8)*

99. $\dfrac{-11}{2,313} + \dfrac{433}{2,313}i$ *(6-8)*   **100.** $\tfrac{1}{3}$ *(5-2)*

CHAPTER 8

EXERCISE 8-1

1. $(2, -3)$ **3.** No solution   **5.** $(1, 3)$ **7.** $(5, -2)$ **9.** $(-3, 4)$ **11.** $(-2, -4)$ **13.** $(1, 4)$

15. $(-2, -3)$ **17.** $(6, 1)$ **19.** $(-2, 3)$ **21.** $(5, -4)$ **23.** $(1, -8)$ **25.** $(p, q) = (-\tfrac{4}{3}, 1)$

27. Infinite number of solutions, $(x, 3x - 9)$ for any real number x **29.** $(-1, -4)$ **31.** $(-2, 5)$ **33.** $(7, -3)$

35. $(-4, -2)$ **37.** $(-2, 3)$ **39.** $(1, -5)$ **41.** $(\tfrac{1}{3}, -2)$

43. Infinite number of solutions, $(x, \tfrac{1}{2}x + 3)$ for any real number x **45.** $(-2, 2)$ **47.** $(1, \tfrac{1}{2})$ **49.** $(-\tfrac{1}{3}, 2)$

51. $(\tfrac{1}{2}, \tfrac{1}{3})$ **53.** $(-\tfrac{1}{4}, \tfrac{2}{3})$ **55.** $(1.1, 0.3)$ **57.** $(8, 6)$ **59.** $(\tfrac{101}{100}, \tfrac{99}{100})$ **61.** No solution   **63.** $(2\tfrac{3}{4}, \tfrac{1}{6})$

65. 1   **67.** Infinite number of solutions   **69.** 1   **71.** No solution   **73.** Infinite number of solutions

EXERCISE 8-2

1. 30 quarters; 70 dimes **3.** 700 \$2 tickets; 2,800 \$4 tickets **5.** 20 dL **7.** 14 nickels; 8 dimes **9.** 52°; 38°
11. 20 liters **13.** 60 dL 20% solution; 30 dL 50% solution **15.** 25 kg \$5 tea; 50 kg \$6.50 tea
17. \$8,000 at 10%, \$12,000 at 15% **19.** 84 $\frac{1}{4}$-lb packages; 60 $\frac{1}{2}$-lb packages
21. 60 mL 80% solution; 40 mL 50% solution **23.** (A) $1\frac{9}{13}$ sec; $7\frac{9}{13}$ sec (B) Approx. 8,462 ft **25.** 3.6 liters
27. Both companies pay \$136 on sales of \$1,700. The straight-commission company pays better for sales over \$1,700;
the other company pays better for sales below \$1,700.
29. $p = 9$ **31.** 42 $\frac{1}{2}$-lb packages; 68 $\frac{1}{4}$-lb packages **33.** 2,800 advance; 1,500 at gate
35. \$40,000 at 4.5%, \$60,000 at 7% **37.** 30 mph; 45 mph **39.** 80; 88

EXERCISE 8-3

1. $(2, -5, 3)$ **3.** $(\frac{3}{2}, -\frac{1}{2}, \frac{1}{2})$ **5.** $(-4, 1, \frac{2}{3})$ **7.** $(-1, -1, -2)$ **9.** $(0, -2, 5)$ **11.** $(2, 0, -1)$
13. $(-1, 2, 0)$ **15.** $(0, 2, -3)$ **17.** $(1, -2, 1)$ **19.** $(3, 0, 1)$ **21.** $(-1, 2, 0)$ **23.** No solution
25. No solution **27.** $(1, 2, -3, 4)$ **29.** $(0, 0, -1, 2)$ **31.** $(18 - z, 14 - z, z + 5, z)$ **33.** $(1, -1, 0, 2)$
35. $\left(\dfrac{2 - 17z}{7}, \dfrac{-13 + 2z}{7}, z\right)$ **37.** $(1 - z, 3, z)$ **39.** $D = -4, E = -4, F = -17)$ **41.** $a = 1, b = -2, c = 3$
43. 1,200 style A; 800 style B; 2,000 style C **45.** 60 g mix A; 50 g mix B; 40 g mix C **47.** $w + y = 80$
$$x + z = 120$$
$$w + x = 140$$
$$y + z = 60$$

49. Oldest press: 12 hr; middle press: 6 hr; newest press: 4 hr

EXERCISE 8-4

1. $\begin{bmatrix} 4 & -6 & | & -8 \\ 1 & -3 & | & 2 \end{bmatrix}$ **3.** $\begin{bmatrix} -4 & 12 & | & -8 \\ 4 & -6 & | & -8 \end{bmatrix}$ **5.** $\begin{bmatrix} 1 & -3 & | & 2 \\ 8 & -12 & | & -16 \end{bmatrix}$ **7.** $\begin{bmatrix} 1 & -3 & | & 2 \\ 0 & 6 & | & -16 \end{bmatrix}$

9. $\begin{bmatrix} 1 & -3 & | & 2 \\ 2 & 0 & | & -12 \end{bmatrix}$ **11.** $\begin{bmatrix} 1 & -3 & | & 2 \\ 3 & -3 & | & -10 \end{bmatrix}$

13. $\begin{bmatrix} 3 & 2 & -1 & | & 5 \\ 2 & -1 & 0 & | & 4 \\ 1 & 0 & 2 & | & 3 \end{bmatrix}$ **15.** $\begin{bmatrix} 1 & 0 & 2 & | & 3 \\ -4 & 2 & 0 & | & -8 \\ 3 & 2 & -1 & | & 5 \end{bmatrix}$ **17.** $\begin{bmatrix} 3 & -1 & 2 & | & 7 \\ 2 & -1 & 0 & | & 4 \\ 3 & 2 & -1 & | & 5 \end{bmatrix}$ **19.** $\begin{bmatrix} 1 & 0 & 2 & | & 3 \\ \frac{7}{2} & 0 & -\frac{1}{2} & | & \frac{13}{2} \\ 3 & 2 & -1 & | & 5 \end{bmatrix}$

21. $(3, 2)$ **23.** $(3, 1)$ **25.** $(2, 1)$ **27.** $(2, 4)$ **29.** $(-8, 5)$ **31.** $(-10, 7)$ **33.** $(-20, 16)$
35. No solution **37.** $(1, 4)$ **39.** Infinitely many solutions: $(2y - 3, y)$ **41.** Infinitely many solutions: $(\frac{1}{2}y + \frac{1}{2}, y)$
43. $(2, -1)$ **45.** $(2, -1)$ **47.** No solution **49.** $\left(\dfrac{5y - 4}{2}, y\right)$ **51.** $(1.1, 0.3)$ **53.** $(-2, 3, 1)$
55. $(0, -2, 2)$ **57.** $(\frac{1}{9}, \frac{61}{27}, -\frac{19}{27})$ **59.** $(0, -11, 7)$ **61.** $(\frac{3}{2}, \frac{1}{2}, \frac{5}{2})$ **63.** $x = 1, y = 0, z = -1, w = 0$
65. $x = 0, y = -1, z = 2, w = 0$

EXERCISE 8-5

1. 1 **3.** 0 **5.** -14 **7.** 2 **9.** -2.6 **11.** 1 **13.** $\frac{1}{12}$ **15.** $\begin{vmatrix} a_{22} & a_{23} \\ a_{32} & a_{33} \end{vmatrix}$ **17.** $\begin{vmatrix} a_{11} & a_{12} \\ a_{31} & a_{32} \end{vmatrix}$

19. $(-1)^{1+1}\begin{vmatrix} a_{22} & a_{23} \\ a_{32} & a_{33} \end{vmatrix}$ **21.** $(-1)^{2+3}\begin{vmatrix} a_{11} & a_{12} \\ a_{31} & a_{32} \end{vmatrix}$ **23.** $\begin{vmatrix} 1 & -2 \\ -4 & 8 \end{vmatrix}$ **25.** $\begin{vmatrix} -2 & 0 \\ 5 & -2 \end{vmatrix}$ **27.** $(-1)^{1+1}\begin{vmatrix} 1 & -2 \\ -4 & 8 \end{vmatrix} = 0$

29. $(-1)^{3+2}\begin{vmatrix} -2 & 0 \\ 5 & -2 \end{vmatrix} = -4$ **31.** $(5, -2)$ **33.** $(1, -1)$ **35.** $(-1, 1)$ **37.** 10 **39.** -21 **41.** -120
43. -40 **45.** -12 **47.** 0 **49.** $(2, -2, -1)$ **51.** $(2, -1, 2)$ **53.** $(2, -3, -1)$ **55.** $(1, -1, 2)$
57. $a = -1, b = 2, c = 0$ **59.** -8 **61.** 48
63. Expand the determinant about the first two rows to obtain $x - 3y + 7 = 0$; then show that the two points satisfy this
linear equation.
65. $\frac{23}{2}$ **69.** $D = 0$; infinitely many solutions

EXERCISE 8-6

1.

3.

5.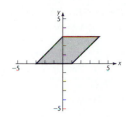

7.

9.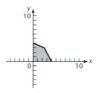

11.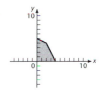

13.

15.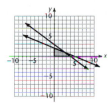

17.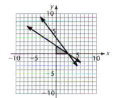

19.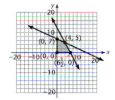

21.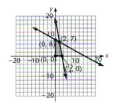

23.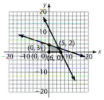

25.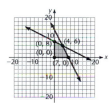

27.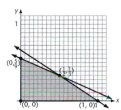

29.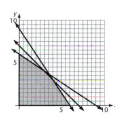

31.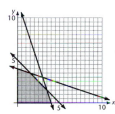

33.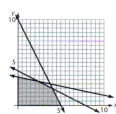

35.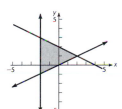

37.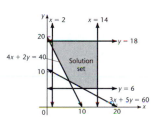

39.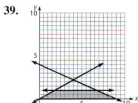

41.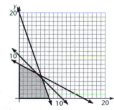

43. $6x + 4y \leq 108$
$x \geq 0$
$y \geq 0$

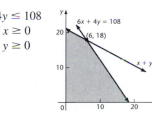

45. 6 trick skis; 18 slalom skis

47. $30x + 10y \geq 360$
$10x + 10y \geq 160$
$10x + 30y \geq 240$
$x \geq 0$
$y \geq 0$

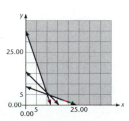

49. 12 oz M; 4 oz N

EXERCISE 8-7

1. $(-3, -4), (3, -4)$ **3.** $(4, 2\sqrt{5}), (4, -2\sqrt{5})$ **5.** $(\frac{1}{2}, 1)$ **7.** $(3, 0)$ **9.** $(4, -2), (-4, 2)$
11. $(0, 1), (\frac{9}{5}, -\frac{8}{5})$ **13.** $(3 - i, 4 - 3i), (3 + i, 4 + 3i)$ **15.** $(2, 1), (2, -1), (-2, 1), (-2, -1)$
17. $(-3, -2), (-3, 2), (3, -2), (3, 2)$

19. 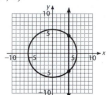   **21.** 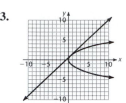   **23.** 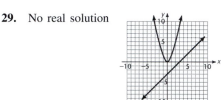   **25.**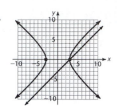

27. 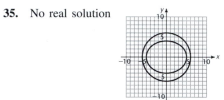   **29.** No real solution   **31.** No real solution

33. No real solution   **35.** No real solution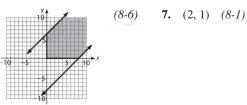

37. $(2 + \sqrt{10}, -2 + \sqrt{10}), (2 - \sqrt{10}, -2 - \sqrt{10})$ **39.** $(-1, 2), (4, 7)$ **41.** $(i, 2i), (i, -2i), (-i, 2i), (-i, -2i)$
43. $(2, 4), (-2, 4), (i\sqrt{5}, -5), (-i\sqrt{5}, 5)$ **45.** $(4, 0), (-3, \sqrt{7}), (-3, -\sqrt{7})$ **47.** 2 by 16 ft
49. $(2\sqrt{2}, \sqrt{2}), (-2\sqrt{2}, -\sqrt{2}), (1, 4), (-1, -4)$ **51.** $(3, 3), (-3, -3), (0, 6), (0, -6)$
53. $(4, 4), (-4, -4), (\frac{4}{5}\sqrt{5}, -\frac{4}{5}\sqrt{5}), (-\frac{4}{5}\sqrt{5}, \frac{4}{5}\sqrt{5})$

CHAPTER 8 REVIEW EXERCISE

1. $(6, 1)$ *(8-1)*   **2.** $(3, 2)$ *(8-1)*   **3.** $(8, 3)$ *(8-1)*
4. *(8-6)*   **5.** *(8-6)*   **6.** *(8-6)*   **7.** $(2, 1)$ *(8-1)*

8. $(5, -3)$ *(8-1)*   **9.** $(2, 1)$ *(8-1)*   **10.** $(-3, 7)$ *(8-1)*   **11.** $(1, 7)$ *(8-1)*   **12.** $(-2, -3)$ *(8-1)*
13. $(-1, 2, 1)$ *(8-3)*   **14.** $(1, -1), (\frac{7}{5}, -\frac{1}{5})$ *(8-6)*   **15.** $(4, 3), (-4, 3), (4, -3), (-4, -3)$ *(8-6)*
16. $(2, 1, -1)$ *(8-3)*   **17.** $(1, 3, 4)$ *(8-3)*   **18.** $(8, \frac{4}{5}, -\frac{4}{5})$ *(8-3)*   **19.** No solution *(8-1)*
20. $(3, 4), (-3, -4)$ *(8-7)*   **21.** $(0, 2), (\sqrt{3}, -1), (-\sqrt{3}, -1)$ *(8-7)*   **22.** $(z, 2 - z, z)$ *(8-3)*

23. $(0, 2, 0)$ *(8-3)*   **24.** $(\frac{16}{7}, \frac{13}{7}, -\frac{12}{7})$ *(8-3)*   **25.** $(6, 0), \left(-7, \frac{\sqrt{13}}{3}\right), \left(-7, \frac{-\sqrt{13}}{3}\right)$ *(8-7)*

26. $\left(5, \frac{i\sqrt{11}}{3}\right), \left(5, \frac{-i\sqrt{11}}{3}\right), \left(-4, \frac{2i\sqrt{5}}{3}\right), \left(-4, \frac{-2i\sqrt{5}}{3}\right)$ *(8-7)*

27. $\left(\frac{3i\sqrt{5}}{10}, \frac{9i\sqrt{5}}{10}\right), \left(\frac{-3i\sqrt{5}}{10}, \frac{-9i\sqrt{5}}{10}\right)$ *(8-7)*   **28.** $(1, -2, 3)$ *(8-3)*   **29.** $(55, -20)$ *(8-1)*

30. $(-48, 34)$ *(8-1)*   **31.** $\left(\frac{3\sqrt{2}}{2}, \frac{3\sqrt{2}}{2}\right), \left(\frac{-3\sqrt{2}}{2}, \frac{-3\sqrt{2}}{2}\right)$ *(8-7)*   **32.** $(3, 36), (-\frac{1}{4}, \frac{1}{4})$ *(8-7)*

33. -2 *(8-5)*   **34.** 0 *(8-5)*   **35.** 30 *(8-5)*   **36.** $\frac{1}{6}$ *(8-5)*

37. $\begin{bmatrix} 4 & 5 & | & 6 \\ 1 & 2 & | & 3 \end{bmatrix}$ *(8-4)*   **38.** $\begin{bmatrix} 1 & 2 & | & 3 \\ 3 & 3 & | & 3 \end{bmatrix}$ *(8-4)*   **39.** $\begin{bmatrix} 1 & 2 & | & 3 \\ 1 & \frac{5}{4} & | & \frac{3}{2} \end{bmatrix}$ *(8-4)*   **40.** $\begin{bmatrix} 3 & \frac{9}{2} & | & 6 \\ 4 & 5 & | & 6 \end{bmatrix}$ *(8-4)*

41. $(-69, 27)$ *(8-5)*   **42.** $(\frac{8}{7}, \frac{1}{7})$ *(8-5)*   **43.** $(2, 1)$ *(8-5)*   **44.** $(46, -18)$ *(8-5)*

45. 14 nickels; 16 dimes *(8-2)*   **46.** $2,000 at 10%; $4,000 at 6% *(8-2)*

47. $3\frac{1}{3}$ lb $2.40 candy; $6\frac{2}{3}$ lb $3 candy *(8-2)*   **48.** 1,000 advance; 2,700 at door *(8-2)*

49. 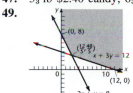 *(8-6)*   **50.** 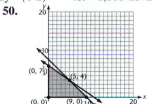 *(8-6)*

51. 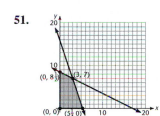 *(8-6)*   **52.** 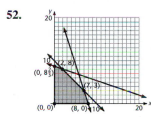 *(8-6)*   **53.** 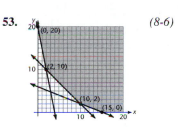 *(8-6)*

54. No solution *(8-1)*   **55.** $(2, -1, 2)$ *(8-3)*   **56.** $(1, 3), (1, -3), (-1, 3), (-1, -3)$ *(8-6)*

57. $(1 + i, 2i), (1 - i, -2i)$ *(8-6)*   **58.** $(1, -5, 3)$ *(8-3)*   **59.** $(\frac{1}{2}, \frac{1}{3}, \frac{1}{4})$ *(8-3)*

60. $(3, 4), (-3, 4), (\sqrt{51}/2, -\frac{7}{2}), (-\sqrt{51}/2, -\frac{7}{2})$ *(8-3)*

61. $(2\sqrt{7}, i\sqrt{3}), (-2\sqrt{7}, i\sqrt{3}), (2\sqrt{7}, -i\sqrt{3}), (-2\sqrt{7}, -i\sqrt{3})$ *(8-7)*

62. $(0, 0), (4, 4), (-2 + 2i\sqrt{3}, 2 - 2i\sqrt{3}), (-2 - 2i\sqrt{3}, 2 + 2i\sqrt{3})$ *(8-7)*

63. $(1, 0), \left(-\dfrac{10}{9}, \dfrac{i\sqrt{19}}{3}\right), \left(-\dfrac{10}{9}, -\dfrac{i\sqrt{19}}{3}\right)$ *(8-7)*   **64.** $(0, 3), (0, -3)$ *(8-7)*   **65.** No real solution *(8-7)*

66. $(-1, 9), (\frac{13}{9}, \frac{169}{9})$ *(8-7)*   **67.** $x = -1, y = 3$ *(8-4)*   **68.** $(-1, -3)$ *(8-4)*   **69.** $(1, 2)$ *(8-4)*

70. $(1, 2)$ *(8-4)*   **71.** $(4, -3, 2)$ *(8-2)*   **72.** $(-1, 0, 1)$ *(8-4)*   **73.** $(\frac{7}{2}, \frac{1}{2}, 0)$ *(8-4)*   **74.** No solution *(8-4)*

75. 0 *(8-5)*   **76.** 0 *(8-5)*   **77.** 3,744 *(8-5)*   **78.** 0 *(8-5)*   **79.** $(4, 5, 6)$ *(8-5)*   **80.** $(3, 4, 5)$ *(8-5)*

81. 6 by 5 cm *(8-1)*   **82.** 30 g 70% solution; 70 g 40% solution *(8-2)*   **83.** $x^2 + y^2 - 4x - 6y - 12 = 0$ *(8-3)*

84. 12 1-ft^3 boxes; 18 1.5-ft^3 boxes *(8-2)*   **85.** No solution *(8-1)*

86. $(0, 2), (0, -2), (i\sqrt{2}, i\sqrt{2}), (-i\sqrt{2}, -i\sqrt{2})$ *(8-6)*

87. $\left(\dfrac{3\sqrt{377}}{13}, \dfrac{8\sqrt{13}}{13}\right), \left(-\dfrac{3\sqrt{377}}{13}, \dfrac{8\sqrt{13}}{13}\right), \left(\dfrac{3\sqrt{377}}{13}, -\dfrac{8\sqrt{13}}{13}\right), \left(-\dfrac{3\sqrt{377}}{13}, -\dfrac{8\sqrt{13}}{13}\right)$ *(8-7)*

88. $\left(\dfrac{12\sqrt{85}}{17}, \dfrac{6\sqrt{119}}{17}\right), \left(-\dfrac{12\sqrt{85}}{17}, \dfrac{6\sqrt{119}}{17}\right), \left(\dfrac{12\sqrt{85}}{17}, -\dfrac{6\sqrt{119}}{17}\right), \left(-\dfrac{12\sqrt{85}}{17}, -\dfrac{6\sqrt{119}}{17}\right)$ *(8-7)*

89. $(5 + 2\sqrt{6}, 5 - 2\sqrt{6}), (5 - 2\sqrt{6}, 5 + 2\sqrt{6})$ *(8-7)*

90. $\left(\sqrt{2 + \sqrt{3}}, \dfrac{1}{\sqrt{2 + \sqrt{3}}}\right), \left(\sqrt{2 - \sqrt{3}}, \dfrac{1}{\sqrt{2 - \sqrt{3}}}\right), \left(-\sqrt{2 + \sqrt{3}}, -\dfrac{1}{\sqrt{2 + \sqrt{3}}}\right), \left(-\sqrt{2 - \sqrt{3}}, -\dfrac{1}{\sqrt{2 - \sqrt{3}}}\right)$ *(8-7)*

91. $\left(\dfrac{3\sqrt{2}}{2}, \dfrac{3\sqrt{2}}{2}\right), \left(-\dfrac{3\sqrt{2}}{2}, \dfrac{3\sqrt{2}}{2}\right), \left(\dfrac{3\sqrt{2}}{2}, -\dfrac{3\sqrt{2}}{2}\right), \left(-\dfrac{3\sqrt{2}}{2}, -\dfrac{3\sqrt{2}}{2}\right)$ *(8-7)*   **92.** No solution *(8-7)*

93. $(-1, -1, 5)$ *(8-3)*   **94.** $(-1, -1, 5)$ *(8-3)*   **95.** $\dfrac{5 + \sqrt{5}}{2}$ and $\dfrac{5 - \sqrt{5}}{2}$ *(8-7)*

96. $y = 2x^2 - \dfrac{x}{2} + \dfrac{1}{4}$ *(8-3)*   **97.** 48 $\frac{1}{2}$-lb packages; 72 $\frac{1}{3}$-lb packages *(8-2)*

98. 40 g mix A; 60 g mix B; 30 g mix C *(8-2)*

99. $0.3x + 0.2y \geq 27$, $0.03x + 0.05y \geq 5.4$, $0.1x + 0.2y \geq 19$, $x \geq 0$, $y \geq 0$ *(8-6)*

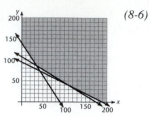

100.   30 g A: 90 g B *(8-6)*

CHAPTER 8 PRACTICE TEST

1. $(-1, 2)$ *(8-1)*   **2.** $(3, -2)$ *(8-1)*   **3.** $(3, 2, 1)$ *(8-3)*   **4.**   No solution   *(8-3)*
5. $(z - 2, 3 - 2z, z)$ *(8-3)*   **6.** $(-1, 2)$ *(8-4)*   **7.** $(\frac{8}{3}, \frac{5}{3}, -1)$ *(8-4)*   **8.** $(1, 1, 1)$ *(8-4)*
9. $(1, 2, -1)$ *(8-5)*   **10.** $(\sqrt{7}, 3), (0, -4), (-\sqrt{7}, 3)$ *(8-7)*   **11.** $(2\sqrt{2}, 2\sqrt{2}), (-2\sqrt{2}, -2\sqrt{2})$ *(8-7)*
12. $\left(\dfrac{10\sqrt{55}}{11}, \dfrac{18i\sqrt{11}}{11}\right), \left(\dfrac{-10\sqrt{55}}{11}, \dfrac{18i\sqrt{11}}{11}\right), \left(\dfrac{10\sqrt{55}}{11}, \dfrac{-18i\sqrt{11}}{11}\right), \left(\dfrac{-10\sqrt{55}}{11}, \dfrac{-18i\sqrt{11}}{11}\right)$ *(8-7)*
13.   *(8-6)*   **14.**   *(8-6)*   **15.**   *(8-6)*

16. $\begin{bmatrix} 1 & -1 & \vline & -3 \\ 3 & 6 & \vline & 9 \end{bmatrix}$ *(8-4)*   **17.**   13   *(8-5)*   **18.** -1 and $\frac{1}{2}$ *(8-7)*
19.   620 general admission; 140 reserved   *(8-2)*   **20.**   80 lb A; 100 lb B *(8-2)*

CHAPTER 9

EXERCISE 9-1

1.   Function   **3.**   Not a function   **5.**   Function   **7.**   Not a function   **9.**   Function   **11.**   Function
13.   Function   **15.**   Not a function   **17.**   Function   **19.**   Function   **21.**   Not a function   **23.**   Not a function
25.   Not a function   **27.**   Function   **29.**   Not a function

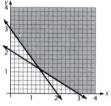

31.   Not a function

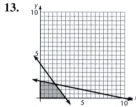

33.   Function, $y = 3x - 1$ **35.**   Function, $y = x^2 - 3x + 1$ **37.**   Not a function, $y = \pm\sqrt{x}$
39.   Not a function, $y = \dfrac{1 \pm \sqrt{1 + 4x}}{2}$ **41.**   Function, $y = x^2 - 3x$ **43.**   Function, $y = \dfrac{x + 1}{x - 1}$
45.   Function, $y = x^2 + 2x + 1$ **47.**   Not a function, $y = \pm\sqrt{x^2 - 1}$

49.   Not a function, $y = \pm\sqrt{9 - \frac{9}{4}x^2}$ **51.**   Function   **53.**   Function   **55.**   Not a function

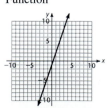
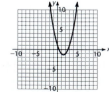
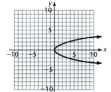

57. Function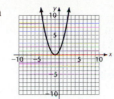

59. Not a function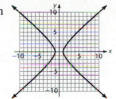

61. Not a function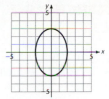

63. Function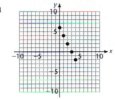

65. Not a function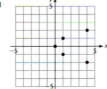

67. Not a function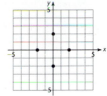

69. Function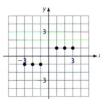

71. R **73.** R **75.** All real numbers except $x = 0$

77. All real numbers except $x = -2$ and $x = 3$ **79.** All real numbers except $x = -4$ and $x = 3$
81. $x \leq 4$ or $(-\infty, 4]$ **83.** $x < -3$ or $x \geq 1$, or $(-\infty, -3) \cup [1, \infty)$ **85.** $x \leq -3$ or $x \geq 2$ or $(-\infty, -3] \cup [2, \infty)$
87. $x \geq 0$ or $(0, \infty)$ **89.** R **91.** Function 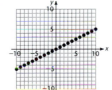   **93.** Not a function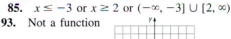

95. Function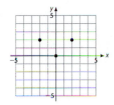

97. Yes; 9. Domain: letters A through Y with Q omitted; Range: integers 2 through 9

99. **(A)** Partial table:

t	d
0	0
1	144
2	256
3	336
4	384
5	400
6	384
7	336
8	256
9	144
10	0

(B) Domain: $0 \leq t \leq 10$
Range: $0 \leq d \leq 400$
(C) The relation is a function.

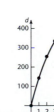

101. $p = -\frac{1}{5}d + 70$; $p = 55$

EXERCISE 9-2

1. 4 **3.** -8 **5.** -2 **7.** -2 **9.** -12 **11.** -6 **13.** 2 **15.** 0 **17.** 42 **19.** $3x + 1$

21. $x^2 + x$ **23.** $-x^2 + 3x - 2$ **25.** -27 **27.** 2 **29.** 6 **31.** 25 **33.** 22 **35.** -91

37. $10 - 2u$ **39.** $3a - 9a^2$ **41.** $2 - 2h$ **43.** -2 **45.** 10 **47.** 48 **49.** 0 **51.** -7

53. $-x^2 + 4x - 2$; all real numbers **55.** 0; all real numbers **57.** $2x - 2x^2$; all real numbers

59. $-x^2 - 2x + 2$; all real numbers **61.** $-x^2 + 4x - 2$; all real numbers **63.** $3x^3 - 5x^2 + 2x$; all real numbers

65. $\dfrac{3x - 2}{x - x^2}$; $x \neq 0, 1$ **67.** $\dfrac{x^2 - x}{3x - 2}$; $x \neq \frac{2}{3}$ **69.** -1; $x \neq 0, 1$ **71.** $53 - 20x$ **73.** $-10x^2 + 10x - 7$

75. $6 - 6x^2$ **77.** $300x^2 - 420x + 147$ **79.** $3x^2 - 6x^3 + 3x^4$ **81.** $-4x^2 + 22x - 30$ **83.** $10x + 10h - 7$

85. $3x^2 + 6xh + 3h^2$ **87.** 10 **89.** $6x + 3h$ **91.** $A(5) = \frac{1}{5}$, $A(0) = -\frac{3}{5}$, $A(-5)$ not defined, $A(x - 5) = \dfrac{x - 8}{x}$

93. Total sales of the two models combined **95.** Total revenue from the sales **97.** $C(x) = 8.6x$

99. $C(F) = \frac{5}{9}(F - 32)$ **101. (A)** 30 mi, 300 mi **(B)** 30

EXERCISE 9-3

1. Slope: 2
y intercept: -4

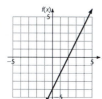

3. Slope: -2
y intercept: 4

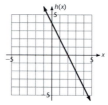

5. Slope: $-\frac{2}{3}$
y intercept: 4

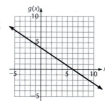

7. Slope: 10
y intercept: 120

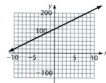

9. Slope: 300
y intercept: $-1,500$

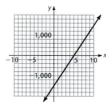

11. Slope: $-10,000$
y intercept: 80,000

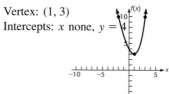

13. Vertex: $(-4, 0)$
Intercepts: $x = -4$, $y = 16$

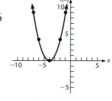

15. Vertex: $(1, 3)$
Intercepts: x none, $y = 4$

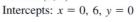

17. Vertex: $(2, 6)$
Intercepts: $x = 2 \pm \sqrt{6}$, $y = 2$

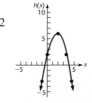

19. Vertex: $(3, 9)$
Intercepts: $x = 0, 6$, $y = 0$

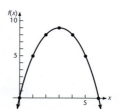

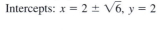

21. Vertex: $(0, -4)$
Intercepts: $x = 2, -2, y = -4$

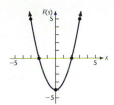

23. Vertex: $(0, 4)$
Intercepts: $x = 2, -2, y = 4$

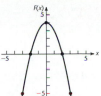

25. Vertex: $(3.5, -2.25)$
Intercepts: $x = 2, 5, y = 10$

27. Vertex: $(1.5, 6.25)$
Intercepts: $x = -1, 4, y = 4$

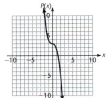

29. Vertex: $(-2, -2)$
Intercepts: $x = 0, -4, y = 0$

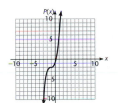

31. Vertex: $(-2, 6)$
Intercepts: $x = -2 \pm \sqrt{3}, y = -2$

33.

35.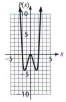

37.

39.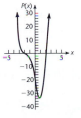

41.

43.

45.

47.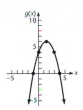

49.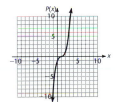

51.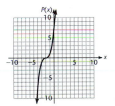

53.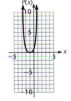

55.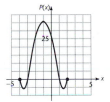

57.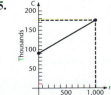

59.

61.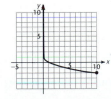

63.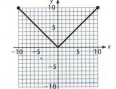

65.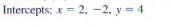

67. **(A)**

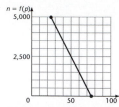

(B) $R(p) = 7,500p - 100p^2$

(C) Largest revenue = \$140,625 when $p = \$37.50$

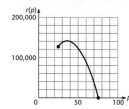

69. **(A)** $A(x) = x(50 - x)$

(B) Domain: $0 < x < 50$

(C)

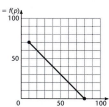

(D) The largest area is 625 in^2 when the pen is 25×25 in

71. **(A)**

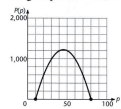

(B) $P(p) = -p^2 + 90p - 800$

(C) Largest profit = \$1,125 when $p = \$45$

73. **(A)** $V(x) = (12 - 2x)(8 - 2x)x$
$= 4x^3 - 40x^2 + 96x$

(B) Domain: $0 < x < 4$ [*Note:* At $x = 0$ and $x = 4$, we have 0 volume.]

(C)

(D) Largest volume: $V(x) \approx V(1.5) \approx 67.5$ in.^{3}; in this case a 1.5 in.-square should be cut from each corner and the dimensions of the box are 5 by 9 by 1.5 in. $\left(\text{The largest volume actually occurs at } x = \dfrac{10 - 2\sqrt{7}}{3}.\right)$

75. (A) $V(x) = 2x^2(108 - 6x) = 216x^2 - 12x^3$

(B) Domain: $0 < x < 18$

(C)

(D) Largest volume: $V(12) = 10{,}368$ in.^{3}; 12 by 24 by 36 in.

77. (A) $Y(x) = (120 + x)(40 - \frac{1}{4}x)$

(B)

(C) Largest yield: 4,900 bushels at $x = 20$

79. (A) $R(x) = (8 + 0.25x)(5{,}000 - 100x)$

(B)

(C) Largest revenue: $42,025 at $x = 9$

81. (A) $R(x) = (420 + 10x)(180 - x)$

(B)

(C) Largest revenue: at $x = 69$, price = $1,110

83. (A) $R(x) = (7 + x)(300 - 20x)$

(B)

(C) Largest revenue: at $x = 4$, price = $11

EXERCISE 9-4

1. Function has an inverse   **3.** Function does not have an inverse

5. $f^{-1} = \{(1, -2), (2, -1), (3, 0), (4, 1), (5, 2)\}$;   Domain: $\{1, 2, 3, 4, 5\}$

7. $f^{-1} = \{(\frac{1}{5}, -2), (\frac{1}{4}, -1), (\frac{1}{3}, 0), (\frac{1}{2}, 1), (1, 2)\}$; Domain: $\{\frac{1}{5}, \frac{1}{4}, \frac{1}{3}, \frac{1}{2}, 1\}$

9. 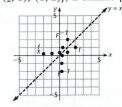   **11.**

13. $f^{-1}(x) = \dfrac{x+2}{3}$; $\frac{1}{3}$, -1, -1 **15.** $f^{-1}(x) = 3x + 6$; 3, -1, -1 **17.** Yes **19.** No **21.** Yes

23. (A) $f^{-1}(x) = \dfrac{x+2}{3}$ (B) $\frac{4}{3}$ (C) 3 **25.** (A) $F^{-1}(x) = 3(x+2)$ (B) 3 (C) 4 **27.** $f^{-1}(x) = (x-1)^2$

29. $f^{-1}(x) = x^2 + 3$ **31.** $f^{-1}(x) = (x+3)^2 - 1$ **33.** $f^{-1}(x) = \dfrac{1-x}{x}$ **35.** $f^{-1}(x) = x^3$

37. $f^{-1}(x) = x^3 + 1$ **39.** $f^{-1}(x) = \sqrt[5]{x+2}$ **41.** $f^{-1}(x) = \dfrac{1}{1-x}$ **43.** $f^{-1}(x) = \dfrac{x+3}{x-1}$ **45.** $f^{-1}(x) = \sqrt{-x}$

67. **69.** **71.** **73.**

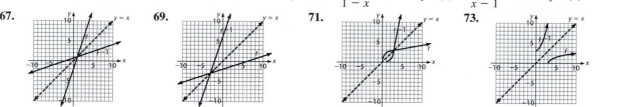

75. $3p - 5 = 3q - 5 \rightarrow 3p = 3q \rightarrow p = q$ **77.** $\sqrt{p+1} = \sqrt{q+1} \rightarrow p + 1 = q + 1 \rightarrow p = q$

79. $\dfrac{1}{p+1} = \dfrac{1}{q+1} \rightarrow p + 1 = q + 1 \rightarrow p = q$

81. **83.** **85.** **87.**

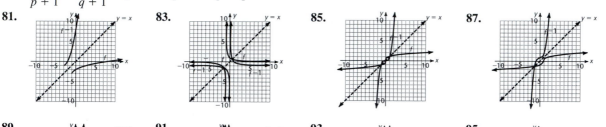

89. **91.** **93.** **95.**

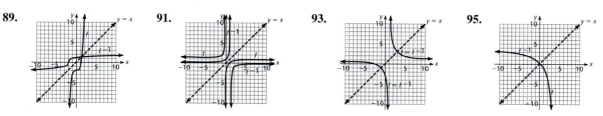

EXERCISE 9-5

1. $F = kv^2$ **3.** $f = k\sqrt{T}$ **5.** $y = k/\sqrt{x}$ **7.** $t = k/T$ **9.** $R = kSTV$ **11.** $V = khr^2$ **13.** 20
15. 12 **17.** 4 **19.** $9\sqrt{3}$ **21.** $U = k(ab/c^3)$ **23.** $L = k(wh^2/l)$ **25.** -12 **27.** 83 lb **29.** 75 mph
31. 20 amp **33.** 480 lb **35.** 20 hr **37.** 31,250 **39.** The new horsepower must be eight times the old.
41. No effect **43.** $t^2 = kd^3$ **45.** 1.47 hr (approx.) **47.** 20 days **49.** Quadrupled **51.** 540 lb
53. (A) $\Delta S = kS$ (B) 10 oz (C) 8 candlepower **55.** 32 times/sec **57.** $N = k(F/d)$ **59.** 1.2 mi/sec
61. 20 days **63.** The volume is increased by a factor of 8.

CHAPTER 9 REVIEW EXERCISE

1. Not a function *(9-1)* **2.** Function *(9-1)* **3.** Function *(9-1)* **4.** Function *(9-1)*
5. Not a function *(9-1)* **6.** Function *(9-1)* **7.** Function *(9-1)* **8.** Not a function *(9-1)*
9. Not a function *(9-1)* **10.** Not a function *(9-1)* **11.** Function *(9-1)* **12.** Function *(9-1)*
13. Domain: 1, 3, 5; Range: 2, 4, 6 *(9-1)* **14.** Domain: -1, 1, 3, 5; Range: 0 *(9-1)*
15. Domain: $x \neq 0$; Range: $y \neq 0$ *(9-1, 9-3)* **16.** Domain: $x > 0$; Range: $y > 0$ *(9-1, 9-3)*
17. Domain: $x > 1$; Range: $y > 0$ *(9-1, 9-3)* **18.** Domain: $x \geq 1$; Range: $y \geq 0$ *(9-1)*
19. Domain: all real numbers; Range: all real numbers *(9-1)* **20.** Domain: $x \neq 1$; Range: $y \neq 0$ *(9-1, 9-3)*
21. 0, 6, 9 *(9-2)* **22.** $-6, 0, -3$ *(9-2)* **23.** $6 - m$, $6 - x - h$ *(9-2)*
24. $c - 2c^2$, $x + h - 2x^2 - 4xh - 2h^2$ *(9-2)* **25.** $6 - 2x^2$ *(9-2)* **26.** $6 - 2x + 2x^2$ *(9-2)*
27. $-2x^2 + 2x - 6$ *(9-2)* **28.** $2x^3 - 13x^2 + 6x$ *(9-2)* **29.** $\dfrac{6-x}{x-2x^2}$ *(9-2)* **30.** $\dfrac{x-2x^2}{6-x}$ *(9-2)*
31. $6 - x + 2x^2$ *(9-2)* **32.** $-2x^2 + 23x - 66$ *(9-2)*

33. Slope = 2
y intercept: −4 *(9-3)*

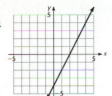

34. *(9-3)*

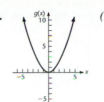

35. *(9-4)*

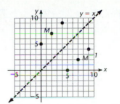

36. Domain: 5, 7, 9; Range: 0, 2, 4 *(9-4)*   **37.** $f^{-1}(x) = \dfrac{x-5}{3}$ *(9-4)*   **38.** $f^{-1}(x) = \dfrac{3-5x}{x}$ *(9-4)*

39. $f^{-1}(x) = \dfrac{x^2-5}{3}$ *(9-4)*   **40.** $f^{-1}(x) = \left(\dfrac{x-5}{3}\right)^2$ *(9-4)*   **41.** $f^{-1}(x) = \dfrac{3}{x-5}$ *(9-4)*   **42.** $m = kn^2$ *(9-5)*

43. $P = \dfrac{k}{Q^3}$ *(9-5)*   **44.** $A = kab$ *(9-5)*   **45.** 45 *(9-5)*   **46.** 16.2 *(9-5)*   **47.** 168 *(9-5)*

48. $C(x) = 200 + 0.05x$ *(9-3)*

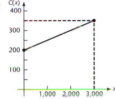

49. $V(t) = 80,000 - 7,500t$ *(9-3)*

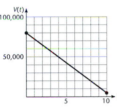

50. No; a function has an inverse only when it is one-to-one, that is, only when every element in the range corresponds to exactly one element in the domain. *(9-4)*

51. $x^2 + x - 6$ *(9-2)*   **52.** $-x^2 - x + 6$ *(9-2)*   **53.** $-x^3 + x^2 + 4x - 4$ *(9-2)*

54. $-x^3 + 2x^2 + 3x - 6$ *(9-2)*   **55.** $-(x+2)$ *(9-2)*   **56.** $\dfrac{-1}{x+2}$ *(9-2)*   **57.** -12 *(9-2)*   **58.** -7 *(9-2)*

59. $-x^2 + 4x$ *(9-2)*   **60.** $2 - x^2$ *(9-2)*   **61.** $-x^4 + 8x^2 - 12$ *(9-2)*   **62.** $x - 4$ *(9-2)*

63. **(A)** $3 + 2h$ **(B)** 2 *(9-2)*

64. Slope: $-\frac{3}{2}$
y intercept: 6 *(9-3)*

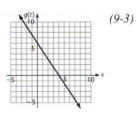

65. Vertex: (2, 1)
Intercepts: x none, y = 5 *(9-3)*

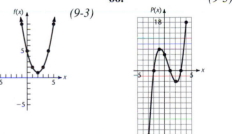

66. *(9-3)*

67. Problem 3 *(9-4)*   **68.** Problem 6 *(9-4)*   **69.** None *(9-4)*

70. Vertex: (3, 144)
Intercepts: $t = 0, 6$, $y = 0$ *(9-3)*

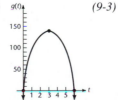

71. $4 \to 1, 5 \to 2, 6 \to 3$ *(9-4)*   **72.** Function does not have an inverse; not one-to-one   *(9-4)*

73. $1 \to 1, -1 \to 2, -3 \to 3$ *(9-4)*   **74.**   *(9-4)*

75. Function does not have an inverse; fails horizontal-line test   *(9-4)*

76. Function does not have an inverse; fails horizontal-line test   *(9-4)*   **77.** $(3, 1), (7, 5), (11, 9)$ *(9-4)*

78. $(3, 1), (1, 3), (7, 5), (5, 7)$ *(9-4)*   **79.** Function does not have an inverse; not one-to-one   *(9-4)*

80. $y = \sqrt[3]{x} - 5$ *(9-4)*   **81.** Function does not have an inverse; graph fails horizontal-line test   *(9-4)*

82. $y = \dfrac{x - 2}{3}$ *(9-4)*   **83.** $(-\infty, -2] \cup (5, \infty)$ *(9-1)*   **84.** All real numbers   *(9-1)*   **85.** $x \neq 5$ *(9-1)*

86. $M^{-1}(x) = 2x - 3$ *(9-4)*   **87.** $y = kx/z$ *(9-5)*   **88.** $y = kx^3/\sqrt{z}$ *(9-5)*   **89.** $A = k\dfrac{ab^3}{c^2}$ *(9-5)*

90. $\frac{4}{3}$ *(9-5)*   **91.** 1 *(9-5)*

92. **(A)** 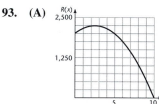   **(B)** $p = \$100$; largest revenue $= f(100) = \$300{,}000$ *(9-3)*

93. **(A)** 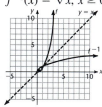   **(B)** $\$2.50$; largest monthly revenue $= \$22{,}500$ *(9-3)*

94. **(A)** $-3 - 4h - h^2$ **(B)** $-4 - h$ *(9-2)*   **95.** **(A)** $\dfrac{1}{2 + h}$ **(B)** $\dfrac{-1}{4 + 2h}$ *(9-2)*

96. **(A)** $f^{-1}(x) = \sqrt{x}, x \geq 0$

(B)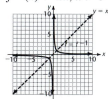

(C) $f^{-1}(9) = 3$; $f^{-1}[f(x)] = x, x \geq 0$ *(9-4)*

97. **(A)** $f^{-1}(x) = 1/x, x \neq 0$

(B)

(C) $f^{-1}(9) = \frac{1}{9}$; $f^{-1}[f(x)] = x, x \neq 0$ *(9-4)*

98.   *(9-4)*   **99.** $t = kwd/P$; $t = 24$ sec   *(9-5)*   **100.** The total force is doubled.   *(9-5)*

CHAPTER 9 PRACTICE TEST

1. Not a function *(9-1)*   **2.** Not a function *(9-1)*   **3.** Function *(9-1)*   **4.** Not a function *(9-1)*
5. 7 *(9-2)*   **6.** $x^2 + 6x + 7$ *(9-2)*   **7.** $x + h + 3$ *(9-2)*   **8.** 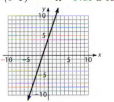 *(9-3)*

9. 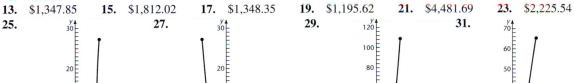 *(9-3)*   **10.** $x \geq 2$ or $x \leq -2$ *(9-1)*   **11.** $x \neq \pm 2$ *(9-1)*

12. Has an inverse: $(0, -1)$, $(1, 0)$, $(2, 1)$ *(9-4)*   **13.** Does not have an inverse; fails horizontal-line test *(9-4)*
14. Does not have an inverse; fails horizontal-line test *(9-4)*   **15.** $(3, 1)$, $(4, 2)$, $(5, 3)$; Domain: 3, 4, 5 *(9-4)*

16. $f^{-1}(x) = (x - 2)^2$; Domain: $x \geq 2$ *(9-4)*   **17.** $f^{-1}(x) = \dfrac{1 + x}{x}$; Domain: $x \neq 0$ *(9-4)*   **18.** 45 *(9-5)*

19. 3.6 *(9-5)*   **20.** 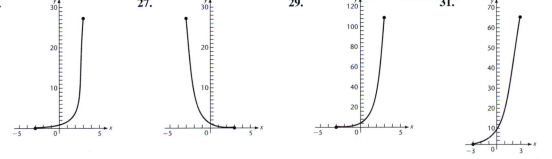 *(9-3)*

Largest profit is \$3,200 when $x = 12$.

CHAPTER 10

EXERCISE 10-1

1. 4.7288 . . .   **3.** 0.1026 . . .   **5.** 7.3890 . . .   **7.** 0.0497 . . .   **9.** 1453.0403 . . .   **11.** 4.1132 . . .
13. \$1,347.85   **15.** \$1,812.02   **17.** \$1,348.35   **19.** \$1,195.62   **21.** \$4,481.69   **23.** \$2,225.54

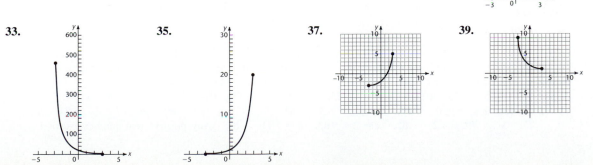

25.   **27.**   **29.**   **31.**

33.   **35.**   **37.**   **39.**

41.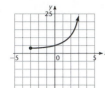

43.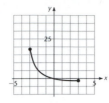

45.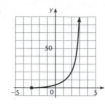

47.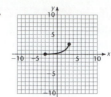

49.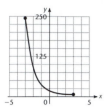

51.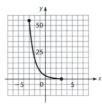

53.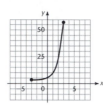

55.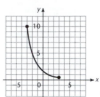

57.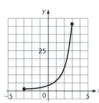

59.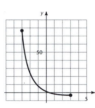

61. $y = 10e^{-0.12x}$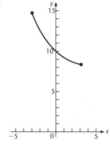

63.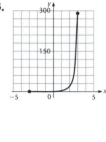

65. $y = 10e^{-x^2}$

67. $y = a(2^x)$

69.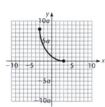

71.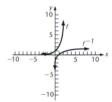

73.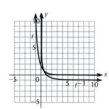

75.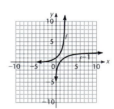

77. $P = 14.7e^{-0.21x}$

79. $A = 100(\frac{1}{2})^{t/28}$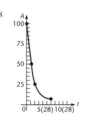

81. $N(i) = 100e^{-0.11(i-1)}$

EXERCISE 10-2

1. 2   **3.** -3 **5.** 1   **7.** $-\frac{1}{2}$ **9.** 2   **11.** $\frac{3}{2}$ **13.** -1 **15.** $-\frac{1}{2}$ **17.** 0   **19.** 5   **21.** -4

23. 2   **25.** 3   **27.** $9 = 3^2$ **29.** $81 = 3^4$ **31.** $1,000 = 10^3$ **33.** $1 = e^0$ **35.** $\log_8 64 = 2$

37. $\log_{10} 10,000 = 4$ **39.** $\log_v u = x$ **41.** $\log_{27} 9 = \frac{2}{3}$ **43.** $x = 4$ **45.** $b = 4$ **47.** $b = 9$

49. $x = 125$ **51.** $x = \sqrt[3]{25}$ **53.** $0.001 = 10^{-3}$ **55.** $3 = 81^{1/4}$ **57.** $16 = (\frac{1}{2})^{-4}$ **59.** $N = a^e$

61. $\log_{10} 0.01 = -2$ **63.** $\log_e 1 = 0$ **65.** $\log_2(\frac{1}{8}) = -3$ **67.** $\log_{81}(\frac{1}{3}) = -\frac{1}{4}$ **69.** $\log_{49} 7 = \frac{1}{2}$ **71.** u

73. $\frac{1}{2}$ **75.** 0   **77.** 3   **79.** $\sqrt{2}$ **81.** $x = 2$ **83.** $b = 100$ **85.** Any positive real number except 1

87. (A)

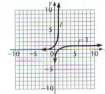

(B) f: domain all real numbers, range $y > 0$; f^{-1}: domain $x > 0$, range all real numbers

(C) $f^{-1}(x) = \log_{10} x$

89. (A)

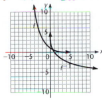

(B) f: domain all real numbers, range $y > 0$; f^{-1}: domain $x > 0$, range all real numbers

(C) $f^{-1}(x) = \log_{1/2} x$

91. Domain of f is the set of all real numbers; range of f is 1. No, f does not have an inverse function. **93.** -3

95. $e^{\ln(1/x)} = \dfrac{1}{x}$ by definition of ln: $e^{-\ln x} = \dfrac{1}{e^{\ln x}} = \dfrac{1}{x}$; therefore, $e^{\ln(1/x)} = e^{-\ln x}$, $\ln e^{\ln(1/x)} = \ln e^{-\ln x}$, and $\ln(1/x) = -\ln x$

EXERCISE 10-3

1. $\log_b u + \log_b v$ **3.** $\log_b A - \log_b B$ **5.** $5 \log_b u$ **7.** $\frac{3}{5} \log_b N$ **9.** $\frac{1}{2} \log_b Q$
11. $\log_b u + \log_b v + \log_b w$ **13.** $\log_b AB$ **15.** $\log_b (X/y)$ **17.** $\log_b (wx/y)$ **19.** 3.40 **21.** -0.920
23. 3.300 **25.** -0.683 **27.** -0.252 **29.** 1.114 **31.** 1.366 **33.** 0.216 **35.** $2 \log_b u + 7 \log_b v$
37. $-\log_b a$ **39.** $\frac{1}{3} \log_b N - 2 \log_b p - 3 \log_b q$ **41.** $\frac{1}{4}(2 \log_b x + 3 \log_b y - \frac{1}{2} \log b_z)$ **43.** $\log_b(x^2/y)$
45. $\log_b(x^3y^2/z^4)$ **47.** $\log_b \sqrt[5]{x^2y^3}$ **49.** $\log_b 3x^2y$ **51.** $\log_b(x/y)^{1/2}$ **53.** $\log_b 2x^3y^4$ **55.** 2.02
57. 0.23 **59.** -0.05 **61.** -0.095 **63.** 0.957 **65.** 1.162 **67.** 8 **69.** $\frac{1}{9}$ **71.** $\frac{27}{64}$
73. $y = cb^{-kt}$
75. Let $u = \log_b M$ and $v = \log_b N$; then $M = b^u$ and $N = b^v$. Thus,

$$\log_b \frac{M}{N} = \log_b \frac{b^u}{b^v} = \log_b b^{u-v} = u - v = \log_b M - \log_b N.$$

77. $MN = b^{\log_b M} b^{\log_b N} = b^{\log_b M + \log_b N}$; hence, by definition of logarithm, $\log_b MN = \log_b M + \log_b N$.

EXERCISE 10-4

1. 4.9177 **3.** -2.8419 **5.** 3.7623 **7.** -2.5128 **9.** Not defined **11.** -2.3010 **13.** -5.1549
15. 4.3429×10^{-7}, that is, 0.0000 to four decimal places **17.** 200,800 **19.** 0.000 664 8 **21.** 47.73
23. 0.6760 **25.** 293.8 **27.** 3.697×10^{-6} **29.** 5.811×10^{-5} **31.** 3.436 **33.** 4.959 **35.** 7.861
37. 3.301 **39.** 3.6776 **41.** -1.6094 **43.** -1.7372 **45.** 2.1110 **47.** -0.6309 **49.** 2.0164
51. -2.000 **53.** **55.** **57.**

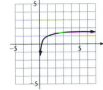 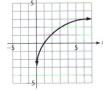 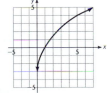

59. 12.725 **61.** -25.715 **63.** 1.1709×10^{32} **65.** 4.2672×10^{-7} **67.**

69.

71.

73.

EXERCISE 10-5

1. 1.46   **3.** 0.321   **5.** 1.29   **7.** 3.50   **9.** 1.80   **11.** 2.07   **13.** 1.43   **15.** -0.719 **17.** 1.03
19. 20   **21.** 5   **23.** $\frac{9}{16}$ **25.** $\sqrt{10}$ **27.** 14.2   **29.** -1.83 **31.** 11.7   **33.** 5   **35.** 1, e^2, e^{-2}
37. e^e **39.** 100, 0.1   **41.** $x = -(1/k)\ln(I/I_0)$ **43.** $I = I_0 10^{N/10}$ **45.** $t = (-L/R)\ln[1 - (RI/E)]$
47. Inequality sign should have been reversed when both sides were multiplied by $\log \frac{1}{2}$, a negative quantity.
49. 6.116 yr   **51.** 28.011 yr   **55.** Approx. 3.8 hr   **57.** 12.8   **59.** Approx. 35 yr   **61.** Approx. 28 yr
63. Approx. 22,689 yr   **65.** Divide both sides by I_0, take logarithms of both sides, and then multiply both sides by 10.
67. 95 ft; 489 ft   **69.** 17.3 weeks

CHAPTER 10 REVIEW EXERCISE

1. $n = \log_{10} m$ *(10-2)*   **2.** $x = 10^y$ *(10-2)*   **3.** 3 *(10-2)*   **4.** $\frac{1}{2}$ *(10-2)*   **5.** $\frac{5}{2}$ *(10-2)*   **6.** -1 *(10-2)*
7. $\frac{1}{2}$ *(10-2)*   **8.** -2 *(10-2)*   **9.** 8 *(10-2)*   **10.** 5 *(10-2)*   **11.** 100 *(10-2)*   **12.** $\frac{1}{8}$ *(10-2)*
13. 9 *(10-2)*   **14.** $\frac{1}{2}$ *(10-2)*   **15.** 1.24 *(10-5)*   **16.** 11.9 *(10-5)*   **17.** 3.32 *(10-5)*   **18.** -2.10 *(10-5)*
19. 900 *(10-3, 10-5)*   **20.** 5 *(10-3, 10-5)*   **21.** 100 *(10-3)*   **22.** 49 *(10-2)*   **23.** $y = e^x$ *(10-2)*
24. $y = \ln x$ *(10-2)*   **25.** -2 *(10-2)*   **26.** 3 *(10-2)*   **27.** -5 *(10-2)*   **28.** $-\frac{1}{2}$ *(10-2)*   **29.** $\frac{1}{3}$ *(10-2)*
30. 64 *(10-2)*   **31.** e *(10-2)*   **32.** 33 *(10-2)*   **33.** 1 *(10-2)*   **34.** $1/\sqrt{2}$ *(10-2)*   **35.** All x *(10-2)*
36. 1 *(10-2)*   **37.** 2.32 *(10-5)*   **38.** 3.92 *(10-5)*   **39.** 92.1 *(10-5)*   **40.** 3.86 *(10-5)*   **41.** -2 *(10-5)*
42. -2.70 *(10-5)*   **43.** 300 *(10-3, 10-5)*   **44.** 2.00 *(10-3, 10-5)*   **45.** 1, 10^3, 10^{-3} *(10-3, 10-5)*
46. 10^e *(10-3, 10-5)*   **47.** 1.948 *(10-4)*   **48.** 2.585 *(10-4)*   **49.** 0.927 *(10-4)*   **50.** 0 *(10-4)*
51.   *(10-1)*   **52.**   *(10-1)*   **53.**   *(10-4)*

54.   *(10-4)*   **55.**   *(10-1)*   **56.** $y = ce^{-5t}$ *(10-5)*   **57.** $y = e^{3x+4}$ *(10-5)*

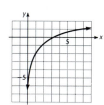
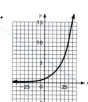
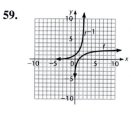
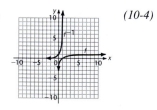

58. $y = \ln(3x + 4)$ *(10-5)*   **59.**   *(10-4)*   **60.**   *(10-4)*

61.   *(10-4)*   **62.**   *(10-4)*   **63.** 10 *(10-5)*   **64.** e^e *(10-5)*

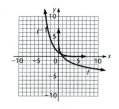
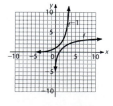

65. $x > 0$ *(10-3)*   **66.** No solution *(10-5)*

67. If $\log_1 x = y$, then we would have to have $1^y = x$; that is, $1 = x$ for arbitrary positive x, which is impossible. *(10-2)*

68. Let $u = \log_b M$ and $v = \log_b N$; then $M = b^u$ and $N = b^v$.
Thus, $\log_b(M/N) = \log_b(b^u/b^v) = \log_b b^{u-v} = u - v = \log_b M - \log_b N$. *(10-3)*

69. \$1,195.62 *(10-1)*   **70.** \$1,491.82 *(10-1)*   **71.** 23.4 yr *(10-5)*   **72.** 23.1 yr *(10-5)*

73. 37,100 yr *(10-5)*   **74.** $I = I_0 e^{-kx}$ *(10-3, 10-5)*   **75.** $n = -\log[1 - (Pi/r)]/\log(1 + i)$ *(10-3, 10-5)*

CHAPTER 10 PRACTICE TEST

1. $\frac{3}{2}$ *(10-2)*   **2.** $\frac{3}{2}$ *(10-2)*   **3.** 1.277 *(10-3)*   **4.** $\log_2 \dfrac{x}{(x+1)^3}$ *(10-3)*   **5.** $\log \dfrac{xz^3}{y^2}$ *(10-3)*

6. 32 *(10-2)*   **7.** $\sqrt{5}$ *(10-2)*   **8.** 100 *(10-2)*   **9.** 5 *(10-5)*   **10.** 10 *(10-2)*   **11.** e *(10-2, 10-5)*

12. 3.3219 *(10-5)*   **13.** 2.3026 *(10-2, 10-5)*   **14.** 374.4665 *(10-5)*   **15.** -2.3979 *(10-2, 10-5)*

16. *(10-4)*   **17.** *(10-4)*

18. f: domain all x, range $y > 0$; f^{-1}: domain $x > 0$, range all y; $f^{-1}(x) = \ln x$ *(10-4)*

19. \$241.17 *(10-4)*   **20.** $t = \dfrac{1}{k} \ln \dfrac{T - M}{C_0}$ *(10-5)*

CUMULATIVE REVIEW EXERCISE, CHAPTERS 1–10

1. -36 *(1-4, 1-5)*   **2.** 125 *(1-6)*   **3.** $12x^2y^2$ *(3-3)*   **4.** $x^3 - 2x^2 + 2x - 4$, $R = 11$ *(3-4)*   **5.** 1 *(6-2)*

6. $-\frac{1}{4}$ *(5-2)*   **7.** $y = 2x - 10$ *(5-2)*   **8.** -4 *(6-4)*   **9.** 76; 2 *(6-2)*   **10.** -2 *(8-5)*   **11.** $10\frac{2}{3}$ *(9-1)*

12. $5 + 4i$ *(6-8)*   **13.** i *(6-8)*   **14.** $(-1, -9)$ *(6-5)*   **15.** 0 *(5-8)*   **16.** $3\left(x + \dfrac{1}{x}\right) > x - 4$ *(1-2)*

17. $R(x) = 3{,}000x - 12x^2$ *(9-2)*   **18.** $f(t) = -8{,}000t + 60{,}000$ *(9-2)*   **19.** $1 - x + y - z$ *(1-3)*

20. $x^3 + x^2 + x - 3$ *(2-1)*   **21.** $3x^2 + 14x + 8$ *(2-2)*   **22.** $4x^2 - 12x + 9$ *(2-2)*   **23.** x^3 *(6-2)*

24. $2x^2y(y^2 + 3x)$ *(2-3)*   **25.** $(x + 4)^2$ *(2-3)*   **26.** $(4x + 1)(x + 4)$ *(2-4, 2-5)*   **27.** $(x - 8)(x + 8)$ *(2-3)*

28. $\dfrac{4xy^2z^3}{6yz^5}$ *(3-1)*   **29.** $\dfrac{x+2}{x+1}$ *(3-1)*   **30.** $\dfrac{xy^5}{z^2}$ *(6-2)*   **31.** 345,600 *(6-3)*   **32.** 3.456×10^{-3} *(6-3)*

33. $4\sqrt{2}$ *(6-6)*   **34.** $\dfrac{2\sqrt{3x}}{3}$ *(6-6)*   **35.** $\dfrac{x^2\sqrt{x}}{3}$ *(6-6)*   **36.** $y = -\frac{3}{4}x + \frac{5}{4}$ *(5-2)*

37. $\begin{bmatrix} 2 & 1 & -1 & \bigm| & 4 \\ 1 & 3 & 2 & \bigm| & 5 \\ -1 & 2 & 3 & \bigm| & 6 \end{bmatrix}$ *(8-4)*   **38.** $\frac{8}{3}$ *(1-7)*   **39.** $\dfrac{3 \pm \sqrt{41}}{4}$ *(7-2)*   **40.** $9, -1$ *(2-7)*   **41.** $(5, -2)$ *(8-1)*

42. $1, \frac{1}{2}, -3$ *(2-7)*   **43.** $2 < x$ or $(2, \infty)$ *(5-3)*   **44.** $-3 \le x \le 7$ or $[-3, 7]$ *(5-4)*   **45.** $2, -1$ *(4-1)*

46. 8 *(7-4)*   **47.** $(0, 2), (2, 0)$ *(7-9)*   **48.** $x = \dfrac{y - 5}{3}$ *(4-4)*   **49.** $(5, -3, 1)$ *(8-3)*   **50.** $3, -3$ *(7-1)*

51.   *(5-1)*   **52.**   *(7-4)*   **53.**   *(9-3)*

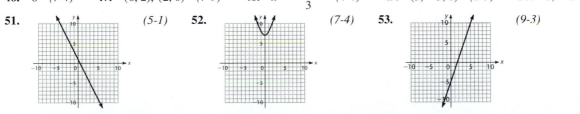

54. 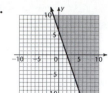 *(5-5)* **55.** 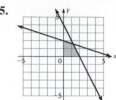 *(8-6)* **56.** 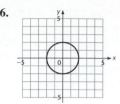 *(7-7)*

57. 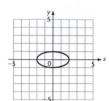 *(7-8)* **58.** 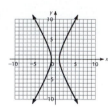 *(7-8)* **59.** 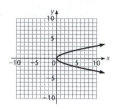 *(7-7)*

60. 9, 15 *(4-1)* **61.** 12 gal *(8-2)* **62.** 4,000 *(4-2)* **63.** \$15,529.69 *(10-1)* **64.** \$15,683.12 *(10-1)*

65. $\frac{1}{3}$ *(9-2)* **66.** $(x + 1)^2(x - 1)^2$ *(3-3)* **67.** Center $(-1, 2)$, radius 3 *(7-7)* **68.** $x + y = 2$ *(5-2)*

69. $f^{-1}(x) = \dfrac{1 + x}{x}$ *(9-4)* **70.** $(x - 4)(x^2 + 4x + 16)$ *(2-6)* **71.** $(x^2 + 8)(x^2 - 8)$ *(2-3)*

72. $(x + 2y)(a + 3b)$ *(2-3)* **73.** $\dfrac{7x + 3}{x(x + 1)}$ *(3-3)* **74.** $\dfrac{(x + 1)(x + 2)}{x + 3}$ *(3-2)* **75.** $\dfrac{1 + 2x}{2x - 1}$ *(3-5)*

76. 0 *(8-5)* **77.** $\dfrac{-5 \pm i\sqrt{59}}{6}$ *(7-2)* **78.** 5 *(7-4)* **79.** -1 *(7-2)* **80.** $x \le -\frac{7}{2}$ or $x \ge \frac{1}{2}$ *(5-4)*

81. 0, 2, 3 *(2-7)* **82.** $(3, -4)$ *(8-2)* **83.** $\dfrac{\ln 5}{3} \approx 0.5365$ *(10-5)* **84.** $10^{1.7} \approx 50.1187$ *(10-5)*

85. $y = \dfrac{2x}{3x - 1}$ *(4-1)* **86.** $b = \frac{1}{2}$ *(10-2)* **87.** 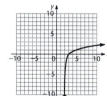 *(8-6)* **88.** 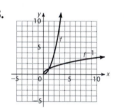 *(9-4)*

89. 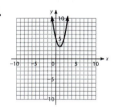 *(7-5, 7-6)* **90.** *(10-4)* **91.** 5 by 8 ft *(7-3)*

92. $A(x) = x(500 - x)$; largest area when $x = 250$ m *(7-5)* **93.** 12 min *(4-3)* **94.** 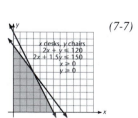 *(7-7)*

95. 54 days *(9-5)* **96.** $(3x - 8)(2x + 9)$ *(2-4, 2-5)* **97.** $(x - 1)(x + 1)(x^2 - x + 1)(x^2 + x + 1)$ *(2-6)*

98. $\dfrac{2x}{x + 2}$ *(3-5)* **99.** 1.75 gal *(8-2)* **100.** 45 mph, 50 mph; 12 hr and 40 min *(4-3)*

CHAPTER 11

EXERCISE 11-1

1. $-1, 0, 1, 2$ **3.** $5, 7, 9, 11$ **5.** $2, 4, 8, 16$ **7.** $0, \frac{1}{3}, \frac{1}{2}, \frac{3}{5}$ **9.** $0, \frac{1}{4}, \frac{8}{27}, \frac{81}{256}$ **11.** $4, -8, 16, -32$

13. $-1, \frac{1}{2}, -\frac{1}{3}, \frac{1}{4}$ **15.** 6 **17.** 39 **19.** 64 **21.** $\frac{99}{101}$ **23.** $(\frac{7}{8})^8 = \dfrac{5,764,801}{16,777,216} \approx 0.3436$ **25.** 256

27. $\frac{1}{20}$ **29.** $S_6 = 1 + \frac{1}{2} + \frac{1}{3} + \frac{1}{4} + \frac{1}{5} + \frac{1}{6}$ **31.** $S_5 = 1 + 2 + 3 + 4 + 5$ **33.** $S_4 = 0 + 1 + 4 + 9$

35. $S_3 = \frac{1}{10} + \frac{1}{100} + \frac{1}{1,000}$ **37.** $S_6 = \frac{1}{5} + \frac{1}{2} + \frac{1}{125} + \frac{1}{625} + \frac{1}{3,125} + \frac{1}{15,625}$ **39.** $S_4 = -1 + 1 - 1 + 1$

41. $S_5 = -1 + 2 - 3 + 4 - 5$ **43.** $1, -4, 9, -16, 25$ **45.** $0.3, 0.33, 0.333, 0.3333, 0.33333$ **47.** $0, 6, 0, 10, 0$

49. $1, 2, 4, 8, 16$ **51.** $7, 3, -1, -5, -9$ **53.** $4, 1, \frac{1}{4}, \frac{1}{16}, \frac{1}{64}$ **55.** $1, 2, 6, 24, 120$ **57.** $1, -1, -4, -8, -13$

59. $x, x^2/2, x^3/6, x^4/24, x^5/120$ **61.** $1, 1, 2, 3, 5$ **63.** $a_n = n + 3$ **65.** $a_n = 3n$ **67.** $a_n = \dfrac{n}{n+1}$

69. $a_n = (-1)^{n+1}$ **71.** $a_n = (-2)^n$ **73.** $a_n = \dfrac{x^n}{n}$ **75.** $-1 + \frac{1}{2} - \frac{1}{3} + \frac{1}{4} - \frac{1}{5}$ **77.** $\frac{4}{1} - \frac{8}{2} + \frac{16}{3} - \frac{32}{4}$

79. $x^2 + \dfrac{x^3}{2} + \dfrac{x^4}{3}$ **81.** $x - \dfrac{x^2}{2} + \dfrac{x^3}{3} - \dfrac{x^4}{4} + \dfrac{x^5}{5}$ **83.** $S_4 = \displaystyle\sum_{k=1}^{4} k^3$ **85.** $S_4 = \displaystyle\sum_{k=1}^{4} k^2$ **87.** $S_5 = \displaystyle\sum_{k=1}^{5} \dfrac{1}{2^k}$

89. $S_n = \displaystyle\sum_{k=1}^{n} \dfrac{1}{k^2}$ **91.** $S_n = \displaystyle\sum_{k=1}^{n} (-1)^{k+1} k^2$ **93.** $2, -4, 8, -16$ **95.** $5, 6, 7, 8$

97. **(A)** $3, 1.833, 1.462, 1.415$ **(B)** Table: $\sqrt{2} = 1.414$ **(C)** $1, 1.5, 1.417, 1.414$ **99.** $1, 1, 2$

EXERCISE 11-2

1. Not arithmetic **3.** $d = -5$ **5.** $d = -6$ **7.** Not arithmetic **9.** $a_2 = -1; a_3 = 3; a_4 = 7$

11. $a_{15} = 67; S_{11} = 242$ **13.** $S_{21} = 861$ **15.** $a_{15} = -21$ **17.** $d = 6; a_{101} = 603$ **19.** $S_{40} = 200$

21. $a_{11} = 2; S_{11} = \frac{77}{6}$ **23.** $a_1 = 1$ **25.** $S_{51} = 4,131$ **27.** $5,050$ **29.** $6,480$ **31.** $5,776$ **33.** $5,490$

35. $4,446$ **37.** $20,727$ **39.** $-1,071$ **41.** $-14,750$ **43.** 35 **45.** 34.5 **47.** $4,446$ **49.** $20,727$

51. $1 + 3 + 5 + \cdots + (2n - 1) = \dfrac{n}{2}[1 + 2n - 1] = \dfrac{n}{2} \cdot 2n = n^2$

53. $3 + 6 + \cdots + 3n = 3(1 + 2 + \cdots + n) = 3\dfrac{n(n+1)}{2}$ **55.** **(A)** 336 ft **(B)** $1,936\text{ ft}$ **(C)** $16t^2$

57.

11 units

EXERCISE 11-3

1. $r = -2$ **3.** Not geometric **5.** Not geometric **7.** Not geometric **9.** $a_2 = 3; a_3 = -\frac{3}{2}; a_4 = \frac{3}{4}$

11. $a_{10} = \frac{1}{243}$ **13.** $S_7 = 3,279$ **15.** $r = \dfrac{1}{\sqrt[5]{100}} \approx 0.398$ **17.** $S_{10} = -1,705$ **19.** $a_2 = 6; a_3 = 4$

21. $S_7 = 547$ **23.** 242 **25.** $13,021$ **27.** $1,365.25$ **29.** $\frac{364}{243} \approx 1.4979$ **31.** $13.579\ 476\ 9$ **33.** $S_\infty = \frac{9}{2}$

35. No sum **37.** $S_\infty = \frac{8}{5}$ **39.** $S_\infty = \frac{4}{3}$ **41.** No sum **43.** 11 **45.** $\sqrt{96} \approx 9.798$

47. $\sqrt{132} \approx 11.4891$ **49.** $\frac{7}{9}$ **51.** $\frac{6}{11}$ **53.** $3\frac{8}{37}$, or $\frac{119}{37}$ **55.** $\dfrac{384,615}{999,999} = \dfrac{5}{13}$ **57.** $\frac{1}{18}$

59. $0.9\overline{999} = \dfrac{9}{10} + \dfrac{9}{100} + \dfrac{9}{1,000} + \cdots = \dfrac{\frac{9}{10}}{1 - \frac{1}{10}} = 1$ **61.** $A = P(1 + r)^n$; approx. 12 yr **63.** $\dfrac{P[(1 + r)^{n+1} - 1]}{r}$

65. 900 revolutions **67.** $28\frac{1}{3}$ ft **69.** $\$4,000,000$

EXERCISE 11-4

1. 720   **3.** 20   **5.** 720   **7.** 15   **9.** 1   **11.** 28   **13.** 9!/8!   **15.** 8!/5!   **17.** 20!/18!
19. 31!/27!   **21.** 126   **23.** 6   **25.** 1   **27.** 2,380   **29.** 630   **31.** 4,960
33. $a^7 + 7a^6b + 21a^5b^2 + 35a^4b^3 + 35a^3b^4 + 21a^2b^5 + 7ab^6 + b^7$
35. $a^{10} + 10a^9b + 45a^8b^2 + 120a^7b^3 + 210a^6b^4 + 252a^5b^5 + 210a^4b^6 + 120a^3b^7 + 45a^2b^8 + 10ab^9 + b^{10}$
37. $u^5 + 5u^4v + 10u^3v^2 + 10u^2v^3 + 5uv^4 + v^5$ **39.** $y^4 - 4y^3 + 6y^2 - 4y + 1$
41. $32x^5 - 80x^4y + 80x^3y^2 - 40x^2y^3 + 10xy^4 - y^5$ **43.** $120a^{14}b^2$ **45.** $6,545a^{32}b^3$ **47.** $5,005u^9v^6$
49. $264m^2n^{10}$ **51.** $924w^6$ **53.** 1.610 51   **55.** 1.104 622 125 411 204 510 01
57. $\binom{n}{r} = \dfrac{n!}{r!(n-r)!} = \dfrac{n!}{(n-r)![n-(n-r)]!} = \binom{n}{n-r}$ **59.** $1 + 6i - 15 - 20i + 15 + 6i - 1 = -8i$

CHAPTER 11 REVIEW EXERCISE

1. Geometric; $r = -\frac{1}{2}$ *(11-3)*   **2.** Arithmetic; $d = 2$ *(11-2)*   **3.** Arithmetic; $d = 3$ *(11-2)*
4. Neither *(11-2, 11-3)*   **5.** Geometric; $r = -2$ *(11-3)*   **6.** Neither *(11-3)*   **7.** Arithmetic; $d = -3$ *(11-3)*
8. Geometric; $r = \frac{4}{3}$ *(11-3)*   **9.** Neither *(11-3)*   **10.** Geometric; $r = \frac{1}{2}$ *(11-3)*   **11.** 5, 7, 9, 11 *(11-1)*
12. 16, 8, 4, 2 *(11-1)*   **13.** $-3, 0, 5, 12$ *(11-1)*   **14.** 2, 6, 12, 20 *(11-1)*   **15.** $-2, 3, -4, 5$ *(11-1)*
16. $1, -\frac{1}{2}, \frac{1}{3}, -\frac{1}{4}$ *(11-1)*   **17.** $-8, -5, -2, 1$ *(11-1)*   **18.** $-1, 2, -4, 8$ *(11-1)*   **19.** 23 *(11-2)*
20. $\frac{1}{32}$ *(11-3)*   **21.** 96 *(11-1)*   **22.** 110 *(11-1)*   **23.** 11 *(11-1)*   **24.** $-\frac{1}{10}$ *(11-1)*   **25.** 19 *(11-1)*
26. 512 *(11-1)*   **27.** 140 *(11-2)*   **28.** $31\frac{31}{32}$ *(11-2)*   **29.** 55 *(11-2)*   **30.** 341 *(11-3)*   **31.** 35 *(11-1)*
32. 70 *(11-1)*   **33.** -4 *(11-1)*   **34.** $\frac{47}{60}$ *(11-1)*   **35.** 720 *(11-4)*   **36.** $20 \cdot 21 \cdot 22 = 9,240$ *(11-4)*
37. 21 *(11-4)*   **38.** 21 *(11-4)*   **39.** 28 *(11-4)*   **40.** 6 *(11-4)*   **41.** 13 *(11-2)*   **42.** 0 *(11-2)*
43. 18 *(11-2)*   **44.** 364.5 *(11-3)*   **45.** 1 *(11-3)*   **46.** $\sqrt{3}$ *(11-3)*   **47.** $\frac{1}{4}$ *(11-2)*   **48.** 20 *(11-2)*
49. $S_4 = \frac{3}{2} + \frac{3}{4} + \frac{3}{8} + \frac{3}{8} = \frac{25}{8}$ *(11-1)*   **50.** $S_4 = \frac{1}{2} - \frac{1}{4} + \frac{1}{8} - \frac{1}{16} = \frac{5}{16}$ *(11-1)*
51. $S_{10} = -6 - 4 - 2 + 0 + 2 + 4 + 6 + 8 + 10 + 12 = 30$ *(11-2)*
52. $S_7 = 8 + 4 + 2 + 1 + \frac{1}{2} + \frac{1}{4} + \frac{1}{8} = 15\frac{7}{8}$ *(11-3)*   **53.** $\dfrac{4,141}{162} \approx 25.56$ *(11-3)*   **54.** $\dfrac{20,475}{256} \approx 79.98$ *(11-3)*
55. 80 *(11-3)*   **56.** $\frac{81}{5}$ *(11-3)*   **57.** $\frac{1}{4}$ *(11-3)*   **58.** $\displaystyle\sum_{k=1}^{n} \frac{(-1)^{k+1}}{3^k}$ *(11-1)*   **59.** $\displaystyle\sum_{k=1}^{17} [10 - 3(k-1)]$ *(11-2)*
60. $\displaystyle\sum_{k=1}^{4} 3(\tfrac{4}{3})^{k-1}$ *(11-3)*   **61.** $\displaystyle\sum_{k=1}^{11} k^2$ *(11-1)*   **62.** $\displaystyle\sum_{k=1}^{6} 10(\tfrac{1}{2})^{k-1}$ *(11-3)*   **63.** 190 *(11-4)*   **64.** 1,820 *(11-4)*
65. 1 *(11-4)*   **66.** 190 *(11-4)*   **67.** 455 *(11-4)*   **68.** $x^5 - 5x^4y + 10x^3y^2 - 10x^2y^3 + 5xy^4 - y^5$ *(11-4)*
69. $x^7 + 7x^6 + 21x^5 + 35x^4 + 35x^3 + 21x^2 + 7x + 1$ *(11-4)*   **70.** $-1,760x^3y^9$ *(11-4)*   **71.** $125,970a^{12}b^8$ *(11-4)*
72. $\frac{8}{11}$ *(11-3)*   **73.** $\dfrac{1,233}{999} = \dfrac{137}{11}$ *(11-3)*   **74.** 1, 8, 28, 56, 70, 56, 28, 8, 1 *(11-4)*
75. 49 $g/2$ ft; 625 $g/2$ ft *(11-2)*

CHAPTER 11 PRACTICE TEST

1. 30 *(11-1)*   **2.** 223 *(11-1)*   **3.** 10 *(11-1)*   **4.** 20 *(11-2)*   **5.** $-\frac{1}{4}$ *(11-3)*
6. Geometric; $r = -\frac{5}{3}$ *(11-3)*   **7.** Neither *(11-2, 11-3)*   **8.** $d = 2$ *(11-2)*   **9.** $r = -\frac{1}{10}$ *(11-3)*
10. $d = \frac{4}{49}$ *(11-2)*   **11.** 258 *(11-2)*   **12.** $\dfrac{2,047}{1,024}$ *(11-3)*   **13.** 2 *(11-3)*   **14.** -16 *(11-2)*
15. 20,100 *(11-2)*   **16.** 636 *(11-4)*   **17.** $\displaystyle\sum_{k=1}^{6} 2(\tfrac{1}{3})^{k-1}$ *(11-3)*
18. $x^6 - 6x^5y + 15x^4y^2 - 20x^3y^3 + 15x^2y^4 - 6xy^5 + y^6$ *(11-4)*
19. $a^5 + 15a^4 + 90a^3 + 270a^2 + 405a + 243$ *(11-4)*   **20.** $31,824a^{11}b^7$ *(11-4)*

APPENDIX A

1. T   **3.** T   **5.** T   **7.** T   **9.** T   **11.** $5 \in P$ **13.** $6 \notin R$ **15.** $P = R$ **17.** $P \neq Q$
19. $\{6, 7, 8, 9\}$ **21.** $\varnothing$ **23.** {Su, M, T, W, Th, F, S}   **25.** $\{a, b, l\}$ **27.** $\varnothing$ **29.** $\{2, 4\}$ **31.** $\varnothing$
33. $\{1, 2, 3, 4, 5, 6, 7, 8\}$ **35.** $\{5, 6, 7, 8\}$ **37.** $\{2, 4, 6, 8\}$ **39.** $\{5, 7\}$ **41.** $\{5, 7\}$ **43.** $\varnothing, \{1\}, \{2\}, \{1, 2\}$
45.

47.

49.

Index

Strategy for Solving Word Problems
(See Section 2-8)

1. Read the problem very carefully—several times if necessary.

2. Write down important facts and relationships on a piece of scratch paper. Draw figures if it is helpful. Write down any formulas that might be relevant.

3. Identify the unknown quantities in terms of a single variable, if possible.

4. Look for key words and relationships in the problem that will lead to an equation involving the variable(s) introduced in step 3.

5. Solve the equation. Write down all the solutions asked for in the original problem.

6. Check the solutions in the original problem.

Metric Units

STANDARD UNITS OF METRIC MEASURE

Meter (m): length (approximately 3.28 ft)

Liter (L): volume (approximately 1.06 qt)

Gram (g): weight (approximately 0.035 oz)

IMPORTANT PREFIXES

kilo ($\times 1,000$) deci $\left(\times \frac{1}{10}\right)$

hecto ($\times 100$) centi $\left(\times \frac{1}{100}\right)$

deka ($\times 10$) milli $\left(\times \frac{1}{1,000}\right)$

ABBREVIATIONS

Length		Volume		Weight	
m	meter	L	liter	g	gram
km	kilometer	kL	kiloliter	kg	kilogram
hm	hectometer	hL	hectoliter	hg	hectogram
dam	dekameter	daL	dekaliter	dag	dekagram
dm	decimeter	dL	deciliter	dg	decigram
cm	centimeter	cL	centiliter	cg	centigram
mm	millimeter	mL	milliliter	mg	milligram

ENGLISH-METRIC CONVERSIONS

Length	Length	Volume (U.S.)	Volume (U.S.)	Weight	Weight
1 in = 2.540 cm	1 cm = 0.3937 in	1 pt = 0.4732 L	1 L = 2.1133 pt	1 oz = 28.35 g	1 g = 0.0353 oz
1 ft = 30.48 cm	1 cm = 0.03281 ft	1 qt = 0.9464 L	1 L = 1.0567 qt	1 lb = 453.6 g	1 g = 0.002205 lb
1 yd = 0.9144 m	1 m = 1.0936 yd	1 gal = 3.785 L	1 L = 0.2642 gal	1 lb = 0.4536 kg	1 kg = 2.205 lb
1 mi = 1.609 km	1 km = 0.6215 mi				